Lecture Notes in Computer Science   11301

Commenced Publication in 1973
Founding and Former Series Editors:
Gerhard Goos, Juris Hartmanis, and Jan van Leeuwen

More information about this series at http://www.springer.com/series/7407

Long Cheng · Andrew Chi Sing Leung
Seiichi Ozawa (Eds.)

Neural Information Processing

25th International Conference, ICONIP 2018
Siem Reap, Cambodia, December 13–16, 2018
Proceedings, Part I

 Springer

Editors
Long Cheng (ID)
The Chinese Academy of Sciences
Beijing, China

Seiichi Ozawa
Kobe University
Kobe, Japan

Andrew Chi Sing Leung
City University of Hong Kong
Kowloon, Hong Kong SAR, China

ISSN 0302-9743   ISSN 1611-3349 (electronic)
Lecture Notes in Computer Science
ISBN 978-3-030-04166-3   ISBN 978-3-030-04167-0 (eBook)
https://doi.org/10.1007/978-3-030-04167-0

Library of Congress Control Number: 2018960916

LNCS Sublibrary: SL1 – Theoretical Computer Science and General Issues

This Springer imprint is published by the registered company Springer Nature Switzerland AG
The registered company address is: Gewerbestrasse 11, 6330 Cham, Switzerland

Preface

The 25th International Conference on Neural Information Processing (ICONIP 2018), the annual conference of the Asia Pacific Neural Network Society (APNNS), was held in Siem Reap, Cambodia, during December 13–16, 2018. The ICONIP conference series started in 1994 in Seoul, which has now become a well-established and high-quality conference on neural networks around the world. Siem Reap is a gateway to Angkor Wat, which is one of the most important archaeological sites in Southeast Asia, the largest religious monument in the world. All participants of ICONIP 2018 had a technically rewarding experience as well as a memorable stay in this great city.

In recent years, the neural network has been significantly advanced with the great developments in neuroscience, computer science, cognitive science, and engineering. Many novel neural information processing techniques have been proposed as the solutions to complex, networked, and information-rich intelligent systems. To disseminate new findings, ICONIP 2018 provided a high-level international forum for scientists, engineers, and educators to present the state of the art of research and applications in all fields regarding neural networks.

With the growing popularity of neural networks in recent years, we have witnessed an increase in the number of submissions and in the quality of submissions. ICONIP 2018 received 575 submissions from 51 countries and regions across six continents. Based on a rigorous peer-review process, where each submission was reviewed by at least three experts, a total of 401 high-quality papers were selected for publication in the prestigious Springer series of *Lecture Notes in Computer Science*. The selected papers cover a wide range of subjects that address the emerging topics of theoretical research, empirical studies, and applications of neural information processing techniques across different domains.

In addition to the contributed papers, the ICONIP 2018 technical program also featured three plenary talks and two invited talks delivered by world-renowned scholars: Prof. Masashi Sugiyama (University of Tokyo and RIKEN Center for Advanced Intelligence Project), Prof. Marios M. Polycarpou (University of Cyprus), Prof. Qing-Long Han (Swinburne University of Technology), Prof. Cesare Alippi (Polytechnic of Milan), and Nikola K. Kasabov (Auckland University of Technology).

We would like to extend our sincere gratitude to all members of the ICONIP 2018 Advisory Committee for their support, the APNNS Governing Board for their guidance, the International Neural Network Society and Japanese Neural Network Society for their technical co-sponsorship, and all members of the Organizing Committee for all their great effort and time in organizing such an event. We would also like to take this opportunity to thank all the Technical Program Committee members and reviewers for their professional reviews that guaranteed the high quality of the conference proceedings. Furthermore, we would like to thank the publisher, Springer, for their sponsorship and cooperation in publishing the conference proceedings in seven volumes of *Lecture Notes in Computer Science*. Finally, we would like to thank all the speakers, authors, reviewers, volunteers, and participants for their contribution and support in making ICONIP 2018 a successful event.

October 2018

Jun Wang
Long Cheng
Andrew Chi Sing Leung
Seiichi Ozawa

ICONIP 2018 Organization

General Chair

Jun Wang
City University of Hong Kong, Hong Kong SAR, China

Advisory Chairs

Akira Hirose
University of Tokyo, Tokyo, Japan

Soo-Young Lee
Korea Advanced Institute of Science and Technology, South Korea

Derong Liu
Institute of Automation, Chinese Academy of Sciences, China

Nikhil R. Pal
Indian Statistics Institute, India

Program Chairs

Long Cheng
Institute of Automation, Chinese Academy of Sciences, China

Andrew C. S. Leung
City University of Hong Kong, Hong Kong SAR, China

Seiichi Ozawa
Kobe University, Japan

Special Sessions Chairs

Shukai Duan
Southwest University, China

Kazushi Ikeda
Nara Institute of Science and Technology, Japan

Qinglai Wei
Institute of Automation, Chinese Academy of Sciences, China

Hiroshi Yamakawa
Dwango Co. Ltd., Japan

Zhihui Zhan
South China University of Technology, China

Tutorial Chairs

Hiroaki Gomi
NTT Communication Science Laboratories, Japan

Takashi Morie
Kyushu Institute of Technology, Japan

Kay Chen Tan
City University of Hong Kong, Hong Kong SAR, China

Dongbin Zhao
Institute of Automation, Chinese Academy of Sciences, China

Publicity Chairs

Zeng-Guang Hou   Institute of Automation, Chinese Academy of Sciences,
                                     China
Tingwen Huang   Texas A&M University at Qatar, Qatar
Chia-Feng Juang   National Chung-Hsing University, Taiwan
Tomohiro Shibata   Kyushu Institute of Technology, Japan

Publication Chairs

Xinyi Le   Shanghai Jiao Tong University, China
Sitian Qin   Harbin Institute of Technology Weihai, China
Zheng Yan   University Technology Sydney, Australia
Shaofu Yang   Southeast University, China

Registration Chairs

Shenshen Gu   Shanghai University, China
Qingshan Liu   Southeast University, China
Ka Chun Wong   City University of Hong Kong,
                                     Hong Kong SAR, China

Conference Secretariat

Ying Qu   Dalian University of Technology, China

Program Committee

Hussein Abbass   University of New South Wales at Canberra, Australia
Choon Ki Ahn   Korea University, South Korea
Igor Aizenberg   Texas A&M University at Texarkana, USA
Shotaro Akaho   National Institute of Advanced Industrial Science
                                     and Technology, Japan
Abdulrazak Alhababi   UNIMAS, Malaysia
Cecilio Angulo   Universitat Politècnica de Catalunya, Spain
Sabri Arik   Istanbul University, Turkey
Mubasher Baig   National University of Computer and Emerging
                                     Sciences Lahore, India
Sang-Woo Ban   Dongguk University, South Korea
Tao Ban   National Institute of Information and Communications
                                     Technology, Japan
Boris Bačić   Auckland University of Technology, New Zealand
Xu Bin   Northwestern Polytechnical University, China
David Bong   Universiti Malaysia Sarawak, Malaysia
Salim Bouzerdoum   University of Wollongong, Australia
Ivo Bukovsky   Czech Technical University, Czech Republic

Ke-Cai Cao	Nanjing University of Posts and Telecommunications, China
Elisa Capecci	Auckland University of Technology, New Zealand
Rapeeporn Chamchong	Mahasarakham University, Thailand
Jonathan Chan	King Mongkut's University of Technology Thonburi, Thailand
Rosa Chan	City University of Hong Kong, Hong Kong SAR, China
Guoqing Chao	East China Normal University, China
He Chen	Nankai University, China
Mou Chen	Nanjing University of Aeronautics and Astronautics, China
Qiong Chen	South China University of Technology, China
Wei-Neng Chen	Sun Yat-Sen University, China
Xiaofeng Chen	Chongqing Jiaotong University, China
Ziran Chen	Bohai University, China
Jian Cheng	Chinese Academy of Sciences, China
Long Cheng	Chinese Academy of Sciences, China
Wu Chengwei	Bohai University, China
Zheru Chi	The Hong Kong Polytechnic University, SAR China
Sung-Bae Cho	Yonsei University, South Korea
Heeyoul Choi	Handong Global University, South Korea
Hyunsoek Choi	Kyungpook National University, South Korea
Supannada Chotipant	King Mongkut's Institute of Technology Ladkrabang, Thailand
Fengyu Cong	Dalian University of Technology, China
Jose Alfredo Ferreira Costa	Federal University of Rio Grande do Norte, Brazil
Ruxandra Liana Costea	Polytechnic University of Bucharest, Romania
Jean-Francois Couchot	University of Franche-Comté, France
Raphaël Couturier	University of Bourgogne Franche-Comté, France
Jisheng Dai	Jiangsu University, China
Justin Dauwels	Massachusetts Institute of Technology, USA
Dehua Zhang	Chinese Academy of Sciences, China
Mingcong Deng	Tokyo University of Agriculture and Technology, Japan
Zhaohong Deng	Jiangnan University, China
Jing Dong	Chinese Academy of Sciences, China
Qiulei Dong	Chinese Academy of Sciences, China
Kenji Doya	Okinawa Institute of Science and Technology, Japan
El-Sayed El-Alfy	King Fahd University of Petroleum and Minerals, Saudi Arabia
Mark Elshaw	Nottingham Trent International College, UK
Peter Erdi	Kalamazoo College, USA
Josafath Israel Espinosa Ramos	Auckland University of Technology, New Zealand
Issam Falih	Paris 13 University, France

Bo Fan	Zhejiang University, China
Yunsheng Fan	Dalian Maritime University, China
Hao Fang	Beijing Institute of Technology, China
Jinchao Feng	Beijing University of Technology, China
Francesco Ferracuti	Università Politecnica delle Marche, Italy
Chun Che Fung	Murdoch University, Australia
Wai-Keung Fung	Robert Gordon University, UK
Tetsuo Furukawa	Kyushu Institute of Technology, Japan
Hao Gao	Nanjing University of Posts and Telecommunications, China
Yabin Gao	Harbin Institute of Technology, China
Yongsheng Gao	Griffith University, Australia
Tom Gedeon	Australian National University, Australia
Ong Sing Goh	Universiti Teknikal Malaysia Melaka, Malaysia
Iqbal Gondal	Federation University Australia, Australia
Yue-Jiao Gong	Sun Yat-sen University, China
Shenshen Gu	Shanghai University, China
Chengan Guo	Dalian University of Technology, China
Ping Guo	Beijing Normal University, China
Shanqing Guo	Shandong University, China
Xiang-Gui Guo	University of Science and Technology Beijing, China
Zhishan Guo	University of Central Florida, USA
Christophe Guyeux	University of Franche-Comte, France
Masafumi Hagiwara	Keio University, Japan
Saman Halgamuge	The University of Melbourne, Australia
Tomoki Hamagami	Yokohama National University, Japan
Cheol Han	Korea University at Sejong, South Korea
Min Han	Dalian University of Technology, China
Takako Hashimoto	Chiba University of Commerce, Japan
Toshiharu Hatanaka	Osaka University, Japan
Wei He	University of Science and Technology Beijing, China
Xing He	Southwest University, China
Xiuyu He	University of Science and Technology Beijing, China
Akira Hirose	The University of Tokyo, Japan
Daniel Ho	City University of Hong Kong, Hong Kong SAR, China
Katsuhiro Honda	Osaka Prefecture University, Japan
Hongyi Li	Bohai University, China
Kazuhiro Hotta	Meijo University, Japan
Jin Hu	Chongqing Jiaotong University, China
Jinglu Hu	Waseda University, Japan
Xiaofang Hu	Southwest University, China
Xiaolin Hu	Tsinghua University, China
He Huang	Soochow University, China
Kaizhu Huang	Xi'an Jiaotong-Liverpool University, China
Long-Ting Huang	Wuhan University of Technology, China

Panfeng Huang	Northwestern Polytechnical University, China
Tingwen Huang	Texas A&M University, USA
Hitoshi Iima	Kyoto Institute of Technology, Japan
Kazushi Ikeda	Nara Institute of Science and Technology, Japan
Hayashi Isao	Kansai University, Japan
Teijiro Isokawa	University of Hyogo, Japan
Piyasak Jeatrakul	Mae Fah Luang University, Thailand
Jin-Tsong Jeng	National Formosa University, Taiwan
Sungmoon Jeong	Kyungpook National University Hospital, South Korea
Danchi Jiang	University of Tasmania, Australia
Min Jiang	Xiamen University, China
Yizhang Jiang	Jiangnan University, China
Xuguo Jiao	Zhejiang University, China
Keisuke Kameyama	University of Tsukuba, Japan
Shunshoku Kanae	Junshin Gakuen University, Japan
Hamid Reza Karimi	Politecnico di Milano, Italy
Nikola Kasabov	Auckland University of Technology, New Zealand
Abbas Khosravi	Deakin University, Australia
Rhee Man Kil	Sungkyunkwan University, South Korea
Daeeun Kim	Yonsei University, South Korea
Sangwook Kim	Kobe University, Japan
Lai Kin	Tunku Abdul Rahman University, Malaysia
Irwin King	The Chinese University of Hong Kong, Hong Kong SAR, China
Yasuharu Koike	Tokyo Institute of Technology, Japan
Ven Jyn Kok	National University of Malaysia, Malaysia
Ghosh Kuntal	Indian Statistical Institute, India
Shuichi Kurogi	Kyushu Institute of Technology, Japan
Susumu Kuroyanagi	Nagoya Institute of Technology, Japan
James Kwok	The Hong Kong University of Science and Technology, SAR China
Edmund Lai	Auckland University of Technology, New Zealand
Kittichai Lavangnananda	King Mongkut's University of Technology Thonburi, Thailand
Xinyi Le	Shanghai Jiao Tong University, China
Minho Lee	Kyungpook National University, South Korea
Nung Kion Lee	University Malaysia Sarawak, Malaysia
Andrew C. S. Leung	City University of Hong Kong, Hong Kong SAR, China
Baoquan Li	Tianjin Polytechnic University, China
Chengdong Li	Shandong Jianzhu University, China
Chuandong Li	Southwest University, China
Dazi Li	Beijing University of Chemical Technology, China
Li Li	Tsinghua University, China
Shengquan Li	Yangzhou University, China

Ya Li	Institute of Automation, Chinese Academy of Sciences, China
Yanan Li	University of Sussex, UK
Yongming Li	Liaoning University of Technology, China
Yuankai Li	University of Science and Technology of China, China
Jie Lian	Dalian University of Technology, China
Hualou Liang	Drexel University, USA
Jinling Liang	Southeast University, China
Xiao Liang	Nankai University, China
Alan Wee-Chung Liew	Griffith University, Australia
Honghai Liu	University of Portsmouth, UK
Huaping Liu	Tsinghua University, China
Huawen Liu	University of Texas at San Antonio, USA
Jing Liu	Chinese Academy of Sciences, China
Ju Liu	Shandong University, China
Qingshan Liu	Huazhong University of Science and Technology, China
Weifeng Liu	China University of Petroleum, China
Weiqiang Liu	Nanjing University of Aeronautics and Astronautics, China
Dome Lohpetch	King Mongkut's University of Technology North Bangoko, Thailand
Hongtao Lu	Shanghai Jiao Tong University, China
Wenlian Lu	Fudan University, China
Yao Lu	Beijing Institute of Technology, China
Jinwen Ma	Peking University, China
Qianli Ma	South China University of Technology, China
Sanparith Marukatat	Thailand's National Electronics and Computer Technology Center, Thailand
Tomasz Maszczyk	Nanyang Technological University, Singapore
Basarab Matei	LIPN Paris Nord University, France
Takashi Matsubara	Kobe University, Japan
Nobuyuki Matsui	University of Hyogo, Japan
P. Meesad	King Mongkut's University of Technology North Bangkok, Thailand
Gaofeng Meng	Chinese Academy of Sciences, China
Daisuke Miyamoto	University of Tokyo, Japan
Kazuteru Miyazaki	National Institution for Academic Degrees and Quality Enhancement of Higher Education, Japan
Seiji Miyoshi	Kansai University, Japan
J. Manuel Moreno	Universitat Politècnica de Catalunya, Spain
Naoki Mori	Osaka Prefecture University, Japan
Yoshitaka Morimura	Kyoto University, Japan
Chaoxu Mu	Tianjin University, China
Kazuyuki Murase	University of Fukui, Japan
Jun Nishii	Yamaguchi University, Japan

Haruhiko Nishimura	University of Hyogo, Japan
Grozavu Nistor	Paris 13 University, France
Yamaguchi Nobuhiko	Saga University, Japan
Stavros Ntalampiras	University of Milan, Italy
Takashi Omori	Tamagawa University, Japan
Toshiaki Omori	Kobe University, Japan
Seiichi Ozawa	Kobe University, Japan
Yingnan Pan	Northeastern University, China
Yunpeng Pan	JD Research Labs, China
Lie Meng Pang	Universiti Malaysia Sarawak, Malaysia
Shaoning Pang	Unitec Institute of Technology, New Zealand
Hyeyoung Park	Kyungpook National University, South Korea
Hyung-Min Park	Sogang University, South Korea
Seong-Bae Park	Kyungpook National University, South Korea
Kitsuchart Pasupa	King Mongkut's Institute of Technology Ladkrabang, Thailand
Yong Peng	Hangzhou Dianzi University, China
Somnuk Phon-Amnuaisuk	Universiti Teknologi Brunei, Brunei
Lukas Pichl	International Christian University, Japan
Geong Sen Poh	National University of Singapore, Singapore
Mahardhika Pratama	Nanyang Technological University, Singapore
Emanuele Principi	Università Politecnica elle Marche, Italy
Dianwei Qian	North China Electric Power University, China
Jiahu Qin	University of Science and Technology of China, China
Sitian Qin	Harbin Institute of Technology at Weihai, China
Mallipeddi Rammohan	Nanyang Technological University, Singapore
Yazhou Ren	University of Science and Technology of China, China
Ko Sakai	University of Tsukuba, Japan
Shunji Satoh	The University of Electro-Communications, Japan
Gerald Schaefer	Loughborough University, UK
Sachin Sen	Unitec Institute of Technology, New Zealand
Hamid Sharifzadeh	Unitec Institute of Technology, New Zealand
Nabin Sharma	University of Technology Sydney, Australia
Yin Sheng	Huazhong University of Science and Technology, China
Jin Shi	Nanjing University, China
Yuhui Shi	Southern University of Science and Technology, China
Hayaru Shouno	The University of Electro-Communications, Japan
Ferdous Sohel	Murdoch University, Australia
Jungsuk Song	Korea Institute of Science and Technology Information, South Korea
Andreas Stafylopatis	National Technical University of Athens, Greece
Jérémie Sublime	ISEP, France
Ponnuthurai Suganthan	Nanyang Technological University, Singapore
Fuchun Sun	Tsinghua University, China
Ning Sun	Nankai University, China

Norikazu Takahashi	Okayama University, Japan
Ken Takiyama	Tokyo University of Agriculture and Technology, Japan
Tomoya Tamei	Kobe University, Japan
Hakaru Tamukoh	Kyushu Institute of Technology, Japan
Choo Jun Tan	Wawasan Open University, Malaysia
Shing Chiang Tan	Multimedia University, Malaysia
Ying Tan	Peking University, China
Gouhei Tanaka	The University of Tokyo, Japan
Ke Tang	Southern University of Science and Technology, China
Xiao-Yu Tang	Zhejiang University, China
Yang Tang	East China University of Science and Technology, China
Qing Tao	Chinese Academy of Sciences, China
Katsumi Tateno	Kyushu Institute of Technology, Japan
Keiji Tatsumi	Osaka University, Japan
Kai Meng Tay	Universiti Malaysia Sarawak, Malaysia
Chee Siong Teh	Universiti Malaysia Sarawak, Malaysia
Andrew Teoh	Yonsei University, South Korea
Arit Thammano	King Mongkut's Institute of Technology Ladkrabang, Thailand
Christos Tjortjis	International Hellenic University, Greece
Shibata Tomohiro	Kyushu Institute of Technology, Japan
Seiki Ubukata	Osaka Prefecture University, Japan
Eiji Uchino	Yamaguchi University, Japan
Wataru Uemura	Ryukoku University, Japan
Michel Verleysen	Universite catholique de Louvain, Belgium
Brijesh Verma	Central Queensland University, Australia
Hiroaki Wagatsuma	Kyushu Institute of Technology, Japan
Nobuhiko Wagatsuma	Tokyo Denki University, Japan
Feng Wan	University of Macau, SAR China
Bin Wang	University of Jinan, China
Dianhui Wang	La Trobe University, Australia
Jing Wang	Beijing University of Chemical Technology, China
Jun-Wei Wang	University of Science and Technology Beijing, China
Junmin Wang	Beijing Institute of Technology, China
Lei Wang	Beihang University, China
Lidan Wang	Southwest University, China
Lipo Wang	Nanyang Technological University, Singapore
Qiu-Feng Wang	Xi'an Jiaotong-Liverpool University, China
Sheng Wang	Henan University, China
Bunthit Watanapa	King Mongkut's University of Technology, Thailand
Saowaluk Watanapa	Thammasat University, Thailand
Qinglai Wei	Chinese Academy of Sciences, China
Wei Wei	Beijing Technology and Business University, China
Yantao Wei	Central China Normal University, China

Guanghui Wen	Southeast University, China
Zhengqi Wen	Chinese Academy of Sciences, China
Hau San Wong	City University of Hong Kong, Hong Kong SAR, China
Kevin Wong	Murdoch University, Australia
P. K. Wong	University of Macau, SAR China
Kuntpong Woraratpanya	King Mongkut's Institute of Technology Chaokuntaharn Ladkrabang, Thailand
Dongrui Wu	Huazhong University of Science and Technology, China
Si Wu	Beijing Normal University, China
Si Wu	South China University of Technology, China
Zhengguang Wu	Zhejiang University, China
Tao Xiang	Chongqing University, China
Chao Xu	Zhejiang University, China
Zenglin Xu	University of Science and Technology of China, China
Zhaowen Xu	Zhejiang University, China
Tetsuya Yagi	Osaka University, Japan
Toshiyuki Yamane	IBM, Japan
Koichiro Yamauchi	Chubu University, Japan
Xiaohui Yan	Nanjing University of Aeronautics and Astronautics, China
Zheng Yan	University of Technology Sydney, Australia
Jinfu Yang	Beijing University of Technology, China
Jun Yang	Southeast University, China
Minghao Yang	Chinese Academy of Sciences, China
Qinmin Yang	Zhejiang University, China
Shaofu Yang	Southeast University, China
Xiong Yang	Tianjin University, China
Yang Yang	Nanjing University of Posts and Telecommunications, China
Yin Yang	Hamad Bin Khalifa University, Qatar
Yiyu Yao	University of Regina, Canada
Jianqiang Yi	Chinese Academy of Sciences, China
Chengpu Yu	Beijing Institute of Technology, China
Wen Yu	CINVESTAV, Mexico
Wenwu Yu	Southeast University, China
Zhaoyuan Yu	Nanjing Normal University, China
Xiaodong Yue	Shanghai University, China
Dan Zhang	Zhejiang University, China
Jie Zhang	Newcastle University, UK
Liqing Zhang	Shanghai Jiao Tong University, China
Nian Zhang	University of the District of Columbia, USA
Tengfei Zhang	Nanjing University of Posts and Telecommunications, China
Tianzhu Zhang	Chinese Academy of Sciences, China

Ying Zhang	Shandong University, China
Zhao Zhang	Soochow University, China
Zhaoxiang Zhang	Chinese Academy of Sciences, China
Dongbin Zhao	Chinese Academy of Sciences, China
Qiangfu Zhao	University of Aizu, Japan
Zhijia Zhao	Guangzhou University, China
Jinghui Zhong	South China University of Technology, China
Qi Zhou	University of Portsmouth, UK
Xiaojun Zhou	Central South University, China
Yingjiang Zhou	Nanjing University of Posts and Telecommunications, China
Haijiang Zhu	Beijing University of Chemical Technology, China
Hu Zhu	Nanjing University of Posts and Telecommunications, China
Lei Zhu	Unitec Institute of Technology, New Zealand
Pengefei Zhu	Tianjin University, China
Yue Zhu	Nanjing University, China
Zongyu Zuo	Beihang University, China

Contents – Part I

Deep Neural Networks

Adaptive Deep Dictionary Learning for MRI Reconstruction 3
  D. John Lewis, Vanika Singhal, and Angshul Majumdar

Deep-PUMR: Deep Positive and Unlabeled Learning
with Manifold Regularization 12
  Xingyu Chen, Fanghui Liu, Enmei Tu, Longbing Cao, and Jie Yang

Viewpoint Estimation for Workpieces with Deep Transfer
Learning from Cold to Hot. 21
  Changsheng Lu, Haotian Wang, Chaochen Gu, Kaijie Wu,
  and Xinping Guan

Co-consistent Regularization with Discriminative Feature
for Zero-Shot Learning 33
  Yanling Tian, Weitong Zhang, Qieshi Zhang, Jun Cheng, Pengyi Hao,
  and Gang Lu

Hybrid Networks: Improving Deep Learning Networks via Integrating
Two Views of Images 46
  Sunny Verma, Wei Liu, Chen Wang, and Liming Zhu

On a Fitting of a Heaviside Function by Deep ReLU Neural Networks 59
  Katsuyuki Hagiwara

Deep Tag Recommendation Based on Discrete Tensor Factorization 70
  Wenwen Ye, Zheng Qin, and Xu Li

Efficient Integer Vector Homomorphic Encryption Using Deep Learning
for Neural Networks 83
  Tianying Xie and Yantao Li

DeepSIC: Deep Semantic Image Compression 96
  Sihui Luo, Yezhou Yang, Yanling Yin, Chengchao Shen,
  Ya Zhao, and Mingli Song

Multi-stage Gradient Compression: Overcoming the Communication
Bottleneck in Distributed Deep Learning 107
  Qu Lu, Wantao Liu, Jizhong Han, and Jinrong Guo

Teach to Hash: A Deep Supervised Hashing Framework
with Data Selection 120
    Xiang Li, Chao Ma, Jie Yang, and Yu Qiao

Multi-view Deep Gaussian Processes 130
    Shiliang Sun and Qiuyang Liu

Deep Collaborative Filtering Combined with High-Level Feature
Generation on Latent Factor Model 140
    Xu Li, Xu Chen, and Zheng Qin

Data Imputation of Wind Turbine Using Generative Adversarial
Nets with Deep Learning Models 152
    Fuming Qu, Jinhai Liu, Xiaowei Hong, and Yu Zhang

A Deep Ensemble Network for Compressed Sensing MRI 162
    Huafeng Wu, Yawen Wu, Liyan Sun, Congbo Cai, Yue Huang,
    and Xinghao Ding

Deep Imitation Learning: The Impact of Depth on Policy Performance 172
    Parham M. Kebria, Abbas Khosravi, Syed Moshfeq Salaken,
    Ibrahim Hossain, H. M. Dipu Kabir, Afsaneh Koohestani,
    Roohallah Alizadehsani, and Saeid Nahavandi

Improving Deep Neural Network Performance with Kernelized
Min-Max Objective 182
    Kai Yao, Kaizhu Huang, Rui Zhang, and Amir Hussain

Understanding Deep Neural Network by Filter Sensitive
Area Generation Network 192
    Yang Qian, Hong Qiao, and Jing Xu

Accelerating Deep Q Network by Weighting Experiences 204
    Kazuhiro Murakami, Koichi Moriyama, Atsuko Mutoh,
    Tohgoroh Matsui, and Nobuhiro Inuzuka

Exploring Deep Learning Architectures Coupled with CRF Based
Prediction for Slot-Filling 214
    Tulika Saha, Sriparna Saha, and Pushpak Bhattacharyya

Domain Adaptation via Identical Distribution Across Models and Tasks 226
    Xuhong Wei, Yefei Chen, and Jianbo Su

A Pointer Network Based Deep Learning Algorithm
for the Max-Cut Problem 238
    Shenshen Gu and Yue Yang

Convolution Neural Networks

Multi-stream with Deep Convolutional Neural Networks for Human Action
Recognition in Videos   251
    Xiao Liu and Xudong Yang

Use 3D Convolutional Neural Network to Inspect Solder Ball Defects   263
    Bing-Jhang Lin, Ting-Chen Tsan, Tzu-Chia Tung, You-Hsien Lee,
    and Chiou-Shann Fuh

User-Invariant Facial Animation with Convolutional Neural Network   275
    Shuiquan Wang, Zhengxin Cheng, Liang Chang, Xuejun Qiao,
    and Fuqing Duan

An Approach for Feature Extraction and Diagnosis of Motor Rotor Bearing
Based on Convolution Neural Network   284
    Hao Wang, Dongsheng Yang, Yongheng Pang, Ting Li, and Bo Hu

Gated Convolutional Networks for Commonsense
Machine Comprehension   297
    Wuya Chen, Xiaojun Quan, and Chengbo Chen

Cursive Scene Text Analysis by Deep Convolutional Linear Pyramids   307
    Saad Bin Ahmed, Saeeda Naz, Muhammad Imran Razzak,
    and Rubiyah Yusof

Human Action Recognition with 3D Convolution Skip-Connections
and RNNs   319
    Jiarong Song, Zhong Yang, Qiuyan Zhang, Ting Fang, Guoxiong Hu,
    Jiaming Han, and Cong Chen

Convolutional Neural Network with Discriminant Criterion
for Input of Each Neuron in Output Layer   332
    Hidenori Ide and Takio Kurita

Proposal of Complex-Valued Convolutional Neural Networks for Similar
Land-Shape Discovery in Interferometric Synthetic Aperture Radar   340
    Yuki Sunaga, Ryo Natsuaki, and Akira Hirose

Feature Learning and Transfer Performance Prediction
for Video Reinforcement Learning Tasks via a Siamese
Convolutional Neural Network   350
    Jinhua Song, Yang Gao, and Hao Wang

Structured Sequence Modeling with Graph Convolutional
Recurrent Networks   362
    Youngjoo Seo, Michaël Defferrard, Pierre Vandergheynst,
    and Xavier Bresson

Part-Level Sketch Segmentation and Labeling Using Dual-CNN   374
 Xianyi Zhu, Yi Xiao, and Yan Zheng

RE-CNN: A Robust Convolutional Neural Networks
for Image Recognition   385
 Zhe Wang, Wenhuan Lu, Yuqing He, Naixue Xiong,
 and Jianguo Wei

MusicCNNs: A New Benchmark on Content-Based
Music Recommendation   394
 Guoqiang Zhong, Haizhen Wang, and Wencong Jiao

Remote Sensing Image Segmentation by Combining Feature Enhanced
with Fully Convolutional Network   406
 Ruiguo Yu, Xuzhou Fu, Han Jiang, Chenhan Wang, Xuewei Li,
 Mankun Zhao, Xiang Ying, and Hongqian Shen

A New LSTM Network Model Combining TextCNN   416
 Xiao Sun, Xiaohu Ma, Zhiwen Ni, and Lina Bian

Self-inhibition Residual Convolutional Networks for Chinese
Sentence Classification   425
 Mengting Xiong, Ruixuan Li, Yuhua Li, and Qi Yang

Recurrent Neural Networks

A Hybrid 2D and 3D Convolution Based Recurrent Network
for Video-Based Person Re-identification   439
 Li Cheng, Xiao-Yuan Jing, Xiaoke Zhu, Fumin Qi, Fei Ma,
 Xiaodong Jia, Liang Yang, and Chunhe Wang

Improving Recurrent Neural Networks with Predictive Propagation
for Sequence Labelling   452
 Son N. Tran, Qing Zhang, Anthony Nguyen, Xuan-Son Vu, and Son Ngo

Design of Synthesizing Multi-valued High-Capacity Auto-associative
Memories Based on Complex-Valued Networks   463
 Chunlin Sha and Hongyong Zhao

Analysis on the Occurrence of Tropical Cyclone in the South Pacific
Region Using Recurrent Neural Network with LSTM   476
 Adarsh Karan Sharma, Vishal Prasad, Roneel Kumar,
 and Anuraganand Sharma

Combining User-Based and Session-Based Recommendations
with Recurrent Neural Networks   487
 Tu Minh Phuong, Tran Cong Thanh, and Ngo Xuan Bach

EMD-Based Recurrent Neural Network with Adaptive Regrouping for Port
Cargo Throughput Prediction 499
   Yan Li, Ryan Wen Liu, Quandang Ma, and Jingxian Liu

Enhancing the Recurrent Neural Networks with Positional Gates
for Sentence Representation 511
   Yang Song, Wenxin Hu, Qin Chen, Qinmin Hu, and Liang He

Spiking Neural Networks

A Visual Recognition Model Based on Hierarchical Feature Extraction
and Multi-layer SNN. 525
   Xiaoliang Xu, Wensi Lu, Qiming Fang, and Yixing Xia

Skewed and Long-Tailed Distributions of Spiking Activity in Coupled
Network Modules with Log-Normal Synaptic Weight Distribution 535
   Sou Nobukawa, Haruhiko Nishimura, and Teruya Yamanishi

Efficient Multi-spike Learning with Tempotron-Like LTP
and PSD-Like LTD. 545
   Qiang Yu, Longbiao Wang, and Jianwu Dang

A Ladder-Type Digital Spiking Neural Network 555
   Hiroaki Uchida and Toshimichi Saito

The Effects of Feedback Signals Mediated by NMDA-Type Synapses
for Modulating Border-Ownership Selective Neurons in Visual Cortex 563
   Nobuhiko Wagatsuma and Hirotoshi Konno

Modelling and Analysis of Temporal Gene Expression Data Using
Spiking Neural Networks. 571
   Durgesh Nandini, Elisa Capecci, Lucien Koefoed, Ibai Laña,
   Gautam Kishore Shahi, and Nikola Kasabov

A Gesture Recognition Method Based on Spiking Neural Networks
for Cognition Development 582
   Dong Niu, Dengju Li, Rui Yan, and Huajin Tang

Delayed Feedback Reservoir Computing with VCSEL. 594
   Jean Benoit Héroux, Naoki Kanazawa, and Daiju Nakano

Modeling the Respiratory Central Pattern Generator
with Resonate-and-Fire Izhikevich-Neurons 603
   Pavel Tolmachev, Rishi R. Dhingra, Michael Pauley,
   Mathias Dutschmann, and Jonathan H. Manton

Proposal of Carrier-Wave Reservoir Computing 616
   Akira Hirose, Gouhei Tanaka, Seiji Takeda, Toshiyuki Yamane,
   Hidetoshi Numata, Naoki Kanazawa, Jean Benoit Heroux,
   Daiju Nakano, and Ryosho Nakane

Spiking Neural Networks for Cancer Gene Expression Time Series
Modelling and Analysis 625
   Jack Dray, Elisa Capecci, and Nikola Kasabov

Dimensionality Reduction by Reservoir Computing and Its Application
to IoT Edge Computing 635
   Toshiyuki Yamane, Hidetoshi Numata, Jean Benoit Héroux,
   Naoki Kanazawa, Seiji Takeda, Gouhei Tanaka, Ryosho Nakane,
   Akira Hirose, and Daiju Nakano

Author Index 645

Deep Neural Networks

Adaptive Deep Dictionary Learning for MRI Reconstruction

D. John Lewis, Vanika Singhal$^{(\boxtimes)}$, and Angshul Majumdar

Indraprastha Institute of Information Technology, Delhi, India
{john16095, vanikas, angshul}@iiitd.ac.in

Abstract. This work addresses the well known problem of reconstructing magnetic resonance images from their partially samples K-space. Compressed sensing (CS) based techniques have been used rampantly for the said problem. Later studies, instead of employing a fixed basis (like DCT, wavelet etc. as used in CS), learnt the basis adaptively from the image itself. Such studies, loosely dubbed as dictionary learning (DL) showed marked improvement over CS. This work proposes deep dictionary learning based inversion. Instead of learning a single level of basis, we learn multiple levels adaptively from the image, while reconstructing it. The results show marked improvement over all previously known techniques.

Keywords: Dictionary learning · Deep learning · Reconstruction

1 Introduction

Magnetic resonance imaging (MRI) is a relatively slow modality (compared to CT or USG). Over the past two decades a lot of research has been carried out to accelerate the data acquisition for MRI. The late 90's and early half of 2000's concentrated on hardware based acceleration techniques. These efforts reached their limits eventually. In the late 2000's, the advent of compressed sensing (CS) ushered in a new era of software (signal processing) based acceleration.

The basic idea in such software based acceleration techniques is to sub-sample the K-space in a physically plausible manner and reconstruct the image from the sub-sampled K-space. CS based techniques exploit the sparsity of the image in a known domain (e.g. wavelet, DCT, finite difference etc.) in order to reconstruct it. Later studies, showed that instead of assuming the sparsity basis, much better results can be obtained if the basis is learnt adaptively from the image during reconstruction.

In recent years, deep learning based techniques have been used for reconstruction. They learn to reconstruct images from a separate set of training images and then assume that the learnt model generalizes to the unknown image. Autoencoders and convolutional neural networks have been used for the purpose.

In essence neither deep learning nor CS is adaptive techniques. The former learns the model (non-adaptively) and applies it to a new image; it does not adapt (learn) from the image it has to reconstruct. Therefore for the image being reconstructed, the model is 'fixed'. CS uses a fixed sparsifying basis, so there is nothing adaptive about it.

© Springer Nature Switzerland AG 2018
L. Cheng et al. (Eds.): ICONIP 2018, LNCS 11301, pp. 3–11, 2018.
https://doi.org/10.1007/978-3-030-04167-0_1

This is the first work that will incorporate deep learning in an adaptive fashion into the reconstruction framework. In dictionary learning, a single layer of basis is learnt. Here we will learn multiple layers of basis – thus making it deep (dictionary learning). These deep dictionaries will be learnt from the image being reconstructed; hence the learning will be adaptive.

2   Literature Review

As has been mentioned before, we are interested in accelerating MRI scan, and therefore need to sub-sample the K-space. The data acquisition model for partially sampled K-space scan can be expressed as follows:

$$y = RFx + \eta, \; \eta \sim N(0, \sigma^2) \tag{1}$$

Where y is the sub-sampled K-space, x is the underlying image to be reconstructed, F is the Fourier operator and R is the restriction operator for partial sampling. The noise η is supposed to be Normally distributed.

The measurement matrix is under-determined. Therefore the inverse problem (1) has infinitely many solutions. CS based techniques such as [1] assume that the signal is sparse in a transform domain (DCT, wavelet, curvelet, finite difference etc.). This transform domain sparsity is exploited in order to reconstruct it. The generic formulation can be given by

$$\min_x \|y - RFx\|_2^2 + \lambda \|\Psi x\|_1 \tag{2}$$

The l_2-norm data fidelity arises from the Gaussian nature of the noise. Here Ψ is the sparsity promoting transform – it can also be a combination of two like wavelet and finite difference [1].

In CS techniques the sparsity promoting transform is fixed. Better results can be obtained if the sparsity basis is learnt from the data. Dictionary learning based methods [2]. The formulation is given by,

$$\min_{x,D,Z} \|y - RFx\|_2^2 + \lambda \left(\sum_i \|P_i x - Dz_i\|_2^2 + \gamma \|z_i\|_1 \right) \tag{3}$$

Here $P_i x$ represents the ith patch of image x; this is represented by a learnt dictionary D and sparse coefficients z_i. This constitutes the dictionary learning term. The data consistency term remains the same from before. Only the measurements (y) and the operator A are known, the rest (image x, dictionary D and coefficients Z) are to be estimated.

Dictionary learning is a synthesis formulation. Its equivalent analysis formulation, known as transform learning, has also been successfully used for MRI reconstruction [3].

$$\min_{x,T,Z} \|y - Ax\|_2^2 + \lambda \left(\sum_i \|TP_ix - z_i\|_2^2 + \gamma \|z_i\|_1 \right) \tag{4}$$

The data consistency term remains the same as before. But in the transform learning term, notice that the learnt transform T operates on the patches P_ix to generate sparse coefficients Z.

In a concise fashion we have discussed the signal processing based techniques used in MRI reconstruction. For a more thorough review the reader can peruse [4]. In recent times, with the advent of deep learning, an alternate approach for reconstruction is being pursued. One approach is based on the autoencoder model.

An autoencoder is a self-supervised neural network where the input and the output are the same. A modified version of it called the denoising autoencoder takes noisy samples as inputs and clean samples at the output. It basically learns to denoise the input. MRI reconstruction has been addressed via the denoising autoencoder framework [5]. A noisy version of the image is obtained from the partial K-space samples via zero-filling ($\hat{x} = (RF)^H y$. This noisy version is input to the autoencoder and the clean version (x) is presented at the output. In the training stage, the autoencoder learns to remove the artifacts and noise from $\hat{x}$ to generate x. In the testing stage, when a new zero-filled image is presented at the input of the autoencoder, it maps it to a clean output.

A similar approach has been tried using the convolutional neural network (CNN) architecture [6]. The basic idea is to pass a corrupted version of a patch from the image through several layers of convolution and pooling. The output of the CNN is a clean version of the patch. Such an approach can only work on toy problems since it assumes that the compressed sensing projection operator can operate on a patch from an image – this is an unrealistic assumption.

3 Proposed Formulation

We are interested in solving the linear inverse problem (1). However for notational ease we replace $A = RF$, leading to: $y = Ax + n$. For the first time, we propose an adaptive deep learning based approach for solving such an inverse problem. It is based on the deep dictionary learning paradigm [7]. In the past it has been used for solving analysis problems in biometrics, remote sensing, biomedical signal analysis etc. But no one tried employing it for solving inverse problems.

In standard (shallow) dictionary learning, each patch (P_ix) of the image is expressed as a product of the learnt basis (D) and sparse coefficients (z_i), i.e. $P_ix = Dz_i \, \forall i$. The entire inversion process consists of ensuring data consistency $\|y - Ax\|_2^2$ and a dictionary learning regularization term $\left(\sum_i \|P_ix - Dz_i\|_2^2 + \gamma \|z_i\|_1 \right)$. The combined formulation has already been mentioned (3).

In this work we extend from single level to multiple layers of (deep) dictionaries. Therefore instead of expressing each patch as a single layer of dictionary and coefficients, we express it in terms of deep dictionaries.

$$P_i x = D_1 \varphi(D_2 \varphi(D_3 z_i)) \, \forall i \tag{5}$$

We are showing the derivation for three layers, but the approach we follow is easily extendable to fewer or more layers.

As in shallow dictionary learning, we will have a data consistency term $\|y - Ax\|_2^2$ and a deep dictionary learning term $\left(\sum_i \|P_i x - D_1 \varphi(D_2 \varphi(D_3 z_i))\|_2^2 + \gamma\|z_i\|_1 \right)$[7]. Putting the two together, we have the final formulation

$$\min_{x, D_1, D_2, D_3, Z} \|y - Ax\|_2^2 + \lambda \left(\sum_i \|P_i x - D_1 \varphi(D_2 \varphi(D_3 z_i))\|_2^2 + \gamma\|z_i\|_1 \right) \tag{6}$$

Here Z is formed by stacking z_i's as columns.

It is worth mentioning here that all prior studies on deep dictionary learning followed a greedy approach while training. Unfortunately, we do not have that liberty here; we need to solve (6) jointly. Therefore, we resort to the variable splitting augmented Lagrangian approach [8]. We substitute: $z_{1,i} = \varphi(D_2 \varphi(D_3 z_i))$ and $z_{2,i} = \varphi(D_3 z_1)$. With these proxies, the augmented Lagrangian [9] formulation is

$$\min_{x, D_1, D_2, D_3, Z, Z_1, Z_2} \|y - Ax\|_2^2 + \lambda \left(\sum_i \|P_i x - D_1 z_{1,i}\|_2^2 + \gamma\|z_i\|_1 \right)$$
$$+ \mu \left(\|Z_1 - \varphi(D_2 Z_2)\|_F^2 + \|Z_2 - \varphi(D_3 Z)\|_F^2 \right) \tag{7}$$

Here Z_1 is formed by stacking the $z_{1,i}$'s as columns and Z_2 is formed by stacking the $z_{2,i}$'s as columns. For more layers, we could have continued the substitutions resulting in an augmented Lagrangian with more terms.

Using alternating direction method of multipliers [10], (7) can be segregated into the following sub-problems.

$$\text{P1}: \min_x \|y - Ax\|_2^2 + \lambda \sum_i \|P_i x - D_1 z_{1,i}\|_2^2 \tag{8}$$

$$\text{P2}: \min_{D_1} \sum_i \|P_i x - D_1 z_{1,i}\|_2^2 \equiv \min_{D_1} \|X - D_1 Z_1\|_F^2 \tag{9}$$

$$\text{P3}: \min_{D_2} \|Z_1 - \varphi(D_2 Z_2)\|_F^2 \equiv \min_{D_2} \|\varphi^{-1}(Z_1) - D_2 Z_2\|_F^2 \tag{10}$$

$$\text{P4}: \min_{D_3} \|Z_2 - \varphi(D_3 Z)\|_F^2 \equiv \min_{D_3} \|\varphi^{-1}(Z_2) - D_3 Z\|_F^2 \tag{11}$$

$$\text{P5}: \min_{Z_1} \|X - D_1 Z_1\|_2^2 + \mu\|Z_1 - \varphi(D_2 Z_2)\|_F^2 \tag{12}$$

Here X is formed by stacking the patches $P_i x$ as columns

$$P6: \min_{Z_2} \|Z_1 - \varphi(D_2 Z_2)\|_F^2 + \|Z_2 - \varphi(D_3 Z_1)\|_F^2 \|\varphi^{-1}(Z_1) - D_2 Z_2\|_F^2 + \|Z_2 - \varphi(D_3 Z_1)\|_F^2 \quad (13)$$

$$P7: \min_{Z} \mu \|Z_2 - \varphi(D_3 Z)\|_F^2 + \lambda\gamma\|Z\|_1 \equiv \|\varphi^{-1}(Z_2) - D_3 Z\|_F^2 + \frac{\lambda\gamma}{\mu}\|Z\|_1 \quad (14)$$

Since the activation functions are unitary (usually sigmoid or tanh) and operate element-wise, they are trivial to invert. Hence we can express them in the equivalent form as and when needed. We find that the sub-problems P1 to P6 are all simple linear least squares problems having a closed form solution. Only P7 is an l_1-norm minimization problem that needs to be solved iteratively. But well known existing algorithms like iterative soft thresholding [11] is there to solve it.

We stop the iterations when they reach a maximum number of iterations or when the objective function reaches a local minimum. Once the algorithm converges, the image is estimated; the other variables are discarded since they are of no use.

Note that in all the substitutions we have used the same multiplicative factor μ. This is because this factor controls the relative importance of each level. Since there is no reason to favor one level over the other, they are kept the same. Moreover, the inputs and the other levels are given equal importance, hence the value of is fixed at unity.

4 Experimental Evaluation

4.1 Dataset Description

All animal experimental procedures were carried out in compliance with the guidelines of the Canadian Council for Animal Care and were approved by the institutional Animal Care Committee. One female Sprague-Dawley rat was anaesthetized and perfused intracardially with phosphate buffered saline for 3 min followed by freshly hydrolysed paraformaldehyde (4%) in 0.1 M sodium phosphate buffer at pH 7.4. The 20 mm spinal cord centred at C5 level was then harvested and post-fixed in the same fixative. MRI experiments were carried out on a 7 T/30 cm bore animal MRI scanner (Bruker, Germany). Single slice multi-echo CPMG sequence was used to acquire fully sampled k-space data from the excised spinal cord sample using a 5 turn, 13 mm inner diameter solenoid coil with 256×256 matrix size, TE/TR = 6.738/1500 ms, 32 echoes, 2.56 cm field-of-view (FOV), 1 mm slice, number of averages (NA) = 8, and the excitation pulse phase cycled between 0° and 180°. This constitutes our In Vivo dataset.

For Ex Vivo experiments, a rectangular coil 22×19 mm was surgically implanted over the lumbar spine (T13/L1) of a female Sprague-Dawley rat as described previously. For MRI experiments, animal was anaesthetized with is fluorine (5% induction, 2% maintenance) mixed with medical air and positioned supine in a specially designed holder. Data was acquired using the same CPMG sequence but with slice thickness of 1.5 mm and in-plane resolution of 117 μm. The slice was positioned at T13/L1 level, and NA = 6.

Undersampled K-space data in the phase encode direction was generated for each set of data for different acceleration factors (2, 4 and 8 which corresponds to 128, 64 and 32 phase encoding lines respectively) from the fully sampled K-space. 33% of the read-out lines were placed around the centre, and the rest randomly distributed in the periphery up to the desired number of phase encoding steps for the prescribed acceleration factor.

4.2 Results

In past studies on dictionary learning (DL) [3] and transform learning (TL) [4] based MRI reconstruction, it has already been shown that they improve over CS based solutions by a large margin. In [6] it was shown that the deep learning based techniques are only comparable to CS (and sometimes worse) but cannot surpass them in reconstruction quality. Therefore, in this work we only compare with TL and DL.

For our proposed technique we need to estimate two parameters λ and γ and one hyper-parameter μ. The hyper-parameter is fixed to unity since we want to give equal importance to all the layers. The parameters were learnt on a separate MRI validation image (not from the datasets used here) for each sub-sampling ratio. The parameters were obtained via grid search. We used patches of size 16×16. The number of dictionary elements was halved in each subsequent level.

In Figs. 1 and 2, we show the results of our proposed approach vis-à-vis existing techniques for 8 times acceleration factor. They show a randomly chosen echo. These figures qualitatively show the superior reconstruction ability of our proposed technique. From the reconstructed images, one can clearly see that we have far less artifacts compared to the others. In fact the reconstructed image from our three layer architecture is visually the same as the ground-truth. This is also discernible from the difference (between ground-truth and reconstructed) images. From our proposed method, they are almost completely dark (especially for our three layer technique), but TL and DL shows visible artifacts.

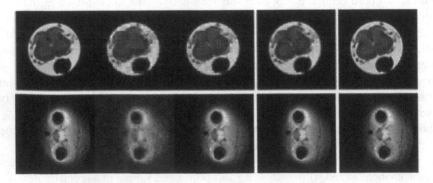

Fig. 1. Reconstructed images. Top – Ex-vivo, bottom – In-vivo. Left to right – original, TL, DL, proposed 2 layer and proposed 3 layer.

Fig. 2. Difference images. Top – Ex-vivo, bottom – In-vivo. Left to right – TL, DL, proposed 2 layer and proposed 3 layer.

In the following table we show the results for Ex Vivo and In Vivo imaging for all the acceleration factors. The standard measured of Signal to Noise Ratio (SNR) is used. We show results from our approach with two and three layers. What is shown here is the mean of the 32 epochs (Table 1).

Table 1. Reconstruction accuracy in SNR (dB)

Method	Ex Vivo imaging			In Vivo imaging		
	128 lines	64 lines	32 lines	128 lines	64 lines	32 lines
DL	24.7	20.6	16.9	23.9	20.2	15.6
TL	24.8	20.9	17.1	24.0	20.2	15.8
Proposed-2 layer	26.4	22.7	20.2	24.5	21.1	17.6
Proposed-3 layer	26.9	23.1	20.9	24.9	21.7	18.0

The results show that our proposed method excels over the state-of-the art by a considerable margin. We have not shown results for 4 layers; this is because the results deteriorate. It owes to the fact that with more layers, there are too many parameters to learn and with limited training data, deeper models overfit which leads to poor results.

5 Conclusion

This is the first work that introduces adaptive deep dictionary learning. A framework for solving inverse problems where multiple layers of dictionaries are learnt in an adaptive fashion from the image it is reconstructing. Results on MRI recovery from partial K-space samples show excellent results – better than all known techniques. In future, we would try to extend this technique to solve problems in other areas, e.g. in parallel MRI and dynamic MRI.

The drawback of our method is that it is considerably slower compared to CS or DL/TL based inversion. Therefore, it cannot be used where reconstruction speed is of essence. However, practically speaking, we are not aware of any situation where this might be an issue.

This is the first work to show how deep dictionary learning can be used for solving inverse problems in medical imaging. In the future, we want to modify this basic approach to solve structured problems like multi-echo MRI [12–14] and parallel MRI [15, 16].

Acknowledgements. We are thankful in part to the Infosys Center for Artificial Intelligence @ IIITD for partial support and in part to 5IOA036 FA23861610004 grant by Air Force Office of Scientific Research (AFOSR), AOARD.

References

1. Lustig, M., Donoho, D., Pauly, J.M.: Sparse MRI: The application of compressed sensing for rapid MR imaging. Magn. Reson. Med. **58**(6), 1182–1195 (2007)
2. Ravishankar, S., Bresler, Y.: MR image reconstruction from highly under sampled k-space data by dictionary learning. IEEE Trans. Med. Imaging **30**(5), 1028–1041 (2011)
3. Ravishankar, S., Bresler, Y.: Sparsifying transform learning for compressed sensing MRI. In: IEEE ISBI, pp. 17–20 (2013)
4. Majumdar, A.: Compressed Sensing for Magnetic Resonance Image Reconstruction. Cambridge University Press, Cambridge (2015)
5. Mehta, J., Majumdar, A.: RODEO: robust DE-aliasing auto encoder for real-time medical image reconstruction. Pattern Recogn. **63**, 499–510 (2017)
6. Kulkarni, K., Lohit, S., Turaga, P., Kerviche, R., Ashok, A.: ReconNet: non-iterative reconstruction of images from compressively sensed measurements. In: IEEE CVPR, pp. 449–458 (2016)
7. Tariyal, S., Majumdar, A., Singh, R., Vatsa, M.: Deep dictionary learning. IEEE Access **4**, 10096–10109 (2016)
8. Combettes, P.L., Pesquet, J.C.: Proximal splitting methods in signal processing. In: Fixed-point Algorithms for Inverse Problems in Science and Engineering, pp. 185–212 (2011)
9. Afonso, M.V., Bioucas-Dias, J.M., Figueiredo, M.A.: An augmented Lagrangian approach to the constrained optimization formulation of imaging inverse problems. IEEE Trans. Image Process. **20**(3), 681–695 (2011)
10. Boyd, S., Parikh, N., Chu, E., Peleato, B., Eckstein, J.: Distributed optimization and statistical learning via the alternating direction method of multipliers. Found. Trends® Mach. Learn. **3**(1), 1–122 (2011)
11. Daubechies, I., Defrise, M., De Mol, C.: An iterative thresholding algorithm for linear inverse problems with a sparsity constraint. Commun. Pure Appl. Math. **57**(11), 1413–1457 (2004)
12. Maggu, J., Singh, P., Majumdar, A.: Multi-echo reconstruction from partial K-space scans via adaptively learnt basis. Magn. Reson. Imaging **45**, 105–112 (2018)
13. Majumdar, A., Ward, R.K.: Joint reconstruction of multiecho MR images using correlated sparsity. Magn. Reson. Imaging **29**(7), 899–906 (2011)

14. Majumdar, A., Ward, R.K.: Accelerating multi-echo T2 weighted MR imaging: analysis prior group-sparse optimization. J. Magn. Reson. **210**(1), 90–97 (2011)
15. Majumdar, A., Ward, R.K.: Calibration-less multi-coil MR image reconstruction. Magn. Reson. Imaging **30**(7), 1032–1045 (2012)
16. Majumdar, A., Ward, R.K.: Nuclear norm-regularized SENSE reconstruction. Magn. Reson. Imaging **30**(2), 213–221 (2012)

Deep-PUMR: Deep Positive and Unlabeled Learning with Manifold Regularization

Xingyu Chen[1], Fanghui Liu[1], Enmei Tu[1], Longbing Cao[2], and Jie Yang[1(✉)]

[1] Institute of Image Processing and Pattern Recognition,
Shanghai Jiao Tong University, Shanghai, China
{slinene,tuen,jieyang}@sjtu.edu.cn, lfhsgre@outlook.com
[2] Advanced Analytics Institute, University of Technology at Sydney,
Ultimo, Australia
longbing.cao@uts.edu.au

Abstract. Training a binary classifier only on positive and unlabeled examples (i.e., the PU learning) is an important yet challenging issue, widely seen in many problems in which it is difficult to obtain negative examples. Existing methods for handling this challenge often perform unsatisfactorily, since they often ignore the relations between positive and unlabeled examples and are also limited to the traditional shallow learning frameworks. Therefore, this work proposes a new approach: *Deep Positive and Unlabeled learning with Manifold Regularization* (Deep-PUMR), which integrates the manifold regularization with deep neural networks to address the above issues with classic PU learning. Deep-PUMR holds two major advantages: (i) Our method exploits the manifold properties of data distribution to capture the relationship of positive and unlabeled examples; (ii) The adopted deep network enables Deep-PUMR with strong learning ability, especially on large-scale datasets. Extensive experiments on five diverse datasets demonstrate that Deep-PUMR achieves the state-of-the-art performance in comparison with classic PU learning algorithms and risk estimators.

Keywords: PU learning · Deep neural network · Manifold learning

1 Introduction

Conventional supervised machine learning algorithms focus on training an accurate classifier based on both positive and negative examples (PN learning). However, in many practical tasks, the explicit negative training examples may not be available. In contrast, the unlabeled examples may be easily collectable which may be either positive or negative. To tackle this situation, *Positive and Unlabeled* learning (PU learning) was proposed in [18], which has been intensively applied to various applications such as retrieval [7], matrix completion [10], and sequential data [16].

© Springer Nature Switzerland AG 2018
L. Cheng et al. (Eds.): ICONIP 2018, LNCS 11301, pp. 12–20, 2018.
https://doi.org/10.1007/978-3-030-04167-0_2

The existing PU learning algorithms consist of two categories in term of how the unlabeled data are handled. The first category tries to distinguish the unlabeled data as positive or negative before training a conventional classifier. For example, Li et al. [17] firstly identify some reliable negative examples from unlabeled data, and then perform ordinary supervised learning. Liu et al. [18] utilize the two-step EM algorithm with the naive Bayesian classification to identify the negative data. Both methods are quite heuristic and ad-hoc, with their performance very sensitive to the quality of the negative data. The second category treats all unlabeled data as negative, and then uses the weighting-based cost-sensitive learning to approximate the real data distribution. For example, Elkan et al. [7] assign small weights to the unlabeled data that are treated as negative (Weighted PU), of which the specific weights have to be carefully estimated via a computationally expensive way. To avoid the empirical weight estimation, Plessis et al. [4] propose the notion of *Unbiased Risk Estimators* and introduce an unbiased non-convex surrogate loss function to accomplish PU learning. However, sometimes the unbiased risk estimators yield negative empirical risks, which is definitely unreasonable. Accordingly, Kiryo et al. [12] propose a new method called *Non-Negative Risk Estimator* to constrain the loss function with a non-negative lower bound, which achieves the current state-of-the-art results on most of the benchmark tasks.

Although massive PU learning algorithms are available, their performance is still far from perfect. The reasons include: (1) Most existing models follow the scheme of risk minimization and then adopt the shallow learning framework, both resulting in highly limited learning capability especially on complex data; and (2) None of them consider the internal geometric relationship between the labeled and unlabeled examples, which is inconsistent with reality hence restricting the determination of the groundtruth labels of unlabeled examples. For instance, the positive data could be close to negative data in the Euclidean space but is far away in the manifold space. To address the above issues, this paper proposes a new method dubbed *Deep Positive and Unlabeled learning with Manifold Regularization* (Deep-PUMR), which builds a deep neural network and also introduces the *manifold regularization* [1] for PU learning. Specifically, we introduce a new risk estimator named *Manifold Non-negative Risk Estimator* (MNRE), which is able to capture the underlying manifold properties hidden in the entire dataset, and embed it into the high-performing convolutional neural network (CNN) ResNet [9]. Substantial experimental results demonstrate that the proposed Deep-PUMR obtains superior performance to the state-of-the-art PU learning methods on a variety of large-scale datasets.

2   Preliminaries

Assume that $\mathcal{X} \subset \mathbb{R}^d$ is a d-dimensional input feature space and $\mathcal{Y} \in \{0,1\}$ is the label space, then n examples can be i.i.d. sampled from the joint distribution $\mathcal{X} \times \mathcal{Y}$ to form the training set $\mathcal{D} = \{(\mathbf{x}_1, y_1), (\mathbf{x}_2, y_2), ..., (\mathbf{x}_n, y_n)\}$ where $\mathbf{x}_i \in \mathcal{X}$ are example features and $y_i \in \{0,1\}$ $(i = 1, 2, \cdots, n)$ are the corresponding labels. Here $y_i = 1$ means that $\mathbf{x}_i$ is positive, while $y_i = 0$ indicates that $\mathbf{x}_i$ is unlabeled. Given $\mathcal{F}$ as the hypothesis space, our goal is to find a classifier $f \in \mathcal{F}$ based on $\mathcal{D}$ such that any unseen test example $\mathbf{x}$ can be assigned correct labels $f(\mathbf{x}) \in \{\pm 1\}$.

2.1 Non-negative PU Learning

We provide a brief description of the Non-negative PU learning algorithm [12] which is useful for our subsequent discussions.

Given a dataset $\mathcal{D}$, $\pi_P = p(y = +1)$ is the prior probability for positive class, and $\pi_N = p(y = -1) = 1 - \pi_P$ is the prior for negative class. Here π_P can be easily estimated according to [19]. Besides, $\mathcal{L}(f(\mathbf{x}), y)$ is the loss function of predicted output $f(\mathbf{x})$ on groundtruth y. Mathematically, we denote $R_P^+(f) = \mathbb{E}_P[\mathcal{L}(f(\mathbf{x}), +1)]$, $R_P^-(f) = \mathbb{E}_P[\mathcal{L}(f(\mathbf{x}), -1)]$, $R_U^-(f) = \mathbb{E}_U[\mathcal{L}(f(\mathbf{x}), -1)]$, where $\mathbb{E}_P = \frac{1}{n_P} \sum_{i=1}^{n_P}$, $\mathbb{E}_U = \frac{1}{n_U} \sum_{i=1}^{n_U}$ with n_P and n_U being the numbers of positive examples and unlabeled ones, respectively. The Non-negative Risk Estimator can be expressed as

$$\check{R}_{PU}(f) = \pi_P R_P^+(f) + \max\left\{0, R_U^-(f) - \pi_P R_P^-(f)\right\}. \tag{1}$$

Note that Eq. (1) is always non-negative, therefore the Non-negative PU (nnPU) estimator can be obtained by minimizing the empirical risk of $\check{R}_{PU}$. It is worth mentioning that the loss function $\mathcal{L}$ has a variety of options, and the possible choices are summarized in [3].

2.2 Manifold Regularization

Manifold regularization is one of the graph-based methods which has been widely adopted in many machine learning techniques such as semi-supervised-learning (SSL) [8], spectral clustering [20], and dimensionality reduction [6]. The basic assumption of manifold regularization is that the data distribution lies on a manifold $\mathcal{M}$ of which the dimension is lower than the normal data space $\mathcal{X}$. In other words, the labels of examples should vary smoothly along the manifold. Under the manifold assumption, the data form a nonlinear manifold $\mathcal{M} \subset \mathcal{X}$ and hence the intrinsic data property can be fully explored. One natural choice for the manifold regularizer is the gradient on the manifold $\nabla_{\mathcal{M}}$ [2], which is able to measure the smoothness of target function. This enables an appropriate form for the manifold regularizer $\|f_I\|$ as

$$\|f_I\|^2 = \int_{\mathbf{x} \in \mathcal{X}} \|\nabla_{\mathcal{M}} f(\mathbf{x})\|^2 dp_{\mathcal{X}}(\mathbf{x}), \tag{2}$$

where $p_{\mathcal{X}}$ is the marginal distribution of $\mathcal{X}$. Many methodologies relying on manifold regularization are developed based on Eq. (2).

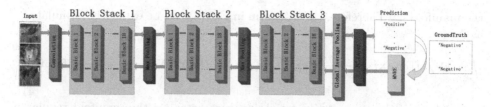

Fig. 1. Deep-PUMR network structure. We use a 152-layer residual network, which contains 3 block stacks. Each block stack consists of 18 basic blocks and each basic block has 2 convolutional layers, 2 activation layers with ReLU and 2 batch normalization layers. A max pooling layer is added after each block stack and the last block stack is followed by a global average pooling step. A fully-connected layer (FC Layer) with the tanh activation function is employed to generate the model prediction, and MNRE acts as the loss function for comparing the model outputs and the groundtruth labels.

3   The Proposed Deep-PUMR Model

3.1   Network Architecture

We use a well-known deep convolutional neural network named "ResNet" [9] for our Deep-PUMR model, which contains 152 layers and has been demonstrated to be effective and scalable to various big data applications. The structure of the proposed Deep-PUMR is shown in Fig. 1, from which we see that our model has a architecture similar to ResNet [9], and the numbers of the feature maps in each block stack are 16, 32 and 64, respectively. Moreover, we develop a novel *Manifold Non-negative Risk Estimator* (MNRE) to simultaneously deal with the positive and unlabeled examples, which will be detailed in the next section.

3.2   Loss Function

Existing PU learning models do not fully consider the relationships between positive and negative examples. We believe that such information is important as it helps to transfer the known labels of positive examples to the unlabeled examples. Therefore, we propose a *Manifold Non-negative Risk Estimator* (MNRE) by integrating the nnPU with a manifold regularizer. Specifically, we assume that the marginal distribution p_x is a compact sub-manifold $\mathcal{M} \subset \mathbb{R}^m$ $(m < d)$, and the manifold regularizer is given by Eq. (2). Hence, the MNRE is defined as follows:

$$\widetilde{R}_{PU}(f) = \pi_P R_P^+(f) + \max\left\{0, R_U^-(f) - \pi_P R_P^-(f)\right\} + \int_{\mathbf{x} \in \mathcal{X}} \|\nabla_{\mathcal{M}} f\|^2 dp_x(\mathbf{x}). \tag{3}$$

However, in most applications, p_x is not known. Hence, we must have an empirical estimation of p_x. It has been shown that the exponential weights for the adjacency graph lead to the convergence of graph Laplace-Beltrami operator on the manifold [2]. Therefore, the last term in Eq. (3) can be explicitly formulated as

$$\int_{\mathbf{x}\in\mathcal{X}}\|\nabla_{\mathcal{M}}f\|dp_x(\mathbf{x}) = \frac{\gamma_I}{(n_P + n_U)^2}\sum_{i,j=1}^{n_P+n_U}(f(\mathbf{x}_i) - f(\mathbf{x}_j))^2\mathbf{W}_{ij} = \frac{\gamma_I}{(n_P+n_U)^2}\mathbf{f}^\top\mathbf{L}\mathbf{f},$$

(4)

in which γ_I is the nonnegative weighting parameter for the manifold regularizer, n_P is the number of labeled positive data, and n_U is the number of unlabeled data, $\mathbf{f} = [f(\mathbf{x}_1), f(\mathbf{x}_2), \ldots, f(\mathbf{x}_{(n_P+n_U)})]^\top$ records the labels of all training examples. The graph Laplacian matrix $\mathbf{L}$ is defined by $\mathbf{L} = \mathbf{D} - \mathbf{W}$. Here, $\mathbf{W}_{ij}$ is the edge weight between $\mathbf{x}_i$ and $\mathbf{x}_j$, which is defined by $\mathbf{W}_{ij} = exp(\frac{-\|\mathbf{x}_i-\mathbf{x}_j\|_2^2}{2\eta^2})$, if $\mathbf{x}_j \in \mathcal{N}_i^k$; $\mathbf{W}_{ij} = 0$, otherwise. $\mathcal{N}_i^k$ represents the k-nearest points to $\mathbf{x}_i$, η is the Gaussian kernel width and $\mathbf{D}$ is the diagonal matrix with the i-th diagonal element $\mathbf{D}_{ii} = \sum_{j=1}^{n_P+n_U}\mathbf{W}_{ij}$. By minimizing Eq. (4), the examples that are strongly connected by an edge will obtain similar labels, and two examples can be assigned with quite different labels if they have very weak connections. As a result, Eq. (3) can be transformed to

$$\widetilde{R}_{PU}(f) = \pi_P R_P^+(f) + \max\left\{0, R_U^-(f) - \pi_P R_P^-(f)\right\} + \frac{\gamma_I}{(n_P + n_U)^2}\mathbf{f}^\top\mathbf{L}\mathbf{f}.$$

(5)

By minimizing the risk estimator $\widetilde{R}_{PU}(f)$, we obtain the optimal classifier $f^*(\mathbf{x})$.

Since our model is built on a deep neural network (i.e., ResNet), an efficient solution is necessary to train the network. A commonly used optimization algorithm for CNN training is the mini-batch Stochastic Gradient Descent (mini-batch SGD) [5]. However, due to the existence of the manifold regularizer in Eq. (4), the loss $\widetilde{R}_{PU}(f)$ is not directly decomposable, which forms the obstacle for utilizing mini-batch SGD. Therefore, we split the manifold regularizer into sub-matrix to facilitate the sequential optimization. Suppose that $\mathcal{D}^i = \{\mathcal{D}_P^i, \mathcal{D}_U^i\}$ is the i-th mini-batch of dataset $\mathcal{D}$ ($i = 0, 1, \cdots, N$) with N being the number of batches. $\mathcal{D}_P^i$ is the labeled positive examples in $\mathcal{D}^i$, and $\mathcal{D}_U^i$ is the unlabeled examples in $\mathcal{D}^i$. $\mathbf{L}^i$ is the sub-Laplacian matrix[1] produced by mini-batch $\mathcal{D}^i$. We define $\mathbf{L}_{ab}^i = \mathbf{L}_{ab}$ if $\mathbf{x}_a, \mathbf{x}_b \in \mathcal{D}^i \subset \mathcal{X}$, and $\mathbf{L}_{ab}^i = 0$ otherwise. Based on above definitions, Eq. (5) can be minimized in sequence as

$$\widetilde{R}_{PU}^i(f) = \pi_P R_P^{i+}(f) + \max\left\{0, R_U^{i-}(f) - \pi_P R_P^{i-}(f)\right\} + \frac{\gamma_I}{(n_P^i + n_U^i)^2}\mathbf{f}^{i\top}\mathbf{L}^i\mathbf{f}^i.$$

(6)

where n_P^i and n_U^i are the size of labeled positive data and unlabeled data in $\mathcal{D}^i$, respectively. The label vector $\mathbf{f}^i = \mathbf{C}^i\mathbf{f}$, where $\mathbf{C}^i$ is the diagonal matrix to pick up the data of mini-batch $\mathcal{D}^i$ from $\mathcal{D}$, $\mathbf{C}_{tt}^i = 1$ if $\mathbf{x}_t \in \mathcal{D}^i$, and $\mathbf{C}_{tt}^i = 0$ otherwise ($t = 0, 1, \cdots, n$). Accordingly, the proposed Deep-PUMR can be successfully minimized via mini-batch SGD. Furthermore, we provide the backward propagation rule of our designed network. The *sigmoid loss* is chosen as the loss

---

[1] $\mathbf{L}^i$ represents the sub-matrix of $\mathbf{L}$ corresponding to $\mathcal{D}^i$. For example, if the i-th mini-batch corresponds to the l_1-th examples to l_2-th examples, $\mathbf{L}^i = \mathbf{L}_{l_1:l_2,l_1:l_2}$ in matlab formation.

Algorithm 1. Framework of Deep-PUMR based on stochastic optimization.

Input:
$\quad$ $\mathcal{D}$, $\pi_P \in (0,1)$, hyperparameters $\alpha \in (0,1)$, $\beta \in (0, \pi_P)$, γ_I, σ and kernel width η.
Output:$\quad$ model parameters θ of optimal classifier.
1: **repeat**
2: $\quad$ Shuffle $\mathcal{D}$ into N mini-batches, $\mathcal{D}^i$ is the i-th mini-batch
3: $\quad$ Calculate the Laplacian matrix $\mathbf{L}$
4: $\quad$ Let Υ be an external mini-batch SGD optimization algorithm such as [12]
5: $\quad$ **for** $i = 1$ to N **do**
6: $\quad\quad$ **if** $R_U^-(f) - \pi_P R_P^-(f) \geq -\beta$ **then**
7: $\quad\quad\quad$ Set the gradient $\nabla_\theta[\pi_P R_P^+(f) + R_U^-(f) - \pi_P R_P^-(f)] + \nabla_{\mathcal{M}}^i$
8: $\quad\quad\quad$ update θ by Eq. (7) and Υ with the current step size ξ
9: $\quad\quad$ **else**
10: $\quad\quad\quad$ Set the gradient $\nabla_\theta[\pi_P R_P^-(f) - R_U^-(f)] + \nabla_{\mathcal{M}}^i$
11: $\quad\quad\quad$ update θ by Eq. (8) and Υ with step size $\alpha \cdot \xi$
12: $\quad\quad$ **end if**
13: $\quad$ **end for**
14: $\quad$ Set $\gamma_I = \sigma \cdot \gamma_I$
15: **until** stopping criterion has been met

---

function $\mathcal{L}(a,b) = \frac{1}{(1+exp(-ab))}$ throughout this paper because it has shown very satisfactory performance in CNN [9]. The backward propagation rule of sigmoid loss is $\frac{\partial \mathcal{L}}{\partial \theta} = \mathcal{L}(1 - \mathcal{L}) \cdot \frac{\partial f}{\partial \theta}$, where θ is the parameter vector to be optimized. The proposed objective function Eq. (6) is summed over all the labeled, unlabeled examples and the example pairs formed by them. Here we detail the backward propagation rule of the i-th mini-batch. To this end, two cases are considered.

(a) If $0 \leq R_U^-(\mathbf{f}_U^i) - \pi_P R_P^-(\mathbf{f}_P^i)$:

$$\mathcal{C}_{MNRE}(\theta) = \frac{\partial \mathbf{f}_P^i}{\partial \theta} \kappa \cdot \left[\mathcal{L}_{\mathbf{f}_P^i}^+ (1 - \mathcal{L}_{\mathbf{f}_P^i}^+) - \mathcal{L}_{\mathbf{f}_P^i}^- (1 - \mathcal{L}_{\mathbf{f}_P^i}^-) \right] + \frac{\partial \mathbf{f}_U^i}{\partial \theta} \mathcal{L}_{\mathbf{f}_U^i}^- (1 - \mathcal{L}_{\mathbf{f}_U^i}^-) + \nu \frac{\partial \mathbf{f}^i}{\partial \theta};$$
$$(7)$$

(b) If $0 > R_U^-(\mathbf{f}_U^i) - \pi_P R_P^-(\mathbf{f}_P^i)$:

$$\mathcal{C}_{MNRE}(\theta) = \frac{\partial \mathbf{f}_P^i}{\partial \theta} \kappa \cdot \mathcal{L}_{\mathbf{f}_P^i}^- (1 - \mathcal{L}_{\mathbf{f}_P^i}^-) - \frac{\partial \mathbf{f}_U^i}{\partial \theta} \mathcal{L}_{\mathbf{f}_U^i}^- (1 - \mathcal{L}_{\mathbf{f}_U^i}^-) + \nu \frac{\partial \mathbf{f}^i}{\partial \theta}, \qquad (8)$$

where $\mathbf{f}_P^i = [f(\mathbf{x}_1), f(\mathbf{x}_2), \ldots, f(\mathbf{x}_{(n_P^i)})]^\top$, and $\mathbf{f}_U^i = [f(\mathbf{x}_1), f(\mathbf{x}_2), \ldots,$ $f(\mathbf{x}_{(n_U^i)})]^\top$, $\mathbf{f}^i = \{\mathbf{f}_P^i, \mathbf{f}_U^i\}$, $\kappa = \pi_P/n_P$, $\nu = \gamma_I/(n_P^i + n_U^i)^2 \sum_{o=1}^{n_P^i + n_U^i} \mathbf{L}^i(o, :)$ and $\mathbf{L}^i(o, :)$ is the sum of the o-th row of $\mathbf{L}^i$, $\mathcal{L}_{\mathbf{f}_P^i}^\pm = \sum \mathcal{L}(f(\mathbf{x}), \pm 1)(\mathbf{x} \in \mathcal{D}_P^i)$ and $\mathcal{L}_{\mathbf{f}_U^i}^\pm = \sum \mathcal{L}(f(\mathbf{x}), \pm 1)(\mathbf{x} \in \mathcal{D}_U^i)$. Based on this, we obtain the gradient of $\frac{\partial \mathcal{C}_{MNRE}}{\partial f}$ by Eqs. (7) and (8), and $\frac{\partial f}{\partial \theta}$ can be optimized by existing SGD-like algorithms [24]. Hence, we can conduct the backward propagation rule of Deep-PUMR by the above schemes.

During the training stage, we set a nonnegative decay factor σ to tune the weighting parameter as $\gamma_I = \sigma * \gamma_I$ after each epoch. The hyperparameters α, β are introduced to regulate the Deep-PUMR model by following [12]. Moreover, a pre-trained layer is applied to generate a proper initial value to improve the convergence speed.

4 Experiments

Baselines. We evaluate the performance of the proposed Deep-PUMR by comparing it to some typical baseline algorithms on five diverse datasets. The adopted baselines include Weighted PU learning (Weighted PU) [7], Biased SVM [15], Unbiased PU learning (uPU) [4] and Non-negative PU learning (nnPU) [12]. In addition, the classical SVM is also compared when it is trained on the real positive and negative data (dubbed "PN SVM"). For the fairness of comparison, all the algorithms are conducted 10 times individually with randomly selected labeled data.

Datasets. Five large-scale datasets are adopted for our experiments and the relevant metadata is shown in Table 1. For MNIST and CIFAR-10, we use the default splits of training and test sets. For the other datasets, we randomly choose 70% examples for training and the remaining 30% are used for testing.

Experimental Settings. Two learning methods are set up as follows: (A) For PU, we set $n_P = 1,000$ and n_U to be the total number of training data; (B) For PN, we set $n_P = 1,000$ and $n_N = (\pi_N/2\pi_P)^2 \cdot n_P$ by following [12]. Besides, based on our parameter sensitivity test, we fix batch size to 500, $\eta \in [0.8, 1.2]$ and $\frac{\gamma_I}{(n_P + n_U)} = 10^{-5}$ on all experiments. We set $\alpha = 0.5$ and $\beta = 0.01$, respectively. Other approaches are optimized accordingly as recommended by their authors.

Table 1. Specification of benchmark datasets.

Dataset	# Training data	# Test data	# Feature	π_P	Positive/negative
MNIST [14]	60, 000	10, 000	784	0.50	Odd/even
CIFAR-10 [13]	50, 000	10, 000	3, 072	0.40	Traffic tools/animals
VOC2012 [22]	15, 585	6, 678	270, 000	0.35	Human/without human
IMDB-WIKI [23]	43, 628	18, 699	10, 000	0.76	Male/female
Kaggle-DRD [11]	24, 588	10, 539	90, 000	0.73	Healthy/sick

Results and Analyses. The experimental results on five datasets are presented in Table 2. We can see that the neural network based algorithms (i.e. nnPU and Deep-PUMR) achieve better results when compared to traditional SVM based PU learning algorithms (i.e. Weighted PU, Biased SVM and uPU). Specifically, our method surpasses the second best method (nnPU) with margins about 3.9%,

Table 2. Means and standard deviations of accuracies on all datasets.

Method	MNIST	CIFAR-10	VOC2012	IMDB-WIKI	Kaggle-DRD
PN SVM	0.897 ± 0.003	0.867 ± 0.008	0.794 ± 0.025	0.879 ± 0.032	0.809 ± 0.023
Weighted PU [7]	0.884 ± 0.024	0.846 ± 0.015	0.783 ± 0.024	0.863 ± 0.021	0.791 ± 0.027
Biased SVM [15]	0.862 ± 0.009	0.834 ± 0.011	0.769 ± 0.019	0.855 ± 0.011	0.779 ± 0.023
uPU [4]	0.891 ± 0.017	0.852 ± 0.030	0.798 ± 0.045	0.871 ± 0.031	0.800 ± 0.016
nnPU [12]	0.934 ± 0.023	0.911 ± 0.024	0.839 ± 0.023	0.884 ± 0.027	0.842 ± 0.031
Deep-PUMR	$\mathbf{0.966 \pm 0.012}$	$\mathbf{0.932 \pm 0.026}$	$\mathbf{0.864 \pm 0.011}$	$\mathbf{0.905 \pm 0.021}$	$\mathbf{0.871 \pm 0.021}$

4.1%, 5.5%, 2.1% and 4.9%, respectively. Such significant improvement demonstrates the superiority of our Deep-PUMR to the existing approaches. This is because the deeper neural network can gradually extract some high-level features to better represent the examples. Meanwhile, our method shows a lower standard deviation than nnPU on most datasets (except CIFAR-10), which indicates that our model is more stable and robust than the existing neural network-based models. Additionally, our method exceeds the PN SVM by 6.9%, 6.5%, 7.0%, 2.6% and 6.2%, respectively, which also verifies that the PU learning may perform potentially better than PN learning as $n_U \to \infty$ [21].

5   Conclusion

This paper proposes a novel approach called *Deep Positive and Unlabeled learning with Manifold Regularization* (Deep-PUMR), which utilizes manifold regularization and deep neural network for PU learning. The contributions of our work include two aspects: one is the designed *Manifold Non-negative Risk Estimator* (MNRE) for relating the positive and unlabeled examples, and the other is the innovation of very deep neural network for PU learning. Moreover, we develop the forward and backward propagation rules for our network based on mini-batch SGD, making our method very effective for handling large-scale datasets. Sufficient experimental results demonstrate that these two contributions are critical for Deep-PUMR to obtain encouraging performance.

Acknowledgments. This research is partly supported by NSFC, China (No: 61572315, 6151101179) and 973 Plan, China (No. 2015CB856004).

References

1. Belkin, M., Niyogi, P., Sindhwani, V.: Manifold regularization: a geometric framework for learning from labeled and unlabeled examples. JMLR **7**(1), 2399–2434 (2006)
2. Belkin, M., Niyogi, P., Sindhwani, V.: On manifold regularization. In: AISTATS, p. 1 (2005)
3. Du Plessis, M., Niu, G., Sugiyama, M.: Convex formulation for learning from positive and unlabeled data. In: ICML, pp. 1386–1394 (2015)

4. Du Plessis, M.C., Niu, G., Sugiyama, M.: Analysis of learning from positive and unlabeled data. In: NIPS, pp. 703–711 (2014)
5. Duchi, J., Hazan, E., Singer, Y.: Adaptive subgradient methods for online learning and stochastic optimization. JMLR **12**(Jul), 2121–2159 (2011)
6. Elhamifar, E., Vidal, R.: Sparse manifold clustering and embedding. In: NIPS, pp. 55–63 (2011)
7. Elkan, C., Noto, K.: Learning classifiers from only positive and unlabeled data. In: KDD, pp. 213–220 (2008)
8. Gong, C., Liu, T., Tao, D., Fu, K., Tu, E., Yang, J.: Deformed graph laplacian for semisupervised learning. TNNLS **26**(10), 2261–2274 (2015)
9. He, K., Zhang, X., Ren, S., Sun, J.: Deep residual learning for image recognition. In: CVPR, pp. 770–778 (2016)
10. Hsieh, C.J., Natarajan, N., Dhillon, I.: PU learning for matrix completion. In: JMLR, pp. 2445–2453 (2015)
11. Jaafar, H.F., Nandi, A.K., Al-Nuaimy, W.: Automated detection and grading of hard exudates from retinal fundus images. In: EUSIPCO, pp. 66–70. IEEE (2011)
12. Kiryo, R., Niu, G., du Plessis, M.C., Sugiyama, M.: Positive-unlabeled learning with non-negative risk estimator. NIPS (2017)
13. Krizhevsky, A., Hinton, G.: Learning multiple layers of features from tiny images (2009)
14. LeCun, Y., Bottou, L., Bengio, Y., Haffner, P.: Gradient-based learning applied to document recognition. Proc. IEEE **86**(11), 2278–2324 (1998)
15. Lee, W.S., Liu, B.: Learning with positive and unlabeled examples using weighted logistic regression. In: ICML, pp. 448–455 (2003)
16. Li, X.L., Yu, P.S., Liu, B., Ng, S.K.: Positive unlabeled learning for data stream classification. In: SDM, pp. 259–270 (2009)
17. Li, X., Liu, B.: Learning to classify texts using positive and unlabeled data. In: IJCAI, pp. 587–592 (2003)
18. Liu, B., Lee, W.S., Yu, P.S., Li, X.: Partially supervised classification of text documents. In: ICML, pp. 387–394 (2002)
19. Menon, A., Van Rooyen, B., Ong, C.S., Williamson, B.: Learning from corrupted binary labels via class-probability estimation. In: ICML, pp. 125–134 (2015)
20. Nadler, B., Lafon, S., Kevrekidis, I., Coifman, R.R.: Diffusion maps, spectral clustering and eigenfunctions of fokker-planck operators. In: NIPS, pp. 955–962 (2006)
21. Niu, G., du Plessis, M.C., Sakai, T., Ma, Y., Sugiyama, M.: Theoretical comparisons of positive-unlabeled learning against positive-negative learning. In: NIPS, pp. 1199–1207 (2016)
22. Redmon, J., Farhadi, A.: Yolo9000: better, faster, stronger. In: CVPR (2016)
23. Rothe, R., Timofte, R., Gool, L.V.: Deep expectation of real and apparent age from a single image without facial landmarks. IJCV **126**, 144–157 (2016)
24. Ruder, S.: An overview of gradient descent optimization algorithms. arXiv (2016)

Viewpoint Estimation for Workpieces with Deep Transfer Learning from Cold to Hot

Changsheng Lu, Haotian Wang, Chaochen Gu$^{(\boxtimes)}$, Kaijie Wu, and Xinping Guan

Key Laboratory of System Control and Information Processing,
Shanghai Jiao Tong University, Shanghai 200240, China
ChangshengLuu@gmail.com, HaotiannWang@gmail.com
{jacygu,kaijiewu,xpguan}@sjtu.edu.cn

Abstract. With the revival of deep neural networks, viewpoint estimation problem can be handled by the learned distinctive features. However, the scarcity and expensiveness of viewpoint annotation for the real-world industrial workpieces impede its progress of application. In this paper, we propose a deep transfer learning method for viewpoint estimation by transferring priori knowledge from labeled synthetic images to unlabeled real images. The synthetic images are rendered from 3D Computer-Aided Design (CAD) models and annotated automatically. To boost the performance of deep transfer network, we design a new two-stage training strategy called cold-to-hot training. At the cold start stage, deep networks are trained for the joint tasks of classification and knowledge transfer in the absence of labels of real images. But after it turns into the hot stage, the pseudo labels of real images are employed for controlling the distributions of input data. The satisfactory experimental results demonstrate the effectiveness of the proposed method in dealing with the viewpoint estimation problem under the scarcity of annotated real workpiece images.

Keywords: Viewpoint estimation · Deep transfer learning
Cold-to-hot training · Workpiece · Synthetic image · CAD model

1 Introduction

Viewpoint estimation is crucial in many tasks such as workpiece grasping in automated assembly pipeline, autonomous car driving [4] and virtual reality. The aim of viewpoint estimation is to determine the orientation of the target object from 2D or 3D data, which is challenging especially when lacking annotated data. Currently, a variety of viewpoint estimation methods have been proposed and the used training data can be generally categorized as: (1) 3D data, e.g. point cloud, Computer-Aided Design (CAD) model; (2) 2D real or synthetic images.

© Springer Nature Switzerland AG 2018
L. Cheng et al. (Eds.): ICONIP 2018, LNCS 11301, pp. 21–32, 2018.
https://doi.org/10.1007/978-3-030-04167-0_3

In consideration of rich information provided by 3D data, there exists large potential to improve the performance of viewpoint estimation. Most of 3D data-driven viewpoint estimation methods rely on traditional features such as 3D Shape Context (3DSC) [5] and signatures of histograms of orientation (SHOT) [27]. Unfortunately, such methods are time-consuming in extracting 3DSC or SHOT features and solving the correspondences for matching. In addition, the performance of estimation is subject to the texture of workpieces, illumination variation, and noise disturbance. With the growing scale of point cloud and CAD model datasets, e.g. ScanNet [19], S3DIS [1], ShapeNet [3], neural networks are developed to consume 3D data directly. Charles R. Qi et al. propose an efficient neural network architecture called PointNet [18] to deal with the irregular and disordered point cloud data and develop its hierarchical version PointNet++ [19] for learning spatial local features. Recently, PointCNN [12], a generalization of the typical convolutional neural network (CNN) has been designed for the point cloud. Both PointNet++ and PointCNN achieve impressive performance on tasks such as 3D object classification and 3D scene segmentation.

Unlike handling the complex 3D point cloud or CAD model data, viewpoint estimation from 2D images is more straightforward. As for the workpieces with simple geometric structures, traditional shape detection methods [10,15] may be efficient. However, CNN [11] is more general and powerful to learn distinct features from images, which leads to many great successful applications such as object detection [21,22], image segmentation [23] and viewpoint estimation [25,30]. To guarantee the superior performance and the generalization ability of CNN, tremendous labeled real images are required. In terms of industrial workpiece viewpoint estimation, it is quite expensive and time-consuming to manually annotate such tremendous real images precisely. To cope with the above issues, a popular solution is to utilize data augmentation techniques. Many studies [17,25,28,30] augment training set with synthetic images by rendering CAD models in different real scenes, lighting conditions and virtual camera views. To generate synthetic images with higher fidelity, feasible techniques include domain randomization [28] and adversarial training strategy [7,24]. However, there still exists a huge domain discrepancy between synthetic images and real images. In this case, transfer learning can be adopted to bridge the gap between synthetic images and real images. The representative methods [6,13,14,29] try to use domain adaption for learning common features between the source domain and target domain. As a result, it enables deep networks to estimate workpiece viewpoints with rich annotated synthetic images and unlabeled (or few labeled) real images.

In the absence of annotated real workpiece images, we propose a deep transfer network equipped with a novel training strategy for viewpoint estimation, as shown in Fig. 1. Firstly, we augment our dataset with synthetic workpiece images by rendering from CAD models in the virtual environment, in which the labels can be automatically generated. Then we take the equal-sized batches of labeled synthetic images and unlabeled real images as input to feature extractor. For the extracted high-level features, Maximum Mean Discrepancy (MMD) [2,13] is

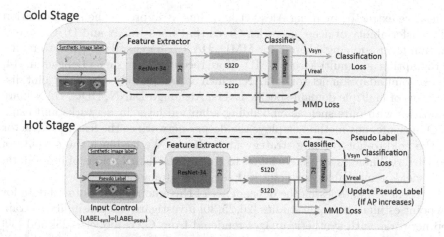

Fig. 1. The framework of the proposed deep transfer network with cold-to-hot training strategy. The high-level features are extracted from randomly selected batches of both synthetic images and real images by feature extractor (Resnet-34 [8]). At cold training stage, the real images have no labels. When the network turns into the hot stage, the pseudo labels are employed to real images, which are used for controlling the distributions of input batches of synthetic and real images to be equivalent. The pseudo labels will be updated once the average precision exceeds the previous one over a validation set which is only composed of real images.

employed for domain adaption. Afterwards, we formulate MMD and geometric aware viewpoint classification loss functions to optimize the deep model. In order to improve the performance of networks, a novel cold-to-hot training strategy is applied. The deep transfer network starts at cold status with unlabeled real workpiece images and is expected to learn priori knowledge from labeled synthetic images. While at the hot stage, the viewpoint estimation of the real image is used as the pseudo label for better controlling the distributions of input data, which organizes the unlabeled real images from disorder to order.

2   Related Work

In terms of the viewpoint estimation problem in unlabeled real workpiece data, a popular solution is to borrow the priori knowledge from relevant synthetic images by transfer learning. Usually, transfer learning aims to remove the discrepancy of the distributions between the source and target data representations in higher space, which enables the model trained on source domain to work on target domain, e.g. fine tuning with pre-trained model [26]. The pioneering work DaNN [6], incorporates MMD [2] in a two-layer network pre-trained with a denoising auto-encoder for objection detection, which reveals that MMD is good at domain adaptation and able to help networks learn domain invariant knowledge effectively. Considering that the depth of DaNN is shallow, [29] deepens the feature extractor by using AlexNet [9]. Thus, it improves the representation and transfer ability of deep networks. Given that both DaNN and DDC use one adaption layer and single kernel for MMD, DAN [13] generalizes them to multi-layer adaption and multi-kernel MMD. More recently, Mingsheng Long et al. [14] propose joint adaption networks (JAN) which approximately aligns the joint distributions of multiple domain-specific layers. Although DaNN, DDC, DAN share similarities with ours such as embedding adaption layer in networks and using MMD measure, our method employs the deeper network as the feature extractor and the cold-to-hot training strategy for training. We feedback the outputs of unlabeled real images as pseudo labels to control the distributions of input data, which yields remarkable viewpoint estimation accuracy boost.

In recent years, to lessen the dependence of the amount of annotated data for viewpoint estimation, some studies [20,25,30] investigate the possibility to train deep networks with synthetic images rendered from CAD models. [25] and [30] propose similar image synthesis pipelines by overlaying images rendered from CAD models on top of real images. [25] builds a neural network architecture by stacking multiple classifiers over a shared feature extractor, while [30] modifies AlexNet for viewpoint estimation. In addition, [20] trains a classifier with synthetic images, and meanwhile, trains another mapping network combining with real images, which tries to map the real images into the feature space of synthetic data. Our method and [20,25,30] all use synthetic images during training deep networks, but they are essentially different. First, the frameworks of deep networks are different. Second, our method integrates transfer learning methodology into deep networks while [20,25,30] not. Third, the training set of our learning model excludes labels of real images while [20,25] use the annotations of both synthetic and real images.

3   Viewpoint Estimation for Workpieces

Generally, viewpoint estimation follows a multi-classification fashion. Thus we discretize the continuous viewpoint space as $\mathcal{V}$, and denote $v \in \mathcal{V}$ as the discretized viewpoint. Our purpose is to build a deep transfer network by transferring priori knowledge from source domain to target domain. The training set are formulated as $\mathcal{T} = \mathcal{T}^s \cup \mathcal{T}^t$, where $\mathcal{T}^s = \{x_i^s, y_i^s\}$ is composed by synthetic images with annotations, and $\mathcal{T}^t = \{x_i^t\}$ is made up of unlabeled real-world workpiece images.

3.1   Deep Transfer Network and Loss Function

In order to sufficiently excavate the deep features of images in $\mathcal{T}$, we adopt ResNet-34 [8] as the feature extractor f. The extracted feature $f(x_i)$ of each image x_i is in fixed length, and assumed to be in feature space $\mathcal{X}$. Then, a classifier g takes high-level feature $f(x_i)$ as input for viewpoint estimation. Despite the simplicity of networks structure, it is competent for joint tasks of viewpoint estimation and knowledge transfer.

For training the deep networks, every two batches, namely B^s and B^t, are randomly sampled from synthetic image set $\mathcal{T}^s$ and real image set $\mathcal{T}^t$ ($B^s \subset \mathcal{T}^s, B^t \subset \mathcal{T}^t$). Then we can acquire their high-level features $f(B^s)$ and $f(B^t)$ by feeding B^s and B^t into feature extractor f simultaneously. For the purpose of empowering the networks to have transfer ability, $f(B^s)$ and $f(B^t)$ are mapped into Reproducing Kernel Hilbert Space $\mathcal{H}$ (RKHS) for the computation of their MMDs. The MMD loss function $\mathcal{L}_{MMD}$ can be formulated as

$$\mathcal{L}_{MMD} = \left\| \frac{1}{|B^s|} \sum_{x_i^s \in B^s} \phi(f(x_i^s)) - \frac{1}{|B^t|} \sum_{x_j^t \in B^t} \phi(f(x_j^t)) \right\|_{\mathcal{H}}^2 \tag{1}$$

where $\phi(\cdot) : \mathcal{X} \to \mathcal{H}$ and $|B^\bullet|$ refers to the number of samples in batch $B^\bullet$. In order to calculate $\mathcal{L}_{MMD}$ efficiently, we introduce kernel trick for the computation of the inner product of vectors in $\mathcal{H}$. In addition, to reduce the risk of selecting inappropriate kernel function k, it is more reasonable to choose multiple weighted kernel functions instead, namely $\phi(\cdot)^T \phi(\cdot) = \sum_u \beta_u k_u(\cdot, \cdot)$, where $\sum_u \beta_u = 1$ and $\beta_u \geq 0$.

Since the annotations of $\mathcal{T}^t$ are unknown, classifier g merely takes $f(B^s)$ as input for training. Meanwhile, object viewpoint can be equivalently considered as the shooting position of the camera on a sphere, which owns geometric significance. Inspired by this observation and combined with cross entropy, the geometric aware viewpoint classification loss function $\mathcal{L}_{CLS}$ can be written as

$$\mathcal{L}_{CLS} = - \sum_{x_i^s \in B^s} \sum_{v \in \mathcal{V}} w(v, y_i^s) y_i^s P_v(x_i^s) \tag{2}$$

where $P_v(x_i^s)$ is the probability of viewpoint prediction of x_i^s for class v ($v \in \mathcal{V}$), and $w(v, y_i^s)$ is the geometric aware weight which is inversely proportional to the geodesic distance between predicted class v and label y_i^s. As a result, we can build the joint loss function $\mathcal{L}$ as

$$\mathcal{L}(\theta_f, \theta_g; \mathcal{T}) = \mathcal{L}_{CLS} + \lambda \mathcal{L}_{MMD} \tag{3}$$

where θ_f and θ_g are the model parameters of feature extractor f and classifier g, respectively. λ is the trade-off parameter between $\mathcal{L}_{CLS}$ and $\mathcal{L}_{MMD}$. Using the proposed novel loss function $\mathcal{L}$, we can train the deep transfer network for joint tasks of knowledge transfer and classification.

3.2   Cold-to-Hot Training

To enhance the transfer ability of networks, we propose a new two-stage training strategy called cold-to-hot training. Recall that, at the beginning of networks training, the real images in $\mathcal{T}^t$ do not have labels. Therefore, we call the period that deep networks simply transfer the priori knowledge from annotated $\mathcal{T}^s$ as cold stage. Once the deep transfer network reaches the bottleneck of average precision (AP), we turn it into hot training stage for moving ahead a further step.

At hot stage, we first utilize the viewpoint prediction $\tilde{y}_i^t$ of the real image x_i^t as its pseudo label. Thus the real image set $\mathcal{T}^t = \{x_i\}$ can be reformulated as $\tilde{\mathcal{T}}^t = \{x_i^t, \tilde{y}_i^t\}$. Instead of utilizing the pseudo labels for training classifier g, we employ the pseudo labels of $\tilde{\mathcal{T}}^t$ for controlling the distribution $\tilde{P}^t(v)$ of batch $\tilde{B}^t$ ($\tilde{B}^t \subset \tilde{\mathcal{T}}^t$) to be equivalent to the distribution $P^s(v)$ of batch B^s ($B^s \subset \mathcal{T}^s$), where $v \in \mathcal{V}$. The concrete implementation of input control can be illustrated as: Randomly select $|B^s|P^s(v)$ real image samples from set $\{x_i^t | (x_i^t, \tilde{y}_i^t) \in \tilde{\mathcal{T}}^t, \tilde{y}_i^t = v\}$ for each $v \in \mathcal{V}$ to form the batch $\tilde{B}^t$ w.r.t. the randomly sampled batch B^s. The difference of distributions of the inputs for deep transfer network between cold and hot training stages can be written as

$$\text{cold} \begin{cases} B^s \sim P^s(v) \\ B^t \sim P^t(v) \\ P^s(v) \neq P^t(v) \end{cases} \Rightarrow \text{hot} \begin{cases} B^s \sim P^s(v) \\ \tilde{B}^t \sim \tilde{P}^t(v) \\ P^s(v) = \tilde{P}^t(v) \end{cases}. \tag{4}$$

The reasons for input control are twofold: (1) The deep transfer network should also work when inputting two identical distribution batches B^s and $\tilde{B}^t$ from source domain and target domain because the expectancy of closer distance between $f(B^s)$ and $f(\tilde{B}^t)$ is reasonable from the perspective of distance measure of MMD. (2) We believe that the deep model could learn more essential features and increase transfer ability if networks are fed with the data that are with higher correlation. Eventually, the deep transfer network will self-update the pseudo labels if the AP surpasses the previous one over a validation set such that the meaningful iterations could organize the real images from disorder to order.

4   Experimental Results

4.1   Experiment Setup

The experiments are divided into three parts. Firstly, we prepare the workpiece dataset composed of labeled synthetic images and unlabeled real images. Then we perform extensive convincing experiments and compare our method with novel methods in the literature. Finally, the learned features are visualized in 2D space to study their distinctiveness.

Workpiece Dataset. The dataset is composed of eight types of industrial workpieces, each type including 12,400 synthetic images and 840 real images. Synthetic images are automatically generated and annotated by the CAD tool 3Ds-Max. As shown in Fig. 2, we build a CAD model in 3Ds-Max and place a virtual camera in the predefined position in the geographic coordinate system. For simplicity, we only choose seven frontal viewpoints (marked by red dots in Fig. 2) in the upper hemisphere which is divided into three different latitudes ($0°$, $45°$,

90°) and four different longitudes (0°, 90°, 180°, 270°). To expand the dataset, we slightly rotate the workpiece and adjust the position of the virtual camera in the vicinity of these predefined positions, and so force, rendering a large number of synthetic images. For the real images, we first fabricate the workpieces in the light of their corresponding CAD models based on 3D printing technology and then capture the images under three different kinds of background from simple to complex. The position of the camera is set according to the counterpart in the virtual coordinate system.

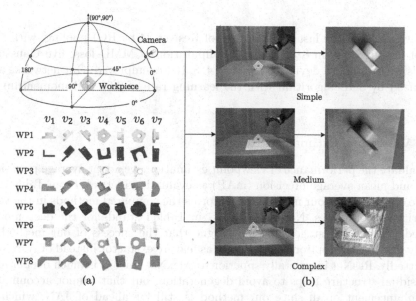

Fig. 2. Schematic diagram of workpiece image generation. (a) The process of synthetic image generation. The virtual camera is placed in the vicinity of red dots, and below are some examples of synthetic images of eight workpieces. (b) The process of real image generation. The workpiece is placed in three different backgrounds, from simple to complex. The right side shows the corresponding images. (Color figure online)

Compared Methods. We compare our method[1] with three popular transfer learning methods for viewpoint estimation: DDC [29], DAN [13], and JAN [14]. Note that both DDC and DAN are based on AlexNet, while JAN uses ResNet-50. DDC is implemented by us based on the original paper, and DAN and JAN are tested using source codes from the authors.

Implementation Details. Our deep transfer network employs ResNet-34 [8] as a feature extractor, whose weights are pretrained on ImageNet. We insert a fully-connected layer and Sigmoid activation functions before the classification layer as an adaptation layer and initialize its parameters with the normal distribution of $\mu = 0$ and $\sigma = 0.01$. Besides, in order to get a wider and fix-sized feature

---

[1] https://github.com/haotian-wang/viewpoint-estimation.

Table 1. Mean average precision (mAP) across all viewpoint classes from workpiece one (WP1) to workpiece eight (WP8). The last column is the average of mAPs.

Method	WP1	WP2	WP3	WP4	WP5	WP6	WP7	WP8	Avg.
DDC	60.0%	46.6%	48.4%	48.3%	53.0%	53.2%	48.4%	79.8%	54.7%
DAN	70.9%	55.4%	49.5%	49.8%	70.2%	65.5%	70.2%	52.9%	60.6%
JAN	78.8%	61.5%	56.3%	66.6%	70.1%	66.0%	**80.9%**	73.2%	60.2%
Ours⁻	87.8%	79.9%	**59.0%**	72.2%	73.6%	67.7%	73.6%	**88.8%**	75.3%
Ours	**91.9%**	**87.4%**	57.8%	**77.5%**	**76.6%**	**69.9%**	74.0%	87.4%	**77.8%**

map, we replaced the last pooling layer of ResNet-34 by ROI pooling, with the size of output fixed on 8×8. In the computation of MMD loss, five Gaussian kernels with standard deviation $\sigma = 1, 2, 4, 8, 16$ are employed. Our networks are optimized by SGD algorithm with the learning rate of 0.01 and momentum of 0.9.

4.2   Viewpoint Estimation Results

To evaluate the performance of viewpoint estimation, we adopt average precision (AP) and mean average precision (mAP) as evaluation metrics. From Table 1, it can be observed that our method outperforms the compared methods in the vast majority of workpieces. Note that Ours⁻ in Table 1 stands for the deep model trained without hot stage. One may argue that the success of our method is merely due to the adoption of ResNet-34 as feature extractor instead of AlexNet. Admittedly, ResNet is generally superior to AlexNet since it is much deeper and its residual structure helps to avoid degeneration, but that cannot account for the enhancement of all since our method is still far ahead of JAN, which is on the basis of ResNet-50. The boosted performance of our method is due to two ways of amelioration: the geometric aware viewpoint classification loss and the cold-to-hot training. Considering the former, the geometric significance is incorporated in the training procedure, and hence the probability of mistakenly dividing two disparate viewpoints into one class has been abated.

In order to investigate the effect of cold-to-hot training in detail, we apply cold-to-hot training strategy to all compared methods (denoted as DDC†, DAN†,

Table 2. Average precision (AP) of each viewpoint class on workpiece one (WP1).

Method	$(90°, 90°)$	$(45°, 180°)$	$(45°, 0°)$	$(0°, 90°)$	$(0°, 180°)$	$(45°, 90°)$	$(0°, 0°)$	mAP
DDC	82.4%	66.0%	**72.9%**	66.8%	41.4%	73.8%	46.0%	60.0%
DAN	98.6%	51.2%	48.6%	97.4%	**56.8%**	90.1%	51.4%	70.9%
JAN	**99.9%**	42.6%	54.9%	100.0%	59.4%	95.2%	44.1%	**78.8%**
DDC†	92.5%	**70.8%**	69.4%	**78.5%**	**69.9%**	71.7%	57.1%	69.3%
DAN†	99.8%	51.3%	**55.3%**	99.5%	51.6%	**96.6%**	53.3%	77.1%
JAN†	**99.9%**	**52.8%**	**58.2%**	99.9%	49.9%	**99.3%**	48.9%	76.9%
Ours⁻	100.0%	99.7%	99.9%	89.5%	**84.7%**	77.8%	**84.7%**	87.8%
Ours	100.0%	100.0%	100.0%	94.6%	83.4%	80.9%	79.9%	91.9%

JAN[†]) and compare APs of each viewpoint class on workpiece one (WP1), as shown in Table 2. It is worth noting that nearly all methods with cold-to-hot training get promoted, most of which can achieve about 5% increase of mAP. The essential reason for such a large enhancement is that cold-to-hot training can better control the distributions of input batches. Even in the most indistinguishable classes, our method can achieve over 79.9% AP (Table 2), which is rightly the reflection of the effectiveness of cold-to-hot training.

4.3   Visualization of Learned Transfer Features

Due to the uninterpretability of deep neural network, a proper method must be used to explore the learned features. The features before the classification layer make up 512D vectors, which are difficult to understand. Thanks to t-SNE [16] algorithm, we can observe the high dimensional features in 2D space. Figure 3 shows the visualization on the test images of WP1.

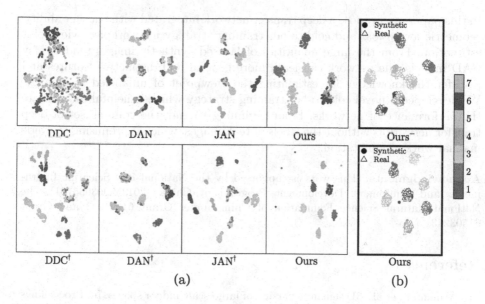

Fig. 3. Visualization of learned features before classification layer on test images of workpiece one (WP1). The points are converted from high dimensional space using t-SNE [16]. The color bar on the right side marks the category of each viewpoint. The test images of (a) are composed with real images while (b) uses both synthetic and real images.

Without cold-to-hot training, we observe that DDC is the worst among the upper four figures of Fig. 3(a), for points with various colors are almost mixed up. DAN and JAN further clarify the boundaries of various clusters, but also remain a great number of mistaken points in each cluster. Clusters generated from our method are most distinct and this exactly accords with the experimental results, revealing that our method shines over the three compared methods.

After applying cold-to-hot training, clusters in all methods become more concentrated and compact. Although cold-to-hot training cannot eliminate all wrong points, it can still improve the classification accuracy by generating clearer boundaries and widening the distances of different clusters (lower four figures of Fig. 3(a)). Moreover, the transfer ability of our deep network gets enhanced after hot stage since the distribution overlap ratio of high-level features between synthetic and real images becomes larger (by comparing Ours⁻ and Ours in Fig. 3(b)). The core of this behavior is that through cold-to-hot training, the distribution of high-level features in target domain has been adjusted closer to the distribution of those in source domain by utilizing pseudo labels, which can reduce the erroneous classification made by the classifier.

5 Conclusion

In this paper, we propose a deep transfer network integrated with transfer ability, geometric aware loss and cold-to-hot training strategy for workpiece viewpoint estimation. From the large quantities of labeled synthetic images rendered by CAD models, the network can learn more general and distinctive features and transfer the knowledge for estimating the viewpoint of unlabeled real image. We developed a novel cold-to-hot training strategy which is helpful to facilitate the performance of networks. From beginning to end, the training set of deep transfer network is without the labels of real images, which is promising to evade manual work of annotation.

Acknowledgments. This work is supported by the National Key Scientific Instruments and Equipment Development Program of China (2013YQ03065101), the National Natural Science Foundation of China under Grant 61521063 and Grant 61503243.

References

1. Armeni, I., et al.: 3D semantic parsing of large-scale indoor spaces. In: Proceedings of the IEEE Conference on Computer Vision and Pattern Recognition, pp. 1534–1543 (2016)
2. Borgwardt, K.M., Gretton, A., Rasch, M.J., Kriegel, H.P., Schölkopf, B., Smola, A.J.: Integrating structured biological data by kernel maximum mean discrepancy. Bioinformatics **22**(14), e49–e57 (2006)
3. Chang, A.X., et al.: ShapeNet: an information-rich 3D model repository. arXiv preprint arXiv:1512.03012 (2015)
4. Chen, T., Lu, S.: Robust vehicle detection and viewpoint estimation with soft discriminative mixture model. IEEE Trans. Circuits Syst. Video Technol. **27**(2), 394–403 (2017)

5. Frome, A., Huber, D., Kolluri, R., Bülow, T., Malik, J.: Recognizing objects in range data using regional point descriptors. In: Pajdla, T., Matas, J. (eds.) ECCV 2004. LNCS, vol. 3023, pp. 224–237. Springer, Heidelberg (2004). https://doi.org/10.1007/978-3-540-24672-5_18

6. Ghifary, M., Kleijn, W.B., Zhang, M.: Domain adaptive neural networks for object recognition. In: Pham, D.-N., Park, S.-B. (eds.) PRICAI 2014. LNCS (LNAI), vol. 8862, pp. 898–904. Springer, Cham (2014). https://doi.org/10.1007/978-3-319-13560-1_76

7. Goodfellow, I., et al.: Generative adversarial nets. In: Advances in Neural Information Processing Systems, pp. 2672–2680 (2014)

8. He, K., Zhang, X., Ren, S., Sun, J.: Deep residual learning for image recognition. In: Proceedings of the IEEE Conference on Computer Vision and Pattern Recognition, pp. 770–778 (2016)

9. Krizhevsky, A., Sutskever, I., Hinton, G.E.: ImageNet classification with deep convolutional neural networks. In: Advances in Neural Information Processing Systems, pp. 1097–1105 (2012)

10. Leavers, V.F.: Shape Detection in Computer Vision Using the Hough Transform. Springer, Heidelberg (1992). https://doi.org/10.1007/978-1-4471-1940-1

11. LeCun, Y., Bengio, Y., Hinton, G.: Deep learning. Nature **521**(7553), 436 (2015)

12. Li, Y., Bu, R., Sun, M., Chen, B.: PointCNN. arXiv preprint arXiv:1801.07791 (2018)

13. Long, M., Cao, Y., Wang, J., Jordan, M.: Learning transferable features with deep adaptation networks. In: International Conference on Machine Learning, pp. 97–105 (2015)

14. Long, M., Zhu, H., Wang, J., Jordan, M.I.: Deep transfer learning with joint adaptation networks. In: International Conference on Machine Learning, pp. 2208–2217 (2017)

15. Lu, C., Xia, S., Huang, W., Shao, M., Fu, Y.: Circle detection by arc-support line segments. In: 2017 IEEE International Conference on Image Processing (ICIP), pp. 76–80. IEEE (2017)

16. Maaten, L.V.D., Hinton, G.: Visualizing data using t-SNE. J. Mach. Learn. Res. **9**(Nov), 2579–2605 (2008)

17. Peng, X., Sun, B., Ali, K., Saenko, K.: Exploring invariances in deep convolutional neural networks using synthetic images. CoRR, abs/1412.7122 2(4) (2014)

18. Qi, C.R., Su, H., Mo, K., Guibas, L.J.: PointNet: deep learning on point sets for 3D classification and segmentation. Proc. Comput. Vis. Pattern Recognit. (CVPR) **1**(2), 4 (2017)

19. Qi, C.R., Yi, L., Su, H., Guibas, L.J.: Pointnet++: deep hierarchical feature learning on point sets in a metric space. In: Advances in Neural Information Processing Systems, pp. 5105–5114 (2017)

20. Rad, M., Oberweger, M., Lepetit, V.: Feature mapping for learning fast and accurate 3D pose inference from synthetic images. arXiv preprint arXiv:1712.03904 (2017)

21. Redmon, J., Farhadi, A.: YOLOv3: an incremental improvement. arXiv preprint arXiv:1804.02767 (2018)

22. Ren, S., He, K., Girshick, R., Sun, J.: Faster R-CNN: towards real-time object detection with region proposal networks. IEEE Trans. Pattern Anal. Mach. Intell. **39**(6), 1137–1149 (2017)

23. Shelhamer, E., Long, J., Darrell, T.: Fully convolutional networks for semantic segmentation. IEEE Trans. Pattern Anal. Mach. Intell. **39**(4), 640–651 (2017)

24. Shrivastava, A., Pfister, T., Tuzel, O., Susskind, J., Wang, W., Webb, R.: Learning from simulated and unsupervised images through adversarial training. In: The IEEE Conference on Computer Vision and Pattern Recognition (CVPR), vol. 3, p. 6 (2017)

25. Su, H., Qi, C.R., Li, Y., Guibas, L.J.: Render for CNN: viewpoint estimation in images using CNNs trained with rendered 3D model views. In: Proceedings of the IEEE International Conference on Computer Vision, pp. 2686–2694 (2015)

26. Tajbakhsh, N., Shin, J.Y., Gurudu, S.R., Hurst, R.T., Kendall, C.B., Gotway, M.B., Liang, J.: Convolutional neural networks for medical image analysis: full training or fine tuning? IEEE Trans. Med. Imaging 35(5), 1299–1312 (2016)

27. Tombari, F., Salti, S., Di Stefano, L.: Unique signatures of histograms for local surface description. In: Daniilidis, K., Maragos, P., Paragios, N. (eds.) ECCV 2010. LNCS, vol. 6313, pp. 356–369. Springer, Heidelberg (2010). https://doi.org/10. 1007/978-3-642-15558-1_26

28. Tremblay, J., et al.: Training deep networks with synthetic data: bridging the reality gap by domain randomization. arXiv preprint arXiv:1804.06516 (2018)

29. Tzeng, E., Hoffman, J., Zhang, N., Saenko, K., Darrell, T.: Deep domain confusion: maximizing for domain invariance. Computer Science (2014)

30. Wang, Y., Li, S., Jia, M., Liang, W.: Viewpoint estimation for objects with convolutional neural network trained on synthetic images. In: Chen, E., Gong, Y., Tie, Y. (eds.) PCM 2016. LNCS, vol. 9917, pp. 169–179. Springer, Cham (2016). https://doi.org/10.1007/978-3-319-48896-7_17

Co-consistent Regularization with Discriminative Feature for Zero-Shot Learning

Yanling Tian[1,2,3], Weitong Zhang[1,2,3], Qieshi Zhang[1,2($\boxtimes$)], Jun Cheng[1,2], Pengyi Hao[4], and Gang Lu[3]

[1] Guangdong Provincial Key Laboratory of Computer Vision and Virtual Reality Technology, Shenzhen Institutes of Advanced Technology, Chinese Academy of Sciences, Shenzhen, China
{yl.tian,wt.zhang,qs.zhang,jun.cheng}@siat.ac.cn
[2] The Chinese University of Hong Kong, Hong Kong, China
[3] School of Computer Science, Shaanxi Normal University, Xi'an, China
goforlg@126.com
[4] Zhejiang University of Technology, Hangzhou, China
haopy@zjut.edu.cn

Abstract. With the development of deep learning, zero-shot learning (ZSL) issues deserve more attention. Due to the problems of projection domain shift and discriminative feature extraction, we propose an end-to-end framework, which is different from traditional ZSL methods in the following two aspects: (1) we use a cascaded network to automatically locate discriminative regions, which can better extract latent features and contribute to the representation of key semantic attributes. (2) our framework achieves mapping in visual-semantic embedding space and calculation procedure of the dot product in deep learning framework. In addition, a joint loss function is designed for the regularization constraint of the whole method and achieves supervised learning, which enhances generalization ability in test set. In this paper, we make some experiments on Animals with Attributes 2 (AwA2), Caltech-UCSD Birds 200-2011 (CUB) and SUN datasets, which achieves better results compared to the state-of-the-art methods.

Keywords: Zero-shot learning (ZSL) · Discriminative region Regularization · Projection domain shift · Supervised learning

1 Introduction

In recent years, deep learning has become a hot topic. Purely supervised learning has achieved amazing results on many tasks [8], but there still exist some limitations: (1) a large number of samples are needed to train a good model, and (2) classifiers trained with a certain dataset can only classify the categories in the dataset and cannot identify other new categories. This model obviously

© Springer Nature Switzerland AG 2018
L. Cheng et al. (Eds.): ICONIP 2018, LNCS 11301, pp. 33–45, 2018.
https://doi.org/10.1007/978-3-030-04167-0_4

does not meet the ultimate aim of artificial intelligence that machines have the ability to identify new categories through reasoning. This is called as zero-shot learning (ZSL) [9,13]. Therefore, ZSL aims to recognize a new object or a new category never seen before.

In ZSL problem, the classifier or model is trained with seen categories in training set and tested with unseen categories in test set which is disjoint with training set. Generally, some descriptors such as object attributes can be represented from all the categories, and we can utilize these descriptors to settle this ZSL problem. As shown in Fig. 1, we get some attribute features from the seen class in the training set, such as "horselike" from donkey, "stripe" from tiger and "black & white" from panda. Then we can predict the unseen class, zebra, from above attribute features by a predictor with zebra descriptors.

Although it can solve a lot of problems in deep learning, there are some difficulty for ZSL problem [8,11]. For example, (1) discriminative feature extraction problem: due to object features are extracted by using pre-trained convolutional neural network (CNN) or hand-crafting, it is difficult to find stable and effective deep-visual features without noise for accurate identification because foreground elements, illumination and view angle contribute to complexity and diversity of images. (2) projection domain shift problem [8]: a projection from a low-level feature space to the semantic representation space is learned from the training set and applied without adaptation to the test set. Because there are disjoint and potentially unrelated classes between training set and test set, the projection functions learned from the training set are biased when applied directly to the test set.

To solve above shortcomings, a co-consistent regularization with discriminative feature framework is proposed which has two contributions:

(1) An iteration network is applied, which can form object-centric regions automatically in different scales and concentrate on learning deep detailed features to avoid redundant calculation and increase the robustness.
(2) An end-to-end network is proposed, which achieves supervised learning and completes the mapping in visual-semantic embedding space and dot product calculation process, which effectively solves the problem of domain shift and enhances the generalization ability of the test set greatly.

2   Related Work

ZSL is widely used in various research areas, such as object recognition [10], natural language processing [3] and video understanding [7]. In particular, there are a lot of work in zero-shot recognition (ZSR), which includes two main tasks: semantic representations and learning an embedding model.

Through the use of semantic representations, the relationship of seen and unseen classes can be made full use of, which can better recognize the unseen classes. There are various kinds of semantic representations such as user-defined

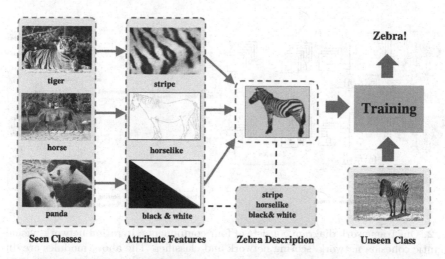

Fig. 1. ZSL aims to give the model ability to recognize the unseen class by simulating human reasoning process. Specifically, using training sets to make the model recognize the test data, although there is no intersection between them. The attributes of the training data provide the semantics for the unseen class, so that the model can predict the test class to solve the ZSL problem.

attributes [5,9], concept ontology [9] and semantic word vectors [23]. Meanwhile, lots of researches have been studied with semantic representations in ZSR. Fouhey *et al.* proposed a method [5] to infer 3D shape attributes from a single image and infer a 3D shape embedding a low dimensional vector representing the 3D shape. Zhang *et al.* proposed an efficient structure prediction algorithm [23] based on test-time adaptation of similarity functions learned using training data. Although all of these methods contributed to the ZSL problem, they did not take discriminative features into account when performing semantic representations. Inspired by [6,18], we propose a cascaded network to automatically locate the discriminative region so as to better extract deep detailed features and contribute to the representation of semantic features.

With the help of semantic representations, ZSR can be realized by learning an embedding model, which aims to establish connections between unseen class and seen class by projecting the low-level features of image close to their corresponding semantic vectors so that new classes can be recognized. Akata *et al.* proposed Attribute Label Embedding (ALE) [1] which utilizes ranking loss to learn a bilinear compatibility function between the image and the attribute space. Akata *et al.* proposed a Structured Joint Embedding (SJE) framework [2] which combines several functions linearly to form a joint embedding space and evaluates four different class embeddings including attributes, word2vec [12], glove [16] and wordnet hierarchy. Recently, Morgado *et al.* proposed a Semantically Consistent Regularization (SCoRe) model [13]. It adds a semantically consistent regularization to make the learned transformation matrix perform better on test set, which are similar to our work. SCoRe model leverages the advantages of

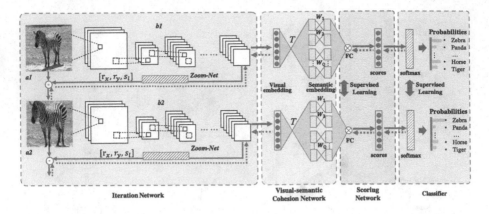

Fig. 2. Our framework diagram consists of four components: iteration network, visual-semantic cohesion network, scoring network and classifier. The above modules are differentiable, so that our method is an end-to-end network.

Recognition using Independent Semantics (RIS) method and Recognition using Semantic Embeddings (RSE) method to choose better semantic score to further classify. Different from this, we propose a visual-semantic cohesion network and a scoring network, which map the discriminative features into visual embedding space and further map into semantic space according to the proposed regularization constraints with supervised manner. Then attributes score can be obtained by scoring network so that unseen class can be represented further. Moreover, it enables to reduce influence of projection domain shift problem and to improve the generalization ability of the test set. Thus, better performance of prediction and classification can be obtained for unseen class.

3   Proposed Method

In this section, we propose an end-to-end framework for ZSL problem. As shown in Fig. 2, there are multiple scales in our framework and for easy to describe, we use two scales as an example. Our framework consists of four components: iteration network, visual-semantic cohesion network, scoring network and classifier. The iteration network manages to locate the pivotal features and regions. The visual-semantic bridging achieves two tasks including mapping the feature representation into visual embedding space and following the proposed regularization constraints for further mapping into semantic space. Through the fully-connected layer, each probability is produced in semantic embedding space. Finally, the classifier makes predictions based on the scores. All modules are differentiable, so that proposed model can be trained in an end-to-end manner.

3.1   Iteration Network

In this section, we aim to automatically locate discriminative regions and further extract deep detailed features in multiple scales. Therefore, we extract the image features ($a1$ to $a2$) through CNNs ($b1$ to $b2$) in Fig. 2 and zoom the discriminative region by Zoom-Net.

Feature Extraction. A full-size image is fed as an input into the pre-trained CNN model to get the convolution feature, which is the start for ZSL problem. Therefore, at each scale, given an input image or a zoom region X, we can obtain region-based feature representation:

$$\varphi(X) = I * X, \tag{1}$$

where I is the parameters of the CNN, "$*$" donates a set of operations of convolution, pooling and activation.

Zoom Discriminative Region. Inspired by [6,18], we suppose there are some discriminative regions which benefit ZSL. In addition, in order to avoid a large amount of redundant computation for global feature and get useful local information, we should obtain discriminative regions which are located to complete feature extraction of the next scale by zooming the pivotal region. For this purpose, the pivotal discrimination region is defined as a square which effectively reduces the parameters in the model and can be represented by three parameters as follows:

$$[r_x, r_y, s_l] = Z(I_z * X). \tag{2}$$

Setting the plane rectangular coordinate system with the origin of original image' upper left, which x-axis and y-axis are defined from left to right and top to bottom. Where r_x, r_y denotes the region's center coordinates, s_l denotes the half of the square's side length. I_z is overall parameters of the Zoom-Net, and $Z(\cdot)$ denotes fully-connected layers which can generate three region parameters.

After the localization of the pivotal region has been generated via Zoom-Net, we will further shrink to the proposed scale with higher resolution to extract more fine-grained features. Similarly, we set the starting point of the upper left corner in input images at each level as the origin of a pixel coordinate system. Accordingly, the coordinates of the top-left (tl) point and bottom-right (br) point of the pivotal discrimination region can be described:

$$r_{x(tl)} = r_x - s_l, r_{y(tl)} = r_y - s_l, r_{x(br)} = r_x + s_l, r_{x(br)} = r_y + s_l. \tag{3}$$

By multiplying the variables of the pivotal discrimination region and the pivotal discrimination region on the previous scale, a further cropping operation can be realized with the responding value of the pivotal feature:

$$X_{zoom} = X \odot [f(x - r_{x(tl)}) - f(x - r_{x(br)})] \cdot [f(y - r_{y(tl)}) - f(y - r_{y(br)})], \tag{4}$$

where $f(\cdot)$ donates sigmoid function, $\odot$ refers to the multiplication between variables. Thus, the pivotal discrimination region X_{zoom} is obtained which can make further feature extraction.

3.2   Visual-Semantic Cohesion Network

For traditional ZSL task, the method based on independent semantics and semantic embedding has a good performance. But they still have their own fundamental flaws and do not effectively combine with deep learning. According to their respective advantages, we make a bridge that each independent semantics learned independently from the given pivotal features and achieves a learnable mapping method to the visual embedding space in deep learning. In this section, we first introduce the implementation details on visual embedding and semantic embedding based on deep learning. Then, the relationship between them and specific process of combination are listed to prove the effectiveness of this task.

Visual Embedding and Semantic Embedding. Inspired from the strategy of sharing the parameters among attributes and common feature extraction mechanism in CNN, the attribute predictor a_k in visual embedding process after iteration network in deep learning can be represented as follows:

$$a_k(x; t_k, \Theta) = \sigma \left(t_k^T \theta(x; \Theta) \right), \tag{5}$$

where t_k is a parameter vector that contains parameter Θ involved. Then, put it through a sigmoid function with feature extractor $\theta(x; \Theta)$. In order to learn the mapping function from sample x to attribute vector s, set the training set and representations as follows:

$$\mathcal{D} = \left\{ (x^{(i)}, s^{(i)})_{i=1}^N \right\}, \tag{6}$$

$$s^{(i)} = \left(s_1^{(i)}, ..., s_Q^{(i)} \right), \tag{7}$$

where Q attribute labels are indicated as $s_Q^{(i)}$. y denotes a specific class and v denotes the semantic vocabulary of x. Loss function can be construed as a cross-entropy loss, $L_s(v, y) = -y \log(v) - (1 - y) \log(1 - v)$. The parameters t_k and Θ can be optimized through a process as follows:

$$\mathcal{R}[a_1, ..., a_Q, \mathcal{D}] = \sum_i \sum_k L_s \left(a_k(x^{(i)}; t_k, \Theta), s_k^{(i)} \right). \tag{8}$$

As for semantic embedding, it is important to balance the semantic space constraints and the conditions for establishing attribute models. Given Q binary attributes and C classes, the each class can transform into a one-hot coding scheme as follows:

$$\phi_k(y) = \begin{cases} 1, & \text{if class } y \text{ contains attribute } k \\ -1, & \text{if class } y \text{ lacks attribute } k \end{cases}, \tag{9}$$

where $\phi(y)$ is a defined mapping from attributes to labels in space for class y, containing Q dimensional vectors. Additionally, adding a fully connected layer for Q and parameters T to ensure implementing in deep learning. Thus, an output obtained as follows:

$$h(x; \mathbf{T}, \Theta) = \Phi^T a(x) = \Phi^T \mathbf{T}^T \theta(x; \Theta), \tag{10}$$

where $\theta(x; \Theta)$ is the feature representation. $\mathbf{T}$ is a learnable mapping matrix between feature and semantic space. A cross-entropy loss between the output in CNN and samples can be denoted as L. Similarly, in Eq. 6, $y^{(i)}$ later introduced is the class of $x^{(i)}$. The parameters $\mathbf{T}$ and θ can be optimized as follows:

$$\mathcal{R}[h, \mathcal{D}] = \sum_i L\left(h(x^{(i)}; \mathbf{T}, \Theta), y^{(i)}\right). \tag{11}$$

Relationship and Complementarity. Comparing both loss function, inter-relationship and respective defects are obvious. The implementation of visual embedding achieves a strong constraint on each $a_k(x)$. As for semantic embedding, though it provides a semantic space for sample s, class c and establishes an abundant bond between attributes, redundant information and space.

Assume that $\mathcal{K}$ denotes semantic space of attribute vectors in training dataset. Accordingly, there is a $\mathcal{M}$ making the equation $\mathcal{U} = \mathcal{M} - \mathcal{K}$ true. $\mathcal{K}$ tend to be much smaller than the $\mathcal{M}$ in semantic space. If the attribute vectors of test dataset are in $\mathcal{U}$, the classifier trained by the training sample cannot recognize them correctly. According to the process of visual embedding, $\mathcal{U}$ becomes small through making $\mathcal{M}$ and $\mathcal{K}$ similar. On the contrary, the method based on semantic embedding cannot deal with the class in $\mathcal{U}$ in the same parameter settings. In summary, they can't perform well when $\mathcal{U}$ is in an extreme situation. But there is also the possibility of complementarity and desirability. To make constraints on independent attributes after CNN into the process of semantic embedding.

3.3   Co-consistent Regularization

Scoring and recognition task can be implemented by the mean of nearest neighbor in semantic embedding space after visual embeddings and semantic embeddings cohesion.

Specifically, given an image, visual embedding can be obtained through visual embedding network. Afterwards, visual embedding is mapped to semantic embedding by visual-semantic cohesion network. Finally, a dot-product computation process between the consistent regularization semantic embeddings and the projected embedding gets the score. Therefore, the score can be denoted as the form of confidence score h in a rough form as follows:

$$h_c(x) = \langle W_c, g(x) \rangle, \tag{12}$$

where $g(\cdot)$ denote a predictor. And w_c is the decision of set for C class through the one-hot counting process. It is a learnable function that to interpret as follows:

$$h^* = \arg\min_h \left(\mathcal{R}_E[h] + \lambda \Omega[h]\right), \tag{13}$$

where $\mathcal{R}_E[h]$ denotes the process that $h^*(\cdot)$ can be optimized by risk function $\mathcal{R}$. And λ is a Lagrange multiplier, usually $\lambda \geq 0$. $\Omega[h]$ is a process of joint constraint and it will be covered later. The predictor g can learn from cross-entropy as a further constraint for semantic embedding, and it has a new classifier representation for combination as follows:

$$h(x; W, \mathbf{T}, \Theta) = W^T f(x) = W^T \mathbf{T}^T \theta(x; \Theta), \tag{14}$$

where W contains the weight of w_c in semantic embedding CNN. With the weight decreases gradually in CNN learning for λ in Eq. 13, a common rule-generalization can shorten as follows:

$$\Omega[W] = -\sum_{c=1}^{C} w_c^T \phi(c). \tag{15}$$

The use of rule-generalization achieves a constraint for output to make it as close as possible to the defined representation of the attribute vector. By integrating new loss function of visual embedding and semantic embedding and adding a rule-generalization, a Co-consistent Regularization (CcR) function is denoted as:

$$\underset{\Theta, \mathbf{T}, W}{\text{minimize}} \sum_i L\left(h(x^{(i)}; W, \mathbf{T}, \Theta), y^{(i)}\right) + \lambda \sum_i \sum_k L_s\left(f_k(x^{(i)}; t_k, \Theta), s^{(i)k}\right) + \beta \Omega[W], \tag{16}$$

where $f_k(\cdot)$ denotes the kth predictor, λ and β are Lagrange multipliers that control the different degree of the constraints above. In this way, the method manages to build an unconstrained semantic space and achieve the constraint on the mapping of sample X to attribute vector to reduce unseen classes through the first two functions. Adjusting the two super parameters can achieve better results than independent methods. CcR needs deep learning to achieve a good effect. To be more specific, $\theta(x; \Theta)$ in Eq. 14 is computed by a CNN. Parameters Θ, W and $\mathbf{T}$ can be learned from Eq. 17 by using multiple semantic state sets. From Eq. 12, the class scores can also be listed as follows:

$$h_c(x) = \sum_k h_c^{(k)}(x) = \sum_k \left\langle W_{s_k^c}^{(k)}, f_k(x) \right\rangle, \tag{17}$$

where s_k^c denotes the state of the kth semantic in class c, $W_{s_k^c}^{(k)}$ the corresponding codeword, and $f_k(\cdot)$ the corresponding subspace of $f(\cdot)$. Semantic predictions are obtained by computing the dot-products. After the scoring network, we apply a softmax to produce the predicted probability vector for training and test classes. Obviously, the final predicted class is one with the highest score.

4 Experiment

4.1 Experimental Setup

Database. We evaluate our method and make comparison with the state-of-the-art methods. The considered benchmark datasets have diverse domains and scales, so that the performance can be validated and illustrated from different aspects. Three datasets are used as follows: Animals with Attributes 2 (AwA2) [21], Caltech-UCSD Birds 200-2011 (CUB) [19] and SUN [14]. Table 1 summarizes their key statistics and the detailed settings.

Table 1. Summary of three datasets.

Dataset	Images	Train/ZS classes	Labeled (per class)
AwA2 [21]	33,322	40/10	750
CUB [19]	11,788	150/50	60
SUN [14]	14,340	645/72	20

Implementation Details. The iteration network is initialized using two learning rates 0.0005 and 0.0001 of pre-trained CNN models on GoogLeNet and VGG-19 to learn $\phi(c)$, respectively. Due to the large images and relative center position in the AwA2 dataset, the iteration network is set in two scales. For dataset CUB and SUN, three scales are contained because of the complex images. 224×224 is adopted in each scale for the input image or zoomed region.

Baselines. To prove the effectiveness of the different components in our method, six baselines are designed to compare with the proposed method.

- Single scale with RIS (S-RIS):
  The iteration network just contains one scale without zoomed operation. Then, features are mapped to attributes through RIS, an authoritative and official approach, achieve classification and recognition by given classifier.

- Single scale with RSE (S-RSE):
  Compared with S-RIS, same settings are adopted for using RSE to replace RIS. Thus, we can observe the difference between them and our framework.

- Single scale with Visual-Semantic Cohesion Network (S-VSC):
  While showing the superiority of iteration network, single scale network can also indicate the superiority of semantic cohesion network (VSC) than anyone of RIS and RSE even under the universal feature extraction method.

- Double scale with RIS (D-RIS):
  Two scales are contained with zoomed mechanism, so that it can prove the effectiveness of the iteration network and RIS in deep learning under the proposed framework.

- Double scale with RSE (D-RSE):
  Make sure whether RSE is valid and better than our combined model.

- Double scale with Visual-Semantic Cohesion Network (D-VSC):
  Compared with the baselines above, the whole iteration network and semantic cohesion network are added into this model. While the classifier to recognize the attributes is different from the proposed method.

4.2   Experimental Results

In this section, we first discuss the gains of superparameters λ and β to prove the effectiveness of CcR function on mean class accuracy (MCA). Then, we evaluate the ZSL performance of our method with modularized baselines and state-of-the-art method on three datasets.

The Gains of Superparameters. At the beginning of the experiments, the loss based on superparameters should be adjusted respectively to ensure other modules work well. We use visual attributes and two scales iteration network on two datasets to measure the profits, and set $w_c = \phi(c)$, $\lambda = 0$. As for the λ, it is evaluated by increasing the value while keeping $\beta = 0$, so that the MCA influenced by L and L_s changes in the limit of $\lambda \to \infty$. On the contrary, the gains of $\Omega[W]$ are measured by increasing β while keeping $\lambda = 0$. Fig. 3 shows a relative improvement in a function of the Lagrange multipliers with a GoogLeNet instead of iteration network. Both results perform over cohesion network with a high profit, which indicates the importance of deep learning frame rather than fixing the parameters. The points with good performance in Fig. 3 note that regularized union constraints for the subnetwork are effective to consistency between the semantic and classification codes. Enough flexibility for learning classification codes superior to normal corresponding semantic method. In all cases, our method is much better than the compared methods proving the significance of modeling attributes and the mapping process. Finally, visual-semantic network and scoring performance are also superior to CNN with single embedding space.

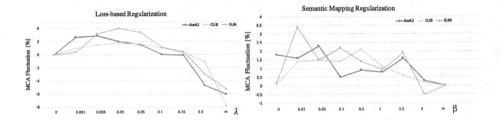

Fig. 3. Gain curves of loss-based and semantic mapping regularization of our method.

Comparison. We evaluate the proposed method with the state-of-the-art methods on AwA2, CUB, and SUN datasets for more comprehensive and detailed analysis. Note that there are many differences between methods, including CNN implementation, training and zero-shot class, and semantic space representation. In order to reduce such differences and perform the effectiveness of our separate modules, we choose [2, 15, 20] as reference to compare with feature extraction and semantic embedding network. Similarly, [4, 10, 17, 22] are chosen to compare with visual-semantic cohesion network. In summary, our method performs well on different datasets with multi-way classification accuracy. Though comparing with the state-of-the-art on GoogLeNet and VGG-19, our method has varying degrees of improvement in GoogLeNet with 1.5% on AwA2, 2.0% on CUB, and 2.8% on SUN, respectively. At the same time achieves a good performance in VGG-19 by 1.8% on AwA2, 1.7% on CUB, and 1.0% on SUN, respectively (Table 2).

Table 2. MCA results compared with different methods using GoogLeNet & VGG-19.

Method	AwA2	CUB	SUN
DAP [10]	57.2 (59.4)	44.5 (43.7)	25.1 (24.4)
ESZSL [17]	75.3 (69.3)	44.0 (43.4)	27.9 (26.1)
SJE [2]	66.7 (65.6)	50.1 (49.3)	25.1 (23.2)
LatEM [20]	71.9 (71.7)	45.5 (42.5)	26.3 (24.5)
JLSE [22]	80.4 (76.8)	42.1 (56.3)	30.2 (28.7)
Low-Rank [4]	82.8 (78.9)	44.8 (43.8)	29.5 (30.5)
JSLA [15]	82.7 (81.0)	52.1 (51.3)	28.3 (29.8)
S-RIS	74.3 (71.2)	42.3 (41.7)	24.6 (25.1)
S-RSE	75.5 (75.1)	43.2 (42.8)	25.2 (24.3)
S-VSC	78.3 (78.0)	45.5 (45.1)	26.2 (25.5)
D-RIS	79.2 (77.5)	47.9 (46.8)	26.5 (25.4)
D-RSE	78.7 (76.3)	49.1 (48.6)	27.1 (26.3)
D-VSC	81.2 (80.1)	51.8 (51.5)	29.4 (28.5)
Our method	**83.2 (82.8)**	**54.1 (53.0)**	**31.1 (30.8)**

5 Conclusions

In this paper, we proposed an end-to-end framework for ZSL problem. Our framework uses a cascaded model to extract the multi-scale discriminative features of the images. In addition, our framework integrates four modules, iteration network, visual-semantic cohesive network, scoring network and the classifier, which enables to train in an end-to-end frame. In this way, the generalization ability of the test set enhances greatly. By doing this, our framework achieved better results on the AwA2, CUB and SUN datasets.

Acknowledgments. This work was supported by National Natural Science Foundation of China (61772508, 61801428, U1713213), National Key R&D Program of China (2017YFB1402100), Zhejiang Provincial Natural Science Foundation (LY18F020034), Natural Science Basic Research Plan in Shaanxi Province of China (2017JM6101, 2017JM6060, 2017JQ6077, 2017JM6103), Guangdong Technology Project (2016B010108010, 2016B010125003, 2017B010110007), CAS Key Technology Talent Program, Shenzhen Engineering Laboratory for 3D Content Generating Technologies ([2017]476), Shenzhen Technology Project (JCYJ 20170413152535587,

JSGG20160331185256983, JSGG20160229115709109), Key Laboratory of Human-Machine Intelligence-Synergy Systems, Shenzhen Institutes of Advanced Technology, CAS (2014DP173025), Fundamental Research Funds for the Central Universities (GK201703060, GK201801004), Teaching Reform and Research Project of Shaanxi Normal University (17JG33).

References

1. Akata, Z., Perronnin, F., Harchaoui, Z., Schmid, C.: Label-embedding for image classification. TPAMI **38**(7), 1425–1438 (2016)
2. Akata, Z., Reed, S., Walter, D., Lee, H., Schiele, B.: Evaluation of output embeddings for fine-grained image classification. In: CVPR, pp. 2927–2936. IEEE Press, Boston (2015)
3. Blitzer, J., Foster, D.P., Kakade, S.M.: Zero-shot domain adaptation: a multiview approach. Technical report, TTI-TR-2009-1. Toyota Technological Institute, Chicago (2009)
4. Ding, Z., Shao, M., Fu, Y.: Low-rank embedded ensemble semantic dictionary for zero-shot learning. In: CVPR, pp. 2050–2058. IEEE Press, Honolulu (2017)
5. Fouhey, D., Gupta, A., Zisserman, A.: From images to 3D shape attributes. TPAMI **1**(1), 1–14 (2017)
6. Fu, J., Zheng, H., Mei, T.: Look closer to see better: recurrent attention convolutional neural network for ne-grained image recognition. In: CVPR, pp. 4476–4484. IEEE Press, Honolulu (2017)
7. Fu, Y., Hospedales, T.M., Xiang, T., Gong, S.: Learning multi-modal latent attributes. TPAMI **36**(2), 303–316 (2014)
8. Fu, Y., Hospedales, T.M., Xiang, T., Gong, S.: Transductive multi-view zero-shot learning. TPAMI **37**(11), 2332–2345 (2015)
9. Lampert, C.H., Nickisch, H., Harmeling, S.: Learning to detect unseen object classes by between-class attribute transfer. In: CVPR, pp. 951–958. IEEE Press, Miami (2009)
10. Lampert, C.H., Nickisch, H., Harmeling, S.: Attribute-based classification for zero-shot visual object categorization. TPAMI **36**(3), 453–465 (2014)
11. Lazaridou, A., Dinu, G., Baroni, M.: Hubness and pollution: delving into class-space mapping for zero-shot learning. In: IJCNLP, pp. 270–280. ACL, Beijing (2015)
12. Mikolov, T., Sutskever, I., Chen, K., Corrado, G.S., Dean, J.: Distributed representations of words and phrases and their compositionality. In: NIPS, pp. 3111–3119. Curran Associates, Long Beach (2013)
13. Morgado, P., Vasconcelos, N.: Semantically consistent regularization for zero-shot recognition. In: CVPR, pp. 10–16. IEEE Press, Honolulu (2017)
14. Patterson, G., Hays, J.: Sun attribute database: discovering, annotating, and recognizing scene attributes. In: CVPR, pp. 2751–2758. IEEE Press, Providence (2012)
15. Peng, P., Tian, Y., Xiang, T., Wang, Y., Pontil, M., Huang, T.: Joint semantic and latent attribute modelling for cross-class transfer learning. TPAMI **40**(7), 1625–1638 (2017)
16. Pennington, J., Socher, R., Manning, C.: Glove: global vectors for word representation. In: EMNLP, pp. 1532–1543. ACL, Doha (2014)
17. Romera-Paredes, B., Torr, P.H.S.: An embarrassingly simple approach to zero-shot learning. Visual Attributes. ACVPR, pp. 11–30. Springer, Cham (2017). https://doi.org/10.1007/978-3-319-50077-5_2

18. Tian, Y., Zhang, W., Zhang, Q., Lu, G., Wu, X.: Selective multi-convolutional region feature extraction based iterative discrimination CNN for fine-grained vehicle model recognition. In: ICPR, pp. 3279–3284. IEEE Press, Beijing (2018)
19. Welinder, P., et al.: Caltech-UCSD birds 200. Technical report CNS-TR-2010-001, California Institute of Technology (CIT) (2010)
20. Xian, Y., Akata, Z., Sharma, G., Nguyen, Q., Hein, M., Schiele, B.: Zero-shot recognition via structured prediction. In: CVPR, pp. 69–77. IEEE Press, Las Vegas (2016)
21. Xian, Y., Lampert, C.H., Schiele, B., Akata, Z.: Zero-shot learning-a comprehensive evaluation of the good, the bad and the ugly. In: CVPR, pp. 3077–3086. IEEE Press, Honolulu (2017)
22. Zhang, Z., Saligrama, V.: Zero-shot learning via joint latent similarity embedding. In: CVPR, pp. 6034–6042. IEEE Press, Las Vegas (2016)
23. Zhang, Z., Saligrama, V.: Zero-shot recognition via structured prediction. In: Leibe, B., Matas, J., Sebe, N., Welling, M. (eds.) ECCV 2016. LNCS, vol. 9911, pp. 533–548. Springer, Cham (2016). https://doi.org/10.1007/978-3-319-46478-7_33

Hybrid Networks: Improving Deep Learning Networks via Integrating Two Views of Images

Sunny Verma[1], Wei Liu[1(✉)], Chen Wang[2], and Liming Zhu[2]

[1] Advanced Analytics Institute, School of Software,
University of Technology Sydney, Sydney, Australia
Sunny.Verma@student.uts.edu.au, Wei.Liu@uts.edu.au
[2] CSIRO, Data61, Sydney, Australia
{Chen.Wang,Liming.Zhu}@data61.csiro.au

Abstract. The *principal component* analysis network (**PCANet**) is an unsupervised parsimonious deep network, utilizing *principal components* as filters in the layers. It creates an amalgamated view of the data by transforming it into column vectors which destroys its spatial structure while obtaining the principal components. In this research, we first propose a tensor-factorization based method referred as the **Tensor Factorization Networks** (**TFNet**). The **TFNet** retains the spatial structure of the data by preserving its individual modes. This presentation provides a minutiae view of the data while extracting matrix factors. However, the above methods are restricted to extract a single representation and thus incurs information loss. To alleviate this information loss with the above methods we propose **Hybrid Network** (**HybridNet**) to simultaneously learn filters from both the views of the data. Comprehensive results on multiple benchmark datasets validate the superiority of integrating both the views of the data in our proposed **HybridNet**.

Keywords: Tensor decomposition · Classification · Feature extraction

1 Introduction

Features extraction is an important operation in the development of classification tasks. This process has matured with a multitude of developments evolving from the machine learning, computer vision, data mining, and signal processing communities [26]. Today, in the era of deep learning, features are extracted by processing the data through multiple stacked convolution layers. The crux of feature extraction with deep architectures is to perform sophisticated operations with multiple layers in a sequential manner [1]. The features obtained by the deep networks promise better feature representations than the conventional shallow networks. However, these networks are trained via stochastic optimization techniques which necessitates multiple flops of the same data to effectively learn its

© Springer Nature Switzerland AG 2018
L. Cheng et al. (Eds.): ICONIP 2018, LNCS 11301, pp. 46–58, 2018.
https://doi.org/10.1007/978-3-030-04167-0_5

representation. This leads to longer training time while obtaining feature representations from the data. Furthermore, the fundamental operations utilized in deep networks are expensive regarding memory and space complexities. This limits the usability of deep architectures on micro devices like cellphones. The current research trend focuses on alleviating the above problem associated with the development of deep architectures [11,14].

PCANet is one such promising architecture: it is an unsupervised deep parsimonious network extracting *principal components* in its cascaded layers [4]. Due to the remarkable performance of **PCANet** on several benchmark face datasets, the network is currently accepted as a simple deep learning baseline for image classification. However, the features extracted by **PCANet** (and its later variant **FANet** [13]) does not achieve similar performance on challenging object recognitions datasets like CIFAR-10 [15]. There is a major reason for this performance degradation: vectorizing image patches (which we call the amalgamated view of the data) while extracting *principal components*. This results in loss of spatial information present in the images. This loss is amplified when one vectorizes an *RGB*-image which incurs the loss of both color and spatial information present in the data. However, this operation is inherent with the *principal components* analysis and necessitates the development of sophisticated techniques to reduce this information loss.

In this paper, we explore the feasibility of reducing the spatial information in **PCANet** by first devising **Tensor Factorization Networks** (**TFNet**). The **TFNet** extracts features from the original multi-mode data (which we call the minutiae view of the data) by utilizing multi-linear algebraic operations in its cascaded deep architecture. Contrary to the **PCANet** the **TFNet** *does not vectorizes the data while learning its convolution filters and hence preserves the spatial information present in the data*. Also, each mode of the multi-mode data is decomposed individually providing several degrees of freedom to the filter learning procedure in the **TFNet**.

We then propose the **Hybrid Network** (**HybridNet**) which integrates the advantages of both the **PCANet** and the **TFNet**. The **HybridNet** utilizes both tensor and matrix decompositions techniques while obtaining features representations from different views of the data. Our hypothesis is that the information from either the amalgamated view or the minutiae view is individually insufficient for classification. Since the information captured from the two views contains complementary information and hence both of them are necessary and their integration can enhance the performance of classification systems. To validate our claims we utilize multiple real world benchmark datasets to extensively evaluate the classification performance of the features obtained through the **PCANet**, the **TFNet**, and the **HybridNet**.

We summarize our contributions in this paper as follows:

- We propose **Tensor Factorized Network** (**TFNet**) which preserves the spatial information present in the data and enables extraction of matrix factors from the minutiae view of the tensorial data.

- We propose **Hybrid Network (HybridNet)** which integrates the filter learning procedure from the amalgamated view and the minutiae view of the data and simultaneously extracts features from them.
- We perform comprehensive evaluations with the features obtained via **PCANet**, **TFNet**, and the **HybridNet** on multiple benchmark real world datasets.

The rest of the paper is organized in the following sections: prior work (i.e. the **PCANet**) and tensor preliminaries is presented in Sect. 2. Our proposed **TFNet** and **HybridNet** are presented in Sects. 3 and 4 respectively. Next we describe the experimental setup, results and discussions in Sect. 5 and finally the conclusions in Sect. 6.

2   Background

2.1   PCANet

The **PCANet**'s 3-layer architecture is summarized in this section. Assume that there are N input training images denoted as $\{I_i\}_{i=1}^N$ of size m×n. Also, assume learning L_1 and L_2 number of filters in the first and the second layer respectively.

The First Layer. The procedure begins by extracting overlapping patches of size $k_1 \times k_2$ around each pixel in the image; the patches from image I_i are denoted as $\mathbf{x}_{i,1}, \mathbf{x}_{i,2}, ..., \mathbf{x}_{i,\tilde{m}\tilde{n}} \in \mathbb{R}^{k_1 k_2}$, where $\tilde{m} = m - \lceil \frac{k_1}{2} \rceil$ and $\tilde{n} = n - \lceil \frac{k_2}{2} \rceil$, where $\lceil z \rceil$ gives the smallest integer greater than or equal to z. Then, the obtained patches are *vectorized* and the mean of the image patches is subtracted from them to obtain the patch matrix as $\boldsymbol{X}_i \in \mathbb{R}^{k_1 k_2 \times \tilde{m}\tilde{n}}$. Obtaining patch representation for all the images one obtains

$$\boldsymbol{X} \in \mathbb{R}^{k_1 k_2 \times N\tilde{m}\tilde{n}} \tag{1}$$

the *PCA* minimizes the reconstruction error with a family of orthonormal filters known as L_1 principal eigenvectors of $\boldsymbol{X}\boldsymbol{X}^T$ calculated as below:

$$\min_{V \in \mathbb{R}^{k_1 k_2 \times L_1}} \| \boldsymbol{X} - VV^T \boldsymbol{X} \|_F, \quad s.t. \quad V^T V = I_{L_1} \tag{2}$$

where I_{L_1} is an identity matrix of size $L_1 \times L_1$, the filters are then expressed as:

$$W^1_{l_{\text{PCANet}}} = mat_{k_1,k_2}(ql(\boldsymbol{X}\boldsymbol{X}^T)) \in \mathbb{R}^{k_1 \times k_2}, \quad l = 1, 2, ..., L_1 \tag{3}$$

where $mat_{k_1,k_2}(v)$ is a function that maps $v \in \mathbb{R}^{k_1 k_2}$ to a matrix $W \in \mathbb{R}^{k_1 \times k_2}$, and $ql(\boldsymbol{X}\boldsymbol{X}^T)$ denotes the l-th principal eigenvector of $\boldsymbol{X}\boldsymbol{X}^T$. Each input image I_i in this layer are then convolved with the L_1 filters obtained as below:

$$I^l_{i_{\text{PCANet}}} = I_i * W^1_{l_{\text{PCANet}}}, \quad i = 1, 2, ..., N, \quad l = 1, 2, ..., L_1 \tag{4}$$

where $*$ denotes the 2D convolution. The boundary of image I_i is zero-padded before convolution to obtain $I^l_{i_{\text{PCANet}}}$ with the same dimensions as in I_i. From Eq. 4, one obtains $N \times L_1$ outputs attributed as the output from the first layer.

The Second Layer. In the second layer, the overlapping patches from the input images in this layer (i.e., $I^l_{i_{PCANet}}$) are collected and then the mean of the patches is subtracted from them. Next the patches are vectorized to obtain the final patch matrix which is factorized to obtain the PCA filters:

$$Y \in \mathbb{R}^{k_1 k_2 \times L_1 N \tilde{m} \tilde{n}} \tag{5}$$

$$W^2_{l_{PCANet}} = mat_{k_1,k_2}(ql(YY^T)) \in \mathbb{R}^{k_1 \times k_2}, \quad l = 1, 2, ..., L_2 \tag{6}$$

we then convolve input images in the this layer with the L_2 filters to obtain the output from this layer and proceed with the next layer of the network

$$O^l_{i_{PCANet}} = I^l_{i_{PCANet}} * W^2_{l_{PCANet}}, \quad i = 1, 2, ..., NL_1, \quad l = 1, 2, ..., L_2 \tag{7}$$

The Output Layer. In the final output layer, the convolution outputs from the previous layers of **PCANet** are combined to obtain the final feature vector. First, each of the real-valued outputs from Eq. 7 are binarized by using a Heaviside function $H(O^l_{i_{PCANet}})$ which converts positive entries to 1 otherwise 0. Then the L_2 outputs in the second layer corresponding to the L_1 outputs in the first layer are combined by summing and multiplying with weights. This converts them back into a single image whose pixel value is in the range $[0, 2^{L_2} - 1]$:

$$\mathcal{I}^l_{i_{PCANet}} = \sum_{l=1}^{L_2} 2^{l-1} H(O^2_{l_{PCANet}}) \tag{8}$$

Then, each of the L_1 images from Eq. 8 are partitioned into B blocks and then for each block a histogram is computed with 2^{L_2} bins. Finally the histograms from B blocks are concatenated and denoted as $Bhist(\mathcal{I}^l_{i_{PCANet}})$. This block-wise encoding process encapsulates the L_1 images from Eq. 8 into a feature vector as:

$$f_{i_{PCANet}} = [Bhist(\mathcal{I}^1_{i_{PCANet}}), ..., Bhist(\mathcal{I}^{L_1}_{i_{PCANet}})]^T \in \mathbb{R}^{(2^{L_2})L_1 B} \tag{9}$$

One can now utilize these feature vectors to perform classification.

2.2   Tensor Preliminaries

Tensors are multi-mode arrays, where the modes (also known as orders) of a tensor are analogous to rows and columns (i.e., the two modes) of a matrix. Vectors are defined as first order tensors denoted as x, whereas matrices are defined as second order tensors denoted as X. Tensors are of order-3 or higher and are denoted as $\mathcal{X}$. Few important tensors operations utilized in this paper are defined below.

Tensor Unfolding: Also known as tensor matriziation, is the way of rearranging the elements of an n-mode tensor $\mathcal{X} \in \mathbb{R}^{i_1 \times i_2 ... \times i_N}$ as a matrix in chosen mode n denoted as $X_{(n)} \in \mathbb{R}^{i_n \times j}$, where $j = i_1 ... \times i_{n-1} \times i_{n+1} ... \times i_N$.

Tensor to matrix multiplication: The n-mode tensor product of matrix $A \in \mathbb{R}^{j \times i_n}$ with tensor $\mathcal{X} \in \mathbb{R}^{i_1 ... \times i_{m-1} \times i_m \times i_{m+1} ... \times i_n}$ is denoted as $\mathcal{X} \times_n A$, which results in another tensor $\hat{\mathcal{X}}$ of size $\mathbb{R}^{i_1 \times i_2 \times i_{n-1} \times j \times i_{n+1} ... \times i_n}$.

Tensor Decomposition. Tensor decomposition is a form of generalized of matrix decomposition for factorizing tensors. The algorithm factorizes an n-mode tensor $\mathcal{X} \in \mathbb{R}^{i_1 \times i_2 \cdots \times i_n}$ into two subcomponents: (1) $\mathcal{G} \in \mathbb{R}^{r_1 \times r_2 \cdots \times r_n}$ which is a lower dimensional tensor called the *core-tensor* and, (2) $\mathbf{U}^{(n)} \in \mathbb{R}^{r_n \times i_n}$ which are matrix factors associated with each mode of the tensor. Entries in the *core tensor* $\mathcal{G}$ represents the level of interaction between different components. By contrast, entries in the factor matrices $\mathbf{U}^{(n)}$ can be thought as the *principal components* associated with the mode-n. This form of tensor factorization falls under the *Tucker* family of tensor decomposition [5]. The original tensor can be reconstructed by taking product of the *core-tensor* with the factor matrices as:

$$\mathcal{G} \times_1 \mathbf{U}^{(1)} \times_2 \mathbf{U}^{(2)} \cdots \times_N \mathbf{U}^{(n)} \approx \mathcal{X} \tag{10}$$

The advantages of *Tucker* based factorization methods has already studied in several domains. In computer vision, [25] applied them to the face recognition problem and popularized them as Tensor faces. In data mining, [20] considered the problem of handwritten digits recognition through tensor factorization. In signal processing, [5,7] considered the problem of brain signal analysis with tucker decomposition.

3   The Tensor Factorization Network (TFNet)

The development of **TFNet** is motivated by the information loss which occurs while vectorizing image-patches in **PCANet**. This transformation is inherent while extracting the *principal components* and incurs the loss of geometric structure present in the data. Furthermore, the vectorization of the data results in high dimensional vectors which generally requires more computational resources. Motivated by the above shortcomings with **PCANet** we propose **TFNet** which is computationally efficient and extracts information while preserving the spatial structure of the data for obtaining feature representation.

3.1   The First Layer

Similar to the first layer in **PCANet**, we collect all overlapping patches of size $k_1 \times k_2$ around each pixel from the image $\{I_i\}$. However, contrary to **PCANet** here the obtained patches forms a 3-mode tensor $\mathcal{X}_i \in \mathbb{R}^{k_1 \times k_2 \times \tilde{m}\tilde{n}}$ instead of a matrix. The mode-1 and mode-2 of this tensor represents the row-space and column-space spanned by the pixels in the image, while the third mode of this tensors represents the total number of image patches and we obtain

$$\mathcal{X} \in \mathbb{R}^{k_1 \times k_2 \times N\tilde{m}\tilde{n}} \tag{11}$$

as our final tensor. We decompose the tensor using our custom-designed *LoMOI* algorithm presented in Algorithm 1 to obtain the factor matrices corresponding to the first two modes, which are later utilized in our tensorial filter generation.

Algorithm 1. Left One Mode Out Orthogonal Iteration (LoMOI)

1: **Input:** n-mode tensor $\mathcal{X} \in \mathbb{R}^{i_1, i_2, \dots, i_n}$; factorization ranks for each mode of the tensor $[r_1 \dots r_{m-1}, r_{m+1} \dots r_n]$, where $r_k \leq i_k \forall\ k \in 1, 2, \dots, n$ and $k \neq m$; factorization error-tolerance ε, and Maximum allowable iterations $= Maxiter$, $m = $ mode to discard while factorizing
2: **for** $i = 1, 2, \dots, n$ and $i \neq m$ **do**
3:     $\mathbf{X}_i \leftarrow$ unfold tensor $\mathcal{X}$ on mode-i
4:     $\mathbf{U}^{(i)} \leftarrow r_i$ left singular vectors of $\mathbf{X}_i$ ▷ extract leading r_i matrix factors
5: $\mathcal{G} \leftarrow \mathcal{X} \times_1 (\mathbf{U}^{(1)})^T \dots \times_{m-1} (\mathbf{U}^{(m-1)})^T \times_{m+1} (\mathbf{U}^{(m+1)})^T \dots \times_n (\mathbf{U}^{(n)})^T$ ▷ Core tensor
6: $\hat{\mathcal{X}} \leftarrow \mathcal{G} \times_1 (\mathbf{U}^{(1)})^T \dots \times_{m-1} (\mathbf{U}^{(m-1)})^T \times_{m+1} (\mathbf{U}^{(m+1)})^T \times_N \mathbf{U}^{(n)}$ ▷ reconstructed tensor obtained by multilinear product of the core-tensor with the factor-matrices; Eq. 10.
7: $loss \leftarrow \|\mathcal{X} - \hat{\mathcal{X}}\|$ ▷ decomposition loss
8: $count \leftarrow 0$
9: **while** $[(loss \geq \varepsilon)\ Or\ (Maxiter \leq count)]$ **do**   ▷ loop until convergence
10:     **for** $i = 1, 2, \dots, n$ and $i \neq m$ **do**
11:         $\mathcal{Y} \leftarrow \mathcal{X} \times_1 (\mathbf{U}^{(1)})^T \dots \times_{(i-1)} (\mathbf{U}^{(i-1)})^T \times_{(i+1)} (\mathbf{U}^{(i+1)})^T \dots \times_n (\mathbf{U}^{(n)})^T$ ▷ obtain the variance in mode-i
12:         $\mathbf{Y}_i \leftarrow$ unfold tensor $\mathcal{Y}$ on mode-i
13:         $\mathbf{U}^{(i)} \leftarrow r_i$ left singular vectors of $\mathbf{Y}_i$
14:     $\mathcal{G} \leftarrow \mathcal{X} \times_1 (\mathbf{U}^{(1)})^T \dots \times_{(m-1)} (\mathbf{U}^{(m-1)})^T \times_{(m+1)} (\mathbf{U}^{(m+1)})^T \dots \times_n (\mathbf{U}^{(n)})^T$
15:     $\hat{\mathcal{X}} \leftarrow \mathcal{G} \times_1 \mathbf{U}^{(1)} \dots \times_{(m-1)} (\mathbf{U}^{(m-1)})^T \times_{(m+1)} (\mathbf{U}^{(m+1)})^T \dots \times_n \mathbf{U}^{(n)}$
16:     $loss \leftarrow \|\mathcal{X} - \hat{\mathcal{X}}\|$
17:     $count \leftarrow count + 1$
18: **Output:** $\hat{\mathcal{X}}$ the reconstructed tensor and $[\mathbf{U}^{(1)} \dots \mathbf{U}^{(m-1)}, \mathbf{U}^{(m+1)} \dots \mathbf{U}^{(n)}]$ the factor matrices

---

$$[\hat{\mathcal{X}}, \mathbf{U}^{(1)}, \mathbf{U}^{(2)}] \leftarrow LoMOI(\mathcal{X}, r_1, r_2) \tag{12}$$

where $\hat{\mathcal{X}} \in \mathbb{R}^{r_1 \times r_2 \times N\tilde{m}\tilde{n}}$, $\mathbf{U}^{(1)} \in \mathbb{R}^{k_1 \times r_1}$, and $\mathbf{U}^{(2)} \in \mathbb{R}^{k_2 \times r_2}$. We discard obtaining the matrix factors from mode-3 i.e. $\mathbf{U}^{(3)}$ of the tensor as the *mode*-3 matricization of tensor $\mathcal{X}$ denoted as $\mathbf{X}_3 \in \mathbb{R}^{N\tilde{m}\tilde{n} \times k_1 \times k_2}$ is equivalent to the transpose of the patches matrix $\mathbf{X}$ defined in Eq. 1 which is not decomposed in the **PCANet** while obtaining their filters. A total of $L_1 = r_1 \times r_2$ filters (equivalent to the number of filters in the **PCANet**) are obtained from the factor matrices $\mathbf{U}^{(1)}$ and $\mathbf{U}^{(2)}$ as:

$$W^1_{l_{\mathbf{TFNet}}} = \mathbf{U}^{(1)}_{(:,i)} \otimes \mathbf{U}^{(2)}_{(:,j)} \in \mathbb{R}^{k_1 \times k_2}, \ i = 1 \dots r_1, \ j = 1 \dots r_2, \ l = 1 \dots L_1 \tag{13}$$

where '$\otimes$' is the outer product between two vectors and $\mathbf{U}^{(m)}_{(:,i)}$ represents 'i^{th}' column of the 'm^{th}' factor matrix. Our filters obtained in Eq. 13 does not require any explicit reshaping as the operation *outer*-product between two vectors naturally results in a matrix. Hence, we can straightforwardly convolve our tensorial filters with the input images to obtain output from the first stage as:

$$I^l_{i_{\mathbf{TFNet}}} = I_i * W^1_{l_{\mathbf{TFNet}}}, \quad i = 1, 2, \dots, N, \quad l = 1, 2, \dots, L_1 \tag{14}$$

When the data are *RGB*-images, every patch $\mathbf{x}_{i,j}$ extracted from image is a 3-order tensor $\mathcal{X} \in \mathbb{R}^{k_1 \times k_2 \times 3}$ (*RowPixels* $\times$ *ColPixels* $\times$ Color). After collecting image patches from the training images, we obtain a 4-mode tensor

$$\mathcal{X} \in \mathbb{R}^{k_1 \times k_2 \times 3 \times N\tilde{m}\tilde{n}} \tag{15}$$

decomposing the above tensor to obtain factor matrices are as follows:

$$[\hat{\mathbf{X}}, \mathbf{U}^{(1)}, \mathbf{U}^{(2)}, \mathbf{U}^{(3)}] \leftarrow LoMOI(\mathbf{X}, r_1, r_2, r_3) \tag{16}$$

$$W^1_{l_{\textbf{TFNet}}} = U^{(1)}_{(:,i)} \otimes U^{(2)}_{(:,j)} \otimes U^{(3)}_{(:,k)} \quad \forall i \in 1...r_1, \ j \in 1...r_2, \ k \in 1...r_3 \tag{17}$$

3.2   The Second Layer

Similar to the first layer, we extract overlapping patches from the input images and then subtract the patch mean from the patches and build a 3-mode tensor denoted as $\mathbf{y}_i$ and then decompose it to obtain our factor matrices as:

$$[\hat{\mathbf{y}}, \mathbf{V}^{(1)}, \mathbf{V}^{(2)}] \leftarrow LoMOI(\mathbf{y}, r_1, r_2) \tag{18}$$

where, $\hat{\mathbf{y}} \in \mathbb{R}^{r_1 \times r_2 \times NL_1 \tilde{m}\tilde{n}}$, $\mathbf{V}^{(1)} \in \mathbb{R}^{k_1 \times r_1}$, and $\mathbf{V}^{(2)} \in \mathbb{R}^{k_2 \times r_2}$. We then generate our tensorial filters from the matrix factors of the first two modes as:

$$W^2_{l_{\textbf{TFNet}}} = \mathbf{V}^{(1)}_{(:,i)} \otimes \mathbf{V}^{(2)}_{(:,j)} \in \mathbb{R}^{k_1 \times k_2}, \quad i = 1..r_1, \ j = 1..,r_2, \ l = 1...L_2 \tag{19}$$

Now, each of the L_1 input images in the first layer are convolved with tensorial filters obtained in the second layer as:

$$O^l_{i_{\textbf{TFNet}}} = I^l_{i_{\textbf{TFNet}}} * W^2_{l_{\textbf{TFNet}}}, \quad l = 1, 2, ..., L_2 \tag{20}$$

The number of output images obtained from this operation is equal to $L_1 \times L_2$. We now utilize the output layer of **PCANet** to obtain our final feature vectors

$$\mathcal{I}^l_{i_{\textbf{TDNet}}} = \sum_{l=1}^{L_2} 2^{l-1} H(O^2_{l_{\textbf{TensorNet}}}) \tag{21}$$

$$f_{i_{\textbf{TFNet}}} = [Bhist(\mathcal{I}^1_{i_{\textbf{TFNet}}}), ..., Bhist(\mathcal{I}^{L_1}_{i_{\textbf{TFNet}}})]^T \in \mathbb{R}^{(2^{L_2})L_1 B} \tag{22}$$

the features vectors obtained in Eq. 22 can now be utilized for classification.

4   The Hybrid Network (HybridNet)

The **PCANet** extracts features from the amalgamated view of the data whereas the **TFNet** extracts features from the minutiae view of the data. Our hypothesis is that both these views are important as they conceal distinct representations of the data and integrating feature representations from these views can enhance the performance of classification systems. Motivated by the above, we propose the **HybridNet** which integrates the filter learning process from the minutiae view and the amalgamated view. We explain the feature extraction procedure in **HybridNet** with the help of Fig. 1.

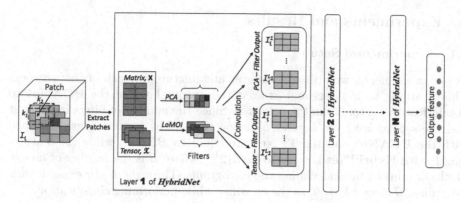

Fig. 1. The proposed hybrid network

4.1 The First Layer

The first layer in **HybridNet** consists of image-patches expressed as both as tensors and matrices. In this way, the first layer in **HybridNet** perceives more diverse information from different views of the data while learning its filters. Further, the filters for the tensorised patches were obtained via *LoMOI*, while the filters for the patch-matrices are obtained via the principal components. Since the first layer of **HybridNet** consists of hybrid filters, the output from this layer is obtained by convolving input images with: (a) the PCA-filters and (b) the tensorial-filters. This injects more diversity to the output from the first layer in **HybridNet** or equivalently to the input of the succeeding layer. Consequently, the covariance matrix in the **HybridNet** captures more variability than that of the covariance matrix obtained in either of the **PCANet** or **TFNet**. Therefore, *the hybrid filters captures more variability in the data which leads to better disentangled representations*. This results in superior performance from the features obtained with the **HybridNet**. Since we obtain L_1-PCA filters from the patch-matrices and L_1-tensor filters from the patch-tensors, a total of $2 \times L_1$ output images are obtained after the convolution of images with these filters.

4.2 The Second Layer

In the second layer of **HybridNet** the filters are learned with the hybrid data obtained from the first layer. Moreover, the output images from this layer are obtained by (a) convolving the L_1 images corresponding to the output from the PCA-filters in the first layer with the -filters in the second layer, and (b) convolving the L_1 images corresponding to the output from the tensor-filters in the first layer with the tensor-filters in the second layer. The number of output images obtained from the second layer produces a total of $2 \times L_1 \times L_2$ outputs. Finally, the outputs from the second layer of **HybridNet** are processed with the same **Output Layer** as in **PCANet** and **TFNet**, to obtain hybrid features.

5   Experiments and Results

5.1   Experimental Setup

In our experiments, we utilized two-layer architecture for each of the networks while learning their filters and utilized the output layer of the **PCANet** to obtain feature vectors from the networks. Since the number of filters in the first and the second layer are L_1 and L_2 respectively. The feature-length obtained with the **PCANet** and the **TFNet** are equal to $BL_1 2^{L_2}$, while the feature-length with **HybridNet** is equal to $2BL_1 2^{L_2}$; where B is the number of image-blocks obtained while calculating the histograms. Throughout our experiments, we utilized *Linear SVM* [9] as the classifier while performing classification.

5.2   Datasets

We utilize the following datasets and hyper-parameters in our experiments:

1. MNIST variations [16], which consists of 28×28 gray scale handwritten digits with controlled factors of variations such as background noise, rotations, background-images etc. Each variation contains $10K$ training and $50K$ testing images. We set, $L_1 = 9$, $L_2 = 8$, $k_1 = k_2 = 7$, with block size $= 7 \times 7$.[1]
2. CUReT texture dataset [24], consisting of 61 classes of image textures where each class has images of the same material with different pose, illumination conditions, specularity, shadowing, and surface normals. A subset of 92 cropped images were taken from each category as in [4,24]. Following the standard procedure in [4], we randomly split the data into train and test set with a split ratio of 50% and classification results are averaged over 10 trails. We set, $L_1 = 9$, $L_2 = 8$, $k_1 = k_2 = 5$, and the block size $= 50 \times 50$.
3. CIFAR-10 [15] consisting of $50K$ training and $10K$ testing images distributed among 10 classes. The RGB images are of dimensions 32×32 and vary significantly in object position, scale, colors, and textures. We vary L_1 as 9 and 27, keep $L_2 = 8$; whereas the patch sizes $k_1 = k_2$ are varied as 5, 7, and 9. Following [4] we also applied spatial pyramid pooling (SPP) [10] to the output layer of **PCANet**, while the block size $= 8 \times 8$. We additionally applied PCA to reduce the dimension of each pooled feature to 100.[2]

5.3   Results and Discussions

Classification errors obtained on handwritten digits variations and texture recognition datasets are reported in Table 1. The hybrid features obtained with our proposed **HybridNet** outperforms the state of the art results on five out of seven MNIST variations dataset. For texture classification the hybrid features achieves the lowest error among the three networks, however on this dataset

---

[1] Overlapping regions between the blocks is equal to half of the block size.
[2] Results does not vary significantly on increasing the projection dimensions.

Table 1. Classification error (%) obtained on MNIST variations and CuReT datasets

L_1	L_2	k_1	k_2	**PCANet** Error (%)	**TFNet** Error (%)	**HybridNet** Error (%)
8	8	5	5	34.80	32.57	**31.39**
27	8	5	5	26.43	29.25	**24.98**
8	8	7	7	39.92	37.19	**35.24**
27	8	7	7	30.08	32.57	**28.53**
8	8	9	9	43.91	39.65	**38.04**
27	8	9	9	33.94	34.79	**31.36**

Methods	Error (%)
Tiled CNN [18]	26.90
K-means [6] (1600 dim.)	22.10
Conv. Maxout [21]	11.68
NIN [17]	**10.41**
PCANet	26.43
TFNet	29.25
HybridNet	24.98

(a) Performance of **PCANet**, **TFNet**, and the **HybridNet** by varying hyperparamters

(b) Benchmark comparisons

Table 2. Classification error (%) obtained on CIFAR-10 without data augmentation

L_1	L_2	k_1	k_2	**PCANet** Error (%)	**TFNet** Error (%)	**HybridNet** Error (%)
8	8	5	5	34.80	32.57	**31.39**
27	8	5	5	26.43	29.25	**24.98**
8	8	7	7	39.92	37.19	**35.24**
27	8	7	7	30.08	32.57	**28.53**
8	8	9	9	43.91	39.65	**38.04**
27	8	9	9	33.94	34.79	**31.36**

Methods	Error (%)
Tiled CNN [18]	26.90
K-means [6] (1600 dim.)	22.10
Conv. Maxout [21]	11.68
NIN [17]	**10.41**
PCANet	26.43
TFNet	29.25
HybridNet	24.98

(a) Performance of **PCANet**, **TFNet**, and the **HybridNet** by varying hyperparamters

(b) Benchmark comparisons

they perform slightly lower than the state of the art. For object recognition the classification errors obtained on CIFAR-10 are reported in Table 2, again the hybrid features achieves the lowest error among the three networks studied in this paper. However on this dataset the performance of hybrid features is **14.57%** lower than state of the art - **NIN** i.e. **10.41%** (without data augmentation) [17]. This is because (a) **NIN** is comparatively deeper and more importantly (b) **NIN** performs supervised nonlinear feature extraction whereas the **HybridNet** performs an unsupervised linear feature extraction. However, the classification errors obtained with **HybridNet** are still promising and can be enhanced with more layers and non-linear operations. Besides, the above we have also evaluated the performance of **PCANet**, **TFNet** and the **HybridNet** by varying training data size on CIFAR and MNIST variation[3] dataset, shown in Fig. 2. The above experiments validates our claim of improving the classification accuracy by integrating the information from two views of the data.

---

[3] Random background, images, and rotation is utilized with block size = 4×4.

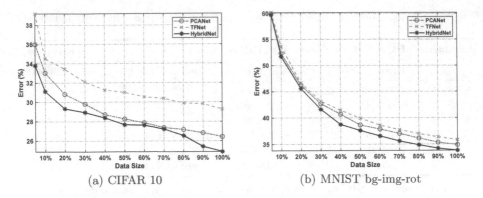

(a) CIFAR 10   (b) MNIST bg-img-rot

Fig. 2. Performance comparison by varying size of the training data

6   Conclusion and Future Work

In this paper we have first introduced **Tensor Factorization Networks** which preserves the spatial structure of the data while extracting features from the minutiae view of the data. Since both the amalgamated view and the minutiae view are individually insufficient feature representations, we propose a hybrid parsimonious network called the **Hybrid Network**. The **Hybrid Network** simultaneously learns its hybrid convolution filters by integrating the two views of the data. The features obtained through hybrid filters enhances the classification performance on several benchmark datasets. The experiments validates the advantages of obtaining superior feature representation by integrating the two views of the data. In future, we plan to study the effects of unaligned images during the filter learning phase in the **Hybrid Network**.

References

1. Bengio, Y., Courville, A., Vincent, P.: Representation learning: a review and new perspectives. IEEE Trans. Pattern Anal. Mach. Intell. **35**(8), 1798–1828 (2013)
2. Broadhurst, R.E.: Statistical estimation of histogram variation for texture classification. In: Proceedings International Workshop on Texture Analysis and Synthesis, pp. 25–30 (2005)
3. Bruna, J., Mallat, S.: Invariant scattering convolution networks. IEEE Trans. Pattern Anal. Mach. Intell. **35**(8), 1872–1886 (2013)
4. Chan, T.H., Jia, K., Gao, S., Lu, J., Zeng, Z., Ma, Y.: PCANet: a simple deep learning baseline for image classification? IEEE Trans. Image Process. **24**(12), 5017–5032 (2015)
5. Cichocki, A., et al.: Tensor decompositions for signal processing applications: from two-way to multiway component analysis. IEEE Sig. Process. Mag. **32**(2), 145–163 (2015)
6. Coates, A., Ng, A., Lee, H.: An analysis of single-layer networks in unsupervised feature learning. In: Proceedings of the Fourteenth International Conference on Artificial Intelligence and Statistics, pp. 215–223 (2011)

7. Cong, F., Lin, Q.H., Kuang, L.D., Gong, X.F., Astikainen, P., Ristaniemi, T.: Tensor decomposition of EEG signals: a brief review. J. Neurosci. Methods **248**, 59–69 (2015)
8. Crosier, M., Griffin, L.D.: Using basic image features for texture classification. Int. J. Comput. Vision **88**(3), 447–460 (2010)
9. Fan, R.E., Chang, K.W., Hsieh, C.J., Wang, X.R., Lin, C.J.: LIBLINEAR: a library for large linear classification. J. Mach. Learn. Res. **9**(Aug), 1871–1874 (2008)
10. Grauman, K., Darrell, T.: The pyramid match kernel: discriminative classification with sets of image features. In: Tenth IEEE International Conference on Computer Vision 2005, ICCV 2005, vol. 2, pp. 1458–1465. IEEE (2005)
11. Han, S., Mao, H., Dally, W.J.: Deep compression: compressing deep neural networks with pruning, trained quantization and huffman coding. In: ICLR (2016)
12. Hayman, E., Caputo, B., Fritz, M., Eklundh, J.-O.: On the significance of real-world conditions for material classification. In: Pajdla, T., Matas, J. (eds.) ECCV 2004. LNCS, vol. 3024, pp. 253–266. Springer, Heidelberg (2004). https://doi.org/10.1007/978-3-540-24673-2_21
13. Huang, J., Yuan, C.: FANet: factor analysis neural network. In: Arik, S., Huang, T., Lai, W.K., Liu, Q. (eds.) ICONIP 2015. LNCS, vol. 9491, pp. 172–181. Springer, Cham (2015). https://doi.org/10.1007/978-3-319-26555-1_20
14. Kossaifi, J., Khanna, A., Lipton, Z., Furlanello, T., Anandkumar, A.: Tensor contraction layers for parsimonious deep nets. In: 2017 IEEE Conference on Computer Vision and Pattern Recognition Workshops (CVPRW), pp. 1940–1946. IEEE (2017)
15. Krizhevsky, A., Hinton, G.: Learning multiple layers of features from tiny images. Master's thesis, Department of Computer Science, University of Toronto (2009)
16. Larochelle, H., Erhan, D., Courville, A., Bergstra, J., Bengio, Y.: An empirical evaluation of deep architectures on problems with many factors of variation. In: Proceedings of the 24th International Conference on Machine Learning, pp. 473–480. ACM (2007)
17. Lin, M., Chen, Q., Yan, S.: Network in network. In: ICLR (2013)
18. Ngiam, J., Chen, Z., Chia, D., Koh, P.W., Le, Q.V., Ng, A.Y.: Tiled convolutional neural networks. In: Advances in Neural Information Processing Systems, pp. 1279–1287 (2010)
19. Rifai, S., Vincent, P., Muller, X., Glorot, X., Bengio, Y.: Contractive auto-encoders: explicit invariance during feature extraction. In: Proceedings of the 28th International Conference on International Conference on Machine Learning, pp. 833–840. Omnipress (2011)
20. Savas, B., Eldén, L.: Handwritten digit classification using higher order singular value decomposition. Pattern Recogn. **40**(3), 993–1003 (2007)
21. Snoek, J., Larochelle, H., Adams, R.P.: Practical Bayesian optimization of machine learning algorithms. In: Advances in Neural Information Processing Systems, pp. 2951–2959 (2012)
22. Sohn, K., Lee, H.: Learning invariant representations with local transformations. In: Proceedings of the 29th International Conference on Machine Learning, pp. 1339–1346 (2012)
23. Sohn, K., Zhou, G., Lee, C., Lee, H.: Learning and selecting features jointly with point-wise gated Boltzmann machines. In: International Conference on Machine Learning, pp. 217–225 (2013)
24. Varma, M., Zisserman, A.: A statistical approach to material classification using image patch exemplars. IEEE Trans. Pattern Anal. Mach. Intell. **31**(11), 2032–2047 (2009)

25. Vasilescu, M.A.O., Terzopoulos, D.: Multilinear image analysis for facial recognition. In: 2002 Proceedings 16th International Conference on Pattern Recognition, vol. 2, pp. 511–514. IEEE (2002)
26. Zheng, L., Yang, Y., Tian, Q.: Sift meets CNN: a decade survey of instance retrieval. IEEE Trans. Pattern Anal. Mach. Intell. 40(5), 1224–1244 (2018)

On a Fitting of a Heaviside Function by Deep ReLU Neural Networks

Katsuyuki Hagiwara[✉]

Faculty of Education, Mie University,
1577 Kurima-Machiya-cho, Tsu 514-8507, Japan
hagi@edu.mie-u.ac.jp

Abstract. A recent research interest on deep neural networks is to understand why deep networks are preferred to shallow networks. In this article, we considered an advantage of a deep structure in realizing a heaviside function in training. This is significant not only as simple classification problems but also as a basis in constructing general non-smooth functions. A heaviside function can be well approximated by a difference of ReLUs if we can set extremely large weight values. However, it is not so easy to attain them in training. We showed that a heaviside function can be well represented without large weight values if we employ a deep structure. We also showed that update terms of weights at input side can be necessarily large if a network is trained to realize a heaviside function. Therefore, apparent acceleration of training is brought about by setting a small learning rate. As a result, we can say that, by employing a deep structure, a good fitting of heaviside function can be obtained within a reasonable training time under a moderate small learning rate. Our results suggest that a deep structure is effective in a practical training that requires a discontinuous output.

Keywords: Deep neural networks · ReLU · Heaviside function

1 Introduction

Deep neural networks are currently the most successful machine learning techniques in many applications and are extensively studied; e.g. [6]. A recent research interest on deep neural networks is to understand why deep networks are preferred to shallow networks. Several studies have been made on this theme in a function approximation problem [4,5,7] and a non-parametric regression problem [3]. [7] and [4] have employed an approach via polynomial expansions for approximating smooth target functions. They have shown that multiplication for constructing polynomial can be approximated by a deep structure that realizes a sawtooth function by ReLUs in [4] or binary expansion by heaviside units in [7]. However, in [5], multiplication is approximated by a linear sum of ReLUs, thus, two layers while the idea of approximating smooth functions is based on [4]. Therefore, the approximation precision may not depend on the depth and

© Springer Nature Switzerland AG 2018
L. Cheng et al. (Eds.): ICONIP 2018, LNCS 11301, pp. 59–69, 2018.
https://doi.org/10.1007/978-3-030-04167-0_6

the benefit of the deep structure seems to be unclear at least in considering smooth target functions. [5] has investigated a complexity (depth and number of weights) of ReLU neural networks in approximating a class of piecewise constant (classifier) functions and piecewise smooth functions that have discontinuity or jump at boundary of a region in the input domain; i.e. it is realized by a multiplication of a smooth function and a indicator function. The important idea in this work is that indicator functions on a set with a smoothly curved boundary can be approximated by a collection of horizon functions. Those can be realized by a ReLU network that is constructed by a concatenation of two networks that approximate a smooth function and a heaviside function. In a context of nonparametric regression problem, [3] has shown that ReLU neural networks have a better generalization performance compared to the other method such as kernel methods when a target function is non-smooth like a piecewise smooth function in [5]. In this work, evaluation of the approximation error is based on [4] and [5]. We intuitively understand that a superposition of smooth functions may not be efficient for approximating a discontinuous functions; e.g. Fourier series [3]. From [5] and [3], a deep structure seems to be efficient when a target function has a discontinuity since a type of layered structure is "needed" for constructing and merging a non-smooth part and smooth part. However, it is not so clear a deeper structure is "preferable"; e.g. a heaviside function that is a basis of constructing non-smooth part can be realized by a network with a single hidden layer. In considering applications of deep neural networks, a discontinuous target function may appear especially in classification problems. It is needed at least for minimizing error on training data since a target label takes a value in $\{0, 1\}$. Actually the most successful applications of deep neural networks may be recognition tasks.

Unlike these works, in this article, we focus on a practical aspect of a realization of discontinuous functions by deep neural networks. We especially consider an advantage of a deep structure in learning of a heaviside function by ReLU networks. Although this setting seems to be restrictive, it is attractive not only in simple classification problems but also in constructing general non-smooth functions as in [3,5]; i.e. discontinuous part of a target function is approximated by a certain transformation of heaviside functions. Our interest is a relationship among weight sizes, the number of layers, fitting accuracy and training speed. Although a relationship among weight sizes, the number of layers and generalization bound has been shortly mentioned in [3], there are no details in their paper. A brief introduction of our work here is as follows. It is well known that a heaviside function can be constructed by a difference of two ReLUs; e.g. [3,5]. Thus, it is realized by a network with a single hidden layer. However, in this case, we need extremely large weight values to approximate a heaviside function well. It takes much time to attain such weight values in training unless we choose a sufficiently large step size; e.g. a large learning rate. However, a large step size may simply lead to instability of training. Such a large step size may not be appropriate for the update of weights that have the other role. We show that this problem is relaxed by just employing a deep structure. In other words, we can obtain a good approximation of a heaviside function without obtaining large weight values by employing a deep structure. More specifically, we give constructions of network to achieve this performance. Based on this result, we furthermore consider an effect of deep structure on gradient-based learning of a heaviside function.

In Sect. 2, we show properties of ReLU and a benefit of deep structure in realizing a heaviside function. In Sect. 3, we consider an effect of deep structure in training. For simplicity, here, we employ a classical back-propagation under a modified ReLU network. We also show a numerical example that supports our idea. Section 4 is devoted for conclusions.

2   Approximation of a Heaviside Function

2.1   Properties of ReLU

We here give some properties of ReLU. Some of them are also in [5]. Activation function of ReLU(Rectified Linear Unit) is defined by

$$\phi(x) := \max(0, x) \tag{1}$$

for $x \in \mathbb{R}$. If $a_1 \in \mathbb{R}$ and $a_2 > 0$ then it is easy to check

$$\phi(a_2\phi(a_1 x)) = \phi(a_2 a_1 x), \tag{2}$$

while $\phi(a_2\phi(a_1 x)) = 0$ for any $x \in \mathbb{R}$ if $a_2 \leq 0$. Therefore, ReLU is closed under a composition if a connection weight is positive. It is easy to check that

$$\phi(x) - \phi(-x) = x \tag{3}$$

holds; i.e. a ReLU network can implement an identity function. By using difference of the two ReLUs (see Fig. 1), we define

$$\sigma(x) := \phi(x + 1/2) - \phi(x - 1/2). \tag{4}$$

We call σ DoReLU(Difference of ReLUs). This can be written as

$$\sigma(x) = \begin{cases} 1 & x > 1/2 \\ x + 1/2 & -1/2 < x \leq 1/2 \\ 0 & x \leq -1/2 \end{cases} \tag{5}$$

It is easy to check

$$\sigma(-x) = 1 - \sigma(x). \tag{6}$$

This implies that $\sigma(-x)$ and $\sigma(x)$ are line symmetry with respect to y-axis. Since $\sigma(x) \geq 0$ for any x, we can write

$$\sigma(x) = \phi(\phi(x + 1/2) - \phi(x - 1/2)). \tag{7}$$

This expression implies that σ is realized by a two layer network of ReLUs (see Fig. 1). We refer to this network as DoReLU network. It is easy to see that

$$\sigma(ax + b) = \phi(\phi(ax + 1/2 + b) - \phi(ax - 1/2 + b)). \tag{8}$$

Note that, for $a > 0$, we have

$$\sigma(ax) = \begin{cases} 1 & x > 1/(2a) \\ ax + 1/2 & -1/(2a) < x \le 1/(2a) \\ 0 & x \le -1/(2a) \end{cases} \tag{9}$$

Therefore, the non-constant region (linear region) becomes narrow and the slope at this region becomes steep if a is large. This fact plays an important role in this article. If $a > 0$ then we also have

$$\sigma(-ax) = 1 - \sigma(ax). \tag{10}$$

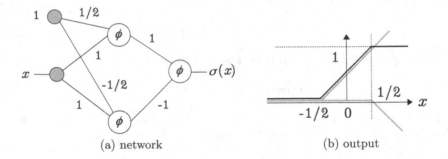

(a) network   (b) output

Fig. 1. DoReLU network.

Let χ_A be an indicator function on a set $A \subseteq \mathbb{R}$; i.e. $\chi_A(x) = 1$ if $x \in A$ and $\chi_A(x) = 0$ if $x \notin A$. For $x \in \mathbb{R}$, we calculate derivatives of σ and ϕ by

$$\sigma'(x) = \chi_{[-1/2, 1/2]}(x) \tag{11}$$
$$\phi'(x) = = \chi_{[0, \infty)}(x). \tag{12}$$

2.2 Approximation by a Simple Network

For $x = (x_1, \ldots, x_d) \in \mathbb{R}^d$, we define

$$h_w(x) = w_1 \cdot x - w_0, \tag{13}$$

where $w_1 = (w_1, \ldots, w_d)$ and $w = (w_1, w_0)$. "." denotes an inner product. We define a heaviside function on $\mathbb{R}^d$ by

$$s_w(x) = \begin{cases} 1 & w_1 \cdot x - w_0 > 0 \\ 0 & w_1 \cdot x - w_0 \le 0 \end{cases}. \tag{14}$$

We consider an approximation of a heaviside function with a parameter vector $w = w^*$; i.e. s_{w^*}. Firstly, we consider a simple network structure:

$$f_w(x) = \sigma(h_w(x)). \tag{15}$$

This is constructed by a DoReLU network, in which we need a modification of Fig. 1 to realize a d-dimensional input. We refer to this network as $\mathcal{N}_0$. If we set $w = aw^*$ for $a \in \mathbb{R}$ then $\lim_{a \to \infty} f_{aw^*}(x) = s_{w^*}(x)$ holds at any x. Thus, a heaviside function is realizable while we need extremely large weight values for a good approximation. In this article, we use the term "approximation" in a somewhat loose sense; e.g. pointwise approximation.

2.3   Approximation by a Deep Structure

In this section, we consider to represent a heaviside function by a deep structure. Although, of course, (15) is the most simple representation, we here consider over-parametrized implementations with a deep layered structure. Firstly, we consider to construct $f(x) = \sigma(ax + b)$ by a deep structure, where $a \in \mathbb{R}$, $b \in \mathbb{R}$ and $x \in I \subset \mathbb{R}$, $I = [x_{\min}, x_{\max}]$. Although there are several implementations, we pick up a natural and economic one here. We define

$$g_1(x) = \phi(a_1 x + b_1) \tag{16}$$
$$g_l(x) = \phi(a_l g_{l-1}(x) + b_l), \ l = 2, \ldots, L, \tag{17}$$

where we assume that $a_l > 0$ and $b_l \in \mathbb{R}$ for all l. We also define $\rho_l = \prod_{j=1}^{l} a_j$. We show that, if $a_1, \ldots, a_L$ is given then we have

$$y_l(x) = \rho_l x + b'_l \tag{18}$$

for all l by choosing appropriate $b_1, \ldots, b_L$, where $b'_1 = b_1$ and $b'_l = a_l b'_{l-1} + b_l$ for $l \geq 2$. If we choose $b_1 \geq -a_1 x_{\min}$ then $a_1 x + b_1 \geq 0$ holds for any $x \in I$, thus, $g_1(x) = a_1 x + b_1$ for any $x \in I$ by the definition of ϕ. Therefore, (18) is true for $l = 1$. We assume that (18) holds for l. If we choose $b_{l+1} > -\rho_{l+1} x_{\min} - a_{l+1} b'_l$ then $\rho_{l+1} x + a_{l+1} b'_l + b_{l+1} \geq 0$ holds for any $x \in I$ and, thus, we have

$$\begin{aligned} g_{l+1}(x) &= \phi(a_{l+1} g_l(x) + b_{l+1}) \\ &= \phi(a_{l+1}(\rho_l x + b'_l) + b_{l+1}) \\ &= \phi(\rho_{l+1} x + a_{l+1} b'_l + b_{l+1}) \\ &= \rho_{l+1} x + b'_{l+1} \end{aligned} \tag{19}$$

by the definition of ϕ, where $b'_{l+1} = a_{l+1} b'_l + b_{l+1}$. Therefore, (18) holds for all l. Under this choice of $b_1, \ldots, b_L$, we have $g_L(x) = \rho_L x + b'_L$. Therefore, by placing a DoReLU network after g_L, a network output is given by

$$f(x) = \sigma(g_L(x) + (b - b'_L)) = \sigma(\rho_L x + b). \tag{20}$$

This implementation requires $L+2$ layer and $L+3$ units. Note that if a DoRELU implements $\sigma(-x)$ in the output then we have $f(x) = 1 - \sigma(\rho_L x + b)$ by (10).

There is an another simple implementation using (2) and (3). By (2), we can have two path that implement $\phi(\rho_L x)$ and $\phi(-\rho_L x)$ in a layered ReLU network if $a_l > 0$ for all l. These outputs are then transformed by a ReLU network constructing an identity function according (3) and we have $\rho_L x$. The network output is obtained by a transformation of it by DoReLU. This requires two units at each layer while we need not biases in intermediate layers. Unlike this, the key point of the above implementation is to use a linear part of ReLU at any layer by choosing appropriate bias values, by which we just need one unit in one layer to realize a linear function. This is possible because we restrict an input range and it may be a natural assumption in applications. By this trick, we may not need an assumption of $a_l > 0$ in the above implementation if we take into account of signs of ρ_l and a_l. However, we omit this proof since it is somewhat complicated and the above fact is enough for the discussions in this article.

We consider a network on $\mathbb{R}^d$:

$$g_1(\boldsymbol{x}) = \phi(a_1 h_w(\boldsymbol{x}) + b_1) \tag{21}$$

$$g_l(\boldsymbol{x}) = \phi(a_l g_{l-1}(\boldsymbol{x}) + b_l), \ l = 2, \ldots, L, \tag{22}$$

and

$$f_{w,a,b,b}(\boldsymbol{x}) = \sigma(g_L(\boldsymbol{x}) + b), \tag{23}$$

where $\boldsymbol{a} = (a_1, \ldots, a_L)$ and $\boldsymbol{b} = (b_1, \ldots, b_L)$. We assume that $|h_w(\boldsymbol{x})| \leq 1$ for any $\boldsymbol{x}$. This enables us to apply the above result by setting $x = h_w(\boldsymbol{x})$. Note that this assumption is not unnatural due to a_1 and b_1; i.e. $a_1 h_w(\boldsymbol{x}) + b_1$ is just a re-parametrization of a usual d-dimensional network input. We refer to this deep network as $\mathcal{N}_L$, where L is the length of $\boldsymbol{a}$. As in the previous result, by choosing an appropriate $\boldsymbol{b}$, we always have

$$f_{w,a,b,0}(\boldsymbol{x}) = \sigma(\rho_L h_w(\boldsymbol{x})) \tag{24}$$

if $a_l > 0$ for any l, in which we set $b = 0$ for simplicity. If $a_l > 1$ then ρ_L can be very large when L is large. If ρ_L is large then $\sigma(\rho_L h_{w^*}(\boldsymbol{x}))$ approximate s_{w^*} well by (9). This implies that a ReLU network that well approximates a heaviside function can be constructed with moderate weight sizes by employing a deep structure; i.e. ρ_L can be very large even when a_l is not so large. This is unlike the implementation by $\mathcal{N}_0$, for which we "need" a very large weight sizes for a good approximation of a heaviside function. It takes much time to attain such weight values in training unless we choose a sufficiently large step size; e.g. a large learning rate. However, a large step size may simply lead to instability of training. Such a large step size may not be appropriate for the update of weights that have the other role. Therefore, we can say that a deep structure is preferable in terms of efficiency in training. Moreover, if the number of layers increases then ρ_L can be very large. Therefore, the depth affects an accuracy of fitting to a heaviside function. Of course, we can construct a better fitting by a linear sum of DoReLU outputs which are linearly dependent. However, it may be weak due to an effect by "sum" not "product". And it is somewhat fruitless, considering a network size. It is natural to consider an implementation of a complex function by a superposition of linearly independent ReLUs since a function that is implemented by a simple deep structure seems to be regular; e.g. a sawtooth function. A general meaningful insight of our result is that a deep structure may help to enhance a discontinuity of a function constructed in input side or to construct a superposition of heaviside functions at output side.

3 Training Issue

3.1 Effect of Deep Structure in Learning

Let $\{(\boldsymbol{x}_i = (x_{i,1}, \ldots, x_{i,d}), y_i), \ i = 1, \ldots, n\}$ be a training data set. For convenience, we define $x_{i,0} = 1$ for any i. Let $\ell_i := \ell(y_i, f(\boldsymbol{x}_i))$ be a loss function of a network output f for the ith training sample. For example, we can set a squared error; i.e. $\ell_i = (y_i - f(\boldsymbol{x}_i))^2$. Although a softmax output is popular for a ReLU network in recognition tasks, for simplicity, we employ DoReLU in the output layer to clip the output; i.e. we consider a unit whose activation function is σ in (4) in the output layer. Then, we can apply the least squares learning in classification problems; e.g. a learning a heaviside function as a special case.

We first consider $\mathcal{N}_0$. We define $f_{w,1,i} := f_w(\boldsymbol{x}_i)$. In a classical batch back-propagation for f_w above, update term is given by

$$\sum_{i=1}^{n} \frac{\partial \ell_i}{\partial w_k} = \sum_{i=1}^{n} \ell_i' \sigma'(h_w(\boldsymbol{x}_i)) x_{i,k}, \quad k = 0, 1, \ldots, d \qquad (25)$$

where σ' is given by (11) and ℓ_i' is a derivative of ℓ_i; e.g. $\ell_i' = -2(y_i - f(\boldsymbol{x}_i))$ in case of squared error. We consider a case where $y_i = s_{w^*}(\boldsymbol{x}_i)$; i.e. training data are samples from a heaviside function. This situation is significant not only as simple classification problems but also as classification by more complex indicator functions since those are shown to be constructed by a transformation of a heaviside function [5]. In this case, $h_w(\boldsymbol{x}_i)$ should go to ∞ in a training process. Therefore, the update term becomes very small at a later stage of training since values of σ' are not zeros at a few samples. Of course, the slow convergence may also be caused by a small error since a heaviside function is realizable.

We next consider $\mathcal{N}_L$. For convenience, it is re-formulated as

$$g_0(\boldsymbol{x}) = \phi(h_w(\boldsymbol{x})) \qquad (26)$$
$$h_l(\boldsymbol{x}) = a_l g_{l-1}(\boldsymbol{x}) + b_l, \ l = 1, \ldots, L+1 \qquad (27)$$
$$g_l(\boldsymbol{x}) = \phi(h_l(\boldsymbol{x})), \ l = 1, \ldots, L \qquad (28)$$
$$g_{L+1}(\boldsymbol{x}) = \sigma(h_{L+1}(\boldsymbol{x})), \qquad (29)$$

where $a_l \in \mathbb{R}$ and $b_l \in \mathbb{R}$ for any l. The network output of $\mathcal{N}_L$ is $g_{L+1}(\boldsymbol{x})$ and, thus, $\ell_i := \ell(y_i, g_{L+1}(\boldsymbol{x}_i))$. For simplicity, we write $g_{l,i} = g_l(\boldsymbol{x}_i)$ and $h_{l,i} = h_l(\boldsymbol{x}_i)$. It is easily derived an update term of a classical back-propagation algorithm for this simple structure and, especially, we have

$$\sum_{i=1}^{n} \frac{\partial \ell_i}{\partial w_k} = \rho_L \sum_{i=1}^{n} \left(\prod_{l=1}^{L} \phi'(h_{l,i}) \right) \ell_i' \sigma'(h_{L+1,i}) x_{i,k}, \quad k = 0, 1, \ldots, d \qquad (30)$$

for renewing w, where σ' and ϕ' are given by (11) and (12) respectively.

We here consider a fitting of samples from a heaviside function; i.e. $y_i = s_{w^*}(x_i)$. For the ϕ', a problem of vanishing gradient may not be applied since we employ ReLU. However, learning speed can be slow, especially in the later stage of training, due to σ' as mentioned in $\mathcal{N}_0$. However, unlike $\mathcal{N}_0$, ρ_L appears in the update term in $\mathcal{N}_L$ that is an over-parametrization case. Since training data are samples from a heaviside function, a network is trained to approximate a heaviside function. Therefore, ρ_L should be large if a network is trained to employ a representation in (24). This causes a large value for the update term. It is possible to lead to an apparent speed up of training if a learning rate is set to a moderate small value while it may induce an oscillation of error (weights) if it is too large. This is known as a problem of exploding gradient as an opposite behavior of vanishing gradient in deep learning problems [1]. The important point is that this situation is possible to automatically occur in fitting a heaviside function. It is important that an acceleration is expected under an appropriate small learning rate that is generally recommended. And the small learning rate may be sufficient for fitting a heaviside function since we do not need large weight values in a deep structure.

3.2   A Simple Numerical Example

We show a numerical example for fitting a heaviside function on $\mathbb{R}^2$. We generate x_i from a uniform distribution on $[-1, 1] \times [-1, 1]$ and $y_i = s_{w^*}(x_i)$ at the x_i, where we set $w^* = (w_1^*, 0)$, $w_1^* = (1, 1)$. We set $n = 50$. We employ a simple batch back propagation since our purpose is to see a behavior of a training process. The number of iteration is $T = 1000$. We conduct 10 trials with different initial values. For simplicity, we set positive initial values for a.

In Fig. 2, we show the resulting error history for the trial that gives the minimum final error. The value in the bracket in Fig. 2 is a learning rate that is defined by η. In Fig. 3, we show the final network outputs of $\mathcal{N}_0$ and $\mathcal{N}_2$, which are trained with a learning rate of $\eta = 0.01$. In Fig. 3, we can see that a separation of $\{0, 1\}$ region is almost perfect for $\mathcal{N}_2$ while there are some intermediate output values between 0 and 1 for $\mathcal{N}_0$. Actually, the reduction of training error is larger and the final error is smaller for $\mathcal{N}_2$ in Fig. 2(a). As shown in Fig. 2(b), we can observe an oscillation at an early stage for $\mathcal{N}_2(\eta = 0.02)$ while we cannot observe for $\mathcal{N}_0(\eta = 0.02)$. This is because that the update term for $\mathcal{N}_2$ is larger than for $\mathcal{N}_0$ as we considered. In Fig. 2(b), we can observe an oscillation for $\mathcal{N}_0(\eta = 0.1)$ while the error of $\mathcal{N}_0$ is still larger than that of $\mathcal{N}_2(\eta = 0.02)$. These results tell us that a learning of large weight values is not easy in a simple training. When we further increase a learning rate for $\mathcal{N}_0$ and set $\eta = 0.5$, we happen to observe the case of zero error (perfect fitting) after a large oscillation. However, this is because the training target and network structure are simple. By taking into account of a large oscillation, such a large learning rate may not be appropriate for a more complex situation.

On the other hand, the resulting weight vector at $\eta = 0.01$ is $(w_0, w_1, w_2) = (0.11, 3.40, 3.38)$ for $\mathcal{N}_0$ and $(w_0, w_1, w_2) = (0.27, 1.73, 1.68)$, $(b_1, a_1, b_2, a_2) =$

$(-0.15, 2.06, -0.59, 2.08)$ for $\mathcal{N}_2$. Obviously, each value of (w_1, w_2, a_1, a_2) in $\mathcal{N}_2$ are smaller than that of (w_1, w_2) in $\mathcal{N}_0$ while the training error of $\mathcal{N}_2$ is smaller than that of $\mathcal{N}_0$. This is consistent with our consideration above since $a_1 a_2 \simeq 4.0$ which is larger than w_1 and w_2 in $\mathcal{N}_0$. If we include w_1 and w_2 into the product in $\mathcal{N}_2$, the total slope is about 7.0 which may be enough for almost perfect fitting.

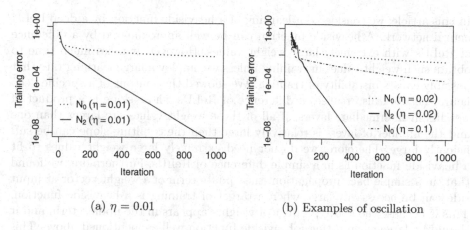

(a) $\eta = 0.01$ (b) Examples of oscillation

Fig. 2. Error history in fitting of a heaviside function.

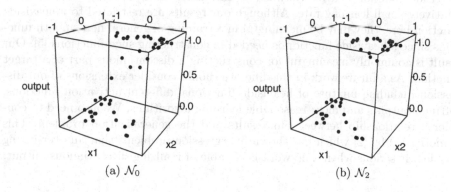

(a) $\mathcal{N}_0$ (b) $\mathcal{N}_2$

Fig. 3. Resulting network output in fitting of a heaviside function ($\eta = 0.01$ for both).

These results tell us that, firstly, it is difficult to attain large weight values for fitting a heaviside function precisely in an actual training unless we employ an extremely large learning rate that possibly causes instability of training. However, secondly, it is solved by employing a deep structure; i.e. we can obtain a precise fitting of a heaviside function within a reasonable training time under a moderate small learning rate. This advantage is supported by a construction of a heaviside function by a deep structure as previously given in this article. Note that the acceleration effect of a deep structure also contributes for sufficiency of a small learning rate. This is also supported by our discussion above.

4   Conclusions and Future Works

In this article, we considered a learning of a heaviside function by a deep ReLU neural network. A heaviside function can be well approximated by a difference of ReLUs with extremely large weight values. However, it takes much time to obtain such weight values in training unless we employ a large learning rate that possibly causes instability of training. We showed that output of a specific deep neural network also realizes a difference of ReLUs whose slope is a product of weights in intermediate layers. If all of those weight values are larger than one and the number of layers is relatively large then the resulting slope can be sufficiently large. Therefore, we do not need extremely large weight values to fit a heaviside function as in a simple difference of ReLUs. Furthermore, we found that, in a simple back propagation, the update term of a weight vector at input side can be necessarily large when a target of training is a heaviside function. This is because the above product of weights appears in the update term and it should be large when fitting a heaviside function well as mentioned above. This brings us an apparent acceleration of learning if we employ a reasonable small learning rate. All these results were verified by a simple numerical experiment. As a result, we showed that, by employing a deep structure, a good fitting of heaviside function can be obtained within a moderate training time under a relatively small learning rate. Although our results are restricted to a heaviside function here, they may be meaningful in a training of general non-smooth functions since a heaviside function is needed in constructing such functions [5]. Our result is commonly meaningful for constructing a discontinuous part of a target function. As a future work in this line, we should consider extensions of our discussion including mixture of heaviside functions, different activation functions, softmax outputs and more reasonable learning algorithms. We also need to consider a relationship between our results and the generalization problem. This is motivated by [2] which has shown in regression problems that an over-fitting problem is serious when a network is capable of realizing discontinuous output.

References

1. Bengio, Y., Simard, P., Frasconi, P.: Learning long-term dependencies with gradient descent is difficult. IEEE Trans. Neural Networks **5**, 157–166 (1994)
2. Hagiwara, K., Fukumizu, K.: Relation between weight size and degree of over-fitting in neural network regression. Neural Netw. **21**, 48–58 (2008)
3. Imaizumi, M., Fukumizu, K.: Deep neural networks learn non-smooth functions effectively. arXiv preprint arXiv:1802.04474 (2018)
4. Liang, S., Srikant, R.: Why deep neural networks for function approximation? arXiv preprint arXiv:1610.04161 (2017)

5. Petersen, P., Voigtlaender, F.: Optimal approximation of piecewise smooth functions using deep ReLU neural networks. arXiv preprint arXiv:1709.05289 (2017)
6. Schmidhuber, J.: Deep learning in neural networks: an overview. Neural Netw. **61**, 85–117 (2015)
7. Yarotsky, D.: Error bounds for approximations with deep ReLU networks. Neural Netw. **94**, 103–114 (2017)

Deep Tag Recommendation Based on Discrete Tensor Factorization

Wenwen Ye[✉], Zheng Qin, and Xu Li

School of Software, Tsinghua University, Beijing, China
{yeww14,qinzh,li-xu16}@mails.tsinghua.edu.cn

Abstract. In the recent years, tag recommendation is becoming more and more popular in both academic and industrial community. Although existing models have obtained great success in terms of enhancing the performance, an important problem has been ignored – the efficiency. To bridge this gap, in this paper, we design a novel discrete tensor factorization model (DTF) to encode user, item, tag into a unified hamming space for fast recommendations. More specifically, we first design a base model to translate the traditional pair-wise interaction tensor factorization (PITF) into its discrete version. Then, to provide our model with the ability to involve content information, we further extend the base model by introducing a deep content extractor for more comprehensive user/item profiling. Extensive experiments on two real-world data sets demonstrate that our model can greatly enhance the efficiency without sacrificing much effectiveness.

Keywords: Recommendation system · Collaborative filtering
Deep learning · Tensor factorization · Discrete

1 Introduction

Tagging is an essential component of the Web 2.0, it allows the user to designate personalized tags on various items with keywords. To automatically suggest users with their favorite tags, personalized tag recommendation is becoming more and more popular in both academic and industrial community. Traditional tag recommender methods mainly are mainly established on the basis of the technology of Tensor factorization (TF), which aims to model user tagging behaviors by capturing the collaborative information among the interactions between users, items and tags.

Although existing tag recommender models have obtained great success in terms of effectiveness, another important aspect has been largely ignored – the efficiency. In the widely applied pair-wise interaction tensor factorization (PITF) method, for an $m \times n \times t$ user-item-tag cube, the space and time complexities are $\mathcal{O}(tr)$ and $\mathcal{O}(mntr + mntr \log k)$, respectively, which may lead to unfavored online efficiency, especially when the scales of the users, items and tags are rapidly growing in real scenarios. Therefore, we'd like to ask *"Can we enhance*

© Springer Nature Switzerland AG 2018
L. Cheng et al. (Eds.): ICONIP 2018, LNCS 11301, pp. 70–82, 2018.
https://doi.org/10.1007/978-3-030-04167-0_7

the online efficiency for tag recommendation without sacrificing much performance?"

In the context of tag recommendation, the online efficiency is mainly restricted by the time cost of the inner production between different latent factors. So the main challenge to answer the above question is how to simplify the inner product operation, while keeping its modeling power. Fortunately, the development of hash technology may shed some lights on this problem. In specific, we can project users, items and tags into a unified hamming space, based on which we optimize them to fit the user tagging behavior. However, the hash optimization problem is NP-hard [5], and the solutions to it fall into two categories: two-stage procedure optimization and direct discrete optimization. The former methods include relaxed optimization stage and binary quantization stage which oversimplify original discrete optimization then lead to a massive quantization loss. The latter directly learn binary codes via discrete optimization. Unfortunately, the basic model of these hashing frameworks is for MF instead of for TF, thus not compatible with the purpose of tag recommendation.

Personal Tag Recommendation. Personalized tag recommendation is an important topic in recommendation systems. Higher-Order-Singular-Value-Decomposition (HOSVD) [13] is a TD model optimized for square-loss where all not observed values are learned as 0. Besides. Tucker decomposition can be used in tag recommendation [1]. And a special case of Tucker decomposition is the canonical decomposition (CD) also known as the parallel factor analysis (PARAFAC) [4]. More specifically, Pairwise Interaction Tensor Factorization (PITF) [11] adapted the Bayesian Personalized Ranking (BPR) [10] framework including the BPR optimization criterion and the BPR learning algorithm from the related field of item recommendation to tag recommendation.

Discrete Optimization. Hashing has been widely shown as an outstanding approach to accelerate similarity search [2,15,18]. [6] mapped user/item latent representation into the Hamming space to obtain hash codes for users and items. Similar to this, [19] followed the idea of Iterative Quantization [3] to generate binary code. According to the analysis of [8] on joint optimizations of quantization losses and intrinsic objective functions [9,12] can demonstrate notable performance gain over the two-stage [14] (real-valued optimization and binary quantization) hashing methods. Differ from two-stage hashing method, discrete learning can learn hash codes by directly solving the discrete optimization problem. Follow this, [15] proposed an efficient discrete CF approach to solve the rating prediction problem in recommendation system. [16] imposed personalized ranking criteria from implicit feedback. Moreover, [7,17] developed an efficient discrete optimization algorithm for Content-aware Matrix Factorization.

Inspired by the above motivations, in this paper, we introduce the hash method into the field of tag recommendation. More specifically, we first propose an efficient tag recommendation framework called Discrete Tensor Factorization (DTF) which combines TF framework and hash techniques to directly learn the binary codes for the user/item/tag vector. Additionally, the balance and uncorrelation constraints are added to learn the compact but informative codes. To make the optimization flexible, we utilize a fast iteratively bit-wise updates optimization algorithm for mixed-integer sub-problems. Besides, we further extend the DTF model by introducing a deep content extractor for more comprehensive user/item profiling as content enhanced model (DTF-CE) to improve the accuracy of recommendation. Finally, we evaluate the proposed framework on several sizes of data sets and proved its superiority to the baselines.

Our contributions are summarized as follows:

1. We first propose to project users, items and tags into a unified hamming space for fast inference in the field of tag recommendation.
2. We design a base model to translate pair-wise interaction tensor factorization model into its discrete version, and further extend this base model by introducing a deep content regularizer to enable our model for capturing more auxiliary information.
3. Extensive experiments on two real-world data sets demonstrate that our model can greatly enhance the efficiency without sacrificing much effectiveness.

2   Preliminaries and Problem Statement

In this section, we will introduce some notations related in this paper and then formulate the problem.

2.1   Notations

Let users, items and tags sets denoted by $U = \{1, 2, \ldots m\}$, $I = \{1, 2, \ldots n\}$, $T = \{1, 2, \ldots t\}$. The interaction sets $S \subseteq U \times I \times T$. For tag recommendation, we recommending a list of tags for a given user-item pair (u, i). Such a combination (u, i) as a post and we define the set of all observed posts $P_s := \{(u, i) | \exists t \in T : (u, i, t) \in S\}$. We follow the BPR criteria for convergence and define the items tagged by user u as $I_u = \{i \in I : (u, i) \in P_s\}$; positive tags for item i by user u as $T_{ui}^+ = \{t \in T : (u, i, t) \in S\}$; and other tags as $T_{ui}^- = T/T_{ui}^+$. The content information denoted as F.

2.2   Problem Statement

Differ from classic item recommendation, tag recommendation aims at recommending tags for items according to users' preferences. And the prediction model is represented through the latent vectors of users, items and tags.

Given the user-item-tag interaction set S. The input is a query post $p := (u, i)$, output is the top-K tags that user u liked for i. We adopt hash techniques to accelerate the traditional tensor factorization model, and work out a power trade-off strategy between efficiency and accuracy.

3 Discrete Tensor Factorization Model

Predictive Function. The predictive function is the key component of a recommendation system. The traditional models for tag recommendation (e.g., PITF) are based on Tensor Factorization which map users, items and tags into the r-dimension real-value latent vectors and the prediction score is measured by the inner product of users, items and tags real-value vectors.

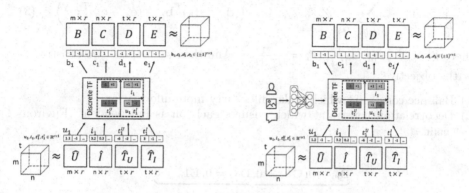

Fig. 1. DTF basic model

Fig. 2. DTF-CE content enhanced model

In this paper, the proposed discrete framework continues to use the same idea of the prediction model but maps them into r-dimension Hamming space, then calculates the prediction score by bits operation instead of the real-value operation according to [19]. The binary codes of users $\hat{U}$, items $\hat{I}$, tags for user $\hat{T_U}$ and tags for item $\hat{T_I}$ are denoted as $\mathbf{B} = [\mathbf{b}_1, ..., \mathbf{b}_m] \in \{\pm 1\}^{r \times m}$, $\mathbf{C} = [\mathbf{c}_1, ..., \mathbf{c}_n] \in \{\pm 1\}^{r \times n}$, $\mathbf{D} = [\mathbf{d}_1, ..., \mathbf{d}_t] \in \{\pm 1\}^{r \times t}$ and $\mathbf{E} = [\mathbf{e}_1, ..., \mathbf{e}_t] \in \{\pm 1\}^{r \times t}$. The prediction score is defined as

$$sim(u, i, t) = \frac{1}{2r} \sum_{k=1}^{r} \mathbb{I}(b_{uk} = d_{tk}) + \frac{1}{2r} \sum_{k=1}^{r} \mathbb{I}(c_{ik} = e_{tk})$$
$$= \frac{1}{2} + \frac{1}{4r} \mathbf{b}_u^T \mathbf{d}_t + \frac{1}{4r} \mathbf{c}_i^T \mathbf{e}_t \tag{1}$$

where $\mathbb{I}(\cdot)$ denotes the indicator function that returns 1 if the true and 0 otherwise and that represents the predicted score of user u over item i for tag t.

Figure 1 shows an overview of the Discrete Tensor Factorization model. It is a personalized ranking approach that adopts the BPR optimization criterion and directly optimizes the personalized ranking objective with the least square loss. To minimize the following least square loss:

$$Loss = \sum_{(u,i,g,h) \in D_S} \frac{1}{|U||I_u||T_{ui}^+||T_{ui}^-|} (1 - (\hat{y}_{uig} - \hat{y}_{uih}))^2 \tag{2}$$

where g and h refer t_A and t_B. We propose to infer pairwise ranking constraints D_S from S like $D_S := \{(u, i, t_A, t_B) : (u, i, t_A) \in S \wedge (u, i, t_B) \notin S\}$. Thus for applying BPR optimization and combine with Eq. (1), we set:

$$\hat{y}_{uigh} = \hat{y}_{uig} - \hat{y}_{uih} = \frac{1}{4r}\mathbf{b}_u^T(\mathbf{d}_g - \mathbf{d}_h) + \frac{1}{4r}\mathbf{c}_i^T(\mathbf{e}_g - \mathbf{e}_h)$$

and combine this loss function with Eq. (2), the DTF objective rewrites as:

$$\underset{B,C,D,E,\Gamma}{\arg\min} \sum_{(u,i,g,h)\in D_S} \mathscr{U}\mathscr{I}_u\mathscr{T}_{ui}\left(4r - \left((\mathbf{d}_g - \mathbf{d}_h)^T\mathbf{b}_u + (\mathbf{e}_g - \mathbf{e}_h)^T\mathbf{c}_i\right)\right)^2 \quad (3)$$

where $\mathscr{U} = \frac{1}{|U|}$, $\mathscr{I}_u = \frac{1}{|I_u|}$, $\mathscr{T}_u = \frac{1}{|T_{ui}^+||T_{ui}^-|}$. And then we propose two constraints for the objective:

(1) Balance constraint which makes bits carry more information.
(2) Decorrelation constraint which ensures each bit is independent. Environment.

$$\underbrace{\mathbf{B}_m = 0, \mathbf{C1}_n = 0, \mathbf{D1}_t = 0, \mathbf{E1}_n = 0}_{\text{Balance Constraints}}$$
$$\underbrace{\mathbf{BB}^T = m\mathbf{I}_r, \mathbf{CC}^T = n\mathbf{I}_r, \mathbf{DD}^T = t\mathbf{I}_r, \mathbf{EE}^T = t\mathbf{I}_r.}_{\text{Decorrelation Constraints}}$$

After we import the two constraints into Eq. (3), this optimization problem becomes more chanllaging since it is NP-hard that involves $\mathcal{O}(2^{(m+n+t)r})$ combinatorial search for the binary codes. According to DCF, we adopt a solution by softening these constrains: import delegate continuous values matrix X, Y, W, Z and minimize the distances with hash codes $\|\mathbf{B} - \mathbf{X}\|_F^2$, $\|\mathbf{C} - \mathbf{Y}\|_F^2$, $\|\mathbf{D} - \mathbf{W}\|_F^2$, $\|\mathbf{E} - \mathbf{Z}\|_F^2$. Note that we can enforce the distances equal 0 by imposing very large tuning parameters. Thus, combining the constrains with Eq. (3), the objective can be equivalently transformed as the following optimization problem:

$$\underset{B,C,D,E,\Gamma}{\arg\min} \sum_{(u,i,g,h)\in D_S} \mathscr{U}\mathscr{I}_u\mathscr{T}_{ui}\left(4r - \left((\mathbf{d}_g - \mathbf{d}_h)^T\mathbf{b}_u + (\mathbf{e}_g - \mathbf{e}_h)^T\mathbf{c}_i\right)\right)^2$$
$$- 2\alpha\, tr(\mathbf{B}^T\mathbf{X}) - 2\beta\,(\mathbf{C}^T\mathbf{Y}) - 2\gamma\, tr(\mathbf{D}^T\mathbf{W}) - 2\zeta\, tr(\mathbf{E}^T\mathbf{Z})$$

$$(4)$$

$$s.t.\, \mathbf{B} \in \{\pm1\}^{r\times m}, \mathbf{C} \in \{\pm1\}^{r\times n}, \mathbf{D}, \mathbf{E} \in \{\pm1\}^{r\times t}$$
$$\mathbf{X1}_m = 0, \mathbf{Y1}_n = 0, \mathbf{W1}_t = 0, \mathbf{Z1}_t = 0$$
$$\mathbf{XX}^T = m\mathbf{I}_r, \mathbf{YY}^T = n\mathbf{I}_r, \mathbf{WW}^T = t\mathbf{I}_r, \mathbf{ZZ}^T = t\mathbf{I}_r$$

where $\mathbf{1}$ refer vectors that all 1, $\mathbf{I}_r$ refer r-dimension identity matrix. $\alpha > 0$, $\beta > 0$, $\gamma > 0$ and $\zeta > 0$ are tuning parameters.

We alternatively solve two kinds of subproblems for DTF: (1) $\mathbf{B}, \mathbf{C}, \mathbf{D}$ and $\mathbf{E}$ will updated by parallel discrete optimization; (2) $\mathbf{X}, \mathbf{Y}, \mathbf{W}$ and $\mathbf{Z}$ will updated by Singular Value Decomposition (SVD).

Model Optimization Update B, fix other: Follow the Eq. (4), we update **B** with fixed **C, D, E, X, Y, W** and **Z**. Since each user is independent, we can update **B** by updating $\mathbf{b}_u$ in parallel, the formulation defined as:

$$
\underset{\mathbf{b}_u \in \{\pm1\}^r}{\arg\min} \; \mathscr{U} \sum_{i \in I_u} \mathscr{I}_u \sum_{g,h \in T} \mathscr{T}_{ui} r_{uig}(1 - r_{uih}) \cdot \left(((\mathbf{d}_g - \mathbf{d}_h)^T \mathbf{b}_u)^2 \right.
$$
$$
\left. + 2(\mathbf{e}_g - \mathbf{e}_h)^T \mathbf{c}_i (\mathbf{d}_g - \mathbf{d}_h)^T \mathbf{b}_u - 8r(\mathbf{d}_g - \mathbf{d}_h)^T \mathbf{b}_u \right) - 2\alpha \mathbf{x}_u^T \mathbf{b}_u
\tag{5}
$$

where r_{uig} represent whether the user u has interaction with tag g over item i, returning 1 if $(u, i, g) \in S$ and 0 otherwise. $r_{uig}(1 - r_{uih}) = 1$ when $(u, i, g, h) \in D_s$. Due to the constraints mentioned above, the optimization is generally NP-hard and we adopt Discrete Coordinate Descent (DCD) to update $\mathbf{b}_u$ bit by bit. Let b_{uk} be the k-th bit of $\mathbf{b}_u$, $\mathbf{b}_u = [\mathbf{b}_{u\bar{k}}^T, b_{uk}]^T$, and discard the irrelevant terms of b_{uk}, the Eq. (7) is redefined as:

$$
\underset{b_{uk} \in \{\pm1\}}{\arg\min} \; -b_{uk}\hat{b}_{uk}
\tag{6}
$$

$$
\hat{b}_{uk} = \mathscr{U} \sum_{i \in I_u} \mathscr{I}_u \sum_{g,h \in T} \mathscr{T}_{ui} r_{uig}(1 - r_{uih}) \left((4r - 2(\mathbf{e}_g - \mathbf{e}_h)^T \mathbf{c}_i) - (\mathbf{d}_{g\bar{k}} - \mathbf{d}_{h\bar{k}})^T \mathbf{b}_{u\bar{k}} \right)
$$

$(d_{gk} - d_{hk}) + \alpha x_{uk}$. Detailed derivation is omitted for the space limit, and the update rule of b_{uk} is

$$
b_{uk} = \begin{cases} sgn(\hat{b}_{uk}) & \hat{b}_{uk} \neq 0 \\ sgn(b_{uk}) & \text{else} \end{cases}
\tag{7}
$$

where $sgn(t) = 1$ if $t \geqslant 0$, otherwise, $sgn(t) = -1$.

Update C, Fix Other: When **B, D, E, X, Y, W** and **Z** are fixed, and discard the irrelevant terms of c_{ik}. Thus the Eq. (4) changed into:

$$
\underset{\mathbf{c}_i \in \{\pm1\}^r}{\arg\min} \; \mathscr{U} \sum_{u \in U} \mathscr{I}_u \sum_{g,h \in T} \mathscr{T}_{ui} r_{uig}(1 - r_{uih})
$$
$$
\left(((\mathbf{e}_g - \mathbf{e}_h)^T \mathbf{c}_i)^2 + 2(\mathbf{d}_g - \mathbf{d}_h)^T \mathbf{b}_u (\mathbf{e}_g - \mathbf{e}_h)^T \mathbf{c}_i \right.
$$
$$
\left. - 8r(\mathbf{e}_g - \mathbf{e}_h)^T \mathbf{c}_i \right) - 2\beta \mathbf{y}_i^T \mathbf{c}_i
\tag{8}
$$

Similar to the **B**-subproblem, we update $\mathbf{c}_i = [\mathbf{c}_{i\bar{k}}^T, c_{ik}]^T$ bit by bit and the objective function as:

$$
\underset{c_{ik} \in \{\pm1\}}{\arg\min} \; -c_{ik}\hat{c}_{ik}
\tag{9}
$$

$$
\hat{c}_{ik} = \mathscr{U} \sum_{u \in U} \mathscr{I}_u \sum_{g,h \in T} \mathscr{T}_{ui} r_{uig}(1 - r_{uih}) \left((4r - 2(\mathbf{d}_g - \mathbf{d}_h)^T \mathbf{b}_u) - (\mathbf{e}_{g\bar{k}} - \mathbf{e}_{h\bar{k}})^T \mathbf{c}_{i\bar{k}} \right)
$$

$(e_{gk} - e_{hk}) + \beta y_{ik}$. Updating c_{ik} when fix $\mathbf{c}_{i\bar{k}}$, the rule of c_{ik} is:

$$
c_{ik} = \begin{cases} sgn(\hat{c}_{ik}) & \hat{c}_{ik} \neq 0 \\ sgn(c_{ik}) & \text{else} \end{cases}
\tag{10}
$$

Update D/E, Fix Other: $\mathbf{D}$ and $\mathbf{E}$ represent hash codes for T_U and T_I. After discarding terms irrelevant $\mathbf{d}_g$ and $\mathbf{e}_g$ separately, the structure of theses two independent objective functions is the same. Thus, we integrate $\mathbf{D}$ and $\mathbf{E}$ subproblems in this section, and the formula is:

$$
\begin{aligned}
&\underset{\mathbf{d}_g \in \{\pm 1\}^r}{\arg\min} \sum_{u \in U} \mathscr{U} \sum_{i \in I_u} \mathscr{I}_u \sum_{h \in T} \mathscr{T}_{ui} r_{uig} (1 - r_{uih}) \cdot \\
&\left((\mathbf{b}_u^T (\mathbf{d}_g - \mathbf{d}_h))^2 + 2\mathbf{c}_i^T (\mathbf{e}_g - \mathbf{e}_h) \mathbf{b}_u^T (\mathbf{d}_g - \mathbf{d}_h) - 8r \mathbf{b}_u^T (\mathbf{d}_g - \mathbf{d}_h) \right) \\
&+ \sum_{u \in U} \mathscr{U} \sum_{i \in I_u} \mathscr{I}_u \sum_{h \in T} \mathscr{T}_{ui} r_{uih} (1 - r_{uig}) \cdot \\
&\left((\mathbf{b}_u^T (\mathbf{d}_h - \mathbf{d}_g))^2 + 2\mathbf{c}_i^T (\mathbf{e}_h - \mathbf{e}_g) \mathbf{b}_u^T (\mathbf{d}_h - \mathbf{d}_g) - 8r \mathbf{b}_u^T (\mathbf{d}_h - \mathbf{d}_g) \right) \\
&- 2\gamma \mathbf{w}_g^T \mathbf{d}_g
\end{aligned}
\tag{11}
$$

Equation (11) is for $\mathbf{D}$ problem. And for $\mathbf{E}$, elements in the formula that need to be replaced are: $\mathbf{b}_u \to \mathbf{c}_i, \mathbf{d}_g \leftrightarrow \mathbf{e}_g, \mathbf{d}_h \leftrightarrow \mathbf{e}_h, \gamma \to \zeta, \mathbf{w}_g \to \mathbf{z}_g$. Similar to the $\mathbf{B}$, $\mathbf{C}$-subproblem, update rules as:

$$
\begin{aligned}
&\underset{d_{gk} \in \{\pm 1\}}{\arg\min} - d_{gk} \hat{d}_{gk} \\
&\underset{e_{gk} \in \{\pm 1\}}{\arg\min} - e_{gk} \hat{e}_{gk}
\end{aligned}
\tag{12}
$$

where

$$
\begin{aligned}
\hat{d}_{gk} = &\underset{\mathbf{d}_g \in \{\pm 1\}^r}{\arg\min} \sum_{u \in U} \mathscr{U} \sum_{i \in I_u} \mathscr{I}_u \sum_{h \in T} \mathscr{T}_{ui} r_{uig} (1 - r_{uih}) \cdot \\
&\left((4r - 2\mathbf{c}_i^T (\mathbf{e}_g - \mathbf{e}_h)) b_{uk} - \mathbf{b}_{u\overline{k}}^T (\mathbf{d}_{g\overline{k}} - \mathbf{d}_{h\overline{k}})^T b_{uk} + d_{hk} \right) \\
&+ \sum_{u \in U} \mathscr{U} \sum_{i \in I_u} \mathscr{I}_u \sum_{h \in T} \mathscr{T}_{ui} r_{uih} (1 - r_{uig}) \cdot \\
&\left((4r - 2\mathbf{c}_i^T (\mathbf{e}_h - \mathbf{e}_g)) b_{uk} - \mathbf{b}_{u\overline{k}}^T (\mathbf{d}_{h\overline{k}} - \mathbf{d}_{g\overline{k}})^T b_{uk} - d_{hk} \right) \\
&+ \gamma w_{gk}
\end{aligned}
\tag{13}
$$

And for $\hat{e}_{gk}$, elements in the formula (13) that need to be replaced are: $\mathbf{c} \leftrightarrow \mathbf{b}, \mathbf{d}_g \leftrightarrow \mathbf{e}_g, \mathbf{d}_h \leftrightarrow \mathbf{e}_h, \gamma \to \zeta, \mathbf{w} \to \mathbf{z}$. Finally, the d_{gk} and e_{gk} update rules:

$$
d_{gk} = \begin{cases} sgn(\hat{d}_{gk}) & \hat{d}_{gk} \neq 0 \\ sgn(d_{gk}) & \text{else} \end{cases}, e_{gk} = \begin{cases} sgn(\hat{e}_{gk}) & \hat{e}_{gk} \neq 0 \\ sgn(e_{gk}) & \text{else} \end{cases}
\tag{14}
$$

Update X/Y/W/Z, Fix Other: $\mathbf{X}$, $\mathbf{Y}$, $\mathbf{W}$ and $\mathbf{Z}$ are the same optimization problem, update one of them when other parameters all fixed. The objective function for $\mathbf{X}$ are defined as:

$$
\underset{\mathbf{X} \in \{\mathbb{R}\}^{r \times m}}{\arg\min} \ tr(\mathbf{B}^T \mathbf{X}), \ s.t. \ \mathbf{X} \mathbf{1}_m = 0, \mathbf{X} \mathbf{X}^T = m \mathbf{I}_r
\tag{15}
$$

We use SVD to tackle the Eq. (14)

$$\mathbf{X} \leftarrow \sqrt{m}[\mathbf{P}_b \; \hat{\mathbf{P}}_b][\mathbf{Q}_b \; \hat{\mathbf{Q}}_b]^T \tag{16}$$

where $\mathbf{P}_b$ and $\hat{\mathbf{P}}_b$ are the left and right singular vectors of the row-centered matrix $\bar{\mathbf{B}}$ and $\bar{B}_{ij} = B_{ij} - \frac{1}{m}\sum_j B_{ij}$, $\hat{\mathbf{P}}_b$ are the left singular vectors corresponding to zero singular values, and $\hat{\mathbf{Q}}_b$ are the obtained by Gram-Schimidt based on $[\hat{\mathbf{Q}}_b \; 1]$. And $\mathbf{Y}$, $\mathbf{W}$ and $\mathbf{Z}$ can refer to $\mathbf{X}$ for analogy. To enhance the performance of recommendation and bring in more content-aware information, we utilizes a deep model (CDN) to extract latent content feature from various data called Discrete Tensor Factorization Content Enhanced model (DTF-CE).

3.1 Content Enhanced Model

Figure 2 shows the overview of the DTF-CE, a discrete hybrid framework that involving content information. This method consists of two parts: First component is DTF model; Second, Content Deep Neural Network (CDN), a neural network used to extract content information. So, the joint optimization objective has two sub-objectives: DTF objective and CDN objective.

$$\underset{B,C,D,E,\Gamma}{\arg\min} \; \lambda \sum_{i=1}^{n} \|CDN(\mathbf{f}_i, \Gamma) - l_i\|_F^2 + Obj(DTF) \tag{17}$$

where $\lambda > 0$ is used to adjust the weight of the two sub-objective. Γ represents the parameters of CDN which is a neural network with H hidden-layers where the biases (first term) and weights (second term) of all *hidden layers* are determined as $[\{\mu\}_{l=1}^{H}, \{\theta\}_{l=1}^{H}] = CDN(\mathbf{F}, \Gamma)$, and F is the input vectors of content information and $\mathbf{L} \in \{\mathbf{B}, \mathbf{C}, \mathbf{D}, \mathbf{E}\}$ is the corresponding hash codes.

Update Γ, Fix Other. The same with other subproblems, $\mathbf{L}$ is fixed and the optimization problem becomes a deep neural model learning problem, the Eq. (16) can be rewritten as:

$$\underset{\Gamma}{\arg\min} \sum_{i=1} \|CDN(\mathbf{f}_i, \Gamma) - \mathbf{l}_i\|^2 \tag{18}$$

All the parameters of CDN are learned by stochastic gradient descent method. We choose *sigmoid* function for all the hidden layers as activation function, and *tanh* function for output since $l_{ik} \in \{\pm 1\}$.

Different from the Original L's Sub-problem. After we introduced CDN into DTF, the original the problem corresponding to $\mathbf{L}$ also changed slightly. We need to add CDN component to both optimization objective function and l_{ik}'s update rule.

$$\lambda \sum_{i=1} \|CDN(\mathbf{f}_i, \Gamma) - \mathbf{l}_i\|^2,$$
$$\frac{\lambda}{2} CDN(\mathbf{f}_i, \Gamma)_k \tag{19}$$

4 Experiments

In this section, we conduct experiments with the aim of answering the following research questions:

RQ1: Is DTF more efficient than the state-of-the-art TF based tag recommendation methods?

RQ2: Is DTF making good trade-off strategies between efficiency and accuracy?

In what follows, we first present the experimental settings, followed by answering the above two research questions.

4.1 Experimental Settings

Data Sets. We experimented with two publicly accessible data sets: MovieLens and Last.fm. The characteristics of the two datasets are summarized in Table 1.

1. **Movielens.** This movie rating data set has been widely applied to the evaluation of collaborative filtering algorithms. It covers the user interaction with tag over item and provides the item link to *themovieDB* for the content information.
2. **Last.fm.** The song data set has been widely used to evaluate tag recommendation. And we take song's description features (e.g. artist, release date, genre...) through last.fm API as the content information.

Table 1. Statistics of the evaluation data sets.

dataset	Users	Items	Tags	Triples	Posts	Sparsity
Moivelens	7,801	19,545	35,169	465,564	174,844	99.99%
Last.fm	2,917	1,853	2,045	219,702	75,565	99.99%

Compared Methods. We experimented with 4 method:

- **HOSVD.** Higher order singular value decomposition is a method for learning a tensor factorization model. HOSVD optimizes for minimal element-wise error and cannot deal with missing values. For tag recommendations, the missing values are usually filled with zeros.
- **BPR-PITF.** This is a Bayesian Personalized Ranking framework based on Pairwise Interaction Tensor Factorization, which is directly optimized by the ranking based evaluation with Bayesian.
- **DTF.** Discrete tensor factorization model proposed in this paper.
- **DTF-CE.** Discrete tensor factorization model with content enhanced (Fig. 3).

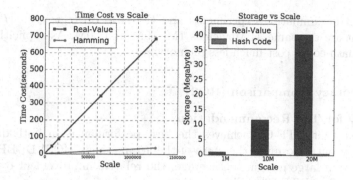

Fig. 3. Efficiency experiments for time and storage. Left: Time. Right: Storage

Parameter Settings and Evaluation. We search α, β, γ and ζ from $\{1e^{-4}, 1e^{-3}, ..., 1e^2\}$, and λ from $\{1e^{-1}, 1, 1e^1\}$ by grid search. For CDN, the hidden layer is $512 \rightarrow 256 \rightarrow 128$. As a result, we set α, β, γ and $\zeta = 1e^{-3}$ and $\lambda = 5$.

We use the common evaluation scheme of F-measure in TopN-lists.

$$F1(S_{test}, N) = \frac{2 \cdot \text{Prec}(S_{test}, N) \cdot \text{Rec}(S_{test}, N)}{\text{Prec}(S_{test}, N) + \text{Rec}(S_{test}, N)}$$

The experiments are repeated 5 times by sampling new training/test sets. We report the average of all the outcomes. The reported f-measure is the f-measure over the average recall and average precision.

4.2 Efficiency Comparison (RQ1)

Efficiency is an important indicator for an online system. We evaluate the efficiency issue on time and storage separately.

Time Efficiency. Assume that there are m users, n items and t tags, the time complexity of predicting top-K tags in real-value system is $\mathcal{O}(mntr + mntr \log k)$ while in hash-based system the time complexity is linearly associated with data size.

We use standard gaussian distribution to generate the user, item and tag real-value feature vectors and hash vectors by *sign* to simulate the computation process. The sizes of $|U| \times |I| \times |T|$ sets in experiment are: 1000, 80000, 160000, 640000, 1250000. The result is shown in the Fig. 1. We can see the that time hash-based cost is much shorter than real-valued. Therefore, hash-based tag recommendation has significant superiority over the real-value tag recommendation in terms of time efficiency.

Storage Efficiency. To save a r-dimension feature vector, we need $32 \times r$ or $64 \times r$ bits in real-value methods $1 \times r$ bit in hash-based methods. We design another set of simulation experiments and the sizes of $|U| \times |I| \times |T|$ sets in experiment are the same as Movielens 1M, 10M and 20M. As explicitly shown in Fig. 1, hash codes cost much less memory.

4.3   Accuracy Comparison (RQ2)

Accuracy for Tag Recommendation
We can see that DTF-CE achieves the best performance on both data sets, significantly outperforming the state-of-the-art methods HOSVD, BPR-PITF and DTF by a large margin (on average, the relative improvement over BPR-PITF 4.21% and DTF 5%) (Figs. 4 and 5).

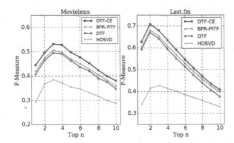

Fig. 4. Comparison with two data sets

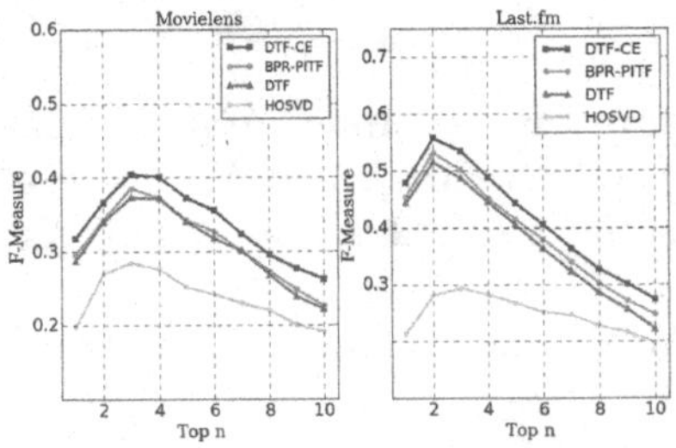

Fig. 5. Comparison with cold-start items

First, we can easily found that methods applied BPR-opt perform better than HOSVD. And by comparing DTF to BPR-PITF, the performance of DTF is very close to the PITF when n is small and becomes worse as n grows larger. Due to the binary codes latent vectors lose some information compared to the real-value vectors, thus it is acceptable and reasonable to have small gaps between DTF and BPR-PITF.

However, from another perspective, the performance of DTF and DTF-CE prove that the Content Enhanced component promotes hash codes' ability to carry more information and improves the performance of recommendation. DTF and DTF-CE are proposed to find some trade-off strategies between efficiency and accuracy.

Accuracy on Cold-Start Recommendation
In our last experiment, we use the item content information as the input of CDN. Moreover, we found that the item content information is useful to tackle the item cold-start problem: when a user wants to tag a new item but there are no historical tags for it. Then we randomly selected the items that are tagged less than 20 times as test set, the random selection is carried out 5 times independently.

We can see that DTF-CE still achieves the best performance on both data sets. The performance of DTF-CE and BPR-PITF illustrate that the Content

Enhanced component enables hash codes to carry more items' information and promotes the performance of cold-start problem. Besides, comparing DT-CE to BPR-PITF, the fact that performance differences between DTF-CE and BPR-PITF increase after 7 (top-N) serves as another evidence for the effectiveness of Content Enhanced component because DT-CE yields more compact yet informative binary codes than BPR-PITF.

5 Conclusion

In this paper, an efficient hashing based tensor factorization framework called Discrete Tensor Factorization (DTF) is proposed to alleviate tag recommendation problem. First, we formulate a predictive function based on hash codes. Second, based on the predictive function, we present the DTF recommendation framework by imposing balance and decorrelation constraints on hash codes and directly learned by several alternative optimization sub-problems. Then, we further extend the DTF model by introducing a deep content extractor for more comprehensive user/item profiling as the content enhanced model (DTF-CE) to improve the accuracy of recommendation. Finally, extensive experiments on two real-world data sets demonstrate that our model can greatly enhance the efficiency without sacrificing much effectiveness and the proposed models provide a good trade-off between recommendation efficiency and accuracy.

References

1. Carroll, J.D., Chang, J.J.: Analysis of individual differences in multidimensional scaling via an n-way generalization of eckart-young decomposition. Psychometrika **35**(3), 283–319 (1970)
2. Das, A.S., Datar, M., Garg, A., Rajaram, S.: Google news personalization: scalable online collaborative filtering. In: Proceedings of the 16th International Conference on World Wide Web, pp. 271–280. ACM (2007)
3. Gong, Y., Lazebnik, S., Gordo, A., Perronnin, F.: Iterative quantization: a procrustean approach to learning binary codes for large-scale image retrieval. IEEE Trans. Pattern Anal. Mach. Intell. **35**(12), 2916–2929 (2013)
4. Harshman, R.A.: Foundations of the PARAFAC procedure: models and conditions for an explanatory multimodal factor analysis (1970)
5. Håstad, J.: Some optimal inapproximability results. J. ACM (JACM) **48**(4), 798–859 (2001)
6. Karatzoglou, A., Smola, A., Weimer, M.: Collaborative filtering on a budget. In: Proceedings of the Thirteenth International Conference on Artificial Intelligence and Statistics, pp. 389–396 (2010)
7. Lian, D., Liu, R., Ge, Y., Zheng, K., Xie, X., Cao, L.: Discrete content-aware matrix factorization. In: Proceedings of the 23rd ACM SIGKDD International Conference on Knowledge Discovery and Data Mining, pp. 325–334. ACM (2017)
8. Liu, W., Mu, C., Kumar, S., Chang, S.F.: Discrete graph hashing. In: Advances in Neural Information Processing Systems, pp. 3419–3427 (2014)
9. Muja, M., Lowe, D.G.: Fast approximate nearest neighbors with automatic algorithm configuration. VISAPP (1) **2**(331–340), 2 (2009)

10. Rendle, S., Freudenthaler, C., Gantner, Z., Schmidt-Thieme, L.: BPR: Bayesian personalized ranking from implicit feedback. In: Proceedings of the Twenty-fifth Conference on Uncertainty in Artificial Intelligence, pp. 452–461. AUAI Press (2009)
11. Rendle, S., Schmidt-Thieme, L.: Pairwise interaction tensor factorization for personalized tag recommendation. In: Proceedings of the Third ACM International Conference on Web Search and Data Mining, pp. 81–90. ACM (2010)
12. Shen, F., Shen, C., Liu, W., Tao Shen, H.: Supervised discrete hashing. In: Proceedings of the IEEE Conference on Computer Vision and Pattern Recognition, pp. 37–45 (2015)
13. Symeonidis, P., Nanopoulos, A., Manolopoulos, Y.: Tag recommendations based on tensor dimensionality reduction. In: Proceedings of the 2008 ACM Conference on Recommender Systems, pp. 43–50. ACM (2008)
14. Wang, J., Kumar, S., Chang, S.F.: Semi-supervised hashing for large-scale search. IEEE Trans. Pattern Anal. Mach. Intell. **34**(12), 2393–2406 (2012)
15. Wang, J., Liu, W., Kumar, S., Chang, S.F.: Learning to hash for indexing big data-a survey. Proc. IEEE **104**(1), 34–57 (2016)
16. Zhang, Y., Lian, D., Yang, G.: Discrete personalized ranking for fast collaborative filtering from implicit feedback. In: AAAI, pp. 1669–1675 (2017)
17. Zhang, Y., Yin, H., Huang, Z., Du, X., Yang, G., Lian, D.: Discrete deep learning for fast content-aware recommendation. In: Proceedings of the Eleventh ACM International Conference on Web Search and Data Mining, pp. 717–726. ACM (2018)
18. Zhang, Z., Wang, Q., Ruan, L., Si, L.: Preference preserving hashing for efficient recommendation. In: Proceedings of the 37th International ACM SIGIR Conference on Research and Development in Information Retrieval, pp. 183–192. ACM (2014)
19. Zhou, K., Zha, H.: Learning binary codes for collaborative filtering. In: Proceedings of the 18th ACM SIGKDD International Conference on Knowledge Discovery and Data Mining, pp. 498–506. ACM (2012)

Efficient Integer Vector Homomorphic Encryption Using Deep Learning for Neural Networks

Tianying Xie[1] and Yantao Li[2(✉)]

[1] College of Computer and Information Sciences, Southwest University,
Chongqing 400715, China
[2] College of Computer Science, Chongqing University, Chongqing 400044, China
yantaoli@cqu.edu.cn

Abstract. Machine learning techniques based on neural networks have achieved significant applications in a wide variety of areas. There is a great risk on disclosing users' privacy when we train a high-performance model with a large number of datasets collected from users without any protection. To protect user privacy, we propose an Efficient Integer Vector Homomorphic Encryption (EIVHE) scheme using deep learning for neural networks. We use EIVHE to encrypt users' datasets, then feed the encrypted datasets into a neural network model, and finally obtain the trained model for neural networks. EIVHE is an innovative bridge between cryptography and deep learning, which aims at protecting users' privacy. The experiments demonstrate that the deep neural networks can be trained by encrypted datasets without privacy leakage, and achieve an accuracy of 89.05% on MNIST. Moreover, this scheme allows us to conduct computation in an efficient and secure way.

Keywords: Deep learning · Neural networks
Homomorphic encryption · MNIST

1 Introduction

Nowadays, neural networks have been applied in a great number of applications, which can be broadly categorized into six fields, including image and object recognition, electronic game, voice generation and recognition, imitation of art and style, prediction, and website design modification [1,2]. With the development of these fields, however, local computing has been found increasingly incapable in limited computing resources. To address computing resource limitation, more and more researchers prefer to use the cloud to do computations. Although the performance of computing has been improved, how cloud platforms guarantee that the security of users' data rises? Therefore, how to protect data privacy and how to ensure its usability are intractable issues in the process of cloud computing. As a key method to solve these issues, homomorphic encryption technology has been widely developed in recent years.

© Springer Nature Switzerland AG 2018
L. Cheng et al. (Eds.): ICONIP 2018, LNCS 11301, pp. 83–95, 2018.
https://doi.org/10.1007/978-3-030-04167-0_8

There are two types of homomorphic encryptions: fully homomorphic encryption (FHE) and somewhat homomorphic encryption (SWHE). For FHE, the homomorphic encryption scheme supports any given function, as long as the function can be described by an algorithm and implemented on a computer. FHE is an efficient scheme, but the computational overhead is high. For SWHE, the homomorphic encryption scheme only supports some specific functions. SWHE is slightly weaker, but the computational overhead is lower and easier to implement. The most popular four algorithms are: efficient homomorphic encryption on integer vectors [3], ring-based fully homomorphic encryption scheme [4], somewhat practical fully homomorphic encryption [5], and fully homomorphic encryption without bootstrapping [6]. However, most of them did not address the privacy of users' data in neural networks.

In this paper, we propose a method that combines state-of-the-art machine learning methods with advanced cryptography [6] and training neural networks within encrypted datasets to protect the dataset privacy for training and testing in neural networks.

The main contributions of this work can be summarized as follows:

- We propose an efficient integer vector homomorphic encryption scheme using deep learning to protect the privacy of the users' data in neural networks.
- To protect the privacy in training and testing images, we exploit the fully homomorphic encryption scheme to encrypt the input data using EIVHE encryption scheme before feeding them into the neural networks.
- We evaluate EIVHE scheme on a standard image classification dataset MNIST, and the experimental results show that the deep neural networks can be trained by encrypted datasets without privacy leakage, and EIVHE achieves an accuracy of 89.05%.

The remainder of this paper is organized as follows: Sect. 2 introduces preliminary knowledge. We detail EIVHE scheme for neural networks in Sect. 3 and describe experiments in Sect. 4. We evaluate the performance of EIVHE in Sect. 5 and conclude this work in Sect. 6.

2   Preliminaries

2.1   Fully Homomorphic Encryption

Homomorphic encryption was first introduced by Rivest, Adleman and Dertouzous shortly after the invention of RSA [8,9]. Homomorphic encryption enables operations on plaintexts to be performed on their respective ciphertexts without disclosing the plaintexts [6]. A homomorphic encryption scheme generally consists of four algorithms: KeyGen, Encrypt, Decrypt, and Evaluate [10,11]. According to the definition of homomorphic encryption, we can intuitively define the four functions as follows:

1. KeyGen. It takes a security parameter w, and produces a secret key s_k and a public key p_k, i.e., $keyGen(w) \rightarrow (p_k, s_k)$.

2. Encrypt. It takes a public key p_k and a plaintext m as inputs, and produces a ciphertext c of m, i.e., $Encrypt(m, p_k) \to c$.
3. Decrypt. It takes the secret key s_k and c as inputs, and produces the plaintext m of c, i.e., $Decrypt(c, s_k) \to m$.
4. Evaluate. It takes the public key p_k, a circuit C and a tuple of ciphertext $(c_1, c_2, \cdots, c_n)$ as inputs, and produces the encrypted result c, i.e., $Decrypt(s_k, c_1, c_2, \cdots, c_n) = f(m_1, m_2, c \cdots, m_n)$, where f is the functionality that we want to perform.

2.2 Concept of Deep Learning

Given a training dataset, the learning task is to determine these weight variables by minimizing a pre-defined cost function, such as the cross-entropy or the squared-error cost function [12]. Deep learning refers to a multi-layer neural network using various machine learning algorithms to solve image, text and other problems of the algorithm set. Deep learning can be classified into neural networks in a broad category, but they are different in implementations. The core of deep learning is feature learning, which aims at obtaining hierarchical feature information through hierarchical networks, thereby solving the important problems that need to artificially design features in the past.

Deep neural networks that are remarkably effective for many machine learning tasks, define parameterized functions from inputs to outputs as compositions of many layers of basic building blocks, such as affine transformations and simple nonlinear functions [15]. By varying parameters of these blocks, we can "train" such a parameterized function with the goal of fitting any given finite set of input/output examples. Neural networks are generally composed of three layers: input layer, hidden layer, and output layer. More specifically, input layer is the interface we feed our dataset into the network. Hidden layer is the processing unit of a model. Output layer shows the results we obtain from the model. Deep learning means the network has many hidden layers. We regard deep learning as a process of constant grinding, where it first defines some standard parameters and then revises them.

3 Efficient Integer Vector Homomorphic Encryption Scheme for Neural Networks

3.1 Efficient Integer Vector Homomorphic Encryption (EIVHE)

We introduce our EIVHE scheme for encrypt datasets, including the encryption scheme and key-switching technique, as illustrated in Algorithm 1.

Encryption Scheme. At first, let $\mathbf{x} \in \mathbb{Z}_p^n$ be an integer vector to encrypt, where n denotes the length of the vector and p indicates the alphabet size. Let

$\mathbf{c} \in \mathbb{Z}_q^{n+1}$ be the ciphertext of $\mathbf{x}$ with length $n+1 > n$ and alphabet size $q >> p$. The secret key is a matrix $\mathbf{S} \in \mathbb{Z}_q^{m \times n}$ and it satisfies

$$\mathbf{Sc} = w\mathbf{x} + \mathbf{e}, \tag{1}$$

where $\mathbf{e} \in \mathbb{Z}^{m \times n}$ indicates a noise matrix, and w is an integer parameter such that $w > 2|\mathbf{e}|$ ($|\mathbf{e}|$ indicates the maximum absolute value of the entries in the noise matrix $\mathbf{e}$). The process of encryption $\mathbf{x}$ is to find a ciphertext $\mathbf{c}$ such that $\mathbf{Sc}$ satisfies Eq. (1). According to Eq. (1), the decryption of ciphertext $\mathbf{c}$ based on the secret key $\mathbf{S}$ can be obtained via computing Eq. (2):

$$\mathbf{x} = \lfloor \frac{\mathbf{Sc}}{w} \rfloor, \tag{2}$$

where $\lfloor \mathbf{a} \rfloor$ indicates rounding $\mathbf{a}$ down to its nearest integer.

Key-Switching Technique. In [13], Brakerski and Vaikuntanathan introduced a very useful re-linearization technique that can switch the secret key in the PVM scheme to any other secret key when they both are vectors. Then, Brakerski, Gentry and Halevi developed a technique to switch two secret keys of matrices [14]. In general, Brakerski and Vaikuntanathan's re-linearization method has two steps, which we apply to switch a secret key $\mathbf{S} \in \mathbb{Z}_q^{m \times n}$ to another secret key $\mathbf{S}' \in \mathbb{Z}_q^{m \times n'}$ via an intermediate key $\mathbf{S}^* \in \mathbb{Z}_q^{m \times nl}$ in three steps, and we get a new ciphertext $\mathbf{c}'$ that still encrypts the same integer vector $\mathbf{x}$.

1. Step 1: $\mathbf{S} \rightarrow \mathbf{S}^*$, indicating that $\mathbf{S}$ is transformed to $\mathbf{S}^*$, such that its corresponding new ciphertext $\mathbf{c}^*$ has a smaller magnitude than $\mathbf{c}$. The goal is to represent each element $\mathbf{c}_i$ in $\mathbf{c}$ with a binary vector (binary representation), hence it results in a new ciphertext $\mathbf{c}^*$ with $|\mathbf{c}^*| = 1$. Assuming that $\mathbf{c}_i = \mathbf{b}_{i0} + \mathbf{b}_{i1}2 + \cdots + \mathbf{b}_{i(l-1)}2^{l-1}$, then we obtain $\mathbf{c}^*$ by expressing each $\mathbf{c}_i$ as $[\mathbf{b}_{i0}, \mathbf{b}_{i1}, \cdots, \mathbf{b}_{i(l-1)}]$. Then, we construct a secret key $\mathbf{S}^* \in 2^{m \times nl}$ such that

$$\mathbf{S}^* \mathbf{c}^* = \mathbf{Sc}, \tag{3}$$

which can be replaced by each $\mathbf{S}_{ij}$ in $\mathbf{S}$ with a vector $[\mathbf{S}_{ij}, \mathbf{S}_{ij}2, \cdots, \mathbf{S}_{ij}2^{l-1}]$.

2. Step 2: $\mathbf{S}^* \rightarrow \mathbf{S}'$, showing that $\mathbf{S}^*$ is transformed to $\mathbf{S}'$. We construct an integer matrix $\mathbf{M} \in \mathbb{Z}^{n' \times nl}$ and a noise matrix $\mathbf{E}$, such that

$$\mathbf{S}' \mathbf{M} = \mathbf{S}^* + \mathbf{E}. \tag{4}$$

Assuming $\mathbf{S}' = [\mathbf{I}, \mathbf{T}]$ with an identity matrix $\mathbf{I}$, $\mathbf{M}$ can be constructed by Eq. (5):

$$\mathbf{M} = \left(\frac{\mathbf{S}^* + \mathbf{E} - \mathbf{T}\mathbf{A}^*}{\mathbf{A}} \right), \tag{5}$$

where $\mathbf{A} \in \mathbb{Z}^{(n'-m) \times nl}$ is a random matrix.

3. Step 3: $\mathbf{S}', \mathbf{M} \rightarrow \mathbf{c}'$, indicating that we obtain a new ciphertext $\mathbf{c}'$ by $\mathbf{M}$, as Eq. (6):

$$\mathbf{c}' = \mathbf{M}\mathbf{c}^*. \tag{6}$$

Algorithm 1. Efficient integer vector homomorphic encryption scheme (EIVHE)

Input: w, x;

Output: c, S;

1: get the row of x m and the column of x n, respectively;
2: generate $key() : S = random.rand(m, n) \times w$;
3: get $T() : T = random.rand(n, 1)$;
4: get l: encode length for a maximum integer in $|x|$;
5: get c^*;
6: for $i = 1$ to m
7:     for $j = 1$ to n
8:         $b = binary\_vector(c[i][j])$;
9:         if $c[i][j] \leq 0$ then $b^* = -1$;
10:        $c^*[i][(j*1) + l - len(b) : (j+1)*l] += b$;
11:    end for
12: end for
13: get S^*;
14: for $i = 1$ to l;
15:    $S^* = S * 2^{l-i-1}$;
16: end for
17: $S' = [I, T]$ and $A = random.rand(1, n*l) * 10$;
18: $E = random.rand(\text{the row of } S^*, \text{the column of } S^*) * l$;
19: $M = \begin{pmatrix} -TA + S^* + E \\ A \end{pmatrix}$;
20: $c' = M * T$;
21: return $c = c'$ and $S = S'$;

3.2 Neural Networks

We construct six hidden layers. To prevent gradient dispersion, we use batch normalization. Activation function is Rectified Linear Unit (ReLu), and output layer is SoftMax. The cost function is cross entropy, and optimization function is Adaptive moment estimation (Adam). These functions are detailed as follows:

Batch Normalization. According to [18], at each Stochastic gradient descent (SGD), the corresponding activation is normalized by mini-batch, so that the mean value of the result (output signal of each dimension) is 0 and the variance is 1.

ReLu. Rectified Linear Unit (ReLu) is an activation function, which is commonly used in artificial neural networks. [7] In general, the linear rectification function refers to the slope function in mathematics, such as:

$$f(X) = max(x, 0). \tag{7}$$

SoftMax. SoftMax is a generalization of logistic function that "squashes" (or maps) a K-dimensional vector Z, which compresses a K-dimensional vector of an arbitrary real number into another K-dimensional real vector, where the values of each element of the vector fall in the range $(0, 1)$ and the sum of theses values comes to 1. SoftMax function has a relationship with logistic function, and is also called normalized exponential function, as shown in Eq. (8):

$$\sigma(\mathbf{Z})_j = \frac{e^{\mathbf{Z}_j}}{\sum_{k=1}^{K} e^{\mathbf{Z}_k}}. \tag{8}$$

Cross Entropy. Cross entropy is a concept of the information theory, which is the earliest used by the information entropy change (relative information entropy, and compression ratio), and then applied in many areas, including communications, error correction circuit, game theory, machine learning [16]. The introduction of cross entropy cost function for neural network is to make up for the defect that the derivative form of sigmoid function is easy to saturate, as shown in Eq. (9):

$$L_H(x, z) = - \sum_{k=1}^{d} x_k log z x + (1 - x_k) log(1 - z_k). \tag{9}$$

Adam. Adaptive moment estimation (Adam) is a first-order gradient-based algorithm, which aims at optimizing random objective function [15]. Based on the first-order and second-order moment estimation of gradient of each parameter via the cost function, Adam dynamically adjusts the learning rate for each parameter [17].

To train the neural networks, we feed **X**, **X_label**, **Y**, and **Y_label** into neural network, where **X** denotes training dataset, **X_label** indicates the label of **X**, **Y** denotes testing dataset, and **Y_label** indicates the label of **Y**. The deep learning algorithm with EIVHE is described in Algorithm 2.

Algorithm 2. Deep learning with EIVHE

Input: **X**, **X_label**, **Y**, **Y_label**;
Output: accuracy;
  1: EIVIIE(**X**), EIVHE(**Y**);
  2: initial weight and parameters of the neural network;
  3: feed training dataset **X** into the neural network;
  4: get mini-batch_X;
  5: forward propagation: computing the cost;
  6: backward propagation: computing gradients;
  7: optimize: Adam;
  8: repeat step 3-7 until the neural network becomes stable;
  9: get deep learning model;
 10: input test dataset **Y** into the neural network;
 11: get accuracy

---

4   Experiments

4.1   Dataset

We conduct experiments based on the standard MNIST dataset for handwritten digit recognition consisting of 60,000 training examples and 10,000 testing examples [12]. Each example is a 28 × 28 gray-level image. That is, one image has 784 feature points before homomorphic encryption, and we get 785 feature points per image using this method. In other words, according to key-switching technique, through matrix transformation, the feature points of one image have changed from 784 to 785. We use a simple forward-propagation neural network with ReLu unit, SoftMax of 10 classes with cross-entropy loss, mini-batch normalization, and the optimization algorithm Adam. Based on MNIST, we conduct experiments on EIVHE.

(a) Cost   (b) Training accuracy   (c) Validation accuracy

Fig. 1. Cost and accuracy of the original dataset

4.2   Baseline Model

Our baseline model uses original MNIST dataset without preprocessing where the input layer has 784 units, six hidden layers are 512, 256, 128, 64, 32, 16 units, respectively, and the output layer has 10 units. As for Adam, we set the parameters as default: $\alpha = 0.001$, $\beta_1 = 0.9$, $\beta_2 = 0.999$ and $\epsilon = 10^{-8}$. Using this method, we can reach an accuracy of 94.87% in about 100 epochs. As shown in Fig. 1(a), the cost of original dataset maintains the downward trend until convergence. Figure 1(b) shows the trend of training accuracy until convergence and the training accuracy reaches up to 94.87%. Figure 1(c) shows the validation accuracy of original testing dataset, and the convergence value is 90.19%. Details are shown in Table 1.

4.3   EIVHE Model

With the same architecture as the baseline model, we test the proposed deep learning with the homomorphic encryption model (EIVHE). The difference is that we use the encrypted dataset to feed the neural network. The original size

Table 1. Details of original dataset

Type	Training accuracy	Validation accuracy	Training time	Epochs
Original dataset	94.87%	90.19%	26.28 m	100

Table 2. Homomorphic encryption time

Dataset	Encryption time	Cost per pixel
Training images	662.77 s	1.41^{-5} s
Testing images	119.98 s	1.53^{-5} s

of one image of MNIST is 28 by 28, and we reshape it as 1 by 784. Thus, training images become a 60000×784 matrix, and testing images become a 10000×784 matrix. In addition, We use One-Hot Encoding to differentiate the training and testing labels. Encryption time and time cost of the two datasets are shown in Table 2.

5　Performance Evaluation

5.1　Accuracy on the Separate Datasets

We encrypt training image datasets and testing image datasets, respectively, by using EIVHE. Then, we feed the encrypted training image datasets into the neural network, where the epoch is set as 100. Under encryption without using the same key, the matrix we obtain is a non-homomorphic matrix that undermines the internal relationship of the datasets to some extent. Then, to solve this issue, we encrypt the two datasets together with the same key as described in Sect. 4.3.

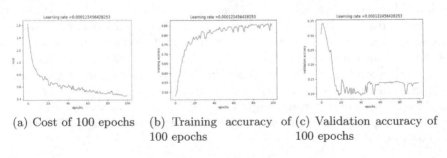

(a) Cost of 100 epochs　(b) Training accuracy of　(c) Validation accuracy of
　　　　　　　　　　100 epochs　　　　　　　100 epochs

Fig. 2. Encrypted dataset

As shown in Fig. 2(a), the cost is much higher than original dataset, and Fig. 2(b) shows the training accuracy of encrypted dataset is 85.97%. The reason for accuracy decrease is that the input dimension of EIVHE is changed from 784 to 785 after encryption. It seems that one fake feature point is artificially added in every input pictures, which can be regarded as a type of noise. However, as shown in Fig. 2(c), we find our validation accuracy presents greatly low. By checking the dataset, we find the fatal point is that we encrypted training and testing datasets, respectively, but each encryption secret key is randomly generated. Details are shown in Table 3.

Table 3. Accuracy of the separate encrypted dataset

Type	Training accuracy	Validation accuracy	Training time	epochs
Encrypted dataset	85.97%	13.54%	28.38 m	100

5.2 Accuracy on the Improved Algorithm

To show a better performance, we improve EIVHE (Algorithm 1) to IEIVHE (Algorithm 3). Thus, we use IEIVHE to encrypt our training and testing datasets together, and then we can get better results. As shown in Fig. 3(a), the cost has declined, 0.3 m lower than original dataset. In Fig. 3(b), the training accuracy has improved to 87.29%, 1.32% higher than original datasets. Figure 3(c) shows that the validation accuracy has improved greatly to 85.04%. Details are shown in Table 4.

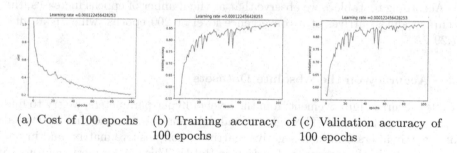

(a) Cost of 100 epochs   (b) Training accuracy of   (c) Validation accuracy of
                          100 epochs   100 epochs

Fig. 3. Improved encrypted datasets (together)

Table 4. Accuracy of the improved algorithm on encrypted dataset

Type	Training accuracy	Validation accuracy	Training time	Epochs
Encrypted dataset	87.29%	85.04%	28.08 m	100

Algorithm 3 . Improved efficient integer vector homomorphic encryption (IEIVHE)

Input: $w, \mathbf{X}, \mathbf{Y}$;
Output: $\mathbf{c_1}, \mathbf{c_2}, \mathbf{S_1}, \mathbf{S_2}$;
 1: $m = \mathbf{X}.shape[0], n = \mathbf{X}.shape[1] = \mathbf{Y}.shape[1], h = \mathbf{Y}.shape[0]$;
 2: $\mathbf{S}$=generate_key(w,n);
 3: $\mathbf{T}$=get_T(n);
 4: $\mathbf{c_1}, \mathbf{S_1}$=encrypt_via_switch($\mathbf{X}$,w,m,n,$\mathbf{T}, \mathbf{S}$);
 5: $\mathbf{c_2}, \mathbf{S_2}$=encrypt_via_switch($\mathbf{Y}$,w,h,n,$\mathbf{T}, \mathbf{S}$);

As for the accuracy improvement, we have done many attempts for hyper-parameter tuning. Unfortunately, there are so many hyper-parameters fixed, which are widely circulated via the industry, so we set the recommended parameters as the default. In addition, there is one parameter that has impact on accuracy of our model: epochs. Following this clue, we vary the number of epochs to test the accuracy improvement. Details are are shown in Table 5.

Table 5. Accuracy of the impact of epochs

Type	Training accuracy	Validation accuracy	Training time	Epochs
Encrypted dataset	87.29%	83.38%	55.59 m	200
	87.28%	83.39%	85.97 m	300

According to Table 5, we observe that as the number of epochs increases, the accuracy almost arrives at a convergence level at 200 epochs, which is around 87.29%.

5.3   Accuracy on the Absolute Datasets

Besides accuracy improvement from the level of hyper-parameters, we try to find a new way to improve it. Through internal data analysis, we find that encrypted data matrix alters from non-negative matrix to non-positive matrix, and in our neural network, the activation function is ReLU. This causes most neurons to be inactive. Hence, a delicate and easy method is devised to solve the issue that takes the absolute value of the encrypted matrix.

As shown in Fig. 4(a), we find the cost is approximately the same. Figure 4(b) shows the improved accuracy, which means our method has effect and has a little bit improvement. Then, as for Fig. 4(c), we can see that the validation accuracy also has been improved. Details are shown in Table 6.

From the above experiments, we have the following observations:

1. The homomorphic encryption protects the privacy of data. Datasets can be encrypted before using them via neural networks, since it takes lots of time

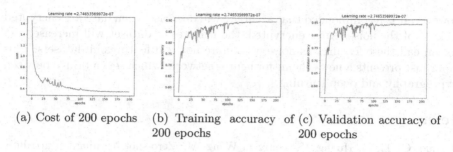

(a) Cost of 200 epochs   (b) Training accuracy of (c) Validation accuracy of
200 epochs   200 epochs

Fig. 4. Encrypted dataset (absolute)

to encrypt the dataset after usage. Our solution just costs a little time to encrypt the training dataset and train our model, and then every image that needs to be encrypted just spends little time to encrypt.
2. When we take the absolute value of the encrypted dataset, the accuracy is improved by 3.08% via the level of data.
3. The training super-parameters have little impact on the model accuracy. As for epochs, we obtain a stable and efficient value 200.

Our algorithm allows for privacy preserving and deep learning. The first initial experiments did not show a significant improvement, but it is interesting to consider more sophisticated schemes for accuracy improvement.

Table 6. Accuracy of the absolute datasets

Type	Training accuracy	Validation accuracy	Training time	Epochs
Encrypted dataset (absolute)	89.05%	84.98%	256.73 m	200

6 Conclusions

We demonstrate the training of deep neural network with encrypted datasets of homomorphic encryption, preserving the users' privacy and avoiding privacy leakage when using neural networks, by computing the entire model with many parameters. In our experiments based on MNIST, we achieve 89.05% training accuracy with encrypted dataset. Our scheme is based on dataset of homomorphic encryption, and after encryption we use encrypted dataset to train our neural network model. Since our approach applies cryptography into deep learning, it can be adapted to many other datasets before using them to feed. When conduct data encryption, others cannot understand data without secret key, and thus it can protect privacy and ensure users to use the neural network safely.

However, it is recommended that accuracy will decrease when data are encrypted because of the noise. After encrypted, the features of dataset will increase, and we can call these features as noise, which are not real features in datasets. This paper just presents a new scheme for neural network, which sets a bridge between cryptography and deep learning.

References

1. Luo, C., Li, Z., Huang, K., Feng, J., Wang, M.: Zero-shot learning via attribute regression and class prototype rectification. IEEE Trans. Image Process. **27**(2), 637–648 (2018)
2. Hu, G., Peng, X., Yang, Y., Hospedales, T.M., Verbeek, J.: Frankenstein: learning deep face representations using small data. IEEE Trans. Image Process. **27**(1), 293–303 (2018)
3. Zhou, H., Wornell, G.: Efficient homomorphic encryption on integer vectors and its applications. In: 2014 Information Theory and Applications Workshop, pp. 1–9. IEEE Press, New York (2014)
4. Bos, J.W., Lauter, K., Loftus, J., Naehrig, M.: Improved security for a ring-based hully homomorphic encryption scheme. In: Stam, M. (ed.) Cryptography and Coding. LNCS, vol. 8308, pp. 45–64. Springer, Heidelberg (2013). https://doi.org/10.1007/978-3-642-45239-0_4
5. Fan, J., Vercauteren, F.: Somewhat practical fully homomorphic encryption. Cryptology ePrint Archive, Report 2012/144 (2012)
6. Brakerski, Z., Gentry, C., Vaikuntanathan, V.: (Leveled) fully homomorphic encryption without bootstrapping. In: Proceedings of the 3rd Innovations in Theoretical Computer Science Conference, pp. 309–325. ACM, New York (2012)
7. Osia, S.A., et al.: A hybrid deep learning architecture for privacy-preserving mobile analytics. arXiv preprint arXiv:1703.02952 (2018)
8. Rivest, R.L., Adleman, L., Dertouzos, M.L.: On data banks and privacy homomorphisms. Found. Secure Comput. **4**(11), 169–180 (1978)
9. Rivest, R.L., Shamir, A., Adleman, L.: A method for obtaining digital signatures and public-key cryptosystems. Commun. ACM **21**(2), 120–126 (1978)
10. Boneh, D., Goh, E.-J., Nissim, K.: Evaluating 2-DNF formulas on ciphertexts. In: Kilian, J. (ed.) TCC 2005. LNCS, vol. 3378, pp. 325–341. Springer, Heidelberg (2005). https://doi.org/10.1007/978-3-540-30576-7_18
11. Gentry, C.: Fully homomorphic encryption using ideal lattices. In: Proceedings of the Forty-first Annual ACM Symposium on Theory of Computing, pp. 169–178. ACM, New York (2009)
12. LeCun, Y., Bottou, L., Bengio, Y., Haffner, P.: Gradient-based learning applied to document recognition. Proc. IEEE **86**(11), 2278–2324 (1998)
13. Brakerski, Z., Vaikuntanathan, V.: Efficient fully homomorphic encryption from (standard) LWE. In: 2011 IEEE 52nd Annual Symposium on Foundations of Computer Science, pp. 97–106. IEEE Press, New York (2011)
14. Brakerski, Z., Gentry, C., Halevi, S.: Packed ciphertexts in LWEbased homomorphic encryption. In: Kurosawa, K., Hanaoka, G. (eds.) Public-Key Cryptography - PKC 2013. LNCS, vol. 7778, pp. 1–13. Springer, Heidelberg (2013). https://doi.org/10.1007/978-3-642-36362-7_1
15. Abadi, M., et al.: Deep learning with differential privacy. In: Proceedings of the 2016 ACM SIGSAC Conference on Computer and Communications Security, pp. 308–318. ACM, New York (2016)

16. Le, T.P., Aono, Y., Hayashi, T., Wang, L., Moriai, S.: Privacy-preserving deep learning via additively homomorphic encryption. IEEE Trans. Inf. Forensics Secur. **13**(5), 1333–1345 (2018)
17. Yang, L., Murmann, B.: Approximate SRAM for energy-efficient, privacy-preserving convolutional neural networks. In: 2017 IEEE Computer Society Annual Symposium on VLSI, pp. 689–694. IEEE Press, New York (2017)
18. Ioffe, S., Szegedy, C.: Batch normalization: accelerating deep network training by reducing internal covariate shift. In: Proceedings of the 32nd International Conference on Machine Learning, pp. 448–456. ACM, New York (2015)

DeepSIC: Deep Semantic Image Compression

Sihui Luo[1], Yezhou Yang[2], Yanling Yin[1], Chengchao Shen[1], Ya Zhao[1], and Mingli Song[1]([✉])

[1] Zhejiang University, Hangzhou 310058, China
{sihuiluo829,yanlingyin,chengchaoshen,yazhao,brooksong}@zju.edu.cn
[2] Arizona State University, Tempe, AZ 85281, USA
yz.yang@asu.edu

Abstract. Incorporating semantic analysis into image compression can significantly reduce the repetitive computation of fundamental semantic analysis in client-side applications such as semantic image retrieval. The same practice also enables the compressed code to carry semantic information of the image during its storage and transmission. In this paper, we propose a Deep Semantic Image Compression (DeepSIC) model to achieve this goal and put forward two novel architectures that aim to reconstruct the compressed image and generate corresponding semantic representations at the same time by a single end-to-end optimized network. The first architecture performs semantic analysis in the encoding process by reserving a portion of the bits from the compressed code to store the semantic representations. The second performs semantic analysis in the decoding step with the feature maps that are embedded in the compressed code. In both architectures, the feature maps are shared by the compression and the semantic analytics modules. Experiments over benchmarking datasets show promising performance of the proposed compression model.

Keywords: Deep image compression · Semantic image compression
End-to-end optimization

1 Introduction

As the era of smart cities and Internet of Things (IoT) unfolds, the increasing number of real-world applications require corresponding image and video transmission services to handle compression and semantic encoding at the same time, hitherto not addressed by the conventional systems. Traditionally, the process of image compression is merely a type of data compression that is applied to digital images to reduce their cost of storage and transmission. Almost all the image compression algorithms stay at the stage of low-level image representation in which the representation is arrays of pixel values.

Standard codecs such as JPEG [16] and JPEG2000 [9] compress images via a pipeline which roughly breaks down into 3 modules: pixel-level transformation,

© Springer Nature Switzerland AG 2018
L. Cheng et al. (Eds.): ICONIP 2018, LNCS 11301, pp. 96–106, 2018.
https://doi.org/10.1007/978-3-030-04167-0_9

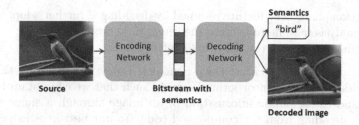

Fig. 1. The general framework of semantic image compression

quantization, and entropy encoding. It is the mainstream pipeline for lossy image compression codec. Although advances in the training of neural networks have helped improving performance in existing codecs of lossy compression, recent learning-based approaches generally follow the same pipeline as well [1,3,10,13–15].

As Rippel and Bourdev [10] point out, generally speaking, image compression is highly related to the procedure of pattern recognition. In other words, if a system can discover the underlying structure of the input, it can eliminate the redundancy and represent the input more succinctly. Recent deep learning based compression approaches discover the structure of images by training a deep compression model and then convert the extracted hidden representation into binary code [1,3,10,13–15]. These important works have achieved significant improvement on the performance of lossy image compression. Nevertheless, these existing DL-based compression codecs, like the conventional codecs, also only compress the images at the pixel level, and do not understand the semantics of the image. Currently, when the client-side applications require the semantic information of an image, they have to first reconstruct the image from the codec and then conduct an additional computing to obtain the semantic information.

Can a system conduct joint-optimization of the objectives for both compression and the semantic analysis? In this paper, we make the first attempt to approach this challenging task which stands between the computer vision and multimedia information processing fields, by introducing the Deep Semantic Image Compression (DeepSIC). Here, our DeepSIC framework aims to encode the semantic information in the codecs, and thus significantly reduces the computational resources needed for the repetitive semantic analysis on the client side.

We depict the DeepSIC framework in Fig. 1, which aims to incorporate the semantic representation within the codec while maintaining the ability to reconstruct visually pleasing images. Two architectures of our proposed DeepSIC framework are given for the joint analysis of pixel information together with the semantic representations for lossy image compression. One is pre-semantic DeepSIC, which integrate the semantic analysis module into the encoder part and reserve several bits in the compressed code to represent the semantics. The other is post-semantic DeepSIC, which only encodes the image features during the encoding process and conducts the semantic analysis process during the

reconstruction phase. The feature retained by decoding is further adopted in the
semantic analysis module to achieve the semantic representation.

In summary, we make the following contributions:

- We propose a concept called Deep Semantic Image Compression that aims
  to provide a semantic incorporated scheme such that we can obtain both the
  visual and the semantic information of an image through a single network
  without decoding from the compressed code. To our best knowledge, this is
  the first work to incorporate semantic analysis task directly within the image
  compression network.
- We put forward two kinds of architecture for the proposed DeepSIC: pre-
  semantic DeepSIC and post-semantic DeepSIC.
- We conduct experiments over multiple datasets to validate our framework
  and compare the two proposed architectures.

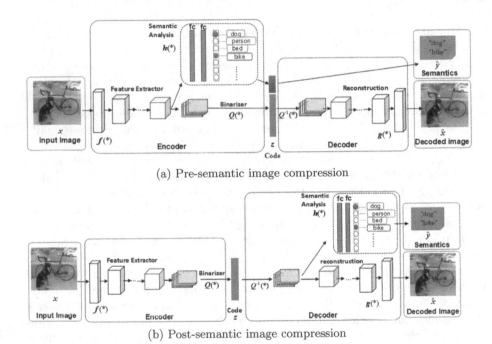

(a) Pre-semantic image compression

(b) Post-semantic image compression

Fig. 2. Two architectures of our semantic compression network: (a) Image compression
with pre semantic analysis module in the encoder; (b) Image compression with post
semantic analysis module in the decoder.

2   Proposed Deep Semantic Image Compression System

We introduce the modules involved in proposed DeepSIC in this section, Fig. 2
shows two architectures of the proposed DeepSIC system. For pre-semantic

DeepSIC, it places the semantic analysis in the encoding process, which is implemented by reserving a portion of the bits in the compressed code to store semantic information. Hence the code innately reflects the semantic information of the image. For post-semantic image compression, the feature retained from decoding is utilized for the semantic analysis module to get the class label and for reconstruction network to synthesize the target image. For both architectures, we extract the feature of the original images, quantize and code them into binary codes for storing or transmitting to the decoder. The reconstruction network then creates an estimate of the original input image based on the received binary code. The compression model is trained under the traditional Rate-Distortion loss [1,13] together with the error rate of the semantic analysis. The specific descriptions of each module are present in the subsequent subsections.

2.1 Feature Extraction

The goal of feature extraction module is to produce the condensed representation of the input image to reduce the redundancy while maintaining the knowledge needed for both the image reconstruction and semantic analysis task. The strategy of reusing the feature throughout the network helps the training of deeper network architectures [4,5,7,11,12]. These CNNs can represent the image better and are demonstrated to be good feature extractors for various computer vision tasks. Our proposed DeepSIC deploys a DenseNet [5] based structure illustrated in Fig. 3.

Given an input image x, the extracted feature is denoted as $f(x)$. Specifically, feature extraction module in the proposed network consists of four stages with skip connections. The first stage is a preprocess convolution block of three 3×3 convolution, the first convolution is of the stride $(2, 2)$ to downsample the input in spatial dimensions. The subsequent three stages are identical that each comprises a dense block and a transition-down block. To be specific, the dense block is formed by 4 densely connected convolutions and the transition down the block is formed by a 1×1 convolution and a $2\times$ downsample layer. Each stage is ended with a batch normalization layer and a Relu activation layer to avoid overfitting and increase the nonlinearity respectively. Each subsequent stages utilizes the output of the previous stage to alleviate the loss of spatial information.

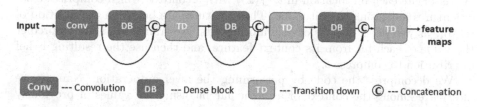

Fig. 3. The four-stage densenet based feature extractor: the first stage is a preprocess convolutional module and each of the subsequent three stage is formed by a dense block and transition-down block.

2.2   Binarizer

The function of a binarizer is converting the extracted tensor $f(x) \in R^{C \times H \times W}$ to the binary compressed code sequence. In proposed compression system, we firstly quantizing the tensor and then further compress the quantized code via a lossless entropy coding scheme. With quantization, the feature tensor is optimally quantized to a lower bit precision B:

$$Q(f(x)) = \frac{1}{2^{B-1}} \left\lceil 2^{B-1} f(x) \right\rceil. \tag{1}$$

The quantization bin B we use here is 6 bit and the $\lceil * \rceil$ operation represent rounding operation. After quantization, the output is converted to a binary tensor. Since the quantization Q is a non-differential function for back-propagation, the marginal density of quantized representation z_i is then given by the training of a continuous and differentiable approximation of expressing the model's distribution Q by a probability density q.

$$Q(f(x)) = \int_{-\frac{1}{2}}^{\frac{1}{2}} q(f(x) + t)d_t \tag{2}$$

And as done in [13], $Q(f(x))$ is upper bounded to

$$-log_2 Q(f(x)) = -log_2 \int_{-\frac{1}{2}}^{\frac{1}{2}} q(f(x) + t)d_t \leqslant \int_{-\frac{1}{2}}^{\frac{1}{2}} -log_2 q(f(x) + t)d_t. \tag{3}$$

However, the entropy of the binary code generated during feature extraction and quantization period are not maximum because the network is not explicitly designed to maximize entropy in its code, and the model does not necessarily exploit visual redundancy over a large spatial extent. We exploit this low entropy by lossless compression via entropy coding, to be specific, we utilize the context-adaptive binary arithmetic coding (CABAC) framework proposed by [8]. Arithmetic entropy encoding schemes are designed to compress discrete-valued data to bit rates closely approaching the entropy of the representation, assuming that the probability model used to design the code approximates the data well. We associate each bit location in $Q(f(x))$ with a context, which comprises a set of features indicating the bit value. These features are based on the position of the bit as well as the values of neighboring bits. We train a classifier to predict the value of each bit from its context feature, and then use the resulting belief distribution to compress b.

We decompress the code by performing the inverse operation. Namely, we interleave among the context of a particular bit using the values of previously decoded bits. The obtained context is employed to retrieve the activation probability of the bit to be decoded. Note that this constrains the context of each bit to only involve features composed of bits already decoded.

2.3   Reconstruction from Features

The module of reconstruction from features mirrors the structure of the feature extraction module, which is four-stage formed as well. Each stage comprises the transpose convolution and upsample layer to gradually recover the spatial dimension of the source image. The output of each previous layer is passed on to the subsequent layer through two paths, one is the deconvolutional network, and the other is a upsampling to target size through interpolation. After reconstruction, we obtain the output decompressed image $\hat{x} = g\left(Q^{-1}(q)\right)$.

Although arithmetic entropy encoding is lossless, the quantization will bring in some loss in accuracy, the result of $Q^{-1}(q)$ is not identical in value with the output of feature extraction $f(x)$.

2.4   Semantic Analysis

We select image classification as the goal of semantic analysis module for experiments in this paper as well. Figure 2 presents the structure of our semantic analysis module $h(*)$. It is position-optional and can be placed in the encoding or decoding process for the two different architectures.

We adjust the learning rate using the related name-value pair arguments when creating the two fully connected layer. Moreover, a softmax function is utilized as the output unit activation function after the last fully connected layer for the multi-class classification. Denote the weight matrix of the two fully connected layer as W_1 and W_2, the bias of them as b_1 and b_2 respectively, the prediction probability distribution can be formulated as:

$$\hat{y} = softmax\left(W_2\left(W_1 f(x) + b_1\right) + b_2\right) \tag{4}$$

Given the ground truth y and the total category number C, we set the cross entropy of the classification results as the semantic analysis loss $\mathcal{L}_{sem}$:

$$\mathcal{L}_{sem} = \sum_i^C y_i log(\hat{y}_i). \tag{5}$$

We note that the input of the semantic analysis module in proposed two architectures is slightly different. The input feature maps of semantic analysis module in pre-semantic DeepSIC are under floating point precision. Differently, the input feature maps of semantic analysis module in post-semantic DeepSIC are under fixed-point precision due to quantization and entropy coding. The performance comparison of these two schemes is present in Sect. 3.

2.5   Joint Training of Compression and Semantic Analysis

We implement end-to-end training for the proposed DeepSIC system, jointly optimize the two constraints of the semantic analysis and image reconstruction modules. Our joint loss is shown in Eq. 6.

$$\mathcal{L} = R + \lambda_1 D + \lambda_2 \mathcal{L}_{sem} \tag{6}$$

Here, λ_1 and λ_2 govern the trade-offs of the three loss terms. One of them is the aforementioned semantic loss $\mathcal{L}_{sem}$, another is the compression rate R measured by the entropy of the quantized code:

$$R = \mathbb{E}\left(-\log_2 Q(f(x))\right). \tag{7}$$

The final loss term is the distortion (or the distance) between the original image and the reconstructed image measured by MSE distance metric:

$$D = \mathbb{E}\left(\|x_i - \hat{x}_i\|_2^2\right). \tag{8}$$

3   Experiment

In this section, we evaluate the performance of the proposed method: DeepSIC on ILSVRC 2012 and Kodak dataset. Firstly, we introduce basic experimental setup, including datasets, evaluation metrics, and implementation details. Then, we report the experimental results over the multiple benchmarking datasets with the aforementioned evaluation metrics. Finally, we analyze the results of our experiments.

3.1   Experimental Setup

Datasets. For training, we jointly optimized the full set of parameters over ILSVRC 2012 classification dataset which consists of 1.2 million images for training, and 50,000 for validation from 1,000 classes. A data augmentation scheme is adopted for training images and randomly crop them to 128×128 patches at training time. Performance tests on Kodak PhotoCD dataset [2] are also present to compare compression quality with other image compression codecs. Kodak is an uncompressed image set with 24 PNG-format images which are popularly used for testing compression performance.

Metric. To assess the visual quality of reconstructed images, we adopt Multi-Scale Structure Similarity (MS-SSIM) [17] for comparing original, uncompressed images to compressed, degraded ones. MS-SSIM [17] is a representative perceptual metric which has been specifically designed to match the human visual system. For image classification task, we utilize the classification accuracy as the metric.

Implementation and Training Details. We conduct training and testing on an NVIDIA Quadro M6000 GPU. All models are trained with 128×128 patches sampled from the ILSVRC 2012 dataset using the Tensorflow API.

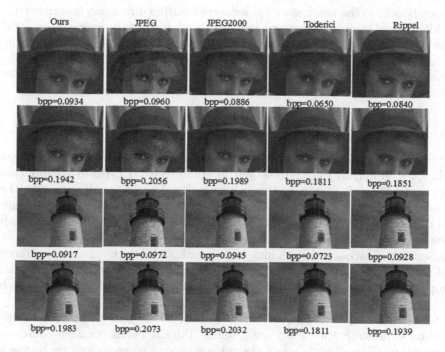

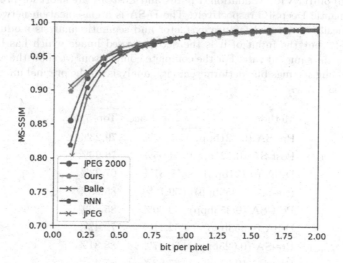

Fig. 4. Examples of reconstructed image parts by different codecs (JPEG, JPEG 2000, ours, Toderici [15] and Rippel [10]) at low bitrate levels.

Fig. 5. Average rate-distortion curves for the images from Kodak dataset. Our DeepSIC is compared with JPEG, JPEG2000, Ballé [1], and Toderici [15]. JPEG and JPEG 2000 results are averaged over images compressed with identical quality settings.

We set $B = 32$ as the batch size. The extracted feature dimension is variable due to different subsample settings to gain variable length of the compressed code. This optimization is performed separately for each weight with the Adam [6] optimizer, yielding separate transforms and marginal probability models. The initial learning rate is set as 0.001, with decaying twice by a factor of 5 during training. We train each model for a total of 8,000,000 iterations.

3.2   Experimental Results

We compare proposed DeepSIC against conventional commercial compression techniques JPEG, JPEG2000, as well as some recent DL-based compression work [1,10,15] with compressed examples in Fig. 4. Additionally, we take the average MS-SSIM [17] as the metric of the distortion between original images and the decoded ones from Kodak PhotoCD dataset, the average rate-distortion curves are shown in Fig. 5. As it is shown in Figs. 4 and 5, though we incorporate semantic analysis task with image compression model, the visual appearance and compression performance of compressed images is better than JPEG and comparable to the state-of-the-art DL-based image compression methods such as [1,10,15].

Comparisons of the semantic analysis result of the two proposed architectures on classification accuracy are given in Table 1. Furthermore, as different compression ratios directly affect the performance of compression, we compare

Table 1. Classification accuracy comparisons over different compression ratios (measured by bpp) on ILSVRC validation. Pre-SA and Post-SA are short for pre-semantic and post-semantic DeepSIC respectively. The D-SA is a classification network which adopts identically structured feature extractor and semantic analysis module as proposed DeepSIC but the input of it is the decompressed image which has been compressed under the same bit rate. For the complete comparison, we show the result with the same training settings but perform semantic analysis on the original uncompressed image in the last row.

Method	Top-1 acc.	Top-5 acc.
Pre-SA (0.121bpp)	55.87%	79.23%
Post-SA (0.121bpp)	54.95%	78.90%
D-SA (0.121bpp)	50.31%	67.23%
Pre-SA (0.353bpp)	59.12%	85.93%
Post-SA (0.353bpp)	59.09%	85.89%
D-SA (0.353bpp)	52.71%	79.42%
Pre-SA (0.625bpp)	68.27%	88.31%
Post-SA (0.625bpp)	67.93%	87.45%
D-SA (0.625bpp)	68.12%	88.33%
Original	70.91%	89.27%

semantic analysis result of the proposed architectures over certain fixed compression ratios in Table 1. It presents the trend of how compression ratio affects the performance of semantic analysis. The placement of semantic analysis module may slightly affect the classification accuracy of semantic inference as it is shown in Table 1 that the Pre-SA does better than Post-SA. The cause of this is that the input of semantic analysis module in Pre-SA is the original feature map and the input of the semantic analysis module in Post-SA is the quantized degraded feature map. The results in Table 1 also show that incorporating semantic directly in compression procedure (Pre-SA and Post-SA) outperforms performing inference on the degraded decoded image (D-SA) as it is commonly done now when we need to perform image understanding task to the compressed images.

Although we incorporate semantic analysis task with image compression model, the visual appearance of compressed images is better than JPEG and comparable to the state-of-the-art DL-based image compression methods such as [1,10,15]. Consistent with the appearance of these example images, we retain the semantic representation of them through a single network without decoding the RGB image from the compressed representation.

4   Conclusion and Future Work

In this paper, we propose an image compression scheme incorporating semantics, which we refer to as Deep Semantic Image Compression (DeepSIC). It aims to decompress the source image and generate corresponding semantic representations at the same time through a single end-to-end optimized compression network. We put forward two architectures of it: pre-semantic DeepSIC and post-semantic DeepSIC. Experiments over Kodak PhotoCD and ILSVRC classification datasets show promising performance of the proposed compression model. Though incorporating semantics analysis task, the proposed DeepSIC yields comparable visual reconstructions while also produce the semantic information of the image without decoding or additional network on image understanding task.

Despite the challenges such as how to efficiently organize semantic representation of multiple objects to explore, deep semantic image compression is an inspiring new direction which breaks through the boundary of multi-media signal processing and pattern recognition. The proposed DeepSIC paves a promising research avenue that we plan to further explore other possible solutions to the aforementioned challenges.

Acknowledgment. This work is supported by National Natural Science Foundation of China (61572428, U1509206), Fundamental Research Funds for the Central Universities (2017FZA5014), National Key Research and Development Program (2016YFB1200203) and Key Research and Development Program of Zhejiang Province (2018C01004).

References

1. Ballé, J., Laparra, V., Simoncelli, E.P.: End-to-end optimized image compression. arXiv preprint arXiv:1611.01704 (2016)
2. Franzen, R.: Kodak lossless true color image suite (1999). http://r0k.us/graphics/kodak
3. Gregor, K., Besse, F., Rezende, D.J., Danihelka, I., Wierstra, D.: Towards conceptual compression. In: Advances in Neural Information Processing Systems, pp. 3549–3557 (2016)
4. He, K., Zhang, X., Ren, S., Sun, J.: Deep residual learning for image recognition. In: IEEE Conference on Computer Vision and Pattern Recognition, pp. 770–778 (2016)
5. Huang, G., Liu, Z., Van Der Maaten, L., Weinberger, K.Q.: Densely connected convolutional networks. In: IEEE Conference on Computer Vision and Pattern Recognition, pp. 2261–2269 (2017)
6. Kingma, D., Ba, J.: Adam: a method for stochastic optimization. arXiv preprint arXiv:1412.6980 (2014)
7. Long, J., Shelhamer, E., Darrell, T.: Fully convolutional networks for semantic segmentation. In: IEEE Conference on Computer Vision and Pattern Recognition, pp. 3431–3440 (2015)
8. Marpe, D., Schwarz, H., Wiegand, T.: Context-based adaptive binary arithmetic coding in the H. 264/AVC video compression standard. IEEE Trans. Circ. Syst. Video Technol. **13**(7), 620–636 (2003)
9. Rabbani, M., Joshi, R.: An overview of the JPEG 2000 still image compression standard. Sig. Process. Image Commun. **17**(1), 3–48 (2002)
10. Rippel, O., Bourdev, L.: Real-time adaptive image compression. In: Proceedings of the 34th International Conference on Machine Learning, vol. 70, pp. 2922–2930. PMLR (2017)
11. Simonyan, K., Zisserman, A.: Very deep convolutional networks for large-scale image recognition. arXiv preprint arXiv:1409.1556 (2014)
12. Szegedy, C., et al.: Going deeper with convolutions. In: IEEE Conference on Computer Vision and Pattern Recognition, pp. 1–9 (2015)
13. Theis, L., Shi, W., Cunningham, A., Huszár, F.: Lossy image compression with compressive autoencoders. arXiv preprint arXiv:1703.00395 (2017)
14. Toderici, G., et al.: Variable rate image compression with recurrent neural networks. arXiv preprint arXiv:1511.06085 (2015)
15. Toderici, G., et al.: Full resolution image compression with recurrent neural networks. In: IEEE Conference on Computer Vision and Pattern Recognition, pp. 5435–5443 (2017)
16. Wallace, G.K.: The JPEG still picture compression standard. IEEE Trans. Consum. Electron. **38**(1), xviii–xxxiv (1992)
17. Wang, Z., Simoncelli, E.P., Bovik, A.C.: Multiscale structural similarity for image quality assessment. In: Thrity-Seventh Asilomar Conference on Signals, Systems Computers, vol. 2, pp. 1398–1402 (2003)

Multi-stage Gradient Compression: Overcoming the Communication Bottleneck in Distributed Deep Learning

Qu Lu[1,2], Wantao Liu[1(✉)], Jizhong Han[1], and Jinrong Guo[1,2]

[1] Institute of Information Engineering, Chinese Academy of Sciences, Beijing, China
{luqu,liuwantao,hanjizhong,guojinrong}@iie.ac.cn
[2] School of Cyber Security, University of Chinese Academy of Sciences, Beijing, China

Abstract. Due to the huge size of deep learning model and the limited bandwidth of network, communication cost has become a salient bottleneck in distributed training. Gradient compression is an effective way to relieve the pressure of bandwidth and increase the scalability of distributed training. In this paper, we propose a novel gradient compression technique, Multi-Stage Gradient Compression (MGC) with Sparsity Automatic Adjustment and Gradient Recession. These techniques divide the whole training process into three stages which fit different compression strategy. To handle error and preserve accuracy, we accumulate the quantization error and sparse gradients locally with momentum correction. Our experiments show that MGC achieves excellent compression ratio up to 3800x without incurring accuracy loss. We compress gradient size of ResNet-50 from 97 MB to 0.03 MB, for AlexNet from 233 MB to 0.06 MB. We even get a better accuracy than baseline on GoogLeNet. Experiments also show the significant scalability of MGC.

Keywords: Gradient compression · Communication optimization
Distributed system · Network bottleneck · Deep learning

1 Introduction

In the past few years, deep learning has experienced explosive development. Deeper and larger models are designed in domains ranging from image processing [6], text classification [10] to translation [14] and many others. The training task of large-scale models has been widely transferred to distributed system because of the reduction on computation time [14]. Stochastic Gradient Descent (SGD) is widely used as optimization method in distributed training. As show in Fig. 1, parameter server collects all gradients from distributed workers, calculates average gradients and updates parameters, then workers pull back the latest parameters and start next iteration. We can reduce the computation time dramatically by increasing the number of workers [16,18]. However, communication time between parameter servers and workers increases because more

© Springer Nature Switzerland AG 2018
L. Cheng et al. (Eds.): ICONIP 2018, LNCS 11301, pp. 107–119, 2018.
https://doi.org/10.1007/978-3-030-04167-0_10

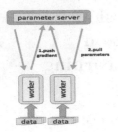
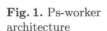

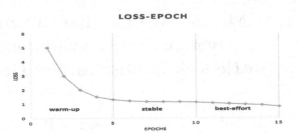

Fig. 1. Ps-worker architecture

Fig. 2. Three different stages in MGC.

gradients need to transmitted on larger cluster. Due to limitation of bandwidth, the reduction on computation time is neutralized by communication cost while increasing the number of workers. Therefore, communication becomes an obvious bottleneck for improving the scalability and performance of distributed machine learning.

Gradient compression has been employed in many work to solve the communication bottleneck [3,4,15]. *Sparsification* and *quantization* are two main approaches of gradient compression. *Sparsification* means only sending a part of gradients and *quantization* quantizes the gradients to low-precision values. We utilize the advantages of these two approaches to further improve compression ratio. The contributions of MGC include: (1) we propose a novel gradient compression strategy which combines quantization and sparsification. (2) we summarize the feature of training and divide the training process into three stages by *Sparsity Automatic Adjustment* and *Gradient Recession*. (3) we propose *Merged Residual Correction* to eliminate the error caused by compression.

The rest of this paper is organized as follows. Section 2 covers the related works of training characteristic and gradient compression strategy. Section 3 describes the motivation of MGC and the algorithm of it. Section 4 shows the compression ratio, convergence and many other performance of MGC. Finally, we draw a conclusion of this work in Sect. 5.

2 Related Works

2.1 Stages in Machine Learning Training

Researchers empirically find the amplitude of variation of model is different in different period of training. He et al. [6] found the huge volatility of weights in the first few epochs and proposed the warmup strategy. Inspired by his work, Goyal et al. [5] let the learning rate gradually ramps up to a large value in warmup stage. Lin et al. [11] followed these work and use the similar method on compression strategy. The growth of learning rate and sparsity of gradient usually achieve a significant convergence by avoiding a sudden variation on weights [5]. Wen et al. [17] found that we can get a better result if we change strategy in good time by stopping quantization.

2.2   Solutions to Communication Bottleneck

A mass of approaches have been proposed to overcome the communication bottleneck of distributed training. One way is updating parameters immediately if fastest node has calculate gradients instead of using strict synchronous stochastic gradient descent as update strategy [7,14]. However, this method may harm convergence and accuracy because of the staleness. Another way is overlapping computation and network communication. In the back-propagation phase, sending gradients asynchronously once all computation of one layer has been finished [18]. The communication time will be covered completely by computation time in the best situation. But this method doesn't match the networks with low computation-to-communication ratio. In addition, we can solve the problem by controlling the size of gradients on the network. Sparsification and quantization are two main approaches to control size of gradients.

Gradient Sparsification. We can reduce the communication overhead by only exchanging part of gradients in a iteration. Strom et al. [16] proposed using a threshold to filter gradients, sending gradients larger than a predefined constant value and gained training speedup around 17x on 20 nodes. However, the threshold is an extra hyper-parameter and it is hard to choose. Ali et al. [1] dropped a fixed ratio of gradients to save 99% exchange on the net. They find this scheme only incurring negligible degradation on BELU score. Aiming to tune compression rate automatically on different deep learning tasks, Chen et al. [3] presented a compression scheme based on localized selection of gradient residues and pushed compression ratio up to 200 x on fc layer without significant loss of top-1 accuracy on ImageNet dataset. Lin et al. [11] found 99.9% of the gradients exchange are redundant and proposed deep gradient compression (DGC) with four methods to preserve accuracy. They claimed DGC achieves a gradient compression ratio up to 600X without losing accuracy. We can see that if we use threshold to filter gradients, the selection of threshold is hard. If we select a fixed ratio of gradients, more computation and some error will be introduced.

Gradient Quantization. Gradients could be represented by lower precision values to reduce the data size of communication. Zhou et al. use bit convolution kernels to accelerate both training and inference which is called DoReFa-Net [19]. DoReFa-Net can achieve comparable prediction accuracy as 32-bit counterparts. Seide et al. [15] proposed quantizing the sub-gradients aggressively, to but one bit per value and got a 10 times speed-up on a 160M-parameter model at a small accuracy loss. Alistarh et al. [2] presented QSGD that allows the user to smoothly trade off communication bandwidth and convergence time and trained the ResNet-152 network to full accuracy on ImageNet 1.8X faster than the full-precision variant. Wen et al. [17] proposed Terngrad requires only three numerical levels which doesn't incur any accuracy loss on most of models and can even get better accuracy. Both of QSGD and Terngrad give the theoretical proof of convergence.

Compared to gradient sparsification, quantization need more computation and it has a limitation on compression ratio. But quantization has a better convergence and achieves lower error [4]. Based on quantization and sparsification, Multi-Stage Gradient Compression employed both of them and changes the compression strategy in different stage.

3 Multi-stage Gradient Compression

3.1 Motivation

Both gradient sparsification and quantization have achieved a good effect on reducing gradient communication time. However, there are still some shortcomings in some aspects.

The quantization compression quantizes gradients to low-precision values, which has a limitation on reduction of the gradient volume [3]. We only get 32x compression ratio even quantizing the gradients to 1-bit value. Opposite to it, gradient sparsification only send gradients larger than threshold. Higher compression ratio is achieved if a larger threshold is chosen. [11] can get up to 600X compression ratio by transmitting only 0.1% of gradients. For a better compression, we combined these two approaches as the major compression strategy of MGC and push the compression ratio up to 3800x (for comparison, the best result of DGC is 597x [11] and TernGrad is 20x [17]). We use sparse update firstly, that means we only send important gradients in a iteration. We send 0.1% of gradients like Lin [11] because of the high compression ratio without losing accuracy obviously. After that the range of gradients is much smaller and we employ ternary quantization proposed by Wen et al. [17] on selected gradients. We will show more details in Sect. 3.2.

Combining the two approaches simply may degenerate accuracy in some situation. To preserve the accuracy, we propose that there are three different stages in training process: warm-up, stable, best-effort. Figure 2 shows that. All weights and biases in deep neural networks will be generated randomly at the beginning of training process. So the network is unstable and parameters change rapidly in the first stage which is called warm-up stage [5]. Over-sparse gradients prolong this period of warm-up because many gradients have been delayed. So we use low sparsity in this stage and propose *sparsity automatic adjustment* to change the sparsity of gradient compression which fits the unstable feature of warm-up stage. It will be explained in Sect. 3.2.

After warm-up stage, the network has been stable already. The change of parameters is much smaller than before, meanwhile, the gradients are monotonous in comparison with warm-up stage [5]. This is a period with stable gradients. We called it stable stage. The gradients is not diverse means it won't make huge error when we use aggressive gradient sparsification and delay the change of parameters [11]. So we employ the compression strategy with highest sparsity (we delay 99.9% of gradients like [11]).

In the late stages of training, the volatility of loss function and weights tends to weeny [17]. As we all know, quantization is a lossy-compression and applying quantization will produce error in most gradient values which means the direction of gradients changed. The training is close to convergence in this stage. But wrong direction may make the model deviates from the optima. We call this period best-effort stage. We hope it will converge to optima more accurate and faster in the final by stopping quantization. *Quantization recession* is proposed to stop quantization in the best-effort period. It will be discussed in Sect. 3.3.

Based on the above idea, we propose MGC (Multi-Stage Gradient Compression) which tunes strategy based on *sparsity automatic adjustment* and *quantization recession*. We also use *merged residual correction* to ensure no loss of accuracy.

Algorithm 1. MGC:distributed SGD training using sparsification and quantization

---
Worker:
    i=1,...,N
1: Input $Z_t(i)$, a part of a mini-batch of training samples Z_t;
2: Compute gradients $g_t(i)$ under $Z_t(i)$, $g_t(i) = g_t(i) + g_{re}(i)$
3: Use Sparsity Automatic Adjustment,get sparsity τ
4: Sparse gradients, $g_{temp}(i) = sparse(g_t(i), \tau)$, get $threshold_\tau$ and residual g_{re}
5: Use Quantization Recession ,get flag useQ
6: **if** useQ = true **then**
7:     Ternarize temp gradients $g_t(i) = ternarize(g_{temp}(i))$
8:     Accumulate gradients residual to $g_{re}(i)$(merged residual)
9: **else**
10:     $g_t(i) = g_{temp}(i)$
11: **end if**
12: Push compressed $g_t(i)$ to the server
13: Use merged residual correction on $g_{re}(i)$
14: Pull average gradients $\overline{g}_t$ from the server
15: Update parameters $w_{t+1} = w_t - \eta \cdot \overline{g}_t$
Server:
    Average gradients $\overline{g}_t = \sum_i g_t(i)/N$

3.2 Multi-stage Gradient Compression

Algorithm 1 formulates the SGD scheme with MGC on t-th iteration. In each iteration, workers compute gradients based on a mini-batch of dataset. Before sending gradients to server, we add two steps (sparsity automatic adjustment and quantization recession). As shown in line 3, sparsity automatic adjustment observes the training and changes the sparsity of our compression algorithm in warm-up stage. Then we employ compression on gradients and accumulate the residual gradients locally. This residual term g_{re} will be added to gradients in next iteration. Here the flag useQ represent whether to use quantization and $threshold_\tau$ is the threshold of gradients in this iteration. At last, workers push compressed gradients to server and employ merged residual correction on local residual gradients. We use sparsity automatic adjustment and quantization recession which divide the training into three stages and tune compression strategy.

As show in Algorithm 1, the major compression strategy in MGC is a combination of sparsification and quantization. Formally, the process of compression is:

$$g_{temp} = sparse(g_t, \tau). \tag{1}$$

where g_t is the gradient vector in t-th iteration. τ is the percentage that we want to reserve locally and the value of τ may be change during the training process. $sparse(\cdot)$ means the process to select gradients, if absolute value larger than $threshold_\tau$, the gradient will be sent.

$$\bar{g}_t = ternarize(g_{temp}) = s_t \cdot sign(g_{temp}) * b_{temp}. \tag{2}$$

Here the ternarize process is proposed by Wen [17]. $sign(\cdot)$ sets each element to its sign. s_t is a scaler that can shrink $sign(g_{temp}) * b_{temp}$ to a much smaller amplitude. $*$ is the Hadamard product. Each element of b_{temp} depends on the corresponding value of g_{temp}:

$$\begin{cases} P(b_{temp,k} = 1|g_t) = |g_{temp,k}|/s_t, & (|g_{temp,k}| = s_t, if |g_{temp,k}| > s_t). \\ P(b_{temp,k} = 0|g_t) = 1 - |g_{temp,k}|/s_t, & (|g_{temp,k}| = s_t, if |g_{temp,k}| > s_t). \end{cases} \tag{3}$$

$b_{temp,k}, g_{temp,k}$ means the k-th element of b_{temp}, g_{temp}. In Terngrad, the scaler s_t is the max value of gradient g_t. We replaced this scaler with the value of $threshold_\tau$ here.

$$s_t \triangleq max(abs(g_t)) \rightarrow s_t \triangleq threshold_{\tau=0.999}. \tag{4}$$

$threshold_{\tau=0.999}$ means the threshold to select top 0.1% gradient. We choose it as scaler for two reasons: (1) Intuitively, range of gradients after sparsification is small, so it won't make a huge error on gradients if we use threshold instead of max value as scaler. (2) Wen introduce a hyper-parameter $c\sigma$ to clip gradients. This hyper-parameter helps to control the magnitude of s_t. If s_t in Eq. (4) is large, most of gradients will probably ternarize to zeros and the convergence of model will be affected [17]. We can achieve the same effect by using $threshold_{\tau=0.999}$ instead of clipping gradients.

Sparsity Automatic Adjustment. We make the sparsity of gradients low in the beginning. The sparsity will gradually raise to a large value to increase the compression rate and accelerate the training. We called sparsity automatic adjustment. A regularization term is usually used in loss function to select feature while this solution helps avoid overfitting in problems. The loss term of regularization will degrade sharply in first few epochs that means weights change frequently and there is a long way to converge. In MGC we change the sparsity value depends on regularization loss term. The sparsity will increase up to max_sparsity (we set the value to 0.999 according to [11]) in order to trade off compression and model accuracy. Algorithm 2 formulates steps of sparsity automatic adjustment, the sparsity will be set 0.8 at first and the value will increase according to volatility of regularization term. When the sparsity has raised to final value or regularization loss doesn't degrade in an epoch, the sparsity will be set to 0.999 finally [11]. We need to end warm-up stage manually because regularization may not always decrease on some datasets. Here we set max_epoch to 5 empirically. As we will see in Sect. 4, sparsity automatic adjustment improve the training performance obviously. This approach helps pushing the compression rate higher without harming accuracy.

Algorithm 2. Sparsity Automatic Adjustment

1: Initialize weights and bias
2: Compute the regularization loss of initial parameters r_0
3: Start next iteration and save the loss of regularization $loss_i$
4: Calculate the average of regularization in one epoch $r_i = \sum_{i=1}^{N} loss_i$
5: Change the sparsity to rate $1 - 0.2^{(r_0/r_i)}$
6: If the sparsity has raised to max_sparsity or regularization loss doesn't degrade in an epoch or epoch larger than max_epoch, set sparsity to max_sparsity.

Quantization Recession. It is easier to understand SGD algorithm by make an analogy between SGD and going down the hill. We walk along the opposite direction of gradient which means the steepest descent way. Intuitively, we can arrive at the destination faster by avoiding the deviation of direction while we are close to minima. Gradient quantization introduce error which will harm convergence greatly in some situations [2]. As we mentioned above, we add a best-effort stage after the stable training stage. In the best-effort stage, we only use gradient sparsification to avoid misguiding the optimized direction. The experiment in Sect. 4 shows that this technique make up the lack of Multi-Stage Gradient Compression and maintain the accuracy.

3.3   Merged Residual Correction

Both of quantization and sparsification yield errors and we merge residual to correct it. In the first step of sparsification we remain all gradients which is not selected to be transmitted immediately. In the second step of quantization we add quantization errors to remained gradients as residual. We will keep residual gradients locally in every worker and add them to new gradients in next iteration. Momentum SGD is be used widely because it can accelerate training, Lin find we can improve the performance of model by using momentum correction [11]. We follow his approach and apply momentum correction to the residual gradient. This approach can improve the model performance because when transmitted gradients being aggregated by considering momentum item, the local residual gradients never be treated with the same way. Asynchronous gradient update has been studied and Mitliagkas et al. [12] proposed a term called implicit momentum to explain why the stalenss of parameter will degrade model performance. Similar to asynchronous distributed machine learning, we delay many gradients to be added to parameters and these gradients will make a contribution after they were transmitted. But implicit momentum will misguide the optimization direction and we need to mitigate staleness by excluding the value of stale gradient. We alleviate staleness by masking threshold value if gradient has been transmitted. We apply this mask simply as follow.

$$u_{t,k} \leftarrow u_{t,k} \odot \neg threshold. \tag{5}$$

$u_{t,k}$ is the local momentum factor on k-th node in t-th iteration.

4   Experiments

4.1   Experiment Settings

We first validate the convergence of MGC on several datasets and show results in Sect. 4.2. Then we compare gradient compression ratio with existed schemes in Sect. 4.3. At the last, the scalability and the performance of MGC are discussed in Sect. 4.4. The experiments are based on TensorFlow. For fair comparison, all hyper-parameters of schemes to be compared are the same unless we point out. We use three well-known datasets here. MNIST is a handwritten digits dataset with 60,000 images, the best-effort stage for MGC is from 6-th epoch of 10 epochs. Cifar10 consists of 60,000 32×32 colour images in 10 classes. ImageNet is a large-scale image dataset of 1000 classes [9]. The best-effort stage for Cifar10 is from 140-th of 200 epochs, for ImageNet it starts from 70-th epoch of 90 epochs. We evaluate gradient compression ratio as follows:

$$GradientCompressionRatio = size[MGC(g_t)]/size(g_t) \tag{6}$$

All experiments run on 4 nodes GPU cluster. Each node has 32 GB RAM and 2 NVIDA Tesla M40s.

4.2   Convergence Research of MGC

We investigate the convergence of MGC on MNIST (LeNet) and Cifar10 (Cifar-Net) dataset. In our experiments, baseline means the original momentum SGD without gradient compression. For MNIST dataset, the mini-batch size is 64 and maximum iteration is 10K. The base learning rate of LeNet is 0.01.

For Cifar10, mini-batch is 128 and maximum iteration is 100K. We set the base LR of CifarNet to 0.002 because it achieves a better result than 0.01. Figures 3 and 4 shows the results of training on LeNet and CifarNet. MGC can converge to the similar accuracy with baseline and if we add sparsity automatic adjustment we will get a better result. The maximum accuracy loss on LeNet

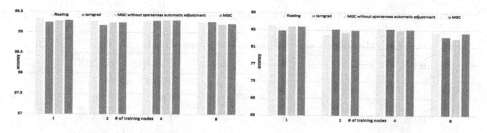

Fig. 3. Accuracy of LeNet   **Fig. 4.** Accuracy of CifarNet

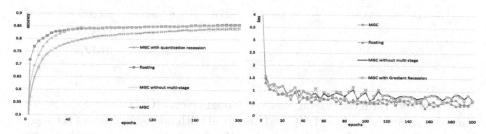

Fig. 5. Accuracy curve of CifarNet   **Fig. 6.** Loss curve of CifarNet

Table 1. Compression ratio and accuracy of TernGrad, DGC, MGC

Model	Method	Top-1	Top-5	Gradient size	Compression ratio
AlexNet	BaseLine	58.17%	80.19%	232.56 MB	-
	TernGrad	57.28% (−0.86%)	80.23% (+0.04%)	29.18 MB	8x
	DGC	58.20% (+0.03%)	80.20% (+0.01%)	0.39 MB	597x
	MGC	57.99% (−0.18%)	80.22% (+0.03%)	0.06 MB	3800x
GoogLeNet	BaseLine	-	87.82%	44 MB	-
	TernGrad	-	85.96% (−1.86%)	5.86 MB	7.5x
	MGC	-	85.81% (−2.01%)	0.04 MB	1100x
ResNet-50	BaseLine	75.96%	92.91%	97.49 MB	-
	DGC	76.15% (+0.19%)	92.97% (+0.06%)	0.35 MB	277x
	MGC	76.00% (+0.04%)	92.52% (−0.39%)	0.03 MB	3200x

is no more than 0.1%. The maximum accuracy loss on CifarNet is 0.6% after we use sparsity automatic adjustment. The accuracy degrades when we use a big mini-batch size. There are many reasons that can explain it. It is generally accepted that larger mini-batch tends to converge to poorer minima [8,13].

4.3 Gradient Compression Ratio and Accuracy on ImageNet

We also test MGC on ImageNet dataset by scaling to a larger and deeper networks such as AlexNet, GoogLeNet, ResNet. Table 1 shows the training result on 4 nodes. We compare the gradient compression ratio with DGC [11] on AlexNet,

Table 2. Accuracy of GoogLeNet on multi-node with different mini-batch size

BaseLR	Mini-batch size	Workers	Iterations	Gradients	Top-5
0.04	128	2	600k	Floating	88.30%
				TernGrad	86.77%
				MGC	87.01%
0.08	256	4	300k	Floating	87.52%
				TernGrad	85.96%
				MGC	87.68%
0.1	512	8	300k	Floating	89.00%
				TernGrad	86.47%
				MGC	87.12%

ResNet-50 and Terngrad [17] on AlexNet, GoogLeNet. MGC make a good performance on compression than DGC and Terngrad with no distinct loss of accuracy. Terngrad didn't give the top-1 accuracy in GoogLeNet, so we only compare the top-5 accuracy in Table 1. For AlexNet, MGC gives 6.4x better compression ratio than DGC and 475x than TernGrad. For ResNet-50, MGC gets a better accuracy than baseline and 11.6x compression ratio than DGC. Table 2 show more details about training GoogLeNet on ImageNet. We use the same hyper-parameters with Terngrad and values are showed in table. We trained all models with the same epochs of dataset. It means that iterations becomes smaller and parameters is updated with lower frequency when we increase the number of workers. So the base learning rate is larger while the mini-batch size is larger. We can see MGC converges to a same level with baseline in accuracy regardless of mini-batch size. MGC improves the accuracy in the large batch-size because sometimes randomness of compressed gradient may change the direction and help it escape from local minima [8]. Compared to Terngrad, MGC achieves excellent compression ratio and even a better accuracy on 4 nodes.

Figures 5 and 6 shows the learning curve of CifarNet on 2 nodes. The learning curve of MGC without multi-stage is slightly worse than the baseline. After we employ Sparsity Automatic Adjustment and Quantization Recession on it, the accuracy and loss curve closely match the baseline. We can observe that the accuracy curve of MGC with Gradient Recession converges to a slight better result than vanilla MGC. Both of Sparsity Automatic Adjustment and Quantization Recession make a contribution to eliminating error and getting a better accuracy.

4.4   Scalability and Speedup Performance

Wen extended a widely accepted CPU-based performance model to evaluate GPU-based deep learning systems [17]. We use the GPU-based performance model to show the speed up of MGC on more nodes. We test AlexNet and

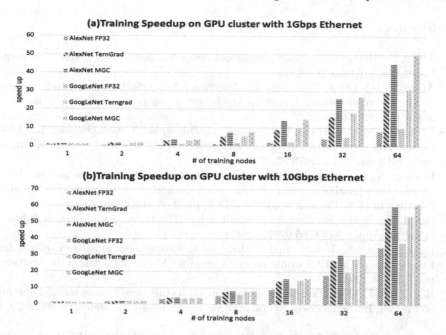

Fig. 7. Compare training speedup with floating (without compression), TernGrad and MGC. Each node has 2 NVIDIA M40 GPUs and one PCI switch.

GoogLeNet on GPU nodes with all-reduce communication model [1]. The bandwidths of cluster are 1 Gbps Ethernet and 10 Gbps Ethernet with PCI switch. Figure 7 shows the comparison of speedup on multi-node and on single node. MGC increases the throughput aggressively on these models. TernGrad only gets 30x speedup on 64 nodes with 1 Gbps while MGC gets 46x (AlexNet). The scalability of MGC is closely linear with 10 Gbps Ethernet. MGC makes a better performance with smaller bandwidth compared with TernGrad and baseline. So we can improve the scalability of distributed training by employing MGC.

5   Conclusion

Multi-Stage Gradient Compression (MGC) achieve 880-3800x better compression ratio than floating gradient without losing accuracy or slowing down the convergence. MGC tune compression strategy by using sparsity automatic adjustment and quantization recession. Merged residual correction is used to preserve accuracy. MGC reduces communication bandwidth and increases the scalability of distributed training.

Acknowledgement. This research is supported by the National Key Research and Development Program of China (No. 2017YFB1010000).

References

1. Aji, A.F., Heafield, K.: Sparse communication for distributed gradient descent. CoRR abs/1704.05021 (2017)
2. Alistarh, D., Li, J., Tomioka, R., Vojnovic, M.: QSGD: randomized quantization for communication-optimal stochastic gradient descent. arXiv preprint arXiv:1610.02132 (2016)
3. Chen, C., Choi, J., Brand, D., Agrawal, A., Zhang, W., Gopalakrishnan, K.: AdaComp : adaptive residual gradient compression for data-parallel distributed training. CoRR abs/1712.02679 (2017)
4. Dryden, N., Moon, T., Jacobs, S.A., Van Essen, B.: Communication quantization for data-parallel training of deep neural networks. In: Workshop on Machine Learning in HPC Environments (MLHPC), pp. 1–8. IEEE (2016)
5. Goyal, P., et al.: Accurate, large minibatch SGD: training imagenet in 1 hour. arXiv preprint arXiv:1706.02677 (2017)
6. He, K., Zhang, X., Ren, S., Sun, J.: Deep residual learning for image recognition. In: Proceedings of the IEEE Conference on Computer Vision and Pattern Recognition, pp. 770–778 (2016)
7. Ho, Q., et al.: More effective distributed ml via a stale synchronous parallel parameter server. In: Advances in Neural Information Processing Systems, pp. 1223–1231 (2013)
8. Keskar, N.S., Mudigere, D., Nocedal, J., Smelyanskiy, M., Tang, P.T.P.: On large-batch training for deep learning: generalization gap and sharp minima. arXiv preprint arXiv:1609.04836 (2016)
9. Krizhevsky, A., Sutskever, I., Hinton, G.E.: Imagenet classification with deep convolutional neural networks. In: Advances in Neural Information Processing Systems, pp. 1097–1105 (2012)
10. Lai, S., Xu, L., Liu, K., Zhao, J.: Recurrent convolutional neural networks for text classification. In: AAAI, vol. 333, pp. 2267–2273 (2015)
11. Lin, Y., Han, S., Mao, H., Wang, Y., Dally, W.J.: Deep gradient compression: reducing the communication bandwidth for distributed training. arXiv preprint arXiv:1712.01887 (2017)
12. Mitliagkas, I., Zhang, C., Hadjis, S., Ré, C.: Asynchrony begets momentum, with an application to deep learning. In: 2016 54th Annual Allerton Conference on Communication, Control, and Computing (Allerton), pp. 997–1004. IEEE (2016)
13. Neelakantan, A., et al.: Adding gradient noise improves learning for very deep networks. arXiv preprint arXiv:1511.06807 (2015)
14. Recht, B., Re, C., Wright, S., Niu, F.: HOGWILD: a lock-free approach to parallelizing stochastic gradient descent. In: Advances in Neural Information Processing Systems, pp. 693–701 (2011)
15. Seide, F., Fu, H., Droppo, J., Li, G., Yu, D.: 1-bit stochastic gradient descent and its application to data-parallel distributed training of speech DNNs. In: Fifteenth Annual Conference of the International Speech Communication Association (2014)
16. Strom, N.: Scalable distributed DNN training using commodity GPU cloud computing. In: Sixteenth Annual Conference of the International Speech Communication Association (2015)
17. Wen, W., et al.: TernGrad: ternary gradients to reduce communication in distributed deep learning. In: Guyon, I., et al. (eds.) Advances in Neural Information Processing Systems, vol. 30, pp. 1509–1519. Curran Associates, Inc. (2017)

18. Zhang, H., et al.: Poseidon: an efficient communication architecture for distributed deep learning on GPU clusters. arXiv preprint (2017)
19. Zhou, S., Wu, Y., Ni, Z., Zhou, X., Wen, H., Zou, Y.: DoReFa-Net: training low bitwidth convolutional neural networks with low bitwidth gradients. arXiv preprint arXiv:1606.06160 (2016)

Teach to Hash: A Deep Supervised Hashing Framework with Data Selection

Xiang Li, Chao Ma, Jie Yang$^{(\boxtimes)}$, and Yu Qiao

Institute of Image Processing and Pattern Recognition,
Shanghai Jiao Tong University, Shanghai, China
{xx.lee,sjtu_machao,jieyang,qiaoyu}@sjtu.edu.cn

Abstract. Recent years have witnessed wide applications of deep learning for large-scale image hashing tasks, as deep hashing algorithms can simultaneously learn feature representations and hash codes in an end-to-end way. However, although these methods have obtained promising results to some extent, they seldom take the effect of different training samples into account and treat all samples equally throughout the training procedure. Therefore, in this paper, we propose a novel deep hashing algorithm dubbed "Teach to Hash" (T2H), which introduces a "teacher" to automatically select the most effective samples for the current training period. To be specific, the "teacher" utilizes two criteria to measure the effectivity of all samples, and iteratively update the training set with the most effective ones. Experimental results on two typical image datasets indicate that the introduced "teacher" can significantly improve the performance of deep hashing framework and the proposed method outperforms the state-of-the-art hashing methods.

Keywords: Deep learning · Data selection · Supervised hashing

1 Introduction

With the explosive growth of image data in real applications, hashing has attracted considerable attention due to its fast query speed and low storage cost. The goal of hashing is to encode the original high-dimensional data into a set of compact binary codes while preserving the similarities between them [17].

Hashing algorithms can be categorized into two parts: data-independent and data-dependent. Data-independent methods (e.g., Locality-sensitive Hashing (LSH) [4]) generally obtain hash functions based on random projections. Unlike data-independent methods, data-dependent methods attempt to generate effective and compact hash codes through a set of training data, and they can be further divided into unsupervised methods [6,20] and supervised methods [2,22]. However, most of these methods cannot well encode the relationship among original data into the learned hash codes because they are based on hand-crafted features, which are independent of the hash function learning procedure and cannot give any feedback to learn better hash codes.

© Springer Nature Switzerland AG 2018
L. Cheng et al. (Eds.): ICONIP 2018, LNCS 11301, pp. 120–129, 2018.
https://doi.org/10.1007/978-3-030-04167-0_11

Recently, the visual recognition for static images has been boosted by deep convolutional neural networks (CNN) [9], which motivated the applications of deep learning in hashing tasks. Xia et al. [21] combined the feature learning and the hash functions learning in a two-stage way. Li et al. [11] developed an end-to-end deep supervised hashing framework and obtained promising results. Lai et al. [10] enforced the network to learn binary-like outputs which can preserve the semantic relations of image-triplets. By coupling the extraction of image feature and the learning of hash codes, these methods all show significantly improved retrieval accuracy.

However, although existing deep hashing algorithms have achieved promising results, these methods treat all the training samples equally throughout the training process. Specifically, during each training epoch, the whole training set is simply shuffled and then randomly divided into several batches. However, it has been studied that the effects of various training samples during learning procedure are quite different [1,5,13]. For example, Gong et al. [5] demonstrated that the label diffusion process could be improved if the labels are propogated from simple to difficult. Loshchilov et al. [13] proposed to help the learning process by focusing on the most relevant samples according to their loss value via an online batch selection algorithm. All these researches illustrate the necessity to perform data sampling in deep hashing frameworks.

Inspired by the human cognitive procedure that people always learn the general outlines firstly and then pay more attention to the most difficult parts for further promotion, we would like to firstly train a model on the whole training set and then select the most effective subset of the training set to enhance the performance of the current model. Based on this consideration, we propose a novel deep supervised hashing framework dubbed "Teach to Hash" (T2H). Specifically, at the beginning of the learning procedure, the deep model is trained on the entire training set until it comes to a rough convergence. After that, we introduce a "teacher" to automatically select a set of *difficult* and *diverse* samples to form the subset which can effectively polish the rough model. Extensive experiments on two real-world datasets (i.e., CIFAR-10 [8] and NUS-WIDE [3]) demonstrate that the introduced "teacher" can boost the performance of deep hashing framework, and the proposed T2H method outperforms the state-of-the-art hashing methods.

2 Methodology

The entire framework of the proposed T2H algorithm is shown in Fig. 1. In this section, we detail the each component of the framework respectively.

2.1 Deep Hashing Network

Suppose $\mathcal{X} = \{\mathbf{x}_i\}_{i=1}^N$ is a data collection, where $\mathbf{x}_i$ represents the raw pixels for the i-th sample. The similarity relationship among the data points is denoted by a set of pairwise labels $\mathcal{S} = \{s_{ij}\}$ with $s_{ij} \in \{0, 1\}$, where $s_{ij} = 1$ means

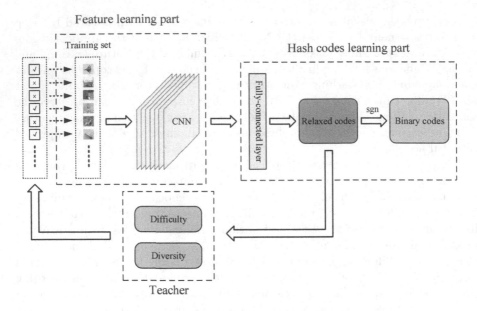

Fig. 1. The framework of the proposed T2H method is composed of two components: the deep hashing network and the teach-to-hash stage. Further, the deep hashing network can be divided into the feature learning part and hash codes learning part.

$\mathbf{x}_i$ and $\mathbf{x}_j$ are similar while $s_{ij} = 0$ means $\mathbf{x}_i$ and $\mathbf{x}_j$ are dissimilar. The goal of supervised hashing is to learn the corresponding hash code $\mathbf{b}_i \in \{0, 1\}^r$ for each sample $\mathbf{x}_i$ with the code length r, and the binary codes $\mathcal{B} = \{\mathbf{b}_i\}_{i=1}^N$ should meanwhile preserve the similarities in $\mathcal{S}$. As discretely learning binary codes is difficult in real cases, a two-stage procedure is employed, where we firstly calculate the relaxed code $\mathbf{u}_i$ for each data point and then quantize it into the binary code through $\mathbf{b}_i = \mathrm{sgn}(\mathbf{u}_i)$.

In this paper, we utilize a general deep hashing framework to perform simultaneous feature learning and hash codes learning (as shown in Fig. 1). Firstly, we employ the widely utilized CNN model, Alexnet [9], to build the feature learning part in our framework. After that, a fully-connected layer is added, whose outputs are regarded as the relaxed codes related to the input points, and the dimension of the output is set to r. In short, the binary code $\mathbf{b}_i$ for each point $\mathbf{x}_i$ can be calculated as:

$$\mathbf{b}_i = \mathrm{sgn}(\mathbf{u}_i) = \mathrm{sgn}(\mathbf{W}^T \phi(\mathbf{x}_i, \theta) + \mathbf{v}), \tag{1}$$

where θ denotes all the parameters of the deep neural network, $\phi(\mathbf{x}_i, \theta)$ denotes the output of the last layer in the feature learning part related to point $\mathbf{x}_i$, $\mathbf{W}$ is the weight matrix, and $\mathbf{v}$ denotes the bias vector.

In order to complete the entire deep hashing framework, we adopt the same object function in [11] as it has shown excellent performance. By taking the negative log-likelihood of the obtained pairwise labels in $\mathcal{S}$, the optimization problem can be defined as:

$$\min_{\mathcal{B},\mathcal{U}} -\sum_{s_{ij}\in\mathcal{S}}(s_{ij}\Theta_{ij}-\log(1+e^{\Theta_{ij}}))+\eta\sum_{i=1}^{N}\|\mathbf{b}_i-\mathbf{u}_i\|_2^2, \qquad (2)$$

where $\Theta_{ij}=\frac{1}{2}\mathbf{u}_i^T\mathbf{u}_j$, $\mathcal{U}=\{\mathbf{u}_i\}_{i=1}^{N}$ and η is the regularization term. The first part of Eq. (2) forces the similar samples to have low Hamming distance in the hash space and the second part can reduce the quantization error. With this loss function, the entire framework can be effectively learned by an alternating optimization method.

2.2 Teach to Hash

In this section, we propose a "teacher" in our deep hashing framework to efficiently select the most effective samples for the current training period. The basic problem is how to measure the effectivity of various samples. On the one hand, as mentioned in Sect. 1, when the deep model converges roughly, those difficult samples may be the most effective ones currently as they help the learning algorithm to decide accurate hash functions. On the other hand, in order to reduce the information redundancy of the training set, the selected samples should be as dissimilar as possible, which means the diversity among samples is supposed to be taken into consideration [7,14]. Inspired by these two aspects, we utilize two criteria to form the "teacher" in our hashing model. As shown in Fig. 1, the "teacher" measures the effectivity of each image through difficulty and diversity by utilizing the corresponding relaxed codes, and decides whether these points are supposed to be employed for training the model (images with ticks indicate that they are selected to be involved in the training process, while images with crosses mean they only take part in the forward propagation in the network.).

Difficulty. Motivated by [19,23], the difficulty of the i-th sample can be judged by the distance to the j-th hyperplane, which can be calculated as:

$$dis(\mathbf{x}_i)_j=\frac{|\mathbf{w}_j^T\phi(\mathbf{x}_i,\theta)+\mathbf{v}_j|}{\|\mathbf{w}_j\|_2}=\frac{|u_i^j|}{\|\mathbf{w}_j\|_2}. \qquad (3)$$

It is clear that the smaller the distance is, the more difficult the i-th sample is. Note that for all the samples, the value of $\|\mathbf{w}_j\|_2$ is the same, so we can approximately employ $dis(\mathbf{x}_i)_j=|u_i^j|$ to measure the distance to the j-th hyperplane. Since there are r hyperplanes, the total difficulty of the i-th sample can be calculated as:

$$c_i=(\sum_{j=1}^{r}dis(\mathbf{x}_i)_j^2)^{\frac{1}{2}}=(\sum_{j=1}^{r}(u_i^j)^2)^{\frac{1}{2}}=\|\mathbf{u}_i\|_2. \qquad (4)$$

Diversity. The diversity of each pair of points is measured according to the pairwise similarities of the involved samples. Motivated by [14], the similarity d_{ij} between the i-th sample and the j-th sample can be measured by the obtained relaxed codes $\mathbf{u}_i$ and $\mathbf{u}_j$, which is defined as:

$$d_{ij} = \frac{\mathbf{u}_i^T \mathbf{u}_j}{\|\mathbf{u}_i\|_2 \|\mathbf{u}_j\|_2}. \tag{5}$$

Here, we adopt d_{ij} instead of directly using the similarity label s_{ij} because it is dynamic changing during the training process and more related to the current model. It is clear that the smaller d_{ij} is, the more dissimilar $\mathbf{x}_i$ and $\mathbf{x}_j$ are.

Overall Objective. In order to select both difficult and diverse samples to form the training subset, the difficulty component related to Eq. (4) and the diversity component related to Eq. (5) should be integrated. As a result, the overall objective can be written as:

$$\min_{\mathcal{X}_s} \sum_{\mathbf{x}_i \in \mathcal{X}_s} c_i + \beta \sum_{\mathbf{x}_i, \mathbf{x}_j \in \mathcal{X}_s} d_{ij}, \tag{6}$$

where $\mathcal{X}_s$ denotes the selected subset, and $\beta > 0$ is the trade-off parameter.

To solve Eq. (6), an indicating vector $\boldsymbol{\mu} \in \{0,1\}^{N \times 1}$ is introduced to formalize the objective, where the i-th element $\mu_i = 1$ indicates $\mathbf{x}_i$ is selected and $\mu_i = 0$ otherwise. Equation (6) can be rewritten as:

$$\min_{\boldsymbol{\mu}} \ \boldsymbol{\mu}^T \mathbf{C}_d + \beta \boldsymbol{\mu}^T \mathbf{D}_s \boldsymbol{\mu}$$
$$s.t. \ \boldsymbol{\mu} \in \{0,1\}^{N \times 1} \tag{7}$$
$$\boldsymbol{\mu}^T \mathbf{1} = L,$$

where $\mathbf{C}_d = [c_1, c_1, \ldots, c_N]^T$ denotes the vector of data difficulty and $\mathbf{D}_s = \{d_{ij}\}^{N \times N}$ denotes the matrix of data diversity, $\mathbf{1}$ is an all one vector, and L is the number of the selected samples.

However, Eq. (7) is still a NP-hard problem because of the integer constraint and is hard to be solved directly. By relaxing the constraint, Eq. (7) can be rewritten as:

$$\min_{\boldsymbol{\mu}} \ \boldsymbol{\mu}^T \mathbf{C}_d + \beta \boldsymbol{\mu}^T \mathbf{D}_s \boldsymbol{\mu}$$
$$s.t. \ 0 \le \mu_i \le 1, \ i = 1, 2, \ldots, N \tag{8}$$
$$\boldsymbol{\mu}^T \mathbf{1} = L.$$

Equation (8) is a typical convex optimization problem and can be solved via Augmented Lagrange Method (ALM). After obtaining the solution of Eq. (8), the "teacher" selects L samples with the largest μ_i for further training. Generally, the selection will start at the T_s-th epoch and every selected subset will be trained for T_i epochs.

The entire training process for T2H is briefly summarized in Algorithm 1.

Algorithm 1. Training procedure for T2H.

Input: Training set $\mathcal{X}$, pairwise labels set $\mathcal{S}$, max epoch T, the epoch to start selection T_s, selection interval T_i, the number of the selected data L, code length r, parameters β, η.

1: Initialize θ, $\mathbf{W}$ and $\mathbf{v}$. Shuffle $\mathcal{X}$ to form the training set $\mathcal{X}_t$.
2: **for** $t = 1$ to T **do**
3:    // Train the deep hashing framework.
4:    Train the entire network via Eq. (2) with the training set $\mathcal{X}_t$;
5:    Calculate $\mathcal{B}$ and $\mathcal{U}$ for the whole set $\mathcal{X}$ via Eq. (1);
6:    // Teach the deep model by data selection.
7:    **if** $t > T_s$ and $(t - T_s) \mod T_i == 0$ **then**
8:       Calculate $\mathbf{C}_d$ via Eq. (4) and $\mathbf{D}_s$ via Eq. (5);
9:       Solve Eq. (8) using ALM and obtain the indicating vector $\boldsymbol{\mu}$;
10:       Select L samples with the largest μ_i to form the new subset $\mathcal{X}_t$;
11:    **end if**
12: **end for**
Output: Binary codes $\mathcal{B}$, parameters $\mathbf{W}$, $\mathbf{v}$ and θ.

---

3   Experiment

In this section, we compare the proposed T2H with several state-of-the-art hashing methods on two typical image datasets. We start by introducing the details of experimental settings, and then perform the comparison results of all these methods. All the experiments are completed on a NVIDIA TITAN XP GPU server.

3.1   Datasets and Settings

We evaluate our method on two image datasets: CIFAR-10 [8] and NUS-WIDE [3]. CIFAR-10 dataset is a single-label dataset which can be divided into ten classes. It has 60,000 color images with the size of 32 × 32 and each image is associated with only one of the ten semantic labels. NUS-WIDE dataset is a multi-label dataset. It contains nearly 270,000 images collected from the Internet and each image belongs to one or multiple classes from 81 classes. As suggested by [10], we only adopt images associated with the 21 most frequent labels. For both two datasets, we randomly sample 500 data from each class to form the initial training set and 100 data from each class to form the test set. Two images are considered similar if they at least share one common semantic label.

To demonstrate the effectiveness of the proposed T2H method, we compare our method with several state-of-the-art algorithms which can be categorized into three classes:

- Non-deep unsupervised methods: spectral hashing (SH) [20] and iterative quantization (ITQ) [6].

- Non-deep supervised methods: sequential projection learning for hashing (SPLH) [18], kernel-based supervised hashing (KSH) [2], latent factor hashing (LFH) [22], fast supervised hashing (FastH) [12], and supervised discrete hashing (SDH) [16].
- Deep supervised hashing methods: convolutional neural network hashing (CNNH) [21], deep neural networks hashing (DNNH) [10], deep hashing network (DHN) [24], and deep pairwise-supervised hashing (DPSH) [11].

For fair comparison, all of the non-deep baselines utilize the deep features extracted by the same framework with our method. For the proposed T2H, we adopt the AlexNet [9] model pre-trained on ImageNet [15] as the initial CNN model and randomly sample from a Gaussian distribution with mean 0 and variance 0.01 for initialization of $\mathbf{W}$ and $\mathbf{v}$. The hyper-parameter η is set to 50, which is demonstrated to achieve satisfactory performance by [11]. The trade-off parameter β is generally set to 0.5 on the both two datasets. We empirically set $T_s = 50$, $L = 3000$ on CIFAR-10, $T_s = 20$, $L = 6000$ on NUS-WIDE, and $T_i = 20$ on both datasets as they show good performance in our experiments. For all these methods, we utilize MAP as the evaluation metric, which is widely used in hashing tasks [17].

Table 1. MAP results on CIFAR-10 and NUS-WIDE. The MAP on NUS-WIDE is calculated based on the top 5000 returned neighbors.

Methods	CIFAR-10				NUS-WIDE			
	12-bits	24-bits	32-bits	48-bits	12-bits	24-bits	32-bits	48-bits
T2H	**0.732**	**0.762**	**0.771**	**0.779**	**0.807**	**0.832**	**0.841**	**0.843**
DPSH	0.706	0.747	0.763	0.772	0.798	0.823	0.830	0.835
DHN	0.555	0.594	0.603	0.621	0.708	0.735	0.748	0.758
DNNH	0.552	0.566	0.558	0.581	0.674	0.697	0.713	0.715
CNNH	0.429	0.511	0.509	0.522	0.611	0.618	0.625	0.608
FastH	0.533	0.607	0.619	0.636	0.779	0.807	0.816	0.825
SDH	0.478	0.557	0.584	0.592	0.780	0.804	0.815	0.824
KSH	0.488	0.539	0.548	0.563	0.768	0.786	0.790	0.799
LFH	0.208	0.242	0.266	0.339	0.695	0.734	0.739	0.759
SPLH	0.299	0.330	0.335	0.330	0.753	0.775	0.783	0.786
ITQ	0.237	0.246	0.255	0.261	0.719	0.739	0.747	0.756
SH	0.183	0.164	0.161	0.161	0.621	0.616	0.615	0.612

3.2   Effect of Teacher

In this section, we perform the comparison between T2H and DPSH to illustrate the effectiveness of the proposed "teacher". As shown in Table. 1, T2H achieves better performance compared with DPSH, which adopts the same object function but simply treats all samples equally throughout the training process. This is because with the teacher's help, the proposed T2H is able to focus on the difficult and diverse samples, which can significantly improve the rough model and lead to better results than the method with simple shuffling.

For further comparison, the training MAP curves of T2H and DPSH are shown in Fig. 2. We can find that after the epoch to start the selection (50 on CIFAR-10 and 20 on NUS-WIDE), the MAP of T2H increases significantly and the learning process becomes more stable than the one without the "teacher". All these experiment results demonstrate the effectiveness of the "teacher".

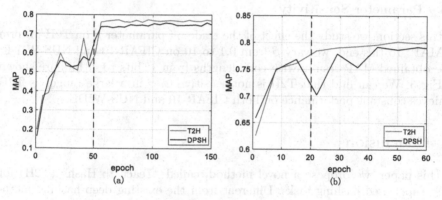

Fig. 2. MAP curves of T2H and DPSH during training process. (a) is on CIFAR-10. (b) is on NUS-WIDE. The imaginary lines indicate the epoch to start the selection.

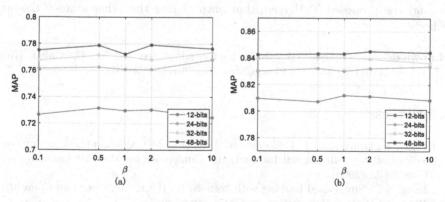

Fig. 3. Parameter sensitivity of β. (a) is on CIFAR-10. (b) is on NUS-WIDE. The MAP on NUS-WIDE is calculated based on the top 5000 returned neighbors.

3.3   Results on CIFAR-10 and NUS-WIDE

The MAP results of all the compared hashing methods with code lengths from 12 to 48 bits are shown in Table. 1. We can find that T2H consistently outperforms the other methods in all cases on both two datasets. In particular, T2H performs better than other non-deep baselines with deep features (i.e., FastH, SDH, KSH, LFH, SPLH, ITQ, SH) and the two-stage deep learning method (i.e., CNNH), because T2H is an end-to-end deep hashing framework which integrates the feature learning part and hash codes learning part. It can also be found that DPSH outperforms DHN and DNNH because of the effective object function.

3.4   Parameter Sensitivity

In this section, we study the effect of the trade-off parameter β on T2H in term of MAP. To this end, we vary β from 0.1 to 10 on CIFAR-10 and NUS-WIDE. The obtained MAP results with code lengths from 12 bits to 48 bits are shown in Fig. 3. We can find that T2H is not sensitive to β in a large range, as T2H achieves constant performance on both CIFAR-10 and NUS-WIDE.

4   Conclusion

In this paper, we propose a novel method, called "Teach to Hash" (T2H) for deep supervised hashing tasks. Different from the existing deep hashing methods that treat all training samples equally throughout the training process, we introduce a "teacher" to automatically "teach" the deep model learning accurate hash functions from the most difficult and diverse samples, which is motivated by the cognitive process of humanity. Experimental results on two real datasets illustrate that the introduced "teacher" can significantly boost the learning process and the proposed T2H algorithm outperforms the other state-of-the-art methods.

Acknowledgement. This research is partly supported by NSFC, China (No: 61572315, 6151101179) and 973 Plan, China (No. 2015CB856004).

References

1. Bengio, Y., Louradour, J., Collobert, R., Weston, J.: Curriculum learning. In: Proceedings of the 26th Annual International Conference on Machine Learning, pp. 41–48. ACM (2009)
2. Chang, S.F.: Supervised hashing with kernels. In: IEEE Conference on Computer Vision and Pattern Recognition, pp. 2074–2081 (2012)
3. Chua, T.S., Tang, J., Hong, R., Li, H., Luo, Z., Zheng, Y.T.: NUS-WIDE: a real-world web image database from national university of Singapore. In: Proceedings of ACM Conference on Image and Video Retrieval (CIVR 2009), Santorini, Greece, 8–10 July 2009

4. Gionis, A., Indyk, P., Motwani, R.: Similarity search in high dimensions via hashing. vol. 8, no. 2, pp. 518–529 (1999)
5. Gong, C., Tao, D., Liu, W., Liu, L., Yang, J.: Label propagation via teaching-to-learn and learning-to-teach. IEEE Trans. Neural Netw. Learn. Syst. **28**(6), 1452–1465 (2017)
6. Gong, Y., Lazebnik, S., Gordo, A., Perronnin, F.: Iterative quantization: a procrustean approach to learning binary codes for large-scale image retrieval. IEEE Trans. Pattern Anal. Mach. Intell. **35**(12), 2916–2929 (2013)
7. Jiang, L., Meng, D., Yu, S.I., Lan, Z., Shan, S., Hauptmann, A.: Self-paced learning with diversity. In: Advances in Neural Information Processing Systems, pp. 2078–2086 (2014)
8. Krizhevsky, A.: Learning multiple layers of features from tiny images (2009)
9. Krizhevsky, A., Sutskever, I., Hinton, G.E.: ImageNet classification with deep convolutional neural networks. In: Advances in Neural Information Processing Systems, pp. 1097–1105 (2012)
10. Lai, H., Pan, Y., Liu, Y., Yan, S.: Simultaneous feature learning and hash coding with deep neural networks, pp. 3270–3278 (2015)
11. Li, W.J., Wang, S., Kang, W.C.: Feature learning based deep supervised hashing with pairwise labels. In: International Joint Conference on Artificial Intelligence, pp. 1711–1717 (2016)
12. Lin, G., Shen, C., Shi, Q., Hengel, A.V.D., Suter, D.: Fast supervised hashing with decision trees for high-dimensional data. In: Computer Vision and Pattern Recognition, pp. 1971–1978 (2014)
13. Loshchilov, I., Hutter, F.: Online batch selection for faster training of neural networks. arXiv preprint arXiv:1511.06343 (2015)
14. Ma, C., Gong, C., Gu, Y., Yang, J., Feng, D.: SHISS: supervised hashing with informative set selection. Pattern Recognit. Lett. **107**, 105–113 (2017)
15. Russakovsky, O., et al.: Imagenet large scale visual recognition challenge. Int. J. Comput. Vis. **115**(3), 211–252 (2015)
16. Shen, F., Shen, C., Liu, W., Shen, H.T.: Supervised discrete hashing. In: Computer Vision and Pattern Recognition, pp. 37–45 (2015)
17. Wang, J., Zhang, T., Song, J., Sebe, N., Shen, H.T.: A survey on learning to hash. IEEE Trans. Pattern Anal. Mach. Intell. **40**(4), 769–790 (2018)
18. Wang, J., Kumar, S., Chang, S.F.: Sequential projection learning for hashing with compact codes. In: International Conference on International Conference on Machine Learning, pp. 1127–1134 (2010)
19. Wang, Q., Si, L., Zhang, Z., Zhang, N.: Active hashing with joint data example and tag selection. In: International ACM SIGIR Conference on Research and Development in Information Retrieval, pp. 405–414 (2014)
20. Weiss, Y., Torralba, A., Fergus, R.: Spectral hashing. In: International Conference on Neural Information Processing Systems, pp. 1753–1760 (2008)
21. Xia, R., Pan, Y., Lai, H., Liu, C., Yan, S.: Supervised hashing for image retrieval via image representation learning. In: AAAI Conference on Artificial Intelligence (2012)
22. Zhang, P., Zhang, W., Li, W.J., Guo, M.: Supervised hashing with latent factor models. In: Proceedings of the 37th International ACM SIGIR Conference on Research & Development in Information Retrieval, pp. 173–182. ACM (2014)
23. Zhen, Y., Yeung, D.Y.: Active hashing and its application to image and text retrieval. Data Min. Knowl. Discov. **26**(2), 255–274 (2013)
24. Zhu, H., Long, M., Wang, J., Cao, Y.: Deep hashing network for efficient similarity retrieval. In: AAAI, pp. 2415–2421 (2016)

Multi-view Deep Gaussian Processes

Shiliang Sun[(✉)] and Qiuyang Liu

Department of Computer Science and Technology, East China Normal University,
3663, North Zhongshan Road, Shanghai 200062, China
slsun@cs.ecnu.edu.cn

Abstract. Deep Gaussian processes (DGPs) have shown their power in
many tasks of machine learning. However, when they deal with multi-
view data, DGPs assume the same modeling depth for different views of
data, which is quite unreasonable because there are usually large diversities among different views. In this paper, we propose the model of
multi-view deep Gaussian processes (MvDGPs), which takes full account
of the characteristics of multi-view data. Combining the advantages of
the DGPs with the multi-view learning, MvDGPs can independently
determine the modeling depths for each view, which is more flexible and
powerful. In contrast with the DGPs, MvDGPs support asymmetrical
modeling depths for different view of data, resulting in better character-
izations of the discrepancies among different views. Experimental results
on multiple multi-view data sets have verified the flexibilities and effectiveness of the proposed model.

Keywords: Multi-view learning · Deep learning · Gaussian process
Unsupervised learning

1 Introduction

Gaussian processes (GPs) [14] have gained tremendous attentions in various
areas of machine learning [6,10,11]. The high popularities are attributed to providing flexible function approximation that can be applied to both supervised
and unsupervised learning tasks [13]. Deep Gaussian processes (DGPs) [5], as
deep generalizations of the GPs, maintain excellent properties of the GPs such
as nonparametric probabilistic nature and well-calibrated predictive uncertainty
estimates. Moreover, due to the hierarchical structure, DGPs show better predictive performance and greater generalization ability [2,9,15].

DGPs have shown effectiveness on many single-view tasks. With the rapid
development of big data applications, more and more data present the multi-view
characteristic [18]. These multi-view data often contains different descriptions on
various perspectives for the same object or concept. On the one hand, as data of
different views represent the same object or concept, there are some connections
among them. On the other hand, they could offer complementary information
since their representations are from different angles or sensors. Therefore, models

© Springer Nature Switzerland AG 2018
L. Cheng et al. (Eds.): ICONIP 2018, LNCS 11301, pp. 130–139, 2018.
https://doi.org/10.1007/978-3-030-04167-0_12

which take full use of the multi-view data are able to provide better performance than models that employ the single-view data in general. Naturally, multi-view data aroused much interest in machine learning communities [16]. Recently, there are an increasing number of algorithms proposed for multi-view learning [21], which can mainly be divided into three major categories [21]: co-training style algorithms [1], co-regularization style algorithms [12,20], and margin-consistency style algorithms [17].

Although the DGPs could take multi-view data as inputs, they model data of different views with the same depth. Since data of different views are based on diversified perspectives, there are often wide variations among them. For instance, a web image could be depicted by its surrounding text and its visual information such as RGB or HSV. Under this circumstance, text view and visual view differ greatly, which is required to be handled by different modeling depths. It is inappropriate to employ the DGPs for this kind of data, which is usually the case in multi-view data.

Therefore, in order to deal properly with multi-view data, we present the multi-view deep Gaussian processes (MvDGPs) model, which is able to employ elastic modeling depths for various views of data. Unlike the DGPs, MvDGPs adequately cover traits of multi-view data, supporting asymmetrical modeling depths. As the MvDGPs use flexible modeling depths for various views of data, they could use the same or different depths to model data of different views, which is much more flexible and reasonable than the same modeling depths in the DGPs. The experiments on the real world multi-view data sets also validate the effectiveness of our proposed MvDGPs, which demonstrates the advantage of the flexible modeling depths.

The rest of this paper is organized as follows. In Sect. 2, we briefly review DGPs. Section 3 introduces the proposed MvDGPs model, covering the principle and the Bayesian inference. Experimental results are provided in Sect. 4. Finally, we conclude this paper and point out the future research directions in Sect. 5.

2   Deep Gaussian Processes

We briefly review the deep Gaussian processes (DGPs) model in this section.

DGPs [4,5] are multi-layer generalizations of the GPs, providing a powerful Bayesian nonparametric tool for probabilistic modeling. As a hierarchical directed graphical model, DGPs are composed of multiple layers of latent variables, and employ a hierarchical structure of GP [14] mappings. Actually, when the depth of the DGPs is one, it is equivalent to a Gaussian process latent variable model (GP-LVM) [19]. Figure 1 depicts the graphical representation of the DGPs. For clarity, we present the DGPs in the unsupervised learning scenario. The right-most layer expresses the observed outputs. Other layers are hidden variables. If there are observed inputs, they are placed on the left-most layer.

Formally, given a data set $D = \{Y\}$, DGPs defines L layers of hidden variables $\{X_l\}_{l=1}^{L}$ ($X_l \in R^{N \times Q_l}$) through the following generative processes:

$$Y = f_1(X_1) + \epsilon_1, \quad \epsilon_1 \sim \mathcal{N}(0, \sigma_1^2 I), \tag{1}$$

$$X_{l-1} = f_l(X_l) + \epsilon_l, \quad \epsilon_l \sim \mathcal{N}(0, \sigma_l^2 I), \quad l = 2, \ldots, L, \tag{2}$$
$$X_L \sim \mathcal{N}(0, I), \tag{3}$$

where f_l samples from a Gaussian process with the covariance function k_l, i.e.,
$f_l(x) \sim \mathcal{GP}(0, k_l(x, x')), l = 1, \ldots, L.$

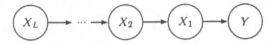

Fig. 1. The graphical model for the DGPs

DGPs can model non-stationarities and sophisticated functions without the use of a non-stationary kernel function because of the successive warping of latent variables through the hierarchy, which makes the DGPs outperform the standard GPs [4]. Exact Bayesian learning in the DGPs is analytically intractable. Approximated inference methods such as variational inference and expectation propagation [3,5] have been developed to address the problem.

Figure 2 demonstrates the graphical model representation of a two-layer DGPs on the multi-view data. It is obviously that the DGPs employ the same depth for data of different views, which is unreasonable when data representations of different views vary enormously. Even data from different views represent the same object, there are often wide variations among them due to their distinct descriptive perspectives. In view of the above problems, DGPs are usually not suitable for the multi-view data.

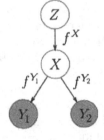

Fig. 2. The graphical model for a two-layer DGPs on multi-view data

Hence, we present a new DGPs for multi-view learning. In contrast to the DGPs, the proposed MvDGPs employ relatively independent depths to model data of different views. If data of different views differ greatly, MvDGPs are able to use asymmetrical depths to model them, which can not be realized by the existing DGPs.

3   Multi-view Deep Gaussian Processes

In this section, we extend the DGPs to the multi-view scenario and propose the model of multi-view deep Gaussian processes (MvDGPs). MvDGPs contain multiple layers of latent variables. GPs are leveraged to model the mapping relationships between the adjacent layers. Combing the peculiarities of multi-view data and advantages of the DGPs, MvDGPs employ the relatively independent modeling depths for different views of data. The flexible modeling depths allow the model to use the same or different depths to model data of different views according to the given multi-view data, which is much more flexible than the same modeling depths in the DGPs. When there are sharp differences among different views of data, MvDGPs allow asymmetrical modeling depths for them, which could model the discrepancies among different views better and turn to learning a more reasonable fusion representation.

3.1   The Proposed Model

Figure 3 illustrates the graphical model representation of the MvDGPs on a two-view data set $D = \{(\boldsymbol{x}_i^{\langle 1 \rangle}, \boldsymbol{x}_i^{\langle 2 \rangle})\}_{i=1}^N$. In the deep architecture of the MvDGPs, all the latent variables are used as the inputs of the variables in the next layer and the outputs of the variables of the previous layer. $\boldsymbol{X}^{\langle 1 \rangle} = [\boldsymbol{x}_1^{\langle 1 \rangle}, \ldots, \boldsymbol{x}_N^{\langle 1 \rangle}]^{\mathbf{T}}$ ($X^{\langle 1 \rangle} \in R^{N \times D_1}$) denote the observation data from the first view, $\boldsymbol{X}^{\langle 2 \rangle} = [\boldsymbol{x}_1^{\langle 2 \rangle}, \ldots, \boldsymbol{x}_N^{\langle 2 \rangle}]^{\mathbf{T}}$ ($X^{\langle 2 \rangle} \in R^{N \times D_2}$) denote the observation data from the second view. $X^{\langle v, h \rangle} \in R^{N \times Q_{v,h}}$ denote the latent variables of the vth view on the hth layer, where $v \in \{1, 2\}$ indexes the views, $h \in \{1, 2, \ldots, H_v\}$ with $H_1 = L, H_2 = R$ indexes the layers. $Xs^{\langle h \rangle} \in R^{N \times Q_h}$ denote the shared latent variables on the hth layer, where $h \in \{1, 2, \ldots, H\}$ indexes the layers.

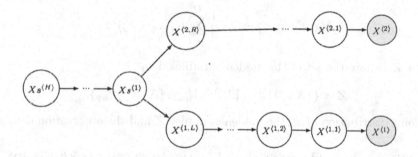

Fig. 3. The graphical model for the MvDGPs

By maintaining the chain of the shared variables and appending more chains of latent variables and latent functions to model the data of new views, the MvDGPs are able to applied to the scenarios where data have more than two views easily. For clarity, we focus on the two view learning tasks to demonstrate the MvDGPs.

Formally, given a two-view data set $D = \{\boldsymbol{X}^{\langle 1 \rangle}, \boldsymbol{X}^{\langle 2 \rangle}\}$, MvDGPs define L layers of the latent variables of the first view $\{X^{\langle 1, h \rangle}\}_{h=1}^L$, R layers of the latent variables of the second view $\{X^{\langle 2, h \rangle}\}_{h=1}^R$, and H layers of the shared latent variables $\{Xs^{\langle h \rangle}\}_{h=1}^H$ through the following generative processes. Denote $H_1 = L$ and $H_2 = R$.

$$Xs^{\langle H \rangle} \sim \mathcal{N}(0, I). \tag{4}$$

$$Xs^{\langle h-1 \rangle} = f_s^{\langle h \rangle}(Xs^{\langle h \rangle}) + \epsilon_s^{\langle h \rangle},$$

$$\epsilon_s^{\langle h \rangle} \sim \mathcal{N}(0, (\sigma_s^{\langle h \rangle})^2 I), \quad h = 2, \dots, H. \tag{5}$$

where $f_s^{\langle h \rangle}$ sample from a Gaussian process with the covariance function $k_s^{\langle h \rangle}$, i.e., $f_s^{\langle h \rangle}(x) \sim \mathcal{GP}(0, k_s^{\langle h \rangle}(x, x'))$, $h = 2, .., H$.

$$X^{\langle v, H_v \rangle} = f^{\langle v, H_v + 1 \rangle}(Xs^{\langle 1 \rangle}) + \epsilon^{\langle v, H_v + 1 \rangle},$$

$$\epsilon^{\langle v, H_v + 1 \rangle} \sim \mathcal{N}(0, (\sigma^{\langle v, H_v + 1 \rangle})^2 I), \quad v \in \{1, 2\}, \tag{6}$$

$$X^{\langle v, h-1 \rangle} = f^{\langle v, h \rangle}(X^{\langle v, h \rangle}) + \epsilon^{\langle v, h \rangle},$$

$$\epsilon^{\langle v, h \rangle} \sim \mathcal{N}(0, (\sigma^{\langle v, h \rangle})^2 I), \quad v \in \{1, 2\}, \quad h = 2, .., H_v, \tag{7}$$

$$X^{\langle v \rangle} = f^{\langle v, 1 \rangle}(X^{\langle v, 1 \rangle}) + \epsilon^{\langle v, 1 \rangle},$$

$$\epsilon^{\langle v, 1 \rangle} \sim \mathcal{N}(0, (\sigma^{\langle v, 1 \rangle})^2 I), \quad v \in \{1, 2\}, \tag{8}$$

where $f^{\langle v, h \rangle}$ sample from a Gaussian process with the covariance function $k^{\langle v, h \rangle}$, i.e., $f^{\langle v, h \rangle}(x) \sim \mathcal{GP}(0, k^{\langle v, h \rangle}(x, x'))$, $v \in \{1, 2\}, h = 1, .., H_v + 1$.

In this paper, we use the automatic relevance determination squared exponential (ARD-SE) kernel as the covariance function:

$$k(x, x') = \sigma_f^2 \exp(-\frac{1}{2} \sum_{q=1}^{Q} \alpha_q (x_q - x_q')^2). \tag{9}$$

Let $\mathbf{Z}$ denote the set of the hidden variables, i.e.,

$$\mathbf{Z} = \{\{Xs^{\langle h \rangle}\}_{h=1}^H, \{X^{\langle 1, h \rangle}\}_{h=1}^L, \{X^{\langle 2, h \rangle}\}_{h=1}^R\}.$$

The joint distribution of all the hidden variables $\mathbf{Z}$ and the observation data is

$$p(X^{\langle 1 \rangle}, X^{\langle 2 \rangle}, \mathbf{Z}) = \prod_{v=1}^{2} \{p(X^{\langle v \rangle}|X^{\langle v, 1 \rangle}) \prod_{h=1}^{H_v - 1} [p(X^{\langle v, h \rangle}|X^{\langle v, h+1 \rangle})]p(X^{\langle v, H_v \rangle}|Xs^{\langle 1 \rangle})\}$$

$$\prod_{h=1}^{H-1} [p(Xs^{\langle h \rangle}|Xs^{\langle v, h+1 \rangle})]p(Xs^{\langle H \rangle}). \tag{10}$$

3.2   Variational Bayesian Training

According to the standard Bayesian training procedure, the optimization of the model evidence is required, i.e., optimization of the $\log p(X^{\langle 1 \rangle}, X^{\langle 2 \rangle})$.

$$\log p(X^{\langle 1 \rangle}, X^{\langle 2 \rangle}) = \log \int_{Z} \prod_{v=1}^{2} \{p(X^{\langle v \rangle}|X^{\langle v, 1 \rangle}) \prod_{h=1}^{H_v - 1} [p(X^{\langle v, h \rangle}|X^{\langle v, h+1 \rangle})]$$

$$p(X^{\langle v, H_v \rangle}|Xs^{\langle 1 \rangle})\} \prod_{h=1}^{H-1} [p(Xs^{\langle h \rangle}|Xs^{\langle v, h+1 \rangle})]p(Xs^{\langle H \rangle})d\mathbf{Z}. \tag{11}$$

In accordance with the standard DGPs, exact Bayesian learning in the MvDGPs is analytically intractable. Here, we take the same variational inference methods used in [5] to address the problem. Specifically, we introduce the corresponding inducing points $\{U^{\langle 1,h\rangle}\}_{h=1}^{L+1}$, $\{U^{\langle 2,h\rangle}\}_{h=1}^{R+1}$, $\{U^{\langle h\rangle}\}_{h=2}^{H}$ for each layer, and assume the following variational posterior distributions for hidden variables:

$$q(f^{\langle v,1\rangle}, U^{\langle v,1\rangle}|X^{\langle v\rangle}) = p(f^{\langle v,1\rangle}|U^{\langle v,1\rangle}, X^{\langle v\rangle})q(U^{\langle v,1\rangle}), \quad v \in \{1,2\}. \tag{12}$$

$$q(f^{\langle v,h\rangle}, U^{\langle v,h\rangle}|X^{\langle v,h-1\rangle}, X^{\langle v,h\rangle}) = p(f^{\langle v,h\rangle}|U^{\langle v,h\rangle}, X^{\langle v,h\rangle})q(U^{\langle v,h\rangle}),$$
$$v \in \{1,2\}, \quad h = 2,\ldots,H_v. \tag{13}$$

$$q(f^{\langle v,H_v+1\rangle}, U^{\langle v,H_v+1\rangle}|X^{\langle v,H_v\rangle}, Xs^{\langle 1\rangle}) = p(f^{\langle v,H_v+1\rangle}|U^{\langle v,H_v+1\rangle}, Xs^{\langle 1\rangle})$$
$$q(U^{\langle v,H_v+1\rangle}), \quad v \in \{1,2\}. \tag{14}$$

$$q(f_s^{\langle h\rangle}, U^{\langle h\rangle}|Xs^{\langle h-1\rangle}, Xs^{\langle h\rangle}) = p(f_s^{\langle h\rangle}|U^{\langle h\rangle}, Xs^{\langle h\rangle})q(U^{\langle h\rangle}),$$
$$h = 2,\ldots,H. \tag{15}$$

$$q(X^{\langle v,h\rangle}) = \prod_{j=1}^{Q_{v,h}} \mathcal{N}(\mu_j^{\langle v,h\rangle}, S_j^{\langle v,h\rangle}), \quad v \in \{1,2\}, \quad h = 1,..,H_v. \tag{16}$$

$$q(Xs^{\langle h\rangle}) = \prod_{j=1}^{Q_h} \mathcal{N}(\mu_j^{\langle h\rangle}, S_j^{\langle h\rangle}), \quad h = 1,..,H. \tag{17}$$

Following the above settings, we can get the following closed-form variational lower bound $\mathcal{L}$ for the model evidence:

$$\mathcal{L} = E_Z \log p(X^{\langle 1\rangle}, X^{\langle 2\rangle}, Z) - E_Z \ln q(Z)$$
$$= \sum_{v=1}^{2}\{g^{\langle v\rangle} + \sum_{h=1}^{H_v} r^{\langle v,h\rangle} + \sum_{h=1}^{H_v+1} \mathcal{H}_{q(X^{\langle v,h\rangle})}\}$$
$$+ \sum_{h=1}^{H-1} r_s^{\langle h\rangle} + \sum_{h=1}^{H-1} \mathcal{H}_{q(X^{\langle h\rangle})} - KL(q(Xs^{\langle H\rangle})\|p(Xs^{\langle H\rangle})), \tag{18}$$

where $\mathcal{H}$ denotes the entropy of a distribution, and $KL(q(Xs^{\langle H\rangle})\|p(Xs^{\langle H\rangle}))$ is the Kullback-Leibler divergence [8] between $q(Xs^{\langle H\rangle}$ and $p(Xs^{\langle H\rangle}))$.

$$g^{\langle v\rangle} = g(X^{\langle v\rangle}, f^{\langle v,1\rangle}, U^{\langle v,1\rangle}, X^{\langle v,1\rangle})$$
$$= \langle \log p(X^{\langle v\rangle}|f^{\langle v,1\rangle}) - \log \frac{p(U^{\langle v,1\rangle})}{q(U^{\langle v,1\rangle})} \rangle_{p(f^{\langle v,1\rangle}|U^{\langle v,1\rangle}, X^{\langle v,1\rangle})q(U^{\langle v,1\rangle})q(X^{\langle v,1\rangle})},$$
$$v \in \{1,2\}. \tag{19}$$

$$r^{\langle v,h-1\rangle} = r(X^{\langle v,h-1\rangle}, f^{\langle v,h\rangle}, U^{\langle v,h\rangle}, X^{\langle v,h\rangle})$$
$$= \langle \log p(X^{\langle v,h-1\rangle}|f^{\langle v,h\rangle})$$
$$- \log \frac{p(U^{\langle v,h\rangle})}{q(U^{\langle v,h\rangle})} \rangle_{p(f^{\langle v,h\rangle}|U^{\langle v,h\rangle}, X^{\langle v,h\rangle})q(U^{\langle v,h\rangle})q(X^{\langle v,h-1\rangle})q(X^{\langle v,h\rangle})},$$
$$v \in \{1,2\}, \quad h = 2,\ldots,H_v. \tag{20}$$

$$
\begin{aligned}
r^{\langle v, H_v \rangle} &= r(X^{\langle v, H_v \rangle}, f^{\langle v, H_v+1 \rangle}, U^{\langle v, H_v+1 \rangle}, Xs^{\langle 1 \rangle}) \\
&= \langle \log p(X^{\langle v, H_v \rangle} | f^{\langle v, H_v+1 \rangle}) \\
&\quad - \log \frac{p(U^{\langle v, H_v+1 \rangle})}{q(U^{\langle v, H_v+1 \rangle})} \rangle_{p(f^{\langle v, H_v+1 \rangle} | U^{\langle v, H_v+1 \rangle}, Xs^{\langle 1 \rangle}) q(U^{\langle v, H_v+1 \rangle}) q(X^{\langle v, H_v \rangle}) q(Xs^{\langle 1 \rangle})}, \\
&\quad v \in \{1, 2\}.
\end{aligned} \tag{21}
$$

$$
\begin{aligned}
r_s^{\langle h-1 \rangle} &= r(Xs^{\langle h-1 \rangle}, f_s^{\langle h \rangle}, U^{\langle h \rangle}, Xs^{\langle h \rangle}) \\
&= \langle \log p(Xs^{\langle h-1 \rangle} | f_s^{\langle h \rangle}) \\
&\quad - \log \frac{p(U^{\langle h \rangle})}{q(U^{\langle h \rangle})} \rangle_{p(f_s^{\langle h \rangle} | U^{\langle h \rangle}, Xs^{\langle h \rangle}) q(U^{\langle h \rangle}) q(Xs^{\langle h-1 \rangle}) q(Xs^{\langle h \rangle})}, \\
&\quad h = 2, \ldots, H.
\end{aligned} \tag{22}
$$

In the above Eqs. (19–22), $\langle \cdot \rangle$ denotes the expectation.

3.3 Parameter Estimation

We have chosen the ARD-SE kernel as the covariance functions in the MvDGPs. The parameters to be optimized in the MvDGPs cover the variational parameters $\{\mu_j^{\langle v,h \rangle}, S_j^{\langle v,h \rangle}\}$ in Eq. (16), $\{\mu_j^{\langle h \rangle}, S_j^{\langle h \rangle}\}$ in Eq. (17) and the model parameters $\{\theta, \sigma, \tilde{X}\}$, where θ denotes the set of all the parameters in the kernel functions, $\sigma = \{\{\sigma_s^{\langle h \rangle}\}_{h=2}^H, \{\sigma^{\langle 1,h \rangle}\}_{h=1}^{L+1}, \{\sigma^{\langle 2,h \rangle}\}_{h=1}^{R+1}\}$, and $\tilde{X}$ represents the set of all the inducing inputs for each layer. Both the model parameters and the variational parameters can be optimized by the gradient descent method. The hyper-parameters of the MvDGPs are L, R, and H, which can be obtained by the grid search strategy.

4 Experiments

In order to evaluate the performance of the proposed MvDGPs, we conduct experiments on multiple multi-view data sets.

4.1 Data Sets

In the experiments, we use the web-page data sets. The web-page data sets are frequently-used multi-view data sets, which are composed of two-view web pages collected from computer science department web sites at four universities: Cornell university, university of Washington, university of Wisconsin, and university of Texas. The two views are the content view and the cite view, where the content view denotes words occurring in a web page and the cite view denotes words appearing in the links pointing to that page. The web pages are classified into five classes: student, project, course, staff and faculty. In the four data sets, we set the category with the greatest size to be the positive class, and all the other categories as the negative class. The data sets are unbalanced, and their sizes range from 187 to 265.

4.2   Setting

We evaluate the MvDGPs as a discriminative model for multi-view data. For the model we trained, the fused representation of the multi-view data was extracted and fed to an SVM classifier [7]. As the left-most layer of the model in Fig. 3 captures the multi-view data better, we employ the mean of the posterior distribution of the hidden variables in the left-most layer as the fused representation of the multi-view data. For comparison, the similar setting is performed on the DGPs.

In order to assess the proposed model adequately, we use two kinds of partitions for the data. We repeat the experiments for all the data sets ten times and record the average accuracies and the corresponding standard deviations for the classification task.

4.3   Results

We present the average accuracies and standard deviations of all the methods on the webpage data sets in Table 1. Figure 4 demonstrates the average accuracies and the corresponding standard deviations of the MvDGPs with different depth settings on the Cornell data set.

Table 1. The average accuracies and standard deviations (%) on real world data sets

Data set	Model			
	70% training set		60% training set	
	DGPs	MvDGPs	DGPs	MvDGPs
Cornell	78.79 ± 6.67	$\mathbf{79.65 \pm 3.01}$	78.21 ± 4.25	$\mathbf{82.05 \pm 4.05}$
Washington	76.81 ± 2.73	$\mathbf{82.75 \pm 2.60}$	77.61 ± 2.25	$\mathbf{85.00 \pm 3.21}$
Wisconsin	78.97 ± 4.05	$\mathbf{86.02 \pm 3.52}$	78.11 ± 3.91	$\mathbf{86.42 \pm 3.63}$
Texas	85.01 ± 4.08	$\mathbf{85.33 \pm 4.14}$	85.30 ± 4.39	$\mathbf{85.55 \pm 3.50}$

It is clearly shown in Table 1 that our proposed MvDGPs are superior to the DGPs on the multi-view data, which means that the MvDGPs are able to capture the fused representation of the multi-view data better, leading to the better performance on the classification tasks. As shown in the Fig. 4, the superior performance of the MvDGPs is attributed to the proper modeling depths for data of different views, which allows asymmetrical modeling depths when data of different views differ greatly. This could not be fulfilled by the DGPs. In consideration of the complexities and diversities among the different views in the multi-view data, the flexible and relatively independent modeling depths make the model more powerful, resulting in leveraging multi-view data more reasonably.

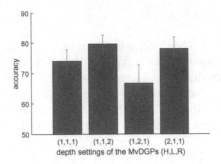

Fig. 4. The average accuracies and standard deviations (%) on the Cornell data set with 70% training set

5   Conclusion

In this paper, we have proposed the MvDGPs for multi-view learning, which extend the DGPs to the scenario of learning with multiple views via flexible modeling depths. As a novel multi-layer hierarchical multi-view framework, MvDGPs take adequate consideration of the characteristics of the multi-view data and the DGPs. The settings of the modeling depths for data of different views are elastic and reasonable, which not only interrelate with each other, but also pay attention to the distinctions among different views. Experimental results on real-word data validate the effectiveness of the proposed MvDGPs.

In the future, it is interesting to consider employing the labels as another view. Through inferring the missing values for the label view in the prediction stage, MvDGPs are able to be extended to the supervised learning task.

Acknowledgments. This work is supported by the National Natural Science Foundation of China under Project 61673179, and Shanghai Knowledge Service Platform Project (No. ZF1213).

References

1. Blum, A., Mitchell, T.: Combining labeled and unlabeled data with co-training. In: Proceedings of the 11th Annual Conference on Computational Learning Theory, pp. 92–100 (1998)
2. Bui, T., Hernández-Lobato, D., Hernandez-Lobato, J., Li, Y., Turner, R.: Deep Gaussian processes for regression using approximate expectation propagation. In: International Conference on Machine Learning, pp. 1472–1481 (2016)
3. Bui, T.D., Hernández-Lobato, J.M., Li, Y., Hernández-Lobato, D., Turner, R.E.: Training deep Gaussian processes using stochastic expectation propagation and probabilistic backpropagation. arXiv preprint arXiv:1511.03405 (2015)
4. Dai, Z., Damianou, A., González, J., Lawrence, N.: Variational auto-encoded deep Gaussian processes. arXiv preprint arXiv:1511.06455 (2015)
5. Damianou, A., Lawrence, N.: Deep Gaussian processes. In: Artificial Intelligence and Statistics, pp. 207–215 (2013)

6. Damianou, A.C., Titsias, M.K., Lawrence, N.D.: Variational Gaussian process dynamical systems. In: Proceedings of the 25th Annual Conference on Neural Information Processing Systems, pp. 2510–2518 (2011)
7. Hearst, M.A., Dumais, S.T., Osuna, E., Platt, J., Scholkopf, B.: Support vector machines. IEEE Intell. Syst. Appl. **13**(4), 18–28 (1998)
8. Joyce, J.M.: Kullback-Leibler divergence. Springer (2011)
9. Kandemir, M.: Asymmetric transfer learning with deep Gaussian processes. In: International Conference on Machine Learning, pp. 730–738 (2015)
10. Krause, A., Guestrin, C.: Nonmyopic active learning of Gaussian processes: an exploration-exploitation approach. In: Proceedings of the 24th International Conference on Machine Learning, pp. 449–456 (2007)
11. Lawrence, N.D., Jordan, M.I.: Semi-supervised learning via Gaussian processes. In: Proceedings of the 18th Annual Conference on Neural Information Processing Systems, pp. 753–760 (2004)
12. Liu, Q., Sun, S.: Multi-view regularized Gaussian processes. In: Kim, J., Shim, K., Cao, L., Lee, J.-G., Lin, X., Moon, Y.-S. (eds.) PAKDD 2017. LNCS (LNAI), vol. 10235, pp. 655–667. Springer, Cham (2017). https://doi.org/10.1007/978-3-319-57529-2_51
13. Quiñonero-Candela, J., Rasmussen, C.E.: A unifying view of sparse approximate Gaussian process regression. J. Mach. Learn. Res. **6**(Dec), 1939–1959 (2005)
14. Rasmussen, C.E., Williams, C.K.I.: Gaussian Processes for Machine Learning. MIT Press, Cambridge (2006)
15. Salimbeni, H., Deisenroth, M.: Doubly stochastic variational inference for deep Gaussian processes. In: Advances in Neural Information Processing Systems, pp. 4591–4602 (2017)
16. Sun, S.: A survey of multi-view machine learning. Neural Comput. Appl. **23**(7–8), 2031–2038 (2013)
17. Sun, S., Chao, G.: Multi-view maximum entropy discrimination. In: IJCAI, pp. 1706–1712 (2013)
18. Sun, S., Shawe-Taylor, J., Mao, L.: PAC-Bayes analysis of multi-view learning. Inf. Fusion **35**, 117–131 (2017)
19. Titsias, M., Lawrence, N.D.: Bayesian Gaussian process latent variable model. In: Proceedings of the Thirteenth International Conference on Artificial Intelligence and Statistics, pp. 844–851 (2010)
20. Yu, S., Krishnapuram, B., Rosales, R., Rao, R.B.: Bayesian co-training. J. Mach. Learn. Res. **12**, 2649–2680 (2011)
21. Zhao, J., Xie, X., Xu, X., Sun, S.: Multi-view learning overview: recent progress and new challenges. Inf. Fusion **38**, 43–54 (2017)

Deep Collaborative Filtering Combined with High-Level Feature Generation on Latent Factor Model

Xu Li, Xu Chen, and Zheng Qin[✉]

School of Software, Tsinghua University, Haidian District, Beijing, China
luxi_li@sina.com, {xu-ch14,qingzh}@tsinghua.edu.cn

Abstract. Recommender System becomes indispensable in the era of information explosion nowadays. Former researchers have noticed the important role of high-level feature playing on semantic factor cases. However, in more common scenes where semantic features cannot be reached, research involving high-level feature on latent factor models is lacking. Analogizing to the idea of the convolutional neural network in image processing, we proposed a Weighted Feature Interaction Network to generate high-level features from the low-level latent factors. An intuitive interpretation is also given to help understand. Then it is integrated into a Deep Collaborative Filtering Model. The results on two real-world datasets show that weighted feature interaction network works and our Deep Collaborative Filtering Model outperforms some conventional and state-of-the-art models. Our work improves the feature representation and recommendation performance on Latent Factor Model.

Keywords: Recommender systems · Latent factor model
Collaborative filtering · Deep neural network · Implicit feedback

1 Introduction

In the era of information explosion, people are facing a variety of commodities in E-commerce, which leads to a difficulty in finding the most appropriate items satisfying their needs. Thus, Recommender System plays an important role.

It is common that the preference of someone is determined by more than one feature. It is obviously reasonable to recommend a college boy a basketball. It is based on the fact that he is a *male* in *age* between 10 to 30, and he's a *student* which implies that he is more likely to access a chance to play. This can be called as a three-order feature or a high-level feature. Some researchers did notice this and have done some works [2,3]. However, for some circumstances when personal information is not collected (e.g. targeted advertising [4]), or when users tend to provide no information in their profile page, we can only make recommendations by latent features based on the user-item interaction (e.g. purchase records, click logs, etc.). Due to the non-interpretability of latent factors, few works have been done on high-level latent features.

© Springer Nature Switzerland AG 2018
L. Cheng et al. (Eds.): ICONIP 2018, LNCS 11301, pp. 140–151, 2018.
https://doi.org/10.1007/978-3-030-04167-0_13

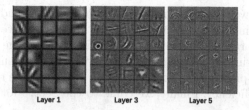

Layer 1   Layer 3   Layer 5

Fig. 1. The feature maps in convolutional layers of different depths showing low-level and high-level features [1]

We deem that the pattern of low-level and high-level features exists in latent factors as well. Just like the images contain low-level details like corners and edge/color conjunctions as well as complex high-level features like outlines, shapes, and textures (Fig. 1). Users also have low-level features like gender, age, profession, etc. and high-level features like dynamic, geek, genteel, etc. Items contain color, size, price, etc. as low-level features and deluxe, functional, succinct, classy, etc. as high-level features.

The learned high-level latent features imply an abstraction of one's preference which is more meaningful. Usually, one who likes doing sports may prefer lively music than classical; one obsessed with masterpieces of Mondrian may prefer pure color plain dresses rather than a flowery one. It is the high-level mental motivation in one's brain maintains this consistency in preference. The questions in psychological tests aren't intended to learn which picture do you like most, but to draw a profile of what a person you are according to your choice (Fig. 2). This shows the importance of high-level tendentiousness underlying the subconsciousness and personality. In this paper, we focused on the high-level feature generation on latent factor model and propose a specific network structure for the issue.

Which one do you like most?

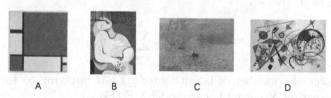

A   B   C   D

Fig. 2. A commonly asked question in psychological testing: which one do you like most? (A. Piet Mondrian, *Composition with Red, Blue, and Yellow* (Neoplasticism); B. Pablo Picasso, *The Dream* (Cubism and Surrealism); C. Claude Monet, *Impression Sunrise* (Impressionism); D. Wassily Kandinsky, *Straight Line* (Abstract Expressionism)) (Color figure online)

In what follows, we first introduce some preliminaries including implicit feedback, latent factor model and neural networks. In Sect. 3, we proposed a new model Deep Collaborative Filtering (DCF), which is a 3-stage model combined with Weighted Feature Interaction Network (WFIN). The surprising effect of DCF with WFIN is presented in Sect. 4. Section 5 shows some related works and discusses their insufficiency. Finally, we come to the Sect. 6 for a conclusion.

2   Preliminaries

2.1   Implicit Feedback

When focusing only on whether there is an interaction between users and items, it is implicit feedback. An user-item interaction matrix $\mathbf{Y}$ consisting of 0s and 1s is used to record that. The two dimensions of $\mathbf{Y}$ are the users and the items. The value of element y_{ui} in $\mathbf{Y}$ is given by Eq. 1:

$$y_{ui} = \begin{cases} 1 & \text{interaction exists between user } u \text{ and item i;} \\ 0 & \text{interaction not observed.} \end{cases} \tag{1}$$

2.2   Latent Factor Model

Collaborative Filtering (CF) is a set of conventional recommender algorithms based on the assumptions that users who like the same items can recommend items to each other [5–7], and items that are liked by the same users can be recommended among these users [8–10]. Latent factor model (LFM) is an implementation on implicit feedback which is based on matrix factorization (MF) of CF [11]. LFM decomposes the user-item interaction matrix into user latent factor matrix $\mathbf{P}$ and item latent factor matrix $\mathbf{Q}$ by singular value decomposition (SVD) [12–14]. Each row $\boldsymbol{p}_u$ denotes the latent factor vector of user u and each column $\boldsymbol{q}_i$ denotes that of item i. Similar users or items have closer distance between their vectors. The multiplication of $\boldsymbol{p}_u$ and $\boldsymbol{q}_i$ should be equal to the real value y_{ui} in $\mathbf{Y}$ (Eq. 2),

$$y_{ui} \approx \hat{y}_{ui} = \boldsymbol{p}_u^{\mathrm{T}} \boldsymbol{q}_i = \sum_{k=1}^{k} \boldsymbol{p}_{uk} \boldsymbol{q}_{ik}, \tag{2}$$

where k denotes the number of latent factors. LFM is trying to simulate the user-item interaction $\mathbf{Y}$ with a linear model MF (Eq. 3),

$$\mathbf{Y} \approx \hat{\mathbf{Y}} = \mathbf{P}^{\mathrm{T}}\mathbf{Q}. \tag{3}$$

2.3   Neural Networks

The linearity limits the expression ability of MF. Deep Neural Networks (DNN) provides a capability of complex nonlinear function fitting. The basic unit of DNN is neuron, which is shown in Fig. 3 and formulated in Eq. 4:

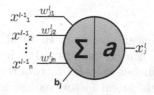

Fig. 3. Structure of a neuron

$$x_j^l = a(\boldsymbol{w}^{\mathrm{T},l}\boldsymbol{x}^{l-1} + b_j) = a(\sum_{i=1}^{n} w_{ji}^l x_{ji}^{l-1} + b_j), \tag{4}$$

where $\boldsymbol{x}^{l-1}$ is the output of the last layer, as well as the input of the current layer l. The connection between the i-th neuron in layer $l-1$ and the j-th neuron in l layer is weighted by $\boldsymbol{w}_{ji}^l$. The activation function $a(\cdot)$ (ReLU or sigmoid) realizes the non-linear capacity.

The output of a training sample will be compared with the ground truth by calculating a log loss. All the weights and biases in the networks either for interaction function fitting or feature engineering, are able to be updated by the back-propagation algorithm. For implicit recommender systems, the goal is to fit the potential user-item interaction function by DNN [15,16]. When a new user or new item comes, the trained model gives a possibility $\hat{y}_{u,i}$ to show how likely the user is to interact with the item.

3 Deep Collaborative Filtering

The Deep Collaborative Filtering model (DCF) is a 3-stage network as shown in Fig. 4. Each stage will be elaborated below in this section.

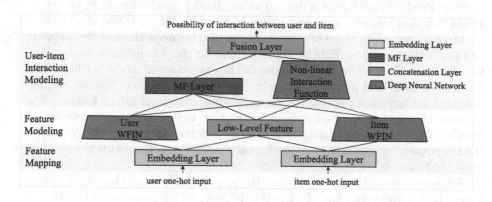

Fig. 4. The structure of DCF

3.1   Feature Mapping

The first stage in DCF is feature mapping. In LFM, embedding method is a fully connected layer used to convert user/item IDs in one-hot codes into dense latent factor vectors. The number of latent factors is set manually, which is 32 in our model. We consider the latent factors as low-level features. Though the latent factors are incomprehensible, every element within a dense vector can be treated as a low-level feature similar to the semantic ones like "ages", "locations", "education backgrounds", etc.

Some researches on semantic cases bind two or more low-level features together to generate a two-order feature (e.g. "age" + "salary"). However, it's hard to guarantee the independence between these features. The overlap on low-level features makes improvement insignificant (aged people tends to sit in higher positions with higher salaries). Thus to make sure an effective progress, extra works on feature engineering and adequate prior knowledge are required. For latent cases where latent factors are generated by embedding, the lookup table of embedding is full rank in the circumstance of big data (large amount of users and items), which means that the n_{factor} vectors of $1 \times |\mathbf{U}|$ in $\mathbf{P}$ or $1 \times |\mathbf{I}|$ in $\mathbf{Q}$, are linear independent. In another word, the n_{factor} latent factors have smaller overlaps and lower correlation. It makes the fusion of latent low-level features more significant.

3.2   Weighted Feature Interaction Network

In the feature modeling stage, we designed the Weighted Feature Interaction Network (WFIN) to generate the high-level feature. So far, the mainstream method of Recommender Systems is focusing on optimizing the fitting of user-item interaction function. Seldom of them consider the feature interaction. The commonly used architecture in recommender systems is a two-stage structure like the left graph in Fig. 5: the low-level latent features are fed into the prediction machine directly. Prediction Machine is defined as any model which aims to fit the interaction function between users and items. It could be MF or DNN, etc.

In DCF, we add a stage of feature interaction between the feature mapping and prediction machine (right in Fig. 5) to mining the relationships among low-level features. Two stacks of fully connected layers are set to model the high-level features of users and items respectively. The value of elements in user outputs indicates his interests towards the corresponding high-level latent feature. The value of elements in item outputs indicates how well it fits the corresponding high-level latent characteristics. The high-level feature vector of the output of WFIN is used by the following prediction machine together with the original low-level features.

Although we may never understand the meaning of low-level latent factors, to have a better understanding of what WFIN does, we introduce a comprehensive instance by analogizing semantic features. Usually people like a movie for a main reason, either the plot or the starring. This can be interpreted by the weights in WFIN. Figure 6 gives a simplified example of WFIN.

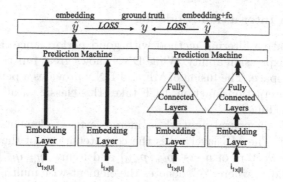

Fig. 5. The architectures of 2-stage model (left) and 3-stage DCF (right)

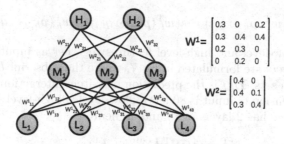

Fig. 6. A snapshot of simplified WFIN

The bottom of WFIN reads low-level features L like "plot", "starring", "publicity" and "public reviews", etc. Through the mid-level feature neurons M, we get the weights of the low-level features on high-level ones like "Soap opera", "Asian drama movie", "American action movie", etc. For this 3-layer WFIN, the contribution of L_i on H_j, noted by $P(L_i \mid H_j)$, is calculated by Eq. 5:

$$P(L_i \mid H_j) = \sigma(\sum_{k=1}^{M\#}(\mathbf{W}_{kj}^2\mathbf{W}_{ik}^1)) = \sigma(\mathbf{W}_{1j}^2\mathbf{W}_{i1}^1 + \mathbf{W}_{2j}^2\mathbf{W}_{i2}^1 + \mathbf{W}_{3j}^2\mathbf{W}_{i3}^1), \quad (5)$$

where σ denotes the softmax function for normalization. For this instance, $H_1 = 0.18L_1 + 0.48L_2 + 0.12L_3 + 0.08L_4$. The proportions of each low-level feature in H_1 are 21%, 56%, 14%, 9% after normalization. This demonstrates that this user likes H_1-type movies ("soap opera"), 56% depends on the feature L_2 ("starring").

The activation functions are identical functions in this instance for the ease of explanation. In practice, WFIN adopts ReLU. It's a 3-layer structure end up with 8 last-layer neurons in our model.

3.3 User-Item Interaction Modeling

MF is a collaborative filtering method with good performance. But the linear nature limits its expression ability. While DNN shows great function fitting ability on non-linear space. The fusion of MF and DNN provides a perfect model for fitting complex interaction function. DCF takes this classic parallel structure as the prediction machine.

To realize MF on neural network, a merge layer of element-wise multiplication is adopted. The MF layer takes the concatenations of the low-level and high-level feature vectors of users $[p_{l,u}, p_{h,u}]$ and items $[q_{l,i}, q_{h,i}]$ as inputs. It is calculated by Eq. 6, where "$\odot$" denotes the element-wise multipication operation of vectors. When the activation function $a(\cdot)$ is set to an element summation function and $\mathbf{w}$ is set to a vector of 1s, Eq. 6 is equal to Eq. 2.

$$o_{ui} = a(\boldsymbol{w}^{\mathrm{T}}(\boldsymbol{p}_u \odot \boldsymbol{q}_i)) = a(\boldsymbol{w}_l^{\mathrm{T}}(\boldsymbol{p}_{l,u} \odot \boldsymbol{q}_{l,i}) + \boldsymbol{w}_h^{\mathrm{T}}(\boldsymbol{p}_{h,u} \odot \boldsymbol{q}_{h,i})). \tag{6}$$

DNN takes low-level and high-level features together as inputs. The operations between layers are formulated in Eq. 7, where the subscript U, I stands for the user and item sets. $\hat{\boldsymbol{Y}}_{U,I}$ is the prediction user-item interaction matrix. Considering the dimension of input is 80 ($2\times$(32low-level+8high-level)), the DNN in prediction machine has 2 layers of 80 and 40 neurons.

$$\mathbf{X^1} = a^1(\mathbf{W^{T,1}X^0_{U,I}} + \boldsymbol{b^1}),$$
$$\mathbf{X^2} = a^2(\mathbf{W^{T,2}X^1_{U,I}} + \boldsymbol{b^2}),$$
$$\cdots \tag{7}$$
$$\mathbf{X^{l-1}} = a^{l-1}(\mathbf{W^{T,l-1}X^{l-2}_{U,I}} + \boldsymbol{b^{l-1}}),$$
$$\hat{\mathbf{Y}}_{U,I} = a^l(\mathbf{W^{T,l}X^{l-1}_{U,I}} + \boldsymbol{b^l}).$$

The fusion of MF layer and DNN is implemented by an 80-neuron layer and fed to one logistic loss function for joint training. Sigmoid is chosen as the activation function to give the possibility of a recommendation.

4 Experiments

4.1 Experimental Settings

Datasets. DCF is tested on two real-world datasets. *MovieLens* records 1,000,209 ratings from 6,040 users on 3,706 movies (Sparsity 95.53%). All the ratings are processed into 1 for implicit feedback. *Pinterest* records whether a user pined a image to his board, with 1,500,809 interactions between 55,187 users and 9,916 images (Sparsity 99.73%). All the statistics above is for the pre-processed datasets. Users with less than 20 interactions are eliminated. The last items which each user interacted with are retained as test data.

Evaluation Protocols. The baselines and DCF return a top-10 ranking list of prospective items for each user according to the possibility prediction. Two commonly-used evaluation protocols are adopted in our paper. *Hit Ratio*(HR) tells whether the last item is included in the top-k recommendation list. While *Normalized Discounted Cumulative Gain*(NDCG) assesses the ranking quality. If the test item appears on the top of the list, it will get a higher NDCG. The performance of a model is the averages of all users for both metrics.

Baselines. We choose MF, DNN, and NeuMF as the baselines. *MF* is a basic collaborative filtering recommender algorithm by SVD. *DNN* uses non-linear neural units to fit complex interaction function. The DNN baseline in this paper has a 5-layer structure with [64, 32, 16, 8, 1] neurons each layer. Embedding layers are put ahead for feature extraction. *NeuMF* is a classic parallel structure which shows state-of-the-art performance. It combines MF and DNN without considering high-level features. To control variables, all the baselines including DCF run on embedding factors of 32 in feature extraction and no regularization against over-fitting is done.

4.2 Performance

Table 1 shows the results of the experiments. DCF with WFIN outperforms the other baselines and reaches a state-of-the-art performance. The HR and NDCG shown in the table are the averages of the best results in 10 independent experiments of 100 iterations. We found that 100 iterations are enough to converge for all models.

Table 1. Comparison of HR and NDCG between DCF and baselines.

Models		MF	DNN	NeuMF	DCF
MovieLens	HR	0.6450	0.6738	0.6845	**0.6924**
	NDCG	0.3728	0.3981	0.4083	**0.4125**
Pinterest	HR	0.8673	0.8576	0.8694	**0.8696**
	NDCG	0.5426	0.5317	0.5450	**0.5480**

We chose the best-performance experiments for each model on two datasets and show them in Fig. 7. It's easy to find that DCF with WFIN performs better than the other three baselines significantly on MovieLens and also defeats them with a smaller gap on Pinterest. WFIN gives high-level features, which makes our proposed model outperform NeuMF, the state-of-the-art one. Faster decreasing on HR and NDCG is observed in Fig. 7(d) and (e) because of over-fitting. The integration of WFIN in DCF involves more parameters, which makes DCF easier to overfit than the others in the absence of regularization and dropout [17]. However, DCF converges faster and reaches a higher spot in the graph and lower loss than the others.

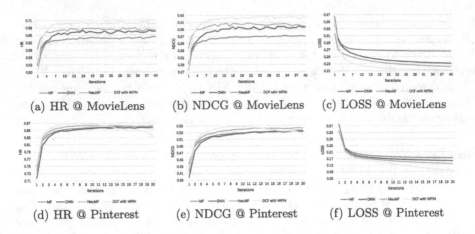

Fig. 7. Performance of DCF with WFIN and baselines

5   Related Work

Many models on explicit feedback and implicit feedback are generic [15,18–20], we mainly focus on the latter in this paper. For the recommendation tasks, the parallel structure of MF and DNN has achieved a strong performance [3,21–24]. MF fits a linear function offering memorization, while DNN learns an arbitrary function to provide generalization. The fusion of these two prediction models is more powerful and expressive.

Expert researchers have done some great jobs on high-level semantic features. Papers [2,3] combined age, the number of app installs, user demographics, device class, etc. into high-order features by particular network structures. Paper [25] proposed a deep knowledge-aware network that fuses semantic-level and knowledge-level as high-level representations on news. Papers [26,27] involved visual feature of aesthetic value or attractiveness. Paper [28] captured user reviews and extract sentiment information as a high-level feature in recommendations. All these researches improves the performance in their specific domain but fail to realize the general conception of high-level feature generation. Meanwhile, it takes strong dependence on manual feature selection and prior knowledge in specific domains.

In the absence of auxiliary semantic features, the common practice is to extract feature vectors from the user-item interaction matrix by LFM [29–31]. In this way, the users and items are vectorized by latent factors. But for most models, these features are fed into the prediction machine directly [22], which lost the information of feature interaction among the latent factors. By feature refactor and interaction, the high-level feature can be generated and make a difference.

6   Conclusions

In this paper, we improved the deep neural network structure for collaborative filtering (DCF) by integrating a weighted feature interaction network (WFIN). The introduction of high-level feature has an intuitive background and could be well interpreted theoretically. Experiments show that WFIN works and DCF with WFIN outperforms the other baselines, including a state-of-the-art model. While the structure of connecting DNN with embeddings is common, the idea of WFIN for high-level feature generation on latent factor model (LFM) is unique and novel and the performance is significantly improved.

In the future, regularization is expected to be introduced to enhance the performance of this model. There are no sufficient works on hyper-parameters, which means that DCF with WFIN still has room to get better. Besides, the DNN in user-item interaction modeling stage could be substituted by LSTM to deal with interaction time series and further provide an efficient online recommendation.

References

1. Zeiler, M.D., Fergus, R.: Visualizing and understanding convolutional networks. In: Fleet, D., Pajdla, T., Schiele, B., Tuytelaars, T. (eds.) ECCV 2014. LNCS, vol. 8689, pp. 818–833. Springer, Cham (2014). https://doi.org/10.1007/978-3-319-10590-1_53
2. Cheng, H., et al.: Wide & deep learning for recommender systems. In: Conference On Recommender Systems, pp. 7–10 (2016)
3. Guo, H., Tang, R., Ye, Y., Li, Z., He, X.: DeepFM: a factorization-machine based neural network for CTR prediction. In: International Joint Conference on Artificial Intelligence, pp. 1725–1731 (2017)
4. Juan, Y., Zhuang, Y., Chin, W.S., Lin, C.J.: Field-aware factorization machines for CTR prediction. In: ACM Conference on Recommender Systems, pp. 43–50 (2016)
5. Zhao, Z.D., Shang, M.S.: User-based collaborative-filtering recommendation algorithms on Hadoop. In: International Conference on Knowledge Discovery and Data Mining, pp. 478–481 (2010)
6. Font, F., Serra, J., Serra, X.: Class-based tag recommendation and user-based evaluation in online audio clip sharing. Knowl. Based Syst. **67**, 131–142 (2014)
7. Tung, W., Chen, Y.: User-based social ranking service design for tagging search and recommendation. J. Converg. Inf. Technol. **6**(10), 385–390 (2011)
8. Jamali, M., Ester, M.: TrustWalker : a random walk model for combining trust-based and item-based recommendation. In: ACM SIGKDD International Conference on Knowledge Discovery and Data Mining, pp. 397–406 (2009)
9. Sarwar, B., Karypis, G., Konstan, J., Riedl, J.: Item-based collaborative filtering recommendation algorithms. In: International Conference on World Wide Web, pp. 285–295 (2001)
10. Li, C., Luo, Z.: A hybrid item-based recommendation algorithm against segment attack in collaborative filtering systems. In: International Conference on Information Management, Innovation Management and Industrial Engineering, pp. 403–406 (2011)

11. Rendle, S.: Factorization machines with libFM. ACM Trans. Intell. Syst. Technol. **3**(3), 1–22 (2012)
12. Koren, Y.: Factorization meets the neighborhood: a multifaceted collaborative filtering model. In: ACM SIGKDD International Conference on Knowledge Discovery and Data Mining, pp. 426–434 (2008)
13. Vozalis, M.G., Margaritis, K.G.: Applying SVD on item-based filtering. In: Proceedings of the International Conference on Intelligent Systems Design and Applications, Isda 2005, pp. 464–469 (2006)
14. Bryt, O., Elad, M.: Compression of facial images using the K-SVD algorithm. Academic Press Inc (2008)
15. Rendle, S., Freudenthaler, C., Gantner, Z., Schmidtthieme, L.: BPR: Bayesian personalized ranking from implicit feedback. In: Uncertainty in Artificial Intelligence, pp. 452–461 (2009)
16. Wang, H., Wang, N., Yeung,D.Y.: Collaborative deep learning for recommender systems. In: Knowledge Discovery and Data Mining, pp. 1235–1244 (2015)
17. Srivastava, N., Hinton, G., Krizhevsky, A., Sutskever, I., Salakhutdinov, R.: Dropout: a simple way to prevent neural networks from overfitting. J. Mach. Learn. Res. **15**(1), 1929–1958 (2014)
18. Ebadi, A.: An intelligent hybrid multi-criteria hotel recommender system using explicit and implicit feedbacks. Ebadi Ashkan **9**, 1431–1441 (2016)
19. Tkalčič, M., Odić, A., Košir, A., Tasič, J.F.: Impact of implicit and explicit affective labeling on a recommender system's performance. In: Ardissono, L., Kuflik, T. (eds.) UMAP 2011. LNCS, vol. 7138, pp. 342–354. Springer, Heidelberg (2012). https://doi.org/10.1007/978-3-642-28509-7_32
20. Li, Q., Zheng, X.: Deep collaborative autoencoder for recommender systems: a unified framework for explicit and implicit feedback (2017). arXiv: Learning
21. Socher, R., Chen, D., Manning, C.D., Ng, A.Y.: Reasoning with neural tensor networks for knowledge base completion. In: International Conference on Neural Information Processing Systems, pp. 926–934 (2013)
22. He, X., Liao, L., Zhang, H., Nie, L., Hu, X., Chua, T.: Neural collaborative filtering. In: International World Wide Web Conferences, pp. 173–182 (2017)
23. He, X., Chua, T.S.: Neural factorization machines for sparse predictive analytics. In: The International ACM SIGIR Conference, pp. 355–364 (2017)
24. Xiao, J., Ye, H., He, X., Zhang, H., Wu, F., Chua, T.: Attentional factorization machines: learning the weight of feature interactions via attention networks. In: International Joint Conference on Artificial Intelligence, pp. 3119–3125 (2017)
25. Wang, H., Zhang, F., Xie, X., Guo,M.: Dkn: deep knowledge-aware network for news recommendation. In: International World Wide Web Conferences, pp. 1835–1844 (2018)
26. Yu, W., Zhang, H., He, X., Chen, X., Xiong, L., Qin, Z.: Aesthetic-based clothing recommendation. In: World Wide Web Conference, pp. 649–658 (2018)
27. Wang, Y., et al.: Telepath: understanding users from a human vision perspective in large-scale recommender systems. In: National Conference on Artificial Intelligence (2018)
28. Lu, Y., Dong, R., Smyth, B.: Coevolutionary recommendation model: mutual learning between ratings and reviews. In: World Wide Web Conference, pp. 773–782 (2018)
29. Cheng, Z., Ding, Y., Zhu, L., Kankanhalli, M.S.: Aspect-aware latent factor model: rating prediction with ratings and reviews. In: International World Wide Web Conferences, pp. 639–648 (2018)

30. Yao, X., Tan, B., Hu, C., Li, W., Xu, Z., Zhang, Z.: Recommend algorithm combined user-user neighborhood approach with latent factor model. In: International Conference on Mechatronics and Intelligent Robotics, pp. 275–280 (2017)

31. He, Z.: Personalized recommendation based on latent factor model and trust of users. Comput. Knowl. Technol. 4, 044 (2016)

Data Imputation of Wind Turbine Using Generative Adversarial Nets with Deep Learning Models

Fuming Qu[1], Jinhai Liu[1(✉)], Xiaowei Hong[1], and Yu Zhang[2]

[1] College of Information Science and Engineering, Northeastern University, Shenyang 110819, China
liujinhai@mail.neu.edu.cn
[2] Datang New Energy Experimental Research Institute, Beijing 100052, China

Abstract. Data missing problem is one of the most important issues in the field of wind turbine (WT). The missing data can lead to many problems that negatively affect the safety of power system and cause economic loss. However, under some complicated conditions, the WT data changes according to different environments, which would reduce the efficiency of some traditional data interpolation methods. In order to solve this problem and improve data interpolation accuracy, this paper proposed a WT data imputation method using generative adversarial nets (GAN) with deep learning models. First, conditional GAN is used as the framework to train the generative network. Then convolutional neural network is applied for both the generative model and the discriminative model. Through the zero-sum game between the two models, the imputation model can be well trained. Due to the deep learning models, the trained data imputation model can effectively recover the data with a few parameters of the input data. A case study based on real WT SCADA data was conducted to verify the proposed method. Two more data imputation methods were used to make the comparison. The experiments results showed that the method proposed in this paper is effective.

Keywords: Wind turbine · Data interpolation
Generative adversarial nets (GAN) · Deep learning

1 Introduction

Data missing problem is one of the most significant issues in the industrial field. Due to such problems as sensor faults, communication faults and data storage failure, there have been frequent occurrences of data missing. Complete data plays an important role in WT's operation, maintenance, and fault detection. In the field of wind turbine practice, some data (wind speed, for example) plays such an important role in WT operation and maintenance that the loss of such data may result in failure of power prediction, affect the normal operation of power grid and cause economic losses. Therefore, data imputation is essential in the daily operation in WT industry. Data imputation methods are developed to deal with the problem of data missing.

© Springer Nature Switzerland AG 2018
L. Cheng et al. (Eds.): ICONIP 2018, LNCS 11301, pp. 152–161, 2018.
https://doi.org/10.1007/978-3-030-04167-0_14

In general, the commonly-used method on data imputation is the model-based method, which finds probabilistic distributions of the data and then recovers the missing data from these distributions. In [1], a wavelet-based time hierarchical Bayesian models were build, which can be used to simultaneously model trend. Fabio Orianiae proposed the direct sampling multiple-point statistical technique as a non-parametric missing data simulator for hydrological flow rate time-series [2]. An auto-regressive and moving average (ARMA) method was used as the model to forecast the wind speed data [3].

With the development of artificial intelligence, data-driven methods for data interpolation, neural network (NN) and support vector regression (SVR) for instance, are often used. Zjavka used polynomial neural networks to forecast the wind speed and realize the data imputation [4]. In [5], Petković proposed the fractal interpolation method, which uses back-propagation neural world and extreme learning machine to compensate the wind speed data set. In [6], a hybrid wind speed forecasting model, which combined fast ensemble empirical mode decomposition, sample entropy, phase space reconstruction and back-propagation neural network with two hidden layers, was proposed to enhance the accuracy of wind speed imputation.

However, under some complicated conditions, the accurate of data imputation is limited. In WT field, the data is changing with different environments, in which case, the accuracy of traditional data imputation methods is limited. A simple model cannot deal with non-linear relationships. Therefore, a non-linear data imputation that can work effectively under complicated conditions is needed.

This paper proposed a data imputation method of WT using generative adversarial nets (GAN) [7] based on deep learning models. GAN is a kind of generative model, which has been successfully applied in generating realistic data in many fields [8, 9]. Based on the sound data generation capabilities of GAN, the data imputation of WT can be well conducted. First, the WT data is reasonably prepared for the model training. Then, deep learning networks are used in the architecture of GAN to deal with non-linear relationships. Finally, the trained generative model can be well applied in WT data imputation. A case study based on real WT data was conducted, and the experiment result indicated that the method proposed in this paper is effective.

The rest of this paper is organized as follows. Section 2 is the description of GAN-based WT data interpolation method. Experiments and comparisons are listed in Sect. 3. Section 4 is the conclusion.

2 GAN Based Data Interpolation Method

2.1 Generative Adversarial Nets

GAN [7] is a framework for estimating a generative model via an adversarial process, in which two models are simultaneously trained: a generative model G and a discriminative model D. The generative model G captures the data distribution and the discriminative model D estimates the probability that a sample came from the real data rather than from G. GAN can be trained to generate more realistic data via the zero-sum game. In other words, D and G play the following two-player min-max game with value function V (G, D):

$$\min_{G} \max_{D} V(D, G) = \min_{G} \max_{D} (E[\log D(x)] + E[\log(1 - D(G(z)))]) \tag{1}$$

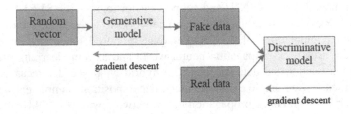

Fig. 1. Schematic diagram of the original GAN.

Figure 1 is the schematic diagram of the original GAN. First, a random vector is transformed into the fake data by the generative model. Then the real data is involved, and combined as the input of the discriminative model. Third, the discriminative model distinguishes the real data from the fake data. The parameters of the generative model and the discriminative model are updated by the gradient descent method in every epoch during the model training. Finally, after the model training, the generative model can be obtained.

2.2   Interpolation Method

However, in the case of data imputation, in order to impute the missing data correctly, the known data is used to recover the missing data, which is a kind of supervised learning. The original GAN is unsupervised learning, so it cannot process the data in a supervised way. Therefore, a supervised GAN is needed.

Conditional GAN is developed as a supervised learning, so that the generated data can be obtained in a supervised way. Conditional GAN is used as the framework of WT data imputation. Moreover, in order to get a better data imputation model, deep learning networks are used in the framework of GAN. In this paper convolutional neural network is chosen.

For the training data, the WT data should be prepared in a proper way to train the deep learning network. In this paper, other known WTs data is used to impute the missing data of a certain WT. The data organization flowchart is shown in Fig. 2.

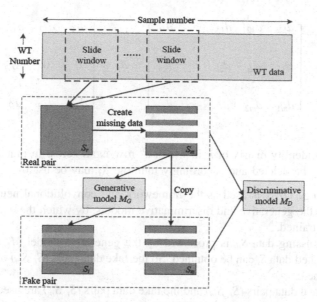

Fig. 2. Flowchart of WT data imputation.

Data Organization. The WT data is collected and organized as the training samples of the data imputation model.

First, in order to provide the training data with more useful information, the collected data from each WT are put together. Suppose d_{ij} is the collected data from i^{th} WT, where i is the WT number and j is the sample number. A training data can be organized as a matrix:

$$D = \begin{bmatrix} d_{11} & d_{12} & \cdots & d_{1n} \\ d_{21} & d_{22} & \cdots & \\ \vdots & \vdots & \vdots & \vdots \\ d_{m1} & d_{m2} & \cdots & d_{mn} \end{bmatrix} \tag{2}$$

where m is the number of WTs and n is the sample numbers of the data.

Second, a slide window δ is constructed to get every training sample. The height of δ is equal to m and the width of δ is equal to n. Then the data samples S_r can be obtained by moving δ from the beginning to the end of the WT data.

Third, the data S_r in some specified rows (the data of specified WTs) are eliminated. The eliminated data S_m are considered as the missing data. So the real data pairs (S_r, S_m) of the training data are obtained.

$$S_m = S_r \times R_i = \begin{bmatrix} d_{11} & d_{12} & d_{13} & \cdots & d_{1n} \\ \vdots & \vdots & \vdots & \vdots & \vdots \\ d_{1i} & d_{2i} & d_{3i} & \cdots & d_{ni} \\ \vdots & \vdots & \vdots & \vdots & \vdots \\ d_{m1} & d_{m2} & d3 & \cdots & d_{mn} \end{bmatrix} \times R_i = \begin{bmatrix} d_{11} & d_{12} & d_{13} & \cdots & d_{1n} \\ \vdots & \vdots & \vdots & \vdots & \vdots \\ 0 & 0 & 0 & 0 & 0 \\ \vdots & \vdots & \vdots & \vdots & \vdots \\ d_{m1} & d_{m2} & d3 & \cdots & d_{mn} \end{bmatrix}$$

$$(3)$$

where R_i is an identity matrix of n in which i^{th} row is all zeros. In practice, multiple rows of S_r may be deleted and accordingly multiple R_i may be used.

Model Training. GAN is used as the framework and convolutional neural networks are selected as the generative and discriminative models, by which the data imputation model can be trained.

First, the missing data S_m is processed by the generative model M_G of GAN, by which the imputed data S_i can be obtained. So the fake data pairs (S_i, S_m) of the training data are obtained.

Then, the real data pairs (S_r, S_m) and the fake data pairs (S_i, S_m) are used as the input of the discriminative model M_D of GAN. M_D tries to identify the real data pair from two data pairs. In order to make the imputed data close to the ground truth, a traditional loss is added in $L(M_G)$ and L1 distance is used to measure the distance of S_r and S_i. The loss function of M_D is:

$$L(M_D) = \mathrm{E}[\log M_D(S_r, S_m)] + \mathrm{E}[\log(1 - M_D(S_i, S_m))] \tag{4}$$

and the loss function of M_G is:

$$L(M_G) = \mathrm{E}[\log(1 - M_D(S_i, S_m))] + \lambda||S_r - S_i||_1 \tag{5}$$

Third, parameters of M_G are fixed, and parameters of M_D are updated according to the result. Accordingly, after the updating parameters of M_D, parameters of M_D are fixed, and the parameters of M_G are updated. The parameter updating of M_D is:

$$\theta^{(n+1)}(M_D) = \theta^{(n)}(M_D) - \eta \nabla_{M_D} L(M_D) \tag{6}$$

and parameter updating of M_G is:

$$\theta^{(n+1)}(M_G) = \theta^{(n)}(M_G) - \eta \nabla_{M_G} L(M_G) \tag{7}$$

where $\theta^{(n+1)}$ is the parameters of $(n+1)^{th}$ iteration of the training and $\theta^{(n+1)}$ is the parameters of nth iteration. ∇ is the gradient decent algorithm and η is the learning rate.

Through the zero-sum game of the two models, the data imputation model can be obtained.

Evaluation Indicators. In order to make a clear comparison, Mean Absolute Error (MAE), Mean Squared Error (MSE) and Normalized Root Mean Squared Error (NRMSE) are selected as the evaluation indicators of the effectiveness of interpolation.

$$MAE = \frac{1}{n} \sum_{t=1}^{n} |y_t - \bar{y}_t|^2 \tag{8}$$

$$MSE = \frac{1}{n} \sqrt{\sum_{t=1}^{n} (y_t - \bar{y}_t)^2} \tag{9}$$

$$NRMSE = \frac{\sqrt{\sum_{t=1}^{n} (y_t - \bar{y}_t)^2}}{\sum_{t=1}^{n} \bar{y}_t^2} \tag{10}$$

where y_t is the imputed data and $\bar{y}_t$ is the real data. n is the length of the missing data.

3 Case Study

In this section, the proposed method is verified by experiments with real WT data. The method proposed by this paper is compared with ARMA method and BPNN method.

3.1 Experiment Settings

A wind farm in northern China is selected for this study. This wind farm kept a good record of the supervisory control and data acquisition (SCADA) data in recent five years. The main components of the WT and its collected variables of the SCADA data are shown in Fig. 3. We collected the SCADA data from 34 WTs of the wind farm, and the wind speed is chosen as the interpolation target in this study.

In this study, the data of some normal WTs were used to impute the data of the data-missing WT (for example, suppose the wind speed data of WT06 is missing, and the wind speed data of other WTs is used to impute the data of WT 06). In order to test the interpolation methods, real missing data was not used in experiments. Instead, the complete wind speed data was used. The wind speed data of some certain WTs were eliminated artificially. Then the interpolation methods were used to recover the eliminated data.

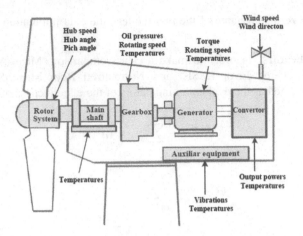

Fig. 3. Main components of WT and main variables of the collected SCADA data.

In this experiment, another two methods were used to make a comparison with the proposed method. The ARMA method was used as a time series method of the regression algorithm, and the back propagation neural network (BPNN) method was chosen as a neuron network method of the machine learning algorithm.

In order to test the data imputation effect of the method, an extreme experiment setting was used. The wind speed data of WT 03, WT 07, WT 011, WT 15, WT 19, WT 23, and WT 27 were kept and wind speed data of the other WTs were all eliminated. The three methods (ARMA method, BPNN method and the proposed method) were conducted to impute the missing data.

3.2 Comparative Experiments

In these experiments, the complete data and the eliminated data were combined as the training data. Another 300 eliminated data was used as the testing data. The data imputation results of WT 12 and WT 18 are shown in Figs. 4 and 5. The three subfigures respectively show the results of the ARMA method, BPNN method and the proposed method.

As can be seen, due to the influence of the environment, the wind speed changes dramatically. Moreover, there are fluctuations in the data. It increases the difficulty of data interpolation. From the data imputation result, the trend of the recovered data is basically correct with the ARMA method and BPNN method. However, in detail, due to data fluctuations and data changing in different environments, the data imputation

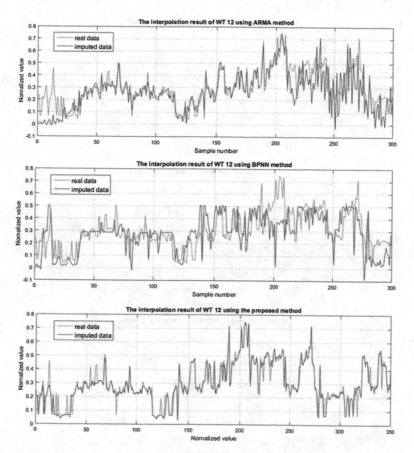

Fig. 4. Data imputation results of WT 12 using ARMA method, BPNN method and the proposed method.

effect is not so good. By contrast, given the same condition, the effect of the proposed method is more accurate in both the data trend and the detail. The result indicates that the proposed method can effectively impute WT wind speed data under such conditons.

The detailed data imputation results of WT 12, WT 18 and WT 24 are listed in Table 1. In can be seen that the proposed method has a good performance in all the three indicators of the data imputation.

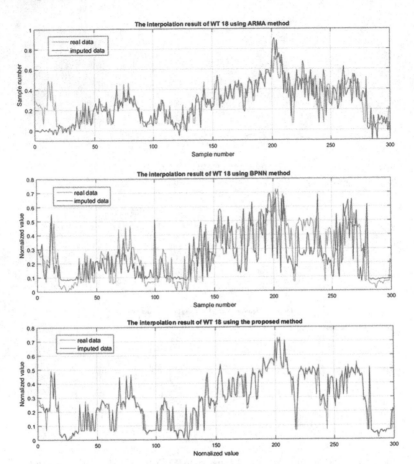

Fig. 5. Data imputation results of WT 18 using ARMA method, BPNN method and the proposed method.

Table 1. Data imputation results.

	ARMA	BPNN	The proposed method
MAE (WT 12)	0.0809	0.1035	0.0180
MSE (WT 12)	0.0084	0.0080	0.0023
NRMSE (WT 12)	0.0246	0.0263	0.0076
MAE (WT 18)	0.0801	0.1452	0.0177
MSE (WT 18)	0.0081	0.0106	0.0023
NRMSE (WT 18)	0.0253	0.0390	0.0084
MAE (WT 24)	0.0927	0.1002	0.0169
MSE (WT 24)	0.0089	0.0079	0.0022
NRMSE (WT 24)	0.0287	0.0277	0.0077

4 Conclusion

Data missing problem is one of the most significant study topics in the field of WT. The collected data is often lost due to connection faults, storage problems, software errors and other reasons. Data missing can lead to many problems, which not only threatens the security of the power system but also affect the economic benefits. In the field of WT, the collected data is changing in different environments, so it is difficult to impute the missing data with high accuracy. In order to solve this problem, this paper proposed a WT data imputation method based on GAN with deep learning models. GAN is used as the architecture, and the zero-sum game between the generative model and discriminative model can make the generator generating data more realistic. Then convolutional neural network is used for both the generative model and the discriminative model. Finally, a case study based on real WT data was conducted. Through the comparison using ARMA method and BPNN method, the effectiveness of the proposed method is proved.

Further research of this study will focus on the training data of the GAN. A good preparation of the training data can ensure the accuracy of the data imputation. More research about this topic will continue to be studied.

References

1. Craigmile, P.F., Guttorp, P.: Space-time modelling of trends in temperature series. J. Time Ser. Anal. **32**(4), 378–395 (2011)
2. Oriani, F., et al.: Missing data simulation inside flow rate time-series using multiple-point statistics. Environ. Model. Softw. **86**(C), 264–276 (2016)
3. Shao, Y., Sun, Y., Liang, L.: Wind speed short-term forecast for wind farms based on ARMA model. Power Syst. Clean Energy **24**(7), 52–55 (2008)
4. Zjavka, L.: Wind speed forecast correction models using polynomial neural networks. Renew. Energy **83**, 998–1006 (2015)
5. Petković, D., et al.: Estimation of fractal representation of wind speed fluctuation by artificial neural network with different training algorithms. Flow Meas. Instrum. **54**, 172–176 (2017)
6. Sun, W., Wang, Y.: Short-term wind speed forecasting based on fast ensemble empirical mode decomposition, phase space reconstruction, sample entropy and improved back-propagation neural network. Energy Convers. Manag. **157**, 1–12 (2018)
7. Goodfellow, I.J., et al.: Generative adversarial nets. In: International Conference on Neural Information Processing Systems, pp. 2672–2680. MIT Press (2014)
8. Chen, Y., et al.: Model-free renewable scenario generation using generative adversarial networks. IEEE Trans. Power Syst. **PP**(99), 1 (2017)
9. Li, J., et al.: WaterGAN: unsupervised generative network to enable real-time color correction of monocular underwater images, p. 99 (2017)

A Deep Ensemble Network
for Compressed Sensing MRI

Huafeng Wu, Yawen Wu, Liyan Sun, Congbo Cai, Yue Huang,
and Xinghao Ding[✉]

Fujian Key Laboratory of Sensing and Computing for Smart City,
School of Information Science and Engineering, Xiamen University,
Xiamen 361005, Fujian, China
dxh@xmu.edu.cn

Abstract. Compressed sensing theory has been proven to accelerate magnetic resonance imaging by measuring less K-space data called CS-MRI. Conventional sparse-optimization based CS-MRI methods lack enough capacity to encode rich patterns within the MR images and the iterative optimization for sparse recovery is often time-consuming. Although the deep convolutional neural network (CNN) models have achieved the state-of-the-art performance on CS-MRI reconstruction recently, the fine structure details can be degraded due to the information loss when the network goes deep. In order to better transfer the information in lossless way, we design *deep ensemble network* (DEN) architecture inspired by the novel interpretation of deep neural network in ensemble respect. The DEN model is formed by cascaded basic blocks for CS-MRI. Within the blocks, information flows forward through different depth. The intermediate outputs reconstructed by each block are merged via 3×3 convolution to generate the final reconstruction result. The experimental results show the proposed DEN model outperforms other state-of-the-art nondeep and deep CS-MRI models.

Keywords: Compressed sensing magnetic resonance imaging
Convolutional neural network · Ensemble learning

1 Introduction

Magnetic resonance imaging (MRI) is a popular medical imaging technique with the advantages of high resolution and low radiation. The major limitation of MRI is the low imaging speed. Therefore, Compressed sensing (CS) technique has been introduced for MRI acceleration by random sub-sampling of k-space (i.e., Fourier space) called compressed sensing magnetic resonance imaging (CS-MRI).

The CS-MRI is a classic inverse imaging problem and can be formulated as the following objective function

$$\hat{x} = \arg\min_{x} \parallel F_u x - y \parallel_2^2 + \sum_i \alpha_i \Psi_i(x). \tag{1}$$

© Springer Nature Switzerland AG 2018
L. Cheng et al. (Eds.): ICONIP 2018, LNCS 11301, pp. 162–171, 2018.
https://doi.org/10.1007/978-3-030-04167-0_15

where $x \in C^{N \times 1}$ is the complex-valued MR image to be reconstructed, $F_u \in C^{M \times N}$ is the under-sampled Fourier matrix and $y \in C^{M \times 1}$ are the k-space data measurements. The first data fidelity term ensures the consistency between the Fourier coefficients of the reconstructed image and the k-space measurements, while the second term called prior regularizes the reconstruction.

In conventional CS-MRI [1] methods, the sparse assumption is utilized to regularize the solution space of the ill-posed problem, while the model of sparseness is implemented in situ, leaving large amount of MRI data overlooked. In addition to this limitation, optimization for the sparse-regularized model is quite time-consuming because it requires tons of iterations, which makes the real-time application difficult. The deep learning models like convolutional neural network (CNN) have shown great advantages in extracting complex patterns within massive image databases. The forward reconstruction can be accomplished in high speed without iterations if the deep model is well-trained. Thus deep neural networks have been introduced into CS-MRI recently.

The deep neural network can extract the features efficiently while leaving fine structural details unprotected, which is more critical in the regression task. Thus more efficient way to transfer information in network needs to be studied. The interpretation of residual learning network in ensemble aspect indicates the ensemble learning strategy can be effectively incorporated into deep neural network to fuse the information flow from different depth of the network.

Motivated by the novel interpretation of residual learning network using ensemble learning theory, we propose a deep ensemble network (DEN) for CS-MRI, where the ensemble strategy is applied in both intra/inter block manner. The contributions of the paper can be summarized as follow:

- We propose a novel strategy to fuse low-layer features into high layers within each block of the proposed model, which makes information flow forward via different depth.
- DEN model is block-formed. We design a framework to merge the intermediate reconstruction of each block through convolution, which can lead to a well utilization of each intermediate reconstruction. We call it inter-block ensemble.
- The proposed DEN model achieves the state-of-the-art performance on two datasets in CS-MRI field with comparable number of parameters.

2   Related Works

2.1   Sparse-Optimization Methods

In the conventional work on CS-MRI, different sparse-based regularizations are proposed in CS-MRI pioneered by SparseMRI [2]. According to CS theory, the MR image can be represented sparsely in transform domains, which is conducive to reducing the required k-space measurement times and shortening the imaging time consumption. Since adaptive transform bases have stronger representation abilities, the variants of wavelet regularization are proposed based on geometric information like PBDW [3] and GBRWT [4]. Adaptive dictionary learning models have also been applied in CS-MRI such as DLMRI [5], BPTV [6] and TLMRI [7], which enable more flexible representations with adaptive sparse modelling. Generally, adaptive priors can capture more structures while non-adaptive ones has higher computational efficiency. The sparse prior can also be incorporated with other popular priors like non-local PANO [8].

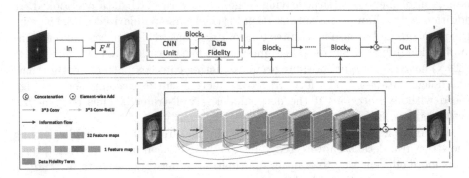

Fig. 1. The proposed Deep Ensemble Network (DEN) architecture. (Color figure online)

2.2 Deep Neural Network Models

Deep convolutional neural network (CNN) has shown great ability to model complex structures and patterns within images. In the recent researches, the deep neural network is also introduced in the field of CS-MRI. Wang et al. [9] firstly used a vanilla CNN model to learn the mapping from a zero-filled input MRI to the fully-sampled output MRI. Lee et al. [10] proposed a modified U-Net architecture to learn the mapping in the residual domain. [11] proposed variational network reconstructions to preserve the natural appearance of MR images. A deep cascade convolutional neural network (DC-CNN) proposed by Schlemper et al. [12] is currently the state-of-the-art CS-MRI inversion technique, unrolling the standard paradigm of CS-MRI into the deep learning architecture.

2.3 Ensemble Learning in Deep Learning

Breiman et al. [13] first well studied the idea of ensemble learning which combines predictors instead of a single predictor. Bagging [14] and boosting [15] are two widely used ensemble techniques. In Bagging, the bootstrap aggregation is utilized to reduce the variance for the strong learners, while in the boosting algorithm, the decisions are made by boosting the capacity of the weak learners.

Ensemble learning has also been widely used in neural networks explicitly and implicitly. As an early application of ensemble learning in neural network, the committee of neural networks are arranged in a simple voting scheme, where the final predictions are output in average in [16].

In [17], the residual networks can be interpreted as a collection of networks with different depth. Those paths representing different networks do not strongly depend on each other, even though they are trained jointly. This phenomenon corresponds to one key property of ensemble learning that the model performance changes smoothly with respect to the number of members. Larsson et al. [18] proposed the FractalNet model where the information goes through different paths in the network, can also be viewed as the ensemble strategy.

3   Method

Motivated by the ensemble learning strategy in deep neural network, we propose a *deep ensemble learning* model shown in Fig. 1 to integrate the information from different depth within the network. We first discuss single basic building block of the network where the intra-block ensemble strategy is utilized, then we will show how the blocks are cascaded and integrated into the final estimation of the reconstruction in inter-block fashion.

3.1   Intra-block Ensemble

The proposed DEN model is made up of basic cascaded building blocks and each block contains a CNN unit and a data fidelity unit.

In the CNN unit of each building block, we design a novel strategy to fuse information from different depth in the network without introducing much parameters. In Fig. 1, for a certain layer in the DEN model, all the features in this layer go through a 3×3 convolution as shown by the red arrow to produce a single feature map colored blue in the figure, which is concatenated to all the subsequent layers. Different from the DenseNet [19] model, the number of parameters of the DEN model only increases linearly instead of quadratically.

Shortcuts strategy was proposed to stabilize the gradient flow in deep residual networks (ResNet) [20]. In the VDSR [21] model, the Global Residual Learning (GRL) was introduced instead of local shortcut. We also adopt the same global residual learning in each CNN unit. As we can see the blue dashed block in Fig. 1, every five convolutional layers include a shortcut as a basic DEN block.

Besides the CNN unit, we also utilize the accurate measurements on the sampled positions in k-space to correct the distortions in the forward pass. We incorporate the data fidelity term into the network design by enforcing greater agreement at the sampled k-space positions with the sampled ones.

Similar to Eq. 1, we solve the following objective function with the data fidelity term

$$\hat{x} = \arg\min_{x} \frac{\lambda}{2} \parallel F_u x - y \parallel_2^2 + \parallel x - x_{in} \parallel_2^2 \tag{2}$$

where x_{in} is the input to the data fidelity term and λ is the regularization parameter. To ensure image consistency between the reconstructed MR image and the measurements, we set λ a large value, e.g. $1e6$. The second term can be viewed as the prior term, where the input image x_{in} is reconstructed by the deep neural network. This term is used in each block in DEN model.

3.2   Inter-block Ensemble

We cascade the basic building blocks and the intermediate reconstructions produced by blocks keeps improved as the network goes deeper. As we mentioned above, the reconstructed MR images of deeper blocks suffer information loss in fine details although with a better reconstruction accuracy. Being Motivated by the ensemble idea, we integrate the intermediate reconstruction of each block to produce the final reconstruction result. The reconstruction of each block is aggregated by concatenation operations. Then the final reconstruction is obtained from the fusion of reconstructions which goes through a 3×3 convolution.

4   Experiments

In this section, we present the experimental results using the training data consisting of 2800 normalized real-value brain MRI. And the testing data consists of 50 brain MRI. We collect these data using previously available data from a 3T MR scanner at Xiamen University. All images are T2 weighted MRI and of size 256×256. Informed consent was obtained from the imaging subject in compliance with the Institutional Review Board policy. Under-sampled k-space measurements are manually obtained via 30% 1D Cartesian sampling mask with random phase encodes as shown in Fig. 2(h).

4.1   Experimental Setup

We train and test the algorithm using Tensorflow [22] for the Python environment on a NVIDIA GeForce GTX 1080 with 8 GB GPU memory. Padding is applied to keep the size of features the same. We use the Xavier method [23] to initialize the network parameters, and we apply ADAM [24] with momentum. The implementation uses the initialized learning rate 0.0005, first-order momentum 0.9 and second momentum 0.99. The weight decay regularization parameter is set to 0.0005. The size of training batch is 4. We report our performance after 24000 training iteration.

4.2   Results

We evaluate the proposed DEN framework using PSNR and SSIM [25] as quantitative image quality assessment measures.

We compare the testing results between Baseline (DC-CNN) and DEN with or without intra/inter-block ensemble, shows in Table 1. Both DEN and DC-CNN are implemented with 3 blocks for convenience, each block contains 5 convolution layers with ReLU activation. We decrease the basic feature maps (the feature

Table 1. Comparison of baseline and DEN with/without intra-block/inter-block ensemble.

Indicators	Method			
	Baseline	Baseline with intra-block	Baseline with inter-block	DEN
PSNR/SSIM	37.67/0.954	37.98/0.958	38.15/0.963	**38.51/0.965**
Param.	**85059**	85620	85087	85648

maps without concatenation) linearly within each block in DEN to keep the total number of network parameters close to DC-CNN for fair comparison. Both intra-block and inter-block ensemble improve performance, while the latter one improves more with fewer parameters. The DEN model increases PSNR by nearly 1 dB.

We compare testing results with other algorithms, PANO, PBDW, TLMRI, GBRWT and DC-CNN. We show the reconstruction results and the corresponding error images of an example from the test data in Fig. 2. A quantitative result is shown in Fig. 3. As is clear in Figs. 2 and 3, DC-CNN and DEN both have the

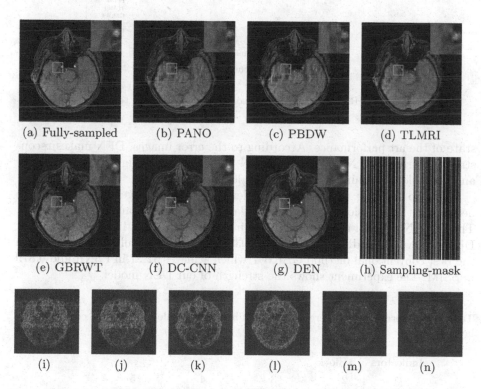

(a) Fully-sampled   (b) PANO   (c) PBDW   (d) TLMRI

(e) GBRWT   (f) DC-CNN   (g) DEN   (h) Sampling-mask

(i)   (j)   (k)   (l)   (m)   (n)

Fig. 2. (b)–(g) is the reconstruction results with local area magnification. (i)–(n) Is the reconstruction error of the above method in order.

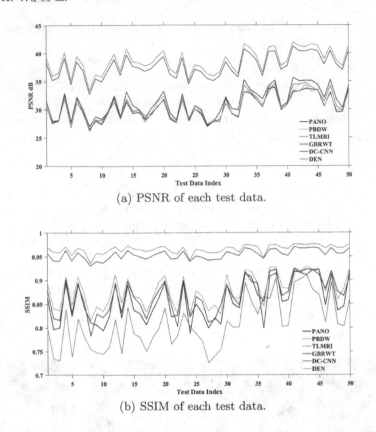

(a) PSNR of each test data.

(b) SSIM of each test data.

Fig. 3. The PSNR (a) and SSIM (b) comparison on the 50 test brain MRI.

state of the art performance. According to the error images, DEN makes reconstruction error less. Note that DC-CNN and DEN are designed with 5 blocks and each block contains 5 convolutional layers.

We also test our model with complex-valued MR image by designing the network with dimensional space twice as large that uses the same strategy as [12]. The DC-CNN model achieves 34.99 dB in PSNR and 0.934 in SSIM, while our DEN achieves 35.36 dB in PSNR and 0.938 in SSIM, the results still outperform the best conventional method GBRWT which achieves 32.27 in PSNR and 0.879 in SSIM. The experiment shows the strength of our DEN model.

Table 2. Performance of different number of blocks, each block contains 5 convolutional layers.

Indicators	Block			
	2	3	4	5
PSNR/SSIM	37.15/0.955	38.90/0.968	39.54/0.970	**39.96/0.974**

Table 3. Performance of different number of convolutional layers within each block in 3-blocks DEN.

Indicators	Convolution			
	4	5	6	7
PSNR/SSIM	38.18/0.963	38.90/0.968	39.03/0.968	**39.04/0.969**

4.3 Discussion on Hyperparameter

Tables 2 and 3 discuss the performance of the number of blocks and convolutional layers separately. As we can see, the performance of the DEN model improves as the number of blocks and convolutional layers increases.

Table 4. Performance on BRATS2015 datasets

Indicators	Methods					
	TLMRI	PANO	GBRWT	PBDW	DCCNN	DEN
PSNR	36.44	40.43	41.45	41.25	42.55	**43.01**
SSIM	0.879	38.15	0.963	0.962	0.970	**0.981**

4.4 Experiments on Another Dataset

Besides the dataset mentioned above, we also test the proposed DEN model on another widely used MRI datasets for MRI brain tumor segmentation (BRATS2015 datasets) which contains 220 high-grade glima (HGG) and 54 low-grade (LGG) patient scans. Only the T_2-weighted MRI is used. We randomly choose 37 high-grade glima (HGG) patient scans for testing the model performance. We use the remaining 183 high-grade glima (HGG) patient scans for training. The averaged objective results are shown in Table 4, we choose 30% 1D Cartesian sampling mask and the deep learning based model contains 5 blocks, each block has 5 convolutional layers.

The extra experiments on the BRATS2015 benchmark validate the state-of-the-art performance of the proposed DEN model on the clinic MRI data with lesions. The DEN model can provide more reliable in the scenarios where the compressed sensing is utilized while the reconstruction accuracy matters.

Table 5. The comparison in runtime (seconds) between different methods.

	TLMRI	GBRWT	PANO	PBDW	DCCNN	DEN
Runtime	127.67	100.60	11.37	68.9	0.03	0.04

4.5   Running Time

We compare the running time of different methods in Table 5. CS-MRI models based on sparse or non-local regularization requires a large number of iterations, resulting slow reconstruction speed. The proposed DEN and DC-CNN model both have 5 blocks and each block contains 5 convolutional layer. Although the running speed of DEN is slower than the other deep-based CS-MRI models, it achieves the state-of-the-art reconstruction accuracy, providing a good balance between running time and reconstruction quality.

5   Conclusion

We have proposed a deep ensemble network for the CS-MRI inversion problem. The network is formed by a series of basic blocks designed to merge the intermediate reconstruction for fully utilizing the merits of each one. Within each block, the CNN unit integrates information from different depths of the network to reduce the loss of information. Experimental results show that the proposed DEN model achieves state-of-the-art performance for CS-MRI while keeping number of free parameter comparable with plain CNN network.

References

1. Lustig, M., Donoho, D.L., Santos, J.M., Pauly, J.M.: Compressed sensing MRI. IEEE Sig. Process. Mag. **25**(2), 72–82 (2008)
2. Lustig, M., Donoho, D., Pauly, J.M.: Sparse MRI: the application of compressed sensing for rapid MR imaging. Magn. Reson. Med.: Official J. Int. Soc. Magn. Reson. Med. **58**(6), 1182–1195 (2007)
3. Qu, X., et al.: Undersampled MRI reconstruction with patch-based directional wavelets. Magn. Reson. Imaging **30**(7), 964–977 (2012)
4. Lai, Z., et al.: Image reconstruction of compressed sensing MRI using graph-based redundant wavelet transform. Med. Image Anal. **27**, 93–104 (2016)
5. Ravishankar, S.: Magnetic resonance image reconstruction from highly undersampled K-space data using dictionary learning (2011)
6. Huang, Y., Paisley, J., Lin, Q., Ding, X., Fu, X., Zhang, X.P.: Bayesian nonparametric dictionary learning for compressed sensing MRI. IEEE Trans. Image Process. **23**(12), 5007–5019 (2014)
7. Ravishankar, S., Bresler, Y.: Sparsifying transform learning for compressed sensing MRI. In: 2013 IEEE 10th International Symposium on Biomedical Imaging (ISBI), pp. 17–20. IEEE (2013)
8. Qu, X., Hou, Y., Lam, F., Guo, D., Zhong, J., Chen, Z.: Magnetic resonance image reconstruction from undersampled measurements using a patch-based non-local operator. Med. Image Anal. **18**(6), 843–856 (2014)
9. Wang, S., et al.: Accelerating magnetic resonance imaging via deep learning. In: 2016 IEEE 13th International Symposium on Biomedical Imaging (ISBI), pp. 514–517. IEEE (2016)
10. Lee, D., Yoo, J., Ye, J.C.: Deep residual learning for compressed sensing MRI. In: 2017 IEEE 14th International Symposium on Biomedical Imaging (ISBI 2017), pp. 15–18. IEEE (2017)

11. Hammernik, K., et al.: Learning a variational network for reconstruction of accelerated MRI data. Magn. Reson. Med. **79**(6), 3055–3071 (2018)
12. Schlemper, J., Caballero, J., Hajnal, J.V., Price, A., Rueckert, D.: A deep cascade of convolutional neural networks for MR image reconstruction. arXiv preprint arXiv:1703.00555 (2017)
13. Breiman, L.: Stacked regressions. Mach. Learn. **24**(1), 49–64 (1996)
14. Breiman, L.: Bagging predictors. Mach. Learn. **24**(2), 123–140 (1996)
15. Freund, Y., Schapire, R.E., et al.: Experiments with a new boosting algorithm. In: ICML, vol. 96, pp. 148–156. Citeseer (1996)
16. Drucker, H., Cortes, C., Jackel, L.D., LeCun, Y., Vapnik, V.: Boosting and other ensemble methods. Neural Comput. **6**(6), 1289–1301 (1994)
17. Veit, A., Wilber, M.J., Belongie, S.: Residual networks behave like ensembles of relatively shallow networks. In: Advances in Neural Information Processing Systems, pp. 550–558 (2016)
18. Larsson, G., Maire, M., Shakhnarovich, G.: FractalNet: ultra-deep neural networks without residuals. arXiv preprint arXiv:1605.07648 (2016)
19. Huang, G., Liu, Z., Van Der Maaten, L., Weinberger, K.Q.: Densely connected convolutional networks. In: CVPR, vol. 1, p. 3 (2017)
20. He, K., Zhang, X., Ren, S., Sun, J.: Deep residual learning for image recognition. In: Proceedings of the IEEE Conference on Computer Vision and Pattern Recognition, pp. 770–778 (2016)
21. Kim, J., Kwon Lee, J., Mu Lee, K.: Accurate image super-resolution using very deep convolutional networks. In: Proceedings of the IEEE Conference on Computer Vision and Pattern Recognition, pp. 1646–1654 (2016)
22. Abadi, M., et al.: TensorFlow: a system for large-scale machine learning. In: OSDI, vol. 16, pp. 265–283 (2016)
23. Glorot, X., Bengio, Y.: Understanding the difficulty of training deep feedforward neural networks. In: Proceedings of the Thirteenth International Conference on Artificial Intelligence and Statistics, pp. 249–256 (2010)
24. Kingma, D.P., Ba, J.: Adam: a method for stochastic optimization. arXiv preprint arXiv:1412.6980 (2014)
25. Wang, Z., Bovik, A.C., Sheikh, H.R., Simoncelli, E.P.: Image quality assessment: from error visibility to structural similarity. IEEE Trans. Image Process. **13**(4), 600–612 (2004)

Deep Imitation Learning: The Impact of Depth on Policy Performance

Parham M. Kebria(✉), Abbas Khosravi, Syed Moshfeq Salaken, Ibrahim Hossain, H. M. Dipu Kabir, Afsaneh Koohestani, Roohallah Alizadehsani, and Saeid Nahavandi

Institute for Intelligent Systems Research and Innovation, Deakin University, Geelong, VIC, Australia
{kebria, abbas.khosravi, syed.salaken, ihossai, dkabir, akoohest, ralizadehsani}@deakin.edu.au

Abstract. This paper investigates the impact of network depth on the performance of imitation learning applied in the development of an end-to-end policy for controlling autonomous cars. The policy generates optimal steering commands from raw images taken from cameras attached to the car in a simulated environment. A convolutional neural network (CNN) is used to find the mapping between inputs (car images) and the desired steering angle. The CNN architecture is modified by changing the number of convolutional layers as well as the filter size. It is observed that the learned policy is capable of driving the car in the autonomous mode purely using visual information. In addition, simulation results indicate that deeper CNNs outperform shallower CNNs for learning and mimicking the human driver's behavior. Surprisingly, the best performance is not achieved by the most complex CNN.

Keywords: Autonomous vehicle · Imitation learning · Simulation
Depth

1 Introduction

Imitation learning has become a hot research topic in the field of robotics and autonomous systems [19]. Also known as *behavioral cloning, apprenticeship learning, or learning from demonstrations*, it offers a framework for acquisition of skills or knowledge from observing a task performed by a human [2]. The agent, robot, or the autonomous system will mimic the same behavior it has observed during demonstration [8]. While its inspiration and basis have stemmed from the field of neuroscience, it has now become a main stream in the field of artificial intelligence. The growth in popularity of imitation learning is due to the increasing demand for agents capable of making intelligent decision and performing realistic actions.

In contrast to imitation learning, reinforcement learning does not require supervision during training. It tries to find the best policy leading to the highest reward through nonstop interactions with the environment. This reward could be the lowest lap time, the lowest energy loss, or the highest game score [28].

© Springer Nature Switzerland AG 2018
L. Cheng et al. (Eds.): ICONIP 2018, LNCS 11301, pp. 172–181, 2018.
https://doi.org/10.1007/978-3-030-04167-0_16

Simplicity and flexibility are promising features of reinforcement learning. However, defining a suitable reward function for enabling agents to learn the desired behavior could be quite challenging [20]. Designing a feasible reward function is even more difficult for novel applications such as high speed racing of agents [15]. Another challenging application is defining a generalizable function capable of proper mapping of perceptual inputs to the cost [10–13, 21, 24, 28]. Last but not least, reinforcement learning is a quite time consuming process which may continue for several weeks. Such a thing is not practical for real world applications where long and repeated experiments as well as agent failure could not be afforded. These issues have contributed to the popularity of imitation learning in different fields, in particular robotics and autonomous systems.

A generic pipeline for imitation learning is shown in Fig. 1. The first step is the data collection which is done through conducting some demonstrations. Recoded data may include information about sensory measurements, environment, and the state of the learner and demonstrator (performer). This data is then processed by a machine to develop a policy for imitating the demonstrated task or behavior. This policy development could be done using raw or extracted features and using a variety of machine learning techniques. Policy creation is often done using a *mapping function*. This function directly approximates the mapping from the agent's state observation to executed actions [2]. A classification or regression policy will be required depending on the output type (class label or a continuous value). The learned policy could then be fine-tuned through inter-actions of the agent with the environment. Policy refinement could be achieved using active learning, reinforcement learning, transfer learning, optimization, and apprenticeship learning [8].

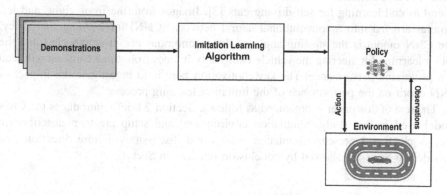

Fig. 1. Imitation learning pipeline.

Some applications of imitation learning have been shown in Fig. 2. These include games [23], vehicle control [9], unmanned aerial vehicles (UAV) flying [6, 16], robotics [27, 29], and autonomous systems [4, 18, 25, 26]. In all those applications, the key requirement of imitation learning is the availability of high quality demonstration [29]. Such records can be obtained from a human driving a car or controlling a UAV [1, 3, 15].

The availability of perfect interfaces and ease of control make it possible to collect data for different scenarios which can be later used for the development of a reliable model.

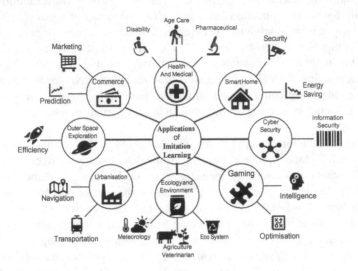

Fig. 2. Imitation learning applications in different fields.

The purpose of this paper is to investigate the impact of deep neural network structure on the performance of imitation learning. The case study here is the development of an autonomous car using a simulation environment. This is based on the idea of end to end learning for self-driving cars [3]. Images from the front, right, and left cameras are fed into a convolutional neural network (CNN) for learning the policy. The CNN output is the steering angle of the autonomous car. It is expected that the policy learning for steering the vehicle using only images from three cameras will lead to a promising performance. The key motivation here is to investigate the impact of CNN depth on the performance of the imitation learning process.

The rest of this paper is organized as follows. Section 2 briefly introduces the CNN model used in this study. Simulation environment and setup are then described in Sect. 3. Section 4 presents simulation results and discussions. Future directions are provided in Sect. 5, followed by conclusion remarks in Sect. 6.

2 Convolutional Neural Network

CNNs are a member of deep neural network family that have been widely used for computer vision and natural language processing in recent years. Their popularity has massively increased since 2012 when AlexNet won the ImageNet large scale visual recognition challenge [14]. It achieved a prediction error of 15.3%, which was more than 10.8% points ahead of the second best solution. The promising performance of CNN for image classification in that competition has been one of the main reasons for the rapid growth of deep learning popularity in the field of machine learning.

Two key characteristics of CNNs are sparse connectivity and shared weights. Deep CNNs employ spatially local correlations by enforcing local connectivity patterns between neurons of adjacent layers. This simply means that the inputs of neurons in layer i are from a subset of units with spatially contiguous receptive fields in layer $i - 1$. Also each filter in a CNN is applied across the entire field. These are enabler for CNN to optimally capture and present local structural information in its inputs. Figure 3 shows the structure of a typical CNN used for the task of image classification. There are four main operations in a typical CNN:

- Convolution: This performs the convolution operation on its inputs. It simply applies a sliding window to recognize particular patterns in different locations. This convolution operation eliminates the need for generating hand crafted features required in traditional machine learning approaches.
- Pooling: This block applies simple max or mean functions to the output of the convolution block. This is performed to reduce the dimensionality of intermediate representations and increase the prediction robustness.
- Activation (nonlinearity): The pooling component is then followed by an activation function. This nonlinearity is key to CNN learning performance.
- Classification (dense layer): The last stage of a CNN is a fully connected multi layer neural network. Neurons in its first layer take input values from every feature map in the final convolution layer. The output of this dense network is the CNN output.

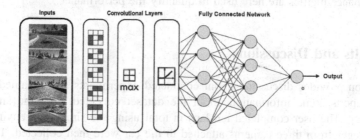

Fig. 3. The structure of a convolutional neural network

The first three blocks together form a convolutional layer. A deep CNN could be built by sequentially adding many convolutional layer. How to determine the optimal number of convolutional layers is more of art than science.

More background information about CNN and their training process using gradient decent method could be found in [5] and [7].

3 Experimental Setup

The simulation environment is a car racing track built using an open source game engine. Multiple snapshots of this simulator taken by different cameras are shown in Fig. 4. This simulation environment provides capabilities for dataset generation for offline supervised model training.

Fig. 4. The simulation environment for data collection and testing the developed policy.

Steps taken during simulation setup and experiments are as follows:

- Use the simulator to collect data: As mentioned in Sect. 1, imitation learning requires data. Here the user drives the car and completes multiple laps. If the user is happy with the driving performance, images from front, left, and right cameras are time-stamped and recorded. Also the steering angle, as the desired output, is saved along those images.
- Build a CNN model: A CNN model to predict the steering angle from three cameras' images is then developed. Gradient descent algorithm is used to tune CNN parameters. This CNN will be used as the policy in the imitation learning pipeline. This is similar to the concept of pixel to action originally proposed in [17] for mastering Atari games.
- Test the model: Once the CNN model is developed, its performance is examined to see how good it generates steering angles similar to the desired ones for the test set. Performance metrics are here used to quantify the performance.

4   Results and Discussions

This section provides discussion about conducted experiments and obtained results. Table 1 reports some information about the dataset collected from the simulation environment. The user completed ten laps in total using the simulation environment. 12,871 images from three cameras attached to the car were then collected. The resolution of these images are 320×160 pixels. This set of images is then split into train (80%) and test (20%) sets. Different metrics including total loss, root mean squared error (RMSE), and mean absolute percentage error (MAPE) are measured for quantifying the imitation learning performance for mimicking the driver's behavior.

Table 1. Datasets used for imitation learning

Item	Description
Number of completed laps	10
Resolution of images	320×160
Total number of images	12871
Number of images used for training	10293 (80%)
Number of images used for testing	2578 (20%)

Information about CNN models used for imitation learning is summarized in Table 2. Model 1 is the simplest model with only one convolutional layer. Model 5 has six convolutional layers and is the most complex one. The fully connected network is the same in all five models with a structure of 10050x10x1. The output of this network is the predicted steering angle based on three images fed to the convolutional layers. The number of epochs and batch size for training are set to 10 and 40, respectively. The learning rate is also 0.0001 in the gradient descent algorithm.

Table 2. The architecture of CNN models used for imitation learning

Model	Convolutional layers	Dense layer
Model 1	24@5x5, 64x@3x3	100x50x10x1
Model 2	24@5x5, 36@5x5, 64x@3x3	100x50x10x1
Model 3	24@5x5, 36@5x5, 48@5x5, 64x@3x3	100x350x310x31
Model 4	24@5x5, 36@5x5, 48@5x5, 64@3x3, 80x@3x3	100x350x310x31
Model 5	24@5x35, 36@5x35, 48@5x35, 64@3x33, 80@3x33, 96@3x33	100x350x310x31
Model 6	24@5x35, 48@5x35, 96@8x38, 64@5x35, 48@3x33	100x350x310x31
Model 7	24@5x35, 36@5x35, 48@5x35, 64@3x33, 64@3x33	100x350x310x31

The profiles of loss function during training for all models are shown in Fig. 5. The best loss function has been achieved by models 1 and 2 within ten epochs. This is because these two models have the least number of parameters, so ten epochs are enough for their training. Also model 5, the most complex model, has converged well after 8 epochs. The same pattern is observed for other models as well. These profiles indicate that ten epochs for training these policy models based on provided visual information has been sufficient, as trained models have shown a satisfactory performance.

Table 3 reports the performance metrics for 7 models applied to the test set. According to these stats, the best performing models are model 4 (evaluated based on loss metric), model 7 (evaluated based on RMSE metric) and model 6 (evaluated based on MAPE metric). The best performing CNNs for learning the driver's behavior are those with at least five convolutional layers. The minimum RMSE is achieved by model 7 with five convolutional layers. The best MAPE is also achieved by model 6 whose convolutional layer complexity gradually decreases. Accordingly, we may conclude that policies developed using deeper CNNs outperform policies developed using shallower CNNs. Of course, this comes with the cost of more computational requirements for policy development. The good news is that online prediction of steering commands is super fast, so more complexity does not cause any issue.

Although we concluded that more depth has a positive impact on the learned policy performance, this positive impact is capped. It is important to note model 5 which is the most complicated network is not the best model based on considered evaluation metrics. This simply indicates that no benefit is gained by increasing the network complexity for this case of imitation learning. Ample care should be exercised to fine tune the network architecture to maximize the policy performance while minimizing

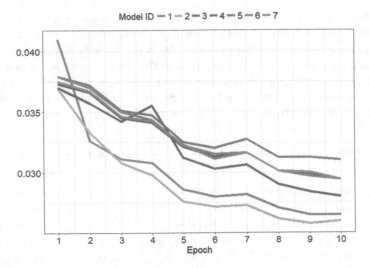

Fig. 5. The profiles of loss function during training for all models.

Table 3. The performance metrics calculated for the test set (autonomous driving)

Model	Loss	RMSE	MAPE
Model 1	0.043	0.067	602.670
Model 2	0.043	0.067	603.220
Model 3	0.042	0.067	626.945
Model 4	**0.040**	0.070	673.917
Model 5	0.041	0.068	625.764
Model 6	0.043	0.068	**580.332**
Model 7	0.042	**0.066**	603.017

resources required for policy development. Accordingly, it may be required to use deep neuroevolution algorithms to optimize the architecture of the CNN model used as policy [16, 22].

Some snapshots of the vehicle driving in autonomous mode using policy 6 are shown in Fig. 6. In the autonomous mode, obtained policy using imitation learning is able to successfully drive the vehicle in different sections of the track. Policies obtained throughout experiments demonstrated varying degrees of success in autonomous driving mode. Their success positively correlates to the policy performance and its convergence profile.

Fig. 6. Autonomous driving using trained models

5 Future Works

The scope of this study will be further extended in future by inclusion of more complicated simulation environment as well as data collection from multiple users. Also it is planned to develop policies using CNNs with more than ten convolutional layers.

6 Conclusion

This paper investigated the problem of imitation learning for a simulated autonomous car. The purpose was to quantify the impact of depth on performance of the developed policy estimator. The data required for imitation learning was collected by completing multiple laps by the user. This data include images from three cameras attached to the front, left, and right side of the car as well as the steering angle. A convolutional neural network is then trained to optimally find the mapping between scene images and the steering angles. Multiple networks with different structure and learning capacity were implemented to quantify depth impact on network performance. The study shows, in general, increasing the depth improves the performance of imitation learning policy.

References

1. Abbeel, P., Coates, A., Ng, A.Y.: Autonomous helicopter aerobatics through apprenticeship learning. Int. J. Robot. Res. **29**(13), 1608–1639 (2010). https://doi.org/10.1177/0278364910371999
2. Argall, B.D., Chernova, S., Veloso, M., Browning, B.: A survey of robot learning from demonstration. Robot. Auton. Syst. **57**(5), 469–483 (2009)
3. Bojarski, M., et al.: End to end learning for self-driving cars (2016)
4. Cardamone, L., Loiacono, D., Lanzi, P.L.: Learning drivers for TORCS through imitation using supervised methods. In: 2009 IEEE Symposium on Computational Intelligence and Games, pp. 148–155 (7–10)
5. Deng, L., Yu, D.: Deep learning: methods and applications. Found. Trends Sig. Process. **7**, 197–387 (2014)
6. Gandhi, D., Pinto, L., Gupta, A.: Learning to fly by crashing (2017)
7. Goodfellow, I., Bengio, Y., Courville, A.: Deep Learning. The MIT Press, Cambridge (2016)
8. Hussein, A., Gaber, M.M., Elyan, E., Jayne, C.: Imitation learning: a survey of learning methods. ACM Comput. Surv. **50**(2), 1–35 (2017). https://doi.org/10.1145/3054912

9. Innocenti, C., Lindn, H., Panahandeh, G., Svensson, L., Mohammadiha, N.: Imitation learning for vision-based lane keeping assistance (2017)
10. Khatami, A., Babaie, M., Khosravi, A., Tizhoosh, H., Nahavandi, S.: Parallel deep solutions for image retrieval from imbalanced medical imaging archives. Appl. Soft Comput. **63**, 197–205 (2018)
11. Khatami, A., Khosravi, A., Nguyen, T., Lim, C.P., Nahavandi, S.: Medical image analysis using wavelet transform and deep belief networks. Expert Syst. Appl. **86**, 190–198 (2017)
12. Khosravi, A., Nahavandi, S., Creighton, D.: A neural network-GARCH-based method for construction of prediction intervals. Electr. Power Syst. Res. **96**, 185–193 (2013)
13. Khosravi, A., Nahavandi, S., Creighton, D.: Quantifying uncertainties of neural network-based electricity price forecasts. Appl. Energy **112**, 120–129 (2013)
14. Krizhevsky, A., Sutskever, I., Hinton, G.E.: Imagenet classification with deep convolutional neural networks. In: Proceedings of the 25th International Conference on Neural Information Processing Systems, vol. 1, pp. 1097–1105. Curran Associates Inc., Lake Tahoe (2012)
15. Li, G., Mueller, M., Casser, V., Smith, N., Michels, D.L., Ghanem, B.: Teaching UAVs to race with observational imitation learning (2018)
16. Müller, M., Casser, V., Smith, N., Michels, D.L., Ghanem, B.: Teaching UAVs to race using Sim4CV (2017)
17. Mnih, V., et al.: Human-level control through deep reinforcement learning. Nature **518**, 529 (2015)
18. Nahavandi, S.: Trusted autonomy between humans and robots: toward human-on-the-loop in robotics and autonomous systems. IEEE Syst. Man, Cybern. Mag. **3**(1), 10–17 (2017)
19. Narayanan, K.K., Posada, L.F., Hoffmann, F., Bertram, T.: Robot programming by demonstration. In: Ando, N., Balakirsky, S., Hemker, T., Reggiani, M., von Stryk, O. (eds.) SIMPAR 2010. LNCS (LNAI), vol. 6472, pp. 288–299. Springer, Heidelberg (2010). https://doi.org/10.1007/978-3-642-17319-6_28
20. Ng, A.Y., Harada, D., Russell, S.J.: Policy invariance under reward transformations: theory and application to reward shaping. In: Proceedings of the Sixteenth International Conference on Machine Learning, pp. 278–287. Morgan Kaufmann Publishers Inc. (1999)
21. Nguyen, T., Khosravi, A., Creighton, D., Nahavandi, S.: Medical data classification using interval type-2 fuzzy logic system and wavelets. Appl. Soft Comput. **30**, 812–822 (2015)
22. Poulsen, A.P., Thorhauge, M., Funch, M.H., Risi, S.: DLNE: a hybridization of deep learning and neuroevolution for visual control. In: 2017 IEEE Conference on Computational Intelligence and Games (CIG), pp. 256–263 (2017)
23. Priesterjahn, S., Kramer, O., Weimer, A., Goebels, A.: Evolution of reactive rules in multi player computer games based on imitation. In: Wang, L., Chen, K., Ong, Y.S. (eds.) ICNC 2005. LNCS, vol. 3611, pp. 744–755. Springer, Heidelberg (2005). https://doi.org/10.1007/11539117_105
24. Salaken, S.M., Khosravi, A., Khatami, A., Nahavandi, S., Hosen, M.A.: Lung cancer classification using deep learned features on low population dataset. In: 2017 IEEE 30th Canadian Conference on Electrical and Computer Engineering (CCECE), pp. 1–5. IEEE (2017)
25. Saleh, K., Hossny, M., Nahavandi, S.: Intent prediction of vulnerable road users from motion trajectories using stacked LSTM network. In: 2017 IEEE 20th International Conference on Intelligent Transportation Systems (ITSC), pp. 327–332. IEEE (2017)
26. Saleh, K., Hossny, M., Nahavandi, S.: Towards trusted autonomous vehicles from vulnerable road users perspective. In: 2017 Annual IEEE International Systems Conference (SysCon), pp. 1–7. IEEE (2017)

27. Saunders, J., Nehaniv, C.L., Dautenhahn, K.: Teaching robots by moulding behavior and scaffolding the environment. In: Proceedings of the 1st ACM SIGCHI/SIGART Conference on Human-Robot Interaction, Salt Lake City, Utah, USA, pp. 118–125. ACM (2006). https://doi.org/10.1145/1121241.1121263
28. Silver, D., Bagnell, J.A., Stentz, A.: Applied imitation learning for autonomous navigation in complex natural terrain. In: Howard, A., Iagnemma, K., Kelly, A. (eds.) Field and Service Robotics, vol. 62, pp. 249–259. Springer, Heidelberg (2010). https://doi.org/10.1007/978-3-642-13408-1_23
29. Zhang, T., et al.: Deep imitation learning for complex manipulation tasks from virtual reality teleoperation (2017)

Improving Deep Neural Network Performance with Kernelized Min-Max Objective

Kai Yao[1], Kaizhu Huang[1(✉)], Rui Zhang[2], and Amir Hussain[3]

[1] Department of EEE, Xi'an Jiaotong-Liverpool University, 111 Ren'ai Rd., Suzhou, Jiangsu, People's Republic of China
Kai.Yao14@student.xjtlu.edu.cn, kaizhu.huang@xjtlu.edu.cn
[2] Department of MS, Xi'an Jiaotong-Liverpool University, 111 Ren'ai Rd., Suzhou, Jiangsu, People's Republic of China
rui.zhang02@xjtlu.edu.cn
[3] Division of Computing Science and Maths, School of Natural Sciences, University of Stirling, Stirling FK9 4LA, UK
ahu@cs.stir.ac.uk

Abstract. In this paper, we present a novel training strategy using kernelized Min-Max objective to enable improved object recognition performance on deep neural networks (DNN), e.g., convolutional neural networks (CNN). Without changing the other part of the original model, the kernelized Min-Max objective works by combining the kernel trick with the Min-Max objective and being embedded into a high layer of the networks in the training phase. The proposed kernelized objective explicitly enforces the learned object feature maps to maintain in a kernel space the least compactness for each category manifold and the biggest margin among different category manifolds. With very few additional computation costs, the proposed strategy can be widely used in different DNN models. Extensive experiments with shallow convolutional neural network model, deep convolutional neural network model, and deep residual neural network model on two benchmark datasets show that the proposed approach outperforms those competitive models.

1 Introduction

In the past several decades, Convolutional Neural Networks (CNN) have played a critical role in solving object recognition problems [10,15]. In the early research of image recognition, one big challenge is how to organize features, for the reason that the features of image data are hard to be extracted by human understanding. Later, CNN is widely considered to be a suited way to deal with this problem. CNN can directly use the original pixels of the image as input, alleviating the heavy demand in doing substantial repetitive and tedious data preprocessing work. Meanwhile, as an end-to-end approach, it does not separate the two processes of feature extraction and classification training because it automatically

© Springer Nature Switzerland AG 2018
L. Cheng et al. (Eds.): ICONIP 2018, LNCS 11301, pp. 182–191, 2018.
https://doi.org/10.1007/978-3-030-04167-0_17

extracts the most effective features during the training process. Because of the rapid progress of modern computing technologies and the availability of massive labelled data, CNN made great achievements in the past few years. Meanwhile, the implements of training strategies are also more critical to improve the CNN performance. Training strategies, such as dropout [14], batch normalization [6], different types of pooling [2] and others [19], can help improve the model performance with minor additional computation costs instead of making the requirements of training process harder.

Invariant features are a major factor for image recognition. Feature vectors learned by a CNN for different input samples with a same label, would be different, which is caused by positional shifts, lighting changes, shapes, angles of view, etc. When these feature vectors are projected into a high dimensional feature space, they could lie possibly at a low-dimensional manifold. Invariance features could be achieved if (1) each manifold (associated one single category) can be made more compact, and (2) the margin between a pair of manifolds can be enlarged. In order to have an insight into this problem, feature vectors in the ideal condition and in the real condition are illustrated in Fig. 1. In the ideal condition as shown in Fig. 1 (a), the within manifold becomes very compact, and the between-manifold margin is enlarged. Unfortunately, many present DNNs, e.g., CNNs, are widely observed to be unable to ensure this property. The two within manifolds shown in Fig. 1 (b) have overlapping parts, and the margin between those two classes becomes blurry.

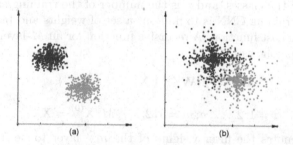

Fig. 1. Feature vectors visualization in (a) ideal condition and (b) real condition. Each color denotes one class and each node denotes one sample feature vector learned by a CNN model (better viewed in color).

To alleviate this problem, [11] proposed to utilise a min-max strategy inspired from Marginal Fisher Analysis (MFA) [17]. This work encourages each object manifold with the same label becomes as inseparable as possible, and it enforces the margin between two object manifolds sharing different labels as large as possible. However, [11] mainly restricted their objective description in the original feature space. Though a so-called kernelization was engaged, they mainly used them for weighting the distance still in the original feature space.

Instead, in this paper, we develop a new truly kernelized min-max strategy. Particularly, the feature space associated with a certain upper layer of DNN is further mapped to a high-dimensional kernel space with the kernel trick [3–5]. The Min-Max objective is then calculated in the kernel space rather than the original feature space. With the kernel trick, no explicit mapping needs to be specified, offering more potentials and flexibility to further improve the performance of DNNs.

Note that, in principle, the kernelized Min-Max Objective can be conveniently used in any layers with its independence of any CNN structures, which means the contributions made here have wide applicability. We have detailed our novel approach and evaluated our methodology extensively with many different types of DNNs in two benchmark datasets. Experimental results show that the novel kernelized Min-Max objective significantly and consistently outperforms the other competitive models.

2   Proposed Method

In this section, we first present the general Min-Max strategy [11] and then describe in details the kernelized Min-Max training.

2.1   General Framework

Let $\{\mathbf{X}_i, c_i\}_{i=1}^{n}$ be the set of input training data, where $\mathbf{X}_i$ denotes the ith raw input data, $c_i \in \{1, 2, \ldots, C\}$ denotes the corresponding ground truth label, C is the number of the classes, and n is the number of the training samples.

The goal of training CNN is to find out a set of weights and biases that can minimize a given cost function. A recursive function for an M-layer CNN model can be defined as follows:

$$\mathbf{X}_i^{(m)} = f(\mathbf{W}^{(m)} * \mathbf{X}_i^{(m-1)} + \mathbf{b}^{(m)}), \tag{1}$$

$$i = 1, 2, \ldots, n; m = 1, 2, \ldots, M; \mathbf{X}_i^{(0)} = \mathbf{X}_i, \tag{2}$$

where $\mathbf{W}^{(m)}$ denotes the filter weights of the mth layer to be learned, $\mathbf{b}^{(m)})$ refers to the corresponding biases, $*$ denotes the convolution operation. $f(\ )$ is an element-wise non-linear activation function such as ReLU, and $\mathbf{X}_i^{(m)}$ represents the features maps generated at layer m for sample $\mathbf{X}_i$.

The CNN model is trained using the cost function as follows:

$$\min_{W} L = \sum_{i=1}^{n} l(\mathbf{W}, \mathbf{X}_i, c_i), \tag{3}$$

where $l(\mathbf{W}, \mathbf{X}_i, c_i)$ is the loss associated with sample $\mathbf{X}_i$.

2.2 Min-Max Strategy

The objective of the new Min-Max strategy is to establish connections between samples by strengthening the compactness of each object manifold and enlarging the between-manifold margin. To better illustrate the idea, we borrow Fig. 2 from [11], where the intrinsic graph illustrates the node adjacency map for the entire manifolds, and the penalty graph provides node adjacency map among the between-manifolds. Then, from the intrinsic graph, the within-manifold compactness can be defined as:

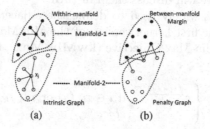

Fig. 2. Illustrations of the adjacency relationship [11].

$$\mathcal{L}_1 = \sum_{i,j=1}^{n} G_{ij}^{(I)} \|x_i - x_j\|^2, \tag{4}$$

$$G_{ij}^{(I)} = \begin{cases} 1 & \text{, if } i \in \tau_{k_1}(j) \text{ or } j \in \tau_{k_1}(i), \\ 0 & \text{, else.} \end{cases} \tag{5}$$

Here x_i is the column expansion of $\mathbf{X}_i^{(k)}$, and $G_{ij}^{(I)}$ indicates the element (i, j) of $\mathbf{G}^{(I)} = (G_{ij}^{(I)})_{n \times n}$, while $\tau_{k_1}(i)$ refers to the index set of the k_1-nearest samples of x_i in the same manifold.

The between-manifold margin can be described as follows:

$$\mathcal{L}_2 = \sum_{i,j=1}^{n} G_{ij}^{(P)} \|x_i - x_j\|^2, \tag{6}$$

$$G_{ij}^{(P)} = \begin{cases} 1 & \text{, if } (i, j) \in \xi_{k_2}(c_i) \text{ or } (i, j) \in \xi_{k_2}(c_j), \\ 0 & \text{, else.} \end{cases} \tag{7}$$

In the above $G_{ij}^{(P)}$ describes element (i, j) of $\mathbf{G}^{(P)} = (G_{ij}^{(P)})_{n \times n}$. Moreover, $\xi_{k_2}(c_i)$ denotes a set, containing the index pairs with the k_2-nearest pairs among the set $\{(i, j)|i \in \pi_c, j \notin \pi_c\}$. Furthermore, π_c means the index set of the sample belonging to the cth manifold.

Consequently, we can write the overall Min-Max objective as :

$$\mathcal{L} = \mathcal{L}_1 - \mathcal{L}_2, \tag{8}$$

The cost function is updated as follows:

$$\min_W L = \sum_{i=1}^{n} l(\mathbf{W}, \mathbf{X}_i, c_i) + \lambda \mathcal{L}(\mathcal{X}^{(k)}, c). \tag{9}$$

2.3   Kernelized Min-Max Objective

In order to further improve the performance, kernel trick is considered to be used in the Min-Max objective. There are two versions of Min-Max objective with kernel trick and we will discuss them in turn.

First, heat kernel is widely used to define the adjacency matrix in Laplacian eigenmaps. For the first kernel version of the Min-Max objective, which is called Kernel Weight Min-Max Objective (**kwMin-Max**), the adjacency matrix is defined as:

$$G_{ij}^{(I)} = \begin{cases} e^{\frac{\|x_i - x_j\|^2}{t}} & \text{, if } i \in \tau_{k_1}(j) \text{ or } j \in \tau_{k_1}(i), \\ 0 & \text{, else.} \end{cases} \tag{10}$$

$$G_{ij}^{(P)} = \begin{cases} e^{\frac{\|x_i - x_j\|^2}{t}} & \text{, if } (i,j) \in \xi_{k_2}(c_i) \text{ or } (i,j) \in \xi_{k_2}(c_j), \\ 0 & \text{, else.} \end{cases} \tag{11}$$

It can be observed that the kw-Min-Max actually weights the Min-Max objective in the original feature space and no transformation was conducted to the kernel space. This is the actually the version proposed in [11], which is truly kernelization.

To combine strictly the kernel trick with the Min-Max strategy, motivated from Kernel Marginal Fisher Analysis (KMFA)[17] and other similar models [16], we derive another strict kernel version for the Min-Max objective, called Kernel Space Min-Max (**ksMin-Max**) to define the similarity in the kernel space between each of the two feature vectors pair. Assume the mapping function Φ is used to remap $\mathbf{X} = [x_1, \ldots, x_n]$ into a kernel space, then we get a new high dimensional feature vector $\mathbf{Z} = [\Phi(x_1), \ldots, \Phi(x_n)] = [z_1, \ldots, z_n]$. Then we can get the strict kernel version of the Min-Max objective. The new kernel version of Min-Max objective can be defined as:

$$\mathcal{L} = \sum_{i,j=1}^{n} G_{ij} \|\Phi(x_i) - \Phi(x_j)\|^2 = \sum_{i,j=1}^{n} G_{ij} \|z_i - z_j\|^2, \tag{12}$$

where $G_{ij} = G_{ij}^{(I)} - G_{ij}^{(P)}$ is the graph matrix. Because the mapping function Φ cannot be written explicitly, the kernel function $K_{ij} = k(x_i, x_j) = \Phi(x_i) \cdot \Phi(x_j)$ is applied. Then the Kernel Space Min-Max Objective can be expressed as follows:

$$\mathcal{L} = \sum_{i,j=1}^{n} G_{ij}(z_i^T z_i - 2 z_i^T z_j + z_j^T z_j) = \sum_{i,j=1}^{n} G_{ij}(K_{ii} + K_{jj} - 2K_{ij}) \tag{13}$$

It is noted again that, in the viewpoint of kernel trick, ksMin-Max projects high level feature map of CNN into a further high dimensional Hilbert kernel space, from which the similarity is calculated. This is significantly from the kw-Min-Max with which the similarity is weighted but still computed in the original feature space. The flexibility of kernel mapping offers the proposed ksMin-Max more potentials in improving the performance of object recognition with DNN.

3 Experiments

3.1 Setting

In order to validate the advantage of the proposed kernelized framework, experimental evaluations will be conducted with three different types of models in different aspects. Cross validation was conducted on each dataset to get the hyper parameters in each model. We then test which layer to be embedded with the Min-Max objective would be most effective. After that, we only use the most effective layer to apply the Min-Max objective and their kernelized version and keep other layers unchanged. For simplicity, we set $k_1 = 5$, $k_2 = 10$ for all the experiments, it is possible to even get better results by tuning k_1 and k_2. $\lambda \in [10^{-5}, 10^{-8}]$ and t will be selected from $\{0.1, 0.5\}$. The Gaussian kernel function $K_{ij} = e^{\frac{\|x_i - x_j\|^2}{2\sigma^2}}$ is used in ksMin-Max. Following many relevant works, we conduct our evaluations on two benchmark data sets CIFAR-10 and CIFAR-100. For comparison, we implement the model [11].

3.2 On Shallow CNN

In this subsection, the CNN quick model (named Quick-CNN) is chosen as the baseline shallow model. The Quick-CNN model contains layers of convolution, pooling, rectified linear unit nonlinearities, and local contrast normalization with a linear classifier on top of it. The Quick-CNN consists of three convolutional layers and one full connected layer, with all weights using the base learning rate and all biases using twice as base learning rate.

The Min-Max strategies including kwMin-Max and ksMin-Max are embedded into the model to find out the effect of embedded position.

The experimental results are reported in Tables 1 and 2. From Table 1 it can be observed that, compared with the baseline Quick-CNN on CIFAR-10, Min-Max objective and its two versions of kernel trick can remarkably decrease the test error rates by 3.93%, 2.23% and 4.44%. Then, in the Table 2 we also list the respective comparison results, the superiority of the proposed ksMin-Max can also be observed. In general, it can be seen that Min-Max objective has a good effect on Quick-CNN model and significantly improves the capability of the model's feature extraction. It can be also seen that the kwMin-Max proposed in [11] actually decreases the performance of Min-Max, showing that the weighted Min-Max may not be good. In comparison, the proposed strict kernlized Min-Max, i.e., the ksMin-Max show its power on further improving the Min-Max on shallow model.

Table 1. Recognition accuracies for CIFAR-10 of various methods

Method	# of param	Error rate(%)
Quick-CNN	0.145M	23.46
Quick-CNN+Min-Max	0.145M	19.53
Quick-CNN+kwMin-Max	0.145M	21.23
Quick-CNN+ksMin-Max	0.145M	**19.02**

Table 2. Recognition accuracies for CIFAR-100 of various methods

Method	# of param	Error rate(%)
Quick-CNN	0.150M	59.07
Quick-CNN+Min-Max	0.150M	55.47
Quick-CNN+kwMin-Max	0.150M	57.06
Quick-CNN+ksMin-Max	0.150M	**55.13**

3.3 On Deep CNN

We examine the performance the Min-Max objectives on the famous deep model, i.e., Network In Network(NIN) [9] which consists of 9 convolution layers and 1 global average pooling layer. The CIFAR-10 and CIFAR-100 datasets are pre-processed as in maxout network [1]. For CIFAR-10, we also applied data augmentation by following [11].

We follow the training procedure used by Krizhevsky et al. [7].

The various performance was reported in Tables 3 and 4 with comparisons against various representative methods, including Stochastic Pooling [18],

Table 3. Recognition accuracies for CIFAR-10 of various methods

Method	#of param	Test Error(%)	# of param	Test error(%)
	Without augmentation		Data augmentation	
Stochastic pooling	——	15.13	——	9.50
CNN+Spearmint	——	14.98	——	9.39
Maxout networks	>5M	11.68	>5M	9.38
Prob. maxout	>5M	11.35	>5M	9.32
NIN	0.97M	10.41	0.97M	8.81
DSN	0.97M	9.78	0.97M	8.22
NIN (my baseline)	0.97M	10.52	0.97M	8.90
NIN+Min-Max	0.97M	9.71	0.97M	7.69
NIN+kwMin-Max	0.97M	10.01	0.97M	8.09
NIN+ksMin-Max	0.97M	**9.40**	0.97M	**7.39**

CNN+Spearmint [12], Maxout Network [1], Prob. Maxout [13] and DSN [8]. As a result, we can see after using Min-Max objective, the NIN model object recognition accuracy is improved over the DSN model and the proposed ksMin-Max have a 1.12% extra promotion compared with the baseline without data augmentation. Again, it is observed that the kwMin-Max objective may decrease the performance compared with the basic Min-Max objective. In comparison, the proposed ksMin-Max objective shows again performance increases consistently than the original Min-Max objectives on all the data sets.

Table 4. Recognition accuracies for CIFAR-100 of various methods

Method	# of param	Test error(%)
Learned pooling	——	43.71
Stochastic pooling	——	42.51
Maxout metworks	>5M	38.57
Prob. maxout	>5M	38.14
Tree based priors	——	36.85
NIN	0.98M	35.68
DSN	0.98M	34.57
NIN (my baseline)	0.98M	35.77
NIN+Min-Max	0.98M	34.14
NIN+kwMin-Max	0.98M	34.99
NIN+ksMin-Max	0.98M	**33.81**

3.4   On Deep Residual Model

We further conduct more studies on the ResNet model trying to figure out the behaviors of the proposed Min-Max objective embedded in extremely deep networks.

The network inputs are 32×32 images, with the per-pixel mean subtracted. Batch normalization and data argumentation are applied but no dropout (Table 5).

Through the table, we can get the following conclusion: On deep residual neural network, the Min-Max objective can get better improvement in the deeper model. The higher level feature maps reflect the affinity for the Min-Max objective. With the growth of layers, improving layer makes the model reach to saturation, while Min-Max objective can continue to improve the object recognition. After using ksMin-Max, the ResNet-56 test error rate reach 6.51%, which is very close to ResNet-110 baseline accuracy 6.43%, while ResNet-56 has 0.85M parameters and ResNet-110 has 1.7M parameters. It is noted that, for conciseness, we only report the performance on CIFAR10, while the conclusion remains the same on the other dataset CIFAR100.

Table 5. Comparison result on CIFAR-10 dataset with different layers ResNet

	Reference(%)	Repeated(%)	Min-Max(%)	ksMin-Max(%)
ResNet-20	8.75	8.13	8.00	7.90
ResNet-32	7.51	7.51	7.31	7.19
ResNet-56	6.97	7.08	6.68	6.51
ResNet-110	6.43	—	—	—

4   Conclusion

In this paper, we aimed to provide insights into the relationship between training samples, and we proposed a new framework able to lift up the performance of DNN by designing a truly kernelized Min-Max strategy. The proposed strategy has the following characteristics: it enforces the learned object feature maps with better within-manifold compactness and between-manifold separability. Experiments with shallow and deep networks on two benchmark datasets showed the effectiveness of the proposed model. We also found that using kernel trick to define distance matrix offers more potentials in improving the learning performance.

Acknowledgments. The work was partially supported by the following: National Natural Science Foundation of China under grant no. 61473236 and 61876155; Natural Science Fund for Colleges and Universities in Jiangsu Province under grant no. 17KJD520010; Suzhou Science and Technology Program under grant no. SYG201712, SZS201613; Jiangsu University Natural Science Research Programme under grant no. 17KJB- 520041; Key Program Special Fund in XJTLU under no. KSF-A-01 and KSF-P-02.

References

1. Goodfellow, I., Wardefarley, D., Mirza, M., Courvile, A., Bengio, Y.: Maxout networks. In: ICML (2013)
2. He, K., Zhang, X., Ren, S., Sun, J.: Spatial pyramid pooling in deep convolutional networks for visual recognition. In: Fleet, D., Pajdla, T., Schiele, B., Tuytelaars, T. (eds.) ECCV 2014. LNCS, vol. 8691, pp. 346–361. Springer, Cham (2014). https://doi.org/10.1007/978-3-319-10578-9_23
3. Huang, K., Yang, H., King, I., Lyu, M.R.: Learning classifiers from imbalanced data based on biased minimax probability machine. Proc. CVPR **2**, 558–563 (2004)
4. Huang, K., Yang, H., King, I., Lyu, M.R.: Machine Learning: Modeling Data Locally and Globally. Springer, Heidelberg (2008). https://doi.org/10.1007/978-3-540-79452-3. ISBN 3-5407-9451-4
5. Huang, K., Yang, H., King, I., Lyu, M.R.: Maxi-min margin machine: learning large margin classifiers globally and locally. IEEE Trans. Neural Netw. **19**, 260–272 (2008)
6. Ioffe, S., Szegedy, C.: Batch normalization: Accelerating deep network training by reducing internal covariate shift. In: ICML (2015)

7. Krizhevsky, A., Sutskever, I., Hinton, G.: ImageNet classification with deep convolutional neural networks. In: NIPS (2012)
8. Lee, C., Xie, S., Gallagher, P., Zhang, Z., Tu, Z.: Deeply-supervised nets. In: NIPS (2014)
9. Lin, M., Chen, Q., Yan, S.: Network in network. In: ICLR (2014)
10. Lyu, C., Huang, K., Liang, H.N.: A unified gradient regularization family for adversarial examples. In: 2015 IEEE International Conference on Data Mining (ICDM), pp. 301–309. IEEE (2015)
11. Shi, W., Gong, Y., Wang, J.: Improving CNN performance with min-max objective. In: International Joint Conference on Artificial Intelligence (2016)
12. Snock, J., Larochelle, H., Adams, R.: Practical Bayesian optimization of machine learning algorithm. In: NIPS (2012)
13. Springenberg, J., Riedmiller, M.: Improving deep neural networks with probabilistic maxout units. In: ICLR (2014)
14. Srivastava, N., Hinton, G., Krizhevsky, A., Sutskever, I., Salakhutdinov, R.: Dropout: a simple way to prevent neural networks from overfitting. In: JMLR (2014)
15. Wang, J., Zhang, B., Sun, Z., Hao, W., Sun, Q.: A novel conjugate gradient method with generalized Armijo search for efficient training of feedforward neural networks. Neurocomputing 275, 308–316 (2018)
16. Xu, B., Huang, K., Liu, C.L.: Maxi-min discriminant analysis via online learning. Neural Netw. 34, 56–64 (2012)
17. Yan, S., Xu, D., Zhang, B., Zhang, H., Yang, Q., Lin, S.: Graph embedding and extensions: a general framework for dimensionality reduction. IEEE Trans. PAMI 29, 40–51 (2007)
18. Zeiler, M., Fergus, R.: Stochastic pooling for regularization of deep convolutional neural networks. In: ICLR (2013)
19. Zhang, S., Huang, K., Zhang, R., Hussain, A.: Learning from few samples with memory network. Cogn. Comput. 10(1), 15–22 (2018)

Understanding Deep Neural Network by Filter Sensitive Area Generation Network

Yang Qian[1,2], Hong Qiao[1,2,3,4(✉)], and Jing Xu[5]

[1] The State Key Lab of Management and Control for Complex Systems,
Institute of Automation, Chinese Academy of Science, Beijing 100190, China
{qianyang2016,hong.qiao}@ia.ac.cn
[2] University of Chinese Academy of Sciences, Beijing, China
[3] CAS Center for Excellence in Brain Science and Intelligence Technology,
Shanghai, China
[4] Cloud Computing Center, Chinese Academy of Sciences, Dongguan, China
[5] Department of Mechanical Engineering, Tsinghua University, Beijing, China
jingxu@tsinghua.edu.cn

Abstract. Deep convolutional networks have recently gained much attention because of their impressive performance on some visual tasks. However, it is still not clear why they achieve such great success. In this paper, a novel approach called Filter Sensitive Area Generation Network (FSAGN), has been proposed to interpret what the convolutional filters have learnt after training CNNs. Given any trained CNN model, the proposed method aims to figure out which object part each filter represents in a high conv-layer, through appropriate input image mask which filters out unrelated area. In order to obtain such a mask, a mask generation network is designed and the corresponding loss function is defined to evaluate the changes of feature maps before and after mask operation. Experiments on multiple datasets and networks show that FSAGN clarifies the knowledge representations of each filter and how small disturbance on specific object parts affects the performance of CNNs.

Keywords: Convolutional neural network · Interpretability
Knowledge representations

1 Introduction

Recent years have seen spectacular improvements in artificial intelligence. Particularly, deep neural networks (DNNs) has achieved superior performance in a variety of visual tasks, such as fine-grained classification [1, 2], object detection [3, 4] and semantic segmentation [5, 6]. Although DNNs outperform previous machine learning techniques on the comparison of accuracy, we still have little knowledge about what they have learnt. When they fail on some cases, it is hard to explain what caused the DNNs to make such decisions. One Pixel Attack cheated the DNN successfully by changing value of a single pixel, which is impossible for human to make such mistakes. This lack of interpretability of DNNs is largely due to the end-to-end structure and learning

© Springer Nature Switzerland AG 2018
L. Cheng et al. (Eds.): ICONIP 2018, LNCS 11301, pp. 192–203, 2018.
https://doi.org/10.1007/978-3-030-04167-0_18

strategy, which lead to the difficulties of understanding the main role of individual neurons during the whole process of completing visual tasks.

Recently, a large number of researchers have realized the necessity of improving interpretability of DNNs and have proposed a variety of models to dig the interpretable knowledge representations learned by DNNs, especially by Convolutional Neural Networks (CNNs). Zeiler et al. [7] examine the pattern of every layer by visualization with a deconvnet and figure out whether a model is truly identifying the location of the object in the image by occluding different portion of the input image and observe the probability of the correct class. This approach finds the occlusion sensitive region of convolutional filters and classifiers, but the size of region is limited to a rectangle and the process is time-consuming. Yosinski et al. [8] visualized filters by finding an image that maximize the activation of this unit via regularized optimization. Much other work tries to leverage heatmaps to understand the decision-making process of networks. An approach called CLEAR [9] is invented to visualize attentive regions of DNNs during the decision-making process. These approaches change the original network structure or learning process more or less and give little insight about what each individual filter has learnt after a network is trained.

In this paper, we mainly focus on the question, which area of the input image does a convolutional filter mainly focuses on? Based on the observation that a specific filter has strong activations for certain parts of the object and keep silent for other areas, we expect to figure out the intrinsic activation mode of some filters and interpret what these filters have remembered after training.

To find out which parts each filter pays attention to automatically and efficiently, we propose a Filter Sensitive Area Generation Network (FSAGN) for generating input image mask to mask unimportant regions in an image. In consideration of sparse activation properties of neural network, we first statistically analyze average activation of every filter and filter out the silent filters. For each active filter, a network is designed to generate a mask of the input image and obtain a new input image by mask operation with the original image. Through a forward propagation, we can get new feature maps. By minimizing the difference between the original and new feature maps, FSAGN converges gradually and finally obtain the power to localize the key part that certain filter represents. Simultaneously, we also adopt an occlusion strategy to generate occlusion sensitive area. After we have a clear insight about which parts each filter focuses on, adversarial samples can be designed to cheat the original network.

The rest of this paper is organized as follows. The proposed framework and design of network are introduced in Sect. 2. Section 3 presents experimental results and corresponding analysis. Sections 4 and 5 make a discussion and conclusion of the paper respectively.

2 Filter Sensitive Area Generation Network

This section describes the proposed network for finding which parts contribute most to the response of certain filter.

2.1 Filters Selection

Thanks to the sparsity of feature maps, only a few filters response strongly to some parts of objects, while others remain inactivated. If we select a filter randomly to generate its concerned part, we may fail because it has small probability to represent specific part of objects theoretically.

In order to find which filters are sensitive for part discovery and are valuable to analyze, we first test all samples and record their activation for every channel. For better measuring the importance of filters, we calculate the sum of each feature map. Then we visualize the response map over all samples and channels, where the vertical axis is channel number, and the horizontal axis is the sample number. Due to the sparse response distributions of CNNs, some filters are always activated for all samples, while others keep silent no matter what images are selected as input.

As is seen in Fig. 1, there are several bright lines in the map, where most areas are dark, which indicates that these filters are potential to have strong response for some specific parts of objects. Therefore, we refer to these filters as the target filters for sensitive area discovery.

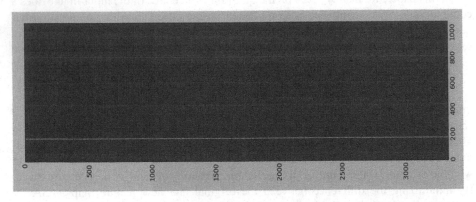

Fig. 1. Average response of each filter in CNN over multiple samples

2.2 Filter Sensitive Area Generation Network

Inspired by the observation that some convolutional filters only response to a small specific area on the input image, which means that when we occlude other areas, the filter activation is not affected dramatically. Zeiler et al. [7] manually adopt a gray rectangle window for occlusion test by sliding window over the whole image to generate an occlusion sensibility map, which is limited for the fixed shape and size of part area and the whole process is time-consuming because of sliding window strategy.

In this section we proposed a Filter Sensitive Area Generation Network (FSAGN) to locate the area that a filter focuses on. The network structure is shown in Fig. 2. Taking feature maps of the last convolutional layer in a trained CNN as input, the FSAGN outputs a mask with the same size as input image through a deconvolution structure [6]. Then the new image generated by mask operation is input to the original network and the new feature maps are obtained too. By comparing the original and the new feature maps, we can evaluate the influence of different regions in images on the response of filters. Two criterions are adopted to define filter sensitive area. When the image except the sensitive area of a filter is set to zero, the response distribution of this filter will keep unchanged compared with the response from the intact image. This strategy is called as filter sensitive area reservation.

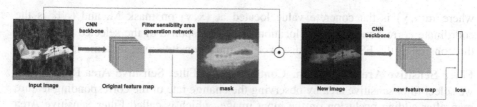

Fig. 2. Framework of FSAGN.

On the contrary, when we occlude the sensitive area on the image, the filter response distribution will change significantly, which is called filter sensitive area occlusion. In either case, the sensitive area should be as small as possible to avoid the area from converging to the whole image.

Next, two methods will be introduced in detail respectively.

Filter Sensitive Area Reservation. For a filter in certain layer, it represents a specific part for some objects, which means activation of the filter mainly originates from a subarea of the whole input image I. Our target is to find a corresponding mask $M \in [0, 1]$ for input image to generate filter sensitive area reservation image I'.

$$I' = I \odot M \tag{1}$$

Given a trained CNN f, the original response of the c-th filter in layer l when inputting the original image is denoted as $r_{l,c} = f(I)[l, c]$. Then the new image I' is fed into the same CNN, and we get the new feature map of the c-th filter in layer l denoted as $r'_{l,c} = f(I)[l, c]$. To find some object parts that contribute the most to the response of specific filter, the optimization goal of FSAGN is to minimize the difference between old feature maps and new feature maps. However, the generated mask is usually sparse. To encourage a compact distribution of mask, we introduce a new constraint in the loss function as follows:

$$Loss(M) = L_{dif}(r, r') + \lambda * L_{area}(M) \tag{2}$$

where L_{dif} and L_{area} represents the feature map difference loss and the mask generation loss respectively. The feature map difference loss is used to describe the difference between the original and new feature maps. In order to focus on the consistency of response distribution rather than the concrete value of activation, two feature maps are firstly normalized to [0, 1], then the feature map difference loss is given by:

$$L_{dif}(r, r') = \|r - r'\|_F \tag{3}$$

To introduce compact distribution constraint on generated mask, $L_{area}(M)$ is formulated as follows [12]:

$$L_{area}(M) = \sum_{(x,y) \in M} m(x, y)\left[(x - t_x)^2 + (y - t_y)^2\right] \tag{4}$$

where $m(x, y)$ is the concrete value located at (x, y) on mask M, and t_x, t_y is the coordinate corresponding to the location of peak response of the selected filter. With the constraint, the FSAGN will discover the most sensitive part to some filters.

Filter Sensitive Area Occlusion. Contrary to the Filter Sensitive Area Reservation, we select filter sensitive area by observing the change rate of the corresponding feature map after adding occlusion on the input image, which is called Filter Sensitive Area Occlusion. Our target is developing a Filter Sensitive Area Generation Network to find some areas in input image so that when these areas are occluded, the response of the related filter changes dramatically. This method helps us better understand what the filters have learnt and which part they focus on.

The optimization function is given as follows:

$$Loss(M) = L_{sim}(r, r') + \lambda L_{area}(M) \tag{5}$$

where L_{sim} and L_{area} represents the feature map similarity loss and the mask generation loss respectively. Different from the Filter Sensitive Area Reservation, the similarity between new feature map and original feature map should be as small as possible.

Herein the activation function is selected as ReLU, thus the feature map is non-negative. When occluding some parts of the object, the new response of this filter will drop rapidly and even decrease to zero. Therefore, the similarity loss is designed as follows:

$$L_{sim}(r, r') = \|r'\|_F \tag{6}$$

Meanwhile, the mask generation loss L_{area} keeps the same as that in Filter Sensitive Area Reservation.

3 Experiments

In this section, we will illustrate the efficiency of Filter Sensitive Area Generation Network and show some examples to figure out which parts the specific filters pay attention to. Experiments were conducted on two public datasets, including MNIST and FGVC-Aircraft [11]. Next, more implementation details and experimental results are explained.

3.1 Implementation Details

Before analyzing the sensitive area for some filters, CNN models for object recognition should be trained first. Specifically, a small-scale convolutional neural network is designed for MNIST classification. It has two convolutional layers and two fully connected layers, taking $28 * 28$ gray images as inputs, as shown in Fig. 3(a). We achieved an accuracy of 99.18% for MNIST datasets. Then we consider the feature maps of the last convolutional layer as reference. A deconvolutional network is adopted as the Filter Sensitive Area Generation Network, as shown in Fig. 3(b). It takes the feature maps of the last convolutional layer as inputs, and adds a sigmoid layer to the output, which generates a single-channel mask $M \in (0, 1)$.

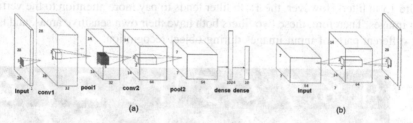

(a)   (b)

Fig. 3. Network structure for MNIST. (a) Network for classification. (b) Network for sensitive area generation.

For FGVC-Aircraft benchmark, a VGG-16 [10] model pre-trained on ImageNet [13] with inputs of size $224 * 224$ are used for better recognition performance, which gained 74% accuracy. We removed the last three fully-connected layers and augmented with a deconvolutional network for filter sensibility area generation. The structure of FSAGN is shown in Fig. 4.

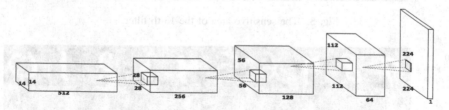

Fig. 4. Structure of FSAGN for VGG-16 trained on FGVC-Aircraft dataset.

When training the whole network, the parameters of basic recognition network remain fixed, with only the Filter Sensitive Area Generation Network updated.

3.2 Experiments on MNIST

Filter Selection. We get a collection of filter response distribution in the last convolutional layer tested on randomly selected 1K samples and plot the filter response diagram, which looks like sparse stripes. Following the method described in Sect. 2.1, the 18-th and 13-th channels are finally chosen as the target filters.

Sensitive Area of the 13-th Filter. We adopt filter sensitive area reservation strategy to generate the sensitive areas for the 13-th filter shown in Fig. 5. The figure shows the original images, original feature maps, generated sensitive areas, new input images after mask operation and new feature maps corresponding to the new image. From the results, some observations can be made: (1) This filter mainly focuses on a small region of the whole image, which means that removing other parts does not have dramatic effects on the activation of this filter. (2) From the similarity of sensitive areas on different samples, the concerned part of the 13-th filter is the slash of handwritten numeral. After training for MNIST recognition, this filter has learned to capture the inclined part of images.

Sensitive Area of the 18-th Filter. The same experimental process is applied on the 18-th filter. Results are shown in Fig. 6. Apparently, we can get similar conclusions with the 13-th filter. However, the 18-th filter tends to pay more attention to the vertical line in images. Therefore, these two filters both have their own sensitive areas, and they detect different parts of input images during object recognition.

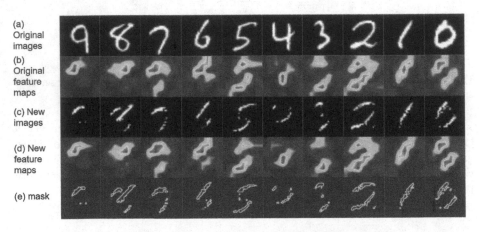

Fig. 5. The sensitive area of the 13-th filter.

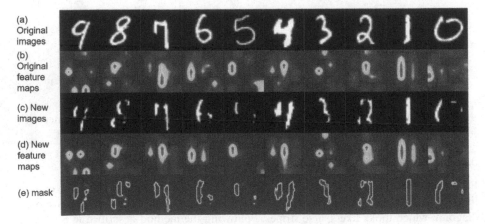

Fig. 6. The sensitive area of the 18-th filter.

3.3   Experiments on FGVC-Aircraft

Filter Selection. We randomly choose 200 aircraft images from FGVC-Aircraft dataset and record feature maps of every channel to form average filter response diagram shown in Fig. 7. We can observe from the diagram that filters which keeping active all the time (bright vertical lines in the image) account for a rather small part of all channels. In the statistical sense, it is consistent with our intuition that the response of CNN is sparse. Following the method described in Sect. 2.1, the 26-th and 262-th filter in the last convolutional layer of the VGG-16 model are selected for finding the key parts that the filters represent.

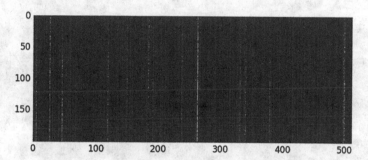

Fig. 7. Response diagram of each filter in VGG-16 for FGVC-aircraft over multiple samples.

Sensitive Area of the 26-th Filter. Firstly, filter sensitive area reservation strategy is adopted to generate the sensitive areas of aircrafts, as shown in Fig. 8. From the result, the key observations are the following: (1) The generated mask can filter out the background and localize the object coarsely, which is unsupervised without any bounding box labels. (2) Occluding most of the background will take little effect on the activation distribution of this filter. (3) We select the region with the biggest value on the mask (red circle on the new images) and find that this filter tends to be the most sensitive to the nose of aircraft. After training on the FGVC-Aircraft dataset, the 26-th filter has remembered the pattern of aircrafts' nose.

Next, occlusion strategy-based experiments are conducted to figure out which area has dramatical effect on the response of the filter when it is occluded. As shown in Fig. 9, some interesting observations are made as following: (1) The occlusion region consists of discrete points and lines rather than a whole continuous area. Although human can still recognize the aircraft after such occlusion, the response of the filter weakens rapidly. (2) The occlusion sensitive area of the filter tends to cover the whole object, which is apparently different from that in reservation strategy.

Sensitive Area of the 262-th Filter. Similar experiments are repeated for the 262-th filter. From the results of sensitive area reservation strategy (see Fig. 10), we observe that: (1) Unsurprisingly, this filter has the same preliminary ability to localize the object without supervision. (2) The main part that the 262-th filter focuses on is the fuselage close to the engine, which is different from the 26-th filter. It implies the diversity of filters and these filters have learnt the key parts of aircraft.

From the results of sensitive area occlusion strategy (see Fig. 11), some unexpected observations are the following: (1) The occlusion region degenerates to multiple parallel vertical lines. This confirms that small disturbance can lead to the network's failure.

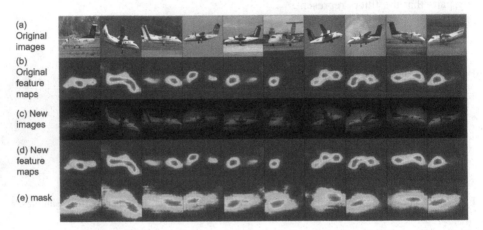

Fig. 8. The sensitive area of the 26-th filter of VGG-16 by reservation strategy.

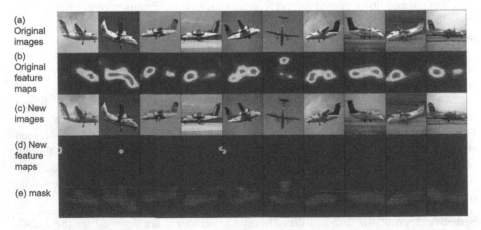

Fig. 9. The sensitive area of the 26-th filter of VGG-16 by occlusion strategy.

Fig. 10. The sensitive area of the 262-th filter of VGG-16 by reservation strategy.

Fig. 11. The sensitive area of the 262-th filter of VGG-16 by occlusion strategy.

4 Discussion

From the experiments on different datasets with different convolution neural networks, some interesting discussions are the following: (1) The activations of filters in CNNs are rather sparse. A small proportion of filters in a layer response strongly, while others keep silent all the time. (2) Each activated filter has a specific response pattern. For simple images and small networks, activation of filters may be sensitive to the vertical line or horizontal line. It indicates that when corresponding parts are occluded, the activation drop rapidly. By contrast, the feature map keeps unchanged when these areas are reserved. For complicated images and large networks, the response pattern of filter in high layer show stronger semantics. For example, a filter can represent the key part of object, like the nose of aircraft. It confirms that deep neural networks learned the key components of objects after training and we can establish a correspondence between parts of objects and filters by the proposed method.

5 Conclusion and Future Work

In this paper, we have proposed a general method to analyze the interpretability of the trained CNNs and better understand what the filters have learnt after training on certain dataset. Based on the observation that some filters could localize the key parts of objects, a Filter Sensitive Area Generation Network is designed and trained to generate the key area that every filter represents. To better describe the correlation between certain filter and the key part, reservation sensibility and occlusion sensibility are proposed respectively. Experiments have shown that the filters response to a certain part of the object and different filters have different fixed response pattern. Besides, small occlusion on the input image will take a significant effect on the activation of filters.

In future work, we will explore classifier sensitive area and make use of this interpretability to generate corresponding adversarial samples or improve the robustness of CNNs by adjusting the sensitive area of filters.

Acknowledgements. This work was supported in part by the National Key Research and Development Program of China (2017YFB1300203), in part by the National Natural Science Foundation of China under Grant 91648205.

References

1. Zhang, N., Donahue, J., Girshick, R., Darrell, T.: Part-based R-CNNs for fine-grained category detection. In: Fleet, D., Pajdla, T., Schiele, B., Tuytelaars, T. (eds.) ECCV 2014. LNCS, vol. 8689, pp. 834–849. Springer, Cham (2014). https://doi.org/10.1007/978-3-319-10590-1_54
2. Zhang, X., Xiong, H., Zhou, W., Tian, Q.: Picking deep filter responses for fine-grained image recognition. In: 2016 IEEE Conference on Computer Vision and Pattern Recognition (CVPR 2016), Las Vegas, NV, pp. 1134–1142 (2016)
3. Ren, S., He, K., Girshick, R., Sun, J.: Faster R-CNN: towards real-time object detection with region proposal networks. In: International Conference on Neural Information Processing Systems (NIPS 2015), vol. 39, pp. 91–99. MIT Press (2015)
4. Redmon, J., Divvala, S., Girshick, R., Farhadi, A.: You only look once: unified, real-time object detection. In: Computer Vision and Pattern Recognition (CVPR 2016), pp. 779–788. IEEE Computer Society (2016)
5. Shelhamer, E., Long, J., Darrell, T.: Fully convolutional networks for semantic segmentation. IEEE Trans. Pattern Anal. Mach. Intell. **39**(4), 640–651 (2017)
6. Noh, H., Hong, S., Han, B.: Learning deconvolution network for semantic segmentation. In: IEEE International Conference on Computer Vision (ICCV 2015), pp. 1520–1528. IEEE Computer Society (2015)

7. Zeiler, Matthew D., Fergus, R.: Visualizing and understanding convolutional networks. In: Fleet, D., Pajdla, T., Schiele, B., Tuytelaars, T. (eds.) ECCV 2014. LNCS, vol. 8689, pp. 818–833. Springer, Cham (2014). https://doi.org/10.1007/978-3-319-10590-1_53
8. Yosinski, J., Clune, J., Nguyen, A., Fuchs, T., Lipson, H.: Understanding neural networks through deep visualization. In: International Conference on Machine Learning — Deep Learning Workshop, pp. 12 (2015)
9. Kumar, D., Wong, A., Taylor, G.W., Kumar, D., Wong, A., Taylor, G.W.: Explaining the unexplained: A CLass-Enhanced Attentive Response (CLEAR) approach to understanding deep neural networks. In: IEEE Conference on Computer Vision and Pattern Recognition Workshops (CVPR 2017), pp. 1686–1694. IEEE (2017)
10. Simonyan, K., Zisserman, A.: Very deep convolutional networks for large-scale image recognition. arXiv preprint arXiv:1409.1556. (2014)
11. Maji, S., Rahtu, E., Kannala, J., Blaschko, M., Vedaldi, A.: Fine-grained visual classification of aircraft. arXiv preprint arXiv:1306.5151. (2013)
12. Zheng, H., Fu, J., Mei, T., Luo, J.: Learning multi-attention convolutional neural network for fine-grained image recognition. In: IEEE International Conference on Computer Vision, pp. 5219–5227. IEEE Computer Society (2017)
13. Deng, J., Dong, W., Socher, R., Li, L.J., Li, K., Li, F.F.: ImageNet: a large-scale hierarchical image database. In: IEEE International Conference on Computer Vision and Pattern Recognition (CVPR 2009), pp. 248–255. IEEE (2009)

Accelerating Deep Q Network by Weighting Experiences

Kazuhiro Murakami[1]([✉]), Koichi Moriyama[1], Atsuko Mutoh[1],
Tohgoroh Matsui[2], and Nobuhiro Inuzuka[1]

[1] Department of Computer Science, Graduate School of Engineering,
Nagoya Institute of Technology, Nagoya, Japan
k.murakami.638@nitech.jp
[2] Department of Clinical Engineering, College of Life and Health Sciences,
Chubu University, Kasugai, Japan

Abstract. Deep Q Network (DQN) is a reinforcement learning methodlogy that uses deep neural networks to approximate the Q-function. Literature reveals that DQN can select better responses than humans. However, DQN requires a lengthy period of time to learn the appropriate actions by using tuples of state, action, reward and next state, called "experience", sampled from its memory. DQN samples them uniformly and randomly, but the experiences are skewed resulting in slow learning because frequent experiences are redundantly sampled but infrequent ones are not. This work mitigates the problem by weighting experiences based on their frequency and manipulating their sampling probability. In a video game environment, the proposed method learned the appropriate responses faster than DQN.

Keywords: Reinforcement learning · Deep learning

1 Introduction

Reinforcement learning that aims to produce optimal behavior under specific circumstances has achieved significant results in fields such as robotic control [1]. Meanwhile, deep learning has also shown remarkable results in fields such as image recognition, natural language processing, and speech recognition. It is good at extracting features from input data and approximating a function very accurately. Furthermore, Deep Q Network (DQN) [2,3] was proposed as a deep reinforcement learning method that is a combination of deep learning and reinforcement learning. DQN stores its "experiences" in a memory and uses them for learning optimal responses in a given environment. In an experiment that used the Atari 2600 [4], DQN learned actions that scored more than advanced players.

However, the learning speed of DQN is very slow. One of the reasons for this is that experiences appear at different rates. DQN learns by using experiences randomly sampled from its memory. Given that experiences are skewed, learning

© Springer Nature Switzerland AG 2018
L. Cheng et al. (Eds.): ICONIP 2018, LNCS 11301, pp. 204–213, 2018.
https://doi.org/10.1007/978-3-030-04167-0_19

from frequent experiences is performed redundantly, whereas infrequent ones cannot be used for learning. However, this problem will be addressed if we can sample the experiences in relation to their appearance rates. Therefore, in this study, we propose a method that estimates the unfamiliarity of each experience by using the outputs of the DQN's middle layer and that weights each experience on the basis of the estimated unfamiliarity.

2   Deep Q Network

Deep Q Network (DQN) [2,3] is a deep neural network whose input is the state and output is the value of each action that can be selected in that state. We will refer to this deep neural network as "Q-network" in this study. The memory that accumulates and stores experiences is called "Replay Memory" and the process of updating the weights of the Q-network using randomly sampled experiences from the Replay Memory is called "Experience Replay". NIPS version DQN [2] algorithm is shown in Fig. 1, where s, a, r, and s' indicate the state, action, reward, and next state, respectively. In addition, f is a flag indicating the end of an episode. The tuples of s, a, r, s', f are called "experiences". t indicates the target variables that we intend to generate from the Q-network and E represents the error between the target variables and outputs of the Q-network.

Nature version DQN [3] also adds two ingenuities to stabilize learning. Note that the target variables t are generated from the current Q-network. Given that the weights of the Q-network are constantly updated, the target variables on a certain state s also regularly changes, causing the destabilization of learning. To address this problem, Nature version DQN introduced another network with specifically the same structure as the Q-network called the "target network". The weights of the target network are not constantly updated but periodically copied from the Q-network. Hence, learning is stabilized using the target variables from the target network.

Furthermore, the difference between the target variables and the actual outputs of the Q-network (called the error signal) is squared when an error is calculated. Thus, the error considerably depends on whether the absolute value of an error signal is larger than 1 or not. Error changes result in large gradient changes, leading to unstable learning. To address this problem, the error signal is clipped as follows:

$$e_i = t_i - Q(s_i, a_i),$$

$$e_i \leftarrow \begin{cases} 1 & (e_i \geq 1), \\ e_i & (-1 < e_i < 1), \\ -1 & (e_i \leq -1), \end{cases}$$

where e_i is the ith error signal and the left arrow indicates the substitution. We use the Nature version DQN and simply call it DQN hereafter.

All experiences that may appear should be sufficiently sampled to ensure that the Q-network converges to the optimal action value function. However,

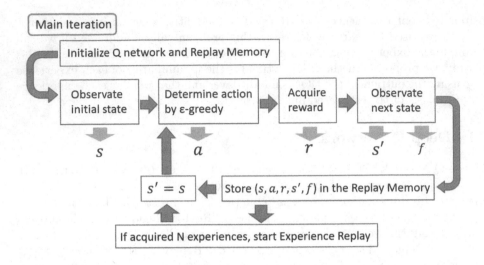

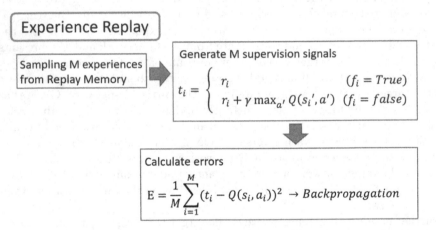

Fig. 1. NIPS version DQN

infrequent experiences have low probability of being sampled. Hence, converging to the optimal action value function takes time.

3   Proposed Method

This work proposes a weighting method by estimating the unfamiliarity of experiences. This unfamiliarity is used to weight experiences. To estimate the unfamiliarity, we introduce a network called "A-network" that predicts the next state s' (shown in vector). Its structure is presented in Fig. 2.

The input of the A-network (shown as $\varphi(s)$ in Fig. 2) is from a middle layer of the Q-network. $\varphi(s)$ shows the novelty of the input state s because

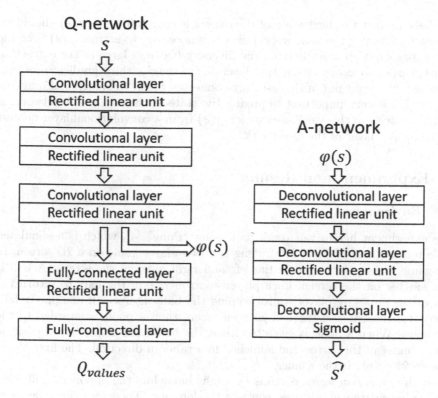

Fig. 2. Proposed network structures

it becomes an unfamiliar value when s is unfamiliar, due to the deep learning characteristic [5]. Consequently, the prediction of s' (i.e., the output of the A-network) considerably deviates. Therefore, if we use the prediction error of the A-network as a weight of an experience, then providing a large weight to an unfamiliar experience that has minimal appearance is possible.

Suppose that the ith experience in the Replay Memory has been sampled. Then, its weight W_i is updated using the following formula:

$$W_i = \sum_j (s'_{i,j} - s^{\hat{}}_{i,j})^2$$

where $s'_{i,j}$ is the jth element of the next state s' contained in the ith experience and $s^{\hat{}}_{i,j}$ is the jth element of the A-network output for the ith experience. Finally, the probability P_k of sampling the kth experience from the Replay Memory is calculated using the weight W_k of each experience as follows:

$$P_k = \frac{W_k}{\sum_n W_n}.$$

Note that repetitive learning using an experience gradually decreases its prediction error and weight of the experience because the A-network will eventually, precisely predict the next state of the experience. In addition, $\varphi(s)$ should be diverse for learning the A-network. Hence it is necessary to extract $\varphi(s)$ from the Q-network's output layer because the diversity becomes large as the extraction position $\varphi(s)$ is closer to the output layer due to the vanishing gradient problem. However, in the output of the last fully-connected layer, the positional information, which is very important to predict the state s' through the A-network, is lost. Therefore, in this work, we extract $\varphi(s)$ from a convolutional layer nearest to the output layer of the Q-network.

4   Experiment and Results

4.1   Environment

The experiment herein was conducted using "Pong" [4], which is a simulated table tennis game of Atari 2600. Pong is a two-player game on a 2D screen. In the game, bars are present on the left and right sides of the screen. A ball is also moving on the screen. Each player controls one of the bars to return the ball to the opposing player whilst keeping the ball onscreen. If one player fails to control the bar and the ball goes offscreen, then a point is awarded to the opponent. When a point is given to either, the ball is set at the initial position at the center of the screen and launched in a random direction. The first player to earn 21 points is the winner.

In this work, an agent controls the right bar while the opponent, following fixed rules written in advance, controls the left one. The agent can take three actions as follows: move up, move down, and stop on the spot. The agent learns its action from rewards given to it, each of which is +1 when it scores a point, while −1 when the opponent scores a point. Otherwise, the score is 0. One episode ends when the winner is declared.

This environment is selected because there is a wide variety of the appearance rate of experiences. Specifically, at the beginning of learning, the agent is expected to keep losing a game at almost no points; hence, the positive-rewarded experiences rarely appear. Meanwhile, the negative-rewarded experiences appear but are much less than the non-rewarded ones. Furthermore, among the non-rewarded experiences, the appearance rates vary because the ball always returns to the initial position after either one earns a point.

In this experiment, we verify that the proposed method will properly weight each experience based on its appearance rate and improve learning speed in this environment.

4.2   Settings

The optimal hyperparameters and the number of the A-network's layers obtained as a result of preliminary experiments are shown in Table 1. The "Mini batch size" M is the number of experiences sampled from the Replay Memory, and the "Replay Memory size" is the maximum number of experiences that can be

Table 1. Hyperparameters

Hyperparameters	Values
Discount rate (γ)	0.99
Initial value of ϵ	1.0
Final value of ϵ	0.1
Mini batch size (M)	32
Replay memory size	100000
Experience replay start size (N)	10000
Update frequency of the target network	10000
Learning rate of the Q-network	0.00025
Learning rate of the A-network	0.00025
The number of the A-network's layers	3

stored in the Replay Memory. The "Experience Replay start size" is the number of experiences necessary to start the Experience Replay. The "Update frequency of the target network" is the number of state transitions that the agent has experienced until the target network is updated. Moreover, all of the output values of the Q-network before learning become 0 by initializing all the weights of the last fully-connected layer to 0. The weights of other layers are initialized using a random number from 0 to 1. The computer we used is composed of two Intel Xeon E5-2650 v4 CPUs and two NVIDIA GeForce GTX 1080 Ti GPUs, with a main memory of 64GB.

4.3 Results

Figures 3 and 4 show the weights of experiences in the Replay Memory after the agent acquired 100,000 experiences, and Fig. 5 shows the weights of experiences in the Replay Memory after the agent acquired 16,000,000 experiences.

As shown in Fig. 3, no positively-rewarded experience is acquired, but four negatively-rewarded ones are present, indicating that the opponent scored four consecutive points, whereas the agent did not acquire a point. In addition, given that the negative-rewarded experiences are arranged at substantially similar intervals, the ball was released from the initial position to the agent's side on three times but the agent could not hit it back. Furthermore, we can read that the weights before a negative-rewarded experience gradually increased but suddenly decreased after it. This result possibly shows a difference among the appearance rates of non-rewarded experiences because the ball was more frequently in the initial position than any other position.

Figure 4 shows that a particularly large weight is given to both the positive-rewarded experience and the experiences leading to it. Meanwhile, Fig. 5 shows that a positive-rewarded experience had a smaller weighting because the experience was sufficiently used for learning and the error of the A-network con-

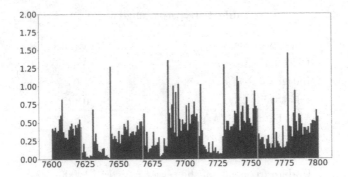

Fig. 3. Weights of experiences from the 7600th to the 7800th: The horizontal axis is the index of experiences in the Replay Memory, which is in chronological order. The vertical axis is the weight of experiences. Blue, orange, and red indicate that the experience with reward 0, −1, and +1, respectively. (Color figure online)

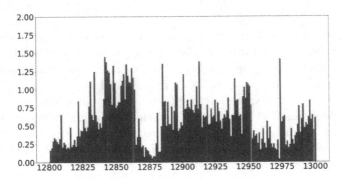

Fig. 4. Weights of experiences from the 12800th to the 13000th: this graph is drawn similar to Fig. 3

sequently decreased. Thus, we know that the proposed method provides the appropriate weightings to the experiences.

Figure 6 shows the learning curves of the scores. As presented in Fig. 6, we confirm the large difference at the beginning of learning. Unlike DQN, the proposed method has acquired higher scores with fewer frames. Moreover, by comparing the learning time until the moving average reaches 0 for the first time, the proposed method was approximately 32% faster than DQN.

Figure 7 shows the learning curves of the average of the maximum Q value. Given that the initial action values are 0 and the learning rate is small, they only change gradually. Therefore, this index should be utilized for measuring the progress of learning. Figure 7 shows that the proposed method immediately obtained a higher average than DQN. The positive-rewarded experiences are repeatedly sampled to ensure that the average of maximum Q value is large, and positive rewards will be transmitted to the previous action values. Therefore,

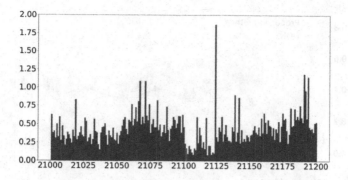

Fig. 5. Weights of experiences from the 15921000th to the 15921200th: this graph is drawn similar to Fig. 3. Note that the horizontal axis shows only the last five digits.

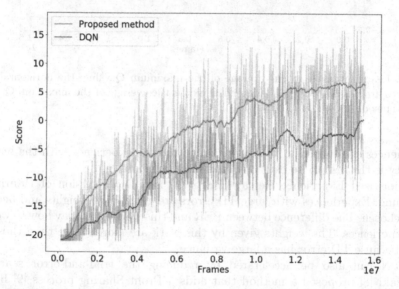

Fig. 6. Learning curves of the scores: the horizontal axis is the total number of game screen frames observed by the agent, whereas the vertical axis is the total reward acquired by the agent in one episode. The blue line shows the result of DQN, and the green line is the result of proposed method. The bold lines are the moving average. (Color figure online)

the proposed method preferentially learns the action value function from the positive-rewarded experiences and the experiences leading to them.

5   Related Works

It is common practice in the field of machine learning to weight the learning data. For example, Prioritized Sweeping [6] is popular in reinforcement learning in which each experience has a priority calculated from TD error. Stored

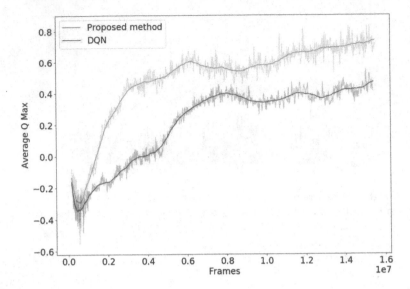

Fig. 7. Learning curves of the average of the maximum Q value: the horizontal axis and lines are similar to Fig. 6. The vertical axis is the average of the maximum Q value in one episode.

experiences are queued in order of priority, and the experience with the highest priority is the first to be extracted and used for learning.

Prioritized Experience Replay (PER) [7] is a DQN version of Prioritized Sweeping. Experiences with large TD errors are given large weights and become more chosen. The difference between PER and this work is the way how to weight the experiences. The weights given by this work are more stable than those by PER because TD error has a large variance.

DQN can also be accelerated by reducing the trial-and-error searches. Miyazaki [8] proposed a method that adds a Profit Sharing process [9] before the Experience Replay. Unlike the Experience Replay, Profit Sharing updates a sequence of action values at once. DQN is accelerated by this updates.

6   Conclusion

In this study, we proposed a weighting method for experiences on the basis of the appearance rate. The method assigned large weights to infrequent experiences, whereas frequent experiences were provided with small weights. On the basis of the experiments in the Pong environment wherein the appearance rate has a large difference, infrequent experiences were provided with large weights. Finally, the proposed method acquired higher performance in fewer frames than DQN.

DQN is effective in many environments as well as Pong. We believe that the proposed method is also versatile, but it is a future work to confirm this belief.

References

1. Abbeel, P., Coates, A., Ng, A.Y.: Autonomous helicopter aerobatics through apprenticeship learning. Int. J. Robot. Res. **29**(13), 1–31 (2010)
2. Mnih, V., et al.: Playing Atari with deep reinforcement learning. arXiv:1312.5602v1 (2013)
3. Mnih, V., et al.: Human-level control through deep reinforcement learning. Nature **518**, 529–533 (2015)
4. Brockman, G., et al.: OpenAI Gym. arXiv:1606.01540 (2016)
5. Rumelhart, D.E., Hinton, G.E., Williams, R.J.: Learning representations by back-propagating errors. Nature **323**, 533–536 (1986)
6. Moore, A.W., Atkeson, C.G.: Prioritized sweeping: reinforcement learning with less data and less time. Mach. Learn. **13**, 103–130 (1993)
7. Schaul, T., Quan, J., Antonoglou, I., Silver, D.: Prioritized experience replay. In: International Conference on Learning Representations (2016)
8. Miyazaki, K.: Exploitation-oriented learning with deep learning introducing profit sharing to a deep Q-network. J. Adv. Comput. Intell. Intell. Inform. **21**(5), 849–855 (2017)
9. Miyazaki, K., Yamamura, M., Kobayashi, S.: A theory of profit sharing in reinforcement learning. J. Jpn. Soc. Artif. Intell. **9**(4), 580–587 (1994). (in Japanese)

Exploring Deep Learning Architectures Coupled with CRF Based Prediction for Slot-Filling

Tulika Saha[✉], Sriparna Saha, and Pushpak Bhattacharyya

Department of Computer Science and Engineering,
Indian Institute of Technology Patna, Dealpur Daulat, India
{sahatulika15,sriparna.saha,pushpakbh}@gmail.com

Abstract. Slot-filling is one of the most crucial module of any dialogue system that focuses on extracting relevant and necessary information from the user utterances. In this paper, we propose variants of Long Short-Term Memory (LSTM) and Gated Recurrent Unit (GRU) models for the task of slot-filling which includes LSTM/GRU networks, Bi-directional LSTM/GRU (Bi-LSTM/GRU) networks, LSTM/GRU-CRF and Bi-LSTM/GRU-CRF networks. Variants of LSTM/GRU is used for discourse modeling i.e., to capture long term dependencies in the input sentences. A Conditional Random Field (CRF) layer is integrated with the above network to capture the sentence level tag information. We show the experimental results of our proposed model on the benchmark Air Travel Information System (ATIS) dataset which indicate that our model performed exceptionally well compared to the state of the art.

Keywords: Dialogue system · Natural language understanding
Slot-filling · LSTM · GRU · CRF

1 Introduction

Natural Language Understanding (NLU) forms one of the most critical module of any dialogue system. Understanding the real intention of the user and to identify the relevant information from the user query - often referred to as *slot-filling*, is fundamental for any human-computer interaction. The NLU module typically consists of the following three tasks: Dialogue Act Classification (DAC), Intent Detection (ID), and slot-filling. With considerable advancement of deep learning (DL) for sentence classification such as DAC [6,7] and ID [2,18], this paper focuses on employing DL based approach to slot filling.

Slot-filling is basically searching of user texts to extract relevant information in order to fill predefined slots in a reference knowledge base [3,17]. Slot-filling is often framed as a sequence labeling task, which maps an observation sequence x = { $x_1, \ldots, x_T$ } to a sequence of labels y = { $y_1, \ldots, y_T$ }, i.e., to acquire the most probable slot sequence given some word sequence. An example of an user

© Springer Nature Switzerland AG 2018
L. Cheng et al. (Eds.): ICONIP 2018, LNCS 11301, pp. 214–225, 2018.
https://doi.org/10.1007/978-3-030-04167-0_20

utterance along with its slot labels are shown in Table 1. The most extensively used idea to solve this problem is the application of Conditional Random Fields (CRFs) [8], where given the input sequence, the probability of a label sequence is computed using an exponential model. Therefore, CRF produces distinct and globally most likely label sequence and it has been applied broadly in [13, 17, 20]. Machine Translation models [10] and Maximum Entropy Markov Models (MEMMs) [16, 17] are some of the other sequence labeling methods that have been studied for this task. The recent growth and success of deep learning has motivated to it being employed for solving the slot-filling task as well. Some of the most notified works include [9, 12, 19, 22, 23] where variations of Recurrent Neural Network (RNN) models have been studied extensively because of their strong potential in modeling temporal dependencies. In this paper, we propose variants of RNN such as LSTM [4], Bi-LSTM, GRU [1] and Bi-GRU to incorporate past and future input features coupled with a CRF layer to model the sentence level tag information; thus, producing state of the art results for the task.

Table 1. An example utterance with its slot

Utterance	Show	me	flights	from	atlanta	to	washington
Slot	O	O	O	O	B-fromloc.city_name	O	B-toloc.city_name

The remaining of the paper is arranged as follows: Sect. 2, presents a brief description of the related works followed by the motivation and contribution of this particular work. The proposed methodology has been discussed in Sect. 3. Section 4 examines the experimental results and its analysis. Lastly, the conclusion and the course for future work are discussed in Sect. 5.

2 Related Works

This section provides a brief description of the works done so far on slot-filling followed by the motivation behind solving this problem.

2.1 Background

Different RNN architectures, including the Jordan-type and Elman-type recurrent networks and their variants were implemented in [12] on the ATIS dataset. They reported a F1-score of 93.98. In another such work, [22] implemented a variation of RNN incorporating context words as features along with some lexical and non-lexical features. They reported a F1-score of 96.60 on the ATIS dataset. In one of the works, [19] proposed a sequential convolution neural network model with previous context words as features and gives attention to current words with its surrounding context. They reported a F1-score of 95.61 on the ATIS dataset. Variants of RNN architecture were presented in [11] that uses the objective function of a CRF, and thereby the RNN parameters are trained based on this objective function, i.e., the whole set of model parameters, including RNN parameters and transition probabilities, are trained jointly. They reported a F1-score of 96.46.

2.2 Motivation and Contribution

Identification of the correct slots can assist an automated system to produce an appropriate response thereby helping the system in resolving the queries of the user. The problem becomes more challenging and difficult when the system needs to handle more realistic, natural utterances expressed in natural language, by a number of speakers. Irrespective of the approach being adopted, the problem is the "naturalness" of the spoken language input. Though RNNs and its variants have been used extensively for slot-filling task but they didn't model label sequence dependencies explicitly. The tokens in a sentence share a dependency with each other in order to capture context information which is addressed using RNN and its variants. Based on this dependency, tags are assigned to each tokens to model this problem. Similarly, the tags assigned to each tokens share dependency with each other which can add valuable information for modeling this sequence labeling problem. Therefore, in this work, we study and assess the effectiveness of using variants of LSTM and GRU networks for slot-filling, with significant attention on modeling label sequence dependencies.

The major contributions of this work are:

- A novel LSTM/GRU network is proposed that takes in past input features coupled with the CRF layer to incorporate the sentence level tag information in order to model label sequence dependencies.
- The proposed model is extended to a Bi-directional LSTM/GRU which incorporates the information from past and future words for prediction along with the CRF layer.
- Experimental analysis of all the models have been presented in detail.

3 Proposed Methodology

In this section the proposed methodology which includes the baseline and proposed models are described in detail.

3.1 Baseline Models

Previously, approaches such as CRF, LSTM and Bi-LSTM have been used to model the task of slot-filling. With the introduction of GRU, it has also found significant attention because of its comparable performance to LSTM. Therefore, we implement each of them as our baseline models to observe their performance and influence.

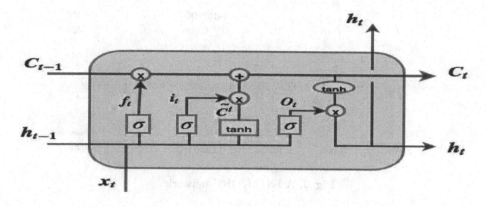

Fig. 1. A LSTM cell

- **Model 1: LSTM Networks.** LSTMs are similar to RNNs with the exception that the updates of the hidden layer in RNNs are changed by purpose-built memory cells in LSTMs. Because of which they are comparatively good in identifying and modeling long range dependencies in input data. A typical LSTM cell is shown in Fig. 1[1]. The working of the LSTM cell is as follows:

$$f_t = \sigma(W_f \cdot [h_{t-1}, x_t] + b_f) \tag{1}$$

$$i_t = \sigma(W_i \cdot [h_{t-1}, x_t] + b_i) \tag{2}$$

$$\tilde{C}_t = tanh(W_C \cdot [h_{t-1}, x_t] + b_C) \tag{3}$$

$$C_t = f_t * C_{t-1} + i_t * \tilde{C}_t \tag{4}$$

$$o_t = \sigma(W_o \cdot [h_{t-1}, x_t] + b_o) \tag{5}$$

$$h_t = o_t * tanh(C_t) \tag{6}$$

where f_t, i_t, o_t are the forget, input and output gate, respectively. C_{t-1}, C_t are cell states at time-step $t-1$ and t, respectively. h_{t-1}, h_t represent hidden state vectors at time-step $t-1$ and t, respectively. W_f, W_i, W_o represent hidden-forget gate, hidden-input gate, hidden-output gate matrix, respectively. Logistic sigmoid function is represented by σ. Figure 2 shows a LSTM based slot-filling model which implements the above mentioned LSTM cell at its core. Pre-trained word embeddings have been used to represent input words as word vectors. The output represents a probability distribution over labels at time t.

- **Model 2: GRU Networks.** GRUs are similar to LSTMs but the key difference is that a LSTM has three gates particularly forget, input and output gates whereas GRU has two gates which are reset and update gates. Analogous to the LSTM unit, the GRU unit also supervises the flow of information, but does so without using a memory unit. It simply unmasks the entire hidden content without any restriction. The performance of GRU is comparable

---

[1]   https://isaacchanghau.github.io/post/lstm-gru-formula/.

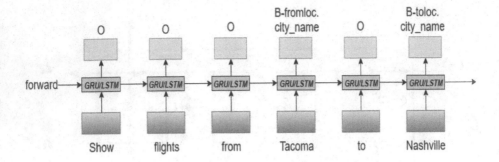

Fig. 2. A LSTM/GRU network

to that of LSTM, but it is computationally more efficient. Figure 3[2] shows a typical GRU cell. The working of the GRU unit is as follows:

$$Z_t = \sigma(W_z \cdot [h_{t-1}, x_t]) \tag{7}$$

$$r_t = \sigma(W_r \cdot [h_{t-1}, x_t]) \tag{8}$$

$$\tilde{h}_t = tanh(W \cdot [r * h_{t-1}, x_t]) \tag{9}$$

$$h_t = (1 - z_t) * h_{t-1} + z_t * \tilde{h}_t \tag{10}$$

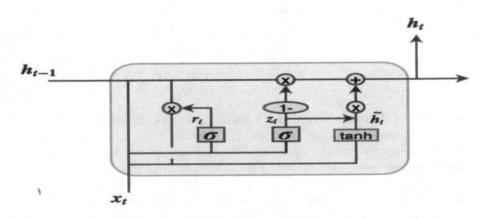

Fig. 3. A GRU cell

where z_t and r_t are update and reset gates, respectively. h_{t-1}, h_t represent hidden state vector at time-step $t-1$ and t, respectively. W_z, W_r represent hidden-update gate, hidden-reset gate matrix, respectively. Logistic sigmoid function is represented by σ . Figure 2 shows a GRU based slot-filling model which implements the above mentioned GRU cell at its core.

---

2   https://isaacchanghau.github.io/post/lstm-gru-formula/.

– **Model 3: Bi-directional LSTM/GRU Networks.** Use of LSTM/GRU units provides access to just past input features. Thus, utilizing a bi-directional LSTM/GRU networks provides access to both past (through forward states) and future (through backward states) input features for a particular time frame. Figure 4 shows a bi-directional LSTM/GRU based slot-filling model.

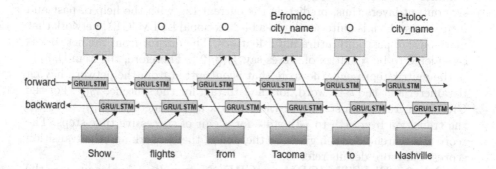

Fig. 4. A bi-directional LSTM/GRU network

– **Model 4: CRF Networks.** A basic CRF model has been implemented with input word and its Part-of-Speech tag[3] as features. Figure 5 shows a CRF based slot-filling model. CRFs work on sentence level rather than individual position; thus, taking the context into account. CRFs, in general have been seen to perform reasonably good for sequence labeling task.

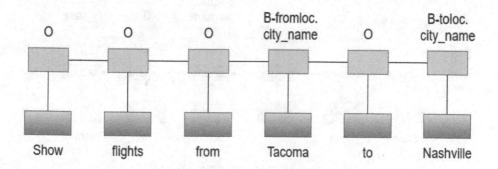

Fig. 5. A CRF network

---

[3] Used Stanford PoS tagger https://nlp.stanford.edu/software/tagger.shtml.

3.2   Proposed Models

- **Model 1: LSTM/GRU - CRF Networks.** This particular approach combines a LSTM/GRU network with a CRF network to obtain a LSTM/GRU-CRF model as shown in Fig. 6. The idea behind such an approach is that this network can then efficiently utilize the past input features because of the presence of the LSTM layer followed by a CRF layer which can then add sentence level tag information. CRF layer is shown by lines that joins successive output layers thus, predicting the current tag with the help of past and future tags which is quite similar to a bi-directional LSTM/GRU network that makes use of past and future input features. The output from the network is considered to be a matrix of scores say $f_\theta([y]_1^K)$. Therefore, the item $[f_\theta]_{(i,k)}$ of the matrix represents the score that is outputted from the network having parameter θ at the k-th word, for the i-th tag, for the sentence $[y]_1^K$. For the CRF layer, there is a state transition matrix as parameters $[A]_{(i,j)}$ to model the transition from i-th to j-th state for a pair of successive time-steps. The score of a sentence is then given by the sum of the network and the transition scores. For more details refer [5,8].
- **Model 2: Bi-LSTM/GRU - CRF Networks.** Analogous to the LSTM/GRU-CRF network, this particular model combines a bi-directional LSTM/GRU network with a CRF model to obtain a Bi-LSTM/GRU-CRF model shown in Fig. 7. Therefore, along with the past input features and sentence level tag information as used in a LSTM/GRU-CRF model, the model utilizes the future input features as well. The training algorithm for the Bi-LSTM/GRU-CRF model is shown in Algorithm 1. For more details of the algorithm, refer [5]. All the proposed models in this paper use a generic Stochastic Gradient Descent forward and backward training method.

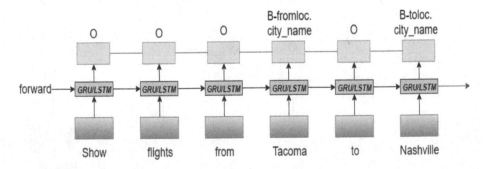

Fig. 6. A LSTM/GRU-CRF model

4   Experimentation, Results and Analysis

This section demonstrates the experimentation, results and analysis of all the proposed approaches. Number of utterances in training, validation and testing

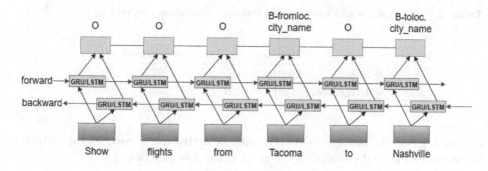

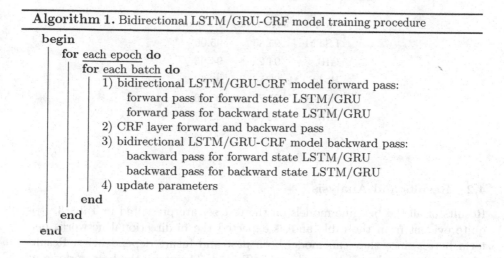

Fig. 7. A Bi-LSTM/GRU-CRF model

Algorithm 1. Bidirectional LSTM/GRU-CRF model training procedure

begin
    for each epoch **do**
        for each batch **do**
            1) bidirectional LSTM/GRU-CRF model forward pass:
                forward pass for forward state LSTM/GRU
                forward pass for backward state LSTM/GRU
            2) CRF layer forward and backward pass
            3) bidirectional LSTM/GRU-CRF model backward pass:
                backward pass for forward state LSTM/GRU
                backward pass for backward state LSTM/GRU
            4) update parameters
        end
    end
end

set for the benchmark ATIS [15] dataset are shown in Table 2. Since the ATIS dataset does not have a standard validation dataset, part of the training data has been used for the validation purpose.

4.1 Experimentation

For implementing the DNN models, Keras[4] has been used. In the input layer, all the unique words of the corpus are given some sequence numbers which are fed to the embedding layer. Pretrained GloVe [14] embedding trained on the CommonCrawl corpus of dimension 300 has been used to represent input words as word vectors. The resultant word embeddings from the input layer are fed to the LSTM/GRU layer for discourse modeling. Number of LSTM/GRU units in a layer is 100. A learning rate of 0.1 is used to train the models. For the baseline models, the number of units in the dense layer is equivalent to the number of

---

[4] https://keras.io/.

Table 2. No. of tokens and sentences in training, validation and testing sets of ATIS dataset

	Train set	Validation set	Test set
# Tokens	47604	8987	9198
# Utterances	4181	797	893

unique tags in the tag-set. Next, the softmax activation is used at the output layer and categorical crossentropy is used as the loss function.

Table 3. Results of all the baseline models

Models	Accuracy	F1-Score
LSTM	95.35	95.06
GRU	94.29	94.95
Bi-LSTM	96.86	96.63
Bi-GRU	95.79	95.56
CRF	72.80	68.91

4.2   Results and Analysis

Results of all the baseline models on the test set are presented in Table 3. It is quite evident from the table and as expected the bi-directional networks perform better since they can model both past and future dependencies. Results of all the proposed models are shown in Table 4. Therefore, the best performing model as seen from the table is that of a Bi-LSTM-CRF model which attains 2% and 26% increments over the corresponding Bi-LSTM and CRF baseline models, respectively in terms of accuracy. We have performed Welch's t-test [21] at 5% significance level and the corresponding results are shown in Table 5. This test signifies that the results produced by all our best performing models are statistically significant.

Table 4. Results of all the proposed models

Models	Accuracy	F1-Score
LSTM-CRF	97.09	96.97
GRU-CRF	96.89	96.38
Bi-LSTM-CRF	98.15	**97.94**
Bi-GRU-CRF	97.71	97.44

Table 5. p-values obtained by Welch's t-test comparing our best performing model with other models

Models	p-values
LSTM-CRF	2.13E−28
GRU-CRF	4.87E−33
Bi-GRU-CRF	3.55E−13

Table 6. Comparison of the proposed approach with the state-of-the-art

Models	F1-Score
RNNs (Mesnil et al. [12])	93.98
RNNs (Yao et al. [22])	96.60
R-CRF (Mesnil et al. [11])	96.46
s-CNN (Vu [19])	95.61
Bi-LSTM-CRF (Our Model)	**97.94**
Bi-GRU-CRF (Our Model)	**97.44**

4.3 Error Analysis

In order to analyze the weakness of the developed model, we have carried out a thorough error analysis of the proposed model. Since the number of unique slot labels in the ATIS corpus is 127, the representation of most of the tags are very less i.e. the dataset is skewed having lesser occurrences of most of the slot labels. This is one of the reasons for the errors. Example utterance such as *"which flights arrive in burbank from las vegas on **saturday april twenty third** in the afternoon"*, here the words marked in bold are wrongly tagged as *"B-arrive_date.day_name"*, *"B-arrive_date.month_name"*, *"B-arrive_date.day_number"*, *"Iarrive_date.day_number"*. It should have been tagged as *"B-depart_date.day_name"*, *"B-depart_date.month_name"*, *"B-depart_date.day_number"*, *"I-depart_date.day_number"*, respectively. Similarly, *"find nonstop flights from salt lake city to new york on **saturday april ninth**"*, have been wrongly tagged as *"B-arrive_date.-day_name"*, *"B-arrive_date.month_name"*, *"B-arrive_date.day_number"* whereas it should have been tagged as *"B-depart_date.day_name"*, *"B-depart_date.month_name"*, *"B-depart_date.day_number"*, respectively. Another such utterance *"does **tacoma airport** offer transportation from the airport to the downtown area"* is wrongly tagged as *"B-toloc.airport_name"*, *"I-toloc.airport_name"* whereas it should have been tagged as *"B-airport_name"*, *"I-airport_name"*, respectively. Mostly, it was found that the errors occurred because the model was not able to distinguish between arrival and departure details.

Comparison with the state-of-the-art approaches. A comparative study has been carried out between our best performing proposed model against the state-of-the-art approaches shown in Table 6. It is evident from the table that our best performing model, Bi-LSTM-CRF and Bi-GRU-CRF, outperformed various state-of-the-art approaches.

5   Conclusions and Future Work

In this paper, various model architectures are proposed for the task of slot-filling to capture the past and future dependencies of the input sentence along with the sentence level tag information. The proposed model outperformed various state-of-the-art approaches on the benchmark ATIS dataset.

In future, we aim to assess the proposed models on datasets belonging to varied domains. Also, we would like to extend our work to investigate different deep learning techniques to increase the accuracy of our model.

References

1. Cho, K., et al.: Learning phrase representations using RNN encoder-decoder for statistical machine translation. arXiv preprint arXiv:1406.1078 (2014)
2. Deng, L., Tur, G., He, X., Hakkani-Tur, D.: Use of kernel deep convex networks and end-to-end learning for spoken language understanding. In: 2012 IEEE Spoken Language Technology Workshop (SLT), pp. 210–215. IEEE (2012)
3. He, Y., Young, S.: A data-driven spoken language understanding system. In: 2003 IEEE Workshop on Automatic Speech Recognition and Understanding, ASRU 2003, pp. 583–588. IEEE (2003)
4. Hochreiter, S., Schmidhuber, J.: Long short-term memory. Neural Comput. **9**(8), 1735–1780 (1997)
5. Huang, Z., Xu, W., Yu, K.: Bidirectional LSTM-CRF models for sequence tagging. arXiv preprint arXiv:1508.01991 (2015)
6. Kalchbrenner, N., Blunsom, P.: Recurrent convolutional neural networks for discourse compositionality. arXiv preprint arXiv:1306.3584 (2013)
7. Khanpour, H., Guntakandla, N., Nielsen, R.: Dialogue act classification in domain-independent conversations using a deep recurrent neural network. In: Proceedings of COLING 2016, the 26th International Conference on Computational Linguistics: Technical Papers, pp. 2012–2021 (2016)
8. Lafferty, J., McCallum, A., Pereira, F.C.: Conditional random fields: probabilistic models for segmenting and labeling sequence data (2001)
9. Liu, B., Lane, I.: Recurrent neural network structured output prediction for spoken language understanding. In: Proceedings of the NIPS Workshop on Machine Learning for Spoken Language Understanding and Interactions (2015)
10. Macherey, K., Och, F.J., Ney, H.: Natural language understanding using statistical machine translation. In: Seventh European Conference on Speech Communication and Technology (2001)
11. Mesnil, G., et al.: Using recurrent neural networks for slot filling in spoken language understanding. IEEE/ACM Trans. Audio Speech Lang. Process. **23**(3), 530–539 (2015)
12. Mesnil, G., He, X., Deng, L., Bengio, Y.: Investigation of recurrent-neural-network architectures and learning methods for spoken language understanding. In: INTER-SPEECH, pp. 3771–3775 (2013)

13. Moschitti, A., Riccardi, G., Raymond, C.: Spoken language understanding with kernels for syntactic/semantic structures. In: IEEE Workshop on Automatic Speech Recognition and Understanding, ASRU, pp. 183–188. IEEE (2007)
14. Pennington, J., Socher, R., Manning, C.: Glove: global vectors for word representation. In: Proceedings of the 2014 Conference on Empirical Methods in Natural Language Processing (EMNLP), pp. 1532–1543 (2014)
15. Price, P.J.: Evaluation of spoken language systems: the ATIS domain. In: Speech and Natural Language: Proceedings of a Workshop Held at Hidden Valley, Pennsylvania, 24–27 June 1990 (1990)
16. Ratnaparkhi, A.: A maximum entropy model for part-of-speech tagging. In: Conference on Empirical Methods in Natural Language Processing (1996)
17. Raymond, C., Riccardi, G.: Generative and discriminative algorithms for spoken language understanding. In: Eighth Annual Conference of the International Speech Communication Association (2007)
18. Tur, G., Deng, L., Hakkani-Tür, D., He, X.: Towards deeper understanding: deep convex networks for semantic utterance classification. In: 2012 IEEE International Conference on Acoustics, Speech and Signal Processing (ICASSP), pp. 5045–5048. IEEE (2012)
19. Vu, N.T.: Sequential convolutional neural networks for slot filling in spoken language understanding. arXiv preprint arXiv:1606.07783 (2016)
20. Wang, Y.Y., Acero, A., Mahajan, M., Lee, J.: Combining statistical and knowledge-based spoken language understanding in conditional models. In: Proceedings of the COLING/ACL on Main Conference Poster Sessions, pp. 882–889. Association for Computational Linguistics (2006)
21. Welch, B.L.: The generalization ofstudent's' problem when several different population variances are involved. Biometrika **34**(1/2), 28–35 (1947)
22. Yao, K., Zweig, G., Hwang, M.Y., Shi, Y., Yu, D.: Recurrent neural networks for language understanding. In: Interspeech, pp. 2524–2528 (2013)
23. Zhang, X., Wang, H.: A joint model of intent determination and slot filling for spoken language understanding. In: IJCAI, pp. 2993–2999 (2016)

Domain Adaptation via Identical Distribution Across Models and Tasks

Xuhong Wei[(✉)], Yefei Chen[(✉)], and Jianbo Su[(✉)]

Shanghai Jiao Tong University, Shanghai, China
{jiushiwolou,marschen,jbsu}@sjtu.edu.cn
http://rcir.sjtu.edu.cn

Abstract. Deep convolution neural network (CNN) models with millions of parameters trained in large-scale datasets make domain adaptation difficult to be realized. In order to be applied for different application scenarios, various light weight network models have been proposed. These models perform well in large-scale datasets but are hard to train from randomly initialized weights when lack of data. Our framework is proposed to connect a pre-trained deep model with a light weight model by enforcing feature distributions of the two models being identical. It is proved in our work that knowledge in source model can be transferred to target light weight model by identical distribution loss. Meanwhile, distribution loss allows training dataset to utilize sparse labeled data in semi-supervised classification task. Moreover, distribution loss can be applied to large amount of unlabeled data from target domain. In the experiments, several standard benchmarks on domain adaptation are evaluated and our work gets state-of-the-art performance.

Keywords: Domain adaptation · Model compression
Identical distribution · Semi-supervised method

1 Introduction

Deep convolution neutral network (CNN) models trained on large-scale dataset have been successfully applied to face recognition such as [24,33]. These deep models perform well on standard benchmarks, but are difficult to use in practical applications directly. Practically, image quality is influenced by different camera specifications and its installation angle, which may not satisfy the requirements of pre-trained models. Moreover, when the model is utilized on mobile devices with limited computational capacity, it is hard to process the image with those deep models within limited time.

Method of fine-tuning [10,31] the pre-trained model with labeled data usually requires hundreds or thousands of samples for each extra category. Generally, fine-tuned model cannot be directly transplanted to devices with weak computational capacity like embedded system. Compressing the models to an adequate size by fine-tuning still requires a large amount of labeled data from

© Springer Nature Switzerland AG 2018
L. Cheng et al. (Eds.): ICONIP 2018, LNCS 11301, pp. 226–237, 2018.
https://doi.org/10.1007/978-3-030-04167-0_21

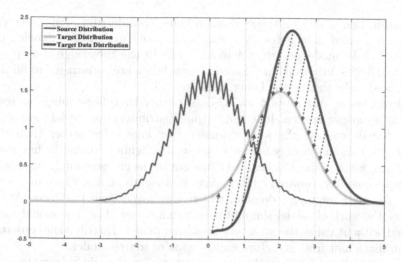

Fig. 1. Source and target domain distribution are assumed to be related, but source model are more complicated (sawtooth). Dataset bias exists in source and target domain. Target data (green) are not in accordance with real target domain distribution (yellow) completely. The red dotted arrow line represents that source domain distribution is utilized to force target data distribution approaching to real target distribution, by fitting the source and target distribution identically. (Color figure online)

target domain during the whole processing. Generally, fine-tuned model cannot be directly transplanted to devices with weak computational capabilities like embedded system. Compressing the deep model to an acceptable size might damage the accuracy of the original model [12], and require lots of labeled data. It is intuitive to think of combining domain adaptation and model compression problem together to improve effect of integration degree about task and model.

In this paper, constraints on distribution of source model and target model are used to realize such idea. Firstly, constraints directly on softmax layer might not be a good method. Data might lie in a manifold which is easy for classification. Various methods are proposed to find such manifold. Recent works [13] point out that triplet loss is more helpful for training and classification rather than softmax, though triplet loss only makes constraints on the relationship of distributions from the same or different classes without knowing the ground-truth of distributions. It implies the fact that distribution contains more information and knowledge of source data, while softmax loss only uses the information of labels. Thus, knowing the complete knowledge of distribution should be more helpful. The general situation of domain adaptation is described as Fig. 1, which exhibits the relationship between source domain, target domain and target data distribution. Even though target data does not share same label space with source, they could also be evaluated by source domain distribution.

Secondly, distribution loss takes advantage of large amount of unlabeled data on target model training. Labelled data might act like anchor points that enforce the classification ability on target domain being considered. Learning distributions on unlabeled data is totally a regression problem. Those unlabeled data helps the concise model on target domain to inherit the knowledge of deep models learned from source domain. Those unlabeled data are encouraged to fill those hollows and make the models being more general.

Thirdly, using distribution constraints means that fine-tuning on source model is no longer needed. It is the output distribution of models learned by source domain contain the key information or knowledge rather than those parameters. Parameters only permute in a small neighbor during the fine-tuning procedures, which make it difficult to climb out of the extreme minimum reached by previous models. However, distribution loss means that a whole new model can be designed for target domain, which may avoid such problems. At last, it makes the work of model simplification much easier. The new model can be designed without using the structure of source model. Distribution constraints perform like a hint from structure and weights of source model.

The method proposed in this paper can realize domain adaptation tasks and model transfer simultaneously. It is based on the idea that the output feature distribution in source domain, contain the key information and knowledge. Knowledge could be inherited by target domain via fitting their feature distribution identically. Nevertheless, constraints on distributions can be applied to large amount of unlabeled data from target domain, which is a good property desired by transfer learning.

The paper is organized as follows. Some related works are reviewed in Sect. 2. And in Sect. 3, the establishment processing of marginal feature distribution constraints are explained in detail. Then three related experiments are utilized to exhibit the effect of our distribution constraints. Finally, the result are completely discussed on our experiment.

2   Related Work

Domain adaption [23] builds frameworks that connect different domains or tasks. Many approaches have been proposed in computer vision [15,26], natural language processing [5], to mitigate the burden of manual label [7,34]. Domain adaptation problems are mainly settled at releasing data shifts in probability distributions of different domains, which can be measured by the subspace in different domain [8,21]. When utilized to domain adaptation, deep neural networks [11,15,22] have better performance than prior shallow transfer learning methods.

For the supervised adaptation scenario, sparse target labeled data is available to train target model. Some approaches are proposed to train a target classifier against source classifier [1,3,32] or learn a feature transformation method to regularize target model simultaneously [16]. To reduce dataset shift influence, some recent work [9,30,31] bridges the source and target in model structure, but the extended structure increases complexity of tasks.

CNN based feature representations have shown better performance in many visual recognition tasks [6,19,27]. In particular, it is efficient to use deep CNN structure that pre-trained on large-scale dataset(ex: ImageNet [25]). Some methods [30,31] share convolution weights from source model to target, but might trap the target model into a situation of local convergence. Inspired by using CNN softmax output layer to distill distributions or model knowledge instead of category labels [2,14], fitting source marginal feature distribution to target domain are proposed in this paper, which does not require any struct connection or parameter sharing in the whole transfer framework.

3   Transfer CNN Distribution Knowledge Across Model and Domain

A complete framework of our approach is shown in Figs. 2 and 3, that utilizes a pre-trained model to transfer distribution knowledge to target model.

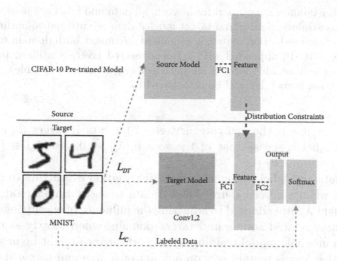

Fig. 2. To avoid source label space and source data disturbance, only target data marginal feature distribution in both model are set to constrain target model training. CIFAR-10 and MNIST are different visual domain dataset. The transfer framework could extract distribution from any well preformed pre-trained source model to assist establishment of target model.

Suppose there are source data $\{x_S, y_S\}$, with the feature representation $f(x; \theta_S)$, and target data $\{x_T, y_T\}$ with feature $g(x; \theta_T)$. Target feature layer is connected with classifier layer θ_C. Our goal is to optimize target model parameter θ_T and θ_C to obtain correct classified result at test time.

For a K categories classifier setting, our model loss function is defined as standard softmax cross entropy loss,

$$L_c = -(\lambda_{sl} \sum_K Y(y_s = k) \log p_k + \sum_K Y(y_t = k) \log p_k), \qquad (1)$$

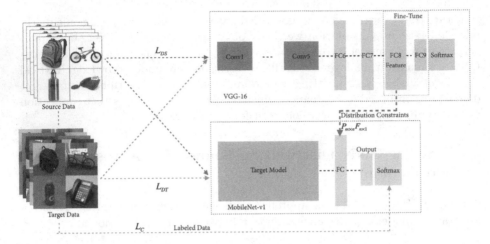

Fig. 3. In Office domain adaptation framework, all data and label are used to support distribution constraints. Source and target labeled data are utilized to optimize target model. A dimension reduction matrix is generated to convert both domain to the same feature space. VGG-16 appended with a full-connected layer, is utilized to fine-tune source domain classifier. MobileNet is assumed as a target tasks acceptable CNN model, where distribution knowledge will be transferred to.

where $Y(\cdot) = 1$ if y is the kth category, else $Y(\cdot) = 0$. If target data does not in the same label space as that of the source, $\lambda_{sl} = 0$, else $\lambda_{sl} \in [1, 0)$, and $p_k = softmax(\theta_C^T g(x; \theta_T))$.

Source data and label will cause target distribution overfitting to the source distribution, when enforced source labeled data to assist target model training process. Weight λ_{sl} are adapted to adjusting the influence of overfitting according to (1). In most related source and target domain, when mostly sharing their labeled data in whole model training, good performance might be arisen.

Distribution loss is mainly rely on amount of source and target data, which is like model probes, regardless of whether they are labeled or not. These data could be regarded as data labeled on feature space by means of distribution. When data transfer knowledge from source model marginal feature distribution to the target, source feature play the role of data label and target feature is desired to fit to the source. To adjust the strength of fitting, source and target data are separated to two independent constituents to build distribution loss.

$$L_{DS} = ||Pf(x_S; \theta_S) - g(x_S; \theta_T)||_2^2, \tag{2}$$

$$L_{DT} = ||Pf(x_T; \theta_S) - g(x_T; \theta_T)||_2^2, \tag{3}$$

$$L_D = \lambda_{ds} L_{DS} + \lambda_{dt} L_{DT}, \tag{4}$$

where P is the dimension reduction matrix that converts source and target feature distribution into the same space. Meanwhile, λ_{ds} and λ_{dt} determine how strongly the distribution constraints influence the optimization process.

Identical distribution can transfer feature knowledge without any labeled data. As is shown in L_D loss, we do not require source and target data contain label information. The data is as similar as a probe that searches knowledge in source model and fits the model distribution to the target. Intuitively, the feature distribution contains plenty of information to describe a model regardless of its whole structure. Therefore, distribution loss is robust on target model structure, and will not be limited by source model structure. In order to minimize the sum of classifier constraints loss (1) and identical distribution loss (4),

$$L = L_c + \lambda_{ds}L_{DS} + \lambda_{dt}L_{DT}. \tag{5}$$

The ideas of domain adaptation via identical distribution loss are universally applicable since it can be applied to almost any couple of source and target CNN structures. By means of distribution fitting, our source domain knowledge can be transferred to generally related target domain.

For the review of this section, we explain the design ideas of our distribution loss. Our novel loss is not limited by target model structure and high correlation of source and target tasks. The hyper-parameters could be adjusted during the training process to conduct target model performance.

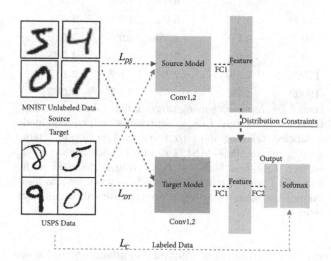

Fig. 4. The red dotted arrow line represents marginal feature distribution. Blue and green dotted line represent our data are evaluated by both models. Feature output layer transfers the distribution knowledge. In our method, convolution layers are not necessarily to shared weights before training. (Color figure online)

4   Evaluation

Firstly, domain knowledge is transferred from MNIST to USPS dataset with the same model structure to evaluate identical distribution constraints. Then, the proposed method is evaluated in a pre-trained CIFAR-10 dataset network when is transferred to a non-related target domain application, MNIST, to observe the ability of feature distribution. Moreover, the framework is evaluated on the Office dataset [26], a standard benchmark dataset for visual domain adaptation.

4.1   Application on Handwritten Digit Dataset

In order to verify the ability of distribution constraints, a CNN network based on LeNet5 [20] has been trained with full MNIST training dataset (Fig. 4). The accuracy of model is 99.2% on MNIST test dataset but 23.1% on that of USPS.

Table 1. MNIST transfer to USPS

USPS Label num	Constraints condition test error rate			
	$L_C(\lambda_{sl} = 0)$	$+L_{DT_l}$	$+L_{DT_{unl}}$	$+L_{DS_{10^4}}$
1000	11.8%	11.2	9.1%	**7.2%**
2000	7.7%	7.4%	**5.0%**	5.3%
3000	6.1%	6.0%	4.7%	**4.5%**
4000	5.4%	5.0%	4.4%	**3.8%**
5000	4.7%	4.3%	3.9%	**3.6%**
10000	3.9%	3.7%	2.9%	**2.7%**
15000	2.8%	**2.6%**	2.8%	2.6%
19000	2.7%	2.5%	2.7%	**2.3%**

Item L_{DT_l} represents using target labeled data to fit distribution, and item $L_{DT_{unl}}$ represents adding remained unlabeled training data to transfer distribution. Item $L_{DS_{10^4}}$ means 10000 MNIST data are randomly sampled to feed L_{DS} loss function. Error rate is declined with increasing labeled number. Meanwhile, when we add loss ingredients stage by stage, the error rate will be mostly reduced.

For the purpose of getting rid of source label advantage in knowledge transfer experiment, λ_{sl} is set to 0. Source unlabeled data are regarded as fitting problems in L_{DS}. Then, the constraints are added stage by stage to evaluate their individual effect.

We randomly extract 10000 samples in MNIST and increased the number of labeled data in USPS to test. The remained data in USPS will be set as target unlabeled data, where $\lambda_{ds} = 5 \times 10^{-4}$, $\lambda_{dt} = 5 \times 10^{-4}$, and learning rate of 0.001.

Compared with signal L_C loss error rate (Table 1), every stage of constraint jointed to distribution loss makes contribution to domain transfer structure (Fig. 4). In the framework, source and target unlabeled data transfer marginal feature distribution which estimated by source model.

4.2 Distribution Loss Applies on Disparate Domains

This experiment is designed to transfer domain knowledge between two disparate label space datasets, which is from CIFAR-10 to MNIST. CIFAR-10 is a 10 classes objection dataset (airplane, dog, ship, . . .), and MNIST is a handwritten digital dataset. 2000 labeled data and 10000 unlabeled data are randomly sampled from MNIST as training subset. During optimization work, MNIST data are projected to CIFAR-10 feature space by a pre-trained source network. The complete transfer framework is in Fig. 2.

Table 2. CIFAR-10 transfer to MNIST

MNIST unlabel num	MNIST labeled num			
	200	1000	5000	10000
1000	**15.3%**	7.2%	4.6%	3%
5000	15.5%	7.0%	4%	2.6%
10000	16.8%	**6.8%**	**3.9%**	**2.2%**
No Transfer	19.9%	14.4%	9.7%	8.3%

As with experiment on MNIST to USPS, source domain distribution constraints observably assist in reducing the error rate in CIFAR-10 to MNIST.

We randomly sample MNIST labeled subset to 4 different size in Table 2, so as unlabeled subset to 3 different size. Identical distribution reduce the error rate when shared knowledge between domain distribution. Parameters $\lambda_{sl} = \lambda_{ds} = 0, \lambda_{dt} = 5 \times 10^{-4}$ and learning rate of 0.001 are set to experiment system.

Although, MNIST and CIFAR-10 is different classification task, the distribution constrains are still enhance the ability of target model obviously. This experiment provides some evidence that distribution do contain transferable knowledge that could be adapted to similar visual domain.

4.3 Adaptation on the Office Dataset

In order to compare our method with other standard benchmarks, we transfer domain knowledge within a standard domain adaptation problem. The Office dataset collect images from three distinct domains, Amazon, DSLR, and Webcam. Amazon, the largest domain, has 2817 labeled images [26]. The 31 categories in the dataset consist of objects commonly encountered in office settings, such as backpack, headphones, and printer.

In all experiments, a VGG-16 [28] model is fine-tuned which is pre-trained on ImageNet [25] dataset. A fc8 layer is added to protect the pre-trained model feature and can enhance the fine-tuned performance. MobileNet [17] is selected as our target domain model, which is proposed recently and has concise CNN structure. Meanwhile, MobileNet weights can be initialized by ImageNet, and

Table 3. Office dataset domain transfer

Method	$A \to W$	$A \to D$	$W \to A$	
DLID [4]	51.9	—	—	
DeCAF$_6$ S+T[6]	80.7 ± 2.3	—	—	
DANN [29]	53.6 ± 0.2	—	—	
SDT [30]	82.8 ± 0.9	**86.1 ± 1.2**	65.2 ± 0.6	
Source VGG S+T	75.0 ± 0.8	80.1 ± 0.2	59.4 ± 0.6	
Target Non-initialized MobileNet S+T	21.5 ± 1.3	19.2 ± 1.8	14.6 ± 1.2	
Target Initialized MobileNet S+T	55.2 ± 1.2	47.1 ± 0.9	37.8 ± 1.4	
Ours: Non-initialized MobileNet	69.9 ± 0.9	72.0 ± 1.1	56.0 ± 1.3	
Ours: Initialized MobileNet	**91.1 ± 0.5**	82.5 ± 0.7	**75.1 ± 0.8**	
Method	$W \to D$	$D \to A$	$D \to W$	Average
DLIDDLID [4]	89.9	—	78.2	—
DeCAF$_6$ S+T [6]	—	—	94.8 ± 1.2	—
DANN [29]	53.6 ± 0.2	—	71.2 ± 0.0	—
SDT [30]	**98.3 ± 0.3**	66.2 ± 0.4	90.8 ± 0.4	<u>82.4</u>
Source VGG S+T	93.7 ± 0.3	60.0 ± 0.4	**95.9 ± 0.6**	<u>77.4</u>
Target Non-initialized MobileNet S+T	49.8 ± 0.9	13.7 ± 1.5	52.3 ± 1.2	28.5
Target Initialized MobileNet S+T	65.8 ± 0.7	38.3 ± 1.7	70.6 ± 0.6	52.5
Ours: Non-initialized MobileNet	90.7 ± 0.6	63.7 ± 1.0	90.9 ± 0.9	73.9
Ours: Initialized MobileNet	95.4 ± 0.9	**74.5 ± 1.1**	95.2 ± 0.5	**85.6**

MobileNet is much conciser than SDT [30] which is based on CaffeNet [18]. Source VGG S+T is taken as source model, which is VGG-16 [28] based. The source model is fine-tuned by source and target data. Our method has top-2 accuracy performance in Office domain adaptation and highest average accuracy rate. Simply fine-tuning a MobileNet cannot reach the original Source VGG S+T accuracy, even though ImageNet per-trained MobileNet weights are shared before training processing. The best result is finally obtained in conciser target model after fitting distribution from source model.

both initialized and non-initialized model are used to evaluate performance in our framework.

We generate subset of each domain by randomly sampling dataset into 5 group splits. Source domain is followed the standard protocol that each split contains 20 examples per category for the Amazon domain, and 8 examples per category for the Webcam and DSLR domains. 3 labeled examples are extracted from each category in the target domain, and the remained data are set as unlabeled samples. Accuracies are reported on the rest unlabeled images by following the standard protocol.

When optimizing the network, our learning rate is set as 0.001, and hyper-parameters are set to $\lambda_{sl} = 1, \lambda_{ds} = \lambda_{dt} = 5 \times 10^{-4}$. The domain adaptation is across tasks and models simultaneously by minimizing sum loss L in (5).

Source VGG S+T is trained by green Dotted box in Fig. 3, which is set as transfer application evaluation criterion. Target initialized and non-initialized MobileNet S+T represents source and target labeled data is directly utilized to train target network.

The source output feature is in 4096 dimensions space. Here, traditional PAC dimension reduction method is used to transfer data from source model feature space to the target one (whose feature dimensions is 1024). However, if the number of source and target data is less than the dimension of the target feature space, the proposed method may not operate normally. Two suggestions should be proposed: 1. Increase number of source and target data; 2. Decrease target feature layer node to a proper number.

The proposed method is to utilize source fine-tuned model to guide the target model training process by identical distribution constraints. We compare the result on 2 target model. One is initialized randomly and the other is initialized by pre-trained model. Both target networks reach the goal that transfer knowledge and train light weight target model simultaneously. Distribution loss not only guarantees the acceptable model accuracy but successfully transfers knowledge to target model. The proposed framework has top-2 accuracy on all subtasks and top-1 performance on average which is shown in Table 3.

5   Discussion and Conclusion

An approach is proposed to transfer domain knowledge via identical distribution loss across models and tasks. Distribution loss is designed in feature space, which is generated by source and target data. When target model is guided by source distribution during training processing, the convergence time and model accuracy are more acceptable. Moreover, large amount of unlabeled data are efficiently involved in the framework to enforce the classification ability, even though they do not share same label space. Target model can be select based on task requirements so that fine-tuning on source model is no longer forced. The model will be trained with a distribution guide but not with a minimum point of source model. Permuting the parameter in target model with distribution guidance provides unlimited possibilities in new target structure. Distribution constraints play a role as weak instruction to support the establishment of target model and have shown its great performance in our experiments.

Acknowledgment. This paper was partially financially supported by National Natural Science Foundation of China under grants 61533012, 91748120 and 61521063.

References

1. Aytar, Y., Zisserman, A.: Tabula rasa: model transfer for object category detection. In: International Conference on Computer Vision, pp. 2252–2259 (2011)
2. Ba, L.J., Caruana, R.: Do deep nets really need to be deep? In: Advances in Neural Information Processing Systems, pp. 2654–2662 (2013)
3. Bergamo, A., Torresani, L.: Exploiting weakly-labeled web images to improve object classification: a domain adaptation approach. In: Neural Information Processing Systems, pp. 181–189 (2010)
4. Chopra, S.: DLID: deep learning for domain adaptation by interpolating between domains. In: ICML Workshop on Challenges in Representation Learning (2013)
5. Collobert, R., Weston, J., Bottou, L., Karlen, M., Kavukcuoglu, K., Kuksa, P.: Natural language processing (almost) from scratch. Mach. Learn. Res. **12**(Aug), 2493–2537 (2011)

6. Donahue, J., et al.: Decaf: a deep convolutional activation feature for generic visual recognition 50(1), I-647 (2013)
7. Duan, L., Tsang, I.W., Xu, D.: Domain transfer multiple kernel learning. Pattern Anal. Mach. Intell. **34**(3), 465–479 (2012)
8. Fernando, B., Habrard, A., Sebban, M., Tuytelaars, T.: Unsupervised visual domain adaptation using subspace alignment. In: International Conference on Computer Vision, pp. 2960–2967 (2013)
9. Ganin, Y., Lempitsky, V.: Unsupervised domain adaptation by backpropagation. In: International Conference on Machine Learning, pp. 1180–1189 (2015)
10. Girshick, R.B., Donahue, J., Darrell, T., Malik, J.: Rich feature hierarchies for accurate object detection and semantic segmentation. In: Computer Vision and Pattern Recognition, pp. 580–587 (2014)
11. Glorot, X., Bordes, A., Bengio, Y.: Domain adaptation for large-scale sentiment classification: a deep learning approach. In: Machine Learning, pp. 513–520 (2011)
12. Gong, Y., Liu, L., Yang, M., Bourdev, L.: Compressing deep convolutional networks using vector quantization. Computer Science (2014)
13. Hermans, A., Beyer, L., Leibe, B.: In defense of the triplet loss for person re-identification. In: Computer Vision and Pattern Recognition (2017)
14. Hinton, G., Vinyals, O., Dean, J.: Distilling the knowledge in a neural network. Comput. Sci. **14**(7), 38–39 (2015)
15. Hoffman, J., et al.: LSDA: large scale detection through adaptation. In: Advances in Neural Information Processing Systems, pp. 3536–3544 (2014)
16. Hoffman, J., Rodner, E., Donahue, J., Darrell, T., Saenko, K.: Efficient learning of domain-invariant image representations. Computer Science (2013)
17. Howard, A.G., et al.: Mobilenets: efficient convolutional neural networks for mobile vision applications. In: Computer Vision and Pattern Recognition (2017)
18. Jia, Y., et al.: Caffe: convolutional architecture for fast feature embedding. In: ACM Multimedia, pp. 675–678 (2014)
19. Krizhevsky, A., Sutskever, I., Hinton, G.E.: ImageNet classification with deep convolutional neural networks. In: Neural Information Processing Systems, pp. 1097–1105 (2012)
20. Lecun, Y., Bottou, L., Bengio, Y., Haffner, P.: Gradient-based learning applied to document recognition. Proc. IEEE **86**(11), 2278–2324 (1998)
21. Mansour, Y., Mohri, M., Rostamizadeh, A.: Domain adaptation: learning bounds and algorithms. arXiv preprint arXiv:0902.3430 (2009)
22. Oquab, M., Bottou, L., Laptev, I., Sivic, J.: Learning and transferring mid-level image representations using convolutional neural networks. In: Computer Vision and Pattern Recognition, pp. 1717–1724 (2014)
23. Pan, S.J., Yang, Q., et al.: A survey on transfer learning. Knowl. Data Eng. **22**(10), 1345–1359 (2010)
24. Parkhi, O.M., Vedaldi, A., Zisserman, A., et al.: Deep face recognition. In: The British Machine Vision Conference, vol. 1, p. 6 (2015)
25. Russakovsky, O., et al.: Imagenet large scale visual recognition challenge. Int. J. Comput. Vis. **115**(3), 211–252 (2015)
26. Saenko, K., Kulis, B., Fritz, M., Darrell, T.: Adapting visual category models to new domains. In: Daniilidis, K., Maragos, P., Paragios, N. (eds.) ECCV 2010. LNCS, vol. 6314, pp. 213–226. Springer, Heidelberg (2010). https://doi.org/10.1007/978-3-642-15561-1_16
27. Sermanet, P., Eigen, D., Zhang, X., Mathieu, M., Fergus, R., Lecun, Y.: Overfeat: integrated recognition, localization and detection using convolutional networks. Eprint Arxiv (2013)

28. Simonyan, K., Zisserman, A.: Very deep convolutional networks for large-scale image recognition. Computer Science (2014)
29. Szegedy, C., Toshev, A., Erhan, D.: Deep neural networks for object detection. In: Advances in Neural Information Processing Systems, vol. 26, pp. 2553–2561 (2013)
30. Tzeng, E., Hoffman, J., Darrell, T., Saenko, K.: Simultaneous deep transfer across domains and tasks. In: IEEE International Conference on Computer Vision, pp. 4068–4076 (2017)
31. Tzeng, E., Hoffman, J., Zhang, N., Saenko, K., Darrell, T.: Deep domain confusion: Maximizing for domain invariance. Computer Science (2014)
32. Yang, J., Yan, R., Hauptmann, A.G.: Adapting SVM classifiers to data with shifted distributions. In: International Conference Data Mining Workshops, pp. 69–76. IEEE (2007)
33. Yi, D., Lei, Z., Liao, S., Li, S.Z.: Learning face representation from scratch. Computer Science (2014)
34. Zhang, K., Schölkopf, B., Muandet, K., Wang, Z.: Domain adaptation under target and conditional shift. In: International Conference on Machine Learning, pp. 819–827 (2013)

A Pointer Network Based Deep Learning Algorithm for the Max-Cut Problem

Shenshen Gu$^{(\boxtimes)}$ and Yue Yang

School of Mechatronic Engineering and Automation, Shanghai University,
Shanghai, China
gushenshen@shu.edu.cn

Abstract. The max-cut problem is one of the classic NP-hard combinatorial optimization problems. In order to solve this problem efficiently, the paper mainly studies the topic of using the pointer network to build a training model to solve the max-cut problem. Then, the network model is trained with supervised learning. The experimental results show that the network trained by this algorithm can obtain the approximate solution to the max-cut problem.

Keywords: Max-cut problem · Pointer network · Supervised learning

1 Introduction

The max-cut problem belongs to the famous twenty-one NP (Nondeterministic Polynomial) problems that Richard M. Karp first proposed [1]. The max-cut problem refers to finding a maximum segmentation for a given directional weighted graph that maximizes the total weights across all edges of these two cut sets [2].

As a typical NP hard problem in combinatorial optimization problem, the max-cut problem has various applications in statistical physics, image processing, communication network design, circuit layout design and other engineering problems. In view of the dual important value of the theory and practice, in the past few decades, researchers have proposed various algorithms to solve the max-cut problem. The algorithm can be divided into two categories, one of which are exact algorithms and the other are heuristic algorithms. The exact algorithms include the enumeration method [3] and the branch and bound method [4] etc. Although the optimal solution to this problem can theoretically be found by an exact algorithm, it is often impossible to achieve it, because the computational time increases exponentially with the increase of the scale of the problem, the search space for the problem also increases rapidly as the scale increases. Even if the current state-of-the-art computer is used for calculation, the time for solving the problem is not tolerable. Therefore, finding an effective approximate heuristic algorithm is of great significance. The effective methods for solving the max-cut problem include immune algorithm, genetic algorithm, greedy algorithm, ant colony algorithm, simulated annealing algorithm, LKH algorithm [5],

© Springer Nature Switzerland AG 2018
L. Cheng et al. (Eds.): ICONIP 2018, LNCS 11301, pp. 238–248, 2018.
https://doi.org/10.1007/978-3-030-04167-0_22

etc. Compared with the exact algorithms, the heuristic algorithms can be applied to solving large-scale problems with thousands or even tens of thousands of variables in a short period of time, so computational efficiency is improved. However, there is a defect that cannot be ignored in the heuristic algorithms, that is, the degree of deviation between the feasible solution and the optimal solution cannot be accurately predicted all the time, and it is easy to fall into a local optimal solution. Therefore, it is of great theoretical significance and application value to study the effective algorithms for solving the max-cut problem.

Recently, deep learning based methods are becoming more and more popular due to the fact that they are capable of discovering their own heuristics based on abundant training data automatically. For this reason, except for the famous application in computer vision, image classification [6] and speech recognition [7], deep learning based methods are now making potential progress in solving various combinatorial optimization problems. Deep learning can represent the categories or features of the data by extracting the underlying features of the combined optimization problem data to form more abstract high-level features, and then using the distributed features of the data. For instance, the famous TSP problem is successfully solved by Oriol Vinyals with RNNs [8]. The quadratic assignment problem is effectively solved by Anto Milan with a data-driven approach [9]. Inspired by these important ideas, a deep learning based method to solve the max-cut problem is proposed in the paper.

The rest of this paper is organized as follows. Section 2 introduces the formulation of the max-cut problem and the architecture of the pointer network. Section 3 explains how to use the pointer network to solve the max-cut problem. Then, Sect. 4 details the experiments and analysis. And finally, the conclusion is given in Sect. 5.

2 Problem Formulation

In this section, the mathematical description of the max-cut problem is first introduced and then the architecture of the pointer network model is described.

2.1 The Max-Cut Problem

$G = (V, E)$ is a graph, where $V = \{1, 2, \ldots n\}$ is vertex set and E is edge set. Suppose that w_{ij} is the weight for edge (i, j) in E. Dividing the vertex set V into two subsets S and S', satisfying $S \cup S' = V$ and $S \cap S' = \varnothing$, then calling S and S' constitute a cut of the graph G. The value of the cut is the number of edges with one end in S and the other end in S', it is calculated by the following equation:

$$cut\,(S, S') = \sum_{\substack{u \in S \\ v \in S'}} w_{uv} \qquad (1)$$

The max-cut problem consists of finding a cut in G with maximum value.

In this paper, we assume that the weight on each edge is one without loss of any generality. Our goal is to find a segment (S, S') of the vertex set V, so that the maximum number of edges is divided (i.e., one vertex of the edge in S and the other vertex in S').

2.2   The Pointer Network

Pointer network is a new type of deep neural architecture combines the popular sequence-to-sequence learning framework [10] with a modified attention mechanism [11] to learn the conditional probability of an output whose values correspond to positions in a given input sequence. It was first proposed by Vinyals et al. [8] to solve TSP problems. The neural network architecture for solving the max-cut problems is shown in Fig. 1. The structure of the pointer network is briefly introduced as follows

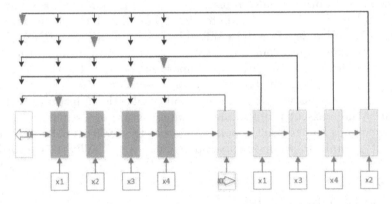

Fig. 1. Architecture of the pointer network (encoder in blue, decoder in yellow) (Color figure online)

The Seq2seq module is mainly composed of an encoder and a decoder. The encoder represents a variable-length input sequence as a vector of fixed dimensions, and the decoder converts this vector into a variable-length output vector. The attention mechanism that connects the encoder and the decoder allows the decoder to query the entire sequence of encoder states, not just the last LSTM cell state. The attention mechanism is actually using a variable-length vector to extract relevant information from the input. It generates corresponding weights for each element of the input sequence, indicating the degree of correlation with the next input of the decoding section. It purposed to tell the decoder network which input parts are more important. This method allows the decoder to focus more on finding useful information in the encoder input sequence that is relevant to the current output, thereby improving the quality of the output.

In the model of this paper, RNN networks are constructed with LSTM units. LSTM is a special recurrent neural network architecture. Compared with feedforward neural networks, RNN has the characteristics of cyclic connections, making it more suitable for the modeling of sequences [12]. The sequence X is fed to the decoder and one element is fed into each time step until the end of the sequence. The end of the sequence is marked with a special end marker. The model then switches to decoding mode, where each time step produces an element in the output sequence of the decoder until the end marker appears. Until this time, the entire process ended.

Each conditional probability of encoder and decoder can be defined as

$$p\left(y_i \,|\, y_1, \ldots, y_{i-1}, X\right) = g\left(y_{i-1}, d_i, c_i\right)$$
$$d_i = h\left(d_{i-1}, y_{i-1}, c_i\right) \tag{2}$$

The c_i vector is calculated as follows:

$$c_i = \sum_{j=1}^{n} \alpha_j^i e_j \tag{3}$$

Where d_i and e_j in (2) and (3) are the hidden states of the decoder and the encoder, respectively, and the weights α_j^i are defined as:

$$\alpha_j^i = \frac{\exp\left(u_j^i\right)}{\sum_{k=1}^{n} \exp\left(u_k^i\right)}$$
$$\left(u_j^i\right) = a\left(d_{i-1}, e_j\right) \tag{4}$$

Among them, a is a feed forward neural network, and the vector u_j^i is called the attention mark of the input sequence element.

Prior to the introduction of the pointer network, there is a problem with the model of Seq2seq combined with the attention mechanism, that is, the output dictionary size of the encoder must depend on the length of the input sequence. Therefore, the pointer network is used to adjust the standard attention mechanism and create a pointer u_j^i to the input sequence element, so that the extra information propagated to the decoder is no longer just the final state of the encoder [13]. Instead, using u_j^i to point to the input sequence element.

$$u_j^i = v^T \tanh\left(W_1 e_j + W_2 d_i\right) \quad j \in (1, \ldots, n)$$
$$p\left(C_i \,|\, C_1, \ldots, C_{i-1}, P\right) = soft\max\left(u^i\right) \tag{5}$$

Where *softmax* normalizes vector u_j^i of length n to make it an output probability distribution on the input dictionary, and v, W_1, W_2 are the parameters that can be learned in the model, and $C = C_1, \ldots, C_m$ is a sequence of m indices.

3   Solving the Max-Cut Problem Using the Pointer Network

3.1   Data Structure of the Max-Cut Problem

In the max-cut problem, our goal is to find a point set S that can make the cut (S, S'), which is the sum of the weights on the edges in E, obtain the maximum value.

Inspired by Vinyals' idea of solving the TSP problem that uses the trained neural network model, input the set of city node coordinates and output the predicted probability distribution of the various nodes of these cities. For the max-cut problem, the input to this network is the weight of the line between the point and the point. The input is represented by a matrix, i.e. w_{ij} represents the weight of the line between point i and point j ($w_{ij} = 1$ or 0, where 1 means there exist a connection between two points and 0 means there is no connection between two points). The output of the network is the segmentation of the vertex set V. The output represents all points in order by 0 and 1, where points marked "1" are placed in one set and points marked "0" are placed in the other set.

For example, let $G = (V, E)$ be an undirected graph with seven vertices. And the weight on edge (i, j) are set to w_{ij} ($w_{ij} = w_{ji}$). This problem can be written as

$$f(x) = x_1x_2 + x_1x_5 + x_2x_5 + x_2x_7 + x_4x_5 + x_4x_7$$

$$x_i \in \{0, 1\}, (i = 1, \ldots, 7) \tag{6}$$

The input weight matrix W is

$$W = \begin{pmatrix} 0 & 1 & 0 & 0 & 1 & 0 & 0 \\ 1 & 0 & 0 & 0 & 1 & 0 & 1 \\ 0 & 0 & 0 & 0 & 0 & 0 & 0 \\ 0 & 0 & 0 & 0 & 1 & 0 & 1 \\ 1 & 1 & 0 & 1 & 0 & 0 & 0 \\ 0 & 0 & 0 & 0 & 0 & 0 & 0 \\ 0 & 1 & 0 & 1 & 0 & 0 & 0 \end{pmatrix} \tag{7}$$

The problem's optimal solution is $x = \begin{pmatrix} 0 & 0 & 1 & 1 & 0 & 1 & 1 \end{pmatrix}^T$. That means vertices x_3, x_4, x_6, and x_7 belong to one set, while vertices x_1, x_2, and x_5 belong to the other set.

3.2   Datasets Generation

Our experiments use the MATLAB program to randomly generate 100 sets of samples as a training set. Each time, ten sets of samples are extracted for training, and 100 sets of samples are generated as a test set.

3.3   Supervised Learning

Supervised learning is a method often used in machine learning. It can learn or create a learning model through training data, and infers the output corresponding to the new given instance input based on this model. The training data consists of the input (usually a vector) of the corresponding problem and the corresponding expected output. The output of a function can be a continuous value (called a regression analysis) or it can be a classification label of a prediction (called a classification). A task of supervised learning is to predict the output of the function corresponding to any possible input value after observing some typical training (input and corresponding expected output). In order to achieve this goal, learners must generalize from existing data to non-observed situations in a "reasonable" manner. This situation is commonly referred to as concept learning in human and animal perceptions.

Supervised learning is, in simple terms, a classification that people often say. Through the existing training samples, which are known data and their corresponding outputs, they are trained to obtain an optimal model. This model is then used to map all inputs to the corresponding outputs and make simple judgments on the outputs to achieve correct classification. So the model has the ability to judge and classify unknown data.

In order to solve a given problem of supervised learning (such as the max-cut problem), the following steps must be considered.

– Initializing Network

Set up hyperparameters, for instance, the number of layers in the neural network, learning rate, the type of neuron activation function, batch size and the method of weight initialization.

– Loading Data

Determine the representation of the input features of the learning function. Convert the input and output data formats to the desired data format.

– Producing a Network Model

Creat a sequence model, attention mechanism functions, loss functions, and optimization functions.

– Training Network

Train the network model and adjust parameters.

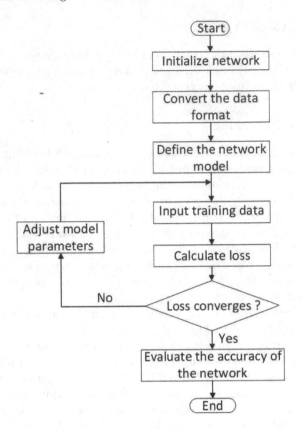

Fig. 2. Scheme of neural network

– Evaluating Network

Use the test set to assess the accuracy of the network on solving the max-cut problem.

The total procedure of the network model is shown in Fig. 2.

4   Experiments Results and Analysis

The pointer network for solving the max-cut problem was implemented with TensorFlow.

In order to validate the pointer network, we performed five experiments. In these experiments, five data sets were generated randomly, with dimensions of 10, 20, 30, 40 and 50. (the dimensions are the number of vertices). Taking the 20-dimensional (see Fig. 3) max-cut problem as an example. The result of 20-dimensional with 1000 training times and 100 training samples is given as follows.

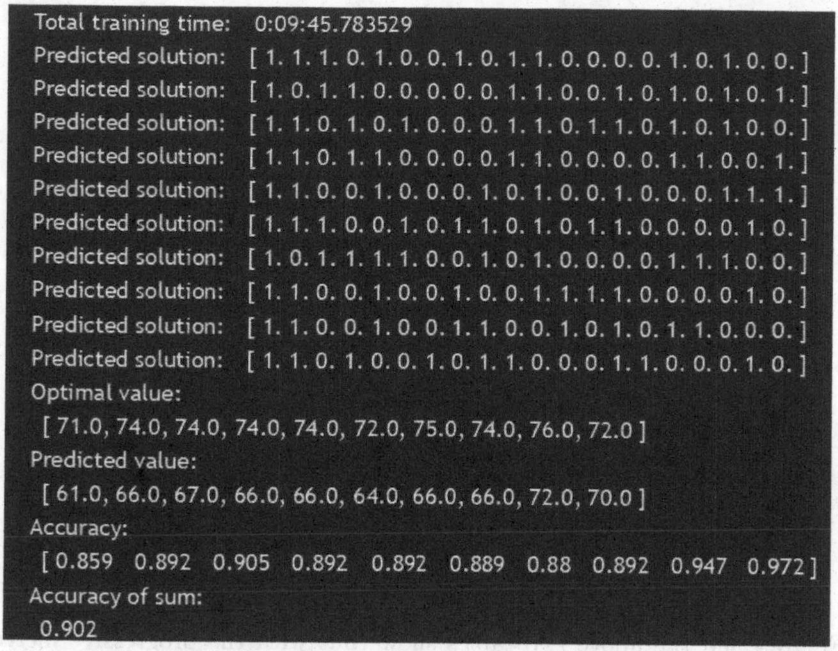

```
Total training time:   0:09:45.783529
Predicted solution:   [ 1. 1. 1. 0. 1. 0. 0. 1. 0. 1. 1. 0. 0. 0. 0. 1. 0. 1. 0. 0. ]
Predicted solution:   [ 1. 0. 1. 1. 0. 0. 0. 0. 0. 1. 1. 0. 0. 1. 0. 1. 0. 1. 0. 1. ]
Predicted solution:   [ 1. 1. 0. 1. 0. 1. 0. 0. 0. 1. 1. 0. 1. 1. 0. 1. 0. 1. 0. 0. ]
Predicted solution:   [ 1. 1. 0. 1. 1. 0. 0. 0. 0. 1. 1. 0. 0. 0. 0. 1. 1. 0. 0. 1. ]
Predicted solution:   [ 1. 1. 0. 0. 1. 0. 0. 0. 1. 0. 1. 0. 0. 1. 0. 0. 0. 1. 1. 1. ]
Predicted solution:   [ 1. 1. 1. 0. 0. 1. 0. 1. 1. 0. 1. 0. 1. 1. 0. 0. 0. 0. 1. 0. ]
Predicted solution:   [ 1. 0. 1. 1. 1. 1. 0. 0. 1. 0. 1. 0. 0. 0. 0. 1. 1. 1. 0. 0. ]
Predicted solution:   [ 1. 1. 0. 0. 1. 0. 0. 1. 0. 0. 1. 1. 1. 1. 0. 0. 0. 0. 1. 0. ]
Predicted solution:   [ 1. 1. 0. 0. 1. 0. 0. 1. 1. 0. 0. 1. 0. 1. 0. 1. 1. 0. 0. 0. ]
Predicted solution:   [ 1. 1. 0. 1. 0. 0. 1. 0. 1. 1. 0. 0. 0. 1. 1. 0. 0. 0. 1. 0. ]
Optimal value:
  [ 71.0, 74.0, 74.0, 74.0, 74.0, 72.0, 75.0, 74.0, 76.0, 72.0 ]
Predicted value:
  [ 61.0, 66.0, 67.0, 66.0, 66.0, 64.0, 66.0, 66.0, 72.0, 70.0 ]
Accuracy:
  [ 0.859   0.892   0.905   0.892   0.892   0.889   0.88   0.892   0.947   0.972 ]
Accuracy of sum:
  0.902
```

Fig. 3. Experimental results of 20-dimensional max-cut problem

Figure 3 shows the training time, the predicted solution to the tested max-cut problem, the optimal value and the predicted value, the accuracy of the network for the ten groups on 20-dimensional max-cut problem. The results in Tables 1 and 2 below were obtained in the same manner as Fig. 3.

As can be seen from the figure, by repeating the training of 10 sets of 20-dimensional data, we obtained the optimal solution to the expected output "Predicted solution", points with "1" means they were placed in one group, and points with "0" means they were placed in the other group. "Optimal value" represents the real optimal value of the input point set, and "Predicted value" represents the predicted optimal value obtained after the input point set was trained by the neural network. "Accuracy" is the ratio of "Predicted value" to "Optimal value", this parameter is used to indicate the quality of the trained model. "Accuracy of sum" is the average of 10 sets of "Accuracy".

Table 1 shows the training time of the max-cut problem model on five different dimensions, t indicates the times of training and s indicates the number of training samples.

Table 2 shows the solution accuracy of the max-cut problem model on five different dimensions with different times of training (t) and the number of training samples (s).

Table 1. Training time of the max-cut problem model

Dimensions	10	20	30	40	50
t = 1000, s = 100	5'16"	9'46"	14'09"	18'42"	23'15"
t = 1000, s = 1000	5'13"	9'36"	14'06"	18'40"	23'13"
t = 2000, s = 100	10'29"	19'31"	28'19"	37'28"	46'21"
t = 2000, s = 1000	10'25"	19'13"	28'14"	37'28"	46'28"

Table 2. Accuracy of the max-cut problem model

Dimensions	10	20	30	40	50
t = 1000, s = 100	95.2%	90.2%	85.2%	81.0%	68.7%
t = 1000, s = 1000	92.0%	91.6%	86.4%	81.3%	79.0%
t = 2000, s = 100	97.5%	88.3%	85.7%	84.3%	73.7%
t = 2000, s = 1000	95.1%	94.2%	86.1%	85.2%	81.4%

Figure 4 and the above two tables show that with the progressive increase of the dimension of the max-cut problem, the time spent on training gradually increases, and the accuracy of the approximate solution decreases. Then, as the training times increase, the quality of the approximate solution is also improved. In addition, The larger the number of training samples, the higher the accuracy of the approximate solution.

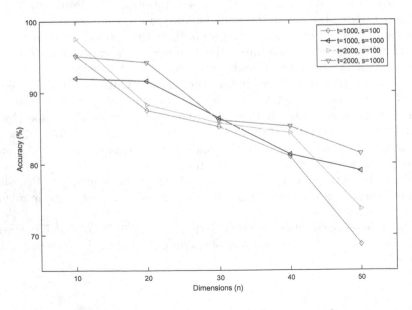

Fig. 4. The accuracy of five dimensions max-cut problem

5   Conclusion

In this paper, we use the pointer network to solve the max-cut problem. The proposed neural network architecture is a variant of the Seq2seq model, which can utilize RNN ordered connections to convey information and allow information to be persisted to predict the final solution. Experiments of the max-cut problem with different dimensions demonstrate that the supervised learning based method can obtain a nice approximate solution. This method greatly reduces the time and cost of calculations compared to conventional algorithms. The experimental results can be said to be very satisfactory, and some of the experimental results even reached the optimal value. The results obtained from the five sets of experiments allow us to see the advantages of solving the combinatorial optimization problem using a pointer network. It indicates that the method has great application potential in exploring combinatorial optimization problems.

Acknowledgments. The work described in the paper was supported by the National Science Foundation of China under Grant 61876105.

References

1. Mehlhorn, K.: NP-completeness. Eatcs Monogr. Theor. Comput. Sci. **5**(3), 359–376 (1984)
2. Bie, T.D., Cristianini, N.: Fast SDP relaxations of graph cut clustering, transduction, and other combinatorial problem. J. Mach. Learn. Res. **7**(3), 1409–1436 (2006)
3. Croce, F.D., Kaminski, M.J., Paschos, V.T.: An exact algorithm for MAX-CUT in sparse graphs. Oper. Res. Lett. **35**(3), 403–408 (2007)
4. Krishnan, K., Mitchell, J.E.: A semidefinite programming based polyhedral cut and price approach for the maxcut problem. Comput. Optim. Appl. **33**(1), 51–71 (2006)
5. Funabiki, N., Kitamichi, J., Nishikawa, S.: An evolutionary neural network algorithm for max cut problems. In: International Conference on Neural Networks, vol. 2, pp. 1260–1265. IEEE (1997)
6. Krizhevsky, A., Sutskever, I., Hinton, G.E.: ImageNet classification with deep convolutional neural networks. In: International Conference on Neural Information Processing Systems, vol. 60, pp. 1097–1105. Curran Associates Inc. (2012)
7. Deng, L., Li, J., Huang, J.T., Yao, K., Yu, D., Seide, F., et al.: Recent advances in deep learning for speech research at Microsoft. In: IEEE International Conference on Acoustics, Speech and Signal Processing, pp. 8604–8608. IEEE (2013)
8. Vinyals, O., Fortunato, M., Jaitly, N.: Pointer networks. In: International Conference on Neural Information Processing Systems. MIT Press (2015)
9. Milan, A., Rezatofighi, S.H., Garg, R., Dick, A., Reid, I.: Data-driven approximations to NP-hard problems (2017)
10. Sutskever, I., Vinyals, O., Le, Q.V.: Sequence to sequence learning with neural networks, vol. 4, pp. 3104–3112 (2014)
11. Bahdanau, D., Cho, K., Bengio, Y.: Neural machine translation by jointly learning to align and translate. Comput. Sci. (2014)

12. Sak, H., Senior, A., Beaufays, F.: Long short-term memory recurrent neural network architectures for large vocabulary speech recognition. Comput. Sci. 338–342 (2014)
13. Acuna-Agost, R., Acuna-Agost, R.: Deep choice model using pointer networks for airline itinerary prediction. In: ACM SIGKDD International Conference on Knowledge Discovery and Data Mining, pp. 1575–1583. ACM (2017)
14. Zhou, M.X.: A benchmark generator for Boolean quadratic programming. Comput. Sci. (2015)
15. Barahona, F., Junger, M., Reinelt, G.: Experiments in quadratic 0–1 programming. Math. Program. **44**(1–3), 127–137 (1989)
16. Gu, S., Hao, T.: A pointer network based deep learning algorithm for 0–1 Knapsack Problem. In: International Conference on Advanced Computational Intelligence (ICACI 2018), pp. 357–361 (2018)
17. Gu, S., Hao, T., Yang, S.: The implementation of a pointer network model for traveling salesman problem on a Xilinx PYNQ board. In: Huang, T., Lv, J., Sun, C., Tuzikov, A.V. (eds.) ISNN 2018. LNCS, vol. 10878, pp. 130–138. Springer, Cham (2018). https://doi.org/10.1007/978-3-319-92537-0_16

Convolution Neural Networks

Multi-stream with Deep Convolutional Neural Networks for Human Action Recognition in Videos

Xiao Liu[✉] and Xudong Yang

National Engineering Laboratory for Integrated Command and Dispatch
Technology, School of Computer Science,
Beijing University of Posts and Telecommunications, Beijing 100000, China
{liu_xiao, xdyang}@bupt.edu.cn

Abstract. Recently, convolutional neural networks (CNNs) have been extensively applied for human action recognition in videos with the fusion of appearance and motion information by two-stream network. However, for human action recognition in videos, the performance over still images recognition is so far away because of difficulty in extracting the temporal information. In this paper, we propose a multi-stream architecture with convolutional neural networks for human action recognition in videos to extract more temporal features. We make the three contributions: (a) we present a multi-stream with 3D and 2D convolutional neural networks by using still RGB frames, dense optical flows and gradient maps as the input of networks separately; (b) we propose a novel 3D convolutional neural network with residual blocks, use deep 2D convolutional neural network as the pre-train network which is added attention blocks to extract the major motion information; (c) we fuse the multi-stream networks by weights not only for networks but also for every action category to take advantage of the optimal performance of each network. Our networks are trained and evaluated on the standard video action benchmarks of UCF-101 and HMDB-51 datasets, and result shows that our method achieves considerable and comparable recognition performance to the state-of-the-art.

Keywords: Action recognition · Multi-stream · 3D CNNs · Attention Category weights

1 Introduction

Recently, convolutional neural networks have been extensively applied for image recognition problems and give the state-of-the-art results on recognition, detection and so on. Inspired by these, many video tasks, such as localizing activities, recognizing actions and understanding videos cause extensive concern of the researchers from all works of life. Most of all, the human action recognition has a wide range of applications including health care, augmented reality game control, security and protection, etc. Specifically, the academics are working hard to improve the performance of human action recognition, while the industrial sectors are committed to making use of these results to make contributions in security and protection.

© Springer Nature Switzerland AG 2018
L. Cheng et al. (Eds.): ICONIP 2018, LNCS 11301, pp. 251–262, 2018.
https://doi.org/10.1007/978-3-030-04167-0_23

There are extensive and significant research efforts in human action recognition. Many approaches, such as two-stream with convolutional neural networks [2], recurrent neural networks (RNNs) such as long short term memory (LSTM) [8] and traditional approaches [1, 28] by using handcrafted features, aims to enable computer automatically recognize action in real world. In addition, the existence of a large number of big datasets including UCF101 [18], HMDB51 [19], Sports-1M [10] and Kinetics-400 [20] greatly improve the development of action recognition.

But for action recognition in videos, there still remains a big gap between videos and images recognition. The complexity of scenes in the videos, low resolution, camera shake and the confusion between sub-action and long-term action, these all have brought considerable difficulties to the recognition tasks. So in this paper, we present a novel approach for recognize action in videos, we propose a 3D CNNs to extract the temporal features and add a network stream using gradient maps to reduce the interference of the complex background. To ensure the performance we also use the RGB images and optical flows to recognize action.

Overall, in this paper, we summarize the three main contributions in our works.

1. We present a multi-stream of 2D and 3D CNNs and use the original RGB frames, dense optical flows and gradient maps as the input of networks. Because of the still RGB frames contain all information of images, dense optical flows contain speed between pixels and the gradient maps are edge-processed images, we can get more appearance and motion information.
2. For 3D CNNs, we propose a novel architecture of network which is added residual blocks to extract more temporal features. For 2D CNNs, we use deeper network and pay more attention to key information. Specifically, before the softmax layer of the network, we add an attention block which is similar to residual block to extract the key motion information.
3. We fuse the streams with network weights and category weights because the different network has different performance for different action categories. We compare several groups of different weights by use results of first evaluation and use the best weights to fuse networks.

The rest of the paper is organized as follows. In Sect. 2 we review the related work on action recognition using both traditional handcrafted features and deep learning architectures. Section 3 introduce the whole multi-stream network's architecture and specify the framework of convolutional neural network, residual block and attention block. Implementation details, the results of experiments which are evaluated on UCF101 and HMDB51 datasets and comparison to the state-of-the-art are given in Sect. 4. In Sect. 5, we give our conclusion and specify the future work.

2   Related Work

Action recognition has been extensively studied in past few years. Previous works related to action recognition in videos fall into two categories [21]: (a) traditional machine learning by using handcrafted features; (b) deep learning using convolutional neural networks to automatically learn the features.

First of all, we introduce the traditional methods which are mainly based on handcrafted features such as Histogram of Oriented Gradients (HOG), Motion Boundary Histogram (MBH), Histogram of Optical Flow (HOF), etc. The typical method is that Wang et al. proposed IDT [26] and DT [1] algorithms, which using HOG, MBH, HOF along improved dense trajectories, then they created a Fisher vector for each video clip and used support vector machine to achieve classification.

The main content of this section will introduce methods based on deep learning. There are three major approaches: two-stream with 2D convolutional neural networks, recurrent neural networks such as LSTM and 3D convolutional neural networks.

Simonyan et al. [2] firstly presented two-stream with convolutional neural networks for action recognition in videos, which use spatial stream to extract the appearance from still RGB frames and temporal stream to extract the motion information from dense optical flows. They fused two streams' softmax scores by averaging scores or training a multi-class linear SVM. Inspired by two-stream, Feichtenhofer et al. [3] used VGG-16 [13] as the basic network to train spatial and temporal streams, and rather than fusing at the softmax layer, the two streams can be fused at any convolution layer without loss of performance. In addition, Wang et al. [4] proposed the Temporal Segment Networks, they use RGB images, optical flows and warped optical flows as input, then they used BN-Inception [25] as basic block.

Ji et al. [5] firstly proposed 3D CNNs and used grayscale image, gradient map and optical flow as five channel image to train the network. But it can only work in a simple scenario just like KTH dataset [24]. So Du T et al. present a new 3D CNNs named C3D [6], which used $3 \times 3 \times 3$ Convolution kernel and 16 frames as input of network. It has a simply architecture and high efficiency.

Recurrent neural networks are often applied for natural language processing due to it contains temporal memory information. LSTM is more sensitive to temporal features in action recognition by avoiding long-term dependency problems. Wu et al. [7] extracted features from frames, optical flows and audio spectrograms by CNNs and LSTM were further adopted on fusion. Donahue et al. [8] presented Long-term Recurrent Convolutional Network (LRCN) model which is end-to-end trainable. Lev et al. [9] used CNNs to extract features, then used RNNs to train fisher vector and reduced dimensionality again, final used SVM to classify actions.

In addition, there are other approaches to explore the action recognition strategies. Karpathy et al. [10] proposed multi-resolution CNNs and different ways of fusion such as late, early or slow fusion. To speeding up the network, they conducted further experiments on training with images of lower resolution while using images of higher resolution to achieve good accuracy. Zhu et al. [11] thought that action may occur sparsely in a few key volumes and training with the key volumes will improve performance. They proposed a deep framework to mini and identify key volumes and used the key volumes to classify action category. Diba et al. [12] presented a new video representation, called temporal linear encoding (TLE) which is embedded inside of CNNs as a new layer so that it can captures the appearance and motion information throughout entire videos. Wang et al. [22] proposed attention on images recognition to focus on key features and inspired of that, Tran et al. [23] proposed an attention to extract temporal features for videos recognition.

Based on these approaches, we propose a novel multi-stream with 3D and 2D convolutional neural networks training on RGB images, optical flows and gradient maps separately. And we fuse these streams by weighted for every action category and every network stream separately.

3 Network Architecture

In this section, we give detailed architectures of 3D and 2D CNNs, residual block, attention block and multi-stream, analyse the performance of different architectures for action recognition. Specifically, we firstly introduce the basic concepts in the framework of 3D CNNs and how to design the architecture of the network and residual block. Then, we explain the 2D CNNs and attention block, and final we give the framework of multi-stream in detail.

3.1   3D CNNs

In 2D CNNs, convolution operation is applied on single-frame image to compute features from the spatial dimension only. Now we can use 3D convolution which operates on continuous multi-frame images to extract the temporal dimension to solve the action recognition problem in videos. So we propose a novel deep convolutional neural network which is achieved by 3D convolution and pooling.

In this section, we mainly introduce the architecture of a novel 3D CNNs in detail. We did several experiments on UCF101 dataset to search for a good architecture. And encouraged by residual framework of Resnet [14], we also use a residual block which is shown in Fig. 1 in our 3D CNNs to obtain more valid features in temporal dimension. In residual block, we use $1 \times 1 \times 1$ and $3 \times 3 \times 3$ filters and add a batch normalization layer after one convolution layer to avoid distribution shifts because the 3D CNNs has large memory capacity. After three convolution layers, we add input layer and the last batch normalization layer and result is processed by Relu activation layer and max pooling layer which use $2 \times 2 \times 2$ pool kernel. After that, the output of residual block is viewed as a layer of the network to continue training.

As we can see in Fig. 2, our proposed architecture of 3D CNNs can be viewed as an extension of the Resnet and VGG-16 network from 2D to 3D. The input data is passed through 12 convolution layers with a certain number of $3 \times 3 \times 3$ convolution kernel and the number is increasing by layers such as 64, 128, etc. And all convolution layers employ padding with "SAME" pattern and one stride to do not change the size of feature maps after convolution layers. The netwrok has 4 pooling layers after one or more convolution layers, and we all use max pooling with $2 \times 2 \times 2$ kernel except for the first pooing layer with $1 \times 2 \times 2$ kernel to ensure the temporal information does not disappear prematurely. In the end of network, we add two fully connected layers, a dropout layer and a softmax loss layer which is to predict action labels.

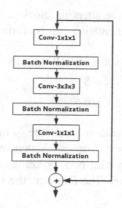

Fig. 1. A residual framework in 3D convolutional neural network.

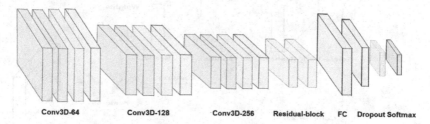

Conv3D-64   Conv3D-128   Conv3D-256   Residual-block   FC   Dropout Softmax

Fig. 2. The architecture of 3D convolutional neural network.

3.2 CNNs and Attention Block

Architecture of basic network is so important that can affect the performance of recognition. For 2D CNNs, we use the models pre-trained on ImageNet Large Scale Visual Recognition Competition. We compared different deep neural networks such as BN-Inception [25], Inception-resnet-v2 [15], Inception-v3 [27] and Resnext-50 [28], and we choose the Inception-v3 because of its high accuracy and not bad efficiency as the pre-train network, and before the fully connected layer, we add an attention block which is similar to residual framework to further extract the key temporal action features and add the fully connected layers and a softmax layer to get result.

For attention block, we want to capture the key objects and movements to classify and recognize the actions. Inspired by Residual Network [22], we design the attention block to extract the key feature by weighted low layers. As shown in Fig. 3, the attention block contains two branches, one weight branch is designed to get every pixel's weight to reduce noise interference while the other feature branch is to get features by convolution operation. Specifically, for weight branch, one convolutional layer with 1×1 filter is designed to get features' weights and the reshape layer is design to adjust to the shape of input of the attention block, softmax layer is to get the weights for every pixel in feature maps. For feature branch, the convolutional layer with 3×3 filter to extract features. After getting the outputs of two branches, we multiply the output of feature branch by the output of weight branch and add itself to get result of the attention block. The attention block can be extended to 3D CNNs and imbedded any layers. The implementation can be formulated in Eq. (1),

$$\delta' = R \circ (\text{softmax} \circ (R \circ (\text{conv}_{1x1} \circ S)))$$
$$\delta'' = \text{conv}_{NxN} \circ S \tag{1}$$
$$\delta = (\delta' \otimes \delta'') \oplus \delta''$$

where the S is the input of the attention block, the $\otimes$ means element-wise multiply by corresponding position, similarly $\oplus$ means element-wise adding too. The conv_{NxN} means the convolution operation with $N \times N$ kernel and conv_{1x1} means the convolution operation with 1×1 kernel. The R means the reshape operation.

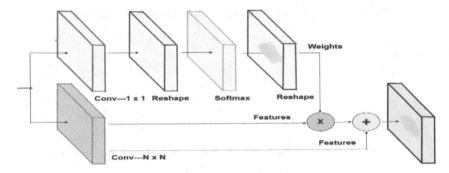

Fig. 3. The framework of attention block.

3.3 Multi-stream

Inspired by the two-stream network, we propose a multi-stream network with architecture of 2D and 3D convolutional neural networks. We use 2D CNNs to train original RGB frames, optical flows and gradient maps, and use 3D CNNs to train original RGB frames and gradient maps separately, and the final result of video-level prediction is fused with weights by the results of the all network streams. The architecture of multi-stream is shown in Fig. 4. As for 2D and 3D CNNs, we have already introduced in the previous section, so the rest of this section we mainly describe how to fuse the five network streams.

After we train all network streams, we fused the results of all streams. Specifically, we assign different weights to not only each action category but also each stream because the streams may have different effect on different categories, so we can maximize the performance of the networks. As shown in Eq. (2), we can firstly get the evaluation result and then we assign every network's weight for each action category according to the proportion of validation accuracy,

$$\vec{w} = \frac{\vec{\Phi}}{\sum_{i=1}^{Dim(\vec{w})} \theta_i} * Dim(\vec{w}) \tag{2}$$

where the $\vec{\Phi}$ is the accuracy vector of first evaluation, and θ_i is the i_{th} value of the $\vec{\Phi}$, $\vec{w}$ is the networks' weights for every action category and $\mathrm{Dim}(\vec{w})$ means the dimension of the $\vec{w}$ where $\mathrm{Dim}(\vec{w})$ is 5.

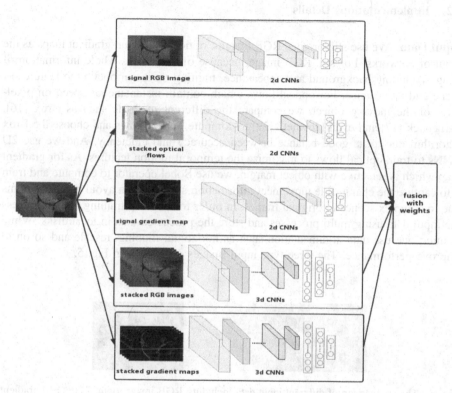

Fig. 4. The framework of proposed multi-stream for action recognition.

4 Experiments

In this section, we will firstly introduce the datasets for experiments, then we give the implementation details in our experiments including the data preprocessing, the setting of hyperparameters and some useful tricks. Nextly, we show the different results of different models and strategies. Finally, we will compare the results with the state-of-the-art.

4.1 Datasets

We evaluate our approach on UCF101 [18] and HMDB51 [19] datasets. UCF101 dataset which is downloaded from YouTube is widely used human action datasets with 13320 videos and includes 101 action classes. Because of its poor lighting and low resolution, HMDB51 is a more challenging dataset which contains 6766 videos and has been annotated for 51 action classes. Both datasets have three train/test splits, so we use the standard splits and show the average accuracy over the three splits in our experiments.

4.2 Implementations Details

Input Data. We use still original RGB images, optical flows and gradient maps as the input of networks. For still RGB image, because of it contains whole information of image including background and appearance, training on it can obtain the features of scene and appearance. For optical flows which contain instantaneous speed on pixel-level of the moving object, we compare the different methods such as Brox [16], Farnebäck [17] and even Flownet2 [29] to compute optical flows and choose the Brox algorithm due to its good balance between accuracy and efficiency. And we use 2D CNNs to train optical flows to maximize the temporal motion features. As for gradient map which is sensitive with object margin, we use Sobel operator to compute and train it to capture the changeable movement information so that it can avoid noise problems due to complex scenes in original frames. In order to speed up training, we preprocess the input data using multi-processes and store the processed data in local disks. While training, we use data argumentation such as horizontal flipping, rescale and so on to improve performance. The three type input images are shown in Fig. 5.

Fig. 5. The comparison of different input data including RGB image, optical flow and gradient map.

3D CNNs. For input data of 3D CNNs, we use a cube of stack of still RGB images and gradient maps. Specifically, all video frames are resized into 112×112 and we take a frame after every constant number in one video and take 20 frames in total. Because of that, we can obtain the entire information of whole video and maximize the extracting feature of long-term action rather than sub-action. After that, we get a cube of size of $D \times H \times W \times C$, which D is the number of frames, H and W are height and width of the frame, C is the number of channels. We train the networks using batch size of 128 clips and using SGD optimizer with initial learning rate of 0.01.

2D CNNs with Attention Block. During training the CNNs, we used the fixed-size 224×224 RGB images, optical flows and gradient maps as the input. We divide the video into five segments and take a random frame every segment for training while testing we divide the video into 25 segments and take the frame in the middle, and take the average result of 25 frames as the final video-level prediction. These tricks also apply to optical flows and gradient maps. We initialize the network weights with pre-trained model from inception-v3 network, use SGD optimizer to learn the parameters. The learning is initialized as 0.1 and batch size is 64.

Multi-stream. We fused the results of all streams. Specifically, we assign different weights to each action category and use the weighted scores to assign weights to each network stream again to maximize the performance of the networks. To speed up training, we train the model by multiple GPUs.

4.3   Evaluation

3D CNNs. We use 3D CNNs to train RGB images and gradient maps separately and compare the different length of frames including 15 and 20. We can see in Tabel 1, and the best result is obtained when using 20 frames.

Table 1. Exploration of different lengths of RGB images and gradient maps for 3D convolutional neural network on the UCF101 dataset.

Length	15frames	20frames
RGB images	82.1%	83.2%
Gradient maps	78.0%	80.1%

2D CNNs. We compared four very deep architectures: BN-Inception, Inception-resnet-v2, Inception-v3 and Resnext-50 network which all trained on same split of dataset. And we found the Inception-v3 achieves the excellent accuracy of 94.1% as shown in Table 2, so we choose it as the basic architecture of 2D CNNs instead of the Inception-resnet-v2 because its low efficiency.

Table 2. The comparison of different pre-train models for 2D convolutional neural network on the UCF101 dataset.

Models	RGB	Optical flows	Gradient maps	Multi-stream (weights)
BN-inception	83.4%	85.7%	80.2%	93.2%
Inception-v3	84.2%	86.3%	82.0%	94.1%
Inception-resnet-v2	84.3%	86.6%	82.2%	94.3%
Resnext-50	82.2%	84.0%	79.6%	92.1%

Table 3. The results of multi-stream with different fusion strategies on the UCF101 dataset.

Strategies	Average	Stream weighted	Class weighted
Multi-stream	92.5%	94.3%	94.8%

Multi-stream. After we train all streams, we fuse all the results when evaluate the model. We compared three strategies to fuse networks. Firstly, we used the average score of all five network stream and the accuracy is 92.5%. Then we traversed the all useful weights for network streams and every action category's weights is same in one network, the weight of 1.2:1.3:0.9:1.1:0.7 can achieve the considerable accuracy. Lastly, we assign weight to each action category because of different stream makes different accuracy to action categories and use the second way of fusion again, the result is shown in Table 3. For example, we use the optical flow stream to predict BodyWeightSquats action, use RGB stream to predict playing JumpingJack action and use gradient map stream to predict GolfSwing action with large weight separately.

Table 4. Comparison of our method based on multi-stream with the state-of-the-art methods on the UCF101 and HMDB51 datasets.

Algorithms	UCF101	HMDB51
iDT [1]	86.4%	61.7%
Two-stream [2]	88.0%	59.4%
TSN [4]	94.2%	69.4%
C3D [6]	85.2%	—
STVLMPF + iDT [31]	94.3%	73.1%
LSTM [32]	93.6%	66.2%
Two-stream I3D (with kinetics-300 k) [33]	98.0%	80.7%
OFF [30]	96.0%	74.2%
Ours	94.8%	71.3%

Comparison with the-State-of-Art. For comparision with the state-of-the-art, we follow the standard evaluation and show the average accuracy over three splits on both UCF101 and HMDB51 datasets. For that we use all discussed 2D and 3D convolutional neural networks and fuse scores of softmax layer with weights. As is shown in Table 4, the multi-stream is comparable to the recent start-of-the-art model.

5   Conclusion

In this paper, we propose a multi-stream network which is fused by 2D and 3D convolutional neural networks for human action recognition in videos. And we use RGB images, optical flows and gradient maps as input of networks to extract more spatiotemporal features. In addition, we also present an attention block to extract key features, and use category and network weights to boost performance of the model. The result shows our method achieves competitive performance to state-of-the-art.

There still remains some problems waiting for us to solve. The first problem to be solved is efficiency because of the complex models and long term training, and we will focus on efficiency in the future work. The other observably question is dataset. Even if the UCF101 and HMDB51 are the most widely used dataset for human action recognition, there are also more challenging datasets which contain more categories and larger amounts of data such as Sports-1 M and Kinetics-400. Therefore, we plan to verify the performance of our approach by testing these more complex datasets.

In future work we hope to train broader action categories on the dataset to obtain more powerful and genetic features, learn the video representation and explore different architecture of network to achieve whole video-level rather than clip-level predictions with high efficiency.

References

1. Wang, H., Schmid, C.: Action recognition with improved trajectories. In: Proceedings of the IEEE International Conference on Computer Vision, vol. 159, pp. 3551–3558. IEEE Press, New York (2013)
2. Simonyan, K., Zisserman, A.: Two-stream convolutional networks for action recognition in videos. In: Advances in Neural Information Processing Systems, pp. 568–576 (2014)
3. Feichtenhofer, C., Pinz, A., Zisserman, A.: Convolutional two-stream network fusion for video action recognition. In: Proceedings of the IEEE Conference on Computer Vision and Pattern Recognition, pp. 1933–1941. IEEE Press, New York (2016)
4. Wang, L., et al.: Temporal segment networks: towards good practices for deep action recognition. In: Leibe, B., Matas, J., Sebe, N., Welling, M. (eds.) ECCV 2016. LNCS, vol. 9912, pp. 20–36. Springer, Cham (2016). https://doi.org/10.1007/978-3-319-46484-8_2
5. Ji, S., Xu, W., Yang, M., Yu, K.: 3D convolutional neural networks for human action recognition. J. IEEE Trans. Pattern Anal. Mach. Intell. 35(1), 221–231 (2013)
6. Tran, D., Bourdev, L., Fergus, R., Torresani, L., Paluri, M.: Learning spatiotemporal features with 3D convolutional networks. In: Proceedings of the IEEE International Conference on Computer Vision, pp. 4489–4497. IEEE Press, New York (2015)
7. Wu, Z., Jiang, Y.G., Wang, X., Ye, H., Xue, X., Wang, J.: Fusing multi-stream deep networks for video classification. J. Comput. Sci. (2015)
8. Donahue, J., et al.: Long-term recurrent convolutional networks for visual recognition and description. In: Proceedings of the IEEE Conference on Computer Vision and Pattern Recognition, pp. 2625–2634. IEEE Press, New York (2015)
9. Lev, G., Sadeh, G., Klein, B., Wolf, L.: RNN fisher vectors for action recognition and image annotation. In: Leibe, B., Matas, J., Sebe, N., Welling, M. (eds.) ECCV 2016. LNCS, vol. 9910, pp. 833–850. Springer, Cham (2016). https://doi.org/10.1007/978-3-319-46466-4_50
10. Karpathy, A., Toderici, G., Shetty, S., Leung, T., Sukthankar, R., Fei-Fei, L.: Large-scale video classification with convolutional neural networks. In: Proceedings of the IEEE Conference on Computer Vision and Pattern Recognition, pp. 1725–1732. IEEE Press, New York (2014)
11. Zhu, W., Hu, J., Sun, G., Cao, X., Qiao, Y.: A key volume mining deep framework for action recognition. In: Proceedings of the IEEE Conference on Computer Vision and Pattern Recognition, pp. 1991–1999. IEEE Press, New York (2016)
12. Diba, A., Sharma, V., Van Gool, L.: Deep temporal linear encoding networks. In: Proceedings of the IEEE Conference on Computer Vision and Pattern Recognition, vol. 1. IEEE Press, New York (2017)
13. Simonyan, K., Zisserman, A.: Very deep convolutional networks for large-scale image recognition. J. Comput. Sci. (2014)
14. He, K., Zhang, X., Ren, S., Sun, J.: Deep residual learning for image recognition. In: Proceedings of the IEEE Conference on Computer Vision and Pattern Recognition, pp. 770–778. IEEE Press, New York (2016)

15. Szegedy, C., Ioffe, S., Vanhoucke, V., Alemi, A.A.: Inception-v4, inception-resnet and the impact of residual connections on learning. In: AAAI, vol. 4, p. 12 (2017)
16. Brox, T., Bruhn, A., Papenberg, N., Weickert, J.: High accuracy optical flow estimation based on a theory for warping. In: Pajdla, T., Matas, J. (eds.) ECCV 2004. LNCS, vol. 3024, pp. 25–36. Springer, Heidelberg (2004). https://doi.org/10.1007/978-3-540-24673-2_3
17. Farnebäck, G.: Two-frame motion estimation based on polynomial expansion. In: Bigun, J., Gustavsson, T. (eds.) SCIA 2003. LNCS, vol. 2749, pp. 363–370. Springer, Heidelberg (2003). https://doi.org/10.1007/3-540-45103-X_50
18. Soomro, K., Zamir, A.R., Shah, M.: UCF101: a dataset of 101 human actions classes from videos in the wild. J. Comput. Sci. (2012)
19. Kuehne, H., Jhuang, H., Garrote, E., Poggio, T., Serre, T.: HMDB: a large video database for human motion recognition. In: Proceedings of the IEEE International Conference on Computer Vision, pp. 2556–2563. IEEE Press, New York (2011)
20. Kay, W., et al.: The kinetics human action video dataset. arXiv preprint arXiv:1705.06950 (2017)
21. Yu, Z., Jiang-Kun, Z., Yi-Ning, W., Bing-Bing, Z.: A review of human action recognition based on deep learning. J. Acta Automiatica Sinica 42(6), 848–857 (2016)
22. Wang, F., et al.: Residual attention network for image classification. arXiv preprint arXiv: 1704.06904 (2017)
23. Tran, A., Cheong, L.F.: Two-stream flow-guided convolutional attention networks for action recognition. arXiv preprint arXiv:1708.09268 (2017)
24. Schuldt, C., Laptev, I., Caputo, B.: Recognizing human actions: a local SVM approach. In: Proceedings of the 17th International Conference on Pattern Recognition, vol. 3, pp. 32–36. IEEE Press, New York (2004)
25. Ioffe, S., Szegedy, C.: Batch normalization: accelerating deep network training by reducing internal covariate shift. arXiv preprint arXiv:1502.03167 (2015)
26. Wang, H., Kläser, A., Schmid, C., Liu, C.L.: Dense trajectories and motion boundary descriptors for action recognition. Int. J. Comput. Vis. 103(1), 60–79 (2013)
27. Szegedy, C., Vanhoucke, V., Ioffe, S., Shlens, J., Wojna, Z.: Rethinking the inception architecture for computer vision. In: Proceedings of the IEEE Conference on Computer Vision and Pattern Recognition, pp. 2818–2826. IEEE Press, New York (2016)
28. Xie, S., Girshick, R., Dollár, P., Tu, Z., He, K.: Aggregated residual transformations for deep neural networks. In: Proceedings of the IEEE Conference on Computer Vision and Pattern Recognition, pp. 5987–5995. IEEE Press, New York (2017)
29. Ilg, E., Mayer, N., Saikia, T., Keuper, M., Dosovitskiy, A., Brox, T.: Flownet 2.0: evolution of optical flow estimation with deep networks. In: Proceedings of the IEEE Conference on Computer Vision and Pattern Recognition, vol. 2, p. 6. IEEE Press, New York (2017)
30. Sun, S., Kuang, Z., Sheng, L., Ouyang, W., Zhang, W.: Optical flow guided feature: a fast and robust motion representation for video action recognition. In: Proceedings of the IEEE Conference on Computer Vision and Pattern Recognition, pp. 1390–1399. IEEE Press, New York (2018)
31. Duta, I.C., Ionescu, B., Aizawa, K., Sebe, N.: Spatio-temporal vector of locally max pooled features for action recognition in videos. In: Proceedings of the IEEE Conference on Computer Vision and Pattern Recognition, pp. 3205–3214. IEEE Press, New York (2017)
32. Sun, L., Jia, K., Chen, K., Yeung, D.Y., Shi, B.E., Savarese, S.: Lattice long short-term memory for human action recognition. In: Proceedings of the IEEE International Conference on Computer Vision, pp. 2166–2175. IEEE Press, New York (2017)
33. Carreira, J., Zisserman, A.: Quo vadis, action recognition? A new model and the kinetics dataset. In: Proceedings of the IEEE Conference on Computer Vision and Pattern Recognition, pp. 4724–4733. IEEE Press, New York (2017)

Use 3D Convolutional Neural Network to Inspect Solder Ball Defects

Bing-Jhang Lin[✉], Ting-Chen Tsan, Tzu-Chia Tung, You-Hsien Lee, and Chiou-Shann Fuh[✉]

Department of Computer Science and Information Engineering,
National Taiwan University, Taipei 10617, Taiwan
lknight8631@gmail.com, fuh@csie.ntu.edu.tw

Abstract. Head-In-Pillow (HIP) is a solder ball defect. The defect can be caused by surface of solder ball oxidation, poor wetting of the solder, or by distortion of the Printed Circuit Board (PCB) by the heat of the soldering process. The current diagnosis of the HIP defects is difficult to find out the problems, and some destructive tests are not recommended for use. In this paper, we use different angles of 2D X-Rays images to reconstruct the 3D PCB volumetric data. We crop the 3D solder balls volumetric data to $46 \times 46 \times 46$ pixels from the 3D PCB model. Because HIP problems do not happen often, we use the data augmentation method to expand our solder ball data. We propose a new 3D Convolutional Neural Network (CNN) to inspect the HIP problems. Our network uses convolutional blocks that consist of different convolutional paths and the dense connectivity method to connect the blocks. The network can learn various features through these convolutional blocks with different convolutional paths. Moreover, the features of each layer will be fully utilized in the following layers by the dense connectivity method, and also avoid some features lost in a deep convolutional path. In the last layer of our network, the global average pooling can let our network process more different sizes of solder ball data, and the normalized same size vector can be used to do the end-to-end learning. Compared with other classic models, our network not only has fewer parameters but also has faster training time.

Keywords: 3D object recognition · 3D CNN · Deep learning
Head-in-Pillow problems

1 Introduction

Head-in-Pillow (HIP) is a solder ball defect. The defect can be caused by surface of solder ball oxidation or poor wetting of the solder, or by distortion of the printed circuit board (PCB) by the heat of the soldering process [6]. The current diagnosis of the HIP defects usually uses 2D X-Ray inspection machine or Burn/In method to filter out boards with HIP. Since most X-Rays can only check from the top to the bottom, it is hard to find out the location of defects. The Burn/In method requires additional costs. In addition, the more reliable methods are destructive tests, but they are not recommended for use.

© Springer Nature Switzerland AG 2018
L. Cheng et al. (Eds.): ICONIP 2018, LNCS 11301, pp. 263–274, 2018.
https://doi.org/10.1007/978-3-030-04167-0_24

It is hard to find out the location of HIP defects from 2D X-Rays images, but we can reconstruct the 3D solder ball model from different angles of 2D X-Rays images. The 3D solder ball model can not only provide more information but also represent the location of HIP defects more clearly. In recent years, Deep Learning has been widely used in many computer vision tasks. In particular, the Convolutional Neural Network (CNN) has an outstanding performance in image object recognition. Different levels of features can be integrated by the deep network structure. Finally, the complex high-level features can be combined with an end-to-end network to predict the result.

In this paper, we reconstruct the 3D PCB model from different 2D X-Ray images (1472 pixels × 1176 pixels × 73 layers), and we crop the 3D solder balls to 46 × 46 × 46 pixels from the 3D PCB model. We propose a 3D CNN to inspect the HIP problems. Our network combines the advantages of many classic CNNs. We design two types of the convolutional blocks composed of many different convolutional paths and use the dense connectivity method to connect these blocks. Because the convolutional block has many different convolutional paths composed of different kernels of convolutions, the network can learn more features from the convolutional block. To process more different sizes of data, we use the global average pooling to adapt these data. In addition, our network is not only faster and more accurate than other classic CNN models, but also makes the speed of convergence faster during the training stage. Because the HIP problems do not happen often, so we use the data augmentation method to expand the 3D solder ball data. In our experiments, our network can achieve excellent results on testing. The size of our network is more refined, and the training and execution time is faster than other models.

2 Related Works

In recent year, many state-of-art CNN models are proposed in the image object recognition. The Visual Geometry Group (VGG) achieved 92.7% top-5 test accuracy in ImageNet in 2014 [8], the ImageNet is a dataset of over 14 million images belonging to 1000 classes. VGG proved that to increase the depth of the network can improve the accuracy of the image recognition, and many papers also used VGG to achieve their recognition tasks [7, 9, 10]. VGG is composed of many consecutive convolutional layers. Therefore, the parameters of the VGG model are not only quite large, but also need to take much time to train this model.

In [11], GoogleNet Inception V1 was the champion on the ImageNet competition in 2014. The Inception V1 block is composed of 1×1, 3×3, and 5×5 convolutions along with a 3×3 max pooling. The network can choose the better features to learn. This architecture allows the model to extract both local feature via smaller convolutions and high abstracted features with larger convolutions. Besides, the 1×1 convolution is a good way to reduce the dimension of the feature maps, and it also represents the raw data well. Because the kernel of the 1×1 convolution has only one parameter, it is easy to scale the dimension of the feature to make it easier to increase the depth of network. Many papers also use the 1×1 convolution to increase the depth of their network [5, 13].

Although increasing the depth of the network can improve the accuracy of the recognition, it is easier for vanishing gradient problem to happen when the network is too deep. This situation will lead to decreased accuracy despite increasing the depth of the network. Simply increasing the depth of the network will not produce very good results, and it will even increase the error. The Deep Residual Neural Network (ResNet) improved the vanishing gradient problem of increasing the depth of the network and made it easier to increase the depth of the network [3]. ResNet proposed a residual method to create a short path to connect earlier layers with later layers. The input of the block not only performs two consecutive convolution processes but also adds to the output in each residual block. This way carries the important information features in previous layer to the later layers, and it also prevents the vanishing gradient problem. Many people do modify the network architecture according to the paper, so that the deep neural network is deeper and more accurate [2, 12].

In 2018, the Densely Connected Convolutional Network (DenseNet) was proposed [4] and used a simple method to increase the depth of the network. This network is one variant of ResNet, but it connects the result of each layer to make the accuracy of the network better. The dense connection makes features of each layer to be fully used, and it also avoids some important feature loss in deep network. The parameters of the network are less than other state-of-art CNNs.

3 Approach

3.1 3D Solder Balls Volumetric Data

The 3D PCB volumetric data are reconstructed by different angles of 2D X-Rays images. In order to obtain clearer 3D PCB volumetric data during the reconstruction, we control the camera to take a 2D X-Rays image every 2.81 degrees. Figure 1(a) shows the 3D PCB volumetric data reconstructed by 128 different angles of 2D X-Rays images, and Fig. 1(b) shows the data reconstructed by 16 different angles of 2D X-Rays images. In contrast to Fig. 1(b) reconstructed result, Fig. 1(a) result is more clear, and background noise is also less.

(a)   (b)

Fig. 1. (a) The 3D PCB model reconstructed with 128 different angles of 2D X-rays images. (b) The 3D PCB model reconstructed with 16 different angles of 2D X-rays images.

The 3D PCB volumetric data size is 1472 × 1176 × 46 pixels. Some solder balls on the edge of PCB will be cut, so we exclude those incomplete solder balls. We receive a total of 277 complete solder balls from Fig. 1(a) PCB. Since the volumetric pattern of solder ball is an elliptical sphere, the middle slice of the volumetric contains the biggest solder ball cross section. However, the maximum bounding box size of the solder ball from the middle slice of the PCB volumetric data is 46 × 46 pixels, so the size of each 3D solder ball is 46 × 46 × 46 pixels. Figure 2(a) shows the volumetric pattern of solder ball. Figure 2(b) shows the solder ball pattern in the middle slice for the solder ball volumetric data.

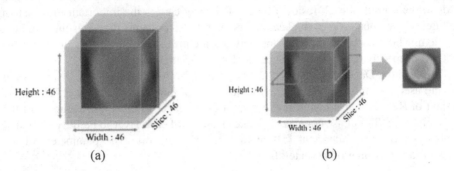

(a)   (b)

Fig. 2. (a) The volumetric pattern of solder ball. (b) Solder ball pattern in the middle slice of the solder ball volumetric data.

In this case, we just receive a total of 277 complete solder balls. There are fewer solder balls with HIP problem. For an image object recognition task, the amount of data is not sufficient, so we also use the Data Augmentation method to obtain more data. The 3D solder ball is a cube-like object whose size is 46 × 46 × 46 pixels. We rotate to expand the data 6 times. In more detail, we have achieved the rotation method by converting between the x, y, and z axes. Figures 3(a) and (b) show the xy plane results. Figures 3(c) and (d) show the yz plane results. Figures 3(e) and (f) show the xz plane results. Besides, we also normalize each solder ball volumetric before inputting to the network for processing.

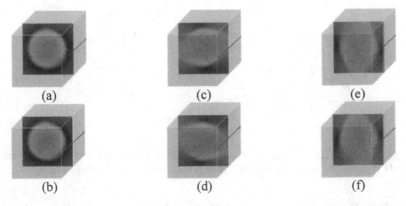

(a)   (c)   (e)

(b)   (d)   (f)

Fig. 3. (a) Combination of positively arranged xy images. (b) Combination of reversed xy images. (c) Combination of positively arranged xz images. (d) Combination of reversed xz images. (e) Combination of positively arranged yz images. (f) Combination of reversed yz images.

3.2   3D Convolutional Neural Network

In this paper, we reference four state-of-art CNN models (VGG, GoogleNet Inception V1, ResNet, DenseNet) in our solder ball HIP problems [3, 4, 8, 11]. These models have a significant impact on the domain about deep learning. We not only convert these four kinds of 2D CNN models into 3D CNN models to inspect our solder ball HIP problems, but also combine the advantages of these models to propose a new model to inspect the problem in our task. We use dense connectivity method to connect our convolutional blocks. Because the feature maps of a dense connection block will have multiple growths, we use the $1 \times 1 \times 1$ convolution with stride 2 to reduce the dimension of the feature maps. The $1 \times 1 \times 1$ convolution can represent the raw data well and compress the features to reduce the dimension, so that the following layer can process these concise features more quickly and efficiently. Besides, the global average pooling is used to normalize the input tensor to the same size, because we want to let our network adapt to the different sizes of the input. Figure 4 shows our 3D CNN structure to inspect the solder ball HIP problems.

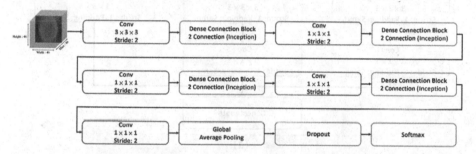

Fig. 4. (a) Our network has four dense connection blocks. The amount of each dense connection block is 2, and the convolutional block type is the "Inception". We add a 3D max polling layer with stride 2 with $2 \times 2 \times 2$ kernel and a 3D convolutional layer with ReLU activation function with $1 \times 1 \times 1$ kernel between the middle dense connection block. Last, the global average pooling layer is used to adapt the different sizes of the input and do the end-to-end learning with the following layers.

Dense Connection Block

In this paper, we propose two types of convolutional blocks for our network. These two blocks all combine the Inception concept [11], and these blocks are composed of many different convolutional paths. The first convolutional block called "Inception Block" has three convolutional paths. The first path includes a 3D max pooling layer with $3 \times 3 \times 3$ kernel and a 3D convolutional layer with $1 \times 1 \times 1$ kernel. The second path includes a 3D average pooling layer with $3 \times 3 \times 3$ kernel and a 3D convolutional layer with $1 \times 1 \times 1$ kernel. The third path includes a single convolutional layer with $3 \times 3 \times 3$ kernel. Last, we concatenate each result of the convolutional path. We also use the residual method [3] in the second convolutional block. This block is called "Inception with Residual Block". This convolutional block has four convolutional paths, and the first three convolutional paths are the same as our "Inception Block". Besides, we add the fourth convolutional path. This added path consists of a 3D convolutional layer with ReLU activation function with $1 \times 1 \times 1$ kernel, a 3D convolutional layer with ReLU activation function with $3 \times 3 \times 3$ kernel, and a 3D convolutional layer with $1 \times 1 \times 1$ kernel. We add the input to the last of the fourth path to achieve the residual method. Finally, we concatenate each result of the convolutional path. The residual method is a good way to prevent vanishing gradient problem, so we can more easily increase the depth of the network. Adding different convolutional paths allows the following layer to learn various features, so the network can choose better features to adapt the data and give us better results on testing data. Figure 5(a) shows our "Inception Block", and Fig. 5(b) shows our "Inception with Residual Block".

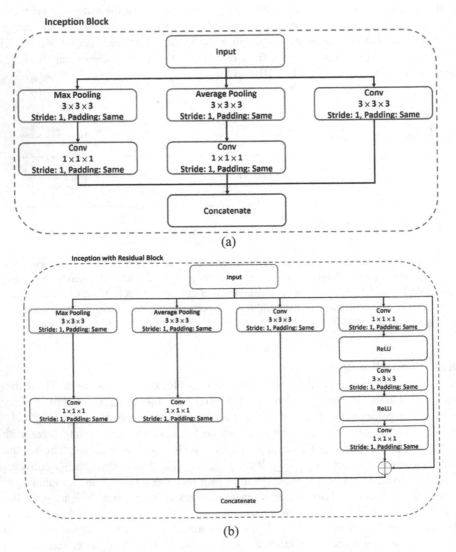

Fig. 5. (a) Our "inception block". (b) our "inception with residual block".

In addition, we use the dense connectivity method [4] to connect our convolutional block. The output of previous layer will be connected with the current output of the convolutional block and do the ReLU activation function. The features of each layer will be fully utilized in the following layers, and also avoid some features lost in a deep convolutional path. Figure 6 shows our dense connection block.

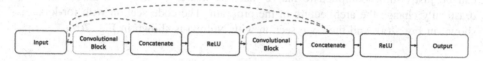

Fig. 6. Our dense connection block. The result of current convolutional block will concatenate with the result of each previous convolutional block, then concatenated result will do the ReLU activation function before inputting the next convolutional block.

4 Experiments

In order to ensure the accuracy and efficiency of our network, we compare our network with other state-of-art CNN models. We also compare computational acceleration to understand what methods can speed up the time of data preprocessing.

4.1 Solder Ball Data

We currently have 277 balls of 3D solder, of which 38 are labeled data and the rest are unlabeled data. The volumetric size of each solder ball is 46 × 46 × 46 pixels. Half of labeled data (19) are defective solder balls, and the other are normal solder balls. We split the labeled data into two parts: training and testing. The training part has 32 solder balls and the testing part has 6 solder balls. Because the defective solder ball do not often appear on the PCB, so that the 239 unlabeled data are temporarily considered as normal solder balls in evaluating stage. In addition to cross comparing the results of different models (VGG, GoogleNet Inception V1, ResNet, DenseNet, Our Network), we also provide the data to the experienced experts to check. We can know which solder ball is defective and the accuracy of the different models. The data are the important part in deep learning. The sampling range is wide enough and the sampling density is dense enough that the results of the model can approach the true distribution. Therefore, before we input the data to our network to inspect the HIP problems, we will augment the data to increase the amount of the labeled data. We rotate to expand the labeled data 6 times, so the training part has 192 (=32 × 6) solder balls, and the testing part has 36 (=6 × 6) solder balls.

4.2 Preprocessing Acceleration Comparison

Although the deep learning tasks are accelerated by the GPU device, the data need to be arranged by the CPU. Therefore, we also pursue the efficiency of CPU computation during preprocessing. Because every pixel in the solder ball volumetric needs to be processed for normalization, our program needs many for-loops to achieve preprocessing. It is a good opportunity to improve the computational efficiency. For speeding up the efficiency, we choose Numba acceleration library [1], and we also use some vectorization techniques to speed up.

Numba can speed up the program with high performance functions written directly in Python. With a few annotations, array-oriented and mathematics-heavy Python code can be just-in-time compiled to native machine instructions, so we do not have to drastically change the architecture of the program. The code section with for-loops is shown in Fig. 7(a), and the code section with Numba is shown in Fig. 7(b).

```
def for_loop_normalization(obj, rs_obj, max_value, min_value):
    # normalization
    for x in range(obj.shape[0]):
        for y in range(obj.shape[1]):
            for z in range(obj.shape[2]):
                rs_obj[x, y, z, 0] = (obj[x, y, z, 0] - min_value) / (max_value - min_value)
    return rs_obj
```
(a)
```
@nb.jit(nopython=True)
def jit_acc_normalization(obj, rs_obj, max_value, min_value):
    # normalization
    for x in range(obj.shape[0]):

        for y in range(obj.shape[1]):

            for z in range(obj.shape[2]):
                rs_obj[x, y, z, 0] = (obj[x, y, z, 0] - min_value) / (max_value - min_value)
    return rs_obj
```
(b)

Fig. 7. (a) The code section with for-loops. (b) The code section with Numba.

The comparisons of normalization of execution times are shown in Table 1. We compare the execution times for these methods for normalization under different quantities of data. In the small data, there is not much difference between Numba and for-loops, but the difference increases as the data increase. Numba is slightly faster than vectorization in performance. Table 1 also shows total execution time of normalization of three methods. We can get approximately about 30 times (= 15.60453/0.56150) acceleration by Numba, and about 50 times (= 15.60453/0.35498) acceleration by vectorization.

Table 1. The comparisons of normalization of execution times with different quantity of data.

Quantity of data	For-loops (sec.)	Numba (sec.)	Vectorization (sec.)
4566 (testing 6 balls)	0.34192	0.32136	0.01253
24352 (training 32 balls)	1.80130	0.02757	0.04110
181879 (testing 239 balls)	13.46030	0.21156	0.30033
Total time	15.60453	0.56150	0.35498

The comparisons of data augmentation of execution times are shown in Table 2. Because we only augment data on labeled data, we only use training and testing data for comparison. Table 2 shows the total execution time of data augmentation of three methods. We can get approximately about 10 times (= 4.03928/0.35697) acceleration by Numba, and about 30 (= 4.03928/0.12433) times acceleration by vectorization.

Table 2. The comparisons of data augmentation of execution times with different quantities of data.

Quantity of data	For-loops (sec.)	Numba (sec.)	Vectorization (sec.)
4566 (testing 6 balls)	0.68535	0.20957	0.01904
24352 (training 32 balls)	3.35342	0.14690	0.10528
Total time	4.03928	0.35697	0.12433

4.3   Experimental Result

We use the Tensorflow 1.5 and CUDA 9.0 to construct our deep learning development environment, and Python 3.6 as our development language. In addition, we use Nvidia GTX 1060 6G as our GPU device, and Intel i7-7700HQ as our CPU processor.

In training stage, we set the batch size to 5 to train models, because the volumetric size is too large that we cannot set higher batch size at one time. The batch size 5 is the most suitable for our development environment and device. Besides, we also discover that some deep networks have to use the lower learning rate for training. Because it has to pass more layers to back-propagate the error, if the learning rate is too high, it may fall into a local optimal solution and achieve poor results. However, the dense connectivity method makes each layer connected to other layers in each dense connection block, so the gradient will not disappear in the deep network. Table 3 shows the information of each model. VGG16 has the largest model size, because it uses many continuous convolutional layers. These layers also have larger amount of the kernels such as 64, 128, 256, and 512. In 3D convolution the $3 \times 3 \times 3$ kernels have 27 parameters that are 3 times more than 2D convolution kernel, so the size of the model parameter also shows significant growth. The time of the training model is proportional to the size of the model. The larger size of model, the more time it takes to train. Table 3 shows the information of each model.

Table 3. The information of each model.

Model	Depth	Training (sec.)	Learning rate	Size of model (KB)
ResNet50	50	1099.3151	0.00001	540568
GoogleNet inception V1	22	1286.3257	0.00001	460280
VGG16	16	5149.8747	0.00001	910582
DenseNet	21	406.5984	0.0001	6945
Our network (inception block)	18	574.8570	0.0001	45318
Our network (inception with residual block)	26	848.4092	0.0001	70449

Table 4 shows the results on testing data. We use the cross entropy as the loss function in this experiment. Our network has the best performance on testing data, and the processing time of the network is shorter. Because the network can learn various features from different kernels of the convolution, it can choose the better feature to adapt the data to achieve better results. GoogleNet Inception V1 also has good accuracy by using different kernels of convolution in network. Besides, the dense connectivity method is a good way to avoid some important features lost in deep network, and the network which uses this way also has good accuracy and speed. Although depth of ResNet50 is very deep, its result is not ideal. Perhaps each residual block using the single shortcut path to connect previous layer with later layer, the network only learns the features from the previous layer, so it is hard to let the network to learn more features to adapt data. VGG16 has decent results, but its size and execution time are quite large compared with other networks.

Table 4. The result on testing data which uses cross entropy as loss function.

Model	Execution time (sec.)	Loss	Accuracy
VGG16	1.52800	1.805191	86%
ResNet50	0.47715	1.939934	83%
DenseNet	0.23764	0.955081	83%
GoogleNet inception V1	0.61803	0.000997	100%
Our network (inception block)	0.30885	0.000093	100%
Our network (inception with residual block)	0.44718	0.000000	100%

Table 5 shows the results on unlabeled data. In the unlabeled data, our network still achieves the best accuracy and faster execution time. The networks use the dense connectivity method will improve the performance. The important features will be preserved by this way, and those features are closer to real data.

Table 5. The results on 239 unlabeled solder balls.

Model	Execution time (sec.)	Accuracy
VGG16	13.86638	84%
ResNet50	5.99745	80%
DenseNet	2.33119	85%
GoogleNet inception V1	6.98257	83%
Our network (inception block)	2.86511	86%
Our network (inception with residual block)	4.29642	87%

When the patterns of these solder balls are skewed near PCB border, we find that the models may misclassify the normal solder balls into defect. The normal solder ball does not have unfused part in the pattern, but we can easily find the overlapping patterns on the defect solder ball in Fig. 8. The skew situation often happens on the PCB border, and the pattern of skew solder ball is blurry which may cause misclassification.

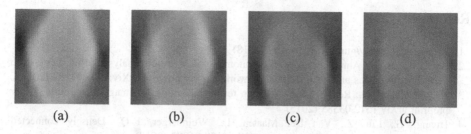

(a)   (b)   (c)   (d)

Fig. 8. (a) Pattern of the normal solder ball. (b) Pattern of the defect solder ball. (c) Skew normal solder ball near PCB border. (d) Skew defect solder ball near PCB border.

5   Conclusions and Future Work

In this paper, we propose a 3D CNN to inspect the HIP problems. Our network combines the advantages of many state-of-art CNN models. We design two types of the convolutional block, and these blocks are composed of many different convolutional paths. Because the network can learn various features from different convolutional paths, so it can choose better features to adapt the data to achieve the better result on testing data. Besides, we reference the dense connectivity method to connect the output of each layer. The previous layers will be connected to the latter layers in each dense connection block, so the features can be fully used in the network. It is a good way to avoid the important features loss in the deep convolutional paths. The global average pooling layer can help our network to adapt many different sizes of data, so our network can use the normalized same size vector to do the end-to-end learning to process more different data. In our experiments, our network has almost the fewest parameters compared with other classical models, and the convergence speed of our network is also the fastest. The accuracy of our network can achieve 100 % on the testing data, and the accuracy of our network is also the best on the unlabeled data.

Data are very important in deep learning. If the sampling range is wide enough and the sampling density is dense enough, then the model can reflect the true distribution of the real situation. Although our network can achieve the astonishing accuracy on the testing data, the amount of the 3D solder balls is not enough. Therefore, we use data augmentation to expand our data before training the network, but these data are still insufficient to summarize the real situation. HIP problems seldom happen. Solder balls on PCB border are often skew and may cause misclassification. In the future, we will not only continue collecting more information on solder balls, but also focus on using those data reconstructed by different angles of the 2D X-Rays images to train the network. We will still endeavor to simplify and accelerate our network to inspect HIP problems.

Acknowledgement. This research was supported by the Ministry of Science and Technology of Taiwan, R.O.C., under Grants MOST 104-2221-E-002-133-MY2 and MOST 106-2221-E-002-220, and by Test Research, Jorgin Technologies, III, Egistec, D8AI, and LVI.

References

1. Numba, A.: https://numba.pydata.org/ (2018)
2. Bi, L., Kim, J., Ahn, E., Feng, D.: Automatic skin lesion analysis using large-scale dermoscopy images and deep residual networks. arXiv preprint arXiv:1703.04197 (2017)
3. He, K., Zhang, X., Ren, S., Sun, J.: Deep residual learning for image recognition. arXiv preprint arXiv:1512.03385 (2015)
4. Huang, G., Liu, Z., Van Der Maaten, L., Weinberger, K.Q.: Densely connected convolutional networks. arXiv preprint arXiv:1608.06993 (2018)
5. Iandola, F.N., Shen, A., Gao, P., Keutzer, K.: DeepLogo: hitting logo recognition with the deep neural network hammer. arXiv preprint arXiv:1510.02131 (2015)
6. Seelig, K.: Head-in-Pillow BGA Defects. AIM Solder, Montreal (2008)
7. Sercu, T., Puhrsch, C., Kingsbury, B., LeCun, Y.: Very deep multilingual convolutional neural networks for LVCSR. arXiv preprint arXiv:1509.08967 (2016)
8. Simonyan, K., Zisserman, Z.: Very deep convolutional networks for large-scale image recognition. arXiv preprint arXiv:1409.1556 (2015)
9. Su, H., Maji, S., Kalogerakis, E., Learned-Miller, E.: Multi-view convolutional neural networks for 3D shape recognition. arXiv:1505.00880 (2015)
10. Sun, Y., Liang, D., Wang, X.G., Tang, X.O.: DeepID3: face recognition with very deep neural networks. arXiv preprint arXiv:1502.00873 (2015)
11. Szegedy, C., et al.: Going deeper with convolutions. arXiv preprint arXiv:1409.4842 (2014)
12. Targ, S., Almeida, D., Lyman, K.: Resnet in Resnet: generalizing residual architectures. arXiv preprint arXiv:1603.08029 (2016)
13. Zhong, Z.Y., Jin, L.W., Xie, Z.C.: High performance offline handwritten chinese character recognition using GoogLeNet and directional feature maps. arXiv preprint arXiv:1505.04925 (2015)

User-Invariant Facial Animation
with Convolutional Neural Network

Shuiquan Wang[1], Zhengxin Cheng[1], Liang Chang[1], Xuejun Qiao[2],
and Fuqing Duan[1(✉)]

[1] College of Information Science and Technology, Beijing Normal University,
Beijing 100875, China
fqduan@bnu.edu.cn
[2] School of Science, Xi'an University of Architecture and Technology,
Xi'an 710055, China

Abstract. In this paper, we propose a robust approach for real-time user-invariant and performance-based face animation system using a single ordinary RGB camera with convolutional neural network (CNN), where the facial expression coefficients are used to drive the avatar. Existing shape regression algorithms usually take a two-step procedure to estimate facial expressions: The first is to estimate the 3D positions of facial landmarks, and the second is computing the head poses and expression coefficients. The proposed method directly regresses the face expression coefficients by using CNN. This single-shot regressor for facial expression coefficients is faster than the state-of-the-art single web camera based face animation system. Moreover, our method can avoid the user-specific 3D blendshapes, and thus it is user-invariant. Three different input size CNN architectures are designed and combined with Smoothed L1 and Gaussian loss functions to regress the expression coefficients. Experiments validate the proposed method.

Keywords: Facial animation · CNN · Face tracking · Expression regression

1 Introduction

Face animation [1] needs to extract facial expression coefficients, and it is different from facial expression recognition, which only needs to classify patterns. The problem of automatically extracting face expression coefficients related attributes from facial images has received increasing attention in computer vision and computer graphics. Techniques based on special equipments (e.g., special facial markers [2], structure light projectors [3] and depth camera [4]) have achieved great success in making films and games, such as Faceshift studio, which is a well-known facial motion capture software. However, these techniques cannot be applied to low cost scenarios since ordinary users don't have the access to use them. Moreover, special facial markers are not convenient to use. Weise et al. [5] developed a real-time face animation system that uses depth and color data from a Microsoft Kinect. They combine the geometry and texture registration with animation priors into a single optimization problem. However, it can only be applied in indoor environments. Many facial animation and tracking systems work

© Springer Nature Switzerland AG 2018
L. Cheng et al. (Eds.): ICONIP 2018, LNCS 11301, pp. 275–283, 2018.
https://doi.org/10.1007/978-3-030-04167-0_25

by tracking appearance features using special equipments [2–4] to animate digital avatars. They have been widely used in film production with high-fidelity animation. Lacking of equipment, these methods cannot be applied. Hence face animation based on a video camera has also been explored. Sauer et al. [6] use the Active Appearance Model and build the mapping from the model pa to the target output. Cao et al. [7] directly regress facial landmarks without using any parametric shape models but minimizing the alignment error over training data in a holistic manner. However, these general models or regressors may not get satisfactory animation or tracking result for images of a specific person of non-frontal faces and various expressions. Cao et al. [1] demonstrated comparable tracking results using a single web camera. They developed a novel user-specific 3D shape regressor, which could regress the 3D position of landmarks directly from 2D video frames. These 3D landmarks are then used to track the facial motion, including the rigid transformation and non-rigid blendshape coefficients, which can be mapped to any digital avatar. However, it is user-variant and the user-specific 3D blendshapes are required.

In this paper, we propose a robust approach for real-time user-invariant and performance-based face animation system using a single ordinary RGB camera with CNN. The main contribution is that we use a single-shot regressor for facial expression coefficients that is faster than the state-of-the-art single web camera based face animation system. The system is user-invariant compared with the user-specific blendshape based method. Three different input size CNN architectures are designed (see Sect. 2.2), and combined with Smoothed L1 and Gaussian loss functions (see Sect. 2.3), to regress the expression coefficients.

2 CNN for Face Animation

CNN has been successfully applied in face detection [8], face recognition [9], age and gender estimation [10] etc. in recent years. Ranjan et al. [11] designed a multi-task learning (MTL) framework and presented a multi-purpose algorithm for simultaneous face detection, face alignment, pose estimation etc. using a single CNN. To our knowledge, it is first to use CNN in face animation from unconstrained photos.

We transform face animation into a regression problem. The pipeline of our system is shown in Fig. 1. The method detects the face firstly, and then facial landmark is detected for face alignment. The two steps can be combined using a multi-task cascaded convolutional networks (MTCNN) for face detection and alignment as proposed by Zhang et al. [8]. Then we use the aligned face image as input of CNN, and use the output of CNN, i.e. facial expression coefficients, to drive the avatar.

The problem of face detection of near frontal faces has been solved well. In this case, face detection is implemented with Opencv. Face alignment is realized using Supervised Descent Method (SDM) [12]. SDM is originally proposed for minimizing a Non-linear Least Squares (NLS) function, and can be applied to face alignment. We give improvement on SDM and make it faster for tracking face from videos.

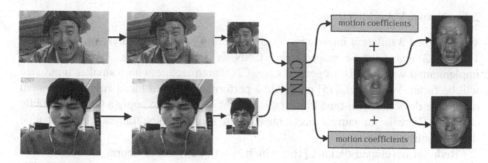

Fig. 1. Pipeline of our system. We use aligned face image as input of CNN, the system outputs motion coefficients.

2.1 Data Preparation

For training CNN, we need a large labeled image set. Gathering a large and labeled training image set for face animation from social image repositories could be time-consuming for manually labeling. In our method, we captured 200 videos with a single web camera from 200 individuals aged 20 – 60 from several ethnicities. And we designed a GUI, as shown in Fig. 2, for labeling the frames from videos. Each frame in video is used as a separate image.

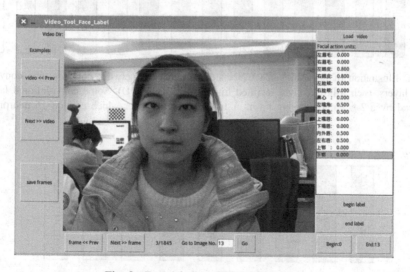

Fig. 2. Data labeling GUI in our system

Different from FACS [14] which uses 46 action units, we define 15 different facial action units. We label the frames with large facial expression variations. For example, we label the coefficient of 15 different facial action units for the 0th and 13th frame images, and then we apply linear interpolation for the 1st to 12th frame to get coefficient of facial action units.

2.2 Network Architecture

We designed 3 different input sizes for 3 different CNN architectures, i.e. 224 * 224, 112 * 112 and 64 * 64 respectively. CNN is time-consuming, especially when implemented with a CPU. In general case, CNN architecture with a smaller input size will be faster. Weng et al. [15] presented a performance-based facial animation system on mobile devices at real-time frame rates. And it is difficult to apply a CNN on mobile devices. There is no current accelerated open source CNN framework on mobile devices until now.

Redmon developed darknet [16], which is an open source neural network framework, the size of tiny dark net is only 4.0 MB, and it is top-1 accuracy is 58.7%. We designed a new network evolved from tiny-darknet as shown in Fig. 3. Our 224 * 224 CNN network comprises nine convolutional layers, four max polling layers and one fully-connected layer with a small number of neurons. For 112 * 112 input size case, we designed 3 convolution layers and 2 fully connected layers, as shown in Fig. 4. Same as the 224 * 224 architecture above, all the max pooling layer size is 2 * 2. As face animation is similar to facial landmark localization, Liang et al. [17] presented a method to regress facial landmark coordinate with a novel CNN. As is shown in Fig. 5, we changed the architecture of the network, especially its fully-connected layer.

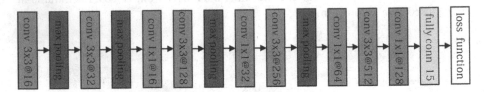

Fig. 3. Illustration of our 244 * 224 CNN architecture. The network contains nine convolutional layers, each followed by a rectified linear operator (ReLU). Some convolutional layers followed by a 2 * 2 max pooling layer. A fully-connected layer which contain 15 neurons is added.

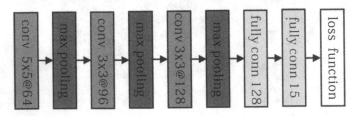

Fig. 4. Illustration of our 112 * 112 CNN architecture. The network contains three convolutional layers, each followed by a ReLU and a 2*2 max pooling layer. Two fully-connected layers are added.

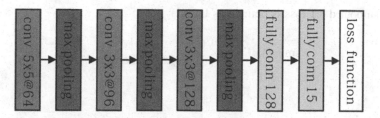

Fig. 5. Illustration of our 64 * 64 CNN architecture. The network contains three convolutional layers, the same as the above 112 * 112 architecture, each convolutional layer is followed by a ReLU and a max pooling layer. Two fully-connected layers are added.

2.3   Loss Function

Facial expression coefficients estimation from a face image is a regression problem. Euclidean loss is used in CNN widely, and it can also be used to solve facial landmark detection problem. Girshick et al. [18] have shown the smoothed L1 loss is less sensitive to outliers than the L2 loss. Ranjan et al. [13] showed that Gaussian loss works better than Euclidean loss for apparent age estimation. However, the gradient of Gaussian loss is close to zero when the predicted expression coefficients is far from the true value (as shown in Fig. 6), which slows the training process. Here, we use a linear combination of smoothed L1 loss and Gaussian loss weighted by λ, as shown in Eq. (1).

$$L_A = (1 - \lambda)\left(1 - \exp\left(-\frac{(P - F)^2}{2\sigma^2}\right)\right) + \lambda \cdot \text{smooth}_{L_1}(P, \ F) \tag{1}$$

Where L_A is the facial expression coefficients loss, P is the predicted coefficients, F is the ground-truth coefficients, and σ is the standard deviation of the facial motion coefficients value as shown in Eq. (1), λ is set to 0.6.

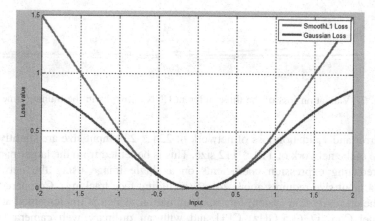

Fig. 6. SmoothL1 loss and Gaussian loss function

The smoothed L1 loss is defined as Eq. (2)

$$\text{smooth}_{L_1}(x) = \begin{cases} 0.5x^2 & \text{if } |x| < 1 \\ |x| - 0.5 & \text{otherwise,} \end{cases} \tag{2}$$

3 Experiments

In our experiments, the resolution of face images used in SDM is 224 * 224. We firstly align the face image with SDM, and CNN is implemented using the Caffe framework. We do not need the pre-trained models for initializing the network, and the weights in all layers are initialized with random values of a zero mean Gaussian distribution. CNN training was performed on an NVIDIA TAINX GPU machine with 1536 CUDA cores.

Figure 7 presents the results for validation loss of our three different CNN networks. We can obtain the mean coefficients error of the three different input size CNN networks. It can be seen that the 112 * 112 network performs best, and the network of 64 * 64 input size performs worse. This may be caused by lacking of the detailed expressional information.

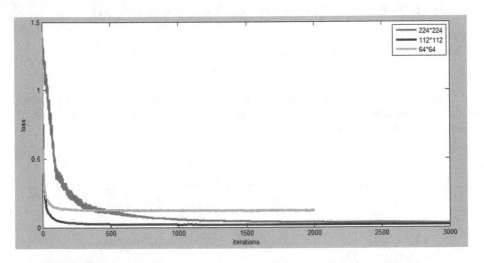

Fig. 7. Validation loss of the three different CNN networks in the training time.

The error and validation loss of network of 224 * 224 input size are slightly larger than those of the network of 112 * 112 size. This is benefited from the larger parameter space. Predicting expression coefficients on a single image using the network of 224 * 224 input size requires about 24 ms (including face tracking). Compared to the state-of-the-art regression system for facial animation, which is implemented at 24 fps on an Intel Core i7 (3.5 GHz) CPU and with an ordinary web camera [1], our

112 * 112 network requires about 20 ms (runs at 50 fps, including face tracking), with NVIDIA GPU. To evaluate the accuracy of our CNN-based method, we captured a face expression video, and manually labeled the 15 expression coefficients, which was used as the ground truth. We compared the estimated expression coefficients to the ground truth in each frame. Figure 8 shows the values of mouth opening degree. It is shown that the expression coefficient estimated by our algorithm (blue curve) is fairly close to manually labeled ground truth (red curve).

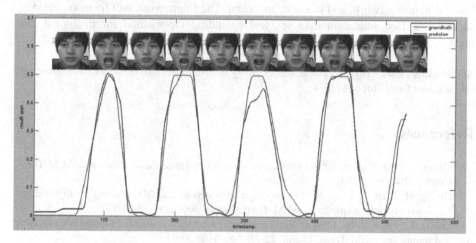

Fig. 8. Comparison of CNN-based regression (in blue) and ground truth manually labeled (in red). (Color figure online)

Finally, we extract expression coefficients and drive the digital avatar by assigning the coefficients of blendshape in MAYA, as shown in Fig. 9 From the eyes and mouth, we can see that the expression of the digital avatar is close to the actor. This validates the proposed method.

Fig. 9. Illustration of our facial animation system using a single camera and MAYA. The camera records 640 * 480 images at 30 fps. Our system extract expression coefficients at over 50 fps on PC with the NVIDIA GPU. We can drive the digital avatar by assigning the coefficients of blendshape in MAYA.

4 Conclusion

Face animation has been widely used in many face related applications. In this paper we propose a CNN-based method for user-invariant face animation. We firstly detect the face on a signal image, and then locate the feature points by supervised descent method SDM. Using the aligned face image as the input of CNN, we can achieve good results for face animation. Different from the state-of-the-art shape algorithms, the proposed method directly regresses the face expression coefficients by using convolutional neural network, and it is user-invariant. The future work will focus on building an accurate face animation data set, and extending the method for unaligned face images.

Acknowledgments. This work was supported by the National Natural Science Foundation of China under Grant No. 61572078.

References

1. Cao, C., Weng, Y., Lin, S.: 3D shape regression for real-time facial animation. ACM Trans. Graph. **32**(4), 96 (2013)
2. Huang, H., Chai, J., Tong, X.: Leveraging motion capture and 3D scanning for high-fidelity facial performance acquisition. ACM Trans. Graph. **30**(4), 76–79 (2011)
3. Zhang, L., Snavely, N., Curless, B.: Spacetime faces: high resolution capture for modeling and animation. ACM Trans. Graph. **23**(3), 546–556 (2008)
4. Bradley, D., Heidrich, W., Popa, T.: High resolution passive facial performance capture. ACM Trans. Graph. **29**(4), 157–166 (2010)
5. Weise, T., Bouaziz, S., Li, H.: Realtime performance-based facial animation. ACM Trans. Graph. **30**(4), 76–79 (2011)
6. Sauer, P., Cootes, T., Taylor, C.: Accurate regression procedures for active appearance models. In: BMVC, vol. 1 no. 6, pp. 681–685 (2011)
7. Cao, C., Weng, Y., Zhou, S.: FaceWarehouse: a 3D facial expression database for visual computing. IEEE Trans. Visual Comput. Graphics **20**(3), 413–425 (2014)
8. Zhang, K., Zhang, Z., Li, Z.: Joint face detection and alignment using multitask cascaded convolutional networks. IEEE Sig. Process. Lett. **23**(10), 1499–1503 (2016)
9. Kang, B.N., Kim, Y., Kim, D.: Deep convolution neural network with stacks of multi-scale convolutional layer block using triplet of faces for face recognition in the wild. In: IEEE International Conference on Systems, Man, and Cybernetics, pp. 4460–4465 (2017)
10. Levi, G., Hassncer, T.: Age and gender classification using convolutional neural networks. In: IEEE Conference on Computer Vision and Pattern Recognition Workshops, pp. 34–42 (2015)
11. Ranjan, R., Sankaranarayanan, S., Castillo, C.D.: An all-in-one convolutional neural network for face analysis. In: 12th IEEE International Conference on Automatic Face and Gesture Recognition, pp. 17–24(2017)
12. Xiong, X., Torre, F.D.L.: Supervised descent method and its applications to face alignment. In: IEEE Conference on Computer Vision and Pattern Recognition, vol. 9, no. 4, pp. 532–539 (2013)

13. Ranjan, R., Zhou, S., Chen, J.C.: Unconstrained age estimation with deep convolutional neural networks. In: IEEE International Conference on Computer Vision Workshop, pp. 351–359 (2015)
14. Ekman, P., Friesen, W.V.: Facial action coding system: a technique for the measurement of facial movement. Rivista Di Psichiatria **47**(2), 126–38 (1978)
15. Weng, Y., Cao, C., Hou, Q.: Real-time facial animation on mobile devices. Graph. Models **76**(3), 172–179 (2013)
16. Redmon, R.: Darknet: open source neural networks in C. http://pjreddie.com/darknet/ (2013–2016)
17. Wu, Y., Hassner, T., Kim, K., et al.: Facial landmark detection with tweaked convolutional neural networks. IEEE Trans. Pattern Anal. Mach. Intell. **99**, 1 (2015)
18. Girshick, R.: Fast R-CNN. In: IEEE International Conference on Computer Vision, pp. 1440–1448 (2015)

An Approach for Feature Extraction and Diagnosis of Motor Rotor Bearing Based on Convolution Neural Network

Hao Wang[1], Dongsheng Yang[1(✉)], Yongheng Pang[1], Ting Li[1],
and Bo Hu[2]

[1] Northeastern University, Shenyang 110819, China
hhao2233@gmail.com, yangdongsheng@mail.neu.edu.cn
[2] State Grid Huludao Electric Power Supply Company, Huludao 125000, China
zwl980517@126.com

Abstract. The traditional rotor bearing fault diagnosis and analysis method is difficult to get the prior knowledge and experience, resulting in the low accuracy of fault diagnosis. In this paper, a method of fault feature extraction and diagnosis of rotor bearing based on convolution neural network is proposed. This method uses the chaotic characteristic of the vibration signal of the rotor bearing, uses the phase space reconstruction method to obtain the embedding dimension as the scale of the convolution neural network input composition, avoid the limitation of traditional frequency analysis method in the process of decomposition and transformation, the fault information can be extracted more comprehensively. In order to make full use of the advantages of the convolution neural network in the field of two-dimensional image analysis and improve the accuracy of the fault diagnosis model, a method of learning input form neural network based on convolution neural network for grayscale graph is proposed. The results of the simulation show the effectiveness of the method.

Keywords: Bearing fault diagnosis · Phase space reconstruction
Convolution neural network · Gray-scale image

1 Introduction

Bearing has always been an important part of modern machinery and equipment. As the rotating bearing is a moving part and generally in a closed working environment, the real time diagnosis of bearing failure has become an important subject. The wear of motor rotor bearing is a continuous process, it is not a "jump" state, so it is necessary to extract both accurate and sufficient fault features to make real time diagnosis for its fault status. The fault signal of the rotor bearing of the motor is not easy to be extracted. Most of the papers are based on the fault status based on the vibration signal. Because of the single signal, how to do a comprehensive, accurate and fast information feature extraction to a single signal has become a subject. This paper also proposes a method of feature extraction and diagnosis for motor rotor bearing based on convolution neural network.

© Springer Nature Switzerland AG 2018
L. Cheng et al. (Eds.): ICONIP 2018, LNCS 11301, pp. 284–296, 2018.
https://doi.org/10.1007/978-3-030-04167-0_26

The most of the methods used for bearing fault diagnosis are decomposed. Fault diagnosis of rolling bearings based on difference spectrum of singular value and stationary subspace analysis by Tang et al. raised the dimension of the signal and decomposed the high dimensional information with the stationary subspace [1]. Zhao et al. used the FFT method to decompose the vibration signal of the bearing [2]. The wavelet analysis method is well received in recent years. The wavelet analysis is widely concerned because it can be used to analyze the time domain frequency signal flexibly and transform it to spectrogram with time characteristic [3–5]. The feature extraction and fault diagnosis of fault signals by decomposition are incomplete, which leads to the limitation of fault information loss.

In recent years, artificial intelligence method has arisen, and many papers have explored the application of AI in bearing fault diagnosis. Jianhu Yuan et al. discussed a method of combining wavelet transform with artificial intelligence method, and formed the intelligent fault diagnosis method of rolling bearing [6]; He and Oh etc. discussed the application of deep learning method in the field of bearing fault diagnosis [7, 8]. Zhang et al. have studied a method of fault feature extraction and diagnosis, which is directly applied to the original vibration signal [9]. Many methods have focused on improving artificial intelligence network to combine the artificial intelligence algorithm and the bearing fault diagnosis. Based on the innovative composition method, this paper makes convolution neural network give full play to its advantages.

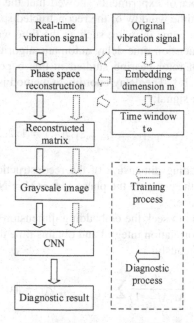

Fig. 1. The overall flow chart

In this paper, we proposed a fault diagnosis method of the rotor bearing described as Fig. 1 above. As you can see, we divided the whole process into two parts, which is the training process and the diagnostic process. When the offline training process is over, the online diagnostic process is relatively simple.

This paper takes into account the chaotic characteristics of rotor bearing vibration signals. The method of phase space reconstruction is innovatively used to reconstruct the grayscale map. And train the convolution neural network by these maps, so as to achieve a comprehensive identification of the fault and make an accurate and rapid diagnosis.

2   Phase Space Reconstruction and Data Processing

2.1   Selection of Time Delay and Embedding Dimension

In the process of phase space reconstruction, the choice of time delay τ and the embedded dimension m of the system is very important, which is related to whether the reconstructed phase space can accurately describe the characteristics of the original vibration signal. There are two main points of view on the choice of time delay τ and the embedded dimension m of the system. The first view is that the selection of m and τ is not related, such as autocorrelation, mutual information and complex autocorrelation methods. This kind of view is based on the precondition of the time series of infinite length and there is no noise interference in the study signal. In most practical applications, there is no guarantee of no noise interference in the signal, so the second viewpoints are produced. The second view holds that the selection of τ and m is interrelated. A large number of experiments showed that the choice of τ and m will have a direct effect on the time window of the reconstructed space. Therefore, the joint algorithm of τ and m were come up with, such as the time window method, the C-C method, the embedded dimension time delay automatic algorithm, and so on [10, 11].

The vibration signal of rotor bearing studied in this paper is chaotic time series signal $x = \{x_i | i = 1, 2, \ldots, N\}$. The delay coordinate method is used to reconstruct the phase space of the chaotic signal:

$$x_i = \left\{ x_i | x_i = \left[x_i, x_{i+\tau}, \ldots, x_{i+(m-1)\tau} \right]^T, i = 1, 2, 3, \ldots, M \right\} \tag{1}$$

Where m is the embedding dimension of the reconstruction matrix. τ is the time delay. M is the number of the points in the phase space ($M = N - (m - 1)\tau$). And N is the length of signal sequence.

We choose C-C method to seek the embedding dimension m and the time delay τ. The C-C method applies correlation integral and obtains time delay τ and time window $t_w (t_w = (m - 1)\tau)$ simultaneously.

$$C(m, N, r, \tau) = \frac{2}{M(M - 1)} \sum_{1 \leq i \leq j \leq M} \theta \left(r - \| x_i - x_j \| \right), r > 0 \tag{2}$$

Where $\theta(x) = 0$, if $x < 0$, $\theta(x) = 1$, if $x \geq 0$; $\| x_i - x_j \|$ represent distance. The correlation integral belongs to the cumulative distribution function, and its significance is the probability that the distance between any two points in the phase space is less than r. the BDS statistic define by

$$S(m, N, r, \tau) = C(m, N, r, \tau) - C^m(m, N, r, \tau) \tag{3}$$

The original vibration signal $x = \{x_i | i = 1, 2, \ldots, N\}$ is decomposed into τ subsequences that do not overlap each other:

$$
\begin{aligned}
x(1) &= \{x_i | i = 1, \tau + 1, \ldots, N - \tau + 1\} \\
x(2) &= \{x_i | i = 2, \tau + 2, \ldots, N - \tau + 2\} \\
&\cdots \\
x(\tau) &= \{x_i | i = \tau, \tau + \tau, \ldots, N - \tau + \tau\}
\end{aligned}
\tag{4}
$$

Where N is an integer multiple of τ. When applying the (4) to (3) to the test statistics, the test statistic is:

$$S_1(m, N, r, \tau) = \frac{1}{\tau} \sum_{s=1}^{\tau} \left[C_s\left(m, \frac{N}{\tau}, r, \tau\right) - C_s^m\left(1, \frac{N}{\tau}, r, \tau\right) \right] \tag{5}$$

When $N \to \infty$,

$$S_1(m, r, \tau) = \frac{1}{\tau} \sum_{s=1}^{\tau} [C_s(m, r, \tau) - C_s^m(1, r, \tau)] \tag{6}$$

When $N \to \infty$, for all r, $S_1(m, r, \tau) = 0$. But in practical application environment, the vibration signal used for fault diagnosis can not be infinite. So, $S_1(m, r, \tau) \neq 0$. In this way, the maximum time interval can take the point when $S_1(m, r, \tau)$ through coordinates (zero crossing) or the minimum difference between all the radius r. At this time, the point in the reconstructed phase space is mostly close to the uniform distribution, and the reconstructed attractor orbit is fully expanded in phase space. Here we choose the maximum and minimum values corresponding to the radius r to define the difference:

$$\Delta S_1(m, \tau) = max\{S_1(m, r_j, \tau, N)\} - min\{S_1(m, r_j, \tau, N)\} \tag{7}$$

$\Delta S_1(m, \tau)$ measures the maximum deviation between all radius r and $\Delta S_1(m, \tau) \sim \tau$. Therefore, the optimal time delay τ is the first minimum point of $\Delta S_1(m, \tau) \sim \tau$, or the first zero point of $\Delta S_1(m, \tau) \sim \tau$.

According to BDS statistical conclusions, $m = 2, 3, 4, 5, r_j = \frac{i\sigma}{2}, i = 1, 2, 3, 4$, where σ is the standard deviation of vibration signal.

$$\overline{S_1}(t) = \frac{1}{16} \sum_{m=1}^{4} \sum_{m=1}^{4} S_1(m, r, \tau) \tag{8}$$

$$\overline{\Delta S_1}(t) = \frac{1}{4} \sum_{m=1}^{4} \Delta S_1(m, \tau) \tag{9}$$

According to the above strategy, finding the global minimum point of $S_{1_{cor}}(t)$ can obtain the optimal delay time window t_w.

This paper proposes to make full use of the C-C method to obtain the time delay τ and the embedding dimension m to preprocess the motor rotor bearing vibration signal. And retain the data characteristics to the greatest degree in the later period when the processed data is given to the convolutional neural network.

2.2 Data Processing and Grayscale Map

We have reconstructed the vibration signal of the bearing and found that the size of the phase space m × M after reconstructing each bearing vibration signal is not same. Therefore, in order to facilitate the convolutional neural network to process information, this paper proposes an approach to expand the reconstructed embedding dimension so that the phase space after reconstruction is consistent in structure. The reason why we adopt this method has the following two advantages:

(1) We perform phase space reconstruction of the rotor bearing vibration signal to form an m-dimensional space. As long as m $\geq$ 2d + 1, the geometry of the dynamic system can be fully opened, where d is the dimension of the attractor in the system, and the condition m $\geq$ 2d + 1 is the sufficient but unnecessary condition of the dynamic system reconstruction. The integer m obtained by the dynamic system reconstruction is called the embedding dimension. This paper applies the method of expanding the reconstructed embedding dimension so that the composition dimension $m' \geq \max\{m\}$ and $m' = N - (m' - 1)\tau$, thus retaining all the chaotic attractors in the reconstruction space. All the features of the system are preserved.

(2) When the convolution neural network identifies the gray scale maps we gave, the best size of the CNN input picture is the same [12], because the final layer needs to be designed as the full connection layer. Based on the above reasons, the former method will adopt the resize method, but this method will cause the reconstructed image to deform, distort and destroy the features of original reconstructed image. In this case, we propose a method to expand the reconstruction and embedding dimension m' which makes $m' = N - (m' - 1)\tau$, and the size of the gray scale maps are consistent. It is convenient for CNN recognition.

To sum up two reasons, while reconstructing the phase space, we will record the size m and τ of each phase space, and make the final grayscale size be $\{m' \times m' | m' = N - (m' - 1)\tau, m' > max\{m\}\}$.

3 Convolutional Neural Network (CNN)

After the above phase space reconstruction, we have reconstructed the vibration signal of the rotor bearing into a grayscale image of $m' \times m'$, but from the perspective of the fault information embodied in the grayscale image, the fault information is very difficult to identified manually. This paper proposes the use of CNN for fault recognition. This recognition method fully exploits the advantages of CNN feature extraction and image recognition, and can fully exploit fault information in reconstructed images [13] (Fig. 2).

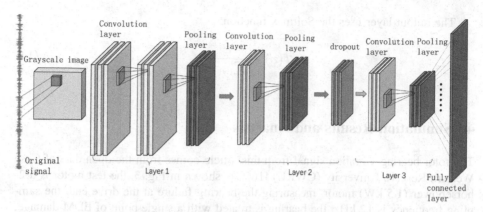

Fig. 2. Convolution neural network (CNN) structure diagram

The convolutional neural network used in this paper has a total of four layers, including three hidden layers and one full-link layer. The first hidden layer consists of two convolution layers and one pooled layer, with convolution kernel sizes of 8 × 8 and 5 × 5. The second and third convolution layers all have one convolution layer and one pooled layer, and the convolution kernel size is 5 × 5. The fourth layer is the output layer, which is a fully connected structure. In order to increase CNN's generalization ability and prevent it from being oversaturated, we added a layer of dropout before the last layer of convolution [14].

In this paper, the CNN model uses the cross-entropy function as the loss function of the estimated distribution and the target distribution. If g(x) is the target distribution and f(x) is the estimated distribution, the cross-entropy between g(x) and f(x) is:

$$\text{loss} = H(f, g) = -\sum_x g(x) \log f(x) \tag{10}$$

This article selects the ReLU function [14] for the activation function:

$$\phi(x) = \max(0, x) \tag{11}$$

The pooling layer uses the max pooling strategy:

$$\frac{\partial g}{\partial x_i} = \begin{cases} 1 & if\ x_i = \max(x) \\ 0 & otherwise \end{cases} \tag{12}$$

The gradient descent update weights process [15] is as follows:

$$W_{ij}^l = W_{ij}^l - \alpha \frac{\partial H}{\partial W_{ij}^l}$$

$$b_i^l = b_i^l - \alpha \frac{\partial H}{\partial b_i^l} \tag{13}$$

The output layer uses the Softmax function:

$$f(z_j) = \frac{e^{z_j}}{\sum_{i=1}^{n} e^{z_j}} \tag{14}$$

4   Simulation Results and Analysis

The rotor bearing vibration signal from this article comes from the open data of Case Western Reserve University (CWRU) [16]. As shown in Fig. 3, the test motor is a 2-horsepower (1.5 kW) motor, measuring the bearing failure at the drive end, the sampling frequency is 12 kHz, the bearing is treated with a single point of EDM damage, the fault is divided into 3 categories: rolling element fault, inner ring fault, outer ring fault.

Fig. 3. Vibration test platform for rotor bearing in CWRU

4.1   Sample Set Preprocessing

Since the original vibration signal has a large amount of data, and the training of neural network requires a large amount of sample, this paper will cut the original vibration signal into a sample set with 400 sample points per sample. The interception step length is 200 per failure type 500 samples. The specific tags are shown in Table 1 below:

Table 1. Classification label of fault sample

Fault type	Scale	Quantity	Lable
Normal	397	500	0
Ball	397	500	1
Inner race	397	500	2
Outer race	397	500	3

The rotor bearing vibration signal in the sample set is shown in Fig. 4 below:

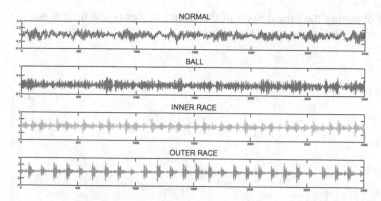

Fig. 4. Raw vibration signal of rotor bearing

From Fig. 4, it can be seen that the vibration signals of the states of the various rotor bearings are too complex and difficult to identify directly by manpower. Therefore, the fault feature extraction and diagnosis method for motor rotor bearings based on CNN proposed in this paper is introduced.

4.2 Selection of Time Delay and Time Window

Using the C-C method described above, we reconstructed the rotor bearing vibration signals in the data set. Firstly, the time delay τ is obtained by formula (7), as shown in Fig. 5 below:

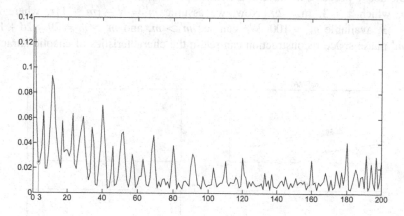

Fig. 5. $\Delta S_1(m, \tau) \sim \tau$ curve

Draw the curve of $\Delta S_1(m, \tau) \sim \tau$, the time t_d is the first minimum point of $\Delta S_1(m, \tau) \sim \tau$ in the image, that is, the time delay of vibration signal of the rotor bearing is $\tau = 3$.

After obtaining the time delay τ of the vibration signal, the time window t_w can be obtained from (9), as shown in Fig. 6 below:

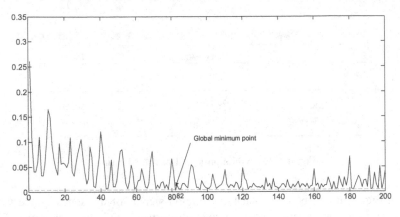

Fig. 6. $S_{1_{cor}}(t) \sim t$ curve

From the above figure, we find the global minimum point t_w in the $S_{1_{cor}}(t) \sim t$ curve, which is the time window $t_w = 82$ for the phase space reconstruction of the rotor bearing vibration signal. By the relation $t_w = (m - 1)\tau$, the reconstructed embedding dimension m = 29 is obtained.

4.3   Composition Reconstruction Grayscale Image

After the above process, we have obtained the phase space reconstruction parameters τ and m, which $\tau = 3$, m = 29. Now, we assume $m' = N - (m' - 1)\tau$, where $N = 397, \tau = 3$, available $m' = 100$. We can see $m' > m$, and $m' > m = 29 \geq 2d + 1$, so using m' phase space reconstruction can retain the characteristics of chaotic attractors.

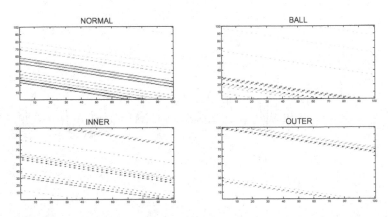

Fig. 7. Reconstructed grayscale map of vibration signal of rotor bearing

Take $N = 397, \tau = 3, m' = 100$. Using (1), the phase space of the rotor bearing vibration signal is reconstructed, and a grayscale image is formed by the phase space reconstruction. Figure 7 shows the four states of the vibration signal.

It can be seen from Fig. 7 that the four states of the vibration signal of the rotor bearing vibration are different, but it is not easy to extract the obvious fault features and it is not easy to find the fault recognition rules. Therefore, we use CNN's image recognition capability to characterize the fault. Identification and diagnosis.

4.4   Result of Fault Diagnosis

We use the reconstructed grayscale image formed by phase space reconstruction as input, and use CNN for fault diagnosis. As shown in Table 2 below, statistics on the diagnostic accuracy of the four kinds of rotor bearing vibration are made in this paper. From the data in the table, it can be seen that the diagnostic accuracy of normal state and rolling element failure can reach 100%, and the accuracy of inner ring failure and outer ring failure can reach 99.4%. It is fully demonstrated that the fault feature extraction and diagnosis method of the motor rotor bearing based on CNN is feasible, and the effect is significant.

Table 2. Fault diagnosis by vibration signal of rotor bearing of motor

Fault type	Quantity	Number of correct	Accuracy
Normal	580	580	1.0000
Ball	580	580	1.0000
Inner race	580	579	0.9965
Outer race	580	578	0.9948
Sum	2320	2317	0.9987

We not only study the fault diagnosis of the rotor bearing using the phase space reconstruction gray image with $\tau = 3$, but we also make fault diagnosis accuracy statistics at $\tau = 1, 2, 4$, and 5. Based on the maximum value of the accuracy of fault diagnosis under different values of τ, we calculated the maximum accuracy of the training of the CNN model and the number of training epoch used to achieve this accuracy. The fact that the accuracy rate is large and the required training period is shorter indicates that the CNN model responds more sensitively and the fault feature extraction is more complete What needs to be explained here is that when $\tau = 1$, it can be obtained from formula (1) that "phase space reconstruction grayscale map" is essentially without phase space reconstruction, in other words, when $\tau = 1$ the grayscale image corresponds to the natural arrangement of the motor rotor bearing original vibration signal into a square form.

Table 3. Accuracy of fault diagnosis under different τ

τ	Scale	Quantity	Max accuracy	Epoch
1	200 × 200	2320	0.9955	6
2	134 × 134	2320	0.9989	10
3	100 × 100	2320	0.9996	9
4	81 × 81	2320	0.9985	5
5	67 × 67	2320	0.9966	4

As shown in Table 3 above, when τ = 3, the fault diagnosis of the motor rotor bearing based on the CNN has the highest accuracy and can be achieved when training the 9th epoch. As shown in Fig. 8 below, the solid red line represents the τ = 3 accuracy curve. When epoch = 3, it achieves a higher accuracy rate. The model training is basically completed and the training speed is faster.

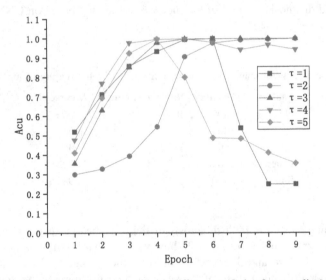

Fig. 8. Model training speed with different τ (Color figure online)

For the phase space reconstruction greyscale map when τ = 3, the CNN can train the model more efficiently. We can interpret it as: When the phase space reconstruction time delay τ = 3, the grayscale image formed by reconstruction retains the complete fault signal characteristics and can extract certain characteristics of the fault information of the rotor bearing vibration signal of the motor. Therefore, the model training can be completed faster when the CNN performs the model training.

5 Conclusion

Finally, at the end of this article, we have three points to summarize.

(1) Taking into account the chaotic characteristics of the rotor bearing vibration signal, this paper proposes the using of phase space reconstruction method to fully open the dynamic system of the motor rotor bearing vibration, so that the fault information of the bearing vibration signal is completely extracted. In the data preprocessing stage, we innovatively avoid the limitations of traditional frequency analysis methods that may cause the loss of fault information in the decomposition transformation process.

(2) In this paper, the vibration signal of the motor rotor bearing after phase space reconstruction is presented in the form of $\tau = 3$, $m' > m = 29 \geq 2d + 1$. On the one hand, the advantages of the phase space reconstruction method are preserved, and the preliminary feature extraction is performed on the fault signal. On the other hand, the vibration signal data can be converted into a grayscale image to fully exploit the CNN. The advantages in the field of image recognition, and can quickly and accurately identify the fault information and make the correct diagnosis.

(3) The advantage of CNN lies in the feature extraction of complex images, thus overcome the disadvantages of artificial fault feature recognition. The innovation of this paper lies in the application of phase-space-reconstructed grayscale image to CNN training and fault diagnosis, which improves the training efficiency of CNN and makes the CNN can complete the training faster and improve the accuracy. Since the reconstructed grayscale image has been subjected to preliminary feature extraction, the number of convolutional kernels can be moderately reduced when constructing a CNN, and even the number of layers of CNN can be reduced. This will also reduce the time for fault diagnosis. Achieve rapid troubleshooting.

References

1. Tang, G.J., Pang, B., Liu, S.K.: Fault diagnosis of rolling bearings based on difference spectrum of singular value and stationary subspace analysis. J. Vib. Shock **34**(11), 83–87 (2015)
2. Zhao, D.Z., Li, J.Y., Cheng, W.D., et al.: Rolling element bearing fault diagnosis based on generalized demodulation algorithm under variable rotational speed. J. Vib. Eng. **2017**(5), 865–873 (2017)
3. Zhang, X.N., Zeng, Q.S., Wan, H.: Bearing fault diagnosis based on improved wavelet denoising and EMD method. Meas. Control Technol. **33**(1), 23–26 (2014)
4. Qin, Y.: A new family of model-based impulsive wavelets and their sparse representation for rolling bearing fault diagnosis. IEEE Trans. Ind. Electron. **65**(3), 2716–2726 (2017)
5. Shao, H., Jiang, H., Li, X., et al.: Intelligent fault diagnosis of rolling bearing using deep wavelet auto-encoder with extreme learning machine. Knowl.-Based Syst. (2017)

6. Yuan, J.H., Han, T., Tang, J., et al.: An approach to intelligent fault diagnosis of rolling bearing using wavelet time-frequency representations and CNN. Mach. Des. Res. **2017**(2), 93–97 (2017)

7. He, M., He, D.: Deep learning based approach for bearing fault diagnosis. IEEE Trans. Ind. Appl. **53**(3), 3057–3065 (2017)

8. Oh, H., Jung, J.H., Jeon, B.C., et al.: Scalable and unsupervised feature engineering using vibration-imaging and deep learning for rotor system diagnosis. IEEE Trans. Ind. Electron. **65**(4), 3539–3549 (2018)

9. Zhang, W., Peng, G., Li, C., et al.: A new deep learning model for fault diagnosis with good anti-noise and domain adaptation ability on raw vibration signals. Sensors **17**(2), 425 (2017)

10. Kim, H.S., Eykholt, R., Salas, J.D.: Nonlinear dynamics, delay times, and embedding windows. Phys. D-Nonlinear Phenom. **127**(1–2), 48–60 (1999)

11. Liu, Y.B., He, B., Liu, F., et al.: Comprehensive recognition of rolling bearing fault pattern and fault degrees based on two-layer similarity in phase space. J. Vib. Shock **36**(4), 178–184 (2017)

12. He, K., Zhang, X., Ren, S., et al.: Spatial pyramid pooling in deep convolutional networks for visual recognition. IEEE Trans. Pattern Anal. Mach. Intell. **37**(9), 1904–1916 (2015)

13. Chang, L., Deng, X.M., Zhou, M.Q., et al.: Convolutional neural networks in image understanding. Acta Autom. Sin. **42**(9), 1300–1312 (2016)

14. Krizhevsky, A., Sutskever, I., Hinton, G.E.: ImageNet classification with deep convolutional neural networks. In: 25th International Conference on Neural Information Processing Systems. Curran Associates Inc., pp. 1097–1105. ACM, Lake Tahoe, Nevada (2012)

15. Lecun, Y., Boser, B., Denker, J.S., et al.: Backpropagation applied to handwritten zip code recognition. Neural Comput. **1**(4), 541–551 (1989)

16. Case western reserve university bearings vibration dataset. http://csegroups.case.edu/bearingdatacenter/home

Gated Convolutional Networks for Commonsense Machine Comprehension

Wuya Chen, Xiaojun Quan[✉], and Chengbo Chen

School of Data and Computer Science, Sun Yat-sen University, Guangzhou, China
{chenwy58,chenchb7}@mail2.sysu.edu.cn, quanxj3@mail.sysu.edu.cn

Abstract. In this paper, we study the problem of commonsense machine comprehension and propose a new model based on convolutional neural networks and Gated Tanh-ReLU Units. The new model, which serves as an alternative to exiting recurrent models, consists of three layers: input layer, gated convolutional layer, and output layer. The input layer produces representations based on various features, such as part-of-speech and relation embeddings. Gated convolutional layer, the key component of our model, extracts n-gram features at different granularities and models the interactions between different texts (questions, answers, and passages). Bilinear interactions are used as output layer to capture the relations among the final expressive representations and to produce the final answers. We evaluate our model on the *SemEval-2018 Machine Comprehension Using Commonsense Knowledge* task. Experimental result shows that our model achieves highly competitive results with the state-of-the-art models but is much faster. To our knowledge, this is the first time a non-recurrent approach gains competitive performance with strong recurrent models for commonsense machine comprehension.

Keywords: Convolutional neural networks · Gated mechanism
Reading comprehension · Commonsense knowledge

1 Introduction

Machine reading comprehension (MRC) aims to teach machines to read and comprehend human languages. MRC is a long-standing objective in natural language processing and it is composed of various subtasks, such as cloze-style reading comprehension [1–3], span-extraction reading comprehension [4,9,17,18], and open-domain reading comprehension [5,19]. One key problem in MRC is for machine to utilize commonsense knowledge for real-life reading comprehension. In *SemEval-2018 Task 11: Machine Comprehension using Commonsense Knowledge* [6], the organizers provide narrative texts about everyday activities and require participants to build systems for answering questions based on such texts. To fulfill this task, it is necessary to introduce external knowledge, such as commonsense knowledge. Wang et al. [8] proposed to use relation embeddings

L. Cheng et al. (Eds.): ICONIP 2018, LNCS 11301, pp. 297–306, 2018.
https://doi.org/10.1007/978-3-030-04167-0_27

gained from ConceptNet [7] as additional features and their proposed model achieved very promising result. We follow this idea but further take the direction of edges in ConceptNet into account, in view of the fact that some relations are designated as symmetric in ConceptNet such as *SimilarTo* while others as asymmetric like *UsedFor*.

Most current works apply recurrent neural networks equipped with attention mechanism for this task. For example, Chen et al. [13] presented a hybrid multi-aspect model with recurrent neural networks combined with a handful of attentions among questions, answers, and passages. The motivation behind this approach is to mimic human intuition in dealing with multiple-choice reading comprehension with multi-aspect attentions. In another representative work, Wang et al. [8] proposed to produce contextual representations for a question, answers and a passage successively using recurrent neural networks and attention in a similar way. However, these models are usually too complicated with very time-consuming training and inference phases as well. Obviously, the high time complexity becomes a bottleneck when applying those models to many NLP problems such as reading comprehension. As an alternative, convolutional architectures are frequently exploited recently. For instance, Dauphin et al. [14] introduced gated convolutional neural networks into language model in their work. Inspired by [8,14], in this paper we proposed a convolutional architecture for text modeling in commonsense machine reading comprehension, which is much more efficient and achieves results very competitive with the state-of-the-art models. In what follows we summarize the main contributions of this work:

- We proposed a fast and effective gated convolutional neural network equipped with attention as a substitution for traditional recurrent architecture.
- This work is the first attempt in MRC at applying the novel Gated Tanh-Relu Units (GTRU) [11], which have been successfully applied to sentiment analysis task but yet to MRC.
- Relation embeddings derived from ConceptNet are introduced as additional input features. The direction of the relationship is also investigated.

2   Task Description

The *SemEval 2018 Task 11* studied in this paper is a competition of machine comprehension using commonsense knowledge. In this task, systems are required to answer multi-choice questions given narrative texts about everyday activity, such as baking a cake, taking a bus, etc. Each question is associated with two candidate answers. A substantial number of questions require inference using commonsense knowledge that beyond the facts mentioned in the texts. Participants are allowed to refer to any form of external resources for such knowledge.

Each example consists of a passage $\{P_i\}_{i=1}^{|P|}$, a question $\{Q_j\}_{j=1}^{|Q|}$, an answer $\{A_k\}_{k=1}^{|A|}$ and a label $y \in \{0, 1\}$. P, Q, A are sequences of word indices, and y indicates whether the candidate is true answer.

3   Model

Our model consists of an input layer, a one-dimension gated convolutional layer, and an output layer. The network architecture is depicted in Fig. 1.

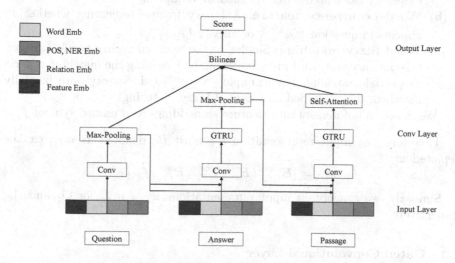

Fig. 1. Neural architecture of our model. Note that we use Conv to denote convolutional neural network, POS to denote part-of-speech, NER to denote named-entity, GTRU to denote Gated Tanh-ReLU Units, and Emb to denote embeddings.

3.1   Input Layer

We intend to utilize multi-aspect information to encode words in texts, questions and answers. For example, the representation of a word P_i in given passages is a concatenation of several vectors:

(1) **Word embeddings**: Pre-trained 300-dimensional *GloVe* vector $\boldsymbol{E}_{P_i}^{glove}$ for word representation.
(2) **Relation embeddings**: Randomly initialized 10-dimensional relation embedding $\boldsymbol{E}_{P_i}^{rel}$. For each word, the relation is determined by querying ConceptNet to check whether there is an edge between P_i and any word in question Q or answer A. If there exists multiple relations, just choose one at random. Furthermore, if relations of ConceptNet's edges are asymmetric, such as *UsedFor* and *HasProperty*, we set their relation embeddings to unidirection. As for the rest of relations, they are initialized as bidirection.
(3) **Part-of-speech and named-entity embeddings**: Randomly initialized 10-dimensional part-of-speech embedding $\boldsymbol{E}_{P_i}^{pos}$ for 49 different types of POS tags, and 8-dimensional named-entity embedding $\boldsymbol{E}_{P_i}^{ner}$ for 18 different types of NER tags. We use small embedding sizes for POS and NER in order to reduce model sizes without much loss of performance.

(4) **Feature embeddings**: We also incorporate several hand-crafted features including logarithmic term frequency, word co-occurrence and word fuzzy matching.

  (a) **Logarithmic term frequency.** The logarithmic transformation of term frequency calculated based on English Wikipedia.

  (b) **Word occurrence feature.** A binary feature indicating whether P_i appears in question $\{Q_j\}_{j=1}^{|Q|}$ or answer $\{A_k\}_{k=1}^{|A|}$.

  (c) **Word fuzzy matching.** Similar to the word co-occurrence feature, we take into account word fuzzy matching by loosening the matching criteria to partial matching. For example, "sweet" and "sweety" are partially matched, thus regarded as a kind of fuzzy matching.

We denote all aforementioned feature embeddings as a unified symbol $\boldsymbol{f}_{P_i}$.

Therefore, the input representation for word P_i, denoted as $\boldsymbol{w}_{P_i}$, can be depicted as

$$\boldsymbol{w}_{P_i} = [\boldsymbol{E}_{P_i}^{glove}; \boldsymbol{E}_{P_i}^{rel}; \boldsymbol{E}_{P_i}^{pos}; \boldsymbol{E}_{P_i}^{ner}; \boldsymbol{f}_{P_i}] \tag{1}$$

Similarly, we can obtain input representations $\boldsymbol{w}_{Q_j}$, $\boldsymbol{w}_{A_k}$ for Q_j and A_k, respectively.

3.2 Gated Convolutional Layer

After the input layer, we obtain the representations $\boldsymbol{w}_{P_i}, \boldsymbol{w}_{Q_j}\ \boldsymbol{w}_{A_k}$ for each token of passage, question, and answer respectively, where $\boldsymbol{w}_{P_i} \in R^{D_P}$, $\boldsymbol{w}_{Q_j} \in R^{D_Q}$, $\boldsymbol{w}_{A_k} \in R^{D_A}$, and D_P, D_Q, D_A denote the dimension size of corresponding embedding vectors.

We first briefly review the vanilla CNN [15] for modeling text representation. For passage $\boldsymbol{w}_P = [\boldsymbol{w}_{P_1}, \boldsymbol{w}_{P_2}, \cdots, \boldsymbol{w}_{P_{|P|}}]$, as we slide the filter across the whole passage, we obtain a sequence of new features $c = [c_1, c_2, ..., c_L]$.

$$c_i = f(\boldsymbol{w}_{P_i:P_{i+k}} * \boldsymbol{w}_c + b_c) \tag{2}$$

where $b_c \in R$ is the bias, f is a non-linear activation function, and $*$ denotes convolution operation. If there are n_k filters of the same width k, the output features form a matrix $\boldsymbol{C} \in R_{n_k X L_k}$.

In order to better model text representation, we choose to use ConvNets and Gated Tanh-ReLU Units (GTRU) [11], which has a novel gate, incorporating the reference vector, to control the flow of information. Specifically, we compute feature c_i as

$$g_i = relu(\boldsymbol{w}_{P_i:P_{i+k}} * \boldsymbol{w}_g + \boldsymbol{V}_g v_g + b_g) \tag{3}$$

$$o_i = tanh(\boldsymbol{w}_{P_i:P_{i+k}} * \boldsymbol{w}_o + b_o) \tag{4}$$

$$c_i = g_i \cdot o_i \tag{5}$$

where v_g is the reference vector. Hence, CNN equipped with GTRU can better extract features regarding the reference information, which is useful for modeling the text related to other reference entities or texts, such as passages of reading comprehension which need to take question into account. We consider it resembles to the unit of match-lstm [4], which incorporates reference vector in the input.

Then, the procedure of gated convolutional layer will be described via three steps. We first map w_Q to contextual features c_Q by a normal convolutional layer with filter sizes 1 and 3, and max-over-time pooling, which gains the most significant features over the whole sentence. After that, we calculate c_A by formula: $f_{GTRU}(w_A, c_Q)$, where f_{GTRU} denotes ConvNets+GTRU and its filter sizes are 1 and 3, followed by a max-over-time pooling. Finally, we obtain c_P using $f_{GTRU}(w_P, c_Q, c_A)$, whose filter sizes are 3, 4, 5, followed by a self-attention layer. The reason we choose multiple filters is that convolutional layers with multiple filters can efficiently extract n-gram features at different granularities on each receptive field. Self-attention [12] is defined as follows:

$$Att_{self}(\{c_i\}_i^{|P|}) = \sum_{i=1}^{|P|} \alpha_i c_i \qquad (6)$$

$$\alpha_i = softmax_i(W_2^T c_i) \qquad (7)$$

The reason we use self-attention rather than max pooling for passage is that the length of passage is usually long, and self-attention can effectively model the long range dependencies between distant words [11,12].

So far, we obtain three final expressive representations of passage, question, answer as c_P, c_Q, c_A respectively.

3.3   Output Layer

The output layer is to connect the three final expressive representations. We choose to use the bilinear interactions:

$$y = \sigma(c_P W_3 c_A + c_Q W_4 c_A) \qquad (8)$$

where y is the final output.

To train the network, cross entropy function is used as the loss function.

4   Experiments and Discussion

4.1   Setup

Dataset. The whole dataset is officially split into training, development and test set, containing 9731, 1411 and 2797 questions, respectively. We follow the preprocessing procedures of [8] and apply $spaCy$[1] to tokenization, part-of-speech tagging and named-entity recognition.

---

[1] https://github.com/explosion/spaCy.

Initialization and Hyperparameters. Our model is implemented based on *PyTorch*[2] and trained on a single Titan Xp GPU. Word embeddings are initialized with 300-dimensional *GloVe* vectors. Moreover, only the word embeddings of the top 10 frequent words are fine-tuned during training. The dropout rates for input embeddings and output layer are set to 0.4 and 0.3 respectively in order to fight overfitting. The number of filters for convolutional layer is 100, following the setting in previous study [16]. We apply *Adamax* [20] optimization approach with an initial learning rate 0.002 which is halved after 10 and 15 epochs. Gradients are clipped to have a maximum L2 norm of 10. Minibatch with batch size 32 is used. Our model converges after 35 epochs.

4.2 Results

Since it has passed the *evaluation phase*[3], we have to evaluate our model in the *post-evaluation phase*. The settings of *post-evaluation phase* are the same as *evaluation phase*. The experimental results and training time are shown in Tables 1 and 2, respectively. We can see that our model achieves comparable result close to TriAN-single, which win the second place in this task, with 0.7 % difference on development dataset and 1.94 % difference on test dataset of post-evaluation phase. However, the training speed of our model is 2.8 times faster than TriAN-single, and our model almost can converge after 35 epochs but TriAN-single needs 50 epochs to converge. The number of model parameters probably accounts for this. Because, our model has 1177263 parameters while TriAN-single has 2038584. Furthermore, the first place model, HFL, is more complicated and should have even more parameters. Note that, although HFL achieves the best ensemble result, its single model gains only 80.94% on test dataset in *evaluation phase*, which underperforms TriAN-single by 1%. Obviously, compared with HFL-single and TriAN-single model, our model is more simple and efficient.

Table 1. Experimental results.

Model	Dev	Test (in *post-evaluation phase*)
Random	50.00%	50.00%
TriAN-RACE	64.78%	61.34%
TriAN-single	83.41% (*our test*)	**80.91%**
HFL-single	**84.48%**	—
Our model (single)	82.71%	78.97%

---

[2] https://pytorch.org/.
[3] https://competitions.codalab.org/competitions/17184#phases.

Table 2. Training time.

Model	Training time(s) of one epoch
TriAN-single	70
our model (single)	**25**

4.3 Ablation Study

As mentioned before, our model takes into account direction of relation embeddings of ConceptNet's edges, which is ignored by previous work [8]. In other words, Wang et al. [8] regard all relations of ConceptNet's edges as bidirection. We perform an ablation study to investigate the effects of this component. As shown in Table 3, the direction of relation embeddings is indeed useful. The reason is that it truly reflects the design principles of the ConceptNet and can distinguish different words more effectively. However, we also understand relation embeddings based on ConceptNet only serve as additional input features and how to combine explicit knowledge with reasoning component is still a open question.

Table 3. Ablation study for direction of relation embeddings of ConceptNet's edges. *w/o* denotes without.

Model	Dev
Our model (single)	**82.71%**
w/o consider direction of relation embeddings	82.12%

4.4 Self-Attention vs Max Pooling

As introduced before, our model is founded fundamentally on convolutional networks. Xue et al. [11] combined gated convolutional networks with max pooling in sentiment analysis task. However, we are aware of the fact that the length of passage in MRC is relatively long so our system maybe can benefit more from self-attention, which can effectively and efficiently capture long range dependencies when the length of sequence is not much longer than the dimension of each token of that sequence. In contrast, when question and answer are both short, max pooling layer is good enough. Therefore, we also conduct a study to compare self-attention layer with max pooling layer. From Table 4, when using self-attention instead of max pooling for passage, we can gain relatively better performance, such as comparing $p\_self + q\_max + c\_max$ with $p\_max + q\_max + c\_max$. In the meantime, comparing $p\_self + q\_max + c\_max$ with $p\_self + q\_self + c\_self$ can confirm our idea that max pooling fit short text processing.

In general, max pooling plays a key role in convolutional networks. It can extract the most significant feature and increase the field of perception. However, self-attention may gain better performance in some cases, such as the processing text is long, as a result of its advantage of modeling long range dependencies.

Table 4. Comparing self-attention with max pooling: p_self denotes self-attention for passage. p_max denotes max pooling for passage. We denote the others in the same way for question and answer.

Model	Dev
our model (single)	**82.71%**
$p\_self + q\_max + c\_max$	**82.71%**
$p\_self + q\_max + c\_self$	81.93%
$p\_self + q\_self + c\_max$	82.14%
$p\_self + q\_self + c\_self$	81.73%
$p\_max + q\_max + c\_max$	81.71%
$p\_max + q\_max + c\_self$	78.17%
$p\_max + q\_self + c\_max$	81.79%
$p\_max + q\_self + c\_self$	79.95%

Table 5. Comparison of different gated units.

Model	Dev
Our model (single)	**82.71%**
GTRU	**82.71%**
GLU	82.21%
GTU	82.50%
GLRU	80.51%

4.5   Gating Mechanisms

In this subsection, we compare GTRU, used in our model, with other possible gate options, i.e., GTU $tanh(X * W + b) \times \sigma(X * W_a + V v_a + b_a)$, GLU $(X * W + b) \times \sigma(X * W_a + V v_a + b_a)$, and GLRU $(X * W + b) \times relu(X * W_a + V v_a + b_a)$ (this gating variant first proposed by us in this paper). From the result in Table 5, we can see the GTRU gains the best performance and GLU or GTU also achieves comparable result, while the GLRU has the worst performance. Maybe the upper bound +1 of sigmoid function, which is not able to amplify the similar features between target vector and reference vector and reduce the different features, accounts for it [11]. Because our model is a shallow architecture, it cannot exploit the advantage of reducing the vanishing gradient problem provided by linear path. Therefore, we cannot be so surprised with the performance of GLRU.

5   Conclusion

Reducing time complexity is a key problem of commonsense machine reading comprehension. To address this, we propose a gated convolutional architecture as an alternative to the typically used recurrent architecture. The architecture includes relation embeddings derived from ConceptNet as additional input features. Moreover, our model also takes into account the direction of relation embeddings which is ignored by previous works. Experimental result proves that the new model is able to achieve competitive results with the state of the art but is much faster.

Acknowledgments. The paper was supported by the Program for Guangdong Introducing Innovative and Enterpreneurial Teams (No. 2017ZT07X355).

References

1. Hill, F., Bordes, A., Chopra, S., Weston, J.: The goldilocks principle: reading children's books with explicit memory representations. arXiv preprint arXiv:1511.02301 (2015)
2. Cui, Y. Liu, T.P., Chen, Z., Wang, S.J., Hu, G.P.: Consensus attention-based neural networks for chinese reading comprehension. arXiv preprint arXiv:1607.02250 (2016)
3. Cui, Y.M., Chen, Z.P., We, S., Wang, S.J., Liu, T., Hu, G.P.: Attention-over-attention neural networks for reading comprehension. In: 55th Annual Meeting of the Association for Computational Linguistics, vol. 1, pp. 593–602. Association for Computational Linguistics, Vancouver (2017)
4. Wang, S.H., Jiang, J.: Machine comprehension using match-LSTM and answer pointer. arXiv preprint arXiv:1608.07905 (2016)
5. Chen, D.Q., Fisch, A., Weston, J., Bordes, A.: Reading Wikipedia to answer open-domain questions. In: 55th Annual Meeting of the Association for Computational Linguistics, vol. 1, pp. 1870–1879. Association for Computational Linguistics, Vancouver (2017)
6. Ostermann, S., Roth, M., Modi, A., Thater, S., Pinkal, M.: SemEval-2018 Task 11: machine comprehension using commonsense knowledge. In: 12th International Workshop on Semantic Evaluation, pp. 747–757, New Orleans (2018)
7. Speer, R., Chin, J., Havasi, C.: Conceptnet 5.5: an open multilingual graph of general knowledge. arXiv preprint arXiv:1612.03975 (2016)
8. Wang, L., Sun, M., Zhao, W.: Yuanfudao at SemEval-2018 Task 11: three-way attention and relational knowledge for commonsense machine comprehension. arXiv preprint arXiv:1803.00191 (2018)
9. Seo, M., Kembhavi, A., Farhadi, A., Hajishirzi, H.: Bidirectional attention flow for machine comprehension. arXiv preprint arXiv:1611.01603 (2016)
10. Wang, W.H., Yang, N., Wei, F.R., Chang, B.B., Zhou, M.: Gated self-matching networks for reading comprehension and question answering. In: 55th Annual Meeting of the Association for Computational Linguistics, vol. 1, pp. 189–198. Association for Computational Linguistics, Vancouver (2017)
11. Xue, W., Li, T.: Aspect based sentiment analysis with gated convolutional networks. In: 56th Annual Meeting of the Association for Computational Linguistics, vol. 1, pp. 2514–2523. Association for Computational Linguistics, Melbourne (2018)

12. Yang, Z., Yang, D., Dyer, C., He, X., Smola, A., Hovy, E.: Hierarchical attention networks for document classification. In: the 2016 Conference of the North American Chapter of the Association for Computational Linguistics: Human Language Technologies, pp. 1480–1489 (2016)
13. Chen, Z., Cui, Y., Ma, W.: HFL-RC system at SemEval-2018 Task 11: hybrid multi-aspects model for commonsense reading comprehension. arXiv preprint arXiv:1803.05655 (2018)
14. Dauphin, Y.N., Fan, A., Auli, M.: Language modeling with gated convolutional networks. arXiv preprint arXiv:1612.08083 (2016)
15. Kalchbrenner, N., Grefenstette, E., Blunsom, P.: A convolutional neural network for modelling sentences. In: 52nd Annual Meeting of the Association for Computational Linguistics, vol. 1, pp. 655–665. Association for Computational Linguistics, Baltimore (2014)
16. Kim, Y.: Convolutional neural networks for sentence classification. In: the 2014 Conference on Empirical Methods in Natural Language Processing (EMNLP), pp. 1746–1751. Association for Computational Linguistics, Doha (2014)
17. Pan, B., Li, H., Zhao, Z., Cao, B., Cai, D., He, X.: Memen: multi-layer embedding with memory networks for machine comprehension. arXiv preprint arXiv:1707.09098 (2017)
18. Wang, Y.Z., et al.: Multi-passage machine reading comprehension with cross-passage answer verification. In: 56th Annual Meeting of the Association for Computational Linguistics, vol. 1, pp. 1918–1927. Association for Computational Linguistics, Melbourne (2018)
19. Wang, S., Yu, M., Jiang, J.: Evidence aggregation for answer re-ranking in open-domain question answering. arXiv preprint arXiv:1711.05116 (2017)
20. Kingma, D.P., Ba, J.: Adam: a method for stochastic optimization. arXiv preprint arXiv:1412.6980 (2014)

Cursive Scene Text Analysis by Deep Convolutional Linear Pyramids

Saad Bin Ahmed[1,2], Saeeda Naz[4], Muhammad Imran Razzak[3(✉)], and Rubiyah Yusof[2]

[1] King Saud bin Abdulaziz University for Health Sciences, Riyadh, Saudi Arabia
ahmedsa@ksau-hs.edu.sa
[2] Malaysia Japan International Institute of Technology (MJIIT),
Universiti Teknologi Malaysia, Kuala-Lumpur, Malaysia
{saad2,rubiyah.kl}@utm.my
[3] University of Technology, Sydney, Australia
imran.razzak@ieee.org
[4] Higher Education Department, Government Post Graduate College No. 01,
Abbottabad, Pakistan
saeedanaz292@gmail.com

Abstract. The camera captured images have various aspects to investigate. Generally, the emphasis of research depends on the interesting regions. Sometimes the focus could be on color segmentation, object detection or scene text analysis. The image analysis, visibility and layout analysis are the tasks easier for humans as suggested by behavioural trait of humans, but in contrast when these same tasks are supposed to perform by machines then it seems to be challenging. The learning machines always learn from the properties associated to provided samples. The numerous approaches are designed in recent years for scene text extraction and recognition and the efforts are underway to improve the accuracy. The convolutional approach provided reasonable results on non-cursive text analysis appeared in natural images. The work presented in this manuscript exploited the strength of linear pyramids by considering each pyramid as a feature of the provided sample. Each pyramid image process through various empirically selected kernels. The performance was investigated by considering Arabic text on each image pyramid of EASTR-42k dataset. The error rate of 0.17% was reported on Arabic scene text recognition.

Keywords: Linear pyramids · Kernels · Feature extraction
Arabic scene text

1 Introduction

The text localization is perceived as a fundamental step in scene text recognition. Although few approaches for text localization have been suggested by pattern recognition and computer vision research communities, more effort is

© Springer Nature Switzerland AG 2018
L. Cheng et al. (Eds.): ICONIP 2018, LNCS 11301, pp. 307–318, 2018.
https://doi.org/10.1007/978-3-030-04167-0_28

still required to define text localization approaches for cursive scene text. The localization techniques are not usually specific to a particular script. One successful approach can be applied on number of scripts having the same nature image. In presented work, the analysis is performed on Arabic cursive script. Later, applied localization technique which has proved encouraging results on EASTR-42k dataset.

In scene text image analysis, features play a prominent role in correct recognition of given scene text. The heterogeneous features increased prospect of obtaining better accuracy. The scene text image contains non-hierarchical features that are extracted from local regions. These features are described as distinctive properties which need to quantize in standard format by using a spatial or statistical techniques. The text represented in scene images contain inter and intra-class variations, this difference is comparatively easier to handle in Latin rather than in Arabic. The characteristics of Arabic and Arabic-like scripts are writing direction (i.e., from right to left), representation of some characters with diacritical marks and appearance of same character with respect to its position makes this script more challenging. Although, recent research on printed and handwritten Arabic-like scripts presented state-of-the-art results as reported by [1–4, 12–15] but scene text recognition of Arabic script is still in its early stages [2]. The scene text recognition techniques fall into texture, component and hybrid based methods [2]. This paper presents texture based methods which help in localizing the text based on filtration techniques.

The background of image is discriminated by the color and lighting effects. The discrimination usually provides a clue about the presence of a text in a particular region. This fact is exploited and established the steps about how this dicriminative property can be helpful in the determination of text localization. This paper proposed a feature extraction technique that helps in Arabic scene text recognition using linear spatial pyramids based on image analysis filters. The Gaussian pyramid is applied for smoothing the images. Every subsequent image passes through various image processing filters. The convolutional approach is applied on image pyramids which helps in developing feature vectors of each image. Every pyramid image is taken into account and passed through the filtration and convolution processes.

The extracted feature vectors were given to Multi-Dimensional Long Short Term Memory (MDLSTM) networks for a purpose to learn the vector sequence. The MDLSTM is an appropriate approach for sequence learning tasks and have been successfully applied on numerous sequence learning cursive text recognition during recent years [1, 3, 13]. The contribution of proposed work is alienated as follows,

1. The linear image pyramid technique has been adapted and applied on localization of Arabic scene text.
2. A novel way is proposed to determine the text location and extract text from given scene image. To accomplish this, the presented work applies image processing filters like laplacian, large blur, small blur, sobel_x, sobel_y and

sharping the image text and convolving all of them with empirically selected kernel.

3. Deep MDLSTM architecture is employed to see the performance of presented method which demonstrated very encouraging results.

Section 2 describe the proposed methodology which further explain the formulation of linear pyramids by image filters. The experimental analysis is elaborated in Sect. 4, whereas, conclusion is summarized in Sect. 5.

2 Related Work

The research on scene text analysis has increased to manifold for since few years. There are numerous techniques proposed for text localization, extraction and recognition from natural images. Text analysis in scene text is not an easy task because it involves a lot of other factors in an image that may create complication in smooth execution of specifically designed techniques. In this section some latest techniques have been compiled that suggest solutions for text localization and extraction from scene image.

Scene text does not have only one type of feature associated with it, instead it has numerous features that may play a conclusive role in correct determination of character/words. Therefore, there is a need to investigate more about the features that may be distinct in nature but helpful in learning and recognition process. One such work is presented by Lazebnik et al. [8]. They proposed a holistic approach for image categorization that uses pyramid matching kernels to demonstrate the performance of their proposed architecture. They evaluated their proposed technique on fifteen categories using Graz dataset [8]. The Graz dataset contains highly discriminative images and their proposed method provides good accuracy on image categorization. Another dataset they evaluated on Caltech-101 wasthe one which contains 31–800 images per category. This dataset is considered as most diverse dataset provided for research purpose. The experimental setting was depicted in their manuscripts [6,21]. They trained 30 images per class and the test set contains 50 images per class. They reported good accuracy by using their dataset as can seen further detail in their paper.

Another presented work is about spatial pyramid matching based on invariant features using sparse coding method presented by [18], they are using max pooling strategy in the histogram on multiple spatial scales to subsume translation and scale invariance. Their work is an extension of the work presented in [8]. Furthermore, they presented the results obtained on Caltech-256, 5 images and TRECVID 2008 surveillance video images. They also compared their results by evaluating their technique on the same datasets as reported in [8]. They implement three variation of spatial pyramid matching (SPM) algorithm. The first variation is SPM with Chi square kernels. Another variation is to use linear kernel on spatial pyramid histogram. The last variation, by using it they reported best accuracy is inclusion of Scale Invariant Feature Transformation (SIFT) features with combination of linear kernel on spatial pyramids. The detail about performed experiments can be found in their respective manuscript.

The spatial pyramid model presents very interesting results by incorporating other techniques for a purpose to enhance the performance. One such technique is presented by [16], they proposed two different models with incorporation of spatial pyramid knowledge. Their first model is an image independent model while the second one adapted the image contents. They used PASCAL VOC 2007, 2008, and 2009 dataset for evaluation of their proposed models. Their average results vary from 62.5%–66.5% as reported.

Although application of spatial linear pyramid is new for scene text recognition problems. SPM has not been applied before for categorization of text related problems. This paper is presenting a novel solution for text localization problem. As this technique has been comparatively applied successfully, on number of object recognition in scene images as can witnessed in [5,6,8,16]. The presented work is exploiting its potential in the field of text localization.

3   Proposed Methodology

The linear pyramids proved very convincing results on image categorization in natural images. The strength of under discussion technique has yet to explore regarding text categorization in natural images. This idea is a focal point of presented research that will try to explore how this technique can be beneficial in determining correct text localization.

This section provides description about the implementation of proposed idea. Figure 1 elaborates proposed idea in detail. Every image is rescaled into standard size. After that linear pyramid's method is applied. The linear pyramid is described in 6 levels as indicated in Fig. 2. The linear pyramid generates 5 images of different resolution. Each image passes through filter pack and the resultant images are converted into grayscale. At the end, all pyramid gray scale images were given to classifier for training purpose. The whole process defined into following subsections.

3.1   Formulation of Linear Spatial Pyramids of Cursive Arabic Scene Text

The image pyramid contemplated as a two dimensional arrays which meant to represent image from smaller to smallest size reducing the image information at each level from base to the top of pyramid. There are numerous ways suggested to define pyramids as explained in [17], but in-practice, procedure to define pyramids is considered a step from base to it's top.

In this work, RGB values are accounted where each cell contains the percentage of R, G, and B colors. The reason for proposing pyramids for text localization is its ability to group the text image/words in an appropriate resolution so it may contribute in depicting feature as a whole in absence of complex computation. In today's era, systems are more intelligent in processing the acquired image, but difficult to understand the content represented in an image. The human perception of recognizing text presented in various fonts regardless of

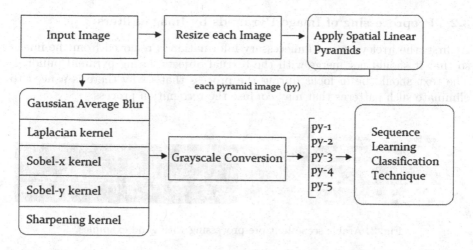

Fig. 1. Flowchart of proposed architecture

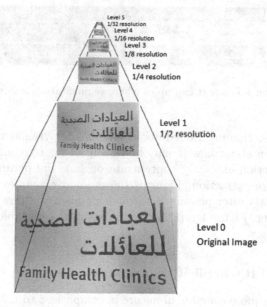

Fig. 2. Visual representation of Linear spatial pyramids with 6 levels.

their sizes inspired the machine models where its realization is implemented by linear pyramids.

The input image is passed through the loop where Gaussian pyramid is applied on each image. Each image is resized to a specific height which should not be greater than 30. The height limit described as the minimum size of the image considered in conducted experiments.

3.2   Preprocessing of Image Pyramids by Image Filters

At first, the irrelevant and unnecessary information is removed from the image so that it should not merge with the textual objects during pyramid building. The text should be in focus during the process that's why there is a need to eliminate such patterns that may confuse the recognition process.

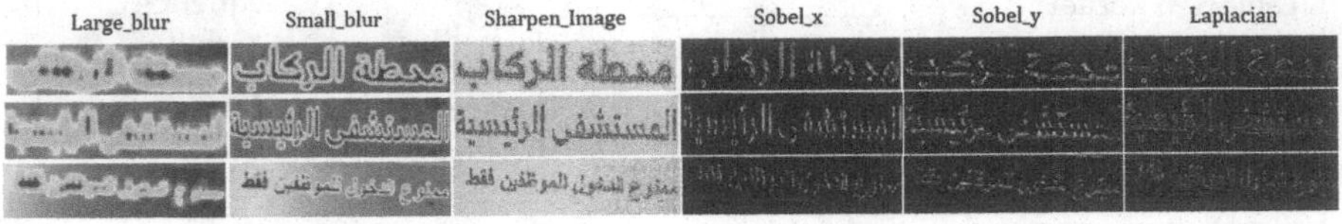

Large_blur	Small_blur	Sharpen_Image	Sobel_x	Sobel_y	Laplacian

Fig. 3. Arabic scene text pre-processing with good examples.

Fig. 4. Arabic scene text examples where segmentation is misclassified.

The previous section mentioned about how image pyramid was constructed, but the description about how it may be beneficial for recognition process not specified. This section explicitly concentrate on how the pyramid images may be used for feature extraction. Figure 3 shows good examples where text has categorized correctly after passing it through number of filters that contribute in feature building. Figure 4 represents misclassified text which degrades the recognition accuracy.

3.3   LSTM and It's Implicit Segmentation

The cursive text is more complex in nature in comparison to Latin. In such text, sequence is very important to know. In the cursive scripts like Arabic where each character has four representation without sharing any similarities among the shapes, the sequence learning is an appropriate choice for learning the temporal sequential behavior. In the scripts where segmentation is tedious to perform, an implicit segmentation is a suitable approach for such problems. In presented work of Arabic or Arabic-like script analysis to date, most authors proposed their methodology by using implicit segmentation which is embedded in their presented technique. One such technique is called Long Short Term Memory (LSTM) network which is a recurrent neural network approach, designed for sequence learning with implicit segmentation.

By implicit approach, the segmentation and recognition of characters are accomplished at the same time. There is a trade-off between the numbers of segments of a given word. The computation time also plays an important role with respect to number of segments. Larger the number of segments of a given word, the more time it requires for computation. In that particular situation, the hypothesis that is made for recognition of a given character increases but at the expense of compromised computation time. On the other hand, less number of segments may produce result in ideal time, but in that specific situation, recognition results may suffer.

Regarding the problems where sequence is important, there is a need to require powerful method that may learn context and make the task easier for recognition. Due to built-in complexities in Arabic scripts, it is usually impossible to segment cursive characters by explicit means. The tentative segmentation of word may predict over segmentation and loop determination problem [7]. In over segmentation, the image may heuristically be over segmented with the horizontal distance in width. The input image is adjusted by x-height which is 60 pixels as shown in Fig. 5.

Fig. 5. Gray scale image of x-height 60 × 1, moving from right to left.

In Arabic or Arabic-like scripts, as mentioned earlier that characters may change its shape according to its position in a word. The traditional segmentation techniques can not apply on Arabic scripts [10,11], for this purpose there is a need to look on implicit segmentation that may follows probabilistic or statistical measure for segmentation of such complex cursive scripts.

4 Experimental Analysis

This section is presenting the experimentation analysis performed by implementing the proposed architecture. Experimental study was carried out by consideration of following points,

1. Each pyramid image participates individually to inquire about which image size of the proposed architecture presented good results.
2. The number of hidden layers contribute in chalking out of the best accuracies.

Following sub section explains about proposed dataset used for experimentation.

4.1   Scene Text Dataset

The dataset has a great importance for evaluating the proposed techniques or methods. The dataset is prepared to demonstrate the performance of presented technique. The dataset for Arabic camera captured images have not been developed yet. The prime concern is to have scene images having Arabic text in focus, segment them and use them to validate the research tasks. The bilingual scripts is common in most of gulf countries which prompts us to present dataset for English as well in addition to Arabic. In dataset, the taken samples tried to cover maximum variability of Arabic scene text data. There are various text font sizes, styles and colors appeared in constraint environment which have been captured for research purpose. The proposed dataset is disintegrated into Arabic and English scene text samples, as indicated in Table 1.

Table 1. EASTR-42K division based on complexity.

EASTR-42K dataset			
Language	Textlines	Words	Characters
Arabic	8,915	10,593	16,000
English	2,601	5,172	7,390

The prime concern of this manuscript is to investigate the performance of Arabic scene text proposed architecture.

Table 2. Experimentation Results reported by considering each Image pyramid.

Pyramid image	Arabic
py-1	77.30 ± 0.45
py-2	73.55 ± 0.63
py-3	68.21 ± 2.17
py-4	56.73 ± 0.59
py-5	49.71 ± 0.81

The best accuracy is computed on py-1. The established hypothesis is that large images have discrete information which posses image features that are used to improve the accuracy.

The experimentation analysis is computed by f-measure on Arabic samples as summarized in Table 3. It has been observed the accuracy gets lower and lower when pyramid gets small. The hypothesis is that large image contains detailed information which contributes to accuracy (Fig. 6).

Table 4 depicts the recognition results reported on Arabic by considering all pyramids of single image as a feature vector. As observed the recognition is

Table 3. F-measure score observed on Arabic samples.

Pyramid images	Precision	Recall	F-measure
py-1	0.82	0.74	0.78
py-2	0.78	0.70	0.74
py-3	0.69	0.58	0.63
py-4	0.60	0.47	0.53
py-5	0.58	0.45	0.51

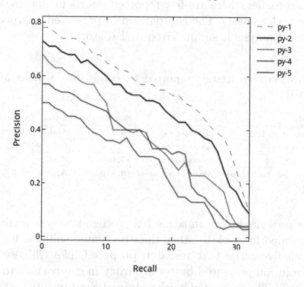

Fig. 6. Precision/Recall curve on Arabic samples.

improved because every time network initial value is guesstimated which may improve the network learning as observed in the table below (Table 2).

Table 4. Experimentation results reported by considering whole image pyramids.

Scripts	Text pyramids
Arabic	82.71 ± 1.13

Table 5, presents the experimental observation calculated on different hidden layer units. As observed, the best accuracy was reported when there are 100 hidden layer units. There is an observation that if reasonable memory units were provided during network learning, it can handle the complex input appropriately.

The reasonable memory unit selection is empirically selected as explained in performed experiments.

Table 5. Experimentation Results reported by considering various hidden memory units in MDLSTM on Arabic data samples.

MDLSTM memory block	20	40	60	80	100
Accuracy		39.68 ± 0.53	48.51 ± 0.22	61.74 ± 1.47	73.42 ± 0.79	82.71 ± 1.13

As mentioned earlier, there are few reported efforts in the research direction of Arabic scene text analysis. The comparison of presented approach with other cursive scene text analysis is summarized in Table 6.

Table 6. Experimentation Results reported by considering various hidden memory units in MDLSTM

Study	Source	No. of Images	Script	Accuracy
Yousfi et al. [20]	Video text	6,532	Arabic	55.03%
Yao et al. [19]	Camera captured	6,532	Chinese	75.0%
Lee et al. [9]	Camera captured	5,000	Korean	88.0%
Proposed technique	**Camera captured**	**10,915**	**Arabic**	**83.0%**

Most of the presented work in scene text recognition was analyzed on Latin script. As mentioned in Table 6, other than presented solution for Arabic scene text, there is cursive scene text research proposed, like Chinese and Korean. The proposed solution presented better accuracy in comparison to other cursive scene text analysis. The presented work obtained good results and reported best accuracy achieved on Arabic scene text recognition.

5   Conclusion

The presented paper has explored the spatial properties associated with an image. The image properties are exploited by considering various representation of scene text image. Each image has passed through empirically selected kernels that aim to enhance features of provided samples. The cursive nature of Arabic script prompted us to employ MDLSTM, as it is the best suited technique for sequence learning tasks [1,3,13]. The dataset was evaluated pyramid wise. The presented technique is relatively a novel idea that has not been used before to address the complexities associated with Arabic scripts. The proposed work indicated best accuracy in comparison to available Arabic scene text recognition results. The evaluation of any technique depends on the dataset in question. One of the possible extension to presented work is to extract statistical features and merge them with presented spatial features for a purpose to perform experiment and evaluate how much merging the said features may impact on the recognition process.

Another important extension of proposed work is to present scale invariant [19] feature extraction technique with the combination of linear pyramids. The invariant technique is applied on each pyramid and extract those points which are in common. This could be an interesting experiment on provided dataset.

Acknowledgement. The authors would like to thank Ministry of Education Malaysia and Universiti Teknologi Malaysia for funding this research project.

References

1. Ahmed, S.B., Naz, S., Razzak, M.I., Rashid, S.F., Afzal, M.Z., Breuel, T.M.: Evaluation of cursive and non-cursive scripts using recurrent neural networks. Neural Comput. Appl. **27**(3), 603–613 (2016)
2. Ahmed, S.B., Naz, S., Razzak, M.I., Yousaf, R.: Deep learning based isolated arabic scene character recognition. In: International Workshop on Arabic Script Analysis and Recognition (ASAR), pp. 46–51. IEEE (2017)
3. Ahmed, S.B., Naz, S., Swati, S., Razzak, M.I.: Handwritten Urdu character recognition using one-dimensional BLSTM classifier. Neural Comput. Appl., pp. 1–9 (2017)
4. Ahmed, S.B., Naz, S., Swati, S., Razzak, M.I., Umar, A.I., Khan, A.A.: UCOM offline dataset-an Urdu handwritten dataset generation. Int. Arab J. Inf. Technol. **14**(2), 239–245 (2017)
5. Gluckman, J.M.: Scale variant image pyramids. In: IEEE Conference on Computer Vision and Pattern Recognition (CVPR), vol. I, pp. 1069–1075 (2006)
6. Grauman, K., Darrell, T.J.: The pyramid match kernel: discriminative classification with sets of image features. In: ICCV, vol. II, pp. 1458–1465 (2005)
7. Graves, A.: Supervised Sequence Labelling with Recurrent Neural Networks. SCI, vol. 385. Springer, Berlin (2012). https://doi.org/10.1007/978-3-642-24797-2
8. Lazebnik, S., Schmid, C., Ponce, J.: Beyond bags of features: spatial pyramid matching for recognizing natural scene categories. In: IEEE Conference on Computer Vision and Pattern Recognition (CVPR), vol. II, pp. 2169–2178 (2006)
9. Lee, S., Cho, M.S., Jung, K., Kim, J.H.: Scene text extraction with edge constraint and text collinearity. In: ICPR, pp. 3983–3986. IEEE Computer Society (2010)
10. Naz, S., Ahmed, S.B., Ahmad, R., Razzak, M.I.: Arabic script based digit recognition systems. In: International Conference on Recent Advances in Computer Systems (RACS), pp. 67–73 (2016)
11. Naz, S., Hayat, K., Razzak, M.I., Anwar, M.W., Madani, S.A., Khan, S.U.: The optical character recognition of Urdu-like cursive scripts. Pattern Recognit. **47**(3), 1229–1248 (2014)
12. Naz, S., Umar, A.I., Shirazi, S.H., Ahmed, S.B., Razzak, M.I., Siddiqi, I.: Segmentation techniques for recognition of arabic-like scripts: a comprehensive survey. Educ. Inf. Technol. **21**(5), 1225–1241 (2016)
13. Naz, S., et al.: Urdu Nastaliq recognition using convolutional-recursive deep learning. Neurocomputing **243**, 80–87 (2017)

14. Naz, S., Umar, A.I., Ahmed, R., Razzak, M.I., Rashid, S.F., Shafait, F.: Urdu Nasta'liq text recognition using implicit segmentation based on multi-dimensional long short term memory neural networks. SpringerPlus **5**(1), 2010 (2016)
15. Razzak, M.I., Anwar, F., Husain, S.A., Belaid, A., Sher, M.: HMM and fuzzy logic: a hybrid approach for online urdu script-based languages' character recognition. Knowl.-Based Syst. **23**(8), 914–923 (2010)
16. Sánchez, J., Perronnin, F., de Campos, T.E.: Modeling the spatial layout of images beyond spatial pyramids. Pattern Recognit. Lett. **33**(16), 2216–2223 (2012)
17. Tan, C.L., Yuan, B., Ang, C.H.: Agent-based text extraction from pyramid images. In: Singh, S. (ed.) International Conference on Advances in Pattern Recognition, pp. 344–352. Springer, London (1999). https://doi.org/10.1007/978-1-4471-0833-7_35
18. Yang, J., Yu, K., Gong, Y., Huang, T.: Linear spatial pyramid matching using sparse coding for image classification. In: IEEE Conference on Computer Vision and Pattern Recognition (CVPR) (2009)
19. Yao, C., Bai, X., Liu, W., Ma, Y., Tu, Z.: Detecting texts of arbitrary orientations in natural images. In: IEEE Conference on Computer Vision and Pattern Recognition (CVPR), pp. 1083–1090. IEEE Computer Society (2012)
20. Yousfi, S., Berrani, S.A., Garcia, C.: ALIF: a dataset for Arabic embedded text recognition in TV broadcast. In: 2015 13th International Conference on Document Analysis and Recognition (ICDAR), pp. 1221–1225. IEEE (2015)
21. Zhang, J., Marszałek, M., Lazebnik, S., Schmid, C.: Local features and kernels for classification of texture and object categories: a comprehensive study. Int. J. Comput. Vis. **73**(2), 213–238 (2007)

Human Action Recognition with 3D Convolution Skip-Connections and RNNs

Jiarong Song[1], Zhong Yang[1(✉)], Qiuyan Zhang[2], Ting Fang[3],
Guoxiong Hu[1,2,3], Jiaming Han[1], and Cong Chen[1]

[1] Nanjing University of Aeronautics and Astronautics, Nanjing 210016, China
jrsnuaa@163.com, YangZhong@nuaa.edu.cn
[2] Electric Power Research Institute of Guizhou Power Grid Co., Ltd.,
Guiyang 550000, Guizhou, China
[3] Hefei University of Technology, Hefei 230009, China

Abstract. This paper proposes a novel network architecture for human action recognition. First, we employ a pre-trained spatio-temporal feature extractor to perform spatio-temporal features extraction on videos. Then, several-level spatio-temporal features are concatenated by 3D convolution skip-connections. Moreover, a batch normalization layer is applied to normalize the concatenated features. Subsequently, we feed these normalized features into a RNN architecture to model temporal dependencies, which enables our network to deal with long-term information. In addition, we divide each video into three parts in which each part is split into non-overlapping 16-frame clips to achieve data augmentation. Finally, the proposed method is evaluated on UCF101 Dataset and is compared with existing excellent methods. Experimental results demonstrate that our method achieves the highest recognition accuracy.

Keywords: Human action recognition · 3D CNNs
3D convolution skip-connections · RNNs · Feature concatenation

1 Introduction

Human action recognition is defined as the process of labeling image sequences with action labels in [15]. Because it has a bright prospect and potential economic value in intelligent video surveillance, human-machine interaction, virtual reality and many other fields, human action recognition has become an essential research in computer vision and has received a significant amount of attention in this community [9,15,17,26].

In this paper, we aim at proposing a new network architecture to achieve human action recognition. Since it is time consuming and easily causing overfitting to train a feature extractor from scratch, we perform transfer learning by employing a pre-trained spatio-temporal feature extractor [22] to address this problem. This feature extractor [22] is well-trained on Sports-1M [7]. Next, in order to improve the recognition performance, skip-connections [3] are used

L. Cheng et al. (Eds.): ICONIP 2018, LNCS 11301, pp. 319–331, 2018.
https://doi.org/10.1007/978-3-030-04167-0_29

to concatenate several-level spatio-temporal features. Moreover, knowing that recurrent neural networks (RNNs) are powerful to capture time dependencies [11], we employ a RNN to model temporal dependencies, which effectively enables our network to deal with long-term information. Furthermore, before feeding these concatenated features into the RNN, a batch normalization (BN) [5] layer is applied to normalize the concatenated features. Lastly, the experiment evaluation is conducted on UCF101 dataset [20]. Experimental results demonstrate that our method achieves the highest recognition accuracy compared with existing excellent methods.

The rest of the paper is organized as follows. Section 2 reviews the related work on human action recognition and points out advantages of our method. Section 3 describes our network architecture and elaborates on the design principle. Section 4 presents experimental results and detailed analysis. Conclusions are provided in Sect. 5.

2 Related Work

In general, human action recognition researches are mainly divided into two categories: (i) based on hand-crafted features methods. (ii) based on deep learning methods. From a hand-crafted perspective, paper [25] proposes a method IDT (improved dense trajectories) which is considered as the state-of-the-art method. In the study by [25], it first cancels out the camera motion from the optical flow. Then, the method encodes motion-based descriptors such as HOF (histograms of optical flow) and MBH (motion boundary histograms) with FV (Fisher vector) [16]. Finally, it uses an SVM (Support Vector Machine) to perform classification on these encoded features. The algorithm IDT with higher-dimensional encodings presented in paper [14] involves three stages: first, pre-process extracted features, IDTs and STIPs (Space-Time Interest Points). Next, three different encoding methods are utilized to encode these features. Lastly, fuse encoded features into a super vector and use an SVM to achieve classification. The two approaches have presented excellent results on human action recognition. However, both of them are strong feature-based methods which need an elaborate feature designing.

Inspired by the remarkable performance of deep learning in still image domain [8], human action recognition relying on deep learning has received lots of interest recently and has achieved significant performances. Paper [7] studies multiple approaches for extending the connectivity of a CNN (Convolutional Neural Network) in time domain to take advantage of local spatio-temporal information. By comparing different architectures, it finds out that the Slow Fusion model performs the best [7]. A two-stream ConvNet architecture is described in [19]. The spatial stream ConvNet operates on individual video frames while the optical flow ConvNet operates on optical flow stacking. Scores from these two ConvNets are then combined by late fusion to get the recognition result [19]. An encode-decode method is presented in [21]. The algorithm first uses an encoder LSTM (Long Short-Term Memory) to map an input sequence into a fixed length representation. Then, decoder LSTMs are used to perform decoding. The result shows this method [21] can improve classification accuracy especially with few training examples. However, it is very difficult to train. A novel 3D CNN model for action recognition is developed in [6]. By performing 3D convolutions kernels, this network [6] is able to extract features from both spatial and temporal dimensions, which significantly benefits the human action recognition on videos. Further researches on the 3D CNN network are investigated in [22]. The paper [22] finally proposes a 3D architecture named C3D with small $3 * 3 * 3$ convolution kernels in all layers. This method [22] performs the best among 3D ConvNets.

Unlike paper [19] which has to compute optical flow in advance and paper [7] easily collapses temporal information by extending only 2D convolution kernels to learn spatio-temporal feature, our approach inspired by C3D [22] is able to extract spatio-temporal features in an end-to-end way. However, while the work [22] only extracts 16-frame clips from a video, we use a strategy that divides each video into three parts in which each part is split into non-overlapping 16-frame clips. This strategy effectively achieves data augmentation to avoid overfitting. Besides, our network is able to take advantage of exceeding 16 frames from videos by this strategy. Moreover, a RNN [11] architecture is employed to deal with long-term information. In addition, we concatenate several-level spatio-temporal features to improve the recognition accuracy.

3 Network Architecture

This section detailly describes our network architecture and discusses the design principle. As shown in Fig. 1, the network mainly consists of four parts: a spatio-temporal feature extractor, 3D convolution skip-connections, a BN layer and a RNN. Since the spatio-temporal feature extractor has been discussed above. In this section, we primarily elaborate on the other three parts.

3.1 3D Convolution Skip-Connections

As a CNN becomes deeper, detailed information is easily discarded. However, low-level information and motion-based features are essential for action recognition on videos [2]. Considering optical flow is useful for action recognition, paper [2] incorporates optical flow to promote the performance of the 3D CNN. However, paper [18] takes a deeper look at the combination of optical flow and action recognition. Finally, they find out that much of optical flow value is not even in the motion per se [18]. Therefore, many approaches incorporating optical flow seldom achieve remarkable promotions.

To improve the action recognition performance, using C3D [22] as our base network, we conduct skip-connections [3,4] to construct our network architecture based on following reasons:

(i) CNNs have been shown to learn powerful and interpretable image features [16].

(ii) Low-level spatio-temporal features extracted by 3D convolution kernels include appearance information and temporal dynamics [24].

(iii) Paper [4] indicates that a layer makes strong use of feature-maps produced its preceding layers before. It means features extracted by early layers are important to global state.

(i) and (ii) demonstrate early layers of C3D can grasp much stronger motion-based features better than optical flow, which offers derailed and useful information to improve classification tasks. (iii) indicates these low-level spatio-temporal features are essential for our recognition tasks in our network. Thus, we employ skip-connections [3,4] to improve the performance of C3D for action recognition in this paper.

As shown in Fig. 1, low-level spatio-temporal features and high-level spatio-temporal features are concatenated. We name this kind of skip-connections as 3D convolution skip-connections in this paper. Results in Sect. 4 demonstrate that the 3D convolution skip-connections method is superior than approaches only incorporating optical flow.

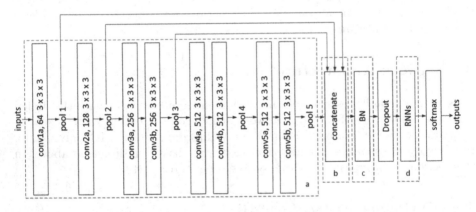

Fig. 1. Network architecture. (a) The spatio-temporal feature extractor is employed from conv1a to pool5 of C3D [5]. The extractor is able to extract spatio-temporal features from inputs. (b) 3D convolution skip-connections are used to concatenate spatio-temporal features from pool1, pool2, pool3 and pool5 layers. (c) A BN layer is applied to normalize the concatenated features to speed up the convergence. (d) A RNN (shown in detail in Fig. 2) is employed to model temporal dependencies to achieve improvements on action recognition.

3.2   The Batch Normalization Layer

We concatenate low-level features with high-level features to form an extremely high dimensional vector in Sect. 3.1 (seen in Fig. 1(b)). Since these high-level features (like pool3 and pool5) are derived from those low-level features (like pool1 and pool2), the features contained in this high dimensional vector are easily correlated. This possibly results in the phenomenon of Internal Covariate Shift [5] causing the risk of divergence. Considering a network training will converge faster if its inputs are whitened [13, 28], we decide to perform normalization on these high dimensional concatenated features.

As the full whitening of the input of each layer is costly and not everywhere differentiable [5], we apply BN [5] which is a differentiable transformation to realize the normalization step by fixing the means and variances of layer inputs. The explicit normalization process is described by (1) [5]. Besides, as explained in [10], (1) can speed up convergence even when the features are not decorrelated, which is suitable to our condition. Considering (1) may change the presentation of a layer [5], additional step scaling and shifting the normalized value is introduced by (2). Step (2) addresses this problem by restoring the representation power of the network [5].

$$\hat{x}^{(k)} = \frac{x^{(k)} - E[x^{(k)}]}{\sqrt{Var[x^{(k)}]}} \tag{1}$$

$$y^{(k)} = \gamma^{(k)}\hat{x}^{(k)} + \beta^{(k)} \tag{2}$$

Suppose $\mathbf{x} = (x^{(1)}, x^{(2)}, \ldots, x^{(n)})$ presents an n-dimensional input for a layer, $x^{(k)}$ presents the activation, the expectation $E[x^{(k)}]$ and variance $Var[x^{(k)}]$ are computed over every mini batch (mini-batches strategy is used in stochastic gradient training), $\gamma^{(k)}$ and $\beta^{(k)}$ are learned along with the original model parameters.

In this work, we add a BN layer between the spatio-temporal feature extractor and the RNN to implement the normalization step. Results in Sect. 4 indicate that the network including the BN layer is significantly more advanced than the architecture without the BN layer.

3.3   The RNN

Human action recognition on video learning tasks require awareness of time. Since RNNs are powerful to capture time dependencies [11], we consider employing RNNs to model temporal dependencies. In this work, according to the average frames of videos on UCF101 dataset, each video is divided into three parts in which each part is split into non-overlapping 16-frame clips. Therefore, we set the time steps to be three. Since the value of time steps is not large, it hardly results in the difficulty of learning long-range dependencies. In this case, it is not difficult to train a simple RNN based on flowing reasons:

(i) The BN layer in Sect. 3.2 allows the input data of the RNN to have the distribution with the mean of zero and the variance of 1. This makes training more efficient and accelerates the convergence of the network.

(ii) The RNN constructed in this paper only contains a single hidden layer. It has less possibility occurring the problem of gradient vanishing while this problem easily occurs in multi-layer network [12].

(iii) Learning with RNNs is considered to be difficult since the gradient vanishing problem and exploding problem occur when backpropagating errors across many time steps [1,11]. However, the value of time steps is set to be 3 in this paper. It means that the maximum time the weight (between the output of a neuron contained in hidden layers and its previous state) multiplied by itself is 3 (demonstrated in (5)) which is small will not making the training too much pressing.

(i) makes the input data are zero-centered, which makes the network easier to train. (ii) and (iii) mean that the initial state (out_{h0}) will not be given too large weight (when $|w_{ii}| > 1$, w_{ii} is the weight between the output of neuron i contained in hidden layers and its previous state) or too small weight (when $|w_{ii}| < 1$) at the last time step. This is demonstrated in (5) and in order to explain clearly, we suppose the activation of the hidden layer is a linear function here. As shown in (5), the maximum time the weight multiplied by itself is 3 which is small, meaning the gradient vanishing or exploding cannot occur easily. Therefor, the RNN including very short time steps is not difficult to train. Thus, we decide to apply simple RNNs. The architecture of the RNN is unfolded in Fig. 2. The detailed process is illustrated in (3) and (4) [11].

$$\mathbf{h}^{(t)} = \sigma(W^{hx}\mathbf{x}^{(t)} + W^{hh}\mathbf{h}^{(t-1)} + \mathbf{b}_h) \qquad t = 1, 2, 3 \tag{3}$$

$$\mathbf{y} = softmax(W^{hx}\mathbf{h}^{(3)} + \mathbf{b}_y) \tag{4}$$

$$\mathbf{h}^{(3)} = W^{hh}[W^{hh}(W^{hh}\mathbf{h}^{(0)} + W^{hx}\mathbf{x}^{(1)}) + W^{hx}\mathbf{x}^{(2)}] + W^{hx}\mathbf{x}^{(3)} \tag{5}$$

In this paper, an input sequence of the RNN can be denoted as $(\mathbf{x}^{(1)}, \mathbf{x}^{(2)}, \mathbf{x}^{(3)})$, where $\mathbf{x}^{(i)}$ presents the ith 16-frame clips features of a video extracted by C3D (out in (1), the output of the part a in Fig. 1), which are flattened into a vector. The output can be denoted as $\mathbf{y}$, a vector presents the probability belonging to each class. $\mathbf{h}^{(t)}$ presents the state that is the output at the time step t of the hidden layer, σ is the sigmoid function. W^{hx} presents the matrix weights between the input and the hidden layer, W^{hh} presents the matrix weights between the hidden layer and its previous state [11]. W^{hy} presents the matrix weights between the hidden layer and the output [11]. The vectors $\mathbf{b}_h$ and $\mathbf{b}_y$ are respectively bias parameters of the hidden layer and the output. $\mathbf{h}^{(0)}$ is the initial state of the hidden layer.

By introducing the RNN architecture, our model is able to carry out human action recognition by taking advantage of exceeding 16 frames. Besides, the RNN allows our network to deal with long-term information compared with C3D [22]. Results in Sect. 4 approve that the method introducing the RNN architecture outperforms C3D.

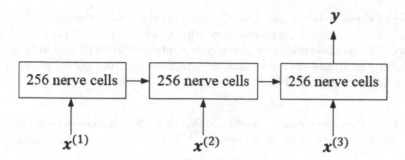

Fig. 2. The RNN architecture. Gives an insight into our RNN architecture. The number of nerve cells contained in the hidden layer is equal to 256. Detailed information is elaborated in Sect. 4.

4 Experiments

4.1 Dataset and Experimental Protocols

The experiment evaluation is conducted on UCF101 dataset [20] which is a standard benchmark for action recognition and is annotated into 101 action classes. It consists of over 13k clips and 27 h of video data. We evaluate our method by calculating the mean accuracy over all splits of UCF101. Some actions under recognition are shown in Fig. 4.

In addition, we divide each video into three parts in which each part is split into non-overlapping 16-frame clips. This strategy effectively achieves data augmentation to avoid overfitting. Besides, we adapt the shape of inputs to our network by this strategy. Moreover, we discard videos containing less than 48 frames.

4.2 Comparison of Different Architectures Variants

Table 1 and Fig. 3 show the comparison within different variants of our model. Table 1 presents the performance of different kinds of feature concatenation and suggests that the pool1&2&3&5 concatenation performs the best. Figure 3 displays the performance of our network with the different number of nerve cells contained in the RNN hidden layer.

It is easy to come to the following conclusions from Fig. 3: (i) the RNN architecture with 512 nerve cells has the highest training accuracy. (ii) the RNN architecture with 256 nerve cells has the highest testing accuracy. (iii) the RNN architecture with 128 nerve cells has both the lowest training accuracy and testing accuracy. The detailed analysis is given as follows.

The increase in the number of nerve cells will increase the model complexity, which enables the model to fit better to training datasets. Therefore, when the number of nerve cells varies from 128 to 512, the training accuracy continuously improves (shown in Fig. 3). However, the augmentation in model complexity sometimes results in the overfitting problem. As indicated in Fig. 3, when the

326   J. Song et al.

number of nerve cells changes from 256 to 512, the training accuracy increases
but the testing accuracy decreases. According to the result in Table 1 and Fig. 3,
we select 256 as the number of nerve cells contained in the RNN hidden layer
(shown in Fig. 2) and decide to concatenate features from pool1, pool2, pool3
and pool5 layers (seen in Fig. 1).

Table 1. Performance comparison within different kinds of feature concatenation. We
use the mean accuracy as the measurement.

Feature concatenation	Accuracy (%)
Without concatenation	82.1
pool1&pool5	86.6
poo2&pool5	86.9
pool3&pool5	86.8
pool1&2&3&5	89.9

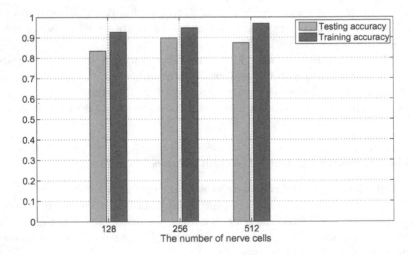

Fig. 3. The mean accuracy across training datasets and testing datasets within the
different number of nerve cells. When the number of nerve cells is set to be 128, the
training accuracy is 92.7% and the testing accuracy is 83.4%. When the number of
nerve cells is equal to 256, the network architecture obtains an accuracy of 94.8% for
training and an accuracy of 89.9% for testing. When the number of nerve cells is set
to be 512, the training accuracy is 96.9% and the testing accuracy is 87.4%.

Fig. 4. Examples of action recognition. Shows some actions under recognition on UCF101 dataset and we list the top 5 action recognition results.

4.3   The Influence of the Batch Normalization Layer

Table 2 compares the mean accuracy between two different network architectures. One architecture includes a BN layer while the other does not include. The result in Table 2 indicates that the architecture including a BN layer achieves a much higher accuracy. In this paper, we concatenate low-level features with high-level features to form an extremely high dimensional vector and feed these concatenated features into RNNs. Since these high-level features are derived from these low-level features, the features contained in this high dimensional vector are easily correlated. Therefore, it is difficult to ensure these concatenated features are orthogonal. Besides, as the network becomes deeper, slight changes in the distributions accumulate. Based on above reasons, small changes to the network parameters will be amplified leading to the change in the distributions. This possibly results in negative consequences on the performance of the network, which slows down the convergence of the network and even leads to the divergence of the network.

However, BN is capable of eliminating the change in the distributions by fixing the distribution of inputs via a normalization step that fixes the means and variances of layer inputs [5]. This can ensure that the distribution of the layer inputs remains stable. By reducing the dependence of gradients on the scale of the parameters or of their initial values [5], we can use higher learning rates which can speed up convergence the network without the risk of divergence. Thus, it is more likely to achieve positive consequences with the BN layer on our recognition tasks.

Table 2. Performance comparison between two different network architectures on UCF101. One architecture includes a BN layer while the other does not include. We use the mean accuracy as the measurement. ·

Network architecture	Nerve cells	Feature concatenation	Accuracy (%)
With batch normalization	256	pool1&2&3&5	89.9
Without batch normalization	256	pool1&2&3&5	63.5

4.4   Comparison with Other Excellent Methods

Table 3 compares the proposed architecture (as shown in Fig. 1) with existing excellent methods on UCF101. The performance comparison is summarized in Table 3. The table is divided into two sets. The first set compares models that use only RGB data as inputs. The second set compares our model against C3D with Optical Flow [2].

As shown in Table 3, it is obvious that our network outperforms other methods by a large margin, where the minimum margin is 2% and the maximum margin is 24.5%. Moreover, our method outperforms Slow Fusion [7] by 24.5% as Slow

Table 3. Mean accuracy on UCF101. The first set takes only RGB frames as inputs. The second set compares our model against C3D with Optical Flow and our method still takes only RGB frames as inputs.

Method	Mean accuracy (%)
IDT+FV [25]	85.9
IDT with higher-dimensional encodings [14]	87.9
Slow Fusion [7]	65.4
C3D [23]	82.3
Spatial stream [19]	72.6
LSTM composite model [21]	75.8
Action transformations [27]	80.8
Our method	89.9
C3D with Optical Flow [2]	83.4
Our method	89.9

Fusion extends only the connectivity of 2D convolution kernels to learn spatio-temporal features. This method easily collapses temporal information. Nevertheless, 3D convolution kernels applied in our method are capable of extracting both spatial and temporal information. The reason our network is superior than C3D [23] is that our approach is able to take advantage of more information. However, C3D only relies on 16 frames to perform recognition tasks. Besides, our network realizes the modeling of temporal dependencies, which allows the network to deal with long-term information. From an incorporating features perspective, our method combines low-level features to improve recognition results. The low-level features extracted by 3D convolution kernels include appearance information and temporal dynamics. Therefore, these features are more powerful and effective than optic flow as much of optical flow value is not even in the motion per se. In this case, it enables our method to outperform C3D with Optical Flow [2].

5   Conclusions

We propose a novel network consisting of a spatio-temporal feature extractor, 3D convolution skip-connections, a BN layer and a RNN architecture. The proposed network outperforms existing excellent methods on UCF101.

In this work, a pre-trained spatio-temporal feature extractor is employed to extract spatio-temporal features from videos in order to avoid the overfitting problem. Then, several-level spatio-temporal features are concatenated by 3D convolution skip-connections to improve the performance of the 3D CNN. Next, we apply a BN layer to speed up the convergence. Moreover, due to the introduced RNN architecture, our method realizes the modeling of temporal dependencies, which enables our network to deal with long-term information.

In addition, a strategy is used to achieve data augmentation in order to avoid overfitting. Besides, this strategy effectively allows our network to take advantage of more information from videos and adapts the shape of inputs to our network. Lastly, experimental results demonstrate that our proposed method achieves a significantly boost on recognition accuracy.

Acknowledgements. This work was supported in part by the National Science Foundation of China (No. 61473144), the Aeronautical Science Foundation of China (Key Laboratory) (No. 20162852031), the Jiangsu Postdoctoral Founding (No. 1402036C), the Special scientific instrument development of Ministry of science, technology of China (No. 2016YFF0103702) and the science and technology project of China Souther Grid Corp (No. 066600KK52170074).

References

1. Bengio, Y., Simard, P., Frasconi, P.: Learning long-term dependencies with gradient descent is difficult. IEEE Trans. Neural Netw. **5**(2), 157–166 (1994)
2. Chavez, K.: 3D convnets with optical flow based regularization
3. He, K., Zhang, X., Ren, S., Sun, J.: Deep residual learning for image recognition. In: Proceedings of the IEEE Conference on Computer Vision and Pattern Recognition, pp. 770–778 (2016)
4. Huang, G., Liu, Z., Weinberger, K.Q., van der Maaten, L.: Densely connected convolutional networks. In: Proceedings of the IEEE Conference on Computer Vision and Pattern Recognition, vol. 1, p. 3 (2017)
5. Ioffe, S., Szegedy, C.: Batch normalization: accelerating deep network training by reducing internal covariate shift. arXiv preprint arXiv:1502.03167 (2015)
6. Ji, S., Xu, W., Yang, M., Yu, K.: 3D convolutional neural networks for human action recognition. IEEE Trans. Pattern Anal. Mach. Intell. **35**(1), 221–231 (2013)
7. Karpathy, A., Toderici, G., Shetty, S., Leung, T., Sukthankar, R., Fei-Fei, L.: Large-scale video classification with convolutional neural networks. In: Proceedings of the IEEE conference on Computer Vision and Pattern Recognition, pp. 1725–1732 (2014)
8. Krizhevsky, A., Sutskever, I., Hinton, G.E.: Imagenet classification with deep convolutional neural networks. In: Advances in Neural Information Processing Systems, pp. 1097–1105 (2012)
9. Lan, Z., Lin, M., Li, X., Hauptmann, A.G., Raj, B.: Beyond Gaussian pyramid: multi-skip feature stacking for action recognition. In: Proceedings of the IEEE conference on Computer Vision and Pattern Recognition, pp. 204–212 (2015)
10. LeCun, Y.A., Bottou, L., Orr, G.B., Müller, K.-R.: Efficient BackProp. In: Montavon, G., Orr, G.B., Müller, K.-R. (eds.) Neural Networks: Tricks of the Trade. LNCS, vol. 7700, pp. 9–48. Springer, Heidelberg (2012). https://doi.org/10.1007/978-3-642-35289-8_3
11. Lipton, Z.C., Berkowitz, J., Elkan, C.: A critical review of recurrent neural networks for sequence learning. arXiv preprint arXiv:1506.00019 (2015)
12. Nielsen, M.A.: Neural networks and deep learning (2015)
13. Orr, G.B., Müller, K.-R. (eds.): Neural Networks: Tricks of the Trade. LNCS, vol. 1524. Springer, Heidelberg (1998). https://doi.org/10.1007/3-540-49430-8
14. Peng, X., Wang, L., Wang, X., Qiao, Y.: Bag of visual words and fusion methods for action recognition: comprehensive study and good practice. Comput. Vis. Image Underst. **150**, 109–125 (2016)

15. Poppe, R.: A survey on vision-based human action recognition. Image Vis. Comput. **28**(6), 976–990 (2010)
16. Sánchez, J., Perronnin, F., Mensink, T., Verbeek, J.: Image classification with the fisher vector: theory and practice. Int. J. Comput. Vis. **105**(3), 222–245 (2013)
17. Schuldt, C., Laptev, I., Caputo, B.: Recognizing human actions: a local SVM approach. In: Proceedings of the 17th International Conference on Pattern Recognition, vol. 3, pp. 32–36. IEEE (2004)
18. Sevilla-Lara, L., Liao, Y., Guney, F., Jampani, V., Geiger, A., Black, M.J.: On the integration of optical flow and action recognition. arXiv preprint arXiv:1712.08416 (2017)
19. Simonyan, K., Zisserman, A.: Two-stream convolutional networks for action recognition in videos. In: Advances in Neural Information Processing Systems, pp. 568–576 (2014)
20. Soomro, K., Zamir, A.R., Shah, M.: UCF101: a dataset of 101 human actions classes from videos in the wild. arXiv preprint arXiv:1212.0402 (2012)
21. Srivastava, N., Mansimov, E., Salakhudinov, R.: Unsupervised learning of video representations using LSTMs. In: International Conference on Machine Learning, pp. 843–852 (2015)
22. Tran, D., Bourdev, L., Fergus, R., Torresani, L., Paluri, M.: Learning spatiotemporal features with 3D convolutional networks. In: 2015 IEEE International Conference on Computer Vision, pp. 4489–4497. IEEE (2015)
23. Tran, D., Bourdev, L.D., Fergus, R., Torresani, L., Paluri, M.: C3D: generic features for video analysis. CoRR, abs/1412.0767 **2**(7), 8 (2014)
24. Tran, D., Ray, J., Shou, Z., Chang, S.F., Paluri, M.: Convnet architecture search for spatiotemporal feature learning. arXiv preprint arXiv:1708.05038 (2017)
25. Wang, H., Schmid, C.: Action recognition with improved trajectories. In: 2013 IEEE International Conference on Computer Vision, pp. 3551–3558. IEEE (2013)
26. Wang, L., Xiong, Y., Wang, Z., Qiao, Y.: Towards good practices for very deep two-stream convnets. arXiv preprint arXiv:1507.02159 (2015)
27. Wang, X., Farhadi, A., Gupta, A.: Actions ~ Transformations. In: Proceedings of the IEEE conference on Computer Vision and Pattern Recognition, pp. 2658–2667 (2016)
28. Wiesler, S., Ney, H.: A convergence analysis of log-linear training. In: Advances in Neural Information Processing Systems, pp. 657–665 (2011)

Convolutional Neural Network with Discriminant Criterion for Input of Each Neuron in Output Layer

Hidenori Ide$^{(\boxtimes)}$ and Takio Kurita

The Department of Information Engineering, Graduate School of Engineering,
Hiroshima University, Kagamiyama, 1-4-1, Higashi-Hiroshima City, Hiroshima, Japan
{hidenori-ide,tkurita}@hiroshima-u.ac.jp

Abstract. Deep convolutional neural network (CNN) performs the state-of-the-art performance in image classification problems. When the neural network is trained for a multi-classes classification problem, each neuron of the output layer of the network is trained to solve the 2 classes classification problem which classifies target class and the rest of the classes. Since the posterior probability of a class can be expressed as soft-max function when the class conditional probabilities of each class are given as the Gaussian distribution with the different means and the same variance, the classifier with soft-max activation function becomes ideal when the input of the neuron in the output layer are given as the Gaussian distributions. Thus it is expected that the former layers in the CNN are constructing the good input for each neuron in the output layer in the learning process. To improve or accelerate the discrimination at the neurons in the output layer, this paper proposes to apply the discriminant criterion to the input value of each neuron in the output layer. The proposed objective function of the deep CNN is given as the sum of the discriminant criterion and the standard cross entropy loss. The results of experiments on MNIST, CIFAR-10 and CIFAR-100 show that the proposed method can improve the performance of classification and classify each class clearly.

1 Introduction

Recently the convolutional neural network (CNN) have achieved a great success in image recognition [1]. After the deep CNN won the ILSVRC 2012 with higher score than the conventional methods, it becomes very popular as the fundamental technique for image classification [1]. To improve the recognition accuracy further, deeper and complex network architectures have been proposed [2–6].

Fisher proposed a linear discriminant analysis to extract categorical dependent feature vector from a given input feature vector [7]. In Fisher's linear discriminant analysis, the discriminant criterion was used to evaluate the separability of the training samples in the mapped feature space. Since then the discriminant criterion has been used in many applications. For example, Otsu

© Springer Nature Switzerland AG 2018
L. Cheng et al. (Eds.): ICONIP 2018, LNCS 11301, pp. 332–339, 2018.
https://doi.org/10.1007/978-3-030-04167-0_30

used the discriminant criterion to select the best threshold for image binarization [8]. The discriminant criterion was used to detect feature points for face recognition [9].

The discriminant criterion or linear discriminant analysis (LDA) also has been applied to the neural networks. Osman et al. showed that minimizing mean squared error at the linear output of neural network was equivalent to minimizing discriminant criterion [10]. Chen et al. proposed combining discriminant criterion with neural network [11]. They designed a decision tree consists of neural networks. They employed discriminant criterion in the intermediate level of the tree for heuristically partitioning a large and complicated task into several smaller and simpler subtasks. The discriminant criterion was also used to define the objective function of the outputs of the deep neural network for person re-identification [12]. Wu et al. extended it by applying the discriminant criterion into fisher networks [13]. They used SVM for person re-identification with feature map generated by neural network with LDA. Conversely Kurita et al. proposed to use the outputs of the trained multi layered Perceptron as the estimations of the posterior probabilities and to construct nonlinear discriminant analysis by using the estimated posterior probabilities [14].

When the neural network is trained for a multi-classes classification problem, each neuron of the output layer of the network is trained to solve the 2 classes classification problem which classifies a target class and the rest of the classes. It is well known that the posterior probability of a class can be expressed as soft-max function when the class conditional probabilities of each class are given as the Gaussian distribution with the different means and the same variance. This means that the classifier with soft-max activation function becomes ideal when the input feature of the neuron in the output layer for the class and the rest of the classes are given as the Gaussian distributions with the different means and the same variance. Thus it is expected that the former layers in the CNN are trying to construct good input feature for the two classes classification problem of each neuron in the output layer in the learning process. In this paper, this phenomenon was experimentally confirmed.

To improve or accelerate the discrimination at the neurons in the output layer, we propose to apply the discriminant criterion of the input values of the neurons in the output layer. Then the objective function is given the sum of the discriminant criterion and the soft-max cross entropy loss function. In this paper we use the discriminant criterion which is defined as the ratio of the within class variance and the total variance.

2   CNN with Discriminant Criterion

It is well known that the posterior probability of a class can be expressed as soft-max function when the class conditional probability of each class is given as the Gaussian distribution with the different means and the same variance.

Let consider the classification problem of K classes $C_1, C_2, \ldots, C_K$ based the features y. The prior probability of k-th class is given by $P(C_k)$ for $k = 1, \ldots, K$.

Assume that the conditional probability distribution of k-th class is also given by Gaussian distribution with the means μ_k and the variance σ^2 as

$$p(y|C_k) = \frac{1}{\sqrt{2\pi\sigma^2}} \exp\left\{-\frac{(y-\mu_k)^2}{2\sigma^2}\right\} \tag{1}$$

where we assume that the variance σ^2 is the same for all classes.

Then the posterior probability of the class C_k is given as

$$
\begin{aligned}
P(C_k|y) &= \frac{P(C_k)p(y|C_k)}{p(y)} \\
&= \frac{P(C_k)p(y|C_k)}{\sum_i P(C_i)p(y|C_i)} \\
&= \frac{\exp\left\{(y-\mu_k)^2\right\}}{\sum_i^K \exp\{(y-\mu_i)^2 + \log\frac{P(C_i)}{P(C_k)}\}} \tag{2} \\
&= \frac{1}{1 + \sum_{i,i\neq k}^{K-1} \exp(\alpha_i)} \tag{3}
\end{aligned}
$$

where $\alpha_i = 2\mu_i y - \mu_i^2 + \log\frac{P(C_i)}{P(C_k)}$.

This means that the classifier with soft-max function becomes ideal when the input feature of the neuron in the output layer for the class and the rest of the classes are given as the Gaussian distributions with the different means and the same variance. Then each neuron of classifier with soft-max is solving two classes classification problem of the target class and the other classes. Thus it is expected that the former layers in the CNN are trying to construct the good input feature for each neuron in the output layer in the learning process.

The left plot of Fig. 1 shows an example of histogram of the input values of a neuron in the output layer of the trained deep CNN. The network was trained by using CIFAR-10 datasets for image classification. Red values are the obtained from the target class samples and blue values are obtained from the other classes samples. It is noticed that each distribution of the target class and the other classes is close to the Gaussian distributions with the different means and the same variance.

To accelerate this process of constructing of two Gaussian distributions, we introduce the discriminant criterion in each neuron of output layer of the deep CNN. The discriminant criterion is used to evaluate the separability of the features from multiple classes [7,8].

By introducing the minimization of the discriminant criterion γ_k for each class, it is expected that CNN with discriminant criterion improves the separability of the two classes, target class and the other classes, at each neuron in the output layer.

We use the discriminant criterion for CNN such as:

$$G = \sum_k^K \gamma_k = \sum_k^K \frac{\sigma_{W_k}^2}{\sigma_{T_k}^2}. \tag{4}$$

$\sigma_{W_k}^2$ and $\sigma_{T_k}^2$ are as follow

$$\sigma_{W_k}^2 = \frac{1}{N} \sum_n^N \left\{ t_{n,k}(\eta_{n,k} - \mu_k)^2 + (1 - t_{n,k})(\eta_{n,k} - \hat{\mu}_k)^2 \right\} \tag{5}$$

$$\sigma_{T_k}^2 = \frac{1}{N} \sum_n^N (\eta_{n,k} - \mu_T)^2. \tag{6}$$

where N is the number of mini-batch, $t_{n,k}$ is the k-th element (k class) of the teacher vector of n-th sample, $\eta_{n,k}$ is the input of k-th neuron at output layer for n-th sample. Also, the number of samples of the k-th class and the other classes are denoted as $N_k = \sum_n^N t_{n,k}$ and $\hat{N}_k = \sum_n^N (1 - t_{n,k})$. Then the means of the k-th class and the other classes are denoted as $\mu_k = \frac{1}{N_k} \sum_n^N t_{n,k}\eta_{n,k}$ and $\hat{\mu}_k = \frac{1}{\hat{N}_k} \sum_n^N (1-t_{n,k})\eta_{n,k}$ respectively. The total mean is defined as $\mu_T = \frac{1}{N} \sum_n^N \eta_{n,k}$.

The objective function of our proposed method is given by the sum of the soft-max cross entropy and the discriminant criterion defined on each neuron of the output layer as

$$E = L + \lambda G \tag{7}$$

where λ is a tuning parameter to control the effect of the discriminant criterion and the cross entropy is defined as

$$L = -\frac{1}{N} \sum_n^N \sum_k^K t_{n,k} \log y_{n,k}. \tag{8}$$

CNN with discriminant criterion minimizes this objective function to update parameter such as

$$w \leftarrow w - \alpha \frac{\partial E}{\partial w} \tag{9}$$

where α is the learning rate of parameters.

3   Experiments

3.1   Datasets

To evaluate the effectiveness of the deep CNN with the discriminant criterion, we have performed experiments using the famous data set CIFAR-10, CIFAR-100 and MNIST. CIFAR-10 and CIFAR-100 are the data sets for object classification and include RGB color images for 10 classes and 100 classes respectively. The size of each image is 32×32 pixels. The number of training and test samples of these data sets are 50000 and 10000. MNIST is the data set for handwritten character recognition and includes grayscale images for 10 classes. The size of the images is 28×28 pixels. The number of training and test samples of this data set is 60000 and 10000.

3.2   Network Architecture

The deep CNN for CIFAR-10 and CIFAR-100 consists of 10 layers, namely 9 convolution layers and one classifier. Max pooling is inserted at third and sixth convolution layers and average pooling is inserted at 9th layer. For each convolution layer, batch normalization is used. We also use dropout for the outputs of max pooling layer. The probability of dropout is set to 0.5. The leaky ReLU is used as the activation function of each neuron in the internal layers. The slope parameter of the leaky ReLU is set to 0.1. The objective function is given in the Eq. 7. For optimization, Adam optimizer is used. The parameters of Adam optimizer are set as $alpah = 0.001$, $beta1 = 0.9$, $beta2 = 0.999$, and $eps = 10^{-8}$.

For MINIS, we made the shallow CNN. The shallow CNN consists of 3 layers with 2 convolution layers and 1 classifier. Batch normalization is inserted at each convolution layer. The activation function, objective function and optimizer are same as deep CNN.

3.3   Classification Performance

We have done the experiments to compare the CNN with the discriminant criterion and the standard CNN with dropout using the data sets CIFAR-10, CIFAR-100 and MNIST. The recognition accuracies for training samples and test samples are shown in Table 1.

From this table, we can find that our method improved the classification accuracy about 1%, about 4.6% and about 0.01% for CIFAR-10, CIFAR-100, and MNIST, respectively. Especially the improvement for CIFAR-100 is large. This is because the test accuracy of CIFAR-100 is far from 100% while the test accuracy of CIFAR-10 and MNIST are close to 100% and there are no room to improve further.

It is noticed that the test accuracy of the CNN with the discriminant criterion is better than the one of the standard CNN.

Table 1. The accuracy for CIFAR-10, CIFAR-100 at 200 epochs and MNIST at 100 epochs.

Dataset	CIFAR-10		CIFAR-100		MNIST	
	Train	Test	Train	Test	Train	Test
Standard CNN with dropout	99.99%	89.82%	99.92%	63.09%	100%	98.72%
CNN with discriminant criterion	100%	**92.04%**	99.95%	**68.24%**	100%	**98.89%**

3.4   Distribution of Input Values

To investigate the effect of the discriminant criterion further, we compared the distributions of the input values of each neuron in the output layer by visualizing the histograms of the input values. Figure 1 shows the histograms of the inputs values of the test samples from CNN. These are obtained by using the trained

CNN for CIFAR-10. The left histogram is obtained by using the standard CNN and the right one is for the CNN with the discriminant criterion. Red bars in histograms shows the target class values of the target class neuron and blue bar is the other classes values of the target class neuron.

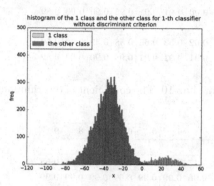 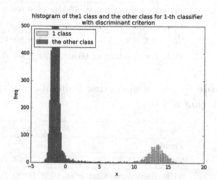

Fig. 1. Histograms of the input values of the neuron in the output layer for class 2. These are obtained by using the test images of CIFAR-10. The left histogram is obtained by using the standard CNN and the right one is for the CNN with discriminant criterion. Red bars in histogram shows the target class values of the target class neuron and blue bar is for the other classes values of the target class neuron. (Color figure online)

From these histograms, we can notice that the CNN is trained to transforms the input images into the two Gaussian distributions of the target class and the rest of the classes. These results give the empirical proof of our conjecture explained in the Sect. 2. We also can notice that the distributions of two classes becomes far if the discriminant criterion is introduced in the objective function.

Tables 2 and 3 shows the discriminant criteria of the input values of each neuron in the output layer. If the value of discriminant criterion is large, two distributions are separated clearly. From these tables we can notice that the values of the discriminant criterion of each neuron are improved very much by introducing the discriminant criterion in the objective function of the CNN training.

3.5 Visualization of Input Values

Finally, we visualized the input values of the neurons of the output layer by applying principal component analysis (PCA). Figure 2 shows the two dimensional PCA scores for CIFAR-10. It is noticed that the clusters of each class obtained by the CNN with the discriminant criterion are more clear than the standard CNN.

Table 2. The values of discriminant criterion in MNIST. The coefficient of discriminant criterion is 0.1.

Class	0	1	2	3	4	5	6	7	8	9
Train										
Normal CNN	0.64	0.59	0.67	0.62	0.53	0.64	0.59	0.48	0.70	0.59
with discriminant criterion	0.93	0.93	0.90	0.90	0.91	0.91	0.93	0.90	0.90	0.89
Test										
Normal CNN	0.64	0.59	0.68	0.62	0.53	0.65	0.58	0.49	0.70	0.59
with discriminant criterion	0.93	0.93	0.90	0.91	0.91	0.91	0.93	0.90	0.90	0.89

Table 3. The values of discriminant criterion in cifar-10. The coefficient of discriminant criterion is 1.0.

Class	0	1	2	3	4	5	6	7	8	9
Train										
Normal CNN	0.54	0.64	0.52	0.47	0.50	0.48	0.56	0.58	0.58	0.58
with discriminant criterion	0.99	0.99	0.98	0.98	0.99	0.99	0.99	0.99	0.99	0.99
Test										
Normal CNN	0.45	0.60	0.41	0.33	0.41	0.37	0.48	0.51	0.53	0.53
with discriminant criterion	0.85	0.92	0.80	0.70	0.85	0.77	0.89	0.89	0.90	0.90

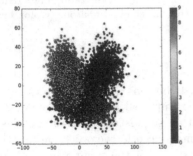 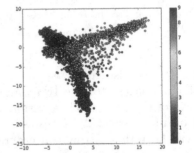

Fig. 2. Visualization of the input values of the neurons in the output layer of the trained CNN for CIFAR-10. The left plot shows the two dimensional PCA scores for the standard CNN and the right is those for the input values obtained by the CNN with the discriminant criterion.

4   Conclusion

In this paper we proposed to apply the discriminant criterion to the input values of the neurons in the output layer. The sum of the soft-max cross entropy loss and the discriminant criterion is defined as the objective function for learning the CNN. Through experiments using CIFAR-10, CIFAR-100, and MNIST data sets, it is confirmed that the proposed method can improve the classification accuracy and separated each class more clearly.

Acknowledgement. This work was partly supported by JSPS KAKENHI Grant Number 16K00239.

References

1. Krizhevsky, A., Sutskever, I., Hinton, G.E.: ImageNet classification with deep convolutional neural networks. In: Proceeding Conference on Neural Information Processing Systems, pp. 1097–1105 (2012)
2. Lin, M., Chen, Q., Yan, S.: Network in network. arXiv **1312**, 4400 (2013)
3. Simonyan, K., Zisserman, A.: Very deep convolutional networks for large scale visual recognition. In: International Conference on Learning Representations (2015)
4. Szegedy, C., et al.: Going deeper with convolutions. In: Proceedings of IEEE Conference on Computer Vision and Pattern Recognition, pp. 1–9 (2015)
5. Szegedy, C., Vanhoucke, V., Ioffe, S., Shlens, J., Wojna, Z.: Rethinking the inception architecture for computer vision. In: Proceedings of IEEE Conference on Computer Vision and Pattern Recognition, pp. 2818–2826 (2016)
6. He, K., Zhang, X., Ren, S., Sun, J.: Deep residual learning for image recognition. In: IEEE Conference on Computer Vision and Pattern Recognition, pp. 770–778 (2016)
7. Fisher, R.A.: The use of multiple measurements in taxonomic problems. Ann. Hum. Genet. **7**(2), 179–188 (1936)
8. Otsu, N.: A threshold selection method from gray-level histograms. IEEE Trans. Syst., Man, Cybern. **9**, 62–66 (1979)
9. Fukui, K., Yamaguchi, O.: Facial feature point extraction method based on combination of shape extraction and pattern matching. Syst. Comput. Jpn. **29**(6), 49–58 (1998)
10. Osman, H., Fahmy, M.M.: On the discriminatory power of adaptive feed-forward layered networks. IEEE Trans. Pattern Anal. Mach. Intell. **16**(8), 837–842 (1994)
11. Chen, K., Xiang, Y., Huisheng, C.: Combining linear discriminant functions with neural networks for supervised learning. Neural Comput. Appl. **6**(1), 19–41 (1997)
12. Dorfer, M., Kelz, R., Widmer, G.: Deep linear discriminant analysis. arXiv **1511**, 04707 (2015)
13. Wu, L., Shen, C., van den Hengel, A.: Deep linear discriminant analysis on fisher networks: a hybrid architecture for person re-identification. Pattern Recognit. **65**, 238–250 (2017)
14. Kurita, T., Asoh, H., Otsu, N.: Nonlinear discriminant features constructed by using outputs of multilayer perceptron. In: Proceedings of International Symposiumon Speech, Image Processing, and Neural Networks, Hong Kong, pp. 417–420 (1994)
15. Giryes, R., Sapiro, G., Bronstein, A.M.: Deep neural networks with random Gaussian weights: a universal classification strategy? IEEE Trans. Signal Process. **64**(13), 3444–3457 (2016)
16. Bishop, C.M.: Pattern Recognition and Machine Learning. Springer, Boston (2006). https://doi.org/10.1007/978-1-4615-7566-5

Proposal of Complex-Valued Convolutional Neural Networks for Similar Land-Shape Discovery in Interferometric Synthetic Aperture Radar

Yuki Sunaga, Ryo Natsuaki, and Akira Hirose[✉]

Department of Electrical Engineering and Information Systems,
The University of Tokyo, Tokyo, Japan
ahirose@ee.t.u-tokyo.ac.jp
http://www.eis.t.u-tokyo.ac.jp/

Abstract. We propose a complex-valued convolutional neural network to extract the areas having land shapes similar to samples in interferometric synthetic aperture radar (InSAR). InSAR extends its application to various earth observations such as volcano monitoring and earthquake damage estimation. Since the amount of data is increasing drastically in these years, it is necessary to structurize them in a big data framework. In this paper, experiments demonstrate that similar small volcanoes are grouped into a single class. We find that the neural network is capable of discovering unidentified lands similar to prepared samples successfully.

Keywords: Interferometric synthetic aperture radar (InSAR)
Feature discovery · Complex-valued neural network (CVNN)

1 Introduction

Synthetic aperture radar (SAR) has been extending its application fields to various earth observations targeting environmental features including diverse surface characteristics such as vegetation, glacier and volcanoes. In order to access SAR data on a database, e.g., Ref. [1], latitude, longitude and period of observation have to be specified. Since the amount of SAR data has been increasing drastically, it will be useful for a user in the future to access SAR data based on the information contained in the data itself, for example, by specifying land shapes that the user wants to obtain. In other words, it is necessary to structurize the data in a big data framework [10].

Previously, we proposed the concept of SAR data structurization in the feature space based on physical observation data. Extracting features from SAR data and classifying land are bases of the structurization. Some papers have proposed related methods based on support vector machine [4], latent dirichlet allocation [20], quaternion neural networks [13,17,18], neural auto-encoders [3,14] and hierarchical self-organizing codebooks [12].

© Springer Nature Switzerland AG 2018
L. Cheng et al. (Eds.): ICONIP 2018, LNCS 11301, pp. 340–349, 2018.
https://doi.org/10.1007/978-3-030-04167-0_31

Since interferometric SAR (InSAR) images have complex values, it is recommended for a system to use complex-valued neural networks [8] that achieve good generalization in complex domain. Suksmono and Hirose extracted features from InSAR images and preliminarily classified land by using a complex-valued self-organizing map, proving that complex-valued neural networks are superior to real-valued ones on InSAR data [19].

Real-valued convolutional neural networks, originating from including [5], has continuously advanced in image recognition area including SAR images. For example, Marmanis et al. classified aerial photographs using a pre-trained deep real-valued convolutional neural network [15]. Learning process sometimes require a large amount of training data. De et al. proposed augmentation of PolSAR training data by rotating observed coherency matrix with uniform steps [2].

In this paper, we propose a complex-valued convolutional neural network. Experiments demonstrate a reasonable land-shape classification successfully. In addition, we find that the network discovers unknown areas similar to a sample in the land shape. For example, we can detect a number of small volcanoes, called Omuroyama's (or scoria cones widely), in InSAR images.

2 Complex-Valued Convolutional Neural Networks

2.1 Construction of Complex-Valued Convolutional Neural Networks

Figure 1 shows the construction of the convolutional neural network we propose here. The basic structure is the same as that of real-valued convolutional neural networks [5]. Single or multiple complex-valued two-dimensional signals (complex-valued images) are fed to the input terminals of the network. As described in detail below, we prepare two complex-valued input images in the following experiments, $^{l=0}\boldsymbol{x}^{\mathrm{ew}}$ and $^{l=0}\boldsymbol{x}^{\mathrm{ns}}$, that have common coordinates. A layer is composed of convolutional and pooling steps, and numbered as l. Input terminal layer is labeled as $l = 0$, while the last layer is $l = L$, which is a fully connected network to work for decision. The total number of layers L is arbitrary.

2.2 Complex-Valued Convolutional Processing

We have single or multiple input signals (images) $^{0}\boldsymbol{x}^{c}$ where c is the image label and 0 means the input layer. The output signal vector of the first convolution layer $^{1}\boldsymbol{v}_{k} = [^{1}v_{kij}]$ is expressed for input signal vectors $^{0}\boldsymbol{x}^{c} = [^{0}x^{c}_{i+p\ j+q}]$ as

$$^{1}\boldsymbol{v} = f(^{1}\boldsymbol{u}), \quad {}^{1}u_{kij} = \sum_{c=1}^{C}\sum_{p=1}^{^{1}P}\sum_{q=1}^{^{1}Q} {}^{1}w^{c}_{kpq} {}^{*}\, {}^{1}x^{c}_{i+p\ j+q} \tag{1}$$

$$f(^{1}u_{kij}) = \tanh(|^{1}u_{kij}|)\exp\left(j\arg(^{1}u_{kij})\right) \tag{2}$$

where $^{1}\mathbf{W} = w^{c}_{kpq}$ is the neural weight connecting an input signal at position (p, q) in image c and a neuron at position (p, q) in convolutional kernel k in

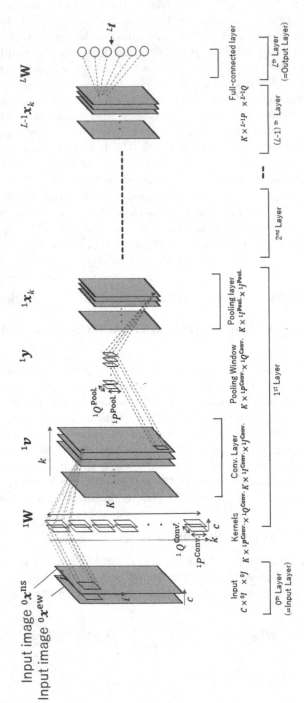

Fig. 1. Construction of the convolutional neural network.

the first layer, and C, ^1P and ^1Q are the numbers of the input images, kernel vertical size and kernel horizontal size, respectively, $f(\cdot)$ is the activation function identical to (4), and $^l\boldsymbol{u}$ is internal state of l-th layer neuron.

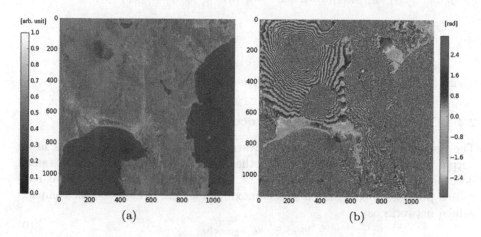

Fig. 2. (a) Amplitude and (b) phase of InSAR interferogram.

2.3   Learning Dynamics

In general in layered complex-valued neural networks, input and output values and connection weights are complex values. The $(l-1)$-th layer output signals $^{l-1}\boldsymbol{x}$ are processed in l-th layer to generate $^l\boldsymbol{x}$ as

$$^l\boldsymbol{x} = f(^l\boldsymbol{u}), \quad ^l\boldsymbol{u} =\,^l\mathbf{W} \,\,^{l-1}\boldsymbol{x} \tag{3}$$

$$f(^lu_m) = \tanh(|^lu_m|) \exp\left(j \arg(^lu_m)\right) \tag{4}$$

where $^l\mathbf{W} = [^lw_{nm}]$ is the connection weight between m $(= iJ + j)$-th neuron in $(l-1)$-th layer and n $(= pQ + q)$-th neuron in l-th layer.

We determine the backpropagation of teacher signals and the dynamics to update connection weights for supervised learning as follows. We calculate teacher signal vectors of each layers $^l\mathbf{t}$ backwards from the output layer. The teacher signal of $(l-1)$-th layer $^{l-1}\mathbf{t}$ is obtained from the teacher signal of l-th layer $^l\mathbf{t}$ as [6–9]

$$^{l-1}\boldsymbol{t} = \left(f\left(^l\boldsymbol{t}^* \,\,^l\mathbf{W}\right)\right)^* \tag{5}$$

Then, $^l\mathbf{W}$ is updated by using $^l\boldsymbol{t}$ as [11,16]

$$
\begin{aligned}
|^lw_{nm}| &\leftarrow |^lw_{nm}| \\
&- \epsilon^{\mathrm{a}}\Big\{(1 - |^lx_n^2|)(|^lx_n| - |^lt_n|\cos(\arg{^lx_n} - \arg{^lt_n}))|^{l-1}x_m|\cos^l\theta_{nm}^{\mathrm{rot}} \\
&- |^lx_n||^lt_n|\sin(\arg{^lx_n} - \arg{^lt_n})\frac{|^{l-1}x_m|}{|^lu_n|}\sin^l\theta_{nm}^{\mathrm{rot}}\Big\}
\end{aligned}
\tag{6}
$$

$$\arg(^l w_{nm}) \leftarrow \arg(^l w_{nm})$$
$$- \epsilon^{\mathrm{p}} \Big\{ (1 - |^l x_n^2|)(|^l x_n| - |^l t_n| \cos(\arg^l x_n - \arg^l t_n))|^{l-1} x_m| \sin^l \theta_{nm}^{\mathrm{rot}}$$
$$+ |^l x_n||^l t_n| \sin(\arg^l x_n - \arg^l t_n) \frac{|^{l-1} x_m|}{|^l u_n|} \cos^l \theta_{nm}^{\mathrm{rot}} \Big\} \tag{7}$$

$$^l \theta_{nm}^{\mathrm{rot}} = \arg(^l x_n) - \arg(^{l-1} x_m) - \arg(^l w_{nm}) \tag{8}$$

where $\epsilon^{\mathrm{a}}, \epsilon^{\mathrm{p}}$ are learning rates of amplitude and phase respectively.

2.4   Complex-Valued Pooling Process

The output of the pooling layer is invariant for small position changes in an output pattern of the former layer. In other words, a pooling layers absorbs small position changes and small rotations of input images. The conventional real-valued pooling process, which selects maximum signal, is modified for complex-valued networks as

$$^l y_{kij} =^l v_{k \, \underset{p,q}{\mathrm{argmax}} \, |^l v_{kpq}|} \tag{9}$$

Here, the output of a pooling layer is a value that has maximum norm in a pooling window. If a local pattern in an input image is similar to a phase pattern in a kernel of a convolutional layer, a norm of output signal becomes large, and then the following pooling layer extracts this signal. This is a phase-centered max-pooling dynamics. That is, when a local pattern in an input image is similar to a pattern in a kernel of the convolutional layer, the signal is passed through the layer.

We propose the following dynamics of the backpropagation in the pooling layer. We make teacher signals backpropagate in order to update only the neurons that pass the signals to the following layer. Specifically, we get teacher signals of the convolution layer $^l \hat{t}$ from teacher signals for the pooling layer $^l t$ as

$$^l \hat{t}_{kpq} = \begin{cases} ^l t_{kij} & \text{if } ^l y_{kij} =^l v_{kpq} \\ ^l v_{kpq} & \text{otherwise} \end{cases} \tag{10}$$

2.5   Decision Network Having Complex-Valued Fully Connected Neurons

For classification, the output layer of the complex-valued convolutional neural network is composed as a full-connected network, just like a usual complex-valued neural network. The input signal of the fully connected layer $^{L-1} x = [^{L-1} x_n]$ is obtained from the preceeding pooling layer as

$$^{L-1} x_{k \times I \times J + i \times J + j} =^{L-1} y_{kij} \tag{11}$$

where $k \times I \times J + i \times J + j (= n)$ represented one dimensionally when the size of the output image of the pooling layer has K images of $I \times J$ pixels.

3 Experiments

3.1 Preprocessing of InSAR Images

We conduct experiments to classify land shapes in an InSAR image by supervised learning in the complex-valued convolutional neural network proposed in Sect. 2. Figure 2 shows an example $1,140 \times 1,138$-pixel InSAR interferogram image of Mt. Fuji and Hakone area obtained by Advanced Land Observing Satellite (ALOS) of Japan Aerospace Exploration Agency (JAXA).

We feed the interferogram to the neural network after the preprocessing described below. First, in order to deal with slope information basically, we generate spatial phase-difference images in east-west and north-south directions as $z_{ij}^{ew} = z_{ij}/z_{i\ j+1}$ and $z_{ij}^{ns} = z_{ij}/z_{i+1\ j}$. Then we apply a 5×5 averaging filter to obtain input images, $[x_{ij}^{ew}]$ and $[x_{ij}^{nw}]$, respectively, as

$$ {}^{0}x_{ij}^{ew} = \sum_{p,q} w_{pq}^{av} \exp\left(j \arg(z_{i+p\ j+q}^{ew})\right) \tag{12} $$

$$ {}^{0}x_{ij}^{ns} = \sum_{p,q} w_{pq}^{av} \exp\left(j \arg(z_{i+p\ j+q}^{ns})\right) \tag{13} $$

with an ordinary conical averaging weighting factor w_{pq}^{av}.

3.2 Parameters in the Neural Network

Figure 1 already presented the structure of the complex-valued convolutional neural network we use here in the following experiments. It consists of a single convolution and pooling layer ($L = 2$) having nine convolution kernels of 27×27 pixels and 2×2-pooling window as well as a full connection layer. The input is 28×28-pixel window images (land pieces) clipped out from $[x_{ij}^{c}]$, that is, $[x_{ij}^{ew}]$ and $[x_{ij}^{ns}]$. The learning rates are $\epsilon^{a} = \epsilon^{p} = 0.01$.

3.3 Teacher Images and the Learning Process

We classify the land pieces into seven classes, namely, north- / east- / south- / west- facing slopes, flatland, sea and *Omuroyama*. Omuroyama is a scoria cone made by a small eruption activity to shape itself into a small volcano.

Table 1. Output teacher values for respective classes.

	North	East	South	West	Flat	Sea	Omuro
L_t	$\begin{bmatrix} 1 \\ -1 \\ -1 \\ -1 \\ -1 \\ -1 \end{bmatrix}$	$\begin{bmatrix} -1 \\ 1 \\ -1 \\ -1 \\ -1 \\ -1 \end{bmatrix}$	$\begin{bmatrix} -1 \\ -1 \\ 1 \\ -1 \\ -1 \\ -1 \end{bmatrix}$	$\begin{bmatrix} -1 \\ -1 \\ -1 \\ 1 \\ -1 \\ -1 \end{bmatrix}$	$\begin{bmatrix} -1 \\ -1 \\ -1 \\ -1 \\ 1 \\ -1 \end{bmatrix}$	$\begin{bmatrix} 0.1 \\ 0.1 \\ 0.1 \\ 0.1 \\ 0.1 \\ 0.1 \end{bmatrix}$	$\begin{bmatrix} -1 \\ -1 \\ -1 \\ -1 \\ -1 \\ 1 \end{bmatrix}$

As a set of teacher input images ${}^{0}x^{c}$, we clip out 600 windows in each category area in $[x_{ij}^{c}]$, resulting in $600 \times 7 = 4,200$ input teachers. The category areas are shown in Fig. 3. Slope teachers are chosen around Mt. Fuji, while Omuroyama teachers are obtained at the Omuroyama located north-west of Mt. Fuji. Since this Omuroyama area is not large enough to yield sufficient

Fig. 3. Regions from which the teacher input images are clipped out.

number of Omuroyama teachers, we cut out not only 300 windows of 28×28 pixels, but also another 300 windows of 35×35 to shrink into 28×28 to generate 600 teachers in total.

Table 1 lists the output teacher values $^L t$ presented to the output layer where the dimension (neuron number) is $\dim {}^L t = 6$. We apply a mini-batch learning process as follows. First, we choose 40 ($\equiv B$) teachers for a single batch at random, calculate the gradients by (6) and (7) for $b = 1, 2, \cdots, B$, and obtain the gradient sum to update the weights in (6) and (7). We name this single process as 1 iteration, and $4,200/40 = 105$ iteration as 1 epoch. The following experiment takes 20 epochs of learning.

We define an error function to monitor the learning process as

$$E = \frac{1}{B} \sum_{b=1}^{B} \sum_{n=1}^{\dim {}^L t} \left| {}^L x_{bn} - {}^L t_{bn} \right|^2 \tag{14}$$

where $^L x_{bn}$ is the output signal of n-th output neuron for teacher input $^0 x_b^c = [^0 x_{ij}^c]((i,j) \in D_b$ (D_b: b-th 28×28-pixel teacher) and $^L t_{bn}$ is the output teacher signal of n-th neuron corresponding to the input teacher $^0 x_b^c$. As the test data for evaluation, we also prepare 4,200 pairs of $^0 x^c$ and $^L t$ separately from the teachers.

After the learning process, we classify the entire interferogram image. We put a 28×28-pixel window to scan the whole image with this window. We calculate the error in (14) for respective $^L t$ values in Table 1 to find the smallest class, to which we decide that the window belongs.

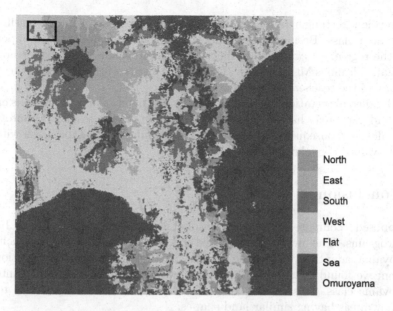

Fig. 4. Classification result obtained by the proposed complex-valued convolutoinal neural network.

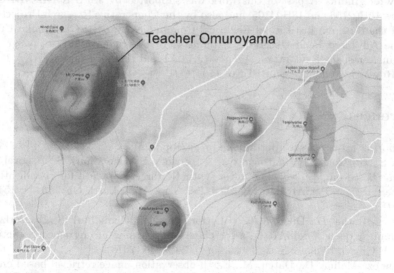

Fig. 5. Land shape map zoom-in on the region indicated by a rectangle in Fig. 4 with red hatches superimposed to present Omuroyama-class areas. (Background map data provided by Zenrin)

4   Results

Figure 4 is the classification result image. We investigate the places assigned to Omuroyama class in detail. Figure 5(a) is a zoom-in on Region marked by an upper-left rectangle in Fig. 4. Red hatches indicate the area classified to Omuroyama class. Besides the teacher Omuroyama located at the upper left inside the region, we can find another small Omuroyama at the center bottom. The small volcano is Mt. Katafuta. Though the size is a little different, the shape is similar to the teacher Omuroyama.

In detailed observation, we find that other red hatches in other regions enable us to find the areas having land shapes similar to the teacher Omuroyama, small volcano. Consequently, we can discover similar land-shape areas with the complex-valued convolutional neural network.

5   Conclusion

We proposed a complex-valued convolutional neural network to deal with InSAR interferograms. The network succeeded in classifying land shapes including Omuroyama adaptively with the results natural for human recognition. In addition, we found that unidentified small volcanos are also grouped into the Omuroyama class. This result showed that the network has the ability to discover new areas having similar land shapes.

Acknowledgment. A part of this work was supported by JSPS KAKENHI Grant Numbers 15H02756 and 18H04105, and also by the Cooperative Research Project Program of the Research Institute of Electrical Communication (RIEC), Tohoku University. The Advanced Land Observing Satellite (ALOS) original data are copyrighted by Japan Aerospace Exploration Agency (JAXA) and provided under JAXA Fourth ALOS Research Announcement PI No. 1154.

References

1. G-portal. https://www.gportal.jaxa.jp/gp/top.html
2. De, S., Bruzzone, L., Bhattacharya, A., Bovolo, F., Chaudhuri, S.: A novel technique based on deep learning and a synthetic target datadata for classification of urban areas in PolSAR data. IEEE J. Sel. Top. Appl. Earth Obs. **11**(1), 154–170 (2018)
3. De, S., Ratha, D., Dikshya, R., Bhattacharya, A., Chaudhuri, S.: Tensorization of multifrequency PolSAR data for classification using an autoencoder network. IEEE Geosci. Remote. Sens. Lett. **15**(4), 542–546 (2018)
4. Espinoza-Molina, D., Datcu, M.: Earth-observation image retrieval based on content, semantics, and metadata. IEEE Trans. Geosci. Remote. Sens. **51**(11), 5145–5159 (2013)
5. Fukushima, K.: Neocognitron: a self-organizing neural network model for a mechanism of pattern recognition unaffected by shift in position. Biol. Cybern. **36**(4), 193–202 (1980)
6. Hirose, A.: Continuous complex-valued back-propagation learning. Electron. Lett. **28**(20), 1854–1855 (1992)
7. Hirose, A.: Applications of complex-valued neural networks to coherent optical computing using phase-sensitive detection scheme. Inf. Sci.-Appl. **2**, 103–117 (1994)

8. Hirose, A.: Complex-Valued Neural Networks, 2nd edn. Springer, Heidelberg (2012). https://doi.org/10.1007/978-3-642-27632-3
9. Hirose, A., Eckmiller, R.: Behavior control of coherent-type neural networks by carrier-frequency modulation. IEEE Trans. Neural Netw. **7**(4), 1032–1034 (1996)
10. Hirose, A., Tsuda, S., Natsuaki, R.: Structurization of synthetic aperture radar information by using neural networks. In: SAR in Big Data Era: Models, Methods and Applications (BIGSARDATA) 2017, pp. 1–4, November 2017. https://doi.org/10.1109/BIGSARDATA.2017.8124936
11. Hirose, A., Yoshida, S.: Generalization characteristics of complex-valued feedforward neural networks in relation to signal coherence. IEEE Trans. Neural Netw. Learn. Syst. **23**, 541–551 (2012)
12. Kim, H., Hirose, A.: Codebook-based hierarchical polarization feature for unsupervised fine land classification using high-resolution PolSAR data. In: International Geoscience and Remote Sensing Symposium (IGARSS) 2018 Valencia. IEEE (to be presented 2018)
13. Kim, H., Hirose, A.: Polarization feature extraction using quaternion neural networks for flexible unsupervised PolSAR land classification. In: International Geoscience and Remote Sensing Symposium (IGARSS) 2018 Valencia, to be presented, July 2018
14. Kim, H., Hirose, A.: Unsupervised fine land classification using quaternion autoencoder-based polarization feature extraction and self-organizing mapping. IEEE Trans. Geosci. Remote. Sens. **56**(3), 1839–1851 (2018)
15. Marmanis, D., Datcu, M., Esch, T., Stilla, U.: Deep learning earth observation classification using imagenet pretrained networks. IEEE Geosci. Remote. Sens. Lett. **13**(1), 105–109 (2016)
16. Oyama, K., Hirose, A.: Adaptive phase-singular-unit restoration with entire-spectrum-processing complex-valued neural networks in interferometric SAR. Electron. Lett. **54**(1), 43–45 (2018)
17. Shang, F., Hirose, A.: Quaternion neural-network-based PolSAR land classification in Poincare-sphere-parameter space. IEEE Trans. Geosci. Remote. Sens. **52**(9), 5693–5703 (2014)
18. Shang, F., Hirose, A.: Averaged-stokes-vector-based polarimetric SAR data interpretation. IEEE Trans. Geosci. Remote. Sens. **53**(8), 4536–4547 (2015)
19. Suksmono, A.B., Hirose, A.: Adaptive complex-amplitude texture classifier that deals with both height and reflectance for interferometric SAR images. IEICE Trans. Electron. **E83–C**(12), 1912–1916 (2000)
20. Tănase, R., Bahmanyar, R., Schwarz, G., Datcu, M.: Discovery of semantic relationships in polsar images using latent Dirichlet allocation. IEEE Geosci. Remote. Sens. Lett. **14**(2), 237–241 (2017)

Feature Learning and Transfer Performance Prediction for Video Reinforcement Learning Tasks via a Siamese Convolutional Neural Network

Jinhua Song, Yang Gao$^{(\boxtimes)}$, and Hao Wang

State Key Laboratory for Novel Software Technology, Collaborative Innovation Center of Novel Software Technology and Industrialization, Nanjing University, Nanjing, China
songjinhua2008@gmail.com, {gaoy,wanghao}@nju.edu.cn

Abstract. In this paper, we handle the negative transfer problem by a deep learning method to predict the transfer performance (positive/negative transfer) between two reinforcement learning tasks. We consider same domain transfer for video reinforcement learning tasks such as video games which can be described as images and perceived by an agent with visual ability. Our method directly trains a neural network from raw task descriptions without other prior knowledge such as models of tasks, target task samples and human experience. The architecture of our neural network consists of two parts: a siamese convolutional neural network to learn the features of each pair of tasks and a softmax layer to predict the binary transfer performance. We conduct extensive experiments in the maze domain and the Ms. PacMan domain to evaluate the performance of our method. The results show the effectiveness and superiority of our method compared with the baseline methods.

Keywords: Transfer learning · Deep neural network
Reinforcement learning task · Transfer performance

1 Introduction

Transfer learning [9,21] is an important learning framework in reinforcement learning (RL), which can reuse the learned knowledge of previously solved tasks (called source tasks) to better solve a new task (called target task). In recent years, lots of transfer learning methods have been studied, which focus on decreasing the learning time and improving the average approximation of the optimal policy. However, if a source task is irrelevant to the target task, *negative transfer* [9,15,21,22] may occur. Negative transfer will cause slower learning speed or even worse performance in the target task.

© Springer Nature Switzerland AG 2018
L. Cheng et al. (Eds.): ICONIP 2018, LNCS 11301, pp. 350–361, 2018.
https://doi.org/10.1007/978-3-030-04167-0_32

Determining when transfer will be useful is an important problem in transfer learning domain. However, most existing methods select a source task by a human (often a domain expert) to guarantee positive transfer. There are only few methods which attempt to autonomously decide if a source task is appropriate for transfer. Some of them identify similar tasks using the models of the tasks [13,18] or samples gathered from the target task [1,10,14]. Others define the relatedness of tasks using some man-made features [7,12,17]. Almost all of the above methods select the most similar source task(s) without a criterion to distinguish positive transfer from negative transfer. They assume that transfer will occur ignoring the scenario where source tasks are all dissimilar to the target task.

To address the above issue, we propose a new method to identify positive transfer for video RL tasks. For some domains, task descriptions can be represented as images. An example is a video game which may have many levels and the differences between the game levels are the environments (i.e., state spaces) that the agent faces. Environmental information is received by the agent as an image (e.g., a map). An autonomous robot with a camera, which can "see" the task scenarios before learning, is another example. The robot understands the tasks when seeing the objects, their locations and other visual information [3]. These tasks are called video RL tasks, and the differences between such tasks can be reflected in the differences of images. We consider a transfer learning scenario where the agent continually solves tasks and needs to decide whether to transfer with only raw task descriptions of new tasks. The tasks faced by the agent are from the same domain, differ in their state spaces and have similar dynamics. Such a setting can be found in many real scenarios such as a robot cleaning different rooms, where the physical rules of the environment and the goals of the agent are the same but the objects in different rooms are quite different.

Our main contribution is the proposed method which employs a deep neural network to solve the problem "to transfer or not to transfer" [15] for video RL tasks since deep learning has been proved as a powerful method to automatically learn features from image data in machine learning [11]. Specifically, we formalise the predicting binary transfer performance problem as a binary classification problem, and adopt a siamese convolutional neural network (CNN) to learn task features from the images and a softmax layer to predict whether transfer is useful between a pair of tasks. Our method only needs the raw task description images which can be acquired by the autonomous agent, thus it can deal with the situations where models and samples of the target tasks as well as man-made features are not available. To evaluate the effectiveness of our method, we conduct extensive experiments in the maze domain and the Ms. PacMan [24] domain. The experimental results show that our method can accurately predict the transfer performance and significantly outperform the baseline methods as well as an existing method using a man-made task feature.

The rest of the paper is organized as follows. In the next section we introduce necessary background for further discussion in subsequent sections. In Sect. 3 we propose our deep learning method which learns an effective classifier to predict the transfer performance of each task pair. We carry out extensive empirical studies in Sect. 4. In particular, we show the effectiveness of our proposed method in the application of the maze domain and the Ms. PacMan domain. Finally, Sect. 5 summarizes contributions of this paper and outlines future research directions.

2   Preliminary

2.1   Reinforcement Learning Tasks

An RL task can be modeled as a Markov decision process: $M = (S, A, P, R)$, where S is the state space; A is the action space; P is the probability transition function; and R is the reward function. The same domain tasks considered in this work have different state spaces but the same action space. The transition and reward functions of the tasks share structural similarity; this means transitions between states are governed by the same environment physics, and rewards are allocated the same way when reaching the same sub-goal.

The goal of an RL agent is to find an optimal policy π^* which maximizes the long-term expected cumulative reward from any initial state $s_0 \in S$. Here, a policy is defined as a mapping, $\pi : S \times A \to [0, 1]$. To learn the optimal policy, many of the RL methods (e.g., temporal difference methods including Q-learning [23] and Sarsa [16]) learn the optimal action-value function (or optimal value function) $Q^*(s, a)$ which estimates the expected cumulative reward of taking action a in state s under policy π^*. In the experiments, Sarsa is used as a representative learning algorithm, while our approach is also suitable for other RL algorithms.

2.2   Transfer Learning Scenario

In this paper, we consider a transfer learning scenario where an agent is facing a potentially infinite random sequence of RL tasks. The agent continually solves tasks and when solving a new task it can reuse the previously learned knowledge to improve the performance on the new task. There are various knowledge transfer methods which reuse different kinds of knowledge. Policy transfer and value function transfer are two of the most commonly used knowledge transfer methods and they have been shown to perform well by initializing the policy or (action-)value function in the target task [20]. For situations where the agent uses a parameterized function approximator, transfer can be achieved by initializing the parameters (e.g., weights of linear methods) in the target task using those learned in a source task.

To evaluate the performance of transfer learning, some measures have been proposed [9,21], including *jumpstart* and *learning speed*. Jumpstart measures the initial performance improvement in the target task. One definition of jumpstart is:

$$jumpstart(m) = R^m_{\pi_t} - R^m_{\pi_0}, \tag{1}$$

where m is the number of episodes for computing jumpstart, π_t is the initial policy with transfer knowledge, π_0 is the initial policy without transfer and R^m_π

is the expected reward after learning for m episodes with policy π. Learning speed measure has been used by a majority of existing literatures. In practice, it can be achieved by *time to threshold*, which computes the reduced learning time needed to achieve a performance threshold in the target task. In this paper, we use the number of the learning episodes to measure time.

Clearly if the source tasks used for transfer learning are dissimilar to the target task, transfer learning will provide no benefit or even cause worse learning performance (negative transfer). Thus it is important to recognize when transfer is useful before actually transferring. In this paper, we focus on learning a model to predict positive/negative transfer between two tasks. When a new task comes, the agent can use the model to acquire a set of tasks which will bring positive transfer. If the task set is not empty, the agent then selects appropriate source tasks from the set for transfer learning.

3 Predicting Transfer Performance

3.1 Problem Formulation and Training Data

Let $\mathcal{T}_{source}$ be a set of source tasks which have been solved and $\mathcal{T}_{target}$ be a set of target tasks which the agent is going to learn. For any task T in $\mathcal{T}_{source}$ or $\mathcal{T}_{target}$, the agent can acquire a description $D \in \mathcal{D}$ of the task, i.e., an image in this paper. However, the agent has no other knowledge of the target task. Given any pair of tasks (T_i, T_j) and their corresponding descriptions (D_i, D_j), where $T_i \in \mathcal{T}_{source}$ and $T_j \in \mathcal{T}_{target}$, our goal is to train a model to predict whether transfer is positive for T_i and T_j. Let $I(T_i, T_j)$ be an indicator which represents the binary performance of transferring the knowledge from T_i to T_j. $I(T_i, T_j) = 1$ indicates positive transfer while $I(T_i, T_j) = -1$ indicates negative transfer. Since I is a binary function, a natural approach is to train a binary classifier model, $h : \mathcal{D} \times \mathcal{D} \to \{-1, 1\}$, to predict its value given the descriptions of the tasks.

For each task pair $T_i, T_i' \in \mathcal{T}_{source}$, we assume that the agent can get their pairwise transfer performance. In practice, the performance data can be acquired by two ways: (1) recording the experiences of previous transfer learning; (2) simulating transfer learning between the already solved tasks. As positive/negative transfer is defined according to a transfer learning measure, we use jumpstart or time to threshold to achieve $I(T_i, T_i')$. Let $I_{ii'} = I(T_i, T_i')$. $I_{ii'}$ is regarded as the label of the task pair (T_i, T_i'), then the set $S = \{(D_i, D_i', I_{ii'}) | D_i, D_i' \in \mathcal{D}, I_{ii'} \in \{-1, 1\}\}$ forms the dataset for learning the model. After the model is learned, the agent can predict I_{ij} given D_i and D_j, where $T_i \in \mathcal{T}_{source}$ and $T_j \in \mathcal{T}_{target}$. As the set $\mathcal{T}_{target}$ may be infinite, the agent can benefit a lot from the learned model by avoiding negative transfer, especially for some real world domains (e.g., Robot learning [5]) where learning in the physical world is costly.

3.2 Predicting Transfer Performance via Deep Neural Network

Our Architecture. In this section, we introduce the details of our classification model. We construct a deep neural network where the input is two images and the output is the binary transfer performance. The architecture is illustrated in Fig. 1. The network consists of two parts: the feature learning part and the classification part. The feature learning part simply learns a function c, which transforms the images into a high-level feature space. Let D_i and D_i' be the input images of the network, then the outputs of the feature learning part are $F_i = c(D_i)$ and $F_i' = c(D_i')$, where $c(\cdot)$ is obtained by a CNN. The CNN in this work has the same architecture as AlexNet. AlexNet contains stacked convolutional layers, max-pooling layers, local normalization layers and fully-connected (FC) layers (more details can be found in [8]). Other CNN architectures can also be adopted, but it is not the focus of our work to study different networks. Hence, we just use one of the effective networks for learning the features. We use a siamese network to simultaneously learn high-level features of the two input images. The siamese network uses the same network c with the same parameters to process the images, thus the learned features are in the same feature space.

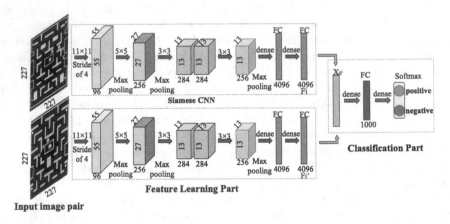

Fig. 1. The network architecture of our proposed model. The size of layers and convolutional filters are shown.

In the classification part, the output vectors of the feature learning part, F_i and F_i', are used as the inputs. We can simply concatenate the two feature vectors to get a new vector $X_{ii'}$. Since the transfer performance of a pair of tasks is related to their differences, an alternative way to obtain $X_{ii'}$ is to subtract the two vectors: $X_{ii'} = F_i - F_i'$. While the concatenated vector maintains more information, the subtracted vector directly provides the differences. We will compare the two vectors in Sect. 4. After $X_{ii'}$ is got, we apply two fully-connected layers. One layer is to map $X_{ii'}$ to a new feature representation $H_{ii'}$, while the other is a softmax layer which contains two units that represent the probabilities of the transfer performance being positive and negative.

Training the Whole Network. We formalise the binary transfer performance prediction problem as a binary classification problem. The training set S is the collection of image pairs labeled as positive transfer and negative transfer. As many classification algorithms, we use cross-entropy loss as classification loss. The optimization objective is to minimize the average loss $\bar{L}$ over all p pairs in the data set: $\bar{L} = -\frac{1}{p} \sum_{i=1}^{p} y_i * \log \hat{y}_i$, where y_i is the true label probability distribution of a pair and $\hat{y}_i$ is the output of the softmax layer. In practice we use a mini-batch method [2] to approximate the objective since the training set may be very large. To train the neural network, we adopt a gradient descent optimization algorithm called Adam (Adaptive Moment Estimation) [6] to update the parameters. Adam can adjust the learning rate adaptively, and it is not easy to plunge into local optima compared with the popular SGD (Stochastic gradient descent) algorithm. As task-level features may be very different from the features of natural images, we train the whole network from scratch instead of using a pre-trained one. The detailed procedure is shown in Algorithm 1.

Algorithm 1. Training the deep neural network

Input: Training set S, batch size m, iterative number I_t, convergence error ε.
Output: Weights and biases of the whole network: (W, b).
1 Initialize W and b.
2 **for** $t = 1, 2, ..., I_t$ **do**
3   Randomly select a batch of m pairs from S:
       $\{(D_{t_1}, D'_{t_1}, I_{t_1 t_1'}), ..., (D_{t_m}, D'_{t_m}, I_{t_m t_m'})\}$.
       // Forward propagation
4   **for** $n = 1, 2, ..., m$ **do**
5   Pass the image pair (D_{t_n}, D'_{t_n}) to the feature leaning part.
6   Get F_n and F'_n, and then compute $X_{nn'}$.
7   Pass $X_{nn'}$ to the classification part, then obtain gradient g_n and
         cross-entropy loss l_n.
       // Back propagation
8   Compute average gradient $\bar{G}_t$ and average loss $\bar{L}_t$
9   Update W and b using Adam algorithm.
10   **if** $(t > 1) \wedge (|\bar{L}_t - \bar{L}_{t-1}| < \varepsilon)$ **then** go to Line 11.
11 **return** (W, b)

---

4 Experimental Evaluation

To empirically evaluate the performance of our proposed method, we conduct extensive experiments in the maze domain and the Ms. PacMan domain.

4.1 Maze Domain

We first consider a simpler domain: the maze domain. The description of a maze can be represented as a map image, thus we can directly apply our method in this domain. An example of the mazes used in our experiment is shown in Fig. 2(a). In a maze, the objective of the learning agent is to find an optimal path to reach the exit from the entrance. In each state, the agent can choose one of the four actions: *up*, *down*, *left*, and *right*. After taking an action, the agent gets to the corresponding neighbor state except when a movement is blocked by an obstacle or the edge of the maze, in which case the agent will stay in its current state. Reward is 100 when arriving at the exit and −1 in other cases.

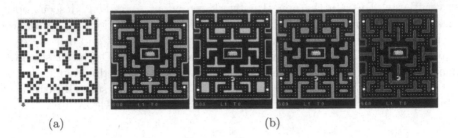

(a)   (b)

Fig. 2. Examples of (a) a 25 × 25 maze and (b) the game Ms. PacMan.

To prepare the training data, we generate 500 mazes, whose size is 25 × 25, with random obstacles but the same entrance and exit (for mazes with different entrances and exits, we can construct a mapping to match their entrances and exits). For each pair of tasks (T_i, T_j), value function transfer which initializes the value function of T_j with that of T_i is performed between them. There are totally 500 × 499 = 249,500 pairs considering the order of tasks. The tasks are learned (with and without transfer) by Sarsa algorithm for 2500 episodes to ensure convergence and the learning processes are repeated for 30 times to get the average learning curves. Then the curves both with and without transfer are used to compute the transfer performance measures. Using which measure to evaluate the transfer performance is determined by the goal of transfer learning. Since the agent can always find the optimal policy in the maze tasks, the goal is often improving the learning speed or achieving a jumpstart improvement. Thus, we use time to threshold and jumpstart measure in this domain. As different measures may bring different labels of the dataset, we respectively obtain labels for the two measures. For time to threshold, we set the threshold to be 1 (the optimal average reward is a bit larger than 1) and compute the reduced number of episodes e for each pair of tasks. Then I_{ij} is got by comparing e with 0:

$$I_{ij} = \begin{cases} 1 & e > 0, \\ -1 & \text{otherwise.} \end{cases} \tag{2}$$

Jumpstart is computed by Eq. (1) where m is set to 1. We do not consider other values of m in the experiments. As the agent keeps learning, the performance after learning for some episodes is not treated as jumpstart. Similar to Eq. (2), I_{ij} is also got by comparing jumpstart with 0.

The datasets that we construct are described in Table 1. 70% of the tasks (350 tasks) form the training task set and the remaining 30% (150 tasks) are used as the test task set. For each measure, we train a network according to Algorithm 1, as different measures may be related to different features. The parameters m, ε and I_t are empirically set to 128, 0.001 and 150, 000, respectively. To find a good implementation of the classification part, we compare the two different ways of computing $X_{ii'}$ (concatenating and subtracting) mentioned in Sect. 3.2, and test two kinds of activation functions for the fully-connected layer which maps $X_{ii'}$ to $H_{ii'}$: one is a ReLU (Rectified Linear Unit) and the other is a linear identical activation function. We use "Con-Linear","Con-ReLU", "Sub-Linear" and"Sub-ReLU" to represent the four combinations.

Table 1. The datasets used in our experiments: the numbers of task pairs in each and the split into positive and negative examples.

Task	Measure	#Pairs	Positive	Negative
Maze	Speed	249,500	210,751	38,749
	Jumpstart	249,500	181,959	67,541
PacMan	Jumpstart	159,600	110,784	48,816

We also compare our method with a baseline method which uses a hand-crafted feature and learns a classifier on the feature. The hand-crafted feature of each maze is got by coding the maze into a $0-1$ matrix where obstacles are coded to 0 and passable blocks are coded to 1. This feature is called code feature. Then the feature of the differences between each pair of tasks is obtained by subtracting two code features. We use 15 classical and commonly used algorithms to build classifiers, containing AdaBoost, Naive Bayes (NB), decision tree (C4.5), random forest, LibSVM and etc which are implemented in Weka [4]. The algorithms are repeated for 30 times and their average performance is acquired. As our goal is to accurately predict the transfer performance, we evaluate the performance of the methods by the classification accuracy.

The results[1] are shown in Table 2. When using concatenated $X_{ii'}$, a linear activation function is better. Conversely, a ReLU is better when using subtracted $X_{ii'}$. The reason can be explained as follows. Some information of the feature vector will be ignored by ReLU. For concatenated $X_{ii'}$, some features of one task may be ignored so that the corresponding differences of the two tasks cannot be acquired. For subtracted $X_{ii'}$, differences have been computed and ReLU can filter some useless features and achieve nonlinearity so as to improve the accuracy. Overall, Con-Linear performs best since concatenated vector maintains more information. Besides, the performance of Con-Linear and Sub-ReLU is close, and they both outperform the baseline method for both measures. This shows the effectiveness of our method which automatically learns task features.

---

[1] Due to the limited space, we only report the results of the best classifier for the baseline method. We do the same in the next subsection.

Table 2. Comparisons of the different methods in the maze domain. For the baseline method, only the results of the best classifiers are presented.

Measure	Con-Linear	Con-ReLU	Sub-Linear	Sub-ReLU	Code+NB
Learning speed	**89.6%**	64.7%	82.4%	**88.7%**	86.1%
Jumpstart	**88.1%**	86.6%	80.3%	**87.8%**	80.9%

4.2 Ms. PacMan Domain

The second experiment is conducted in a more complex domain: a famous video game called Ms. PacMan. As shown in Fig. 2(b), the games are played in several different mazes. Similar to the maze domain, the description of a game can also be represented as a image and the differences between any two games can be extracted from the descriptions. The action space of the Ms. PacMan consists of four actions, *up*, *down*, *left*, and *right*. The goal of the PacMan is to get as many scores as possible by eating edible items such as pills and avoiding ghosts. The scores of eating a pill and a power pill are 10 and 50, respectively. When the PacMan eats a power pill, all ghosts will be edible during a period of time and the scores of eating a first ghost is 200 and will be doubled for each additional ghost that is eaten. When all pills and power pills are eaten, a ghost catches the PacMan or the time limit (2000 time steps) is reached, the game is over.

We generate different Ms. PacMan games by varying 4 game parameters, which are described in Table 3. We totally get 400 different variations of the game. The ghosts behave as "standard ghosts", which usually chase the PacMan and run away when edible. The PacMan learns to play the games from scratch for 2500 episodes by a function approximation method, Sarsa(λ) implemented in [19] which uses 7 heavily engineered distance-related features. We also implement value function transfer for each pair of tasks, which transfers the parameters learned in a source task to initialize the function's parameters in the target task. To compute a reliable estimate for each point of the performance curve, the agent repeats to perform each policy for many times. Specifically, the agent freezes its policy after each episode and plays the game for additional 10 times to compute the average score. The learning process is also repeated for 10 times. Then the agent needs to learn thousands of million episodes for transfer learning between all pairs of tasks. A learning episode (i.e., playing a full game) typically costs $0.5 - 1$ s on our computers, thus all the learning processes in this experiment may cost huge amount of time. Therefore, we use the jumpstart measure as it only requires learning on the target task for 1 episode after transfer.

The constructed dataset is described in Table 1. We implement Con-Linear, which performs best in the previous experiment, on the dataset. Other settings are the same as the previous section. We compare our method with 3 baseline methods using different hand-crafted features: SIFT (scale-invariant feature transform), HOG (histogram of oriented gradient) and LBP (local binary patterns), which are widely used in image processing domain. We also compare our method with the most similar related work proposed in [17], which cre-

Table 3. The parameters of different variations of the Ms. PacMan game

Parameter	Description
Number of ghosts	The number varies from 1 to 4
Maze	There are 4 different mazes, shown in Fig. 2(b)
Power pills	Originally there are 4 power pills
	Delete 1 to generate a new game
Pills	Originally the maze is full of pills. Evenly divide the maze into 4 parts, then delete the pills in 1 part to generate a new game

ated the task feature for PacMan tasks (we call it Sinapov feature) by humans. We compute the training features of the methods using the equation in [17]: $F_{ii'}^{train} = \frac{F_i - F_i'}{\max\{F_i, \xi\}}$, where F_i and F_i' are two feature vectors and ξ is a small number used to avoid being divided by 0. Sinapov feature is used to train the 15 classifiers as the previous experiment does. Since the dimensions of the three features (SIFT, HOG and LBP) are high, we use 6 different types of incremental learning methods implemented in Weka to train classifiers for them. The methods are NB, kNN, RILB (Raced Incremental Logit Boost), DMNB (Discriminative Multinomial Naive Bayes), SPegasos (Primal Estimated sub-GrAdient SOlver for SVM) and LWL (Locally weighted learning). The algorithms are also run for 30 times to compute their average classification accuracies.

Table 4. Comparisons of the different methods in the Ms. PacMan domain.

Method	Sinapov+C4.5	SIFT+DMNB	LBP+DMNB	HOG+DMNB	Con-Linear
Accuracy	81.9%	76.8%	70.0%	70.8%	**96.3%**

The results are shown in Table 4. As expected, the method using the task-related feature defined by experienced humans performs better than those using simply image-related hand-crafted features. Overall, our proposed method significantly outperforms all other methods. Its accuracy on the test set is as high as 96.3% and higher than the second best one by more than 10%. The results show that the features learned from images by our deep learning method are better than hand-crafted features for a specific transfer learning measure, so as to achieve better classification accuracy.

5 Conclusion

In this paper, we propose an effective method to handle the problem "to transfer or not to transfer". Our method uses image descriptions to train a deep neural network to predict the transfer performance between a source task and a target task without other prior knowledge. Our neural network contains two parts:

the feature learning part which uses a siamese CNN to learn high-level features of a pair of images and the classification part which realizes the classification function. Extensive experiments are conducted in the maze domain and the Ms. PacMan domain. Results show that our method can accurately predict the transfer performance and significantly outperform other methods.

In the future, we plan to design more appropriate networks which can better catch task features so as to achieve higher accuracy. In addition, we hope to generalize our method to predict continuous transfer performance which can be used to choose the best source task. Our main idea is to automatically encode the raw task descriptions and then find their differences, thus we hope to generalize our method to apply to tasks with other task descriptions.

Acknowledgements. This work was supported by the National Key R&D Program of China [2017YFB0702600, 2017YFB0702601] and the National Natural Science Foundation of China [grant numbers 61432008, U1435214, 61503178].

References

1. Ammar, H.B., et al.: An automated measure of MDP similarity for transfer in reinforcement learning. In: Workshop at the 28th AAAI Conference on Artificial Intelligence (2014)
2. Bengio, Y.: Practical recommendations for gradient-based training of deep architectures. In: Montavon, G., Orr, G.B., Müller, K.-R. (eds.) Neural Networks: Tricks of the Trade. LNCS, vol. 7700, pp. 437–478. Springer, Heidelberg (2012). https://doi.org/10.1007/978-3-642-35289-8_26
3. Fitzgerald, T., Goel, A., Thomaz, A.: Human-robot co-creativity: task transfer on a spectrum of similarity. In: Proceedings of 8th International Conference on Computational Creativity (2017)
4. Frank, E., Hall, M.A., Witten, I.H.: The WEKA workbench. In: Online Appendix for "Data Mining: Practical Machine Learning Tools and Techniques", 4th edn. Morgan Kaufmann, Los Altos (2016)
5. Hanna, J.P., Stone, P.: Grounded action transformation for robot learning in simulation. In: Proceedings of the 31st AAAI Conference on Artificial Intelligence, pp. 3834–3840. AAAI Press (2017)
6. Kingma, D.P., Ba, J.: Adam: A method for stochastic optimization. CoRR abs/1412.6980 (2014)
7. Konidaris, G., Scheidwasser, I., Barto, A.G.: Transfer in reinforcement learning via shared features. J. Mach. Learn. Res. **13**(1), 1333–1371 (2012)
8. Krizhevsky, A., Sutskever, I., Hinton, G.E.: Imagenet classification with deep convolutional neural networks. In: Advances in Neural Information Processing Systems, pp. 1106–1114 (2012)
9. Lazaric, A.: Transfer in reinforcement learning: a framework and a survey. In: Wiering, M., van Otterlo, M. (eds.) Reinforcement Learning: State-of-the-Art. Adaptation, Learning, and Optimization, vol. 12, pp. 143–173. Springer, Heidelberg (2012). https://doi.org/10.1007/978-3-642-27645-3_5
10. Lazaric, A., Restelli, M., Bonarini, A.: Transfer of samples in batch reinforcement learning. In: Proceedings of the 25th International Conference on Machine Learning, pp. 544–551. ACM (2008)

11. Lecun, Y., Bengio, Y., Hinton, G.: Deep learning. Nature **521**(7553), 436–444 (2015)
12. Mehta, N., Natarajan, S., Tadepalli, P., Fern, A.: Transfer in variable-reward hierarchical reinforcement learning. Mach. Learn. **73**(3), 289–312 (2008)
13. Mousavi, A., Araabi, B.N., Ahmadabadi, M.N.: Context transfer in reinforcement learning using action-value functions. Comput. Intell. Neurosci. **2014**, 428567 (2014)
14. Pan, J., Wang, X., Cheng, Y., Cao, G.: Multi-source transfer ELM-based Q learning. Neurocomputing **137**, 57–64 (2014)
15. Rosenstein, M.T., Marx, Z., Kaelbling, L.P., Dietterich, T.G.: To transfer or not to transfer. In: NIPS Workshop on Inductive Transfer: 10 Years Later (2005)
16. Rummery, G.A., Niranjan, M.: On-line q-learning using connectionist systems. Technical report, University of Cambridge (1994)
17. Sinapov, J., Narvekar, S., Leonetti, M., Stone, P.: Learning inter-task transferability in the absence of target task samples. In: Proceedings of the 14th International Conference on Autonomous Agents and Multiagent Systems, pp. 725–733. ACM (2015)
18. Song, J., Gao, Y., Wang, H., An, B.: Measuring the distance between finite Markov decision processes. In: Proceedings of the 15th International Conference on Autonomous Agents and Multiagent Systems, pp. 468–476. ACM (2016)
19. Taylor, M.E., Carboni, N., Fachantidis, A., Vlahavas, I.P., Torrey, L.: Reinforcement learning agents providing advice in complex video games. Connect. Sci. **26**(1), 45–63 (2014)
20. Taylor, M.E., Stone, P.: Behavior transfer for value-function-based reinforcement learning. In: Proceedings of the 4th International Joint Conference on Autonomous Agents and Multiagent Systems, pp. 53–59. ACM (2005)
21. Taylor, M.E., Stone, P.: Transfer learning for reinforcement learning domains: a survey. J. Mach. Learn. Res. **10**, 1633–1685 (2009)
22. Taylor, M.E., Stone, P.: An introduction to intertask transfer for reinforcement learning. AI Mag. **32**(1), 15 (2011)
23. Watkins, C., Dayan, P.: Q-learning. Mach. Learn. **8**(3–4), 279–292 (1992)
24. Wikipedia: Ms. PacMan (2018). https://en.wikipedia.org/wiki/Ms._Pac-Man

Structured Sequence Modeling with Graph Convolutional Recurrent Networks

Youngjoo Seo[1], Michaël Defferrard[1], Pierre Vandergheynst[1(✉)],
and Xavier Bresson[2]

[1] Signal Processing Laboratory 2, EPFL, Lausanne, Switzerland
{youngjoo.seo,michael.defferrard,pierre.vandergheynst}@epfl.ch
[2] SCSE, Nanyang Technological University, Singapore, Singapore
xbresson@ntu.edu.sg

Abstract. This paper introduces Graph Convolutional Recurrent Network (GCRN), a deep learning model able to predict structured sequences of data. Precisely, GCRN is a generalization of classical recurrent neural networks (RNN) to data structured by an arbitrary graph. The structured sequences can represent series of frames in videos, spatio-temporal measurements on a network of sensors, or random walks on a vocabulary graph for natural language modeling. The proposed model combines convolutional neural networks (CNN) on graphs to identify spatial structures and RNN to find dynamic patterns. We study two possible architectures of GCRN, and apply the models to two practical problems: predicting moving MNIST data, and modeling natural language with the Penn Treebank dataset. Experiments show that exploiting simultaneously graph spatial and dynamic information about data can improve both precision and learning speed.

Keywords: Graph neural networks · Recurrent neural networks
Language modeling

1 Introduction

Many real-world data can be cast as structured sequences, with spatio-temporal sequences being a special case. A well-studied example of spatio-temporal data are videos, where succeeding frames share temporal and spatial structures. Many works, such as [6,10,22], leveraged a combination of CNN and RNN to exploit such spatial and temporal regularities. Their models are able to process possibly time-varying visual inputs for variable-length prediction. These neural network architectures consist of combining a CNN for visual feature extraction followed by a RNN for sequence learning. Such architectures have been successfully used for video activity recognition, image captioning and video description.

More recently, interest has grown in properly fusing the CNN and RNN models for spatio-temporal sequence modeling. Inspired by language modeling, Ranzato et al. [17] proposed a model to represent complex deformations and

© Springer Nature Switzerland AG 2018
L. Cheng et al. (Eds.): ICONIP 2018, LNCS 11301, pp. 362–373, 2018.
https://doi.org/10.1007/978-3-030-04167-0_33

motion patterns by discovering both spatial and temporal correlations. They showed that prediction of the next video frame and interpolation of intermediate frames can be achieved by building a RNN-based language model on the visual words obtained by quantizing the image patches. Their highest-performing model, recursive CNN (rCNN), uses convolutions for both inputs and states. Shi et al. [23] then proposed the convolutional LSTM network (convLSTM), a recurrent model for spatio-temporal sequence modeling which uses 2D-grid convolution to leverage the spatial correlations in input data. They successfully applied their model to the prediction of the evolution of radar echo maps for precipitation nowcasting.

The spatial structure of many important problems may however not be as simple as regular grids. For instance, the data measured from meteorological stations lie on a irregular grid, i.e. a network of heterogeneous spatial distribution of stations [25]. More challenging, the spatial structure of data may not even be spatial, as it is the case for social or biological networks [5]. Eventually, the interpretation that sentences can be regarded as random walks on vocabulary graphs, a view popularized by [15], allows us to cast language analysis problems as graph-structured sequence models.

This work leverages on the recent models of [4,17,23] to design the GCRN model for modeling and predicting time-varying graph-based data. The core idea is to merge CNN for graph-structured data and RNN to identify meaningful spatial structures and dynamic patterns simultaneously. We investigate two different types of GCRN architecture by combining graph convolution and recurrent networks on two application: natural language modeling on Penn Treebank [24] and spatio-temporal sequence modeling on moving-MNIST [23]. First, in an application of graph-level classification, i.e. the words are represented as vocabulary graph, we predict the next word by mapping a label to the graph. Here, the graph convolution is used as a diffusion filter on the graph that indicates possible candidates of next word. And the recurrent networks learn the path walks of the words on the vocabulary graph. Second, in an application of node-level prediction, i.e. weather forecasting on meteorological stations, we can predict the next value on each of the nodes. In this case, the hidden states of the recurrent network should keep the same topology of the graph that can both consider spatial and time relation per each of node on the graph. The graph actually gives a prior knowledge of the structural relationship between the nodes, so that we can take benefits on both precision and learning speed.

2 Preliminaries

2.1 Structured Sequence Modeling

Sequence modeling is the problem of predicting the most likely future length-K sequence given the previous J observations:

$$\hat{x}_{t+1}, \ldots, \hat{x}_{t+K} = \underset{x_{t+1},\ldots,x_{t+K}}{\arg\max} \; P(x_{t+1}, \ldots, x_{t+K} | x_{t-J+1}, \ldots, x_t), \qquad (1)$$

where $x_t \in \mathbf{D}$ is an observation at time t and $\mathbf{D}$ denotes the domain of the observed features. The archetypal application being the n-gram language model (with $n = J + 1$), where $P(x_{t+1}|x_{t-J+1}, \ldots, x_t)$ models the probability of word x_{t+1} to appear conditioned on the past J words in the sentence [7].

In this paper, we are interested in special structured sequences, i.e. sequences where features of the observations x_t are not independent but linked by pairwise relationships. Such relationships are universally modeled by weighted graphs.

Data x_t can be viewed as a graph signal, i.e. a signal defined on an undirected and weighted graph $\mathcal{G} = (\mathcal{V}, \mathcal{E}, A)$, where $\mathcal{V}$ is a finite set of $|\mathcal{V}| = n$ vertices, $\mathcal{E}$ is a set of edges and $A \in \mathbb{R}^{n \times n}$ is a weighted adjacency matrix encoding the connection weight between two vertices. A signal $x_t : \mathcal{V} \to \mathbb{R}^{d_x}$ defined on the nodes of the graph may be regarded as a matrix $x_t \in \mathbb{R}^{n \times d_x}$ whose column i is the d_x-dimensional value of x_t at the ith node. While the number of free variables in a structured sequence of length K is in principle $\mathcal{O}(n^K d_x{}^K)$, we seek to exploit the structure of the space of possible predictions to reduce the dimensionality and hence make those problems more tractable.

2.2   Convolutional Neural Networks on Graphs

Generalizing convolutional neural networks (CNNs) to arbitrary graphs is a recent area of interest. Two approaches have been explored in the literature: (i) a generalization of the spatial definition of a convolution [14,16] and (ii), a multiplication in the graph Fourier domain by the way of the convolution theorem [1,4]. Masci et al. [14] introduced a spatial generalization of CNNs to 3D meshes. The authors used geodesic polar coordinates to define convolution operations on mesh patches, and formulated a deep learning architecture which allows comparison across different meshes. Hence, this method is tailored to manifolds and is not directly generalizable to arbitrary graphs. Niepert et al. [16] proposed a spatial approach which may be decomposed in three steps: (i) select a node, (ii) construct its neighborhood and (iii) normalize the selected sub-graph, i.e. order the neighboring nodes. The extracted patches are then fed into a conventional 1D Euclidean CNN. As graphs generally do not possess a natural ordering (temporal, spatial or otherwise), a labeling procedure should be used to impose it. Bruna et al. [1] were the first to introduce the spectral framework described below in the context of graph CNNs. The major drawback of this method is its $\mathcal{O}(n^2)$ complexity, which was overcome with the technique of [4], which offers a linear complexity $\mathcal{O}(|\mathcal{E}|)$ and provides strictly localized filters. Kipf et al. [11] took a first-order approximation of the spectral filters proposed by Defferrard et al. [4] and successfully used it for semi-supervised classification of nodes. While we focus on the framework introduced by [4], the proposed model is agnostic to the choice of the graph convolution operator $*_{\mathcal{G}}$.

As it is difficult to express a meaningful translation operator in the vertex domain [1,16], Defferrard et al. [4] chose a spectral formulation for the convolution operator on graph $*_{\mathcal{G}}$. By this definition, a graph signal $x \in \mathbb{R}^{n \times d_x}$ is filtered by a non-parametric kernel $g_\theta(\Lambda) = \mathrm{diag}(\theta)$, where $\theta \in \mathbb{R}^n$ is a vector of Fourier coefficients, as

$$y = g_\theta *_\mathcal{G} x = g_\theta(L)x = g_\theta(U\Lambda U^T)x = Ug_\theta(\Lambda)U^T x \in \mathbb{R}^{n \times d_x}, \qquad (2)$$

where $U \in \mathbb{R}^{n \times n}$ is the matrix of eigenvectors and $\Lambda \in \mathbb{R}^{n \times n}$ the diagonal matrix of eigenvalues of the normalized graph Laplacian $L = I_n - D^{-1/2}AD^{-1/2} = U\Lambda U^T \in \mathbb{R}^{n \times n}$, where I_n is the identity matrix and $D \in \mathbb{R}^{n \times n}$ is the diagonal degree matrix with $D_{ii} = \sum_j A_{ij}$ [3]. Note that the signal x is filtered by g_θ with an element-wise multiplication of its graph Fourier transform $U^T x$ with g_θ [19]. Evaluating (2) is however expensive, as the multiplication with U is $\mathcal{O}(n^2)$. Furthermore, computing the eigendecomposition of L might be prohibitively expensive for large graphs. To circumvent this problem, [4] parametrizes g_θ as a truncated expansion, up to order $K - 1$, of Chebyshev polynomials T_k such that

$$g_\theta(\Lambda) = \sum_{k=0}^{K-1} \theta_k T_k(\tilde{\Lambda}), \qquad (3)$$

where the parameter $\theta \in \mathbb{R}^K$ is a vector of Chebyshev coefficients and $T_k(\tilde{\Lambda}) \in \mathbb{R}^{n \times n}$ is the Chebyshev polynomial of order k evaluated at $\tilde{\Lambda} = 2\Lambda/\lambda_{max} - I_n$. The graph filtering operation can then be written as

$$y = g_\theta *_\mathcal{G} x = g_\theta(L)x = \sum_{k=0}^{K-1} \theta_k T_k(\tilde{L})x, \qquad (4)$$

where $T_k(\tilde{L}) \in \mathbb{R}^{n \times n}$ is the Chebyshev polynomial of order k evaluated at the scaled Laplacian $\tilde{L} = 2L/\lambda_{max} - I_n$. Using the stable recurrence relation $T_k(x) = 2xT_{k-1}(x) - T_{k-2}(x)$ with $T_0 = 1$ and $T_1 = x$, one can evaluate (4) in $\mathcal{O}(K|\mathcal{E}|)$ operations, i.e. linearly with the number of edges. Note that as the filtering operation (4) is an order K polynomial of the Laplacian, it is K-localized and depends only on nodes that are at maximum K hops away from the central node, the K-neighborhood. The reader is referred to [4] for details and an in-depth discussion.

3   Related Works

Shi *et al.* [23] introduced a model for regular grid-structured sequences, which can be seen as a special case of the proposed model where the graph is an image grid where the nodes are well ordered. Their model uses convolutional kernels W, instead of classical fully-connected matrix multiplication:

$$
\begin{aligned}
i_t &= \sigma(W_{xi} * x_t + W_{hi} * h_{t-1} + w_{ci} \odot c_{t-1} + b_i), \\
f_t &= \sigma(W_{xf} * x_t + W_{hf} * h_{t-1} + w_{cf} \odot c_{t-1} + b_f), \\
c_t &= f_t \odot c_{t-1} + i_t \odot \tanh(W_{xc} * x_t + W_{hc} * h_{t-1} + b_c), \qquad (5) \\
o_t &= \sigma(W_{xo} * x_t + W_{ho} * h_{t-1} + w_{co} \odot c_t + b_o), \\
h_t &= o_t \odot \tanh(c_t),
\end{aligned}
$$

where $i, f, o \in [0,1]^{d_h}$ are the input, forget and output gates in LSTM [8], $\odot$ denotes the Hadamard product, $\sigma(\cdot)$ the sigmoid function, and $*$ denotes the 2D convolution by a set of kernels. In their setting, the input tensor $x_t \in \mathbb{R}^{n_r \times n_c \times d_x}$ is the observation of d_x measurements at time t of a dynamical system over a spatial region represented by a grid of n_r rows and n_c columns. The model holds spatially distributed hidden and cell states of size d_h given by the tensors $c_t, h_t \in \mathbb{R}^{n_r \times n_c \times d_h}$. The size m of the convolutional kernels $W_h. \in \mathbb{R}^{m \times m \times d_h \times d_h}$ and $W_x. \in \mathbb{R}^{m \times m \times d_h \times d_x}$ determines the number of parameters, which is independent of the grid size $n_r \times n_c$. Earlier, Ranzato et $al.$ [17] proposed a similar RNN variation which uses convolutional layers instead of fully connected layers. The hidden state at time t is given by

$$h_t = \tanh(\sigma(W_{x2} * \sigma(W_{x1} * x_t)) + \sigma(W_h * h_{t-1})), \tag{6}$$

where the convolutional kernels $W_h \in \mathbb{R}^{d_h \times d_h}$ are restricted to filters of size 1×1 (effectively a fully connected layer shared across all spatial locations).

Observing that natural language exhibits syntactic properties that naturally combine words into phrases, Tai et $al.$ [21] proposed a model for tree-structured topologies, where each LSTM has access to the states of its children. They obtained state-of-the-art results on semantic relatedness and sentiment classification. Liang et $al.$ [13] followed up and proposed a variant on graphs. Their sophisticated network architecture obtained state-of-the-art results for semantic object parsing on four datasets. In those models, the states are gathered from the neighborhood by way of a weighted sum with trainable weight matrices. Those weights are however not shared across the graph, which would otherwise have required some ordering of the nodes, alike any other spatial definition of graph convolution. Moreover, their formulations are limited to the one-neighborhood of the current node, with equal weight given to each neighbor.

Motivated by spatio-temporal problems like modeling human motion and object interactions, Jain et $al.$ [9] developed a method to cast a spatio-temporal graph as a rich RNN mixture which essentially associates a RNN to each node and edge. Again, the communication is limited to directly connected nodes and edges.

The closest model to our work is probably the one proposed by Li et $al.$ [12], which showed stat-of-the-art performance on a problem from program verification. Whereas they use the iterative procedure of the Graph Neural Networks (GNNs) model introduced by [18] to propagate node representations until convergence, we instead use the graph CNN introduced by [4] to diffuse information across the nodes. While their motivations are quite different, those models are related by the fact that a spectral filter defined as a polynomial of order K can be implemented as a K-layer GNN[1].

---

[1] The basic idea is to set the transition function as a diffusion and the output function such as to realize the polynomial recurrence, then stack K of those. See [4] for details.

4   Proposed GCRN Models

We propose two GCRN architectures that are quite natural, and investigate their performances in real-world applications in Sect. 5.

Model 1. The most straightforward definition is to stack a graph CNN, defined as (4), for feature extraction based on the LSTM:

$$
\begin{aligned}
x_t^{\text{CNN}} &= \text{CNN}_{\mathcal{G}}(x_t) \\
i_t &= \sigma(W_{xi} x_t^{\text{CNN}} + W_{hi} h_{t-1} + w_{ci} \odot c_{t-1} + b_i), \\
f_t &= \sigma(W_{xf} x_t^{\text{CNN}} + W_{hf} h_{t-1} + w_{cf} \odot c_{t-1} + b_f), \\
c_t &= f_t \odot c_{t-1} + i_t \odot \tanh(W_{xc} x_t^{\text{CNN}} + W_{hc} h_{t-1} + b_c), \\
o_t &= \sigma(W_{xo} x_t^{\text{CNN}} + W_{ho} h_{t-1} + w_{co} \odot c_t + b_o), \\
h_t &= o_t \odot \tanh(c_t).
\end{aligned}
\tag{7}
$$

In that setting, the input matrix $x_t \in \mathbb{R}^{n \times d_x}$ represent the observation of d_x measurements at time t of a dynamical system over a network whose organization is given by a graph $\mathcal{G}$. x_t^{CNN} is the output of the graph CNN. For a proof of concept, we simply choose here $x_t^{\text{CNN}} = W^{\text{CNN}} *_{\mathcal{G}} x_t$, where $W^{\text{CNN}} \in \mathbb{R}^{K \times d_x \times d_x}$ are the Chebyshev coefficients for the graph convolutional kernels of support K. Though graph CNN can give output as same structure as input $x_t^{\text{CNN}} \in \mathbb{R}^{n \times d_x}$, but we feed the graph CNN output as fully connected layer with $x_t^{\text{CNN}} \in \mathbb{R}^{n d_x \times 1}$ as feature for LSTM. The model also holds spatially distributed hidden and cell states of size d_h given by the matrices $c_t, h_t \in \mathbb{R}^{d_h}$. Peepholes are controlled by $w_{c.} \in \mathbb{R}^{d_h}$. The weights $W_{h.} \in \mathbb{R}^{d_h \times d_h}$ and $W_{x.} \in \mathbb{R}^{d_h \times n d_x}$ are the parameters of the fully connected layers. An architecture such as (7) may be enough to capture the data distribution by exploiting local stationarity and compositionality properties as well as the dynamic properties.

Model 2. To generalize the convLSTM model (5) to graphs we replace the Euclidean 2D convolution $*$ by the graph convolution $*_{\mathcal{G}}$:

$$
\begin{aligned}
i_t &= \sigma(W_{xi} *_{\mathcal{G}} x_t + W_{hi} *_{\mathcal{G}} h_{t-1} + w_{ci} \odot c_{t-1} + b_i), \\
f_t &= \sigma(W_{xf} *_{\mathcal{G}} x_t + W_{hf} *_{\mathcal{G}} h_{t-1} + w_{cf} \odot c_{t-1} + b_f), \\
c_t &= f_t \odot c_{t-1} + i_t \odot \tanh(W_{xc} *_{\mathcal{G}} x_t + W_{hc} *_{\mathcal{G}} h_{t-1} + b_c), \\
o_t &= \sigma(W_{xo} *_{\mathcal{G}} x_t + W_{ho} *_{\mathcal{G}} h_{t-1} + w_{co} \odot c_t + b_o), \\
h_t &= o_t \odot \tanh(c_t).
\end{aligned}
\tag{8}
$$

In that setting, the support K of the graph convolutional kernels defined by the Chebyshev coefficients $W_{h.} \in \mathbb{R}^{K \times d_h \times d_h}$ and $W_{x.} \in \mathbb{R}^{K \times d_h \times d_x}$ determines the number of parameters, which is independent of the number of nodes n. To keep the notation simple, we write $W_{xi} *_{\mathcal{G}} x_t$ to mean a graph convolution of x_t with $d_h d_x$ filters which are functions of the graph Laplacian L parametrized by K Chebyshev coefficients, as noted in (3) and (4). In a distributed computing setting, K controls the communication overhead, i.e. the number of nodes any given node i should exchange with in order to compute its local states.

As demonstrated by [23], structure-aware LSTM cells can be stacked and used as sequence-to-sequence models using an architecture composed of an encoder, which processes the input sequence, and a decoder, which generates an output sequence. A standard practice for machine translation using RNNs [2,20].

5 Experiments

5.1 Spatio-Temporal Sequence Modeling on Moving-MNIST

For this synthetic experiment, we use the moving-MNIST dataset generated by [23]. All sequences are 20 frames long (10 frames as input and 10 frames for prediction) and contain two handwritten digits bouncing inside a 64×64 patch. Following their experimental setup, all models are trained by minimizing the binary cross-entropy loss using back-propagation through time (BPTT) and RMSProp with a learning rate of 10^{-3} and a decay rate of 0.9. We choose the best model with early-stopping on validation set. All implementations are based on their Theano code and dataset.[2] The adjacency matrix A is constructed as a k-nearest-neighbor (knn) graph with Euclidean distance and Gaussian kernel between pixel locations. For a fair comparison with [23] defined in (5), all GCRN experiments are conducted with Model 2 defined in (8), which is the same architecture with the 2D convolution $*$ replaced by a graph convolution $*_\mathcal{G}$. To further explore the impact of the isotropic property of our filters, we generated a variant of the moving MNIST dataset where digits are also rotating (see Fig. 2).

Table 1. Comparison between models. Runtime is the time spent per each mini-batch in seconds. Test cross-entropies correspond to moving MNIST, and rotating and moving MNIST. LSTM+GCNN is Model 2 defined in (8). Cross-entropy of FC-LSTM is taken from [23].

Architecture	Structure	Filter size	Parameters	Runtime	Test(w/o Rot)	Test(Rot)
FC-LSTM	N/A	N/A	$142,667,776$	N/A	4832	-
LSTM+CNN	N/A	5×5	$13,524,496$	2.10	3851	4339
LSTM+CNN	N/A	9×9	$43,802,128$	6.10	3903	4208
LSTM+GCNN	$knn = 8$	$K = 3$	$1,629,712$	0.82	3866	4367
LSTM+GCNN	$knn = 8$	$K = 5$	$2,711,056$	1.24	3495	3932
LSTM+GCNN	$knn = 8$	$K = 7$	$3,792,400$	1.61	**3400**	**3803**
LSTM+GCNN	$knn = 8$	$K = 9$	$4,873,744$	2.15	3395	3814
LSTM+GCNN	$knn = 4$	$K = 7$	$3,792,400$	1.61	3446	3844
LSTM+GCNN	$knn = 16$	$K = 7$	$3,792,400$	1.61	3578	3963

Table 1 shows the performance of various models: (i) the baseline fully-connected LSTM (FC-LSTM) from [23], (ii) the 1-layer LSTM+CNN from [23]

---

[2] http://www.wanghao.in/code/SPARNN-release.zip.

with different filter sizes, and (iii) the proposed LSTM+graph CNN(GCNN) defined in (8) with different supports K. These results show the ability of the proposed method to capture spatio-temporal structures. Perhaps surprisingly, GCNNs can offer better performance than regular CNNs, even when the domain is a 2D grid and the data is images, the problem CNNs were initially developed for. The explanation is to be found in the differences between 2D filters and spectral graph filters. While a spectral filter of support $K = 3$ corresponds to the reach of a patch of size 5×5, the difference resides in the isotropic nature of the former and the number of parameters: $K = 3$ for the former and $5^2 = 25$ for the later. Table 1 indeed shows that LSTM+CNN(5×5) rivals LSTM+GCNN with $K = 3$. However, when increasing the filter size to 9×9 or $K = 5$, the GCNN variant clearly outperforms the CNN variant. This experiment demonstrates that graph spectral filters can obtain superior performance on regular domains with much less parameters thanks to their isotropic nature, a controversial property. Indeed, as the nodes are not ordered, there is no notion of an edge going up, down, on the right or on the left. All edges are treated equally, inducing some sort of rotation invariance. Additionally, Table 1 shows that the computational complexity of each model is linear with the filter size, and Fig. 1 shows the learning dynamic of some of the models.

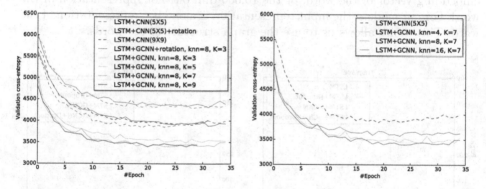

Fig. 1. Cross-entropy on validation set: Left: performance of graph CNN with various filter support K. Right: performance w.r.t. graph construction.

5.2 Natural Language Modeling on Penn Treebank

The Penn Treebank dataset has 1,036,580 words. It was pre-processed in [24] and split[3] into a training set of 929k words, a validation set of 73k words, and a test set of 82k words. The size of the vocabulary of this corpus is 10,000. We use the gensim library[4] to compute a word2vec model [15] for embedding the words of

---

[3] https://github.com/wojzaremba/lstm.

[4] https://radimrehurek.com/gensim/models/word2vec.html.

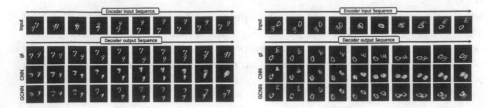

Fig. 2. Qualitative results for moving MNIST, and rotating and moving MNIST. First row is the input sequence, second the ground truth, and third and fourth are the predictions of the LSTM+CNN(5×5) and LSTM+GCNN($knn = 8, K = 7$).

the dictionary in a 200-dimensional space. Then we build the adjacency matrix of the word embedding using a 4-nearest neighbor graph with cosine distance. We used the hyperparameters of the small configuration given by the code[5] based on [24]: the size of the data mini-batch is 20, the number of temporal steps to unroll is 20, the dimension of the hidden state is 200. The global learning rate is 1.0 and the norm of the gradient is bounded by 5. The learning decay function is selected to be $0.5^{\max(0, \#\text{epoch}-4)}$. All experiments have 13 epochs, and dropout value is 0.75. For [24], the input representation x_t can be either the 200-dim embedding vector of the word, or the 10,000-dim one-hot representation of the word. For our models, the input representation is a one-hot representation of the word. This choice allows us to use the graph structure of the words.

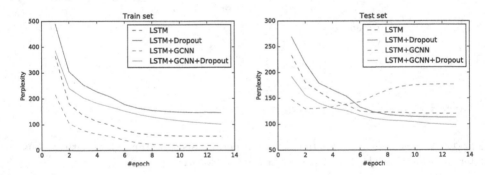

Fig. 3. Learning dynamic of LSTM with and without graph structure and dropout regularization.

Table 2 reports the final train and test perplexity values for each investigated model and Fig. 3 plots the perplexity value vs. the number of epochs for the train and test sets with and without dropout regularization. Numerical experiments show:

---

5 https://github.com/tensorflow/tensorflow/blob/master/tensorflow/models/rnn/ptb/ptb_word_lm.py.

Table 2. Comparison of models in terms of perplexity. [24] code (See footnote 5) is ran as benchmark algorithm. The original [24] code used as input representation for x_t the 200-dim embedding representation of words, computed here by the gensim library (See footnote 3). As our model runs on the 10,000-dim one-hot representation of words, we also ran [24] code on this representation. We re-implemented [24] code with the same architecture and hyperparameters. We remind that GCRN-M1 refers to GCRN Model 1 defined in (7).

Architecture	Representation	Parameters	Train perplexity	Test perplexity
Zaremba *et al.* [24] code (See footnote 5)	embedding	681,800	36.96	117.29
Zaremba *et al.* [24] code (See footnote 5)	one-hot	34,011,600	53.89	118.82
LSTM	embedding	681,800	48.38	120.90
LSTM	one-hot	34,011,600	54.41	120.16
LSTM, dropout	one-hot	34,011,600	145.59	112.98
GCRN-M1	one-hot	42,011,602	18.49	177.14
GCRN-M1, dropout	one-hot	42,011,602	114.29	**98.67**

1. Given the same experimental conditions in terms of architecture and *no* dropout regularization, the standalone model of LSTM is more accurate than LSTM using the spatial graph information (120.16 vs. 177.14), extracted by graph CNN with the GCRN architecture of Model 1, Eq. (7).
2. However, using dropout regularization, the graph LSTM model overcomes the standalone LSTM with perplexity values 98.67 vs. 112.98.
3. The use of spatial graph information found by graph CNN speeds up the learning process, and overfits the training dataset in the absence of dropout regularization. The graph structure likely acts a constraint on the learning system that is forced to move in the space of language topics.
4. We performed the same experiments with LSTM and Model 2 defined in (8). Model 1 significantly outperformed Model 2, and Model 2 did worse than standalone LSTM. This bad performance may be the result of the large increase of dimensionality in Model 2, as the dimension of the hidden and cell states changes from 200 to 10,000, the size of the vocabulary. A solution would be to downsize the data dimensionality, as done in [23] in the case of image data.

6   Conclusion and Future Work

This work aims at learning spatio-temporal structures from graph-structured and time-varying data. In this context, the main challenge is to identify the best possible architecture that combines simultaneously recurrent neural networks like vanilla RNN, LSTM or GRU with convolutional neural networks for graph-structured data. We have investigated here two architectures, one using a stack of CNN and RNN (Model 1), and one using convLSTM that considers convolutions instead of fully connected operations in the RNN definition (Model 2). We have then considered two applications: video prediction and natural language modeling. Model 2 has shown good performances in the case of video prediction, by improving the results of [23]. Model 1 has also provided promising performances in the case of language modeling, particularly in terms of learning speed. It has been shown that (i) isotropic filters, maybe surprisingly, can outperform classical 2D filters on images while requiring much less parameters, and (ii) that graphs coupled with graph CNN and RNN are a versatile way of introducing and exploiting side-information, e.g. the semantic of words, by structuring a data matrix.

Future work will investigate applications to data naturally structured as dynamic graph signals, for instance fMRI and sensor networks. The graph CNN model we have used is rotationally-invariant and such spatial property seems quite attractive in real situations where motion is beyond translation. We will also investigate how to benefit of the fast learning property of our system to speed up language modeling models. Eventually, it will be interesting to analyze the underlying dynamical property of generic RNN architectures in the case of graphs. Graph structures may introduce stability to RNN systems, and prevent them to express unstable dynamic behaviors.

Acknowledgment. This research was supported in part by the European Union's H2020 Framework Programme (H2020-MSCA-ITN-2014) under grant No. 642685 MacSeNet, and Nvidia equipment grant. And XB is supported in part by NRF Fellowship NRFF2017-10.

References

1. Bruna, J., Zaremba, W., Szlam, A., LeCun, Y.: Spectral networks and locally connected networks on graphs. arXiv preprint arXiv:1312.6203 (2013)
2. Cho, K., et al.: Learning phrase representations using RNN encoder-decoder for statistical machine translation. arXiv preprint arXiv:1406.1078 (2014)
3. Chung, F.R., Graham, F.C.: Spectral Graph Theory, no. 92. American Mathematical Society, Providence (1997)
4. Defferrard, M., Bresson, X., Vandergheynst, P.: Convolutional neural networks on graphs with fast localized spectral filtering. In: Advances in Neural Information Processing Systems, pp. 3844–3852 (2016)
5. Doborjeh, Z.G., Kasabov, N., Doborjeh, M.G., Sumich, A.: Modelling peri-perceptual brain processes in a deep learning spiking neural network architecture. Sci. Rep. 8(1), 8912 (2018)
6. Donahue, J., et al.: Long-term recurrent convolutional networks for visual recognition and description. In: Proceedings of the IEEE Conference on Computer Vision and Pattern Recognition, pp. 2625–2634 (2015)
7. Graves, A.: Generating sequences with recurrent neural networks. arXiv preprint arXiv:1308.0850 (2013)
8. Hochreiter, S., Schmidhuber, J.: Long short-term memory. Neural Comput. 9(8), 1735–1780 (1997)
9. Jain, A., Zamir, A.R., Savarese, S., Saxena, A.: Structural-RNN: deep learning on spatio-temporal graphs. In: Proceedings of the IEEE Conference on Computer Vision and Pattern Recognition, pp. 5308–5317 (2016)

10. Karpathy, A., Fei-Fei, L.: Deep visual-semantic alignments for generating image descriptions. In: Proceedings of the IEEE Conference on Computer Vision and Pattern Recognition, pp. 3128–3137 (2015)
11. Kipf, T.N., Welling, M.: Semi-supervised classification with graph convolutional networks. arXiv preprint arXiv:1609.02907 (2016)
12. Li, Y., Tarlow, D., Brockschmidt, M., Zemel, R.: Gated graph sequence neural networks. arXiv preprint arXiv:1511.05493 (2015)
13. Liang, X., Shen, X., Feng, J., Lin, L., Yan, S.: Semantic object parsing with graph LSTM. In: Leibe, B., Matas, J., Sebe, N., Welling, M. (eds.) ECCV 2016. LNCS, vol. 9905, pp. 125–143. Springer, Cham (2016). https://doi.org/10.1007/978-3-319-46448-0_8
14. Masci, J., Boscaini, D., Bronstein, M., Vandergheynst, P.: Geodesic convolutional neural networks on riemannian manifolds. In: Proceedings of the IEEE International Conference on Computer Vision Workshops, pp. 37–45 (2015)
15. Mikolov, T., Chen, K., Corrado, G., Dean, J.: Efficient estimation of word representations in vector space. arXiv preprint arXiv:1301.3781 (2013)
16. Niepert, M., Ahmed, M., Kutzkov, K.: Learning convolutional neural networks for graphs. In: International Conference on Machine Learning, pp. 2014–2023 (2016)
17. Ranzato, M., Szlam, A., Bruna, J., Mathieu, M., Collobert, R., Chopra, S.: Video (language) modeling: a baseline for generative models of natural videos. arXiv preprint arXiv:1412.6604 (2014)
18. Scarselli, F., Gori, M., Tsoi, A.C., Hagenbuchner, M., Monfardini, G.: The graph neural network model. IEEE Trans. Neural Netw. **20**(1), 61–80 (2009)
19. Shuman, D.I., Narang, S.K., Frossard, P., Ortega, A., Vandergheynst, P.: The emerging field of signal processing on graphs: extending high-dimensional data analysis to networks and other irregular domains. IEEE Signal Process. Mag. **30**(3), 83–98 (2013)
20. Sutskever, I., Vinyals, O., Le, Q.V.: Sequence to sequence learning with neural networks. In: Advances in Neural Information Processing Systems, pp. 3104–3112 (2014)
21. Tai, K.S., Socher, R., Manning, C.D.: Improved semantic representations from tree-structured long short-term memory networks. arXiv preprint arXiv:1503.00075 (2015)
22. Vinyals, O., Toshev, A., Bengio, S., Erhan, D.: Show and tell: a neural image caption generator. In: Proceedings of the IEEE Conference on Computer Vision and Pattern Recognition, pp. 3156–3164 (2015)
23. Shi, X., Chen, Z., Wang, H., Yeung, D.Y., Wong, W.K., Woo, W.C.: Convolutional LSTM network: a machine learning approach for precipitation nowcasting. In: Advances in Neural Information Processing Systems, pp. 802–810 (2015)
24. Zaremba, W., Sutskever, I., Vinyals, O.: Recurrent neural network regularization. arXiv preprint arXiv:1409.2329 (2014)
25. Ziat, A., Delasalles, E., Denoyer, L., Gallinari, P.: Spatio-temporal neural networks for space-time series forecasting and relations discovery. In: 2017 IEEE International Conference on Data Mining (ICDM), pp. 705–714. IEEE (2017)

Part-Level Sketch Segmentation
and Labeling Using Dual-CNN

Xianyi Zhu[1], Yi Xiao[1(✉)], and Yan Zheng[2]

[1] College of Computer Science and Electronic Engineering, Hunan University,
Changsha 410082, People's Republic of China
yixiao_csee@hnu.edu.cn
[2] College of Electrical and Information Engineering, Hunan University,
Changsha 410082, People's Republic of China

Abstract. Part-level sketch segmentation and labeling refers to segment an object sketch to semantic component parts. It is a hard task since sketches carry much fewer features than natural images. Inspired by the neural networks used in sketch classification, which shows the performance of the network is significantly affected by the kernel size, we propose a dual-convolutional neural network (CNN) method to tackle automatic sketch segmentation and labeling. The dual-CNN model contains two CNNs, one with large-size convolutional kernels to process long sketches, the other with small-size kernels to work on short ones. Both CNNs have three convolutional layers and three fully connection layers. Except for the first convolutional layer, the rest configurations of these two CNNs are same. To further enhance the performance of the method, we model position and orientation as a triple-channel input of our networks by fusing the minimal oriented rectangle bounding boxes (MORBB) of stroke and its host sketch as masks. Extensive experimental results verify our method and demonstrate that our approach outperforms state of the art.

Keywords: Part-level sketch segmentation · Sketch labeling
Stroke classification · Dual convolutional neural networks

1 Introduction

Freehand sketch is an essential way to depict real entities. Correct semantic understanding of sketches is beneficial for designing user-friendly interfaces. Many researchers have studied sketch understanding, including sketch classification [3,14,27,33], sketch-based photo retrieval [6,25], sketch-based modeling [21], sketch-based animation [17], etc.

Segmenting a sketch into semantic component objects (see Fig. 1) is a challenging sketch understanding task for computer. A sketch can only provide limited information due to its simple drawn strokes without color and texture. Moreover, strokes of the same semantic label can have huge shape differences. Only

© Springer Nature Switzerland AG 2018
L. Cheng et al. (Eds.): ICONIP 2018, LNCS 11301, pp. 374–384, 2018.
https://doi.org/10.1007/978-3-030-04167-0_34

two works [9, 28] have investigated this problem so far. Huang et al. [9] first introduce this problem and give an interactive solution via registering sketches with labeled 3D meshes. Their approach is not suitable for abstract sketches which do not have corresponding 3D models. In addition, because existing algorithms for automatic viewpoint estimation are not accurate enough, manual interactive operations are necessary to conduct an optimal registration. Schneider and Tuytelaars [28] propose a fully automatic method that is appropriate even for abstract sketches. They adopt handcrafted SIFT feature [16] encoded by Fisher Vectors (FV) [24] to classify strokes and used CRFs [12] to improve the results by creating graphs on the relation to strokes. Their method is constrained by the representation ability of SIFT. In contrast, features learned by deep neural networks are more powerful than the handcrafted features. And the two works take much time to derive segmentation and labeling results. Methods of faster speed and higher accuracy are still required.

Fig. 1. The results are acquired automatically using our method, which are trained on a small training set of 15 to 20 labeled sketches for each category

This problem can be considered as stroke classification. We design two CNNs: the one with large kernels and large pooling is dedicated to long strokes, inspired a refined designed network SketchANet [33] for sketch classification, because those strokes can be regarded as a kind of simple and sparse sketch; the other one with small kernels is dedicated to short strokes. These CNNs can extract discriminable features of stroke effectively. Besides learned features, the position and orientation of stroke should also be taken into account. We explicitly model position and orientation as a triple-channel input that is helpful for CNNs to perceive.

The major contributions of this paper are summarized as follows:

1. We propose a dual-CNN method for automatic sketch segmentation and labeling, which improves performance compared with a single-CNN method.
2. We propose to model position and orientation as a triple-channel input by fusing the MORBBs of sketch and inner stroke as masks, which further improves the performances.
3. Extensive experimental results show that our method outperforms the state-of-the-art methods both on accuracy and runtime.

2   Related Works

We review two most related areas in this section: sketch classification and sketch segmentation and labeling.

Sketch Classification: Early works utilize handcrafted features for sketch classification since they regard the sketches as texture images. Eitz et al. [3] are the first to address object sketch classification: they adopt bag of features [20] to encode dense SIFT features extracted from sketch patches. After that, Schneider and Tuytelaars [27] leverage more powerful FV [24] to encode dense SIFT. Li et al. [14] explore to fuse multiple different features using multiple-kernel learning. Compared with the handcrafted features, the features acquired from deep convolutional neural networks are more powerful. Yu et al. [33,34] design a refined convolutional network architecture for sketch classification, which first outperforms the recognition accuracy of human. Recently, researchers [7,26] consider a sketch as a set of sequential strokes, in which deep features of strokes are processed by the recurrent neural network (RNN) models [2,18,22]. Jia et al. [10] combine deep feature extracted by [33] with shape context feature [1] encoded by local-constraint linear coding [31] as the input of RNN.

Sketch Segmentation and Labeling: A few studies have been conducted for sketch segmentation. Noris et al. [19] use a Markov random field model to segment sketches, but this model requires users to provide auxiliary scribbles. Sun et al. [30] propose to extract objects from scenes with drawing sequence, using clip-art images as prior knowledge. The two works are less related to ours. Huang et al. [9] propose to segment a sketch and label its strokes from a given category by using a part-assembly approach, which matches parts of the sketch to parts of 3D meshes and performs a global optimization afterward. Unfortunately, some abstract sketches may not have corresponding 3D models. Moreover, some sketches need to be manually aligned to the corresponding 3D model. Schneider and Tuytelaars [28] proposed a CRF model to segment and label sketches automatically without user's input, in which the unary classifier uses SIFT feature encoded by FV. Compared with deep convolutional features, handcrafted features are not strong enough. The two works [9,28] are highly related to our method, but the runtime of both two is long which makes their methods not suitable for real-time applications.

3   Proposed Method

We assume that the category of the sketch is known which is the same as [9,28]. Each freehand sketch is composed of various strokes, which can be easily indexed in stroke domain. We traverse each stroke and generate its corresponding rasterized image. Then the sketch segmentation and labeling problem can be considered as stroke classification. To tackle this problem, We design two CNNs, one for long strokes and the other one for short strokes. Each CNN has the same configuration except kernel size in the first convolutional layer. Considering the label of a stroke also depends on its position and orientation, we particularly model those properties as a triple-channel input. The overview of the proposed method is illustrated in Fig. 2.

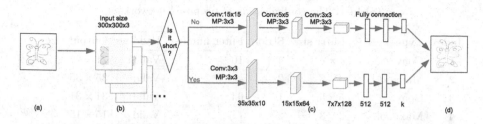

Fig. 2. The overview of proposed method. (a) a sketch (b) 3-channel input images, and each input image is composed of the sketch MORBB mask image (light green), the stroke MORBB mask image (light red) and the stroke image (c) two CNNs have the same architecture except the first convolutional layer (d) the result in which the same labels shaded by a color (Color figure online)

3.1 Modelling Postion and Orientation with Triple Channels

The strokes with similar shapes may belong to different labels when their positions and orientations in a sketch are not same. For instance (the flower in Fig. 1), the sizes and shapes of core, petals and leaves are similar, but the positions and orientations of them are quite different. Meanwhile, the label of a stroke also depends on its host sketch.

We observe that a correct way to determine the orientation and position of sketch or stroke is to judge the orientation and position of its MORBB. Therefore, we explicitly fuse MORBBs into stroke image as masks (see (a) to (b) in Fig. 2) by constructing a triple-channel input: stroke's MORBB mask image (light red), host sketch's MORBB mask image (light green) and stroke image. Specifically, the intensity of masks and stroke is 1, and the rest of intensity is 0. To compute MORBBs, we employ Freeman and Shapira's method [4], which proves the theorem that the smallest-area enclosing rectangle of a polygon has a side collinear with one of the edges of its convex hull. Therefore, the MORBB candidates can be reduced to the number of convex hull edges. We firstly compute the convex hull of sketch or stroke, and then enumerate all MORBB candidates to find the one with the smallest area. The MORBBs are helpful for the convolutional networks to perceive position and orientation and to learn to extract effective features.

3.2 Dual Convolutional Neural Networks

Stroke is similar to sketch as they are sparse lines. From this mind, we construct a CNN with 15×15 kernels in the first convolutional layer (denoted by "N-c15"), inspired by SketchANet [33]. However, we observe that N-c15 may not simultaneously extract effective features of both short and long strokes well. To solve this, we employ another CNN with 3×3 convolutional kernels in the first layer (denoted by "N-c3") to process short strokes exclusively. The rest configuration of N-c3 is same as N-c15, and "N-c3c15" denotes the dual-CNN model. Both CNNs consist of three convolutional layers and three fully connection layers, each with ReLU units. To remain short strokes, the input size of our networks is 300×300. The detailed architecture is listed in Table 1.

Table 1. The configuration of proposed networks

Index	Type	Filter size	Stride	Filter number	Padding	Output size
1	Conv	$c \times c$	4	10	Valid	71×71
2	Maxpool	3×3	2	-	Valid	35×35
3	Conv	5×5	1	64	Valid	31×31
4	Maxpool	3×3	2	-	Valid	15×15
5	Conv	3×3	1	128	Valid	13×13
6	Maxpool	3×3	2	-	Same	7×7
7	Conv (FC)	1×1	-	512	-	1×1
8	Dropout(0.7)	-	-	-	-	1×1
9	Conv (FC)	1×1	-	512	-	1×1
10	Dropout(0.7)	-	-	-	-	1×1
11	Conv (FC)	1×1	-	k	-	1×1

The c can be 3 in N-c3 and 15 in N-c15.

The N-c15 has following two main features different from normal CNNs: **larger first-layer kernels** and **larger pooling**. Many well-known CNNs are equipped with small-size convolutional kernels in the first layer. Early AlexNet [11] exploits 11×11, and later VGG [29] uses 3×3, and ResNet [8] employs 7×7. As [33] explained, because sketches provide none of color and texture information, larger convolutional kernels are indispensable for extracting more structured contexts rather than texture information. Therefore, we adopt 15×15 convolutional kernels in the first layer to acquire structured context such as shape, position and orientation. Most recent CNNs employ 2×2 max pooling with stride 2 to remain some spatial invariance in subsampling [29]. As [33] reported, the 3×3 pooling to provide overlapping areas [11] is more useful to improve performance for sketch recognition, so we take advantage of this pooling in our model.

The learning objective is formulated as a k-class classification problem. For samples $\{x_i\}$ in a batch, we use the softmax cross-entropy loss function, given by $L = - \sum_{i=1}^{n} \sum_{j=1}^{k} y_{ij} \log(y'_{ij})$, where y'_{ij} is the softmax result of the output of single CNN, and y'_{ij} indicates the probability that sample x_i belongs to label j, the notation y_{ij} denotes the ground-truth probability that x_i belongs to label j. For the dual-network method, the two CNNs are trained individually, and then they are gathered during forward-propagation.

We define a threshold θ to distinguish short and long strokes for the dual-CNN method. A short stroke is one that has a rectangular bounding box with height h and width w which are less than θ (namely, $h < \theta$ and $w < \theta$), otherwise it is long. To compute θ, we firstly calculate histogram of maximum value between h and w of each stroke, and then enumerate threshold θ to minimize the intra-class variance using OTSU [23]. Note that single-network method excludes this process.

4   Experiments and Results

4.1   Dataset and Setting

There are two datasets for evaluating the method, the Huang dataset [9] and the Schneider dataset [28] (the subset of TU-Berlin dataset [3]). The Huang dataset consists of 10 classes, and each class has 30 realistic sketches which are associated with 3D meshes. The Schneider dataset consists of 6 classes, and each class has 20 abstract sketches simulating what lay users would draw. To yield segmentation results of all sketches, as done as [28], Huang dataset is uniformly divided into three subsets, two for training and one for testing, and Schneider dataset into four subsets, three for training and one for testing.

As in [9,28], two classical metrics are used to evaluate the accuracy: **pixel metric**, the fraction of pixels that are assigned with the correct labels; and **component metric**, the fraction of correctly labeled original strokes or segmented strokes.

We implement the proposed method using the TensorFlow framework with Python API, and we used the stochastic gradient descent method [13] to train network models. The learning ratio is set to 0.0001, and the number of iterations is 500. We run all our experiments on a desktop PC with an i7-4790k CPU and a GTX1080ti GPU. We rotate the stroke images by $\pm 15, \pm 10, \pm 5°$ to augment the training data.

4.2   Effect of Individual Components

Effect of Kernel Size in the First Layer. Five sizes of kernels are substituted in the first layer. We compare their effects using single networks on Huang dataset, and the results are listed in Table 2. N-c3 and N-c7 with small-sized kernels perform well on component metric in category *Cdlbrm*, *Fourleg* and *Rifle* since these sketches consist of many short strokes. On average, N-c3 performs well on component metric, and N-c15 with large-sized kernels performs well on pixel metric.

Effect of Triple-Channel Modelling and 3×3 Pooling. We choose the single network N-c20 to conduct this experiment on Huang dataset by replacing 3×3 pooling with 2×2 pooling and assigning stroke image without MORBB masks as input. As Table 3 listed, both the results of two operations are inferior.

Effect of Dual Neural Networks. We combine N-c3 with N-c7, N-c11, N-c15 and N-c20 to construct dual networks. We also gather five single networks above to construct penta networks ("N-cs") that employ multi-threshold method [15] to determine which network for inference. We conduct this experiment on Huang dataset, and the results are listed in Table 4. On average, N-c3c15 performs well on both pixel metric and component metric. Comparing with the single network method (Table 2), N-c3c15 outperforms both N-c3 and N-c15 on average.

Table 2. Effect of conv kernel size in first layer, accuracy (%)

	Pixel metric					Component metric				
	N-c3	N-c7	N-c11	N-c15	N-c20	N-c3	N-c7	N-c11	N-c15	N-c20
Airplane	74.1	72.5	74.5	74.5	**76.5**	57.0	56.1	57.0	56.7	**60.0**
Bicycle	74.4	75.6	71.4	72.8	**77.2**	64.3	64.5	59.2	62.2	**65.6**
Cdlbrm	67.3	**68.4**	66.4	66.0	67.2	**70.3**	69.9	66.9	66.6	66.9
Chair	57.3	59.0	**66.8**	**66.8**	**66.8**	53.8	54.4	58.7	57.6	**61.0**
Fourleg	82.9	81.5	81.9	**83.5**	82.7	**77.9**	75.8	75.8	77.0	77.0
Human	**75.7**	73.2	74.1	74.7	73.8	69.3	66.6	69.1	**70.0**	68.9
Lamp	88.5	88.5	85.7	**88.6**	84.6	83.2	82.6	82.6	**85.0**	79.6
Rifle	**59.5**	52.7	50.1	53.2	51.2	**58.6**	54.9	50.4	51.2	52.5
Table	77.1	70.8	79.1	**80.8**	76.2	68.2	64.8	69.3	**70.9**	67.0
Vase	86.5	86.0	87.3	**88.8**	88.2	81.2	81.7	**84.4**	83.9	82.6
Average	74.3	72.8	73.7	**75.0**	74.4	**68.4**	67.1	67.3	68.1	68.1

Table 3. Effect of triple-channel modelling and 3×3 pooling, accuracy (%)

	Pixel metric			Component metric		
	Pooling 2×2	No mask	N-c20	Pooling 2×2	No mask	N-c20
Airplane	63.8	69.3	**76.5**	53.0	52.4	**60.0**
Bicycle	68.6	75.9	**77.2**	57.8	63.0	**65.6**
Cdlbrm	**67.6**	65.8	67.2	66.2	65.5	**66.9**
Chair	63.2	62.3	**66.8**	59.3	54.9	**61.0**
Fourleg	**82.9**	80.3	82.7	76.7	74.8	**77.0**
Human	72.4	71.7	**73.8**	66.1	66.4	**68.9**
Lamp	**87.8**	84.3	84.6	**83.8**	77.8	79.6
Rifle	48.5	49.1	**51.2**	47.0	50.1	**52.5**
Table	71.9	72.3	**76.2**	61.3	63.1	**67.0**
Vase	87.2	86.7	**88.2**	**83.0**	82.6	82.6
Average	71.4	71.8	**74.4**	65.4	65.1	**68.1**

4.3 Comparison with State of the Art and CNN Baseline

We compare the proposed method with state of the art and CNN baseline: [9,28], [9]-A (the automatic version of [9]) and fine-tuned SketchANet [33] (CNN baseline).

Huang Dataset. Since both [9] and [28] label segmented strokes but we label original strokes, the comparison on the pixel metric is more credible. As we can see (Table 5), N-c3c15 outperforms [9]-A on most classes except class *Rifle*, and

Table 4. Results by dual-net and penta-net methods, accuracy (%)

	Pixel metric					Component metric				
	N-c3c7	N-c3c11	N-c3c15	N-c3c20	N-cs	N-c3c7	N-c3c11	N-c3c15	N-c3c20	N-cs
Airplane	73.4	76.2	76.2	75.5	**76.8**	57.0	58.5	58.2	58.2	**59.1**
Bicycle	74.8	74.2	74.1	**75.3**	72.7	64.1	**64.3**	64.1	**64.3**	63.6
Cdlbrm	**70.1**	68.1	68.5	67.9	67.5	**71.6**	70.6	70.9	70.6	69.9
Chair	59.7	**66.7**	66.3	62.8	66.2	55.5	58.7	**59.0**	57.6	58.7
Fourleg	83.3	82.9	**84.6**	83.0	82.9	78.5	77.9	**79.1**	77.9	77.6
Human	**75.8**	74.8	75.5	75.5	74.9	**69.3**	68.9	69.1	69.1	68.6
Lamp	**89.1**	85.5	86.7	88.3	86.1	**83.8**	81.4	82.0	83.2	81.4
Rifle	56.6	57.2	**58.2**	57.0	54.0	57.3	57.8	**58.0**	**58.0**	55.9
Table	71.3	78.2	**79.0**	77.2	76.1	65.5	69.0	**69.7**	69.0	67.0
Vase	86.0	86.2	88.3	**88.7**	88.2	81.2	81.7	82.6	**83.0**	**83.0**
Average	74.0	75.0	**75.7**	75.1	74.5	68.4	68.9	**69.3**	69.1	68.5

Best automatic results are in boldface. "N-cs" means N-c3c7c11c15c20.

Table 5. Results on Huang dataset, accuracy (%)

	Pixel metric							Component metric						
	[9]	[9]-A	[28]	[33]	N-c3	N-c15	N-c3c15	[9]	[9]-A	[28]	[33]	N-c3	N-c15	N-c3c15
Airplane	82.4	74.0	55.1	68.6	74.1	74.5	**76.2**	66.2	55.8	48.7	57.0	57.0	56.7	**58.2**
Bicycle	78.2	72.6	**79.7**	75.4	74.4	72.8	74.1	66.4	58.3	**68.6**	65.7	64.3	62.2	64.1
Cdlbrm	72.7	59.0	**72.0**	61.2	67.3	66.0	68.5	56.7	47.1	66.2	67.2	70.3	66.6	**70.9**
Chair	76.5	52.6	66.5	60.1	57.3	**66.8**	66.3	63.1	42.4	**61.6**	54.9	53.8	57.6	59.0
Fourleg	80.2	77.9	81.5	82.6	82.9	83.5	**84.6**	67.2	64.4	74.2	75.8	77.9	77.0	**79.1**
Human	79.1	62.5	69.7	72.1	**75.7**	74.7	75.5	64.0	47.2	63.1	69.5	69.3	**70.0**	69.1
Lamp	92.1	82.5	82.9	73.1	88.5	**88.6**	86.7	89.3	77.6	77.2	71.9	83.2	**85.0**	82.0
Rifle	75.9	66.9	**67.8**	51.7	59.5	53.2	58.2	62.2	51.5	**65.1**	56.5	58.6	51.2	58.0
Table	79.1	67.9	74.5	71.8	77.1	**80.8**	79.0	69.0	56.7	65.6	63.2	68.2	**70.9**	69.7
Vase	71.9	63.2	83.3	83.5	86.5	**88.8**	88.3	63.1	51.8	79.1	82.6	81.2	**83.9**	82.6
Average	78.8	67.9	73.3	70.0	74.3	75.0	**75.7**	66.7	55.3	66.9	66.4	68.4	68.1	**69.3**

Best automatic results are in boldface.

it outperforms [28] on most classes except class *Bicycle*, *Cdlbrm* and *Rifle*. For class *Bicycle* and *Cdlbrm*, our results are near to [28]. For class *Rifle*, our results are not good. We observed the length of strokes is short and their location is close in class *Rifle*, so our CNNs can not extract effective features. On average, N-c3, N-c15 and N-c3c15 all outperform [9]-A, [28] and [33].

Schneider Dataset. The results are listed in Table 6. Since these sketches do not have corresponding 3D mesh models, [9] and [9]-A can't work. For the same reason, pixel metric is more credible than component metric. N-c3c15 outperforms [28] and [33] on average. Comparing with single neural network N-c3 and N-c15, the results produced by N-c3c15 are improved on average.

Table 6. Results on Schneider dataset, accuracy (%)

	Metric	Airplane	Butterfly	Face	Flower	Pineapple	Snowman	Ave.
N-c3	Pixel metric	80.7	86.8	89.0	71.3	95.4	**85.7**	84.8
N-c15		80.9	86.8	90.5	78.3	96.2	81.4	85.7
N-c3c15		**81.6**	**87.5**	**90.9**	**78.8**	**96.9**	81.3	**86.2**
[28]		76.4	77.7	88.9	74.5	**96.9**	85.2	83.3
[33]		70.2	87.0	86.6	70.0	95.6	83.3	82.3
N-c3	Compo. metric	72.6	81.5	79.9	60.4	94.0	76.5	77.5
N-c15		71.9	82.2	81.4	64.0	95.4	73.7	78.1
N-c3c15		75.1	82.2	81.0	66.9	94.4	73.5	78.8
[28]		76.2	78.0	86.0	73.0	96.1	81.7	81.8
[33]		67.2	80.8	79.1	65.0	96.4	76.7	77.7

Best automatic results on pixel metric are in boldface.

4.4   Runtime Efficiency

On Huang dataset, the training time of N-c15 is about 3 min on average for one category, while N-c3 is about 2 min, and the testing time for one sketch is 0.02 s at most. As reported in [28], for [28], automatically training the model for one category takes several hours, and testing a sketch takes on average 2 to 3 min in unoptimized MATLAB implementation; for [9], a 30-minute global optimization step is followed by a 10-minute local optimization.

5   Discussion

The input sketches are clean and contain well-defined strokes, which are same as the assumptions [9]. A clean sketch contains no extra dots or strokes that belong to unseen classes. A well-defined stroke is drawn as a single curve rather than a bunch of nearly overlapping curves.

The results of sketch segmentation and labeling can benefit many applications, such as sketch-based 3D modeling by part assembly [32], generating 3D views from 2D sketches [5], and categorization task [27].

6   Conclusion

We have proposed a dual-CNN method for automatic sketch segmentation and labeling. Two CNNs with similar structure are designed to tackle stroke classification. To integrate the information of position and orientation, the MORBBs of sketch and stroke are treated as masks and fused with black-white stroke image. Experimental results verify our method and demonstrate that our method outperforms state-of-the-art methods over 2.5% when evaluated in existing datasets. Future work includes the design of a unified model to avoid the requirement of training CNNs for each category and to improve performance on the sketches with dense strokes.

Acknowledgements. The work is supported by the National Key Research & Development Program of China (Grant Num.:2018YFB0203904), NSFC from PRC (Grant Num.:61872137, 61502158, 61803150), Hunan NSF (Grant Num.: 2017JJ3042, 2018JJ3067), and China Postdoctoral Foundation (Grant Num.: 2016M590740).

References

1. Belongie, S.J., Malik, J., Puzicha, J.: Shape matching and object recognition using shape contexts. IEEE Trans. Pattern Anal. Mach. Intell. **24**(4), 509–522 (2002)
2. Chung, J., Gülçehre, Ç., Cho, K., Bengio, Y.: Gated feedback recurrent neural networks. In: Bach, F.R., Blei, D.M. (eds.) ICML 2015. PMLR, vol. 37, pp. 2067–2075. MIT Press, Cambridge (2015)
3. Eitz, M., Hays, J., Alexa, M.: How do humans sketch objects? ACM Trans. Graph. **31**(4), 44:1–44:10 (2012)
4. Freeman, H., Shapira, R.: Determining the minimum-area encasing rectangle for an arbitrary closed curve. Commun. ACM **18**(7), 409–413 (1975)
5. Furusawa, C., Fukusato, T., Okada, N., Hirai, T., Morishima, S.: Quasi 3D rotation for hand-drawn characters. In: SIGGRAPH 2014, Posters Proceedings, p. 12:1. ACM Press, New York (2014)
6. Galea, C., Farrugia, R.A.: Forensic face photo-sketch recognition using a deep learning-based architecture. IEEE Sig. Process. Lett. **24**(11), 1586–1590 (2017)
7. He, J., Wu, X., Jiang, Y., Zhao, B., Peng, Q.: Sketch recognition with deep visual-sequential fusion model. In: Liu, Q., et al. (eds.) ACM Multimedia 2017, pp. 448–456. ACM Press, New York (2017)
8. He, K., Zhang, X., Ren, S., Sun, J.: Deep residual learning for image recognition. In: CVPR 2016, pp. 770–778. IEEE Press, New York (2016)
9. Huang, Z., Fu, H., Lau, R.W.: Data-driven segmentation and labeling of freehand sketches. ACM Trans. Graph. **33**(6), 175:1–175:10 (2014)
10. Jia, Q., Yu, M., Fan, X., Li, H.: Sequential dual deep learning with shape and texture features for sketch recognition. CoRR abs/1708.02716 (2017). http://arxiv.org/abs/1708.02716
11. Krizhevsky, A., Sutskever, I., Hinton, G.E.: Imagenet classification with deep convolutional neural networks. Commun. ACM **60**(6), 84–90 (2017)
12. Lafferty, J.D., McCallum, A., Pereira, F.C.N.: Conditional random fields: probabilistic models for segmenting and labeling sequence data. In: Brodley, C.E., Danyluk, A.P. (eds.) ICML 2001, pp. 282–289. Morgan Kaufmann, San Francisco (2001)
13. Léon, B.: Large-scale machine learning with stochastic gradient descent. In: Lechevallier, Y., Saporta, G. (eds.) COMPSTAT 2010, pp. 177–186. Springer, Heidelberg (2010). https://doi.org/10.1007/978-3-7908-2604-3_16
14. Li, Y., Hospedales, T.M., Song, Y., Gong, S.: Free-hand sketch recognition by multi-kernel feature learning. Comput. Vis. Image Underst. **137**, 1–11 (2015)
15. Liao, P., Chen, T., Chung, P.: A fast algorithm for multilevel thresholding. J. Inf. Sci. Eng. **17**(5), 713–727 (2001)
16. Lowe, D.G.: Object recognition from local scale-invariant features. In: ICCV 1999, pp. 1150–1157. IEEE Press, New York (1999)
17. Mao, C., Qin, S.F., Wright, D.K.: A sketch-based gesture interface for rough 3D stick figure animation. In: Jorge, J.A.P., Igarashi, T. (eds.) Sketch Based Interfaces and Modeling 2005, pp. 175–183. Eurographics Association, Geneva (2005)

18. Mikolov, T., Karafiát, M., Burget, L., Cernocký, J., Khudanpur, S.: Recurrent neural network based language model. In: Kobayashi, T., Hirose, K., Nakamura, S. (eds.) INTERSPEECH 2010, pp. 1045–1048. ISCA Press, Singapore (2010)

19. Noris, G., et al.: Smart scribbles for sketch segmentation. Comput. Graph. Forum **31**(8), 2516–2527 (2012)

20. Nowak, E., Jurie, F., Triggs, B.: Sampling strategies for bag-of-features image classification. In: Leonardis, A., Bischof, H., Pinz, A. (eds.) ECCV 2006. LNCS, vol. 3954, pp. 490–503. Springer, Heidelberg (2006). https://doi.org/10.1007/11744085_38

21. Olsen, L., Samavati, F.F., Sousa, M.C., Jorge, J.A.: Sketch-based modeling: a survey. Comput. Graph. **33**(1), 85–103 (2009)

22. van den Oord, A., Kalchbrenner, N., Kavukcuoglu, K.: Pixel recurrent neural networks. In: Balcan, M., Weinberger, K.Q. (eds.) ICML 2016. PMLR, vol. 48, pp. 1747–1756. MIT Press, Cambridge (2016)

23. Otsu, N.: A threshold selection method from gray-level histograms. IEEE Trans. Syst. Man Cybern. **9**(1), 62–66 (1979)

24. Sánchez, J., Perronnin, F., Mensink, T., Verbeek, J.J.: Image classification with the fisher vector: theory and practice. Int. J. Comput. Vis. **105**(3), 222–245 (2013)

25. Sangkloy, P., Burnell, N., Ham, C., Hays, J.: The sketchy database: learning to retrieve badly drawn bunnies. ACM Trans. Graph. **35**(4), 119:1–119:12 (2016)

26. Sarvadevabhatla, R.K., Kundu, J., Babu, R.V.: Enabling my robot to play pictionary: recurrent neural networks for sketch recognition. In: Hanjalic, A., et al. (eds.) ACM Multimedia 2016, pp. 247–251. ACM Press, New York (2016)

27. Schneider, R.G., Tuytelaars, T.: Sketch classification and classification-driven analysis using fisher vectors. ACM Trans. Graph. **33**(6), 174:1–174:9 (2014)

28. Schneider, R.G., Tuytelaars, T.: Example-based sketch segmentation and labeling using CRFs. ACM Trans. Graph. **35**(5), 151:1–151:9 (2016)

29. Simonyan, K., Zisserman, A.: Very deep convolutional networks for large-scale image recognition. CoRR abs/1409.1556 (2014). http://arxiv.org/abs/1409.1556

30. Sun, Z., Wang, C., Zhang, L., Zhang, L.: Free hand-drawn sketch segmentation. In: Fitzgibbon, A., Lazebnik, S., Perona, P., Sato, Y., Schmid, C. (eds.) ECCV 2012. LNCS, vol. 7572, pp. 626–639. Springer, Heidelberg (2012). https://doi.org/10.1007/978-3-642-33718-5_45

31. Wang, J., Yang, J., Yu, K., Lv, F., Huang, T.S., Gong, Y.: Locality-constrained linear coding for image classification. In: CVPR 2010, pp. 3360–3367. IEEE Press, New York (2010)

32. Xie, X., et al.: Sketch-to-design: context-based part assembly. Comput. Graph. Forum **32**(8), 233–245 (2013)

33. Yu, Q., Yang, Y., Liu, F., Song, Y., Xiang, T., Hospedales, T.M.: Sketch-a-net: a deep neural network that beats humans. Int. J. Comput. Vis. **122**(3), 411–425 (2017)

34. Yu, Q., Yang, Y., Song, Y., Xiang, T., Hospedales, T.M.: Sketch-a-net that beats humans. In: Xie, X., Jones, M.W., Tam, G.K.L. (eds.) BMVC 2015, pp. 7.1–7.12. BMVA Press, London (2015)

RE-CNN: A Robust Convolutional Neural Networks for Image Recognition

Zhe Wang[1], Wenhuan Lu[1(✉)], Yuqing He[2], Naixue Xiong[3], and Jianguo Wei[1]

[1] School of Computer Software, Tianjin University, Tianjin, China
wenhuan@tju.edu.cn
[2] School of Electrical and Information Engineering, Tianjin University,
Tianjin, China
[3] School of Computer Science, Tianjin University, Tianjin, China

Abstract. Recent years we have witnessed revolutionary changes, essentially caused by deep learning and Convolutional Neural Networks (CNN). The performance of image recognition by convolutional neural networks has been substantially boosted. Despite the greater success, the selection of the convolution kernel and the strategy of the pooling layer that only consider the local region and ignore the global region remain several major challenges. These problems may lead to a high correlation between the extracted features and the appearance of the over-fitting. To address the problem, in this paper, a novel and robust method to learn a removal correlation CNN (RE-CNN) model is proposed. This model is achieved by introducing and learning removal correlation layers on the basis of the existing high-capacity CNN architectures. Specifically, the removal correlation layer is trained by the reconstructed CNN features (in this paper, the CNN features are outputs of the layer before classifier layer) using canonical correlation analysis (CCA). The original CNN features are projected into a subspace where the reconstructed CNN features are not correlated. Our extensive experiments on MNIST and LFW datasets demonstrate that the proposed RE-CNN model can improve the recognition capabilities of many existing high-capacity CNN architectures.

Keywords: Image recognition · Convolutional neural networks
Removal correlation · Canonical correlation analysis

1 Introduction

Image recognition is a delicate task for machine. At present, in the filed of Artificial Intelligence (AI) [1], image recognition is still a hot issue because of its promising application for pilotless automobile, robot building and so on. For automatic image classification, feature extraction from the image is the most important step. In order to more effectively represent images, many approaches of feature extraction have been proposed. The features extracted by these approaches can be categorized as man-made feature, and the most widely

© Springer Nature Switzerland AG 2018
L. Cheng et al. (Eds.): ICONIP 2018, LNCS 11301, pp. 385–393, 2018.
https://doi.org/10.1007/978-3-030-04167-0_35

used features are Histogram of Oriented Gradient [2] (HOG) and Local Binary Patterns [3] (LBP). These features are useful for the image recognition [4] on small-scale datasets, but for large-scale datasets, it is difficult to find the proper features that contain enough high-level information. It hinders the performance of the classification accuracy.

In recent years, Deep Neural Networks (DNNS) are considered to be the most mainstream way to solve problems in a wide range of areas. In the systems with the imitation of the mammalian vision, deep convolutional neural networks [5] have become the most suitable approaches of feature extractions for most computer vision problems. As a generic feature extractor, CNNs have been challenging the best classification accuracy in many large-scale datasets, avoiding the disadvantages of man-made feature. With a significant increasing in computing power, deeper and more powerful network structures are generated, such as VGGNet [6], and LBCNN [7].

However, none of these structures take into account the redundant information among the features learned by CNN, which may weaken the performance of the network structures. Many previous studies [8,9] show that the Canonical Correlation Analysis [10] (CCA) can retain the maximum of the high-level information and remove the redundant information. Now the existing CNN models and CCA model are combined in this paper to form the RE-CNN model, which not only can retain the traditional high-capability CNN structures, but also remove the redundant information from the boundedness of the CNN structures. The rest of this paper is organized as follows: in Sect. 2, CCA model is reviewed briefly; in Sect. 3, detailed introduction of the proposed RE-CNN model is given; in Sect. 4, the results of different existing CNN models and RE-CNN models on MNIST and LFW are compared. Finally, Sect. 5 gives the conclusion.

2   Related Work

In this section, CCA model is reviewed briefly. Consider two random variables $x \in R^{D_x}$ and $y \in R^{D_y}$ with zero mean. Simultaneously, let the sets $S_x = \{x_1, x_2, ..., x_n\}$ and $S_y = \{y_1, y_2, ..., y_n\}$ be paired. The goal of CCA model is to find a new coordinate for x by choosing a direction $w \in R^{D_x}$ and similarly for y by choosing a direction $v \in R^{D_y}$, such that the correlation between the projection of S_x and S_y on w and v is maximized,

$$p = \max \frac{w^T C_{xy} v}{\sqrt{w^T C_{xx} w} \sqrt{v^T C_{yy} v}} \tag{1}$$

where p is the correlation, $C_{xx} = E[XX^T] = \frac{1}{n} \sum_{i=1}^{n} x_i x_i^T$ and $C_{yy} = E[YY^T] = \frac{1}{n} \sum_{i=1}^{n} y_i y_i^T$ are the within-set covariance matrices, $C_{xy} = E[XY^T] = \frac{1}{n} \sum_{i=1}^{n} x_i y_i^T$ is the between-set covariance matrix. E denoting the empirical expectation. The problem can be reduced to a generalized eigenvalue problem, where w corresponds to the top eigenvector:

$$C_{xx}^{-1} C_{xy} C_{yy}^{-1} C_{yx} w = \lambda^2 w \tag{2}$$

The asymptotic time complexity for CCA model is $O(nd^2) + O(d^3)$ where $d = \max(D_x, D_y)$; $O(nd^2)$ for computing the covariance matrices and $O(d^3)$ for matrix multiplication, inverse and eigenvalue decomposition.

3   The Proposed RE-CNN Model

The ultimate goal of the proposed method is to learn the uncorrelated features in the CNN features models, and improve the performance of image recognition. The whole model framework can be seen in Fig. 1. To be specific, the removal correlation layer is trained by introducing and learning the removal correlation layers on the basis of the existing high-capacity CNN architectures. The removal correlation layer is trained by the reconstructed CNN features (in this paper, the CNN features are outputs of the layer before classifier layer) using canonical correlation analysis (CCA). The original CNN features are projected into a subspace where the reconstructed CNN features are not correlated. In this way, the redundant information in each feature are removed; hence the classification ability can be improved.

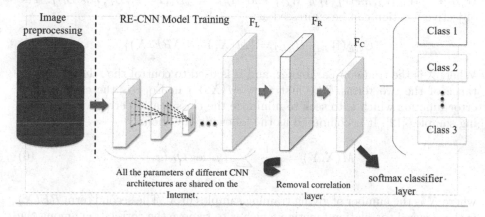

Fig. 1. The framework of the proposed RE-CNN training. There are two steps. The first step is the image preprocessing, it contains the data augmentation. In this paper, we do not care it. In the second step, a new removal correlation layer is learnt and added to the existing high-capacity CNN architectures, in order to train a higher-capacity model.

Model Training: In Fig. 1, in order to remove the correlation among CNN features, a new removal correlation fully layer F_R is added. It uses the output of layer F_L (L is the number of network layers except for the classifier layer) as the input. Compared with the traditional CNN model, in order to enforce that there is no correlation among the learnt RE-CNN features, the proposed RE-CNN model is trained via imposing the CCA criterion on the CNN features.

To reduce the training cost and to avoid the over-fitting, the weights and the biases of $F_1, F_2,...,F_L$ are denoted by $W_1, W_2, ..., W_l$ and $B_1, B_2, ..., B_L$. They are pre-trained on imageset [11]. Then they will be adapted to the recognition tasks (e.g., LFW). After that, the weights and biases can be transferred to the proposed RE-CNN model.

Given a set of initial training samples $X = \{x_1, x_2,\}$. For a training sample $x_i \in X$, let $O_L(x_i)$ be the output of the layer F_L, $O_R(x_i)$ be the output of the layer F_R, $O_C(x_i)$ be the output of the classifier layer F_C, (W_R, B_R) be the new parameters of the layer F_R, and (W_C, B_C) be the new parameters of the layer F_C. $O_L(x_i)_k$ represents the first k features as the outputs of the layer F_L. Finally, the $O_R(x_i)$ and $O_C(x_i)$ can be computed by using the following equations:

$$O_R(x_i) = f(W_R O_L(x_i) + B_R) \tag{3}$$

$$O_C(x_i) = \varphi(W_C O_R(x_k) + B_C) \tag{4}$$

Our object is to train the RE-CNN model with the presentation of high-level features, and to minimize the misclassification on the training dataset. For this purpose, the following objective function is proposed to learn the parameters of $W_{RE} = \{W_1, W_2, ..., W_L, W_R, W_C\}$ and $B_{RE} = \{B_1, B_2, ..., B_L, B_R, B_C\}$. The objective function can be expressed as:

$$G_{RE}(W_{RE}, B_{RE}) = \{M(X, Y) + \lambda RE(X)\} \tag{5}$$

Where λ is the trade-off parameter, and it is used to control the relative importance of the two terms. The first term $M(X, Y)$ in Eq. 5 is the classification error function which is to seek to minimize the classification error for the training samples [12]. It is computed as the following equation:

$$M(X, Y) = -\frac{1}{N} \sum_{x_i \in X} \langle y_i, \log O_C(x_i) \rangle \tag{6}$$

where N is the number of giving training samples in X. The second term $RE(X)$ is the removal correlation function which is to remove the correlation among the CNN features. Then, the CNN features are divided into 10 groups randomly. The first k groups are selected as the first input of CCA and the remaining groups are the second input of CCA. For convenience, x_1 and x_2 are used to represent the two parts of CNN features. The function is defined as following:

$$RE(X) = \max \frac{\alpha^T S_{x_1 x_2} \beta}{\sqrt{\alpha^T S_{x_1 x_1} \alpha \beta^T S_{x_2 x_2} \beta}}$$
$$s.t. \alpha^T S_{x_1 x_1} \alpha = 1, \beta^T S_{x_2 x_2} \beta = 1, \tag{7}$$
$$\alpha_i x_1 x_2{}^T \beta_j = 0, \text{for} \, i \neq j.$$

where $S_{x_1 x_1}$ and $S_{x_2 x_2}$ are, respectively, within-set covariance matrices of the two parts of CNN features, and $S_{x_1 x_2}$ is the between-set covariance matrix between the two parts of CNN features. Here, we assume that $S_{x_1 x_1}$ and $S_{x_2 x_2}$ are non-singular. In this way, the remove function can further produce multiple pairs of projection directions $\{(\alpha_i, \beta_i)\}_{i=1}^{d}$, where d is the smaller one in the two parts of CNN features. In the proposed RE-CNN model, we define $W_R = \alpha \& \beta$, and $B_R = 0$. Then the output of the layer F_R can be computed as:

$$O_R(x_i) = f(\alpha O_L(x_i)_{1\_k}) \tag{8}$$

Or it can be computed as:

$$O_R(x_i) = f(\beta O_L(x_i)_{k+1\_10}) \tag{9}$$

Thus, by incorporating Eqs. 6 and 7 into Eq. 5, the following decorrelation objective function of RE-CNN model can be formed as:

$$
\begin{aligned}
&G_{RE}(W_{RE}, B_{RE}) = \\
&\min\left\{ -\frac{1}{N} \sum_{x_i \in X} \langle y_i, \log O_C(x_i) \rangle - \frac{\lambda \alpha^T S_{x_1 x_2} \beta}{\sqrt{\alpha^T S_{x_1 x_1} \alpha \beta^T S_{x_2 x_2} \beta}} \right\} \\
&s.t. \alpha^T S_{x_1 x_1} \alpha = 1, \beta^T S_{x_2 x_2} \beta = 1, \\
&\quad \alpha_i x_1 x_2^T \beta_j = 0, \text{for} i \neq j.
\end{aligned}
\tag{10}
$$

In this way, the new removal correlation objective function not only minimizes the classification loss, but also imposes the CCA model so that the learnt CNN features are not redundant.

4   Experiments

In order to better compare the RE-CNN model with the traditional CNN models and validate the effectiveness of the RE-CNN model, some complex CNN based structures (VGG-Net [6], LBCNN [7] et al.) are chosen to train the MNSIT [5] and LFW [13]. The necessary implementation details are given in Sect. 4.1, and the split parameter k is discussed in Sect. 4.2. The results of different CNN models are given in Sect. 4.3 to verify the effectiveness of the proposed model.

4.1   Implementation Details

RE-LeNet-5: Firstly, the LeNet-5 is selected as the base structure, and renamed to RE-LeNet-5. The model consists of 3 convolutional layers, interlaced with sub-sampling layers, max pooling layers, one fully connected layer and one gaussian connected layer. The first convolutional layer has relatively 6 large 5×5 kernels. Then, a 2×2 average pooling layer is used to subsample. The second convolutional layer has 16 large 5×5 kernels corresponding to 16 feature maps. Then, the subsampling layer is consistent with the previous one. The third convolutional layer has 160 large 5×5 kernels corresponding to 160 feature maps. The activate function of the whole model is sigmoid.

RE-VGGNet: Recently, VGGNet has get groundbreaking success in image classification tasks on the Image Large Scale Visual Recognition Challenge

(ILSVRC). We use the VGGNet as the base structure. It can be called RE-VGGNet, which is almost identical to the model described in VGGNet. The output dimension of the fully connected layer is 512.

RE-CNN-Centre: In order to make the RE-CNN model be high-capacity, we use the CNN-centre model in [14,15] as the base structure. It can be called RE-CNN-centre. The output dimension of the fully connected layer is 512.

RE-LBCNN: To reduce the computational complexity of RE-CNN, we choose the LBCNN [7] as the base structure. We call it RE-LBCNN.

All convolutional layers are optimized by using stochastic gradient decent with momentum [16]. The base learning rate is initially set to 0.001 and the mini-batch size is 200.

4.2   Discussion About the Split Parameter k

For the split parameter k, respectively, is selected from $\{1, 2, ..., 10\}$. Experiments are conducted to investigate the sensitiveness of the parameter.

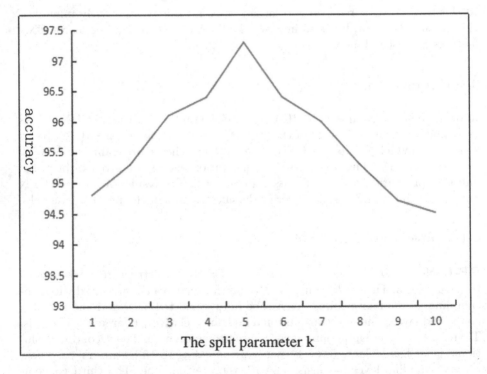

Fig. 2. The accuracy on noise MNIST varies with split parameter k.

Here, the AlexNet [17] is used as the base CNN structure, and the experiment dataset is MNIST. From Fig. 2, it can be seen that the model can get the best recognition accuracy when the k is fixed in 5. To our best knowledge, RE-CNN uses the CCA model to make the whole model extract more advanced features. When the numbers of the input features are the same, RE-CNN model can get the more high-level common features than other models.

4.3   Results

Noise MNIST: The MNIST database of handwritten digits [18] has a training set of 60,000 examples, and a test set of 10,000 examples. It is a subset of a larger set which is available from NIST. It is a good database for people who want to try learning techniques and pattern recognition methods on real-world data while spending minimal efforts on preprocessing and formatting. In order to better compare the advantages and the disadvantages of each method, we randomly add salt-and-pepper noise to MNIST.

LFW Dataset: The LFW dataset consists of 13,233 face images from 5,749 different identities. The pose and the expressions of these identities in the images have large variations. We test on 60,00 face pairs.

Table 1. The accuracy of the image recognition on Noise MNIST and LFW

Methods	Noise MNIST	LFW
LeNet-5	90.1	89.3
RE-LeNet	**92.3**	**91.6**
VggNet-16	95.5	98.0
RE-VggNet-16	**97.7**	**98.4**
Cnn-centre	95.8	97.2
RE-Cnn-centre	**98.1**	**97.6**
LBCNN	94.3	96.7
RE-LBCNN	**96.4**	**96.9**

From the Table 1, it can be seen that the RE-CNN model always get the better recognition in Noise MNIST and LFW datasets than existing CNN structures. For our best known, the CNN features have been improved.

5   Conclusion

Motived by the CCA, it can retain the maximum high-level information and remove redundant information. We combine the existing CNN models and CCA model to form the RE-CNN model. The proposed RE-CNN model not only can retain the traditional high-capability CNN structures, but also remove the redundant information from the boundedness of the CNN structure. From the experiments, the RE-CNN model can get the power feature from the original images. The effective results show that the proposed RE-CNN model performs beyond the traditional CNN models and always get the better accuracy in image recognition. In further work, other defects in the selection of the convolution kernel and the strategy of the pooling layer will be considered and improved. We will focus on using the traditional feature to supervised the learning of CNN features, this can further improve the power of CNN features.

References

1. Russell, S.J., Norvig, P.: Artificial intelligence: a modern approach. Appl. Mech. Mater. **263**(5), 2829–2833 (2010)
2. Dalal, N., Triggs, B.: Histograms of oriented gradients for human detection. In: IEEE Computer Society Conference on Computer Vision and Pattern Recognition, CVPR 2005, vol. 1, pp. 886–893. IEEE (2005)
3. Ojala, T., Pietikainen, M., Maenpaa, T.: Multiresolution gray-scale and rotation invariant texture classification with local binary patterns. IEEE Trans. Pattern Anal. Mach. Intell. **24**(7), 971–987 (2002)
4. Russakovsky, O., Deng, J., Su, H., Krause, J., Satheesh, S., Ma, S., Huang, Z., Karpathy, A., Khosla, A., Bernstein, M., Berg, A.C., Fei-Fei, L.: ImageNet large scale visual recognition challenge. Int. J. Comput. Vis. **115**(3), 211–252 (2015)
5. Lecun, Y.: Gradient-based learning applied to document recognition. Intelligent Signal Processing pp. 306–351 (2001)
6. Simonyan, K., Zisserman, A.: Very deep convolutional networks for large-scale image recognition. In: International Conference on Learning Representations (2015)
7. Juefei-Xu, F., Boddeti, V.N., Savvides, M.: Local binary convolutional neural networks. In: 2017 IEEE Conference on Computer Vision and Pattern Recognition (CVPR), pp. 4284–4293 (2017)
8. Hardoon, D.R., Szedmak, S.R., Shawe-taylor, J.R.: Canonical correlation analysis: an overview with application to learning methods. Neural Comput. **16**(12), 2639–2664 (2004)
9. Andrew, G., Arora, R., Bilmes, J.A., Livescu, K.: Deep canonical correlation analysis. In: Proceedings of The 30th International Conference on Machine Learning, pp. 1247–1255 (2013)
10. Hotelling, H.: Relations between two sets of variates. Biometrika **28**, 321–377 (1936)
11. Deng, J., Dong, W., Socher, R., Li, L.J., Li, K., Fei-Fei, L.: ImageNet: a large-scale hierarchical image database. In: IEEE Conference on Computer Vision and Pattern Recognition, CVPR 2009, pp. 248–255. IEEE (2009)
12. Cheng, G., Zhou, P., Han, J.: RIFD-CNN: rotation-invariant and fisher discriminative convolutional neural networks for object detection. In: IEEE Conference on Computer Vision and Pattern Recognition, pp. 2884–2893 (2016)
13. Huang, G.B., Mattar, M., Berg, T., Learned-Miller, E.: Labeled faces in the wild: a database for studying face recognition in unconstrained environments (2008)
14. Wen, Y., Zhang, K., Li, Z., Qiao, Y.: A discriminative feature learning approach for deep face recognition. In: Leibe, B., Matas, J., Sebe, N., Welling, M. (eds.) ECCV 2016. LNCS, vol. 9911, pp. 499–515. Springer, Cham (2016). https://doi.org/10.1007/978-3-319-46478-7_31

15. Hu, J., Lu, J., Yuan, J., Tan, Y.-P.: Large margin multi-metric learning for face and kinship verification in the wild. In: Cremers, D., Reid, I., Saito, H., Yang, M.-H. (eds.) ACCV 2014. LNCS, vol. 9005, pp. 252–267. Springer, Cham (2015). https://doi.org/10.1007/978-3-319-16811-1_17
16. Liu, B., Wang, M., Foroosh, H., Tappen, M., Penksy, M.: Sparse convolutional neural networks. In: Computer Vision and Pattern Recognition, pp. 806–814 (2015)
17. Krizhevsky, A., Sutskever, I., Hinton, G.E.: Imagenet classification with deep convolutional neural networks. Adv. Neural Inf. Process. Syst. **25**, 1097–1105 (2012)
18. Villa, A.E.P., Masulli, P., Pons Rivero, A.J. (eds.): ICANN 2016. LNCS, vol. 9887. Springer, Cham (2016). https://doi.org/10.1007/978-3-319-44781-0

MusicCNNs: A New Benchmark on Content-Based Music Recommendation

Guoqiang Zhong[✉], Haizhen Wang, and Wencong Jiao

Department of Computer Science and Technology, Ocean University of China,
238 Songling Road, Qingdao 266100, China
gqzhong@ouc.edu.cn

Abstract. In this paper, we propose a new deep convolutional neural network for content-based music recommendation, and call it MusicC-NNs. To learn effective representations of the music segments, we have collected a data set including 600,000+ songs, where each song has been split into about 20 music segments. Furthermore, the music segments are converted to "images" using the Fourier transformation, so that they can be easily fed into MusicCNNs. On this collected data set, we compared MusicCNNs with other existing methods for content-based music recommendation. Experimental results show that MusicCNNs can generally deliver more accurate recommendations than the compared methods. Therefore, along with the collected data set, MusicCNNs can be considered as a new benchmark for content-based music recommendation.

Keywords: Recommendation systems · Music recommendation
Content-based recommendation · Deep learning
Music convolutional neural networks

1 Introduction

In recent years, with the rapid development and popularization of multimedia technology, digital music has become an important component of people's lives, and has fundamentally influenced and changed people's multimedia consumption habits.

Currently, online music stores and streaming services, such as Google Play, iTunes, Spotify and Windows Media Player, have brought great convenience to people's lives. A new problem becomes more and more important, that is, how to accurately recommend songs according to customers' preferences. There have been many studies on the recommendation systems, however, due to the particularity of music, such as the genres, style, rhythm, and geographical characteristics, the music recommendation problem is particularly complicated. It also becomes fairly difficult when doing personalized recommendations for someone at a certain moment, because his feelings and the change in his environment may affect his preferences at that time.

The current music recommendation methods can be summarized as follows.

© Springer Nature Switzerland AG 2018
L. Cheng et al. (Eds.): ICONIP 2018, LNCS 11301, pp. 394–405, 2018.
https://doi.org/10.1007/978-3-030-04167-0_36

Collaborative Filtering. Herlocker et al. introduced the collaborative filtering algorithm for music recommendation [6]. The user firstly scores each item for evaluation. By calculating similarity between different user ratings, the nearest neighbor can be found, and accordingly the recommendation is generated based on the evaluation of the nearest neighbor. Abhinandan et al. use collaborative filtering algorithm for personalized recommendations [4]. However, these approaches suffer from the clod-start problem: when new songs appear, there is no historical data available, so that the collaborative filtering algorithm may fail because of the lack of scoring data.

Content-Based Recommendation. Music contents, such as rhythm, lyric, melody and so on, are very important for music recommendation tasks. However, existing content-based recommendation systems do not perform well enough, because most of them only extract traditional audio content features, such as Mel-frequency cepstral coefficients (MFCC), then use these features to predict user preferences, such as the work of [2,15,17].

In this paper, we propose a new deep learning model, music convolutional neural networks (MusicCNNs), for content-based music recommendation. As it is content-based, this method can effectively avoids the cold-start problem. With MusicCNNs, deep representations of the music segment can be learned. Hence, it generally performs better than existing music recommendation approaches. Moreover, we have collected a music dataset containing approximately 600,000 songs. It provides an excellent resource for music recommendation research. With this data set, we prove that our model is superior to the existing content-based music recommendation models. Therefore, this paper provides a new benchmark for future content-based music recommendation research.

The data and codes used in this paper can be freely downloaded from GitHub with address https://github.com/AdolfKing/MusicCNNs.

2   Related Work

In the following, we mainly review the music recommendation methods published in recent years.

Collaborative Filtering (CF). Collaborative filtering algorithms recommend songs to users based on the information of other users who has the same preferences. Li et al. [9] proposed a mixture-rank matrix approximation (MRMA) method, in which user-item ratings can be characterized by a mixture of LRMA models with different ranks. Koren et al. [8] made a good summary of performing the most advanced collaborative filtering methods which is based on matrix decomposition (MF). However, there is a cold-start problem in the collaborative filtering algorithm, which is ineffective when recommending new or unpopular songs. Our proposed method, MusicCNNs, is based on learned deep representations of the music segments, such that the cold-start problem is well avoided.

Content-Based Recommendation. It is feasible to recommend songs with similar audio content to users according to their preferences. In [14], Soleymani et al. propose a content-based music recommendation system which is based on a set of attributes derived from psychological studies of music preference. Five attributes, Mellow, Unpretentious, Sophisticated, Intense and Contemporary (MUSIC), describe the underlying factors of music preference compared to music genre. In [13], a method which uses a latent factor model for recommendation was proposed. McFee et al. [10] proposed a method by learning from a sample of collaborative filter data to optimize content-based similarity, and they used the typical bag-of-words approach when extracting audio features. However, these methods have limited effects on the extraction of music features. The difference between our proposed model and these previous work is that our model can extract deep music features, which has great advantage over traditional hand-crafted ones.

Hybrid Methods. Yoshii et al. [17] firstly combined CF and content-based methods in music recommendations. In this work, the hybrid method can directly represent substantial (unobservable) user preferences as a set of latent variables introduced in a Bayesian network. Probabilistic relations over users, ratings and contents are statistically estimated. [5] explored the usage of temporal context and session diversity in session-based collaborative filtering techniques for music recommendation. In addition, Wang et al. [16] used a new model based on a deep belief network and a probabilistic graphical model for simultaneous feature extraction and recommendation. However, most of the existing hybrid methods only exploited hand-crafted features. Alternatively, our proposed method directly learns the effective representations of the music.

3   The Collected Data Set

The music data set was collected from the JunoDownload website, with a total of more than 600,000 songs. For each song, the information we collect includes: artist, release_data, track_name, release_label and so on. The most important thing is that we can get the most original audio content information. For the convenience of the experiments, we randomly selected 9 different music genres from the data set, and randomly selected 1000 songs for each genre, to construct the training and test sets. The 9 genres were: Disco, Breakbeat, Euro Dance, Deep House, Downtempo, Dubstep/Grime, Drum And Bass, Dancehall/Ragga, and Electro House.

In order to extract the content features of the audios, we preprocessed the data set, and converted the audios into spectral images.

First of all, we used the Fourier transform [1] to convert MP3 format audios to "images", which was a spectrogram. Each spectrogram is a visualization of a spectrum of sound over time, and the intensity of the color on the image represents the amplitude of the sound at that frequency. In this paper, we chose to create a monochromatic spectrum. Figure 1 shows a converted spectrum from a Breakbeat type of music. The abscissa represents the time and the ordinate represents the frequency of the sound.

In the second step, we cut the spectrum of each music images into segments to facilitate the following representation learning. Each audio file is about 2 min. We cut it with about 5 s for each music segment, so that each audio file corresponds to about 23 images. Then, we normalize the images to the size 256×256. The cut images are shown in Fig. 2.

The above preprocessing operation was performed on all the audio files. Finally, we obtained about 195000 spectrograms of the randomly selected 9000 songs.

Fig. 1. A Fourier-transformed spectrogram from a breakout-type audio file.

Fig. 2. After cutting the spectrogram, the size of each image is 256×256.

4 Music Convolution Neural Networks (MusicCNNs)

In this section, we describe the proposed Music Convolution Neural Networks (MusicCNNs) in detail, including their structure, the used activation functions and regularization techniques. Here, we first introduce our motivation to design the MusicCNNs model for content-based music recommendation.

For music recommendation, the representations of music segments are critical. In order to learn effective representations of the music segments, we try to design deep learning models and use them for music recommendation. In general, deep learning models can learn abstract high-level representations or features by combining low-level features to discover distributed representations of data. In recent years, deep learning has developed rapidly, especially the convolutional neural network models, which have been successfully applied in many fields, such as computer vision, speech recognition and natural language processing. Therefore, we attempt to convert the music information into "images" and apply deep CNNs to learn their intrinsic representations. For concreteness, the deep learning models may bring us some advantages for music recommendation as below.

1. The collected data set is large enough for the effective training of MusicCNNs.
2. Using effective activation functions, such as Rectified Linear Units (ReLu) and exponential linear units (ELU), MusicCNNs can converge fast and slow down the gradient vanishing problem.
3. With GPU parallelization and acceleration, MusicCNNs can be trained in a reasonable amount of time.

In the area of music recommendation, some researchers have exploited the idea of deep learning [7,13]. However, these work did not really take the advantages of deep learning for image classification. Therefore, in this work, we convert the music segments into spectrum images and then use the proposed deep CNN model, MusicCNNs, to extract the effective features of the spectrum images.

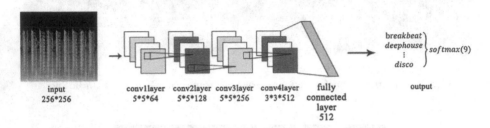

input 256*256   conv1layer 5*5*64   conv2layer 5*5*128   conv3layer 5*5*256   conv4layer 3*3*512   fully connected layer 512   output

Fig. 3. The architecture of MusicCNNs.

4.1   The Architecture of MusicCNNs

The architecture of MusicCNNs is illustrated in Fig. 3. It contains four convolutional layers, where each convolutional layer is followed by a pooling layer, one fully-connected layer and one softmax regression layer. As below, we describe the structure of MusicCNNs in detail.

We suppose that the inputs are spectral images with size 256×256. The kernel size in the first convolutional layer is 5×5, of a total number 64, with a step size of 2 pixels. The second convolutional layer uses 128 kernels of size $5 \times 5 \times 64$. The third convolutional layer has 256 convolutional kernels of size $5 \times 5 \times 128$. The fourth layer has 512 convolutional kernels of size $3 \times 3 \times 256$. Next, the fully connected layer has 512 neurons. For each convolutional layer, a 2×2 max-pooling layer is followed. For each convolutional layer and the fully connected layer, the ELU activation functions are used. Finally, we use the softmax classification at the last layer. In this model, we adopt the dropout technique for model regularization.

In this network model, we use the Exponential Linear Units (ELU) function [3] as the nonlinear activation function, which is shown in Fig. 4. Its advantages are as follows: for the ReLU activation function $max(0, x)$, its output value has no negative value, so that the average value of the output will be greater than 0. When the mean value of the activation value is not 0, it will cause a bias to the next layer, and if the activation values are not offset from each other (i.e.

the mean is non-zero), it will result in a bias shift in the next level of activation. With such stacking, the more cells there are, the greater the bias shift will be. Compared to ReLU, ELU can take a negative value, which makes the cell activation mean can be close to 0, similar to the effect of batch normalization but only requiring lower computational complexity. ELU has soft saturation when the input takes a small value, which improves the robustness to noise. Hence, we use the ELU activation function at every level.

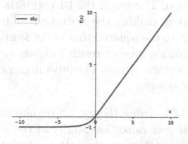

Fig. 4. The curve of ELU activation function.

In order to demonstrate that the ELU activation function in the MusicCNNs model is superior to the ReLU activation function, we visualize the accuracy obtained by them on the validation set, as shown in Fig. 5. We can see that, for the MusicCNNs model, the effect of using the ELU activation function is much better than that of using the ReLU activation function.

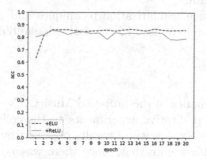

Fig. 5. The effect of using the ReLU and ELU activation function in MusicCNNs, where the abscissa represents the number of epoches and the ordinate represents the accuracy of the model on the validation set.

4.2   Regularization Methods

For preventing the network models from overfitting, one approach is to increase the amount of training data, and the other is to use regularization techniques.

Intuitively, regularization is to make network learn smaller weights while remaining constant in other aspects. In other words, regularization can be regarded as a method that can compromise small weights and minimize the original cost function. The commonly used regularization techniques include L1 regularization, L2 regularization, dropout and so on.

L1 regularization refers to the sum of the absolute values of each element in the vector, also called "lasso regularization". L1 regularization is an important method in machine learning. In the learning process of lasso problems, it is used to optimize the cost function. Therefore, the L1 regularization adds the L1 norm restricts to the cost function, making the learning result to be sparse.

L2 regularization refers to the squared root of the sum of the squared elements of the vector. L2 regularization cannot result in sparsity learning, but can help to avoid over-fitting to a certain extent. The advantages of the L2 regularization include the following two aspects:

1. From the perspective of learning theory, L2 regularization can prevent over-fitting and enhance the generalization ability of the model.
2. In the viewpoint of optimization or numerical calculation, the L2 norm helps to solve the problem of matrix inversion when the condition number is not good.

When training a neural network, dropout sets the probability of eliminating nodes in the neural network. Suppose each node in each layer is set by tossing the coin, the probability that each node can be retained and eliminated is 0.5. This paper is to use this method to set the probability. After setting the node probabilities, we eliminate some nodes and then delete the connections coming in and out of that node. Finally, we get a network with fewer nodes and smaller scale during model training.

Considering the above regularization techniques and the structural characteristics of our model, we have used L2 regularization and dropout in this model.

5   Experiments

To evaluate the performance of the proposed MusicCNNs model, we conducted both quantitative and qualitative experiments on the collected data set. In the following, we report the experimental results obtained by MusicCNNs and the compared methods. Please note that, since there was no available audio data sets for music recommendation, we only used the collected data set.

5.1   Quantitative Evaluation

In order to demonstrate the superiority of MusicCNNs on extracting music content features and content-based music recommendation, we compared it with other existing models. First of all, we divided the pre-processed dataset into three parts: training, validation and test sets according to the ratio of 0.65:0.25:0.1. The models compared with MusicCNNs were as follows.

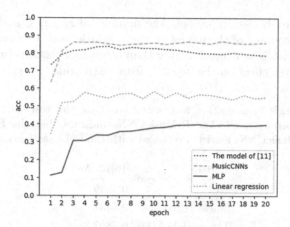

Fig. 6. The effect of MusicCNNs and the compared methods on the extraction of musical features, where the abscissa represents the number of epochs and the ordinate represents the accuracy obtained by the methods on the validation set.

A Multilayer Perceptron (MLP). Due to the input form of the MLP structure, the MLP model cannot directly process our audio content features. Therefore, we used an image descriptor Local Binary Pattern (LBP) to process our audio features into a one-dimensional vector and then used the MLP model for processing. Local Binary Pattern (LBP) is an operator used to describe local texture features of images. It has significant advantages such as rotation invariance and grayscale invariance. It was first proposed by Ojala et al. [12] in 1994 for texture feature extraction. For the LBP feature vector extraction, we first needed to calculate the LBP model for each pixel in the image. And then, we calculated the LBP feature value histogram for each cell, and then normalized the histogram. Finally, the statistical histogram of each cell was connected into a feature vector, which was the LBP texture feature vector of the entire map.

Linear Regression. The network was trained on the same training set as the MLP model.

The Model of [11]. A CNN architecture trained on our training set.

In order to evaluate the performance of MusicCNNs on data representation learning, we made an assessment of the use of MusicCNNs and the compared models to classify 9 classes of music. Using the classification accuracy of the model as an indicator to judge the effect of the model on the extraction of music content features. For this experiment, all the above models were trained on the same dataset. See Sect. 3 for the data preprocessing. Here our training set contains a total of 127,000 spectrograms, the validation set contains approximately 47,000 spectrograms, and the rest is used as a test set. In the training process, the batch size was set to 128 and 20 epochs were used. In order to facilitate the observation of the differences between the above models for music feature extraction, we first compared their performance on the validation set, as shown in the Fig. 6, the abscissa represents the number of epoches, and the ordinate represents the accuracy of the model on the validation set. Obviously, the accuracy rate of MusicCNNs is much higher than other models. It is proved that our model has a better effect on the music feature extraction.

Table 1. The music classification was tested on the entire test set. The MLP is a multilayer perceptron model. The R-MusicCNNs indicate that the ELU activation function in the MusicCNNs model is replaced with the ReLU activation function.

Model	Top-3 Acc
MLP	0.3749
Linear regression	0.5437
The model of [11]	0.7807
R-MusicCNNs	0.7961
MusicCNNs	**0.8196**

To further demonstrate the advantage of MusicCNNs on music recommendation, we used the top-3 accuracy to measure the recommendation accuracy, where **a recommendation is successful if the top-3 candidates include a song belonging to the same class of the query music**. Here, each music segment was represented by the activation outputs of MusicCNNs at the last second layer, while each song was represented by the mean of its segments. The recommendation was based on a nearest neighbor search with the cosine distance. As shown in Table 1, we can see that MusicCNNs is about 4 percent higher than the model of [11], and about 30% higher than the linear regression method, about 40% higher than MLP. Hence, MusicCNNs are superior to other previous models. To further prove the effect of the ELU activation function used in this model, we compared it with R-MusicCNNs (Only replaced the ELU activation function in the MusicCNNs model with the ReLu activation function, leaving the rest architecture unchanged). As shown in Table 1, MusicCNNs obtained the accuracy about 2% higher than that of R-MusicCNNs. Therefore, we can conclude that MusicCNNs is very suitable for the music feature extraction and music recommendation.

5.2   Qualitative Evaluation

How to evaluate the recommendation system is a very complicated issue. The indicator of accuracy does not sufficiently prove the rationality of these recommendations. To determine this point, we also conducted some qualitative experiments on the collected data set. For each song, we searched for similar songs by calculating the cosine similarity between them. Table 2 shows the recommendation of MusicCNNs and the compared 4 models for the same query. Due to space limitation, we only list one query song's recommendation. Here, we named the song for recommendation as a query song. From the table, we can see that the songs recommended by MusicCNNs not only belong to the same parent_genre as the query song, but also they were released by different artists, so that the recommendation is more accurate and diverse. Whereas the songs recommended by other models deviate the query song's parent_genre, and some of the five songs were released by the same artist. The R-MusicCNNs model's recommendation results were very similar to our model, but it still slightly worse than ours. Therefore, we can prove that, for music recommendation, MusicCNNs are not only more precise but also more diverse than other compared models.

Table 2. We used MusicCNNs and other 4 compared models to make recommendations for the same query music. The recommendations are shown in the table below. Due to space limitation, we only show the recommendations for one song. Among the labels of the music, taking "ALEN OVERSOUND - Rise - Deep House" as an example, ALEN OVERSOUND represents the release_artist, Rise represents the release_title, and Deep House represents the parent_genre.

Model	ALEN OVERSOUND - Rise - Deep House
MLP	ALEXANDRA - Marimba Soul - Deep House
	DUB KILLER - Send Me - Dubstep/Grime
	ALLIED/MARKOMAN/TSUNAMI - Concept - Drum And Bass
	ALEXANDRA - Dare Me (remixes) - Electro House
	ALEXANDRA - It's A Friday Night - Euro Dance
Linear regression	MICHAEL DREAM - Feel Like Dancing - Euro Dance
	MICHAEL DREAM - Devil's Eyes - Electro House
	MICHAEL DREAM - Concept - Drum And Bass
	DR KMER feat JIMMY ROLAND - Got To Be Real - Electro House
	DAMIANDEBASS - Still Gone Rock - Breakbeat
The model of [11]	ALPHA INTERNATIONAL - Get Down EP - Disco
	JAY DEEP - High Blown Departure - Deep House
	JAY DEEP - Bang - Disco
	ALFA PAARE - Patti EP - Disco
	CHAPPELL - U Just Have To Deal - Deep House
R-MusicCNNs	DIRTY CULTURE - Sweet Songs EP - Deep House
	DOS AMIGOS - Borriquito (El Axito Del Verano)- Euro Dance
	DADAISMUS - Goldhawk Grooves EP - Deep House
	DARBUKA TRIBE - Deep In Afrika - Deep House
	DADAISMUS - Come With Me - Deep House
MusicCNNs	DIRTY CULTURE - Sweet Songs EP - Deep House
	DOS AMIGOS - Borriquito (El Axito Del Verano) - Deep House
	4BEATCLUB - Goldhawk Grooves EP - Deep House
	DARBUKA TRIBE - Deep In Afrika - Euro Dance
	DADAISMUS - Come With Me - Deep House

6   Conclusion

In this paper, we propose a new deep learning model Music Convolutional Neural Network (MusicCNNs) for music representation learning and content-based music recommendations. Meanwhile, we collected a music data set containing about 600,000 songs. On this music data set, we evaluated the performance of MusicCNNs. Experimental results show that MusicCNNs perform better than other compared models for music representation learning. Unlike previous collaborative filtering algorithms, MusicCNNs have no the cold start problem, while it generally delivers better recommendation results than content-based recommendation methods.

Built upon the collected data set, this paper provides a new benchmark for content-based music recommendations and demonstrates that the latest development of deep learning is very beneficial to music representation learning in music recommendation systems, especially the convolutional neural network. In future work, we would like to exploit other effective deep learning models for content-based music recommendation tasks, and test the deep learning models on large scale of data with multi-modal distribution.

Acknowledgment. This work was supported by the National Key R&D Program of China under Grant 2016YFC1401004, the National Natural Science Foundation of China (NSFC) under Grant No. 41706010, the Science and Technology Program of Qingdao under Grant No. 17-3-3-20-nsh, the CERNET Innovation Project under Grant No. NGII20170416, and the Fundamental Research Funds for the Central Universities of China.

References

1. Bracewell, R.N., Bracewell, R.N.: The Fourier Transform and Its Applications, vol. 31999. McGraw-Hill, New York (1986)
2. Cano, P., Koppenberger, M., Wack, N.: Content-based music audio recommendation. In: ACM MM, pp. 211–212 (2005)
3. Clevert, D.A., Unterthiner, T., Hochreiter, S.: Fast and accurate deep network learning by exponential linear units (ELUs). arXiv preprint arXiv:1511.07289 (2015)
4. Das, A., Datar, M., Garg, A., Rajaram, S.: Google news personalization: scalable online collaborative filtering. In: WWW, pp. 271–280 (2007)
5. Dias, R., Fonseca, M.J.: Improving music recommendation in session-based collaborative filtering by using temporal context. In: ICTAI, pp. 783–788. IEEE (2013)
6. Herlocker, J.L., Konstan, J.A., Borchers, A., Riedl, J.: An algorithmic framework for performing collaborative filtering. In: SIGIR Forum, vol. 51, no. 2, pp. 227–234 (2017)
7. Humphrey, E.J., Bello, J.P., LeCun, Y.: Feature learning and deep architectures: new directions for music informatics. J. Intell. Inf. Syst. **41**(3), 461–481 (2013)
8. Koren, Y., Bell, R.M., Volinsky, C.: Matrix factorization techniques for recommender systems. IEEE Comput. **42**(8), 30–37 (2009)
9. Li, D., Chen, C., Liu, W., Lu, T., Gu, N., Chu, S.M.: Mixture-rank matrix approximation for collaborative filtering. In: NIPS, pp. 477–485 (2017)

10. McFee, B., Barrington, L., Lanckriet, G.R.G.: Learning content similarity for music recommendation. CoRR abs/1105.2344 (2011)
11. Murray, M.: Building a music recommender with deep learning. http://mattmurray.net/building-a-music-recommender-with-deep-learning
12. Ojala, T., Pietikainen, M., Maenpaa, T.: Multiresolution gray-scale and rotation invariant texture classification with local binary patterns. IEEE Trans. Pattern Anal. Mach. Intell. **24**(7), 971–987 (2002)
13. van den Oord, A., Dieleman, S., Schrauwen, B.: Deep content-based music recommendation. In: NIPS, pp. 2643–2651 (2013)
14. Soleymani, M., Aljanaki, A., Wiering, F., Veltkamp, R.C.: Content-based music recommendation using underlying music preference structure. In: ICME, pp. 1–6. IEEE (2015)
15. Wang, X., Rosenblum, D., Wang, Y.: Context-aware mobile music recommendation for daily activities. In: ACM MM, pp. 99–108. ACM (2012)
16. Wang, X., Wang, Y.: Improving content-based and hybrid music recommendation using deep learning. In: ACM MM, pp. 627–636. ACM (2014)
17. Yoshii, K., Goto, M., Komatani, K., Ogata, T., Okuno, H.G.: Hybrid collaborative and content-based music recommendation using probabilistic model with latent user preferences. In: ISMIR, vol. 6, p. 7th (2006)

Remote Sensing Image Segmentation by Combining Feature Enhanced with Fully Convolutional Network

Ruiguo Yu[1], Xuzhou Fu[1], Han Jiang[2], Chenhan Wang[2], Xuewei Li[1(✉)],
Mankun Zhao[1], Xiang Ying[3], and Hongqian Shen[1]

[1] School of Computer Science and Technology, Tianjin University, Tianjin, China
{rgyu,fuxuzhou,lixuewei,zmk,hongqianshen}@tju.edu.cn
[2] Beijing AXIS Technology Company Limited, Beijing, China
{hahn,gabriel}@signcl.com
[3] School of Computer Software, Tianjin University, Tianjin, China
xiang.ying@tju.edu.cn

Abstract. The main idea of this paper is the 25-categories classification task of remote sensing satellite image which is provided by Beijing AXIS Technology Company Limited, and proposes new methods based on fully convolutional network (FCN) and image processing. This method utilized image processing to realize color mapping and feature enhanced of remote sensing satellite image. Consider the influence of equipment and scene shooting environment, there are differences in color performance between remote sensing images, we use color mapping to improve color consistency. Aiming at the disadvantage of FCN has lower sensitivity to details, we add edge information into image as an important signal and expand the image into a five-dimensional one. Then the classification results will be attained through 25-categories classification according to FCN model. The experiment result showed the method is able to enhance the accuracy of FCN model classification to some extent.

Keywords: Remote sensing image · Image segmentation
Feature enhanced · Fully convolutional network

1 Introduction

Satellites that used to observe the earth such as NOAA, MODIS, Landsat TM can take many pictures which have diversity types, resolution, spectral resolution and time resolution. These remote sensing images are widely used in agricultural research, environmental research, city regional classification, as well as analyzing and processing study between natural resources and human activities. So it is particularly necessary to segment the semantic information from remote sensing images. Remote sensing image segmentation refers to process the image and extract the target, which mainly take advantage of images characteristics information to segment images as ground, houses, farmland, vehicles, ground

© Springer Nature Switzerland AG 2018
L. Cheng et al. (Eds.): ICONIP 2018, LNCS 11301, pp. 406–415, 2018.
https://doi.org/10.1007/978-3-030-04167-0_37

vegetation, ocean ice and atmospheric clouds. The features of remote sensing images usually show as high grey grade, large quantity of information data, obscurity boundary, complex target structure. Manually delineated is not only consuming manpower and materials but also existing mistakes. Hence the intelligent segmentation processing of remote sensing images has become the one of the main research directions. This paper plans to use fully convolutional network (FCN) to 25-categories classification remote sensing images, which is more difficult and more challenging than existing classification works.

Till now, the methods of remote sensing images segmentation are traditional image segmentation methods, support vector machine, fuzzy classification, artificial neural network, convolutional neural network, fully convolutional neural network, etc. Among these methods, FCN is able to train end-to-end, pixels-to-pixels on remote sensing images segmentation. Therefore, with the gradual enhancement of image resolution, FCN can extract deeper semantic features in high dimensional representation. Besides, based on the requirement of network structure, CNN requires input images have fixed size while FCN can accept images without size limitations with high efficiency. So far, based on FCN, 12-categories classification and 16-catagories classification of remote sensing images have already realized. This paper plans to tackle following problems.

1. This paper will use FCN to increase the accuracy by judging the abstract features of every pixel.
2. With the development of high resolution satellite image technology, the ground objects information is clear to recognize. Based on surface information, semantic information can be acquired by comprehension, integration and abstraction. And this paper uses FCN to 25-categories classification satellite images to satisfy the practical needs.
3. This paper uses the methods of images processing to transfer color mapping and feature enhanced of remote sensing images. And using FCN model to process remote sensing images to accomplish the task of 25-categories classification.

However, almost all of data set has it own limitations, such as small scale, less number of images, diverse images, low accuracy and so on. These limitations severely restricted the development of new methods, especially methods based on deep learning. In 2017 Li et al. have solved classification problems of supervised remote sensing images by transfer learning. Hence this paper uses the transfer learning to learn the features of the different scene categories in the natural image and pre-training for the classifier. The rest is organized as follows. Section 2 introduces related work of satellite images segmentation and systematically summarize previous research results. Section 3 introduces the main methods of fully convolutional network. In Sect. 4, we discuss the results, performance, and comparison with the results of related techniques in the literature. The discussion, future works and summary are given in Sect. 5.

2 Related Work

Traditional images segmentation includes edge image segmentation [11], regional growth method image segmentation [4], clustering image segmentation [14,15] and threshold image segmentation [3]. Due to remote sensing images characterized high grey grade, large quantity of information data and complex target structure, it will be inefficient to use exhaustion method to choose threshold selection. So maximum likelihood or genetic algorithm can be utilized to improve the efficiency of threshold selection. Besides, the results of traditional segmentation have problems of limited applicability, over segmentation, poor edge smoothness and low accuracy of segmentation.

With the rapid development of computer technology, numbers of new methods such as artificial neural network, support vector machine, fuzzy classification can be applied to classification remote sensing images. Award [1] and Miller et al. [9] use artificial neural network to classify remote sensing data, and to analyze by combining texture information of the image. Mercier et al. [8] use SVM to classify and study remote sensing images. And they considered that SVM is superior than classic supervision classification algorithm and proposed some improved kernel functions to consider the spectral similarity between support vectors, and reduced misinformation caused by traditional kernels. Fuzzy clustering is an important tool for unsupervised classification of remote sensing satellite images. Mukhopadhyay et al. [10] purposed fuzzy clustering based on simulated annealing, which combined with support vector machine to improve its performance. Yuan et al. [17] applied multitask joint sparse representation into high spectrum images classification, which not only reduced spectrum redundancy but also kept the necessary correlation of spectrum during classification. And it also gradually optimized spatial correlation, dramatically enhanced the accuracy and robustness.

Due to its prominent ability of generalization and features of translation, rotation, invariance of local deformation and so on, convolutional neural network is widely used in images segmentation. CNN is a deep learning model of biology and multi-layer with higher recognition rate and more extensive practicality. In 2015, Papandreou et al. [12] used deep convolutional neural network (DCNN) model to develop expectation-maximization EM method for training the semantic image segmentation model. However, when the CNN is applied to image segmentation, each pixel needs to use its surrounding pixel patch as input to CNN, resulting in large storage cost and low computation efficiency. In addition, the CNN requires the input image to have a fixed size. Aiming at this problem, Long et al. [7] put forward the Fully Convolutional Networks (FCN) in 2015 and use the convolutional layer instead of the fully connected layer in standard CNN, maintaining the two-dimensional structure of the image at the same time. FCN has two obvious advantages. (1) It can accept any size of input images without requiring training-images and testing-images to have the identical size; (2) because it avoids the problem of repeat storage and calculation convolution due to the use of patch, it greatly shortens the time of learning and inference. With the engaging of the research, Fu et al. [5] proposed an improved

FCN model to classify the remote sensing images into 12 categories. By using Atrous convolution to increase the density of output diagrams, this method can classify the high-resolution remote sensing images accurately. However, the main limitation is that it requires a large number of high quality Ground Truth tags for model training, which relies on professional experience and amount of manual work. Jiao et al. [6] purpose a new framework of remote sensing image segmentation, named deep multiscale spatialCspectral feature extraction algorithm, which focuses on learning effective discriminant features for remote sensing images and classifies them into 16 categories.

3   Main Idea and Methods

3.1   Data Pretreatment

Remote sensing images refer to pictures or photos that recorded electromagnetic wave of ground objects, which can be mainly classified aerospace pictures and satellite pictures. The styles of remote sensing images differs due to the use of different devices and the change of weather and environment. In this paper, the classification of remote sensing images is about classify the type of ground. In practical, the most common images are types of intertwined ground. To enhance the experiment and eliminate noise interference, the pretreatment is necessary.

Color Mapping. Remote sensing images are acquired by different devices in different weather and environment. So there are more disparities among different images. To eliminate the influence of environment, color transfer is used to train and predict data to reduce the inconsistent by photographing, and unify the color style of the images. According to in-correlation of every channel of $l\alpha\beta$ color space, Reinhard et al. raised a group color transfer formula that in accordance with all color component, which is realized color transfer between color images in a large extent. The basic idea is determining a linear transformation based on statistical analysis of the colored image, which enables source and target images have same mean and variance has in $l\alpha\beta$ space. Remote sensing image itself is a four-channel image, and each image represents information with different color wave band. Although the correlation among each channel is relative strong, to cater to the necessity, linear transformation has conducted on each channel through Reinhard algorithm. If A represents the mean of source image channel a, n_a represents standard deviation of source image channel a, A' represents the mean of target image channel a; n'_a represents the standard deviation of target image channel a. The pixel relation of source image between target image can be expressed as:

$$X = \frac{(x - A) \times n_a}{n'_a} + A' \tag{1}$$

x represents the pixel value of source image. Some of the results of color mapping are show in Fig. 1.

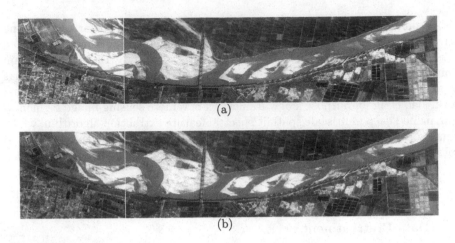

(a)

(b)

Fig. 1. (a) Represent the origin image (b) Represent the image after color mapping.

Feature Enhanced. According to pooling in neural network, it is unavoidable to loss image information, which results in low clarity of edge information after segregating images. To solve this problem, edge detection algorithm is introduced to improve segmentation results. Edge information of images are mainly focused on high frequency stage. The fact of sharpening image or edge detecting is high frequency filter. Differentiation is to calculate the rate of change the signal, with the role of strengthening high-frequency components. In space operations, the sharpening of an image is to differenticate. Due to the signal of digital image is discrete, the differential operation becomes the calculation of difference or the gradient. There are many edge detection (gradient) operators in image processing, which usually include ordinary first difference, Robert operator (cross difference) [13], Sobel operator [16] and so on. The Laplace operator (second difference) is based on the zero crossing point detection. Edge image is acquired by calculating gradient and setting threshold. And canny algorithm [2] is the main method to get the edge image, Canny Edge Detection algorithm is a multi-stage algorithm. The Canny algorithm consists of four stages. 1. Noise Reduction, This step is similar to the Laplacian of Gaussian operator with a Gaussian filter, whose purpose is to eliminate noise, because the noise in the image is also a high-frequency information; 2. Finding Intensity Gradient of the Image. The edges of the image can be pointed to different directions, so the classical Canny algorithm uses four gradient operators to calculate the gradient of the horizontal, vertical and diagonal direction respectively; 3. Non-maximum Suppression. Non-maximum Suppression is an edge thinning method; 4. Hysteresis Thresholding. Use double threshold and lag boundary tracking to improve the accuracy of the algorithm. In Noise Reduction, a big Gaussian filter size can be selected to weaken the edge information in the region of the image. It is more conducive to segregate the image. Some of the results of feature enhanced are show in Fig. 2.

3.2   FCN

As with traditional supervised learning, FCN is pre-training in the beginning and then classify images.

Architecture. FCN transforms the fully connected layers into convolutional layers enables the classical classification CNN. The picture as Fig. 3.: In classical CNN structure the first five layers are convolutional layers, the size of convolutional kernel (the number of channels, the width, and the height) is (64, 3, 3), (128, 3, 3), (256, 3, 3), (512, 3, 3), (512, 3, 3) separately. The 6th and 7th layers are one dimension vector with the length of 4096, the 8 layer is one dimension vector with the length of 25, separately corresponding to 25 probability region. FCN represents the 6, 7, 8 layer as convolution layer, the size of convolutional kernels is (4096, 1, 1), (4096, 1, 1), (25, 1, 1) separately. All layers are convolutional layer, so it also known as fully convolutional network. After many convolutional operations, the resolution is relatively small, FCN will upsample image and recover the original image. In the process of upsampling, deconvolution is used in FCN. However, the stride is 32 pixels in FCN-32s structure, whose segmentation detail is coarse. So Long [7]combines the final prediction layer with lower layers.The pool5 layer is 2x upsampled and forms a skip connection with the pool4 layer, we represent it as deconv2 layer in Fig. 3. Through the fusion of deconv2 layer and pool4 layer, the final net FCN-8s is obtained. And refined image is able to acquire. The process of convolution and deconvolution are show in Fig. 3.

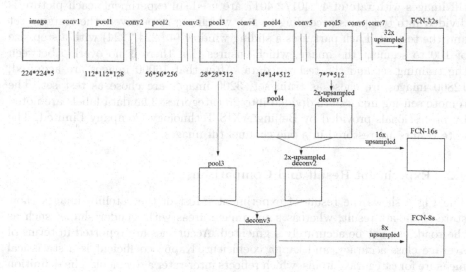

Fig. 2. Architecture of FCN.

Classification Using the Trained Network. By using FCN trained model, the probability of each pixel in the image will be given, then the final classification of each pixel will be acquired through softmax function.

FCN Advantage and Disadvantage. Compared with traditional CNN for image segmentation, FCN has two obvious advantages: one is it can accept any size of input images instead of requiring all experiment images have identical size; the other one is more efficient due to avoid repeat storage and computational convolution brought by using image patch. Meanwhile, the disadvantages of FCN is also obvious: one is the result of up sampling is not fine enough. Although the result of FCN-8s is better than FCN-32s, the results of upsampling is relatively blur, and it is not sensitive to reflect details of images; the other one is the classification of each pixel does not fully take the relations between each pixel into consideration and it neglects spatial regularization steps used in the segmentation methods of pixel classification, so it lacks spatial consistency. Considering that FCN is not sensitive to details, we add the edge information of the image as an important information to the image, and expand the image into a 5 channels image, thus improving the processing edge details of FCN.

4   Experiment

4.1   Data

Data of this paper provided by Beijing AXIS Technology Company Limited. 10 tiff images with rate of $4 * 4017 * 4017$ are used. In experiment, each picture is divided into two parts according to 8:2, which are respectively the training set and the test set. Each part uses a sliding window ($4 * 224 * 224$) with a step size of 100 to segment the image, which ensures that there is no overlap between the training set and the test set data. After that 16000 images are acquired, 12800 images are chose as train set, 3200 images are chose as test set. The remote sensing images are divided into 25 categories. The data labels are noted by professionals provided by Beijing AXIS Technology Company Limited. The note results are restored in a one channel tiff images.

4.2   Experiment Result and Comparison

The Fig. 4 shows the result of experiment. Most of the satellite images show state-of-the-art result, whereas some image areas with slender shape, such as the road, can not be accurately segmented. Accuracies are reported in terms of average class accuracy, and kappa coefficient. Kappa coefficient is a statistical measure for categorical items, which reflects inter-rater agreement. The definition of kappa coefficient is:

$$k = \frac{p_o - p_e}{1 - p_e} \tag{2}$$

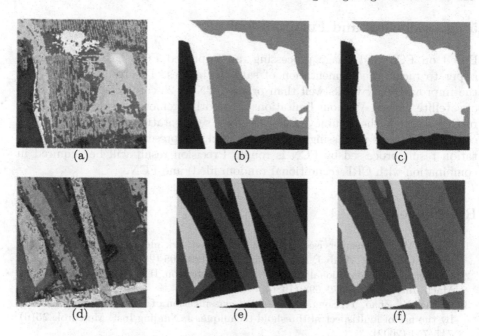

Fig. 3. (a)(d) Represent the origin image (b)(e) Represent the annotation. (c)(f) Represent the result.

p_o represents the overall classification accuracy, which is divided by the sum of the number of samples correctly classified by the total number of samples. N_{k1} and N_{k2} represent the number of predict pixels and the number of real pixels for each category, respectively. N is the total number of samples, that is, the number of pixels.

$$p_e = \frac{1}{N^2} \Sigma_{k=1}^{25} n_{k1} n_{k2} \tag{3}$$

Table 1. Compare with other algorithms

Method	Accuracy	Kappa
FCN	0.727	0.476
Our method	0.733	0.485

The Table 1 shows that compared with original FCN, our method has high accuracy and kappa coefficient. The result showed that the revise of FCN is useful, but the effect is not obvious. If image processing is utilized, detail processing of FCN can be improved.

5   Conclusion and Future Work

Based on FCN and image processing, an improved FCN has raised in this paper to tackle the segmentation of satellite images. The result showed that the improved FCN works well than origin FCN in 25-categories segmentation of satellite images. Without limitations of shooting remote sensing images after further research, the accuracy of 25-categories segmentation can be improved by combining images processing and FCN model. At present, the image segmentation result processed by FCN is rough. Precision result will be acquired in combination with CRF (conditional random filed) and FCN.

References

1. Awad, M.: An unsupervised artificial neural network method for satellite image segmentation. Int. Arab J. Inf. Technol. **7**(2), 199–205 (2010)
2. Canny, J.: A computational approach to edge detection. IEEE Trans. Pattern Anal. Mach. Intell. **11**(6), 679–698 (1986)
3. Cheng, Y., Zhao, Y., Song, G.U.: Cloud classification of GMS-5 satellite imagery by the use of multispectral threshold technique. J. Nanjing Inst. Meteorol. **25**(6), 747–754 (2002)
4. Frauman, E., Wolff, E.: Segmentation of very high spatial resolution satellite images in urban areas for segments-based classification. In: Proceedings for 3rd International Symposium Remote Sensing and Data Fusion Over Urban Areas, Tempe, Arizona (2005)
5. Fu, G., Liu, C., Zhou, R., Sun, T., Zhang, Q.: Classification for high resolution remote sensing imagery using a fully convolutional network. Rem. Sens. **9**(5), 498–498 (2017)
6. Jiao, L., Liang, M., Chen, H., Yang, S., Liu, H., Cao, X.: Deep fully convolutional network-based spatial distribution prediction for hyperspectral image classification. IEEE Trans. Geosci. Rem. Sens. **55**(10), 5585–5599 (2017)
7. Long, J., Shelhamer, E., Darrell, T.: Fully convolutional networks for semantic segmentation. In: Proceedings of the IEEE Conference On Computer Vision and Pattern Recognition, pp. 3431–3440 (2015)
8. Mercier, G., Lennon, M.: Support vector machines for hyperspectral image classification with spectral-based kernels. In: 2003 Proceedings of IEEE International Conference on Geoscience and Remote Sensing Symposium, IGARSS 2003, vol. 1, pp. 288–290. IEEE (2003)
9. Miller, D.M., Kaminsky, E.J., Rana, S.: Neural network classification of remote-sensing data. Comput. Geosci. **21**(3), 377–386 (1995)
10. Mukhopadhyay, A., Maulik, U.: Unsupervised satellite image segmentation by combining SA based fuzzy clustering with support vector machine. In: 2009 Seventh International Conference on Advances in Pattern Recognition, ICAPR 2009, pp. 381–384. IEEE (2009)
11. Munandar, T., Suhendar, A., Abdullah, A., Rohendi, D., et al.: Satellite image edge detection for population distribution pattern identification using levelset with morphological filtering process. In: IOP Conference Series: Materials Science and Engineering. vol. 180, pp. 012064–012064. IOP Publishing (2017)

12. Papandreou, G., Chen, L.C., Murphy, K.P., Yuille, A.L.: Weakly-and semi-supervised learning of a deep convolutional network for semantic image segmentation. In: Proceedings of the IEEE International Conference on Computer Vision, pp. 1742–1750 (2015)
13. Roberts, L.G.: Machine perception of three-dimensional solids. Ph.D. thesis, Massachusetts Institute of Technology (1963)
14. Saha, I., Maulik, U., Bandyopadhyay, S., Plewczynski, D.: SVMeFC: SVM ensemble fuzzy clustering for satellite image segmentation. IEEE Geosci. Rem. Sens. Lett. 9(1), 52–55 (2012)
15. Sharma, S., Buddhiraju, K.M., Banerjee, B.: An ant colony optimization based inter domain cluster mapping for domain adaptation in remote sensing. In: 2014 IEEE International Conference on Geoscience and Remote Sensing Symposium (IGARSS), pp. 2158–2161. IEEE (2014)
16. Sobel, I., Feldman, G.: A 3×3 isotropic gradient operator for image processing. A Talk at the Stanford Artificial Project, pp. 271–272 (1968)
17. Yuan, Y., Lin, J., Wang, Q.: Hyperspectral image classification via multitask joint sparse representation and stepwise MRF optimization. IEEE Trans. Cybern. 46(12), 2966–2977 (2016)

A New LSTM Network Model Combining TextCNN

Xiao Sun[1,2], Xiaohu Ma[1,2(✉)], Zhiwen Ni[1,2], and Lina Bian[1,2]

[1] School of Computer Science and Technology, Soochow University,
Suzhou 215006, China
xhma@suda.edu.cn
[2] Collaborative Innovation Center of Novel Software Technology
and Industrialization, Nanjing 210023, China

Abstract. The development of computer communication technology has brought massive amounts of spam texts. Spammers use a variety of textual means to avoid detection of spam texts, which has brought challenges to spam text filtering technology. Deep neural network has superior performance in feature representation and feature extraction. TextCNN based on convolutional neural network can extract the local feature representation of sentences, but ignore the successive relationship between words. The LSTM based on the recurrent neural network takes into account the sequential relationship between words, but it is not as good as TextCNN in representation of local features. We propose an algorithm that combines the TextCNN and LSTM network called TC-LSTM to implement spam text filtering, and compare the Precision, Recall and F-measure indicators with the traditional TextCNN and LSTM on two datasets. Experiments show that our TC-LSTM algorithm is superior to the traditional TextCNN and LSTM networks in spam text filtering.

Keywords: Deep neural network · Network fusion
Spam text filtering · Deep learning

1 Introduction

With the rapid development of technologies such as cloud computing, big data and the Internet of Things, various applications on the Internet are also complex and diverse [10]. Large amounts, real-time, high-speed and a wide variety of spam text appear. Spam text not only takes up a lot of computing and communication resources, but also has a serious impact on human life, which has become one of the most important problems to be solved.

The job of spam text filtering is nothing. However, manually removing harmful, unwanted and offensive messages before they are delivered to a user costs much time and resources. It is usually to use machine learning methods like Logistic Regression to classify these spam text. Recently, deep learning methods make great progress in computer vision and natural language processing.

© Springer Nature Switzerland AG 2018
L. Cheng et al. (Eds.): ICONIP 2018, LNCS 11301, pp. 416–424, 2018.
https://doi.org/10.1007/978-3-030-04167-0_38

The Convolutional Neural Network was originated from the processing of computer vision and achieved great success [15]. Later research show that the algorithm has a good effect on the field of natural language processing [2]. The neural network uses a convolution operation proposed by LeCun et al. [7] to convolve the data to obtain local features. TextCNN extracts the local keyword features of the sentence through a convolutional neural network, which is proved efficient for the spam text filtering in our later experiment. However, this method ignores the correlation between sample data. Hochreiter et al. [4] proposed a model using long-short-term memory cells to memorize correlations between samples to achieve better classification results.

Our contributions are as follow:

(1) We proposed a model combining TextCNN and LSTM, which called TC-LSTM for spam text filtering. TC-LSTM extracts both keywords and successive relationship features which contributes to classification because it has both the structure of TextCNN and LSTM.
(2) We compared TC-LSTM with TextCNN and LSTM in two spam text datasets of Precision, Recall and F-measure values. Experiments show that our proposed TC-LSTM outperforms the other two networks.

1.1 Task Definition

The spam text we define is what with sensitive content and marketing information. Sensitive content includes sensitive political tendencies, verbal abuse, pornography and other unhealthy information. Marketing information refers to text information related to the marketing of a company.

To solve the problem, we should first implement the two key points below: (1) Artificial interference. In order to evade filtering of keyword matching, spammers usually deformation keywords to unrelated formation. For instance, sentence like {call 19933oo8834 to order} is distortion of sentence {call 19933008834 to order}. (2) Context based classification. Some spam texts do not have obvious keyword features nor distortion of keywords. Classification of spam text like this needs a content-based method.

Given a sentence: $s = \{w_1, w_2, ...w_n\}$, where w_i is a word of a sentence. Each sentence s have a label y. Where $y = 0$ means s is a positive sample which has no spam information. Where $y = 1$ means s is a negative sample which is spam text. What we need to do is to give each sentence a correct label in the datasets.

2   Background

2.1   Word Vectors

Deep learning models have achieved remarkable results in computer vision and natural language processing. When comes to natural language processing, deep learning methods have involved learning word vector representations [1,11]. Much work is focused on classification with word vectors [3]. Word vectors, wherein words are projected from a sparse, 1-of-V encoding (V is the vocabulary size) onto a lower dimension vector space via a hidden layer. In such dense representations, semantically close words are likewise close in lower dimensional vector space.

2.2 TextCNN

Convolutional neural networks (CNN) utilize layers with convolving filters that are applied to local features. Originally invented for computer vision, CNN models have subsequently been shown to be effective for NLP and achieved excellent results in semantic parsing [14], search query retrieval [12], sentence modeling [5], and other traditional NLP tasks.

Figure 1 is the architecture of TextCNN [6], which is used for sentence feature extraction. Let $x_i \in \mathbb{R}^k$ be the k-dimensional word vector of corresponding to the i-th word in the sentences. A sentence of length n is represented as:

$$\mathbf{x}_{1:n} = \mathbf{x}_1 \oplus \mathbf{x}_2 \oplus ... \oplus \mathbf{x}_n \tag{1}$$

Where $\oplus$ is the concatenate operation. $\mathbf{x}_{i:i+j}$ indicates $\mathbf{x}_i, \mathbf{x}_{i+1}, ...\mathbf{x}_{i+j}$. A convolution operation means to apply $\mathbf{W} \in \mathbb{R}^{hk}$ to a window of h words to produce a new feature. For example:

$$c_i = f(\mathbf{W} \cdot \mathbf{x}_{i:i+h-1} + b) \tag{2}$$

Eq. (2) create a new feature c_i. Where $b \in \mathbb{R}$ is a bias and f is a non-linear function. The filter slide with a window of size h on the sentence to create a feature map:

$$\mathbf{c} = [c_1, c_2, ..., c_{n-h+1}] \tag{3}$$

Where $\mathbf{c} \in \mathbb{R}^{n-h+1}$. Then we apply a max-pooling operation to the feature map to get the max value. Figure 1 describes the process of feature extraction of a convolution operation. Each convolution operates on word vectors of a sentence and the results are pooled lately. We use the Max-pooling method in the pooling step.

2.3 Long Short Term Memory

Recurrent Neural Networks were proposed in the 1980's but have just been recently gaining popularity from advances to the networks designs and increased computational power from graphic processing units. They're especially efficient with sequential data because each neuron or unit can use its internal memory to maintain information about the previous input. Recurrent Neural Network can learn the long-term dependencies of words in sentences [8,9,13,16], and have achieved remarkable results in natural language processing based on serialization problems [8].

$$h_t = \begin{cases} 0, & t = 0; \\ f(h_{t-1}, x_t), & \text{otherwise.} \end{cases} \tag{4}$$

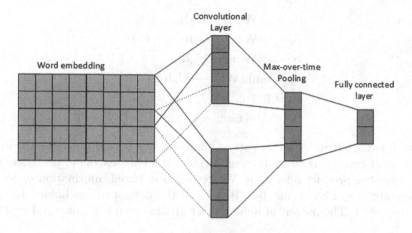

Fig. 1. The architecture of TextCNN

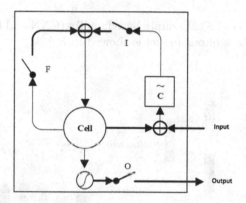

Fig. 2. The architecture of LSTM

Equation (4) describes the process of RNN hidden layer propagation. Where x_t represents the input of the hidden layer at time step t. h_t is the output of the hidden layer at time step t.

Since the RNN usually suffers from vanishing or exploding gradient prob when training long sequences, a solution is to use the Long-Short-Time-Memory [4] (LSTM) unit. The LSTM unit uses an isolation memory units which is updates only when necessary. As shown in Fig. 2, LSTM uses a memory unit to save the important information of an input sentence and abandon the useless information at the same time. A LSTM unit includes a Cell and 3 gate. Where F is forget gate, I is input gate, O is output gate. The propagation of LSTM is as (5)–(10):

$$\mathbf{f}_t = \sigma(\mathbf{W}_f\mathbf{x}_t + \mathbf{U}_f\mathbf{h}_{t-1} + b_f), \tag{5}$$

$$\mathbf{i}_t = \sigma(\mathbf{W}_i\mathbf{x}_t + \mathbf{U}_i\mathbf{h}_{t-1} + b_f), \tag{6}$$

$$\mathbf{o}_t = \sigma(\mathbf{W}_o\mathbf{x}_t + \mathbf{U}_o\mathbf{h}_{t-1} + b_o), \tag{7}$$

$$\widetilde{\mathbf{c}}_t = \tanh(\mathbf{W}_c\mathbf{x}_t + \mathbf{U}_c\mathbf{h}_{t-1}), \tag{8}$$

$$\mathbf{c}_t = \mathbf{f}_t \odot \mathbf{c}_{t-1} + \mathbf{i}_t \odot \widetilde{\mathbf{c}}_t, \tag{9}$$

$$\mathbf{h}_t = \mathbf{o}_t \odot \tanh(\mathbf{c}_t) \tag{10}$$

Where $\mathbf{f}_t$ is the output of forget gate at time step t in LSTM unit. $\mathbf{i}_t$ is the input of the input gate. $\mathbf{o}_t$ is the output of output gate. σ is a sigmoid function. $\odot$ is a element-wise product operation. We get context based information $\widetilde{\mathbf{c}}_t$ by (8) and update $\mathbf{c}_t$ in LSTM unit by (9). $\mathbf{h}_{t-1}$ is the output of the hidden layer at time step $t-1$. The output of hidden layer at time step t is computed by (10).

3   TC-LSTM

We proposed model TC-LSTM simply based on TextCNN and LSTM mentioned above. The architecture of our model is showed in Fig. 3.

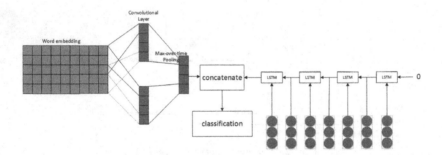

Fig. 3. The architecture of TC-LSTM

TC-LSTM model uses TextCNN to extract keyword features from sentences, and uses LSTM to get correlation between words in a sentence. TextCNN obtains a local feature of a sentence after Max-overtime-pooling, which may be the keyword feature that affects sentence classification. The LSTM in TC-LSTM extracts the serialization features of the sentence, which preserves the sequential relationship of a sentence and can characterize the logical semantics. We combine the keyword features with the successive relationship features to a comprehensive feature. We think this kind of feature is more effective than using only one of the two network structures.

In the task of spam text filtering, obvious sensitive words can usually determine the classification result of a sentence. The TextCNN part of TC-LSTM can extract the local keyword features which is obvious sensitive words mentioned above and the LSTM part obtain the logistic semantic sequential relationship of the sentence. Our TC-LSTM can be used to classify the sentences whose keywords are not significant. We apply the new features to classifier to get a better classification result.

3.1 Feature Combination

We use independent sentences as input to word2vec to get word vectors. These word vectors are applied to TextCNN and LSTM. Make the output of convolution filter in TextCNN be $\mathbf{c} = f(\mathbf{W} \cdot r + b)$ where $\mathbf{W}$ is the convolution weight, r is the word vectors and b is a bias. Then the output of TextCNN is $\tilde{c} = max\{\mathbf{c}\}$. Let the output of LSTM be $\tilde{s} = f(r)$, where f is the function of LSTM unit. Then the feature combination can be defined as (11).

$$e = \tilde{s} \oplus \tilde{c} \tag{11}$$

4   Experiments and Analysis

We use softmax function to access multi-layer perceptron of (11). To prove the effectiveness of our model, We experimented on two datasets compared with TextCNN and LSTM.

Dataset 1 is what we scraped from the Internet. It contains 2888 pornography, political sensitive or abuse sentences and 2888 normal sentences. Dataset 2 is from Sohu Content Identification Algorithm Contest, 2018. The details of the datasets are in Table 1. We use logistic regression with TFIDF to encode documents as baseline. To get vectors from text, we use word2vec to initialize every word. The words which are not in word2vec vocabulary are set 0. The parameters in TextCNN are listed in Table 2:

Table 1. Experiment datas

DataSet	Positives	Negatives
Dataset1	2888	2888
Dataset2	638665	831450

We compared our TC-LSTM with three models in two datasets to verify the effectiveness:

(1) LR: We use Logistic Regression as baseline with TFIDF to get vectors of sentences.
(2) TextCNN: We use different size of filters to exact keyword features of each sentence. The disadvantage is that it is hard to identify sentences with inconspicuous keywords.

Table 2. Parameters in TextCNN

Parameters	Value
Convolution size	2,3,4
Number of filters	128
Dropout	0.3

(3) LSTM: The sentence serialization vector is input into each time step, and output the sentence context relationship features. The disadvantage is that it works not well when sentence is too long and too complicated.

4.1   Experiment Results

Our experiment is based on 4 models on 2 datasets mentioned above. The results are as below in Table 3.

Table 3. The experiment results of TC-LSTM compared with 3 models on 2 DataSets

Model name	Evaluation	DataSet1	DataSet2
LR	P	0.955	0.702
	R	0.920	0.617
	F	0.937	0.657
TextCNN	P	0.955	0.673
	R	0.952	0.705
	F	0.954	0.688
LSTM	P	0.956	0.715
	R	0.965	0.684
	F	0.961	0.699
TC-LSTM	P	0.965	0.697
	R	0.955	0.719
	F	0.975	0.707

Where P is Precision, R is Recall, F is F-measure. It is obvious in Table 3 that TC-LSTM outperforms LR, TextCNN and LSTM on F-measure. Specifically, when comes to DataSet1, TC-LSTM is 2.1% higher than TextCNN, 3.8% higher than LR and 1.4% higher than LSTM. When comes to DataSet2, as the scale of data is up to million, F-measure is lower in DataSet2 than in DataSet1. Specifically, TC-LSTM is 5% higher than LR, 1.9% higher than TextCNN and 0.8% higher than LSTM.

We can obviously see the improvement of TC-LSTM compared with LR, TextCNN, LSTM on DataSet1 in Fig. 4 and on DataSet2 in Fig. 5.

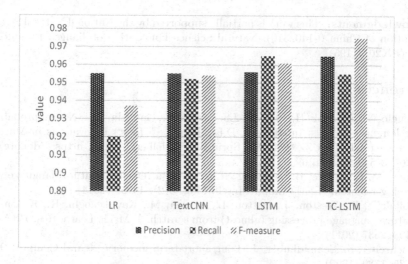

Fig. 4. The experiment on DataSet1

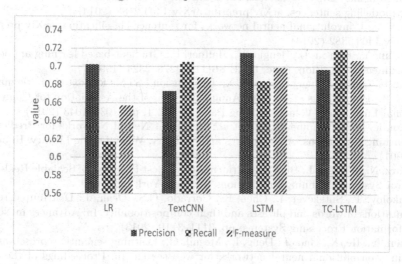

Fig. 5. The experiment on DataSet2

5 Conclusion

We proposed a neural network called TC-LSTM combining TextCNN and LSTM
for spam text filtering. We chose logistic regression as the baseline, together
with TextCNN and LSTM for comparison. The experiment results show that
our algorithm has achieved better performance for spam text filtering, which
is because TC-LSTM has both the advantage of TextCNN and LSTM. TC-
LSTM can both capture the keywords of the spam text and obtain the successive
relationship between words in a sentence. In addition, as LSTM takes too long to
train an epoch, our further work will be focused on reducing the training time.

Acknowledgments. This work is partially supported by the National Natural Science Foundation of China (61402310). Natural Science Foundation of Jiangsu Province of China (BK20141195).

References

1. Bengio, Y., Schwenk, H., Senécal, J.S., Morin, F., Gauvain, J.L.: Neural probabilistic language models. In: Holmes, D.E., Jain, L.C. (eds.) Innovations in Machine Learning. STUDFUZZ, vol. 194. Springer, Heidelberg (2006). https://doi.org/10.1007/3-540-33486-6_6
2. Cambria, E., White, B.: Jumping NLP curves: a review of natural language processing research. IEEE Comput. Intell. Mag. **9**(2), 48–57 (2014)
3. Collobert, R., Weston, J., Bottou, L., Karlen, M., Kavukcuoglu, K., Kuksa, P.: Natural language processing (almost) from scratch. J. Mach. Learn. Res. **12**(Aug), 2493–2537 (2011)
4. Hochreiter, S., Schmidhuber, J.: Long short-term memory. Neural Comput. **9**(8), 1735–1780 (1997)
5. Kalchbrenner, N., Grefenstette, E., Blunsom, P.: A convolutional neural network for modelling sentences. arXiv preprint arXiv:1404.2188 (2014)
6. Kim, Y.: Convolutional neural networks for sentence classification. arXiv preprint arXiv:1408.5882 (2014)
7. LeCun, Y., Bottou, L., Bengio, Y., Haffner, P.: Gradient-based learning applied to document recognition. Proc. IEEE **86**(11), 2278–2324 (1998)
8. Liu, P., Qiu, X., Chen, J., Huang, X.: Deep fusion LSTMs for text semantic matching. In: Proceedings of the 54th Annual Meeting of the Association for Computational Linguistics (Volume 1: Long Papers), vol. 1, pp. 1034–1043 (2016)
9. Mandic, D.P., Chambers, J.A., et al.: Recurrent Neural Networks for Prediction: Learning Algorithms, Architectures and Stability. Wiley Online Library, Hoboken (2001)
10. Marz, N., Warren, J.: Big Data: Principles and Best Practices of Scalable Real-time Data Systems. Manning Publications Co, New York (2015)
11. Mikolov, T., Sutskever, I., Chen, K., Corrado, G.S., Dean, J.: Distributed representations of words and phrases and their compositionality. In: Advances in Neural Information Processing Systems, pp. 3111–3119 (2013)
12. Shen, Y., He, X., Gao, J., Deng, L., Mesnil, G.: Learning semantic representations using convolutional neural networks for web search. In: Proceedings of the 23rd International Conference on World Wide Web, pp. 373–374. ACM (2014)
13. Wang, X., Liu, Y., Chengjie, S., Wang, B., Wang, X.: Predicting polarities of tweets by composing word embeddings with long short-term memory. In: Proceedings of the 53rd Annual Meeting of the Association for Computational Linguistics and the 7th International Joint Conference on Natural Language Processing (Volume 1: Long Papers), vol. 1, pp. 1343–1353 (2015)
14. Yih, W., He, X., Meek, C.: Semantic parsing for single-relation question answering. In: Proceedings of the 52nd Annual Meeting of the Association for Computational Linguistics (Volume 2: Short Papers), vol. 2, pp. 643–648 (2014)
15. Yu, K., Jia, L., Chen, Y., Xu, W.: Deep learning: yesterday, today, and tomorrow. J. Comput. Res. Dev. **20**(6), 1349 (2013)
16. Zhou, X., Wan, X., Xiao, J.: Attention-based LSTM network for cross-lingual sentiment classification. In: Proceedings of the 2016 Conference on Empirical Methods in Natural Language Processing, pp. 247–256 (2016)

Self-inhibition Residual Convolutional Networks for Chinese Sentence Classification

Mengting Xiong, Ruixuan Li[✉], Yuhua Li, and Qi Yang

School of Computer Science and Technology,
Huazhong University of Science and Tecnnology, Wuhan 430074, China
{mengtingxiong,rxli,idcliyuhua,ayang7}@hust.edu.cn

Abstract. Convolutional network has become a dominant approach in many Natural Language Processing (NLP) tasks. However, these networks are pretty shallow and simple so they are not able to capture the hierarchical feature of text. In addition, text preprocessing of those models in Chinese are quite rough, which leads to the loss of rich semantic information. In this paper, we explore deep convolutional networks for Chinese sentence classification and present a new model named Self-Inhibition Residual Convolutional Network (SIRCNN). This model employs extra Chinese character information and replaces convolutional block with self-inhibiting residual convolutional block to improve performance of deep network. It is one of the few explorations which use deep convolutional network in various text classification tasks. Experiments show that our model can achieve state-of-the-art accuracy on three different datasets with a better convergence rate.

Keywords: Self-inhibiting residual · Text classification
Character embedding

1 Introduction

Deep learning approaches in Natural Language Processing (NLP) obtain more attention in academia and industry [2]. There are various NLP tasks, in which text classification is one of the most important ones. For this task, Recursive neural networks (RNN) and convolutional neural network (CNN) based architectures have become standard baselines.

Liu et al. [11] proposed an RNN model which has the ability to capture the contextual information. RNN takes sequences as the input data. Advanced model LSTM [3,16] can also be used to learn text representations and modeling. A big disadvantage of these models is that the training process cannot be parallelized. Thus CNN related methods are more popular in academic and industrial fields.

Convolutional kernels in CNN become the detector of a particular n-gram feature. Kim [7] firstly developed a CNN architecture for classification tasks,

© Springer Nature Switzerland AG 2018
L. Cheng et al. (Eds.): ICONIP 2018, LNCS 11301, pp. 425–436, 2018.
https://doi.org/10.1007/978-3-030-04167-0_39

which is an unbiased model. Kalchbrenner et al. [6] proposed a dynamic convolution neural network for semantic modeling of sentences. They first designed a dynamic k-max pooling strategy to replace the max-pooling strategy in the original CNN. Yang et al. [17] used the hierarchical network structure to fit the structure of the document, and combined the attention mechanism to achieve good results. Johnson and Zhang [5] discussed the effect of word order on text classification.

The convolutional neural networks for computer vision tasks tend to be deeper because deep structures are good at learning hierarchical feature representation from simple edge to shape in images [4]. By contrast, deep network is barely used for text tasks though texts have similar compositional structure, where characters form words, words form phrases, and phrases form sentences, etc. At present, a few articles [10,14] study the performance of deep convolutional networks in text categorization. Schwenk et al. [14] proposed a deep network that borrows ideas from ResNet and VGG networks, which uses only convolution and pooling operations. The network reached 29 layers and performed well on large datasets. Le et al. [10] studied the effect of network layers on text categorization, and they used DenseNet as a typical method of deep networks.

Most methods of RNN or CNN based architectures take words as their input of the network, which benefit from the success of distributed representation [12]. This is called, word embedding in which *word2vec* became a basic step of almost all methods. With the distributed representation of words, we can convert texts into an image-like format. There are several works [8,20] dealing with characters rather than words. For a different language, the definition of the character could be totally different. However, the current preprocessing of Chinese in the literature does not take this into consideration. We carefully study the divergences between Chinese and English languages, and enrich the model input according to the characteristics of Chinese.

In summary, our model aims at extracting compositional information by deepening the CNN model. To solve the problem of information loss in transforming Chinese to alphabet representation and the training problem of deep network, we propose a new model called Self-Inhibition Residual Convolutional Network (SIRCNN) that has two innovative points.

- We redefine the meaning of the characters, and combine words with characters as the input of SIRCNN. This approach enriches the input features of the model, and considers the linguistic features.
- To tackle the training conundrum of deep network, we design an improved residual block, which has a better effect than the traditional residual structure. Experiments show that this structure can achieve better results than the original convolutional network, which is also the best result of deep network in the field of text classification.

2   SIRCNN Architecture

2.1   Architecture Input

In Chinese, rich semantic and syntax information is lost in the process of converting the original word into *Pinyin* representation: (1) Chinese is a very complex language, and there are still some problems in Chinese word segmentation, like ambiguity and new words recognition; (2) Many words or characters have the same Pinyin representation (as shown in Table 1), e.g.: 石狮 (stone lion) and 实施 (implement) have the same character-level representation 'shi2shi1', but completely different meanings and part of speech (POS) information. Different words have the same character-level representation, which will make their meanings confusing.

Table 1. Different representation for an example sentence.

Type	Text
text	北京2008奥运会火炬接力经过世界五大洲21个城市
meaning	Beijing 2008 Olympic Games torch relay through 21 cities on five continents
Pinyin	be3i ji1ng 2008 a4o yu4n hui4 huo3 ju4 jie1 li4 ji1ng guo4 shi4 jie4 wu3 da4 zho1u 21 ge4 che2ng shi4
word	北京 2008 奥运会 火炬 接力 经过 世界 五大洲 21 个 城市
character	北京 2008 奥 运 会 火 炬 接 力 经 过 世 界 五 大 洲 21 个 城 市

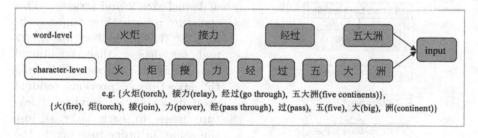

Fig. 1. Relationship of character and word in Chinese for a text sample.

Glyph is a typography term indicating a readable character for the purposes of writing. CJK[1] languages consist of characters that are rich in their topological form, where strokes and parts could represent semantic meaning, and make character glyphs a potentially feasible encoding solution [19]. Thus, we redefine the meaning of Chinese character as unconnected squares glyphs (see Table 1).

Figure 1 demonstrates a more natural segmentation of the Chinese text. Instead of using *pypinyin*[2] package to translate Chinese text into Pinyin representation and then to sequences of "abc ... yz", we use characters redefined as

[1] https://en.wikipedia.org/wiki/CJK_characters.
[2] https://pypi.org/project/pypinyin/.

input. Chinese characters can capture sematic and morphological level information of Chinese text. If not indicated, the characters mentioned below refer to the definition in this subsection. This approach considers more language-specific features and gives the network more textual features that allow the network to start learning from a more substantive level.

2.2 Overall Architecture

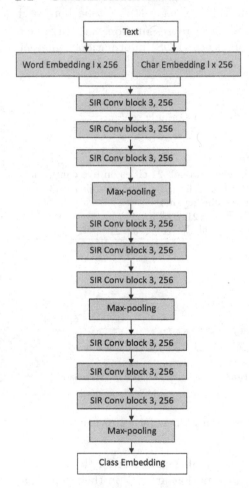

Fig. 2. Overall architecture.

The overall architecture of Self-Inhibition Residual Convolutional Network (SIRCNN) is shown in Fig. 2. The text will be preprocessed in two steps. First, the text is split into words. Second, The text is split into characters. We use *word2vec* to train character and word embeddings. These steps generate two 2D tensors of size (1,256), where l is the length of one sentence and 256 is the dimension of word and character embeddings.

Our network begins with three self-inhibiting residual convolutional blocks, then a max-pooling layer of kernel size 2 and stride 1. This structure repeats three times, making the layer number up to 12 and far deeper than traditional convolutional networks. Following the structure of previous residual deep network, the number of feature maps in each layer is just the same, in order to generate the same output. https://v2.overleaf. com/project/5b2f35b65b43cf4aa2ba 8518 Especially we define the kernel size of last max-pooling layer as 6, and stride 2 to get the important feature in a larger receptive field. This will halve the length of text to get higher level representation feature. Last, we use a simple fully connected classifier to get the final classification result.

The network structure is simple and parallelizable. This structure is currently the result of our choice after many attempts. Most classification networks are very shallow. The deepest network VDCNN [14] needs to reach 29 layers to achieve competitive results than classic approaches. However, our network only needs 12 layers to get better result than shallow networks or ResNet-like deep networks. Thus, both the training time and the number of hyperparameters are greatly reduced.

2.3   Self-inhibiting Residual Convolutional Block

The traditional convolution of one-dimensional involves feature maps sliding over sequences. Feature maps can detect features at different positions. Supposing a map $W \in R^{h \times d}$ of height h, a convolution operation on h consecutive word or character vectors (dimension m) outputs the *feature*

$$c_t = ReLU(W \cdot X_{t:t+h-1} + b) \tag{1}$$

where $X_{t:t+h-1} \in R^{h \times d}$ is the matrix whose i^{th} row is $x_i \in R^d$, and $b \in R$ is a bias. The symbol $\cdot$ refers to the dot product and ReLU is the element-wise rectified linear unit function.

We perform convolution operations with k different feature maps, and denote the resulting features as $c_t \in R^n$, each of whose dimensions comes from a distinct map. Repeating the convolution operations for each window of h consecutive words in the short text, we obtain $c_{1:l-h+1}$, where l is the length of the text.

In short, denoting the desired underlying mapping of each traditional convolution block as $y = H(x, w_h)$, where $x \in R^{h \times d}$ is the input matrix, w_h are the parameters of the mapping and $y \in R^{1:l-h+1}$ is the output vector of convolution block.

Deeper layers and less receptive field can improve the accuracy of network classification. CNN can extract the feature at different level. Moreover, the deeper the network is, the more semantic information extracted features have. At the same time, network training becomes more and more difficult. Highway Network [15] is a network framework that addresses deep network training difficulties. Highway Network is defined as Formula 2

$$y = H(x, w_h) = F(x, w_f)T(x, w_t) + xC(x, w_c) \tag{2}$$

T is the transform gate and C is the carry gate, which are non-linear activation functions. These gates can control how much input and output pass through the layers, where w_t, w_c are the parameters of these transformations.

We set T = 1 and C = 1, which means letting input and output pass through, such that

$$H(x, w_h) = F(x, w_f) + x \tag{3}$$

This is the core formula of ResNet [4]. ResNet lets the convolutional layer to fit $F(x, w_f)$ instead of $H(x, w_h)$, and make a shortcut by adding x with the output, which makes the whole mapping became $F(x, w_f) + x$. To some extent, ResNet can be seen as a variant of Highway Network.

These strategies aim at solving training problems of deep network, which to some extent draws on the design idea of LSTM. Self-inhibiting residual convolutional block (see Fig. 3) is inspired by ResNet and Highway Network in image classification. We experiment with the subtraction operation. We let the input x pass directly, and then suppress x by the output value of the layer. Supposing the mapping we desire is $H(x, w_h)$, the non-linear layers are fitting to $F(x, w_f)$. A shortcut of x subtracts the output of convolutional layer, then the whole mapping becomes

$$H(x, w_h) = x - F(x, w_f) \tag{4}$$

The operation $x - F(x, w_f)$ is performed by a shortcut connection and element-wise subtraction. $F(x, w_f)$ is similar to a differential amplifier, which focuses on learning the difference between $F(x, w_f)$ and x, and more sensitive to the variation of output of each layer. The differential amplifier makes the optimization problem simple and the update speed of all parameters changes faster. The operations introduce no other parameters or computation complexity. The whole network can still be trained by end to end backpropagation. We call it self-inhibiting residual convolutional.

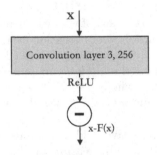

Fig. 3. Self-inhibiting residual convolutional block.

Since the residual convolutional block needs to maintain the input and output in the same dimension, we use only convolutions of size 3, with padding in both sides and followed by a Batch Normalization and a ReLU activation. The length of c in each layer is $l + 2 - 3 + 1 = l$. Stacking of convolutions with size 3 is able to capture the dependence of long-distance, which can learn to combine those minimal size convolutional blocks to get feature size of 4 or 5.

3   Experiments

3.1   Data and Tasks

We use five available datasets, which cover most of the application scenarios.

Chat[3]: This is an intention prediction dataset [18]. Human may have various intents, for example, chit-chatting, booking air tickets, etc. Therefore, after

---

[3] https://github.com/HITlilingzhi/SMP2017ECDT-DATA.

Table 2. Datasets overview

Dataset	Train	Test	Classes	$|w|$	$|c|$	$|v|$
Chat	3025	757	31	450	783	1131
Hotel	8000	2000	2	7732	2525	9390
tc-corpus-answer	7865	1967	20	91816	5665	93623
Sogou	14329	3582	9	83128	5371	85337
THUNews	930730	232683	14	400209	6479	401327

receiving an input message from a user, the first step is to classify the user intent into a specific domain for further processing.

Hotel: This is a sentiment analysis dataset, including positive and negative reviews towards the hotel rooms.

tc-corpus-answer[4]: This is a topic classification dataset, which includes articles of 20 classes. The category contains history, art, computer, literature, education, philosophy and so on.

Sogou: This is a news dataset collected in 2006, including IT, education, culture and so on. It's worth mentioning that this dataset is larger and more complex than dataset called Sogou appears in other paper.

THUCNews[5]: This is a Sina RSS feed dataset, including fashion, lottery, game news and so on.

Table 2 is the overview of these five datasets. $|w|$ means the number of words in each dataset. $|c|$ means the number of characters we redefined. $|v|$ is the number of union set of words and characters.

3.2 Experiment Settings

These five datasets are preprocessed as follows. Jieba[6] Chinese segmentation system is employed to segment texts into words, and texts are simply split into characters. We adopt accuracy as metric. Dimensions of word and character embeddings are $|D| = 256$. For the unknown words we initialize a vector of uniform distribution $[-0.25, 0.25]$. The word vector is fine-tuned along with other parameters during training. The initial learning rates for all models are 0.001, and the learning rate decayed to 0.0009 after four epochs. There is no data enhancement for all models or any other thesaurus for corpus. Adam [9] optimizer is used for all models.

For the reason that CNN-based network requires fixed-length input, generally we fix it to 256 (except a very small Chat corpus). After concatenation is done, input shape of each text becomes 512 × 256. We just delete extra words or characters of those sentences longer than 256 (15 words and 25 characters for Chat corpus).

---

[4] http://www.nlpir.org/download/tc-corpus-answer.rar.

[5] http://thuctc.thunlp.org/message.

[6] https://pypi.org/project/jieba/.

3.3 Comparison of Methods

Based on the direction of improvement, we selected the following representative methods for comparison. For the sake of fairness, all preprocessing steps for all models are the same. We do not perform the step of removing stop words.

CNN: We choose the classic shallow network for text [7]. The network consists of a convolutional layer, a pooling layer, and a full connection layer. It concatenates outputs of different kernel size (3,4,5).

bi-LSTM: We select a typical and widely-used RNN method, bi-LSTM [1]. The basic idea of bi-LSTM is to train one LSTM model forward and one model backward using the same training sequence. Then the outputs of the two models are linearly combined to achieve complete reliance on all contextual information for each node in the sequence.

VDCNN: Schwenk et al. [14] proposed a deep network for text classification, inspirited by ResNet and VGG in computer vision. We experiment with depths 9, 29 because the best results are all observed at depths 29 in the original paper. Network input and parameters are consistent with the original paper.

3.4 Result and Discussion

We evaluate the impact of several tentative configurations in the model based on the final results (see Table 3). Overall, our self-inhibiting residual convolution network has achieved the best performance on most datasets. On Sogou and THUCNews corpus, the training speed of SIRCNN is much faster than that of bi-LSTM, while ensuring a competitive accuracy: a slightly inferior of 0.3%–0.5% absolute.

Table 3. Accuracy of all models on 5 datasets.

Corpus	Chat	Hotel	tc-corpus-answer	Sogou	THUCNews
CNN+random	0.8626	0.9230	0.9064	0.8595	-
CNN+word2vec	-	0.9040	0.9237	0.8724	0.9514
bi-LSTM+random	0.8190	0.9085	-	-	-
bi-LSTM+word2vec	-	-	0.9227	**0.8746**	**0.9618**
VDCNN(9)+random	0.6908	0.8935	-	-	-
VDCNN(9)+word2vec	-	-	0.9008	0.8383	0.9572
VDCNN(29)+random	0.6829	0.8780	-	-	-
VDCNN(29)+word2vec	-	-	0.8678	0.7769	0.9545
SIRCNN+random	**0.9089**	**0.9230**	-	-	-
SIRCNN+word2vec	0.8428	-	**0.9242**	0.8710	0.9564

Word and Character Joint Training Improves Performance. As shown in Fig. 4, the result of word and character joint training is much better than the

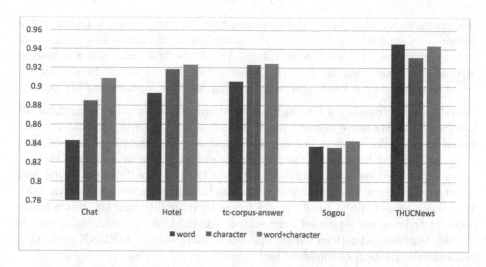

Fig. 4. Accuracy of different input (word vs. character vs. word + character) on SIR-CNN.

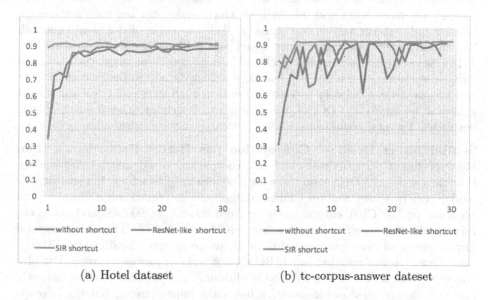

(a) Hotel dataset   (b) tc-corpus-answer dateset

Fig. 5. Accuracy varies with number of iterations without shortcut vs. ResNet-like shortcut vs. SIR shortcut on Hotel corpus and tc-corpus-answer corpus. This figure shows the effect of different shortcut on the convergence rate of the model.

result of word training alone. Since the length of sentence in Chat corpus and Hotel corpus is pretty short, we can see that the joint training of characters and words is very effective. For four datasets, word and character joint training reduce the error rate by 0.61%–6.61%. Characters bring more prior information, which is a good supplement to short sentences. For THUCNews corpus, we ascribe this result to its abundant word dictionary, which almost covers its character dictionary (see Table 2). Word and character joint training cannot bring more features than word in this specific corpus. This is the first time such a definition of characters has been used in the paper. This definition can be similar for languages that do not have inter-word spacing, such as Japanese and Burmese.

Word2vec is Not Good Enough for Sentiment Classification and Small Corpus. It makes sense because word2vec is not able to distinguish those words which have similar context but opposite meaning. This drawback inspires lots of works [13]; [16] to enhance the performance of word2vec in sentiment classification. The word vectors pre-trained with word2vec in small dataset (Chat and Hotel copus) is even less effective than random initialization vectors (see Table 3). We choose some trained word vectors to see their quality and the result is not good. In order to get a trained word vector, it is the best to use a similar corpus of field. For those situations without proper domain corpus, SIRCNN performed random initialization can achieve better result.

Larger Dataset Needs Deeper Convolutional Network. As we can see, the deeper the network is, the more hyper-parameters need to be trained. Large dataset has more texts and information which makes the network get enough trained and perform well on test datasets. Well-trained artificial neural network has an extremely fast convergence speed and a high degree of accuracy. The THUCNews dataset's samples reach millions of orders of magnitude and the number of words reaches 400,000, making it a huge dataset in the field of text. On this largest dataset, all deep models outperform shallow model CNN and have a gain of at least 0.51%. VDCNN, which performs poorly on other datasets, yields better results, again confirming deeper networks outperform shallow networks.

Self-inhibiting Residual Convolution has Better Performance than ResNet-like Convolution. Fig. 5 is a comparison of the results. We can see that the results of VDCNN on these datasets are unsatisfactory. In contrast, our model performs very well on Chinese datasets. The remarkable results are the accuracy on the Chat corpus increases from 69.08% to 90.89% and the accuracy on the Hotel corpus increases from 89.35% to 92.30%. This is a successful improvement of deep convolutional in language-specific classification. At the same time, we can see that our SIRCNN not only improves accuracy but also converges very fast (see Fig. 5). The traditional deep network converges slowly and turbulently. ResNet-like network has some improvement, but the effect is not very satisfactory, gradually stabilized after 20 epochs. However, our SIRCNN can achieve good results on the validation set only in the early phases of the run.

4   Conclusion

In this paper we combine word with Chinese characters, which are redefined as input of the network, and propose a new deep architecture SIRCNN which replaces convolutional block with self-inhibiting residual convolutional block which we design for releasing the training conundrum of deep networks. Our model takes into account that text also has structured information, from characters, words, phrases to sentences. Experiment shows SIRCNN have the natural advantage of capturing this hierarchical information and long-distance dependency and maintain high classification accuracy in five datasets. Besides, our deep network SIRCNN outperforms the deep model VDCNN both in accuracy and the training time, which benefits from its better convergence rate.

This is a meaningful exploration of deep networks in the field of text categorization for specific language. We extend the input features of the network for Chinese without inter-language spacing. The Research on those languages such as Japanese needs more investigation, and will be our future direction.

Acknowledgement. This work is supported by the National Key Research and Development Program of China under grants 2016QY01W0202 and 2016YFB0800402, National Natural Science Foundation of China under grants 61572221, U1401258, 61433006 and 61502185, Major Projects of the National Social Science Foundation under grant 16ZDA092, Science and Technology Support Program of Hubei Province under grant 2015AAA013, and Science and Technology Program of Guangdong Province under grant 2014B010111007.

References

1. Cai, R., Zhang, X., Wang, H.: Bidirectional recurrent convolutional neural network for relation classification. In: Proceedings of the 54th Annual Meeting of the Association for Computational Linguistics. ACL, Berlin (2016)
2. Collobert, R., Weston, J., Bottou, L., Karlen, M., Kavukcuoglu, K., Kuksa, P.P.: Natural language processing (almost) from scratch. J. Mach. Learn. Res. **12**, 2493–2537 (2011)
3. Graves, A.: Long short-term memory. In: Supervised Sequence Labelling with Recurrent Neural Networks. SCI, vol. 385, pp. 37–45. Springer, Heidelberg (2012). https://doi.org/10.1007/978-3-642-24797-2_4
4. He, K., Zhang, X., Ren, S., Sun, J.: Deep residual learning for image recognition. In: 2016 IEEE Conference on Computer Vision and Pattern Recognition, pp. 770–778. IEEE Computer Society, Las Vegas (2016)
5. Johnson, R., Zhang, T.: Effective use of word order for text categorization with convolutional neural networks. NAACL HLT 2015. In: The 2015 Conference of the North American Chapter of the Association for Computational Linguistics: Human Language Technologies, pp. 103–112. NAACL, Denver (2015)
6. Kalchbrenner, N., Grefenstette, E., Blunsom, P.: A convolutional neural network for modelling sentences. In: Proceedings of the 52nd Annual Meeting of the Association for Computational Linguistics. pp. 655–665. ACL, Baltimore (2014)
7. Kim, Y.: Convolutional neural networks for sentence classification. In: Proceedings of the 2014 Conference on Empirical Methods in Natural Language Processing, pp. 1746–1751. ACL, Doha (2014)
8. Kim, Y., Jernite, Y., Sontag, D., Rush, A.M.: Character-aware neural language models. arXiv preprint. arXiv:1508.06615 (2015)
9. Kingma, D.P., Ba, J.: Adam: a method for stochastic optimization. arXiv preprint. arXiv:1412.6980 (2014)
10. Le, H.T., Cerisara, C., Denis, A.: Do convolutional networks need to be deep for text classification? arXiv preprint. arXiv:1707.04108 (2017)

11. Liu, P., Qiu, X., Huang, X.: Recurrent neural network for text classification with multi-task learning. In: Proceedings of the Twenty-Fifth International Joint Conference on Artificial Intelligence, pp. 2873–2879. IJCAI/AAAI Press, New York (2016)

12. Mikolov, T., Sutskever, I., Chen, K., Corrado, G.S., Dean, J.: Distributed representations of words and phrases and their compositionality. In: Advances in Neural Information Processing Systems 26: 27th Annual Conference on Neural Information Processing Systems 2013, pp. 3111–3119. NIPS, Lake Tahoe (2013)

13. dos Santos, C.N., Gatti, M.: Deep convolutional neural networks for sentiment analysis of short texts. COLING 2014. 25th International Conference on Computational Linguistics, Proceedings of the Conference: Technical Papers, pp. 69–78. ACL, Dublin (2014)

14. Schwenk, H., Barrault, L., Conneau, A., LeCun, Y.: Very deep convolutional networks for text classification. In: Proceedings of the 15th Conference of the European Chapter of the Association for Computational Linguistics, pp. 1107–1116. ACL, Valencia (2017)

15. Srivastava, R.K., Greff, K., Schmidhuber, J.: Highway networks. arXiv preprint. arXiv:1505.00387 (2015)

16. Wang, X., Liu, Y., Sun, C., Wang, B., Wang, X.: Predicting polarities of tweets by composing word embeddings with long short-term memory. In: Proceedings of the 53rd Annual Meeting of the Association for Computational Linguistics and the 7th International Joint Conference on Natural Language Processing of the Asian Federation of Natural Language Processing, pp. 1343–1353. ACL, Beijing (2015)

17. Yang, Z., Yang, D., Dyer, C., He, X., Smola, A.J., Hovy, E.H.: Hierarchical attention networks for document classification. In: The 2016 Conference of the North American Chapter of the Association for Computational Linguistics: Human Language Technologies, NAACL HLT 2016, pp. 1480–1489. NAACL, San Diego (2016)

18. Zhang, W., Chen, Z., Che, W., Hu, G., Liu, T.: The first evaluation of Chinese human-computer dialogue technology. arXiv preprint. arXiv:1709.10217 (2017)

19. Zhang, X., LeCun, Y.: Which encoding is the best for text classification in Chinese, English, Japanese and Korean? arXiv preprint. arXiv:1708.02657 (2017)

20. Zhang, X., Zhao, J.J., LeCun, Y.: Character-level convolutional networks for text classification. In: Advances in Neural Information Processing Systems 28: Annual Conference on Neural Information Processing Systems 2015, pp. 649–657. NIPS, Montreal (2015)

Recurrent Neural Networks

A Hybrid 2D and 3D Convolution Based Recurrent Network for Video-Based Person Re-identification

Li Cheng[1], Xiao-Yuan Jing[1,2(✉)], Xiaoke Zhu[1,3], Fumin Qi[1], Fei Ma[1], Xiaodong Jia[1], Liang Yang[1], and Chunhe Wang[1]

[1] School of Computer, Wuhan University, Wuhan, China
cjackl@126.com, jingxy_2000@126.com, qfm120@163.com, mafei0603@163.com, jxdshimon@gmail.com, 1256951152@qq.com, 1173538302@qq.com
[2] College of Automation, Nanjing University of Posts and Telecommunications, Nanjing, China
[3] School of Computer and Information Engineering, Henan University, Kaifeng, China
henuzxk@163.com

Abstract. Video-based person re-identification (re-id), which aims to match people through videos captured by non-overlapping camera views, has attracted lots of research interest recently. In this paper, we propose a novel hybrid 2D and 3D convolution based recurrent neural network for video-based person re-id task, which can simultaneously make use of the local short-term fast-varying motion information and the global long-term spatial and temporal information. Specifically, the 3D convolutional module is able to explore the local short-term fast-varying motion information, while the recurrent layer performed can learn global long-term spatial and temporal information. We evaluate the proposed hybrid neural network on the publicly available PRID 2011, iLIDS-VID and MARS multi-shot pedestrian re-identification datasets, and the experiment results demonstrate the effectiveness of our approach on the task of video-based person re-id.

Keywords: 3D convolution
Short-term fast-varying motion information
Spatial and temporal information

1 Introduction

Person re-identification (re-id) is an important task in automated video surveillance and forensics, which aims to recognize an individual in a large set of candidates captured by different non-overlapping cameras. Since there usually exist visual ambiguity and spatio-temporal uncertainty in person's appearances across different cameras (which is usually caused by some external factors e.g., changes in lightness, viewpoint and resolution), person re-id is a challenging task in practice [14, 18, 28].

© Springer Nature Switzerland AG 2018
L. Cheng et al. (Eds.): ICONIP 2018, LNCS 11301, pp. 439–451, 2018.
https://doi.org/10.1007/978-3-030-04167-0_40

Person re-id techniques can be categorized into two main groups, single-shot methods and multiple-shot methods [4]. The single-shot methods try to associate pairs of images, each containing one instance of an individual. Most of existing methods can be classified as the single-shot methods [2, 13]. For example, a semi-coupled low-rank discriminant dictionary learning (SLD^{2}L) method is developed in [13] for image-based super-resolution person re-identification, which aims to transform the feature of LR image into discriminative HR feature. To match individual images of the same person captured by different non-overlapping camera views against significant and unknown cross-view feature distortion, the CRAFT framework [2] performs cross-view adaptation by automatically measuring camera correlation from cross-view visual data distribution.

The multiple-shot methods extract features from multiple images of the same person to achieve a robust representation of the person. A significant amount of works has gone into the problem of multiple-shot person re-id over the years [1, 21]. In [1], a set of frames of an individual were condensed into a highly informative signature, called the Histogram Plus Epitome (HPE), which incorporates complementary global and local statistical descriptions of the human appearance. Visual-spatial saliency, which represents the visual and spatial relationship among small regions segmented from multiple pedestrian images, is incorporated in region-based matching to improve the performance of person re-id [21]. Video-based person re-id methods are some special multiple-shot methods which require the multiple images of the same person to be a period of continuous video frames or a video clip [26]. Given a video clip of a person captured by one camera (probe person), video-based person re-id tries to find the corresponding person among a video gallery of people captured by other cameras in the surveillance systems. In this paper, we focus on the video-based person re-id problem.

1.1   Motivation

In general, there are two kinds of spatial and temporal information contained in a video clip of one walking person: (1) Global long-term spatial and temporal information; (2) Local short-term fast-varying motion information. The global long-term spatial and temporal information refers to the global long-term motion mode (e.g., speed and gait analysis) which is more abstract than local short-term fast-varying information. While the local short-term fast-varying motion information refers to the quick movements which occur in the partial limbs in a short time (e.g., optical flow and micro gestures). These movements always exist in multiple adjacent frames and can be obtained from detailed (raw) frames [11]. In practice, each pedestrian usually has some unique local short-term fast-varying motions, and thus making full use of these motion information is helpful to improving the discriminability and robustness of the features extracted from pedestrian videos. However, most existing video-based person re-id methods mainly focus on capturing the long-term spatial and temporal information, and ignore the local short-term fast-varying motion information, which will limit the person re-id performance of these methods.

Motivated by the above analysis, we intend to design an approach, which can simultaneously use the local short-term fast-varying motion information and global long-term spatial and temporal information contained in the person videos such that the person re-id performance can be further improved.

1.2 Contribution

Overall, the contributions of this study are mainly in three aspects:

(1) We design a hybrid 2D and 3D convolution based recurrent network (HCRN) for the video-based person re-id task. Specifically, HCRN simultaneously makes use of the local short-term fast-varying motion information and the global long-term spatial and temporal information.
(2) We introduce 3D convolutional operation to capture the local short-term fast-varying motion information contained in multiple adjacent frames of the pedestrian videos. To the best of our knowledge, this is the first work introducing 3D convolutional operation for the video-based person re-id task.
(3) We evaluate the performance of our approach on the public iLIDS-VID, PRID 2011 and MARS pedestrian sequence datasets. Extensive experimental results demonstrate the effectiveness of the proposed approach.

The rest of this paper is organized as follows. The next section briefly reviews the most recent and related developments with this work. Details of the proposed hybrid 2D and 3D convolutional and recurrent network are described in Sect. 3. Experimental results are provided in Sect. 4 to show the accuracy and applicability of the proposed approach. Finally, some concluding remarks are given in Sect. 5.

2   Related Works

In this section, we briefly review two types of works that are related to our approach: (1) Recurrent neural networks, (2) 3D convolutional networks.

2.1   Recurrent Neural Networks

Recurrent neural networks (RNNs) are a powerful family of feedforward neural networks that can model global long-term temporal dependencies contained in inputs which consist of sequences of points that are not independent. There have been a number of works attempt to learn global long-term temporal dependencies contained in the input sequence to address different problems. A hierarchical recurrent neural network is proposed in [24] to capture long-term temporal information for tackling the video captioning problem. Recently, some works [16,29] apply recurrent neural network to extract spatio-temporal features from pedestrian videos for person re-id task.

Although recurrent neural networks have been widely used in many computer vision tasks, they mainly focus on learning global long-term temporal dependencies and ignore the local short-term fast-varying motion features (information). We are the first one which simultaneously use local short-term fast-varying motion information and global long-term spatial and temporal information for the video-based person re-id task.

2.2 3D Convolutional Networks

Deep learning technique has been successfully applied in many areas of computer vision, such as object detection [6], terrain perception [25] and face recognition [17]. Specially, 3D convolutional operation can extract spatial and temporal information from sequence data (e.g. video data) which is very useful for sequence data based recognition targets. Several 3D CNN models are developed in [5, 12] to capture the motion information encoded in multiple adjacent frames for the action recognition problem. Authors in [11] designed a bidirectional recurrent convolutional network based on 3D convolution to capture local short-term fast-varying motion information contained in local adjacent frames for the video super-resolution task.

The major differences between our approach and these methods are twofold: (1) These methods apply 3D convolution to address the action recognition and video super-resolution tasks, while our approach employs the 3D convolution to solve the video-based person re-identification task. (2) In these methods, researchers mainly focus on local short-term fast-varying motion information encoded in multiple adjacent frames. Different from these methods, our approach not only utilizes the local short-term fast-varying motion information, but also can make use of the global long-term information existed in the whole video clips.

3 The Proposed HCRN Network

A diagram of the proposed HCRN network is shown in Fig. 1. The HCRN network consists of a 3D convolutional module, a 2D ResBlock module and a recurrent layer. Specially, we first perform three 3D convolutional layers (3D convolutional module) on raw frames to capture local short-term fast-varying motion information encoded in multiple adjacent frames. Then feature maps produced by 3D convolutional module will be processed by 2D ResBlock module. The 2D ResBlock module consists of three 2D ResBlock block units, which is used to explore high-level feature vectors for each frame. To further explore the global long-term temporal information contained in pedestrian video, we apply a recurrent layer (RNN) to the feature vectors which are produced by the 2D ResBlock module. A temporal pooling layer is adopted at the end of the RNN layer, such that feature vectors for all time-steps are aggregated to give a single feature vector which represents the whole sequence. Finally, we use the 3D convolutional module, 2D ResBlock module, RNN layer and temporal pooling layer as a feature extractor and adopt two loss functions including hinge embedding loss and cross-entropy loss to train the feature extractor in the Siamese architecture. In the following section we will give the details of each component in the proposed hybrid 2D and 3D convolution based recurrent network.

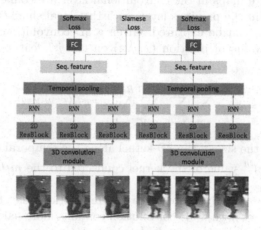

Fig. 1. Architecture of the proposed HCRN network

3.1 Input and Data Augmentation

Let $x_{(i)}^A = x_{(i,1)} \ldots x_{(i,T)}$ be a video sequence, of length T, corresponding to the ith person, where A means the full sequence. Similar to [16], we train the network using a Siamese architecture. For each epoch in the training phase, the input of the Siamese network is a pair of video sequences, $(x_{(i)}, x_{(j)})$, where $x_{(i)}$ and $x_{(j)}$ are the randomly selected subset of 16 consecutive frames over the full sequence $x_{(i)}^A$ and $x_{(j)}^A$, respectively. Note that i and j may refer to the same or different person at each epoch. Specially, when $i = j$, the video sequence pair should be selected from the video clips that are captured from the same person by using two different cameras. When $i \neq j$, the video sequence pair can be selected from video clips captured by the same or different cameras of different persons.

To increase the diversity of the available datasets, we apply several data augmentation methods including randomly mirror all the frames contained in a video clip, and randomly change the brightness, contrast and saturation of each frame in the training phase. In the testing phase, we simply extract feature vector for each video clip (full sequence) from the raw video without any data augmentation.

3.2 3D Convolutional Module

3D convolutional operation has been demonstrated to be a powerful technique for capturing local short-term fast-varying motion information from video [12].

This motivate us to integrate three 3D convolutional layers at the head of the proposed HCRN network to capture local short-term fast-varying motion information (features) encoded in multiple adjacent frames. To perform a 3D convolution operation, we should first stack multiple contiguous frames together to form a cube, then a 3D kernel will be applied to convolve with the cube. In this way, the feature maps in the convolutional layer are connected to multiple contiguous frames in the previous layer, such that local short-term fast-varying motion information can be captured. Given a 3D convolutional operation, we can calculate the value of position (x, y, z) on the jth feature map in the ith layer as follows:

$$v_{ij}^{xyz} = b_{ij} + \sum_{m} \sum_{p=0}^{P_i-1} \sum_{q=0}^{Q_i-1} \sum_{r=0}^{R_i-1} w_{ijm}^{pqr} v_{(i-1)m}^{(x+p)(y+q)(z+r)} \tag{1}$$

where R_i refers to the size of the 3D kernel along the temporal dimension, w_{ijm}^{pqr} denotes the $(p, q, r)^{th}$ value of the kernel connected to the mth feature map in the previous layer.

The reason why we only adopt three 3D convolutional layers at the head of our proposed network is that 3D convolution contains more parameters than 2D convolution which require large-scale dataset to train the network.

3.3   2D ResBlock Module

A typical residual block is showed in Fig. 2(a). The core idea of the residual block is the "shortcut connection" which can be formulated as $F(x) + x$. Several works [8,9,19] have demonstrated that it is easier to optimize the residual mapping than to optimize the original, unreferenced mapping, and can greatly improve the network's ability of feature extraction. Figure 2(b) is a diagram of the 2D ResBlock unit which we propose for the person re-id task. In each 2D ResBlock unit, we first stack five typical residual block as we have showed in Fig. 2(a). Then a max-pooling operation is applied to the feature maps which is produced by the residual blocks, such that the dimension of the feature maps can be reduced. Finally, we adopt a 1D dropout layer at the end of the 2D ResBlock unit to avoid over-fitting problem, which is the main difference between the common residual networks and the proposed 2D ResBlock unit. We stack three 2D ResBlock units in our 2D ResBlock module.

3.4   RNN

Let $c_{(i)} = c_{(i,1)}...c_{(i,T)}$ be the output of the 2D ResBlock module corresponding to the input of $x_{(i)}$. The RNN [16] can learn the global long-term spatial and temporal information existed in $x_{(i)}$ on the following operations:

$$o_{(i,t)} = W_k c_{(i,t)} + W_l r_{(i,t-1)} \tag{2}$$

$$r_t = \tanh\left(o_{(i,t)}\right). \tag{3}$$

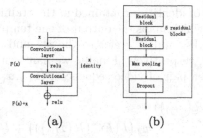

Fig. 2. (a) A typical residual block, (b) Our 2D ResBlock unit

The output, $o_{(i,t)}$, at time step t is a linear combination of the vectors, $c_{(i,t)}$ denotes the output of 2D ResBlock module at time-step t, and, $r_{(i,t-1)}$ is used to remember the information on the RNN's state at the previous time-step.

Then a mean-pooling operation is performed on the $o_{(i)} = o_{(i,1)}...o_{(i,T)}$ over the temporal dimension, such that a single feature vector v_i representing the person's appearance averaged over the video clip can be produced. The mean-pooling operation can be formulated as follows:

$$v_i = \frac{1}{T} \sum_{t=1}^{T} o_{(i,t)}. \tag{4}$$

3.5 Joint Loss Function

The proposed HCRN network illustrated in Fig. 1 is a Siamese network architecture [7]. It consists of two feature extractors with identical weights which we showed above. Given a pair of video sequences $(x_{(i)}, x_{(j)})$, we can get feature vectors $f_i = R(x_{(i)})$ and $f_j = R(x_{(j)})$, respectively, through the feature extractor. Then the Siamese network training objective function of the feature vectors (f_i, f_j) can be written as follows:

$$S(f_i, f_j) = \begin{cases} \frac{1}{2}\|f_i - f_j\|^2 & i = j \\ \frac{1}{2}[\max(m - \|f_i - f_j\|, 0)]^2 & i \neq j \end{cases}, \tag{5}$$

where d is the margin, which means that if a negative pair (f_i, x_j) is already separated by d, then there is no penalty for that pair and $S(f_i, f_j) = 0$. We set the margin d to 2 in our experiments.

Similar to the approach suggested in [16], we also apply the standard cross-entropy loss to optimize the feature extractor network. A cross-entropy loss can be formulated as follows:

$$I(fs_i) = \frac{exp(W_p fs_i)}{\sum_{q=1}^{Q} exp(W_q fs_i)} \tag{6}$$

$$fs_i = FC(f_i), \tag{7}$$

where Q is the number of identities contained in the training set, FC represents a fully connected layer which maps the output of the temporal pooling layer into the classification space, W_p and W_q refer to the pth and q^{th} column of W, the softmax weight matrix, respectively. Finally, we can write the overall training objective G for the given pair of video sequences, $\left(x_{(i)}, x_{(j)}\right)$ as follows:

$$G = w_1 S \left(R\left(x_{(i)}\right), R\left(x_{(j)}\right)\right) + \tag{8}$$
$$w_2 \left(I \left(FC \left(R\left(x_{(i)}\right)\right)\right) + I \left(FC \left(R\left(x_{(i)}\right)\right)\right)\right),$$

where w_1 and w_2 are the weight for the corresponding loss function, which we set both of them to 1 in our experiments.

4 Experimental Results

4.1 Datasets

The PRID 2011 person sequence dataset [10] was captured by two disjoint cameras (Cam-A and Cam-B) in an outdoor street scenario with clean background and rare occlusions. 385 and 749 person sequences were recorded in Cam-A and Cam-B, respectively. Among all persons, only 200 persons were captured in both Cam-A and Cam-B. In our experiments, only these 200 persons who appear in both cameras were considered. The iLIDS-VID dataset [20] consists of 600 video sequences for 300 randomly sampled people with one pair of sequences for each person, which is created based on two non-overlapping camera views at a crowed airport arrival hall under a CCTV network. The MARS dataset [27] is a large-scale video re-id dataset containing 1,261 identities in over 20,000 video sequences. This dataset was collected by six near-synchronized cameras placed in the campus of Tsinghua university, and each identity was captured by at least two cameras.

4.2 Experimental Settings

We follow the evaluation protocol in [16] for both iLIDS-VID and PRID 2011 datasets. In particular, we randomly split all sequence pairs into two sets of equal size, with one for training and the other for testing. Then we further select sequences from the first camera in the testing set to form the probe set, and those from the other camera are used as the gallery set. While for MARS dataset, we follow the evaluation protocol in [22]. We first randomly chose two camera viewpoints of the same person, then set one of them as gallery set and the other as probe set. We employ the standard cumulated matching characteristics (CMC) curve as our evaluation metric for all three datasets, and report the rank-k average matching rates of 10 trials with different train/test splits.

4.3   Compared Methods

To evaluate the proposed HCRN network, we compared it against eight video-based person re-id methods including **DVR** [20], **STFV3D** and its enhancement method **STFV3D+KISSME** [15], **TDL** [23], **SI^{2}DL** [30], **RCN** [16], **TSS** [3] and **ASTPN** [22]. Experiment details will be presented in the following sections.

4.4   Comparison with State-of-the-Art Methods

We compare the proposed HCRN network against these eight video-based person re-id methods mentioned above on iLIDS-VID, PRID 2011 and MARS datasets in Table 1. One can observe that HCRN network always outperforms all the compared video-based person re-id methods on the three datasets. For example, when compare to the second best ASTPN model, the rank-1 matching rates are improved by 10.8% ((68.7 − 62.0)/62.0), 2.6% ((80.0 − 78)/78) and 6.8% ((47 − 44)/44) on iLIDS-VID, PRID 2011 and MARS datasets, respectively. Note that all the RCN, TSS, ASTPN and the proposed HCRN network use Siamese architecture, while the proposed HCRN network is the only one which don't use optical flow features, but with similar performance. The possible reason is that the 3D convolutional module can explore the motion information contained in multiple adjacent frames which play the same role as optical flow features. Among the eight compared methods, the RCN and ASTPN are the most similar methods to the proposed HCRN. The major differences between HCRN and these two methods are two-fold: (1) We apply a 3D convolutional module at the head of the network to explore motion information which is contained in multiple adjacent frames. However, these methods haven't used 3D convolution technique; (2) We adopt a deep residual network (2D ResBlock module) instead of shallow network used in these methods. Several works [8,9,19] have demonstrated that the deep residual architecture is a powerful architecture for

Table 1. Top r ranked matching rates (%) on iLIDS-VID, PRID 2011 and MARS datasets

Method/Rank	iLIDS-VID				PRID 2011				MARS			
	r=1	r=5	r=10	r=20	r=1	r=5	r=10	r=20	r=1	r=5	r=10	r=20
Ours	**68.7**	**90.0**	**96.3**	**98.7**	**80.0**	94.0	97.0	**99.0**	**47.0**	**72.0**	**77.0**	**85.0**
RCN	58.0	84.0	91.0	96.0	70.0	90.0	95.0	97.0	40.0	64.0	70.0	77.0
TSS	60.0	86.0	93.0	97.0	78.0	94.0	97.0	99.0				
ASTPN	62.0	86.0	94.0	98.0	77.0	95.0	**99.0**	99.0	44.0	70.0	74.0	81.0
TDL	56.3	87.6	95.6	98.3	56.7	80.0	87.6	93.6	37.4	62.5	67.7	76.3
SI^{2}DL	48.7	81.1	89.2	97.3	76.7	**95.6**	96.7	98.9	31.1	61.4	68.5	74.5
STFV3D	37.0	64.3	77.0	86.9	42.1	71.9	84.4	91.6	25.5	49.8	59.3	69.9
STFV3D + KISSME	49.7	78.3	84.7	91.7	66.2	87.3	88.4	89.4	31.7	58.6	62.2	73.1
DVR	23.3	42.4	55.3	68.4	28.9	55.3	65.5	82.8	15.9	31.5	41.6	50.4

extracting discriminative features. We have experimentally verified the effects
of 3D convolutional module and 2D ResBlock module. The experimental results
show that each module has played the expected role. Due to limited space, the
related experimental results are not reported in this paper. Overall, the CMC
performance improvements on three datasets demonstrate that HCRN network
can extract more robust and discriminative features (information) than all the
other compared methods.

4.5   Cross Dataset Testing

The generalization capability of person re-id methods always can be estimated by
cross dataset testing. Based on the three datasets, we conducted two sets of cross
dataset testing experiments where these two large and diverse datasets including
iLIDS-VID and MARS were used for training, and testing were performed on
50% of the PRID 2011 dataset. It is evident from Table 2 that the CMC scores of
the proposed method (HCRN) always slightly exceeds that of all compared methods. For instance, when the HCRN is trained on MARS, the proposed method
achieves approx. 15.4% $((30.0 - 26.0)/26.0)$ performance advantage, at rank-1,
over RCN. When trained on iLIDS-VID, the proposed method achieves approx.
6.7% $((32.0 - 30.0)/30.0)$ performance advantage, at rank-1, over ASTPN.

Table 2. Cross-dataset testing accuracy in terms of top r ranked matching rates (%):
trained on MARS and iLIDS-VID, then tested on PRID 2011

Trained on	Method/Rank	$r=1$	$r=5$	$r=10$	$r=20$
MARS	HCRN	30.0	62.0	70.0	79.0
	RCN	26.0	57.0	68.0	78.0
iLIDS-VID	HCRN	32.0	61.0	73.0	86.0
	RCN	28.0	57.0	69.0	81.0
	ASTPN	30.0	58.0	71.0	85.0

5   Conclusion

In this paper, we develop a new hybrid 2D and 3D convolution based recurrent
network for video based person re-id task. The use of 3D convolution layer allows
us to explore the local short-term fast-varying motion information contained in
multiple adjacent frames, while three 2D ResBlock units of each followed by a
dropout layer further extract high-level information from each frame. Finally,
the global long-term spatial and temporal information contained in the whole
videos are learned by an RNN layer. Experiment results on three public video-
based person re-id datasets show that the proposed hybrid network surpass any
other methods in the video-based person re-id literature.

Acknowledgments. The authors would like to thank the editors and anonymous reviewers for their constructive comments and suggestions. This work was supported by NSFC-Key Project of General Technology Fundamental Research United Fund No. U1736211, the National Key Research and Development Program of China under Grant No.2017YFB0202001, the National Nature Science Foundation of China under Grant Nos. 61672208, U1504611,41571417, the Natural Science Foundation Key Project for Innovation Group of Hubei Province under Grant No.2018CFA024, the Science and Technique Development Program of Henan under Grant Nos. 172102210186, 182102311066, the Medical Education Research Project of Henan No. Wjlx2016095.

References

1. Bazzani, L., Cristani, M., Perina, A., Murino, V.: Multiple-shot person re-identification by chromatic and epitomic analyses. Pattern Recogn. Lett. **29**(1), 898–903 (2008)
2. Chen, Y., Zhu, X., Zheng, W., Lai, J.: Person re-identification by camera correlation aware feature augmentation. IEEE Trans. Pattern Anal. Mach. Intell. **40**(2), 392–408 (2018)
3. Chung, D., Tahboub, K., Delp, E.J.: A two stream siamese convolutional neural network for person re-identification. In: International Conference on Computer Vision, ICCV, pp. 1992–2000. IEEE Computer Society (2017)
4. Farenzena, M., Bazzani, L., Perina, A., Murino, V., Cristani, M.: Person re-identification by symmetry-driven accumulation of local features. In: Computer Vision and Pattern Recognition, CVPR, pp. 2360–2367. IEEE Computer Society (2010)
5. Feichtenhofer, C., Pinz, A., Zisserman, A.: Convolutional two-stream network fusion for video action recognition. In: Computer Vision and Pattern Recognition, CVPR, pp. 1933–1941. IEEE Computer Society (2016)
6. Girshick, R.B., Donahue, J., Darrell, T., Malik, J.: Rich feature hierarchies for accurate object detection and semantic segmentation. In: Computer Vision and Pattern Recognition, CVPR, pp. 580–587. IEEE Computer Society (2014)
7. Hadsell, R., Chopra, S., LeCun, Y.: Dimensionality reduction by learning an invariant mapping. In: Computer Vision and Pattern Recognition, CVPR, pp. 1735–1742 (2006)
8. He, K., Zhang, X., Ren, S., Sun, J.: Deep residual learning for image recognition. In: Computer Vision and Pattern Recognition, CVPR, pp. 770–778. IEEE Computer Society (2016)
9. He, K., Zhang, X., Ren, S., Sun, J.: Identity mappings in deep residual networks. In: Leibe, B., Matas, J., Sebe, N., Welling, M. (eds.) ECCV 2016. LNCS, vol. 9908, pp. 630–645. Springer, Cham (2016). https://doi.org/10.1007/978-3-319-46493-0_38
10. Hirzer, M., Beleznai, C., Roth, P.M., Bischof, H.: Person re-identification by descriptive and discriminative classification. In: Heyden, A., Kahl, F. (eds.) SCIA 2011. LNCS, vol. 6688, pp. 91–102. Springer, Heidelberg (2011). https://doi.org/10.1007/978-3-642-21227-7_9
11. Huang, Y., Wang, W., Wang, L.: Video super-resolution via bidirectional recurrent convolutional networks. IEEE Trans. Pattern Anal. Mach. Intell. **40**(4), 1015–1028 (2018)
12. Ji, S., Xu, W., Yang, M., Yu, K.: 3D convolutional neural networks for human action recognition. IEEE Trans. Pattern Anal. Mach. Intell. **35**(1), 221–231 (2013)

13. Jing, X.Y., et al.: Super-resolution person re-identification with semi-coupled low-rank discriminant dictionary learning. In: Computer Vision and Pattern Recognition, CVPR, pp. 695–704. IEEE Computer Society (2015)
14. Li, S., Shao, M., Fu, Y.: Person re-identification by cross-view multi-level dictionary learning. IEEE Trans. Pattern Anal. Mach. Intell. (2017)
15. Liu, K., Ma, B., Zhang, W., Huang, R.: A spatio-temporal appearance representation for viceo-based pedestrian re-identification. In: International Conference on Computer Vision, ICCV, pp. 3810–3818. IEEE Computer Society (2015)
16. McLaughlin, N., del Rincón, J.M., Miller, P.C.: Recurrent convolutional network for video-based person re-identification. In: Computer Vision and Pattern Recognition, CVPR, pp. 1325–1334. IEEE Computer Society (2016)
17. Schroff, F., Kalenichenko, D., Philbin, J.: Facenet: a unified embedding for face recognition and clustering. In: Computer Vision and Pattern Recognition, CVPR, pp. 815–823. IEEE Computer Society (2015)
18. Su, C., Yang, F., Zhang, S., Tian, Q., Davis, L.S., Gao, W.: Multi-task learning with low rank attribute embedding for person re-identification. In: IEEE International Conference on Computer Vision, ICCV. pp. 3739–3747. IEEE Computer Society (2015)
19. Szegedy, C., Ioffe, S., Vanhoucke, V., Alemi, A.A.: Inception-v4, inception-resnet and the impact of residual connections on learning. In: Proceedings of the Thirty-First AAAI Conference on Artificial Intelligence, pp. 4278–4284. AAAI Press (2017)
20. Wang, T., Gong, S., Zhu, X., Wang, S.: Person re-identification by video ranking. In: Fleet, D., Pajdla, T., Schiele, B., Tuytelaars, T. (eds.) ECCV 2014. LNCS, vol. 8692, pp. 688–703. Springer, Cham (2014). https://doi.org/10.1007/978-3-319-10593-2_45
21. Xie, Y., Yu, H., Gong, X., Dong, Z., Gao, Y.: Learning visual-spatial saliency for multiple-shot person re-identification. IEEE Sig. Process. Lett. **22**(11), 1854–1858 (2015)
22. Xu, S., Cheng, Y., Gu, K., Yang, Y., Chang, S., Zhou, P.: Jointly attentive spatial-temporal pooling networks for video-based person re-identification. In: International Conference on Computer Vision, ICCV, pp. 4743–4752. IEEE Computer Society (2017)
23. You, J., Wu, A., Li, X., Zheng, W.: Top-push video-based person re-identification. In: Computer Vision and Pattern Recognition, CVPR, pp. 1345–1353. IEEE Computer Society (2016)
24. Yu, H., Wang, J., Huang, Z., Yang, Y., Xu, W.: Video paragraph captioning using hierarchical recurrent neural networks. In: Computer Vision and Pattern Recognition, CVPR, pp. 4584–4593. IEEE Computer Society (2016)
25. Zhang, W., Chen, Q., Zhang, W., He, X.: Video paragraph captioning using hierarchical recurrent neural networks. Neurocomputing **275**, 781–787 (2018)
26. Zhang, W., Yu, X., He, X.: Learning bidirectional temporal cues for video-based person re-identification. IEEE Trans. Circuits Syst. Video Technol. **28**(10), 2768–2776 (2018)
27. Zheng, L., et al.: MARS: a video benchmark for large-scale person re-identification. In: Leibe, B., Matas, J., Sebe, N., Welling, M. (eds.) ECCV 2016. LNCS, vol. 9910, pp. 868–884. Springer, Cham (2016). https://doi.org/10.1007/978-3-319-46466-4_52
28. Zheng, L., Wang, S., Tian, L., He, F., Liu, Z., Tian, Q.: Query-adaptive late fusion for image search and person re-identification. In: Computer Vision and Pattern Recognition, CVPR, pp. 1741–1750 (2015)

29. Zhou, Z., Huang, Y., Wang, W., Wang, L., Tan, T.: See the forest for the trees: joint spatial and temporal recurrent neural networks for video-based person re-identification. In: Computer Vision and Pattern Recognition, CVPR, pp. 6776–6785. IEEE Computer Society (2017)
30. Zhu, X., Jing, X., Wu, F., Feng, H.: Video-based person re-identification by simultaneously learning intra-video and inter-video distance metrics. In: Proceedings of the Twenty-Fifth International Joint Conference on Artificial Intelligence, IJCAI, pp. 3552–3559. IJCAI/AAAI Press (2016)

Improving Recurrent Neural Networks with Predictive Propagation for Sequence Labelling

Son N. Tran[1]([✉]), Qing Zhang[1], Anthony Nguyen[1], Xuan-Son Vu[2], and Son Ngo[3]

[1] The Australian E-Health Research Centre, CSIRO, Brisbane, QLD 4026, Australia
{son.tran,qing.zhang,anthony.nguyen}@csiro.au
[2] Department of Computing Science, Umeå University, Umeå, Sweden
sonvx@cs.umu.se
[3] Department of Computer Science, FPT University, Hanoi, Vietnam
sonnt69@fe.edu.vn

Abstract. Recurrent neural networks (RNNs) is a useful tool for sequence labelling tasks in natural language processing. Although in practice RNNs suffer a problem of vanishing/exploding gradient, their compactness still offers efficiency and make them less prone to overfitting. In this paper we show that by propagating the prediction of previous labels we can improve the performance of RNNs while keeping the number of parameters in RNNs unchanged and adding only one more step for inference. As a result, the models are still more compact and efficient than other models with complex memory gates. In the experiment, we evaluate the idea on optical character recognition and Chunking which achieve promising results.

Keywords: Natural language processing · Recurrent neural networks Sequence labelling

1 Introduction

Sequence labelling is a machine learning method which has been widely used for natural language processing tasks. In early work, these tasks attract the use of dynamic Bayesian models such as Hidden Markov Models (HMMs) [21] and Conditional Random Fields (CRFs) [17]. An advantage of such models is the ability to learn relationships between sequence labels, which is useful for temporal reasoning. Recently recurrent neural networks (RNNs) have become a central tool for sequence labelling. An RNN is constructed by rolling an artificial neural network with one hidden layer over time. The hidden layer is connected to itself through recurrent weights. A main advantage of RNNs is the ability to learn temporal representations from data using recurrent hidden layers. Different from dynamic Bayesian models, RNNs assume that the class labels in a sequence are

© Springer Nature Switzerland AG 2018
L. Cheng et al. (Eds.): ICONIP 2018, LNCS 11301, pp. 452–462, 2018.
https://doi.org/10.1007/978-3-030-04167-0_41

independent, given the sequence inputs. This makes inference easier, but with the sacrifice of valuable information: the temporal dependencies of sequence labels.

However, as a type of deep architecture, RNNs suffer the problem of vanishing/exploding gradient which necessitates the use of complex memory gates such as Long-short term memory (LSTM) [14] and Gated recurrent unit (GRU) [3]. However, in many cases, especially when efficiency and compactness are of vital importance, RNNs are more desirable. For example, one would choose an RNN over LSTM for memory-limited devices such as mobile phones and smart sensors if its performance is acceptable. Also, the complexity of LSTM and its variants make it prone to overfitting when training data is small. This situation is very common with natural language processing because of the difficulty in labelling ground truth for large number of sequences. Therefore, it would be useful for an improved version of RNNs which can perform comparably well in comparison to complex models such as GRU and LSTM while keeping the number of parameters remain small in size.

In this paper we show that by propagating the prediction of previous labels we can improve the performance of RNNs. This means that we can keep the number of parameters in RNNs unchanged while adding only one more step for inference. Therefore, the models are still more compact and efficient than other models with complex memory gates.

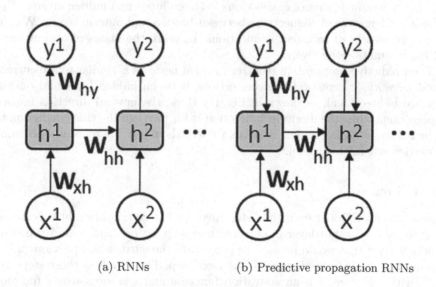

(a) RNNs   (b) Predictive propagation RNNs

Fig. 1. RNNs and ppRNNs

In the experiments, we evaluate RNNs with predictive propagation (ppRNNs) on two sequence labelling tasks: OCR and Chunking. The results show that in most cases predictive propagation help improving the performance of RNNs. We also evaluate the effect of the data sizes on ppRNNs using a POS tagging dataset. We find that ppRNNs achieve higher accuracy than RNNs, GRUs and LSTMs when small training samples are used. Finally, a synthetic dataset is used to compare the computational time needed for training and inference with ppRNNs to that of GRU and LSTM. It shows that due to the additional step for propagating predictions ppRNNs are more computationally expensive than GRU and LSTMS when the dimension of the output layer is very large, e.g. 400 in the experiment. Fortunately, in practice there are many applications that do not require such high number of classes.

The remainder of the paper is organised as follows. In the next section, we review the literature related to this work. Section 2 introduces the idea of predictive propagation in recurrent neural networks. In Sect. 3, we perform the empirical evaluations. Finally, Sect. 4 concludes the paper and discusses the future extensions.

2   Recurrent Neural Networks with Predictive Propagation

2.1   Graphical Structure

A recurrent neural network for sequence labelling is illustrated in Fig. 1a where $\mathbf{W}_{xh}$ is the weight matrix of connections between input and hidden layers; $\mathbf{W}_{hy}$ is the weight matrix of connections between hidden and output layers; $\mathbf{W}_{hh}$ is the weight matrix of recurrent connections; $\mathbf{b}$, $\mathbf{c}$ are the biases of output units and hidden units respectively.

The predictive propagation recurrent neural network is similar to a recurrent neural network, except that the connections between hidden layer and output layer are bidirectional, as shown in Fig. 1b. Here, the upward direction is used for prediction while the downward direction is for propagating that prediction to the next time step. In the next sections we will show how inference and learning are carried out in this model.

2.2   Inference

Inference in ppRNNs at each time step involves the computation of two states: the state of the output layer in current time step for prediction, and the state of hidden layer that would be used to propagate the current and previous information, including the prediction, to the next step. The details of those steps are in Algorithm 1. Here, f is an activation function and s is the softmax function $s(\mathbf{x})_i = \frac{\exp(x_i)}{\sum_{i'} \exp(x_i)}$.

2.3   Learning

We train the RNNs using each sample at a time. In particular for each training pair $\mathbf{x}^{1:T}, \mathbf{y}^{1:T}$ from the dataset we infer $\tilde{\mathbf{y}}^{1:T}$ using the Algorithm 1 and update

---

Algorithm 1. Predictive propagation

Data: Input: $\mathbf{x}^{1:T}$
Result: Output: $\mathbf{o}^{1:T}$
  for $t = 1 : T$ **do**
    $\tilde{\mathbf{h}}^t = f(\mathbf{x}^{t\top}\mathbf{W}_{xh} + \mathbf{h}^{t-1\top}\mathbf{W}_{hh} + \mathbf{c}^\top)$
    $\tilde{\mathbf{y}}^t = s(\tilde{\mathbf{h}}^{t\top}\mathbf{W}_{hy} + \mathbf{b}^\top)$
    $o^t = \arg\max_k \tilde{y}_k^t$
    $\mathbf{h}^t = f(\mathbf{x}^{t\top}\mathbf{W} + \tilde{\mathbf{y}}^{t\top}\mathbf{W}_{hy}^\top + \mathbf{h}^{t-1\top}\mathbf{W}_{hh} + \mathbf{c}^\top)$
  end

---

the parameters by minimising the cross entropy:

$$\mathcal{C} = -\frac{1}{T}\sum_{t=1}^{T}\sum_{l=1}^{L}[y_l^t \log \tilde{y}_l^t + (1 - y_l^t)\log(1 - \tilde{y}_l^t)] \tag{1}$$

where L is the number of classes. A RNN trained by this method is denoted as ppRNN$_p$. Alternatively, with the availability of the true labels, we can use another method to infer $\tilde{\mathbf{y}}^{1:T}$ by replacing $\tilde{\mathbf{y}}^t$ in the last expression in Algorithm 1 with $\mathbf{y}^t$. The state of hidden layer at time t then becomes $\mathbf{h}^t = g(\mathbf{x}^{t\top}\mathbf{W} + \mathbf{y}^{t\top}\mathbf{W}_{hy}^\top + \mathbf{h}^{t-1\top}\mathbf{W}_{hh} + \mathbf{c}^\top)$. We denote a ppRNN using this type of inference for learning as ppRNN$_g$.

3 Experiments

3.1 OCR

The MIT OCR dataset[1] is a widely used benchmark for evaluating sequence labelling algorithms [25]. We use two popular partitions from [19] and [1,9]. In the former, called here "ms" for model selection, the data is partitioned into 10 groups, each consisting of a training, validation, and test set. We select models based on performance on the validation sets and report their average accuracy on the test sets. In the latter, here called "cv" for cross-validation, the data is divided into 10 folds in the usual way but without model selection. Each fold in the "cv" partition has ~6000 training samples, about 10 times larger than each fold in "ms".

Effect of Learning Methods. First, we use the "ms" partition which consists of ten folds, each has ~600, ~100, and ~5400 samples for training, validation, and testing respectively. With such a small number of samples for training we can anticipate the issue of overfitting for large models. Besides, selection of learning methods is also very important. Therefore, we start with testing RNNs with

---

[1] http://www.seas.upenn.edu/~taskar/ocr/.

different number of hidden units: $hidNum = \{100, 500, 1000, 2000, 5000, 10000\}$; and learning methods: $lrn = \{sgd$: vanilla stochastic gradient descent , $adagrad$ [10], $rmsprop$, $adam$ [16]$\}$. For each pair of ($hidNum,lrn$), we select the best learning rate and activation function based on the performance of validation sets. For early stopping, if performance on a validation set does not improve for 20 epochs then we stop the learning and use the best trained model for testing. As we can see from Fig. 2 the performances of RNNs on the test set drop significantly when the number of hidden units is too high. This makes sense because such overfitting phenomenon is more likely to happen when the models become more complex, i.e. have large number parameters. Overall, adaptive learning method $adam$, which is recently introduced, is shown better than the other methods tested here. $adam$ also offers another advantage is that it is fast and works very well with sparse data. Different initial learning rates can be used for $adam$ to produce good results but we found that normally 0.001 is good for all cases. In terms of activation function, surprisingly it seems that using $sigmoid$ for RNNs performs better than $tanh$ despite a study in deep feed-forward neural networks suggesting that the former may suffer from a gradient saturation issue [12].

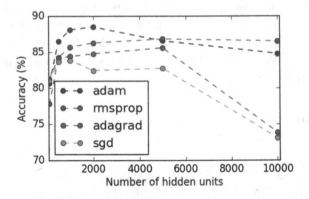

Fig. 2. Performance of RNNs on OCR test set with different learning algorithms and number of hidden units

Overall Results. We compare the performance of SCRBM on the above sequence labelling task with the models: Multiclass support vector machines (SVM-$multiclass$) [6], Structured support vector machines (SVM-$struct$) [26], Max-margin Markov network (M3N) [25], Averaged Perceptron [4], Search-based structured prediction model, a.k.a SEARN [7], Conditional random field (CRF) [17,20], Hidden Markov model (HMM) [21], LogitBoost [11], TreeCRF [8], RTD-RBM [2] (modified to use the inference algorithm proposed in this paper so it works with a sequence labelling task), and state-of-the-art models:

Table 1. Average test set accuracy (%) of various models on the OCR sequence labelling task; "ms" dataset uses model selection and "cv" uses cross-validation without model selection.

Model	ms	cv
ppRNN$_g$	88.54	95.20
ppRNN$_p$	**88.60**	95.54
RNN$_{sigmoid+adam}$	88.44	**95.64**
LSTM	88.28	95.24
GRU	85.77	93.62
RTDRBM	84.54	-
Neural CRF^{CML}	-	95.56
Neural CRF^{LM}	-	95.44
SLE	79.42	-
GBCRF	-	95.36
TreeCRF	-	93.01
LogitBoost	-	90.33
RNN$_{tanh+sgd}$	77.08	86.67
M3N	74.92	86.54
Perceptron	73.60	-
SEARN	72.98	-
SVM$_{multiclass}$	71.46	-
SVM$_{struct}$	78.84	-
HMM	76.30	-
CRF	67.70	85.80

- Structured learning ensemble (SLE) [19]: An optimised ensemble of 7 effective models: SVM-*multiclass*, SVM-*struct*, M3N, Perceptron, SEARN, CRF and HMM.
- Neural CRF [9]: A combination of CRF and deep networks.
- Gradient boosting CRF (GBCRF) [1]: CRF trained by a novel gradient boosting algorithm.

For the "ms" partition, to determine the best model for the task, a grid search was carried out where the number of hidden units, learning rate, activation function are selected similarly as in Sect. 3.1. For the "cv" partition, since model selection is not possible, we use 2000 hidden units and set the initial learning rate for *adam* [16] as 0.001. Each model is trained in 30 epochs.

Table 1 summarises the results and shows that with a right choice of activation function (*sigmoid*) and learning method (*adam*) RNNs can perform very well in this dataset. Although improvement has not been seen in "cv" partition ppRNNs achieve higher accuracy than other methods in "ms" partition where the number of training samples is smaller. The results also imply that propagating prediction probabilities ($ppRNN_p$) seems slightly better than propagating ground truth ($ppRNN_g$), during the learning (Table 2).

Table 2. Features extraction for Conll2000 Chunking task (see [23])

$w_{t-\delta} = w$
w_t matches [A-Z][a-z]+
w_t matches [A-Z]
w_t matches [A-Z]+
w_t matches [A-Z]+[a-z]+[A-Z]+[a-z]
w_t matches .*[0-9].*
w_t appears in list of first names, last names, company names, days, months, or geographic entities
w_t is contained in a lexicon of words with POS T (from Brill tagger)
$T_t = T$
$q_k(x, t + \delta)$ for all k and $\delta \in [-3, 3]$

3.2   CoNLL 2000 Chunking

The CoNLL 2000 shared task is a benchmark data for sequence labelling with a focus on Chunking. The dataset consists of 8,936 and 2012 sentences for training and testing respectively. For this data we use the binary features from [23].

In this experiment, we use 50000 most common features from the training set. The motivation behind the selection of this type of features over word2vec features is to exploit the efficiency of RNNs. Although word2vec features are smaller in terms of size, generic approaches such as RNN, GRU, LSTM do not perform well, and therefore more complex variants, i.e. biLSTM [13], bi-LSTM-CRF [15] and CNN-biLSTM-CRF [18] are needed when using word2vec features as their sole input. Here, the compactness and efficiency of RNNs make it easier to work with such highly dimensional hand-crafted features, which in this case perform better than word2vec features .

Since the dataset only includes training and testing samples we do not perform model selection and early stopping. Instead, we set the number of hidden units to be 500 and the number of training epochs to be 50. Such hyperparameters are chosen based on the capacity of our computers used in this experiment. We also use *adam* to take the advantage of sparsity of the features, the initial learning rate for it is 0.001.

As we can see from Table 3, the features are so effective that even a simple RNN can give a better result than many other approaches using different

Table 3. F1 score in CoNLL 2000 chunking data.

Model	F1 score
ppRNN$_g$	**95.319**
ppRNN$_p$	**95.302**
LSTM	95.118
GRU	94.719
RNN	95.085
Suzuki et. al. [24]	95.15
Huan et. al. [15]	94.46
Sun et. al. [22]	94.34
Collobert et. al. [5]	94.32
Tsuruoka et. al. [27]	93.81

features. Again, we show that predictive propagation can help to improve the performance. It is also worth noting that the ppRNNs in this case only needs ~4 h to train while the *GRU* and *LSTM* take more than ~10 h, using NVIDIA Quadro P1000. Although the other models we compare to in Table 3 are applicable to the hand-crafted features above, in this experiment we show that RNNs can perform very well while being very efficient.

Table 4. Test accuracy of ppRNNs, RNN, GRU and LSTM on POS dataset with different number of training samples.

Model	500	1000	2000	4000	8000
ppRNN$_g$	**88.703**	**90.847**	91.757	92.122	93.336
ppRNN$_p$	88.646	90.767	91.781	92.244	**93.419**
RNN	88.293	90.755	**91.808**	92.303	93.380
LSTM	88.529	90.585	91.606	**92.445**	**93.414**
GRU	88.571	90.526	91.169	92.276	93.046

3.3   POS Tagging: Effect of Training Size

In this experiment we use a POS tagging dataset with different training sizes of 500, 1000, 4000, 8000 samples to compare the effectiveness of ppRNNs with RNN, GRU, and LSTM. The samples are obtained from Penn Treebank 2002 dataset. Model selection and early stopping are done by leaving out 10% of the training set for validation. The test set consists of ~1600 samples and is applied to all cases.

Similar to the Chunking task above, we also use binary features for each token in POS task. The results in Table 4 show that ppRNNs perform better than the others when the training data is small.

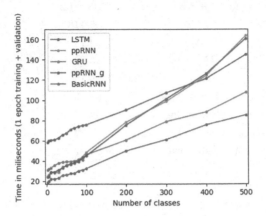

Fig. 3. Computational time.

3.4 Computational Time

Finally, in order to compare the efficiency of ppRNNs over other types of RNNs we use a synthetic dataset of 1000 training samples and 1000 validation samples. Each sample has the length of 100 and the input's dimension is 1000. We train each model in one epoch and evaluate it on the validation set. All RNNs have 100 hidden units. Since the computation in ppRNNs requires the propagation of the prediction back to the hidden layer we test the computational time of the models on the data with different number of classes using a PC with 4 Intel Core^{TM} i5-6500 CPUs @ 3.20 Hz and 32 GiB RAM. From Fig. 3 we can see that ppRNN$_g$ and ppRNN$_p$ takes similar time to learn and infer, and also that when the number of labels grows larger, i.e. ~400, ppRNNs become slower than LSTM. However, it is not common to see such a large number of labels in sequence labelling in practice.

4 Conclusion

We have shown an empirical study on propagating prediction of output layer in RNNs for sequence labelling. In most cases, it helps improve the performance of RNNs while maintaining the compactness. The computational time of the proposed models is practically efficient in comparison with other RNNs having complex memory gates.

References

1. Chen, T., Singh, S., Taskar, B., Guestrin, C.: Efficient second-order gradient boosting for conditional random fields. In: 18th International Conference on Artificial Intelligence and Statistics, vol. 38, pp. 147–155. PMLR, San Diego (2015)
2. Cherla, S., Tran, S.N., d'Garcez, A., Weyde, T.: Discriminative learning and inference in the recurrent temporal RBM for melody modelling. In: 2015 International Joint Conference on Neural Networks (IJCNN), pp. 1–8 (2015)
3. Cho, K., et al.: Learning phrase representations using RNN encoder-decoder for statistical machine translation. In: Conference on Empirical Methods in Natural Language Processing, pp. 1724–1734 (2014)
4. Collins, M.: Discriminative training methods for hidden Markov models: theory and experiments with perceptron algorithms. In: ACL-2002 Conference on Empirical Methods in Natural Language Processing, vol. 10, pp. 1–8. Association for Computational Linguistics, Stroudsburg (2002)
5. Collobert, R., Weston, J., Bottou, L., Karlen, M., Kavukcuoglu, K., Kuksa, P.: Natural language processing (almost) from scratch. J. Mach. Learn. Res. **12**, 2493–2537 (2011)
6. Crammer, K., Singer, Y.: On the algorithmic implementation of multiclass kernel-based vector machines. J. Mach. Learn. Res. **2**, 265–292 (2002)
7. Daumé III, H., Langford, J., Marcu, D.: Search-based structured prediction. Mach. Learn. **75**(3), 297–325 (2009)
8. Dietterich, T.G., Hao, G., Ashenfelter, A.: Gradient tree boosting for training conditional random fields. J. Mach. Learn. Res. **9**(2), 2113–2139 (2008)
9. Do, T., Artieres, T.: Neural conditional random fields. In: 13th International Conference on Artificial Intelligence and Statistics, vol. 9, pp. 177–184. PMLR, Sardinia (2010)
10. Duchi, J., Hazan, E., Singer, Y.: Adaptive subgradient methods for online learning and stochastic optimization. J. Mach. Learn. Res. **12**, 2121–2159 (2011)
11. Friedman, J., Hastie, T., Tibshirani, R.: Additive logistic regression: a statistical view of boosting. Ann. Stat. **28**, 337–407 (2000)
12. Glorot, X., Bengio, Y.: Understanding the difficulty of training deep feedforward neural networks. In: 13th International Conference on Artificial Intelligence and Statistics, vol. 9, pp. 249–256. PMLR, Sardinia (2010)
13. Graves, A., Schmidhuber, J.: Framewise phoneme classification with bidirectional LSTM networks. In: 2005 IEEE International Joint Conference on Neural Networks, Montreal, Quebec, Canada, vol. 4, pp. 2047–2052 (2005)
14. Hochreiter, S., Schmidhuber, J.: Long short-term memory. Neural Comput. **9**(8), 1735–1780 (1997)
15. Huang, Z., Xu, W., Yu, K.: Bidirectional LSTM-CRF models for sequence tagging. CoRR abs/1508.01991 (2015)
16. Kingma, D.P., Ba, J.: Adam: a method for stochastic optimization. CoRR abs/1412.6980 (2014)
17. Lafferty, J.D., McCallum, A., Pereira, F.C.N.: Conditional random fields: probabilistic models for segmenting and labeling sequence data. In: 18th International Conference on Machine Learning, pp. 282–289. Morgan Kaufmann Publishers Inc., San Francisco (2001)
18. Ma, X., Hovy, E.: End-to-end sequence labeling via bi-directional LSTM-CNNS-CRF. In: 54th Annual Meeting of the Association for Computational Linguistics, pp. 1064–1074. Association for Computational Linguistics (2016)

19. Nguyen, N., Guo, Y.: Comparisons of sequence labeling algorithms and extensions. In: 24th International Conference on Machine Learning, pp. 681–688. ACM, New York (2007)
20. Peng, F., McCallum, A.: Information extraction from research papers using conditional random fields. Inf. Process. Manag. **42**(4), 963–979 (2006)
21. Rabiner, L.R.: A tutorial on hidden Markov models and selected applications in speech recognition. In: Readings in Speech Recognition, pp. 267–296. Elsevier, San Francisco (1990)
22. Sun, X., Morency, L.P., Okanohara, D., Tsujii, J.: Modeling latent-dynamic in shallow parsing: a latent conditional model with improved inference. In: 22nd International Conference on Computational Linguistics, pp. 841–848. Association for Computational Linguistics, Stroudsburg (2008)
23. Sutton, C., McCallum, A., Rohanimanesh, K.: Dynamic conditional random fields: factorized probabilistic models for labeling and segmenting sequence data. J. Mach. Learn. Res. **8**, 693–723 (2007)
24. Suzuki, J., Isozaki, H.: Semi-supervised sequential labeling and segmentation using giga-word scale unlabeled data. In: ACL-2008: HLT, pp. 665–673. The Association for Computer Linguistics (2008)
25. Taskar, B., Guestrin, C., Koller, D.: Max-margin Markov networks. In: Advances in Neural Information Processing Systems, vol. 16, p. 25 (2004)
26. Tsochantaridis, I., Joachims, T., Hofmann, T., Altun, Y.: Large margin methods for structured and interdependent output variables. J. Mach. Learn. Res. **6**, 1453–1484 (2005)
27. Tsuruoka, Y., Miyao, Y., Kazama, J.: Learning with lookahead: can history-based models rival globally optimized models? In: 15th Conference on Computational Natural Language Learning, pp. 238–246. Association for Computational Linguistics, Stroudsburg (2011)

Design of Synthesizing Multi-valued High-Capacity Auto-associative Memories Based on Complex-Valued Networks

Chunlin Sha and Hongyong Zhao[✉]

Department of Mathematics, Nanjing University of Aeronautics and Astronautics, Nanjing 210016, China
Hyzhao1967@126.com

Abstract. This paper presents a novel design method which is aimed to synthesize arbitrary multi-valued auto-associative memories via complex-valued neural networks. Globally exponential stable criteria are obtained to guarantee that the unique storage prototype can be retrieved. The proposed procedure enables auto-associative memories to be synthesized by satisfying the constraints of inequalities rather than the learning procedure. The main emphasis of the research presented here is on multi-valued high-capacity auto-associative memories via complex-valued networks. The designed auto-associative memories with $(2r+2)^n$ high memory capacities are robust with respect to design parameter selection and extend the scope of application of complex-valued neural networks. The approach of external inputs via complex-valued neural networks avoids spurious equilibria and retrieves the stored patters accurately. Some applicable experiments are given to illustrate the effectiveness and superiority.

Keywords: Multi-valued associative memories · Network dynamics Design methods · Real-imaginary-type activation · External inputs

1 Introduction

The association is viewed as one of most interesting features of human brain [1]. We can easily associate the human faces with names, recognize people even if they get old. This cognitive function of the brain is called associative memory. It refers to the memory retrieval of a stimulus or behavior in relation to the presentation of an associated stimulus or response. Associative memories have become one of important research areas, and are widely applied to image process, pattern recognition and others [2–4].

To implement associative memories, the most interesting and efficient method is the associative neural network. Associative memory network is a brain-style

---

H. Zhao—This work was supported by National Natural Science Foundation of China (Grant nos. 11571170 and 11501290).

L. Cheng et al. (Eds.): ICONIP 2018, LNCS 11301, pp. 463–475, 2018.
https://doi.org/10.1007/978-3-030-04167-0_42

neural network, and it can store and recall patterns based on data contents rather than data addresses. Whenever an associative memory retrieves a previously stored pattern that closely resembles the recalling input pattern, then it is called an auto-associative memory, otherwise it is called hero-associative memory.

The research subjects of associative memories perhaps stemmed back from the linear associator in 1970s [7]. It is a type of one-layer feedforward network and can be used for synthesizing an auto-associator or hero-associator. However, there are two imperfections in the designed associator. (i) Prototype storage patterns are mutually orthogonal; (ii) The corresponding prototype patterns cannot be recalled accurately when the input probe is noisy or corrupted.

In 1997, Grassi [5] designed associative memories via two-dimensional cell recurrent neural networks (CNNs) described by the nonlinear difference equations. Since then, a lot of achievements [7–12] were scored by using the real-valued neural networks (RVNNs) for synthesizing bi-valued associate memories with external input. In [7], a discrete-time RVNN with high storage capacity was designed to synthesize associative memories. The proposed procedure enables both auto-associative and hero-associative memories to be retrievable accurately. Zeng et al. [8] developed a continuous-time RVNN for associative memories characterized by space-invariant cloning templates. The design parameters of RVNN are obtained by solving a set of linear inequalities and retrieval probes are fed from external inputs instead of initial states. An associative memories model [9] was designed based on stability of continuous-time RVNNs. The conservatism is relaxed and feasible range of the parameter values is extended. Two procedures were designed for synthesizing auto-associative and hero-associative memory [10] with mixed delays separately. The bias vectors can be acquired with large ranges. Zhou et al. [11] proposed a unified associative memory via continuous RVNNs. The robustness of the associative memory in terms of parameters is studied and examined in detail. Sha et al. [12] proposed a bidirectional auto-associative memory RVNN and a new ring RVNN for synthesizing associative memories. The proposed networks are robust with respect to design parameter selection and enable multiple prototype patterns to be retrieved simultaneously.

In the recent years, it was recognized that the researches on multi-valued associative memories via neural networks are important valuable both in theory and in practice. Unfortunately, compared with binary associative memories, the research achievements on multi-valued associative memories with external input are quite rare. Moreover, if binary associative networks are used for multi-valued associative problems, then the structure of networks will become more complex, and the associative performance will be reduced serious [13,14]. Furthermore, it is can effectively avoid some spurious patterns if associative memories via neural networks are dependent on external input [7]. Therefore, it is essential to study multi-valued associative memories with the external input other than the initial state.

As is well known, compared with the RVNNs, activation functions, connection weights and the states of complex-valued neural networks (CVNNs) are all complex-valued. Hence, CVNNs can be viewed as an extension of RVNNs

[15]. Due to its practical applications of CVNNs in physical systems for processing quantum waves, electromagnetic, ultrasonic and light and so on, CVNNs become a hot research spot [17]. It has been shown that CVNNs make it possible to solve some problems that cannot be solved by real-valued models, such as the XOR problem and the detection of symmetry problem which reveals the potent computational power of CVNNs [16]. However, less attention has been paid to multi-valued auto-associative memories via CVNNs with external inputs [21].

Motivated by these works in [10,11], we are to investigate a arbitrary multi-valued associative memory model based on CVNNs with external input. In Sect. 2, the CVNNs model for associative memories is proposed. In Sect. 3, stability analysis and design procedures are established for the proposed CVNNs. In Sect. 4, several numerical simulations are given to demonstrate the effectiveness of the proposed CVNNs. Finally, Sect. 5 concludes this paper.

2   Problem of Descriptions and Preliminaries

In this paper, the steady state of neurons are designed as complex numbers. In order to depict the design problem appropriately, we introduce notions as follows. Write $\{u_1+iu_1, \cdots, u_1+iu_{r+1}, \cdots, u_{r+1}+iu_1, \cdots, u_{r+1}+iu_{r+1}\}^n$ for the set of n-dimensional complex-valued vectors, i.e., $\{u_1+iu_1, \cdots, u_1+iu_{r+1}, \cdots, u_{r+1}+iu_1, \cdots, u_{r+1}+iu_{r+1}\}^n = \{v \in C^n | v = (v_1^R + iv_1^I, \cdots, v_n^R + iv_n^I)^T, v_k^R$ and $v_k^I \in \{u_1, u_2, \cdots, u_{r+1}\}, k = 1, 2, \cdots, n\}$, where i is the imaginary unit.

Problem of Descriptions. Given $p(p < min((r+1)^m, (r+1)^n))$ complex-valued vectors $v^{(1)}, v^{(2)}, \cdots, v^{(p)}$, where $v^{(l)} = (v_1^{R(l)} + iv_1^{I(l)}, \cdots, v_n^{R(l)} + iv_n^{I(l)})^T, v_k^{R(l)}$ and $v_k^{I(l)} \in \{u_1, u_2, \cdots, u_{r+1}\}^n, k = 1, 2, \cdots, n, l = 1, 2, \cdots, p$, design an auto-associative memory via CVNNs such that if $(v_1^{R(l)}+iv_1^{I(l)}, v_2^{R(l)} + iv_2^{I(l)}, \cdots, v_n^{R(l)}+iv_n^{I(l)})^T$ is input into auto-associative memories as a probe, then the output of complex-valued networks converges to the corresponding pattern $(v_1^{R(l)} + iv_1^{I(l)}, v_2^{R(l)} + iv_2^{I(l)}, \cdots, v_n^{R(l)} + iv_n^{I(l)})^T, l = 1, 2, \cdots, p$.

For auto-associative memories, we consider the following continuous-time complex-valued delayed neural networks described by equations of the form

$$\begin{cases} \frac{dz_k(t)}{dt} = -d_k z_k(t) + \sum_{j=1}^{n} [a_{kj} f_j(z_j(t)) \\ \qquad\qquad + b_{kj} f_j(z_j(t - \tau_{kj}(t)))] + v_k, \ t \geq 0, \\ Z_k(t) = f(z_k(t)), \ t \geq 0, \end{cases} \tag{1}$$

with initial conditions

$$z_k(t) = \varphi_k(t), \quad t \in [-\tau, 0],$$

where $z(t) = (z_1(t), z_2(t), \cdots, z_n(t))^T \in C^n$ is the neuron state vector, non-negative delay $\tau_{kj}(t)$ corresponds to finite speed of axonal signal transmission with $\tau_{kj}(t) \leq \tau$, $D = diag(d_1, d_2, \cdots, d_n)$, $d_k \in R$ with $d_k > 0$ represents the self-feedback connection weigh, $A = [a_{kj}]_{n \times n} \in C^{n \times n}$, $B = [b_{kj}]_{n \times n} \in C^{n \times n}$, a_{kj} and b_{kj} stand for the strengths of the connection weight matrix and the asynchronously delayed connection weight among the circuit neurons, respectively, $v = (v_1, v_2, \cdots, v_n)^T \in C^n$, v_k is input vector, $X = (X_1, X_2, \cdots, X_n)^T \in C^n$, X_k is output vector, $\phi(t) = (\phi_1(t), \phi_2(t), \cdots, \phi_n(t))^T$, $\phi_k(t)$ is initial state of the network, $f(z(t)) = (f_1(z_1(t)), f_2(z_2(t)), \cdots, f_n(z_n(t)))^T$, f_k is the real-imaginary-type activation function which can be can be expressed by its real and imaginary parts with

$$f_k(z) = f_k^R(Re(z)) + if_k^I(Im(z)),$$

where

$$f_k^R(Re(z)) = \begin{cases} u_1, & Re(z) \in (-\infty, p_1], \\ \frac{u_2-u_1}{q_1-p_1}(Re(z) - p_1) + u_1, & Re(z) \in (p_1, q_1], \\ u_2, & Re(z) \in (q_1 p_2], \\ \frac{u_3-u_2}{q_2-p_2}(Re(z) - p_2) + u_2, & Re(z) \in (p_2, q_2], \\ u_3, & Re(z) \in (q_2, p_3], \\ \cdots, & \cdots \\ \frac{u_{r+1}-u_r}{q_r-p_r}(Re(z) - p_r) + u_r, & Re(z) \in (p_r, q_r], \\ u_{r+1}, & Re(z) \in (q_r, +\infty), \end{cases}$$

and

$$f_k^I(Im(z)) = \begin{cases} u_1, & Im(z) \in (-\infty, p_1], \\ \frac{u_2-u_1}{q_1-p_1}(Im(z) - p_1) + u_1, & Im(z) \in (p_1, q_1], \\ u_2, & Im(z) \in (q_1 p_2], \\ \frac{u_3-u_2}{q_2-p_2}(Im(z) - p_2) + u_2, & Im(z) \in (p_2, q_2], \\ u_3, & Im(z) \in (q_2, p_3], \\ \cdots, & \cdots \\ \frac{u_{r+1}-u_r}{q_r-p_r}(Im(z) - p_r) + u_r, & Im(z) \in (p_r, q_r], \\ u_{r+1}, & Im(z) \in (q_r, +\infty), \end{cases}$$

with $k \in J = \{1, 2, \cdots, n\}$, $0 \leq u_1 \leq u_2 \leq \cdots \leq u_{r+1}$ and $-\infty < p_1 < q_1 < p_2 < q_2 < \cdots < p_r < q_r < +\infty$ shown in Fig. 1 with $r = 3$.

Definition 1. CVNNs (1) is an auto-associative memory when $u^{(l)} = v^{(l)}$, $l = 1, \cdots, p$.

Assumption 1. (H_1) $d_k > a_{kk}^R l$, where $l = \max\{\frac{u_2-u_1}{q_1-p_1}, \frac{u_3-u_2}{q_2-p_2}, \cdots, \frac{u_{r+1}-u_r}{q_r-p_r}\}$, $\frac{u_2-u_1}{q_1-p_1}, \cdots$ and $\frac{u_{r+1}-u_r}{q_r-p_r}$ are slopes of the different segments.

Denote $z_k = x_k + iy_k$ with $x_k, y_k \in R$, by separating the state, the connection weight, the activation function and the external input, then the network (1) can

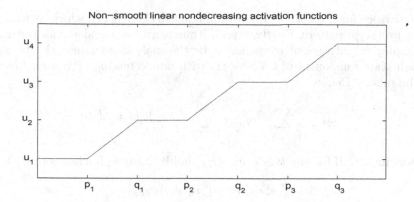

Fig. 1. Piecewise linear nondecreasing activation function

be rewritten in the equivalent form as follows:

$$\begin{cases} \frac{dx_k(t)}{dt} = -d_k x_k(t) + \sum_{j=1}^{n} [a_{kj}^R f_j^R(x_j(t)) - a_{kj}^I f_j^I(y_j(t)) \\ \qquad + b_{kj}^R f_j^R(x_j(t-\tau(t))) - b_{kj}^I f_j^I(y_j(t-\tau(t)))] + v_k^R, \ t \geq 0, \\ \frac{dy_k(t)}{dt} = -d_k y_k(t) + \sum_{j=1}^{n} [a_{kj}^R f_j^I(y_j(t)) + a_{kj}^I f_j^R(x_j(t)) \\ \qquad + b_{kj}^R f_j^I(y_j(t-\tau(t))) + b_{kj}^I f_j^R(x_j(t-\tau(t)))] + v_k^I, \ t \geq 0, \\ X_k(t) = f(x_k(t)), \ t \geq 0, \\ Y_k(t) = f(y_k(t)), \ t \geq 0. \end{cases} \qquad (2)$$

Accordingly, the initial conditions is

$$\begin{cases} x_k(t) = Re(\varphi_k(t)), \quad t \in [-\tau, 0], \\ y_k(t) = Im(\varphi_k(t)), \quad t \in [-\tau, 0], \end{cases}$$

where a_{ij}^R, b_{ij}^R and v_i^R are the real parts of a_{ij}, b_{ij} and v_i respectively, a_{ij}^I, b_{ij}^I and v_i^I are the imaginary parts of a_{ij}, b_{ij} and v_i respectively.

Definition 2. The equilibrium point $z_{xy}^* = \{x_1^*, x_2^*, \cdots, x_n^*, y_1^*, y_2^* \cdots, y_n^*\}$ of neural network (2) is globally exponentially stable if there exist positive numbers $\gamma > 0$ and $\lambda > 0$, such that

$$\|z_{xy}(t) - z_{xy}^*\| \leq \gamma e^{-\lambda t} \|\varphi(t) - z_{xy}^*\|_\tau, \ \forall t \geq 0,$$

where $\|\varphi_{xy}(t) - z_{xy}^*\|_\tau = \sup_{-\tau \leq t \leq 0} \|\varphi(t)_{xy} - z_{xy}^*\|$ and $\varphi(t)_{xy} = (Re(\varphi_1(t)), \cdots,$ $Re(\varphi_n(t)), Im(\varphi_1(t)), \cdots, Im(\varphi_n(t)))$.

3   Stability Analysis and Design Procedures

3.1   Existence and Uniqueness of the Equilibrium Point

The purpose of this paper is to recall the correct desired patterns by the designed CVNNs. In order to achieve this goal, the desired output pattern is designed as the activation function of the equilibrium point in CVNNs (1), which corresponds to the prototype pattern. Firstly, several lemmas will be considered for the design of complex-valued associative memories. By the analysis of lemmas, the dynamics of each state component of CVNNs (1) with nondecreasing activation function can be tacked. Denote

$$\Delta_k = \sum_{j=1, j\neq k}^{n} |a_{kj}^R| + \sum_{j=1}^{n} [|a_{kj}^I| + |b_{kj}^R| + |b_{kj}^I|].$$

Theorem 3.1. If for any $k \in J$ and (H_1) holds. Assume furthermore that

$$v_k^R < d_k p_1 - a_{kk}^R u_1 - \Delta_k u_{r+1}, \tag{3}$$

then there exists $T_1 \geq 0$ such that $x_k(t)$ of CVNNs (1) is located in $(-\infty, p_1)$ for all $t \geq T_1$.

Theorem 3.2. If for any $k \in J$ and (H_1) holds. Assume furthermore that

$$q_{s_1} d_k - u_{s_1+1} a_{kk}^R + \Delta_k u_{s_1+1} < v_k^R < p_{s_1+1} d_k - u_{s_1+1} a_{kk}^R - \Delta_k u_{r+1}, \tag{4}$$

then there exists $T_2 \geq 0$ such that $x_k(t)$ of CVNNs (1) is located in (q_{s_1}, p_{s_1+1}) for all $t \geq T_2$, where $s_1 = 1, 2, \cdots, r-1$.

Theorem 3.3. If for any $k \in J$ and (H_1) holds. Assume furthermore that

$$v_k^R > d_k p_r + a_{kk}^R u_{r+1} + \Delta_k u_{r+1}, \tag{5}$$

then there exists $T_3 \geq 0$ such that $x_k(t)$ of CVNNs (1) is located in $(q_r, +\infty)$ for all $t \geq T_3$.

Theorem 3.4. If for any $k \in J$ and (H_1) holds. Assume furthermore that

$$v_k^I < d_k p_1 - a_{kk}^R u_1 - \Delta_k u_{r+1}, \tag{6}$$

then there exists $T_4 \geq 0$ such that $y_k(t)$ of CVNNs (1) is located in $(-\infty, p_1)$ for all $t \geq T_4$.

Theorem 3.5. If for any $k \in J$ and (H_1) holds. Assume furthermore that

$$q_{s_2} d_k - u_{s_2+1} a_{kk}^R + \Delta_k u_{s_2+1} < v_k^I < p_{s_2+1} d_k - u_{s_2+1} a_{kk}^R - \Delta_k u_{r+1}, \tag{7}$$

then there exists $T_5 \geq 0$ such that $y_k(t)$ of CVNNs (1) is located in (q_{s_2}, p_{s_2+1}) for all $t \geq T_5$, where $s_2 = 1, 2, \cdots, r-1$.

Theorem 3.6. If for any $k \in J$ and (H_1) holds. Assume furthermore that

$$v_k^I > d_k p_r + a_{kk}^R u_{r+1} + \Delta_k u_{r+1}, \tag{8}$$

then there exists $T_6 \geq 0$ such that $y_k(t)$ of CVNNs (1) is located in $(q_r, +\infty)$ for all $t \geq T_6$.

For the sake of convenience, the following symbols are denoted.

$$N_{11} = \{k | k \in J, v_k^R < d_k p_1 - a_{kk}^R u_1 - \Delta_k u_{r+1}, v_k^I < d_k p_1 - a_{kk}^R u_1 - \Delta_k u_{r+1}\},$$

$$N_{12} = \{k | k \in J, q_1 d_k - u_2 a_{kk}^R + \Delta_k u_{r+1} < v_k^R < p_2 d_k - u_2 a_{kk}^R - \Delta_k u_{r+1}, v_k^I$$
$$< d_k p_1 - a_{kk}^R u_1 - \Delta_k u_{r+1}\},$$

$$\cdots \qquad \cdots$$

$$N_{s_1 s_2} = \{k | k \in J,$$
$$q_{s_1} d_k - u_{s_1+1} a_{kk} + \Delta_k u_{r+1} < v_k^R < p_{s_1+1} d_k - u_{s_1+1} a_{kk} - \Delta_k u_{r+1},$$
$$q_{s_1} d_k - u_{s_1+1} a_{kk} + \Delta_k u_{r+1} < v_k^I < p_{s_1+1} d_k - u_{s_1+1} a_{kk} - \Delta_k u_{r+1},$$

$$\cdots \qquad \cdots$$

$$N_{r+1 \ r+1} = \{k | k \in J, v_k^R > d_k p_r + a_{kk}^R u_{r+1} + \Delta_k u_{r+1},$$
$$v_k^I > d_k p_r + a_{kk}^R u_{r+1} + \Delta_k u_{r+1}\},$$

$$\Omega = \prod_{k \in N_{11}} \{z_k = x_k + iy_k | x_k \in (-\infty, p_1), y_k \in (-\infty, p_1)\} \times$$

$$\prod_{k \in N_{12}} \{z_k = x_k + iy_k | x_k \in (-\infty, p_1), y_k \in (q_1, p_2)\}$$

$$\times \quad \cdots \quad \times$$

$$\prod_{i \in N_{r+1 r+1}} \{z_k = x_k + iy_k | x_k \in (q_r, +\infty), y_k \in (q_r, +\infty)\},$$

Obviously, $N_{i_1 i_2} \bigcap N_{j_1 j_2} = \emptyset$ $(i_1, i_2, j_1, j_2 = 1, \cdots, r+1, i_1 i_2 \neq j_1 j_2)$.

Theorem 3.7. If (H_1) and $N_{11} \cup N_{12} \cup \cdots \cup N_{r+1 r+1} = J$ holds, then there exists a unique equilibrium point for CVNNs (1).

3.2   Stability Analysis and Design Procedures

The design approach of auto-associative memories is dependent the analysis of globally exponential stability. The main aim of this subsection is to select the appropriate parameters $d_k, a_{kj}^R, a_{kj}^I, b_{kj}^R$ and b_{kj}^I. Therefore, we will first determine the globally exponential stability criterion CVNNs (1).

In order to discuss the globally exponential stability of CVNNs (1), we rewrite it as follows

$$\begin{cases} \frac{dRe(u_k(t))}{dt} = -d_k Re(u_k(t)) + \sum_{j=1}^{n} \left[a_{kj}^R F_j^R(Re(u_j(t))) - a_{kj}^I F_j^I(Im(u_j(t))) + \right. \\ \qquad \left. b_{kj}^R F_j^R(Re(u_j(t - \tau_{kj}(t)))) - b_{kj}^I F_j^I(Im(u_j(t - \tau_{kj}(t)))) \right], \ t \geq 0, \\ \frac{dIm(u_k(t))}{dt} = -d_k Im(u_k(t)) + \sum_{j=1}^{n} \left[a_{kj}^R F_j^I(Im(u_j(t))) + a_{kj}^I F_j^R(Re(u_j(t))) + \right. \\ \qquad \left. b_{kj}^R f_j^I(Re(u_j(t - \tau_{kj}(t)))) + b_{kj}^I F_j^R(Im(u_j(t - \tau_{kj}(t)))) \right], \ t \geq 0, \\ Re(U_k(t)) = f(Re(u_k(t))), \ t \geq 0, \\ Im(U_k(t)) = f(Im(u_k(t))), \ t \geq 0. \end{cases}$$

$$(9)$$

with initial conditions

$$\begin{cases} Re(u_k(t)) = Re(\psi_k(t)), & t \in [-\tau, 0], \\ Im(u_k(t)) = Im(\psi_k(t)), & t \in [-\tau, 0], \end{cases}$$

where $u_k(t) = z_k(t) - z_k^*, \psi_k(t) = \varphi_k(t) - z_k^*, F_j^R(Re(u_k(t))) = f_j^R(Re(u_k(t)) + x_k^*) - f_k^R(x_k^*)$ and $F_j^I(Im(u_k(t))) = f_j^I(Im(u_k(t)) + x_k^*) - f_k^R(x_k^*), k \in J$.

We further assumption

(H_2) $d_k > \sum\limits_{j=1}^{n} l(|a_{kj}^R| + |a_{kj}^I| + |b_{kj}^R| + |b_{kj}^I|).$

Theorem 3.8. If $N_1 \cup N_2 \cup N_3 \cup N_4 = J$, (H_1) and (H_2) hold, then there exits an exponentially stable equilibrium point of CVNNs (1), and stable output state components belong to $\{u_1 + iu_1, \cdots, u_1 + iu_{r+1}, \cdots, u_{r+1} + iu_1, \cdots, u_{r+1} + iu_{r+1}\}^n$.

Remark 1. In Ref. [9], Han et al. designed the associative memories which stated that the initial states were fixed at zero. Whereas, this limitation is not unnecessary in our paper. Obviously, our approach is less conservative than [9].

Theorem 3.9. If $N_{11} \cup N_{12} \cup \cdots \cup N_{r+1r+1} = J$, (H_1) and (H_2) holds, then CVNNs (1) has at least $(r+1)^{2n}$ storage capacity.

When $y_k = 0, a_{kj} = b_{kj} = 0, f_j^I(\cdot) = 0$ and $I_k^I = 0, k = 1, 2, \cdots, n$, then activation function $f_j^R(Re(z))$ becomes following action function

$$f_j^R(Re(z)) = \begin{cases} -q, & Re(z) \in (-\infty, -p], \\ \frac{q}{p} Re(z), & Re(z) \in (-p, p], \\ q, & Re(z) \in (p, +\infty). \end{cases} \tag{10}$$

Denote

$$\hat{N}_1 = \{k | k \in J, v_k^{(l)} < -|pd_i - qa_{ii}| - \sum_{j=1, j\neq i}^{n} q|a_{ij}| - \sum_{j=1}^{n} q|b_{ij}|\},$$

$$\hat{N}_2 = \{k | k \in J, v_k^{(l)} > |pd_i - qa_{ii}| + \sum_{j=1, j\neq i}^{n} q|a_{ij}| + \sum_{j=1}^{n} q|b_{ij}|\}.$$

We further assumption

(H_3) $d_k > \sum\limits_{j=1}^{n} (|a_{kj}^R| + |b_{kj}^R|).$

Based on Lemmas 1 and 2 in [11] and Theorem 3.8 in this paper, we give the following Corollary.

Corollary 3.1. If $\hat{N}_1 \cup \hat{N}_2 = J$ and (H_3) hold, then there exits an exponentially stable equilibrium point of CVNNs (1) with activation function (10), and stable output state components belong to $\{-q, q\}^n$.

Remark 2. In [5,8], authors designed the associative memories under the assumption that self-inhibition was restricted to be equal to 1. While in our paper, self-inhibition can be chosen arbitrarily. Therefore, our works can be viewed as the improvement of [5,8] in some way.

3.3 The Design Procedure of Auto-Associative Memories

To realize auto-associative memories, a design procedure is given based on the above-mentioned theories about how to obtain parameters.

(i) Denote p desired patterns which need to be memorized as $v^{(1)}, v^{(2)}, \cdots, v^{(p)}$ with $v^{(l)} = (v_1^{R^{(l)}} + iv_1^{I^{(l)}}, \cdots, v_n^{R^{(l)}} + iv_n^{I^{(l)}})^T$, $v_k^{R^{(l)}}$ and $v_k^{I^{(l)}} \in \{u_1, u_2, \cdots, u_{r+1}\}^n$ for $k = 1, 2, \cdots, n$.

(ii) Determine initial values $d_k, a_{kj}^R, a_{kj}^I, b_{kj}^R$ and b_{kj}^I such that $N_{11} \cup \cdots \cup N_{1r+1} \cdots \cup N_{r+1r+1} = \{k|k \in J\}$ and (H_1) hold.

(iii) Synthesize CVNNs (1) with the parameters $d_k, a_{kj}^R, a_{kj}^I, b_{kj}^R$ and b_{kj}^I.

(iv) When external input pattern $v^{(l)}$ is fed to CVNNs (1), then desired output pattern can be retrieved.

Remark 3. Compared with binary associative memories via neural networks with the standard activation function in [7–12], we present an arbitrary muti-valued associative memory with a general activation function, which can be applied to the images of associative memories. Therefore, it is improvement and extension of bipolar associative memories in [7–12].

4 Numerical Solutions

In this section, two numerical examples are given to illustrate the superiority of CVNNs (1). The simulations are implemented by the Matlab R2016a(win32) on 2.40 GHz PC.

Example 1. The purpose of the experiments is to verify the correctness of three-valued auto-associative memories via CVNNs (1). Each stored pattern contains 3×4 units showed in Fig. 2 (blue pixel $= -1$, red pixed $= 0$, while pixed $= 1$).

For pattern C, the vector of input test probe is $v^{(1)} = [-1, 0, -1, 0, 1, 1, -1, 1, 1, 0, -1, 0]^T$. And for pattern T, the vector of input test probe is $v^{(2)} = [-1, 0, -1, 1, -1, 1, 1, 0, 1, 1, -1, 1]^T$.

According to the previous results of theories and the design procedure, 12 three-valued neurons are applied to CVNNs (1) and the following parameters are given: $r = 2, u_1 = -1, u_0 = 0, u_1 = 1, p_1 = -2, p_2 = 1, q_1 = -1, q_2 = 2, d_i = 0.3, \tau_{ij} = sin(t - 0.5)$, $a_{ij} = \begin{cases} 0.01, & i = j \\ 0.04, & i \neq j \end{cases}$, $b_{ij} = 0.01, i, j \in J$. From the numerical experimental results as seen in Figs. 3 and 4, it can be seen that CVNNs (1) converge to correct stored patterns.

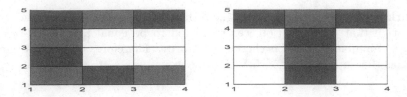

Fig. 2. Two stored patterns in Example 1. (Color figure online)

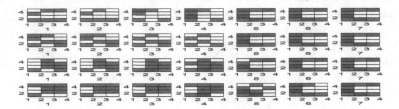

Fig. 3. Recall process with probe $u^{(1)}$ under four random initial states.

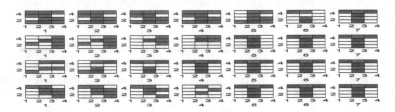

Fig. 4. Recall process with probe $u^{(2)}$ under four random initial states.

Example 2. In this example, we use the design procedure of auto-associative memory to recall the images. Georgia-Tech face database and the annotation can be found in http://www.anefian.com/research/face_reco.htm. Stored patters are given by 640×480 pixel images showed in Fig. 5, each pixel represents one of 256 grey scale values.

For pattern $v^{(1)}$, the vector of input test probe is $v^{(1)} = [74, 110, 110, 105, 110, 106, 105, 102, 151, 118, \cdots, 221, 220, 221, 221, 222, 223, 223, 218, 219, 218, \cdots, 95, 93, 90, 81, 53, 34, 28, 23, 22, 24, 20]^T$.

For pattern $v^{(2)}$, the vector of input test probe is $v^{(2)} = [37, 48, 50, 50, 55, 54, 49, 52, 55, 60, \cdots, 150, 152, 156, 160, 165, 170, 173, 175, 184, 185, \cdots, 25, 26, 25, 26, 27, 27, 27, 26, 25, 24]^T$.

According to the previous results of theories and the design procedure, 153600 mlti-valued neurons are applied to CVNNs (1) and the following parameters are given: $\tau_{kj} = sin(2t + 1), b_{kj} = 10^{-6} + 10^{-6}i, a_{kj} = \begin{cases} 0.15 + 10^{-6}i, & k = j \\ 10^{-6} + 10^{-6}i, & k \neq j \end{cases}, k, j = 1, \cdots, n.$ The simulation results are shown in Fig. 6, in which random initial states and perturbed input patterns are arranged.

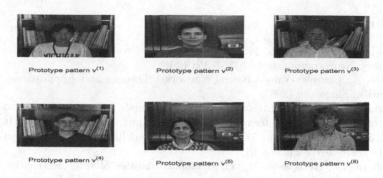

Fig. 5. Four stored patterns in Example 2.

It can be seen from Fig. 6 that CVNNs (1) converge to correct stored patterns. Similarly, other prototype patterns can be recalled accurately.

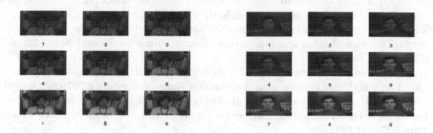

Fig. 6. Recall process input probe $u^{(l)}$ under a random initial state, $l = 1, 2$.

5   Conclusions

In this paper, a new design procedure is presented for multi-valued associative memories via CVNNs models. The proposed model for multi-valued patterns with a piecewise linear nondecreasing activation function generalizes the associative memories in [1,11,13]. Globally exponentially stable criteria are established for the CVNNs (1). The memorized pattern depends on the external input which leads prototype model to be recalled accurately. And our method can recall multi-valued patterns applied to auto-associative memories of coloured letters and face images.

Acknowledgement. The authors would like to thank the anonymous referees and editors for their helpful suggestions, which have improved the quality of this paper. This work was supported by National Natural Science Foundation of China (Grant nos. 11571170 and 11501290).

References

1. Aghajari, Z., Teshnehlab, M., Motlagh, M.: A novel chaotic hetero-associative memory. Neurocomputing **167**, 352–358 (2015)
2. Hirose, A.: Complex-Valued Neural Networks: Theories and Applications. World Scientific, Singapore (2003)
3. Suzuki, Y., Kitahara, M., Kobayashi, M.: Dynamic complex-valued associative memory with strong bias terms. In: Lu, B.-L., Zhang, L., Kwok, J. (eds.) ICONIP 2011. LNCS, vol. 7062, pp. 509–518. Springer, Heidelberg (2011). https://doi.org/10.1007/978-3-642-24955-6_61
4. Kitahara, M., Kobayashi, M.: Fundamental abilities of rotor associative memory. In: 9th IEEE International Conference on Computer and Information Science (ICIS), pp. 497–502 (2010)
5. Grassi, G.: A new approach to design cellular neural networks for associative memories. IEEE Trans. Circuits Syst. I: Fundam. Theory Appl. **44**, 835–838 (1997)
6. Grassi, G.: On discrete-time cellular neural networks for associative memories. IEEE Trans. Circuits Syst. I: Fundam. Theory Appl. **48**, 107–111 (2001)
7. Zeng, Z., Wang, J.: Design and analysis of high-capacity associative memories based on a class of discrete-time recurrent neural networks. IEEE Trans. Syst. Man Cybern Part B Cybern. **38**, 1525–1536 (2008)
8. Zeng, Z., Wang, J.: Associative memories based on continuous-time cellular neural networks designed using space-invariant cloning templates. Neural Netw. **22**, 651–657 (2009)
9. Han, Q., Liao, X., Huang, T., et al.: Analysis and design of associative memories based on stability of cellular neural networks. Neurocomputing **97**, 192–200 (2012)
10. Zhang, H., Huang, Y., Wang, B., et al.: Design and analysis of associative memories based on external inputs of delayed recurrent neural networks. Neurocomputing **136**, 337–344 (2014)
11. Zhou, C., Zeng, X., Yu, J., et al.: A unified associative memory model based on external inputs of continuous recurrent neural networks. Neurocomputing **186**, 44–53 (2016)
12. Sha, C., Zhao, H.: Design and analysis of associative memories based on external inputs of continuous bidirectional associative networks. Neurocomputing **266**, 433–444 (2017)
13. Xiu, C., Liu, C., Cheng, Y.: Associative memory network and its hardware design. Neurocomputing **158**, 204–209 (2015)
14. Chartier, S., Proulx, R.: NDRAM: nonlinear dynamic recurrent associative memory for learning bipolar and nonbipolar correlated patterns. IEEE Trans. Neural Netw. **16**(6), 1393–1400 (2005)
15. Chen, X., Zhao, Z., Song, Q., et al.: Multistability of complex-valued neural networks with time-varying delays. Appl. Math. Comput. **294**, 18–35 (2017)
16. Song, Q., Yan, H., Zhao, Z., et al.: Global exponential stability of impulsive complex-valued neural networks with both asynchronous time-varying and continuously distributed delays. Neural Netw. **81**, 1–10 (2016)
17. Zhao, Z., Song, Q., Zhao, Y.: Stability of complex-valued neural networks with two additive time-varying delay components. In: Cong, F., Leung, A., Wei, Q. (eds.) ISNN 2017. LNCS, vol. 10261, pp. 564–571. Springer, Cham (2017). https://doi.org/10.1007/978-3-319-59072-1_66
18. Zhou, C., Zeng, X., Luo, C., et al.: A new local bipolar autoassociative memory based on external inputs of discrete recurrent neural networks with time delay. IEEE Trans. Neural Netw. Learn. Syst. **28**(11), 2479–2489 (2016)

19. Zheng, P., Tang, W., Zhang, J.: Efficient continuous-time asymmetric Hopfield networks for memory retrieval. Neural Comput. **22**, 1597–1614 (2010)
20. Zheng, P.: Threshold complex-valued neural associative memory. IEEE Trans. Neural Netw. Learn. Syst. **25**, 1714–1718 (2014)
21. Huang, Y., Wang, X., Long, H., et al.: Synthesization of high-capacity auto-associative memories using complex-valued neural networks. Chin. Phys. B **25**, 120701 (2016)
22. Nie, X., Zheng, W., Cao, J.: Multistability of memristive Cohen-Grossberg neural networks with non-monotonic piecewise linear activation functions and time-varying delays. Neural Netw. **71**, 27–36 (2015)
23. Wang, L.: Dynamical analysis on the multistability of high-order neural networks. Neurocomputing **110**, 137–144 (2013)
24. Zeidler, E.: Nonlinear Functional Analysis and its Applications I: Fixed-Point Theorems. Springer, Heidelberg (1986)

Analysis on the Occurrence of Tropical Cyclone in the South Pacific Region Using Recurrent Neural Network with LSTM

Adarsh Karan Sharma, Vishal Prasad, Roneel Kumar,
and Anuraganand Sharma[✉]

The University of the South Pacific, Suva, Fiji
{adarsh.sharma, sharma_au}@usp.ac.fj,
vishalaprasad@gmail.com, roneelkumar808@gmail.com

Abstract. Weather prediction over the years has been a challenge for the meteorological centers in the South Pacific region. This paper presents Recurrent Neural Network (RNN) Architecture with Long Short Term Memory (LSTM) times-series weather data for prediction. From the gathered dataset, the Sea Surface Temperature (SST) is studied since it is known to be the foundation of the cyclone formation. This paper focuses on two scenarios. The first part is predicting upcoming SST using dataset from January 2013 to December 2017. The second part is taking out data of two different cyclones and predicting the SST for the next 14 days. Once the SST prediction is made, the predicted SST is compared with SST in the dataset for those 14 days. The main aim of this paper is to predict the SST using RNN and LSTM to anticipate the occurrence of tropical cyclones. The paper will focus on the reason for this study, a discussion of the model used, how the cyclones are formed, regarding the current threshold, the analysis of the dataset and lastly, the results from the experiment carried out.

Keywords: Time series data · Recurrent neural network
Long short-term memory · Artificial neural networks · Deep learning
Sea surface temperature · Weather prediction

1 Introduction

The weather pattern impacts the life of billions of people every day. The weather forecasting came to practice with the early civilizations and was based on observing frequent astronomical and meteorological events [4]. Weather forecasting nowadays are mainly done on computer-based models where multiple atmospheric factors are taken into consideration [5]. Even though these multiple atmospheric factors are investigated, the important aspect required by human input is to choose the best prediction model for forecasting and analyzing weather pattern and also distinguish which model is better and worse. The weather data collected, is one of the forms of Big data. To processes this Big data, extensive computational power is required to solve all the equations to get the weather forecasting results. At this instance the forecasts seem to become less accurate and computation becomes expensive.

© Springer Nature Switzerland AG 2018
L. Cheng et al. (Eds.): ICONIP 2018, LNCS 11301, pp. 476–486, 2018.
https://doi.org/10.1007/978-3-030-04167-0_43

The advancement in machine learning and artificial intelligence, have introduced neural networks to solve many big data problems. In this paper, the power of neural networks is used for prediction of tropical cyclones. This paper mainly focuses on cyclone prediction in the South Pacific region. There are many factors involved in weather prediction as mentioned earlier and one of the very influential one is the Sea Surface Temperature also known as SST. The SST is an important parameter in the energy balance system of the earth's surface and a critical indicator to measure SST heat [6]. The prediction of SST has become a major of area of research now, since many application domains such as ocean weather and climate forecasting, offshore activities like fishing and mining and ocean environment protection rely on this element.

This paper focuses on predicting SST using the means of Long Short-Term Memory (LSTM). LSTM is a building unit for layers of a Recurrent Neural Network (RNN), which is a class of artificial neural network where connections between units form a directed cycle [7]. Section 2 describes the RNN model with LSTM briefly. Section 3 gives the details of factors that cause formation of cyclone supported by illustrating historical data of tropical cyclones in the South Pacific region. Section 4 discusses experimental results and Sect. 5 concludes the paper by summarizing the results and proposing some further extensions to the research.

2 Model

The models used in this paper are Recurrent Neural Network (RNN) and Long Short Term Memory (LSTM) as pointed out above. There are some other methods used by researchers to determine the SST: linear regression, Support Vector Machines [23], and Artificial Neural Network [24]. The reasons for choosing RNN with LSTM is that when training an RNN model with gradient based optimization techniques and time series data only, the model encounters a Vanishing Gradient Problem [8].

2.1 Recurrent Neural Networks

The RNN Architecture is a natural generalization of feedforward neural networks to sequences, RNNs are networks with loops in them, which results in information persistence [9]. RNN helps to retain the state and the connection between the inputs for an arbitrarily long context window [22]. RNN is a model that can selectively pass information across sequence steps, while processing sequential data one element at a time [22]. However, there is a downfall of using this Neural Network which is known as the Gradient Vanishing Problem.

2.2 Problem of Long-Term Dependencies

When making predictions using the weather dataset, the data is analyzed from past time to present times and this makes it easy to recognize the general pattern. However, as the gap increases between past and the present, simple RNN's fails to learn and keep track of the inputs. To overcome this issue, LSTM Neural networks is introduced.

2.3   Long Short-Term Memory

Hochreiter and Schmidhuber introduced LSTMs as a type of RNN capable of learning long-term dependencies [7]. LSTM units (or blocks) are a building unit for layers of a RNN. A RNN composed of LSTM units is often called an LSTM network. A common LSTM unit usually composed of a cell, an input gate, an output gate and a forget gate. Figure 1 explains it more thoroughly.

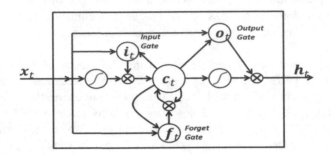

Fig. 1. A simple LSTM gate with input, output and forget gates

2.4   Structure of the Model Used

The model used in this paper is one-to-one sequential LSTM model that simply consists of linear stack of layers.

The number of LSTM input is 100 and input shape of the array is defined accordingly. The model defines the Dense as 1 which means number of outputs is one and the model uses activation function as Sigmoid. The model computes Root Mean Square Error as loss and uses 'adam' as an optimizer.

This research uses Python to develop the LSTM model. Also, the research uses technologies such as Keras and Tensorflow, which are one of the most widely used RNN libraries for Python programming [1].

3   Factors Causing Formation of Cyclones

3.1   The Current Threshold

There are mainly six factors that contribute to cyclone genesis [14]. Sea Surface Temperature (SST) is said to be one of the factors that help in causing the cyclones to form in a region [15]. This also specifies that if the SST is meeting its threshold and a tropical depression is quite near, the region can attract that tropical depression in that area. If the increasing rate of SST and other factors influencing the formation of cyclones remain same, SST can be used as a fuel in Tropical Cyclone genesis and intensification [11, 15].

According to the studies carried out by Palmen [12], the SST threshold of about 26 °C–27 °C degrees can be vital for cyclone formation. This threshold has been in the books for around 70 years without being tested. Gray [13] in his literature concluded that the ideal SST to form a cyclone is 26.5 °C degrees. However, some variations have also occurred whereby a small percentage of (less than 1%) [3] Of cyclones has taken place in situations where the SST is below the threshold mentioned (2–3° cooler) [10].

Moving on, the link between the warming up of the ocean and the tropical cyclone activity has been long and thus, this is very much required in the formation and intensification of cyclone in a region. The thermodynamic states [2] and the increasing SST in a region have increased the occurrence of cyclone. Gray [13] in his research mentioned that 26.5 °C is where the sea gets warm and formation of cumulus clouds takes place. The warm air rises while the cool air at the top rushes down to take the place of the risen air [16]. This causes a bit of depression in the atmosphere and with a spin (Coriolis Force, which is also a vital element in formation of cyclone) [17] can cause a Tropical Depression [TD] to Occur. With an even more intense SST [11], the TD can intensify and turn into a more destructive Tropical Cyclone.

3.2 Analysis of the Dataset

Our SST data are hourly recorded and are available from 1993 to 2018. The dataset was provided by the Fiji Meteorological Center [18]. Our primary region of interest in this analysis was the South Pacific countries and mainly to places where a cyclone was formed or where a country was hit by a cyclone.

The analysis was done by taking each year of a country and seeing the SST of the period when a cyclone was formed. As stated in the above-mentioned literatures, the cyclone formation takes place when the SST is above 26.5 °C. But for the dataset we analyzed, majority of the SST for a given day was above the threshold, that is, even in non-cyclonic season. The average SST for a given month would be around 26.5 °C.

Table 1 quite evidently shows that the mean temperatures for most of the months are more than the threshold. Even in the non-cyclone season (April–November, for the most parts of the South Pacific) the average SST is either more than the threshold or quite near to the threshold.

A new threshold was needed for the algorithm which the team developed. SST during the occurrence of major cyclones in the South Pacific region between 2010 and 2017 was analyzed. It was obvious that the current threshold mentioned, that is, 26.5 ° C will (more likely) cause an occurrence of cyclone. But, the analysis shows that for a better prediction, a new threshold should be learnt.

Table 2 markedly shows that when these major cyclones had occurred, the SST was somewhere greater than 29 °C (averaged). Figure 2 shows the average SST (Data from Table 2) at which cyclone was formed.

The current threshold of 26.5 °C was proposed almost 70 years ago [10]. The atmospheric condition compared to what it is today is quite different. The pollution, growth of population and the emission of greenhouse gasses in the atmosphere has led the earth's atmosphere to warm up, causing a drastic change in the current climate condition as compared to 50–70 years back. Due to the increase in the overall warmth of the earth, the SST has also increased incrementally [16].

Greenhouse gases have eventually increased the SST, which in return increased the cyclone intensity and occurrence as discussed by some scientists [19]. The warmth of the sea greatly depends on its atmospheric condition. And while the atmosphere is

Table 1. This table shows an average SST (Monthly) of 3 countries for the years 2015–2017

Fiji – Monthly Average SST				Vanuatu – Monthly Average SST				Solomon Island – Monthly Average SST		
	2017	2016	2015		2017	2016	2015		2017	2016	2015
Jan	30.47	29.95	29.97	Jan	29.15	28.93	28.35	Jan	30.36	30.11	29.93
Feb	30.10	30.99	31.07	Feb	29.29	30.06	29.21	Feb	30.26	29.28	29.82
Mar	30.10	30.64	30.39	Mar	29.30	29.30	28.14	Mar	30.49	29.29	29.45
Apr	30.17	29.50	29.22	Apr	27.83	29.30	27.61	Apr	30.41	29.23	29.57
May	29.10	28.67	28.12	May	27.03	27.74	26.96	May	30.23	29.60	29.05
Jun	27.96	27.51	26.99	Jun	26.45	26.60	25.72	Jun	29.47	28.99	28.57
Jul	27.01	26.68	26.12	Jul	25.78	25.95	24.78	Jul	28.65	28.54	28.66
Aug	26.51	26.53	25.56	Aug	25.53	25.55	24.60	Aug	28.49	28.60	28.42
Sep	27.75	26.80	25.90	Sep	26.03	25.95	24.87	Sep	28.71	28.84	28.43
Oct	28.46	28.21	26.93	Oct	26.73	26.48	25.28	Oct	29.14	29.53	28.61
Nov	29.22	29.16	28.86	Nov	27.07	27.53	26.71	Nov	30.30	29.99	29.54
Dec	29.58	29.60	29.30	Dec	28.04	28.04	28.02	Dec	30.17	30.10	29.93

Table 2. Table showing average SST for the cyclone period of the countries affected from 2010–2017

Average SST for the Cyclone Period								
Cyclone Name	Date	Affected Countries						
		Fiji	Vanuatu	Solomon Island	Tuvalu	PNG	Samoa	Tonga
Tomas	Mar 9 -17, 2010	28.49						
Ului	Mar 9 – Mar 21, 2010		28.20	28.81				
Yasi	Jan 26 – Feb 7, 2011	31.50	28.69	29.78	N/A	31.50		
Evan	Dec 9 – Dec 19, 2012	29.51					28.68	
Ian	Jan 2 – Jan 15, 2014	30.60						N/A
Ula	Dec 26 – Jan 12, 2016	29.11	28.67		29.65		29.31	26.67
Winston	Feb 7 – Feb 25, 2016	31.76	30.27					26.94

warming up, SST also increases rapidly. SST is said to be the "fuel" [20] which ignites and maintains the cyclone in a region. If the level of fuel remains quite high (keeping in consideration that all other factors remain satisfied) [21], the thermodynamic state of the region will increase causing a disturbance to occur. Similar study was conducted in [23] where LSTM was used to predict the SST. Since SST plays a major part in this prediction of cyclones in the Pacific region, this article displayed some convincing results which helped us in our experiment.

Thus, the model RNN that we are going to use would need this threshold to predict the occurrence of cyclone in the South Pacific region. This piece of data could be quite useful and be quite efficient and precise.

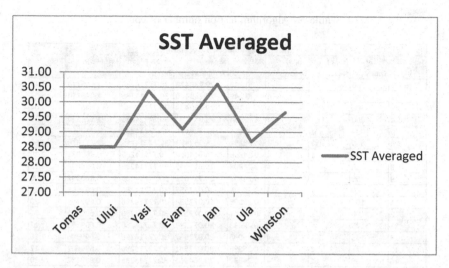

Fig. 2. Figure showing the averaged SST for the past 7 cyclones that hit the Pacific Ocean

4   Analysis and Results

4.1   Testing the Algorithm

The research was basically done to predict cyclone formation. As explained in Sect. 3, SST data has been utilized to meet the objectives. The LSTM network was tested with various approaches to get the best results from the dataset. A research by Dr. Jason Brownlee [23] on how to tune LSTM Hyper-parameters with Keras for Time Series Forecasting data was used as the basis of this experiment. In his research, 1 neuron, 4 batch sizes and 1000 epoch were used. Further in the experiment, we tried to tune our algorithm further by adjusting other hyper-parameters. For this experiment, three major parameters were used, mainly the neuron, batch size and epochs. For each experiment, we changed one parameter while the other two remained the same. Table 3 shows the experiment where the number of epochs was changed while neurons and batch size remained the same.

The objective was to get the best neuron, batch size and epoch. The way we measured the efficiency of our algorithm was by getting the difference mean temperature, which is the actual minus the mean temperature calculated by the algorithm.

Likewise, in Table 4 other parameters remained same expect neuron. For the number of epochs, we used the result obtained from Table 3. Table 5 depicts the experiment made by changing the batch size while the neuron and the epoch remained the same (values calculated from the previous experiments shown in Tables 3 and 4).

It was best noted that through the experiment carried that the correct parameter values for our prediction would be: Batch size = 100, Number of Epoch = 100 and Neurons = 1. This combination not only proved in terms of time but efficiency, less difference of mean temperature, less RMSE and greater accuracy.

Table 3. Algorithm test on number of epoch

Test No	No. of Epoch	No. of Neurons	No. of Batches	RMSE	Mean Temperature	Actual Mean Temp	Diff Mean Temp	Time
				Test the algorithm on No. of Epoch				
1	500	1	4	0.249	29.730429	30.77714286	1.046714	4.41
2	500	1	4	0.249	29.730429	30.77714286	1.046714	5.07
3	500	1	4	0.249	29.730433	30.77714286	1.046710	5.04
4	1000	1	4	0.25154	29.653294	30.77714286	1.123849	10.29
5	1000	1	4	0.25154	29.559475	30.77714286	1.217668	11.52
6	1000	1	4	0.25154	29.559473	30.77714286	1.217670	13.56
7	250	1	4	0.24808	29.914062	30.77714286	0.863081	3.20
8	250	1	4	0.24808	29.914045	30.77714286	0.863098	3.12
9	250	1	4	0.24808	29.914051	30.77714286	0.863092	3.05
10	100	1	4	0.25098	30.077555	30.77714286	0.699588	1.24
11	100	1	4	0.25098	30.077534	30.77714286	0.699609	1.20
12	100	1	4	0.25098	30.077572	30.77714286	0.699571	1.21
13	175	1	4	0.25098	30.077572	30.77714286	0.699571	1.22
14	175	1	4	0.25098	30.077557	30.77714286	0.699586	1.18
15	175	1	4	0.25098	30.077541	30.77714286	0.699602	1.17
16	200	1	4	0.24837	29.964961	30.77714286	0.812182	2.30
17	200	1	4	0.24837	29.96494	30.77714286	0.812203	2.22
18	200	1	4	0.24837	29.964941	30.77714286	0.812202	2.24

Table 4. Algorithm test on number of neuron

Test No	No. of Neurons	No. of Epoch	No. of Batches	RMSE	Mean Temperature	Actual Mean Temp	Diff Mean Temp	Time
				Test the algorithm on No. of Neuron				
1	1	100	4	0.250982	30.077572	30.77714	0.699571	1.21
2	1	100	4	0.250983	30.077572	30.77714	0.699571	1.24
3	1	100	4	0.250982	30.077557	30.77714	0.699586	1.23
4	20	100	4	0.268	28.682894	30.77714	2.094249	1.20
5	20	100	4	0.268	28.689993	30.77714	2.087150	1.25
6	20	100	4	0.268	28.689912	30.77714	2.087231	1.20
7	10	100	4	0.277	28.322474	30.77714	2.454669	1.18
8	10	100	4	0.276	28.333355	30.77714	2.443788	1.17
9	10	100	4	0.277	28.33255	30.77714	2.444593	1.20
10	2	100	4	0.243	29.643412	30.77714	1.133731	1.20
11	2	100	4	0.244	29.643422	30.77714	1.133721	1.21
12	2	100	4	0.241	29.643399	30.77714	1.133744	1.20

4.2   Results

Algorithm was trained using the SST data ranging from Jan 2013 to Dec 2017. Objective was to test how well the system trains the algorithm to predict future temperatures. Figure 3 shows the results that were obtained, that is the prediction done by the algorithm against the actual SST. This prediction is done for the period 1st January 2018 to 13th January 2018.

The experiment above was done to see how the algorithm was behaving with the data input. The prediction made by the algorithm was quite accurate against the actuals. Then further experiments were made where the data was taken before a major cyclone hit an area.

Table 5. Algorithm test on batch size

Test the algorithm on No. of Batch								
Test No	No. of Batch	No. of Epoch	No. of Neurons	RMSE	Mean Temperature	Actual Mean Temp	Diff Mean Temp	Time
1	4	100	1	0.250982	30.077572	30.77714	0.699571	1.21
2	4	100	1	0.250983	30.077572	30.77714	0.699571	1.24
3	4	100	1	0.250982	30.077557	30.77714	0.699586	1.23
4	1	100	1	0.282	29.502321	30.77714	1.274822	5.00
5	1	100	1	0.282	29.50245	30.77714	1.274693	5.05
6	1	100	1	0.282	29.502489	30.77714	1.274654	5.10
7	10	100	1	0.257	30.1999482	30.77714	0.577195	0.46
8	10	100	1	0.257	30.19948	30.77714	0.577663	0.48
9	10	100	1	0.257	30.199484	30.77714	0.577659	0.50
10	50	100	1	0.342	29.10429	30.77714	1.672853	0.20
11	50	100	1	0.342	29.104315	30.77714	1.672828	0.22
12	50	100	1	0.342	29.104315	30.77714	1.672828	0.20
13	25	100	1	0.271	30.178583	30.77714	0.598560	0.20
14	25	100	1	0.271	30.178566	30.77714	0.598577	0.21
15	25	100	1	0.272	30.178588	30.77714	0.598555	0.18
16	100	100	100	0.25	30.171635	30.77714	0.605508	0.14
17	100	100	100	0.25	30.17165	30.77714	0.605493	0.17
18	100	100	100	0.25	30.17164	30.77714	0.605503	0.15

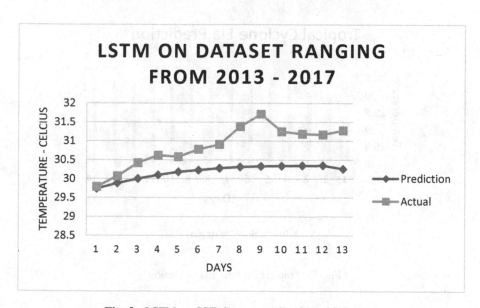

Fig. 3. LSTM on SST dataset ranging from 2013–2017

Since the algorithm was able to predict at a near perfect rate, we tried to test it to see if it would have predicted past years cyclones. For this scenario, we took two cyclones (Tropical Cyclone Winston and Ela) and checked the formation date. TC Winston was formed near Vanuatu and started weakening after hitting Fiji Islands. We took the data for Vanuatu and Fiji. Since we were going to predict the formation of Cyclone Winston, the data taken was from Jan 2013 until 23rd Jan 2016. (7th Feb 2016 was TC

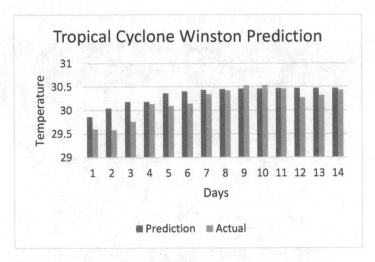

Fig. 4. Tropical cyclone Winston prediction

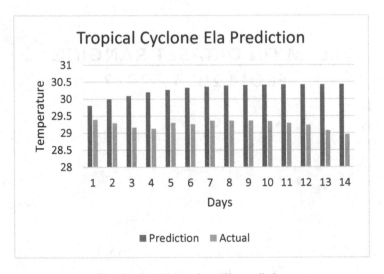

Fig. 5. Tropical cyclone Ela prediction

Winston Formation Date). The algorithm will be set as such, that it will predict future SST for the next 14 days. Figures 4 and 5 shows the actual and predicted values for the next 14 days. The algorithm has predicted that there will be a major tropical depression in the next 14 days. Both the figures quite evidently show that the algorithm can predict the SST quite competently. This proves that LSTM can be used to predict future values quite efficiently.

It was also noted that the prediction goes quite stagnant as you go deeper in predicting the values. One reason could be that the algorithm might not be able to make a relationship between the newer predicted data with the older dataset. One solution could be is increasing our dataset and checking for anomalies that could hinder the prediction, thus making prediction more efficient and much closer to the actual.

5   Conclusion

The major objectives of this paper were accomplished which were to study and formulate the SST of a time series weather data, make cyclone predictions based on the SST and propose a LSTM model for cyclone predictions. The dataset we used showed us that the current threshold can be used predict occurrence of a cyclone. But for the dataset we used, it proved otherwise. For that, we had to move our threshold to 29 °C, to see whether we were able to predict the cyclone correctly or not. The proposed network model utilizes LSTM layer to model the time series data, and fully-connected layer to map the output of LSTM layer to a final prediction. The results of this paper show that a LSTM network can be used in prediction of cyclones with very less RMSE. Python environment together with *Keras* and *Tensorflow* libraries are much more reliable system for compiling and training time series data models. Hence the Artificial Intelligence based system can be efficient enough to make cyclone predictions. Future work would be study on the influence of global warming in cyclone formation based on SST, consequently, forecasting the possible changes in threshold temperature.

References

1. Brownlee, J.: Time Series Prediction with LSTM Recurrent Neural Networks in Python with Keras.   https://machinelearningmastery.com/time-series-prediction-lstm-recurrent-neural-networks-python-keras/. Accessed 01 May 2018
2. Raymond, D., Sessions, S., Carillo, C.: Thermodynamics of tropical cyclogenesis in the northwest Pacific. J. Geophys. Res. **116**(1), 1–18 (2011)
3. Dare, R., McBride, J.: The threshold sea surface temperature condition for tropical cyclogenesis. J. Clim. **24**(1), 4570–4576 (2011)
4. Zaytar, M., Amrani, C.: Sequence to sequence weather forecasting with long short-term memory recurrent neural networks. Int. J. Comput. Appl. **143**(11), 1–5 (2016)
5. Bell, I., Wilson, J.: Visualising the atmosphere in motion. In: Bureau of Meteorology Training Centre, Melbourne, pp. 1–4 (1995)
6. Patil, K., Deo, M., Ravichandran, M.: Prediction of sea surface temperature by combining numerical and neural techniques. J. Atmos. Ocean. Technol. **33**(1), 1715–1726 (2016)
7. Hochreiter, S., Schmidhuber, J.: Long short-term memory. Neural Comput. **9**(8), 1735–1780 (1997)
8. Hochreiter, S.: The vanishing gradient problem during learning recurrent neural nets and problem solutions. Int. J. Uncertain. Fuzziness Knowl.-Based Syst. **6**(2), 1047–1116 (1998)
9. Glorot, X., Bengio, Y.: Understanding the difficulty of training deep feedforward neural networks. In: 13th International Conference on Artificial Intelligence and Statistics, Italy, pp. 249–256 (2010)
10. Tory, K., Dare, R.: Sea surface temperature thresholds for tropical cyclone formation. Centre for Australian Weather and Climate Research, Melbourne, pp. 8171–8183 (2015)

11. Arora, K., Dash, P.: Towards dependence of tropical cyclone intensity on sea surface temperature and its response in a warming world. Climate **4**(30), 1–19 (2016)
12. Palmén, E.H.: On the formation and structure of tropical cyclones. Geophysica **3**(1), 26–38 (1948)
13. Gray, W.M.: Global view of the origin of tropical disturbances and storms. Mon. Weather Rev. **96**(10), 669–700 (1968)
14. Montgomery, M.T.: Recent advances in tropical cyclogenesis. In: Mohanty, U.C., Gopalakrishnan, S.G. (eds.) Advanced Numerical Modeling and Data Assimilation Techniques for Tropical Cyclone Prediction, pp. 561–587. Springer, Dordrecht (2016). https://doi.org/10.5822/978-94-024-0896-6_22
15. ScienceDaile Homepage. www.sciencedaily.com/releases/2015/03/150317162146.htm. Accessed 01 May 2018
16. Morrison, S.: Climate Change Opens the Door to More Intense Tropical Storms. https://ohiostate.pressbooks.pub/sciencebites/chapter/climate-change-opens-the-door-to-more-intense-tropical-storms/. Accessed 02 May 2018
17. Lian, X., Chan, J.: The effects of the full coriolis force on the structure and motion of a tropical cyclone. Part I: effects due to vertical motion. J. Atmos. Sci. **62**(1), 3825–3830 (2005)
18. Australian Government – Bureau of Meteorology. http://www.bom.gov.au/oceanography/projects/spslcmp/data/index.shtml. Accessed 15 Apr 2018
19. Michaels, P., Knappenberger, P., Davis, R.: Sea-surface temperatures and tropical cyclones in the Atlantic basin. Geophys. Res. Lett. **3**(1), 1–4 (2006)
20. Landsea, C.: Climate Variability of Tropical Cyclones: Past, Present and Future. http://www.aoml.noaa.gov/hrd/Landsea/climvari/. Accessed 10 May 2018
21. Ekwurzel, B.: Hurricane Watch Checklist: Four Factors that Strengthen Tropical Cyclones, Union of Concerned Scientists. https://blog.ucsusa.org/brenda-ekwurzel/hurricane-watch-checklist-four-factors-that-strengthen-and-four-that-weaken-tropical-cyclones. Accessed 12 May 2018
22. Lipton, Z., Berkowitz, J., Elkan, C.: A Critical Review of Recurrent Neural Networks for Sequence Learning. arXiv preprint arXiv:1506.00019, pp. 1–38 (2015)
23. Zhang, Q., Wang, H., Dong, J., Zhong, G.: Prediction of sea surface temperature using long short-term memory. IEEE Geosci. Remote Sens. Lett. **14**(10), 1745–1749 (2017)
24. Gandhi, A., D'souza, S., Arjun, N.: Prediction of sea surface temperature using artificial neural network. Int. J. Remote Sens. **39**(12), 4214–4231 (2018)

Combining User-Based and Session-Based Recommendations with Recurrent Neural Networks

Tu Minh Phuong[1(✉)], Tran Cong Thanh[1], and Ngo Xuan Bach[1,2]

[1] Posts and Telecommunications Institute of Technology, Hanoi, Vietnam
{phuongtm, bachnx}@ptit.edu.vn,
thanh.ptit.96@gmail.com
[2] FPT Software, Hanoi, Vietnam

Abstract. Recommender systems generate recommendations based on user profiles, which consist of past interactions of users with items. When user profiles are not available, session-based recommendation can be used instead to make predictions based on sequences of user clicks within short sessions. Although each approach can be used separately, it is desired to utilize both user profiles and session information, and other information such as context, when those are available. In this paper, we propose a Recurrent Neural Networks (RNNs) based method that combines different types of information to generate recommendations. Specifically, we learn user and item representations from user-item interaction data and explore a new type of RNN cells to combine global user embeddings with sequential behavior within each session to generate next item recommendations. The proposed model uses an attention mechanism to adaptively regulate the contributions of different input components based on specific situations. The model can be extended to incorporate other input, such as contextual information. Experimental results on two real-world datasets show that our method outperforms state-of-the-art baselines that use only user or session information.

Keywords: Recommender systems · Session-based recommendation
Recurrent neural networks · Next item recommendation

1 Introduction

Traditional recommendation algorithms rely on user historic data to generate personalized recommendations about products or items that a user is likely interested in [2]. Collaborative filtering (CF) [8], the most widely used recommendation approach, uses past user-item interactions to create a low rank representation for each user which is then matched against item representations to calculate recommendation scores (the model-based CF [11]) or to find users with similar preferences (the neighborhood-based CF [16]). The CF based recommender systems have proved useful as they can provide accurate and personalized recommendations in many domains.

An important condition for CF algorithms to work is the availability of informative user profiles, which implies that each user has enough interactions recorded and the user

© Springer Nature Switzerland AG 2018
L. Cheng et al. (Eds.): ICONIP 2018, LNCS 11301, pp. 487–498, 2018.
https://doi.org/10.1007/978-3-030-04167-0_44

identity is visible for each interaction event. In many online systems such as e-commerce websites or music online service, this condition is not satisfied, either because users are not required to authenticate and/or they are newcomers. To make predictions in such settings session-based recommendation has recently been proposed [10]. A session is a sequence of user interactions (such as clicks) that occur within a given time frame and are separated from other sessions by time intervals of sufficient length. Examples of sessions are a sequence of items clicked by a user on an e-commerce website within a day or a list of songs a user listens to in an evening. Using sequential patterns of session events, session-based systems can produce relatively accurate recommendations, outperforming other approaches such as item-based methods in many domains [7, 10]. However, because users are assumed to be anonymous, session-based recommendations cannot take into account user-specific preferences and therefore are not personalized.

In practice, there are many cases where the systems have access to both session information and user identifiers, for example via explicit authentication or other identifiers such as mobile device IDs. In these settings, using user-only or session-only methods does not allow utilizing all available information, which may be valuable to predict user intents. Specifically, session-based methods cannot model global patterns of user preferences, which are preserved from session to session, while CF methods cannot take into account the user intent specific for each session and the sequential nature of consuming certain types of items such as movie series. Simple modifications, for example, by concatenating sessions belonging to the same user, still ignore part of information and do not give the best results [14].

In this paper, we propose a method to combine user-based CF with session-based recommendation based on RNNs. The method learns representations for users, items, and other context information (if available). These representations are passed to RNNs that model sequential events within sessions. To account for importance of different information sources that vary from situation to situation we explore a new type of RNN cells with an attention mechanism. This makes the method easily extendable to incorporate additional context information, and able to deal with situations when the user is new and does not have enough past interactions. We describe a procedure to prepare training data, which is specially designed to take more training signals from session data.

We evaluate the proposed method on two real-world datasets: one is a public dataset in the music domain, and one proprietary dataset from a video-on-demand domain. The experimental results show clear superiority of our method over plain session-based algorithms, CF algorithms, and a model using concatenated sessions.

2   Related Work

Collaborative Filtering. Collaborative filtering (CF) methods such as matrix factorization [11] and neighborhood-based methods [5, 8] are widely used for building recommender systems based on user-item interaction data. Matrix factorization decomposes the sparse matrix of user-item interactions to represent users and items by latent factor vectors. The recommendation problem is then considered as a matrix completion task, where a missing value is the inner product of the respective user and item vectors. Neighborhood based methods utilize user-item interactions to compute the similarities between users or between items. A neighborhood of an active user (in used-based recommendations) or item neighborhoods (in item-based recommendations) are then exploited to generate recommendations. Content features can also be included to deal with the new user problem [13], or different methods can be combined to reduce the negative effect of data sparseness [1].

Recommendations for Sequential Data. Conventional CF methods can not model the temporal dynamics that naturally exist in user behavior. For instance, series movies are typically consumed one episode after another. Therefore, several methods have been proposed to explicitly incorporate temporal information when generating recommendations, which adapt sequence-based methods from other domains. Among methods of this kind, Markov models [7, 15] and RNNs [6] based recommendations are the most successful. By learning a transition graph over actions of users, Markov chains [7] can model sequential behavior and predict the next action based on the last ones. Donkers et al. [6] present a RNN architecture to model user-item interactions arranged in sequences. They introduce a new type of Gated Recurrent Unit cells that combine user and item vectors for updating RNN hidden states. Our method is similar to [6] with two main differences: (1) we propose a new type of attentive gate which requires less parameters but be easily extended to incorporate more input components; and (2) we adopt a training procedure specially designed to work with session data.

Session-Based Recommendations. Emerged recently, this kind of systems has attracted a great research interest because session-based settings are common in practice. The pioneer work by Hidasi et al. [10] is the first introduction to session-based recommendation which presents a nice application of RNNs for modeling short sessions of user interactions. Since then, RNNs remain the most popular architectures for systems of this kind but with a number of improvements and modifications. Tan et al. [17] present an improved procedure for training data augmentation and selection. Hidasi et al. [9] extend RNNs to incorporate content features of session events. Li et al. [12] propose a mechanism to take into account the user's purpose within a session. Quadrana et al. [14] propose a hierarchical architecture to incorporate user identifiers when those are available. A deviation from the mainstream is the work by Tuan et al. [18], who propose a Convolutional Neural Networks (CNNs) architecture to model both item content and session events. The use of CNNs has two advantages: CNNs are very powerful tool to represent content features while are faster and easier to train than RNNs. Our method in this work is similar to [15] in that it takes into account user identifiers but out model can shift the focus between user and item component depending on a specific session event.

3 Methods

In this section we describe the proposed RNNs based model for combining user-related and session information. We also explain the training procedure and how training data are prepared.

In a typical session-based setting, the system takes as input current session events in forms of clicked items' IDs up to the current time point and outputs, for each available item, a score indicating how likely the item will be clicked next. The top-scored items are then recommended to the user. We consider the settings in which each session is also associated with a user's ID, and possibly the timestamp of each event. Formally, let $[e_1, e_2, \ldots, e_L](L \geq 2)$ denote a session of L clicks by user u where $e_i \in P(1 \leq i \leq L)$ is the index of the clicked item (we omit index u from e_i to simplify the notation). Here P denotes the set of all items. For any prefix of the session $[e_1, e_2, \ldots, e_l](1 \leq l < L)$, the recommendation task is to predict the next click by estimating, for each item $p \in P$, a recommendation score y_i indicating the likelihood of being clicked next. In many cases, we are also given context times of the clicks, which are denoted by $[c_1, c_2, \ldots, c_L]$. We seek to find a model able to take into account all this information when making recommendations.

Figure 1 shows the general architecture of the proposed method, which consists of two main components: an embedding layer for the explicit representation of users, items, and time context, and an RNN to model click sequences. We propose a new type of RNN cells: Gated Recurrent Units (GRUs) with an attention mechanism, which we call *Attentive GRU* (AGRU), to adaptively combine different types of input signals. We describe the proposed model in more detail in the subsequent sections.

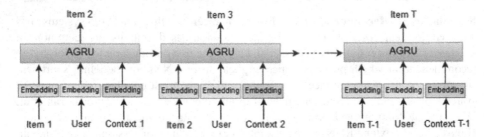

Fig. 1. General architecture of the proposed method. AGRU is GRU with additional attentive gate to regulate the contributions of input components.

3.1   User and Item Embedding

Existing session-based recommendation RNN architectures mainly use one-hot vector to represent input to RNNs [10, 14]. For the recommender system domain, however, it has been proven useful to embed users and items in spaces of lower dimensions so that users with similar tastes are close in the embedded space [11, 19]. Similarly, items appeared in the same context or consumed by the same set of users are also spatially close in the embedded space. For a user performing multiple sessions, the embedding provides a convenient way to represent global preference structures across sessions.

Let $\mathbf{u}_O \in \Re^{|U|}$, $\mathbf{p}_O \in \Re^{|P|}$, and $\mathbf{c}_O \in \Re^{|C|}$ be the one-hot vector representations of user u, item p and context c respectively, where U, P, and C denote the sets of users, items and contexts. We consider embedding matrices $\mathbf{E}_U \in \Re^{n_U \times |U|}$, $\mathbf{E}_P \in \Re^{n_P \times |P|}$, and

$\mathbf{E}_C \in \Re^{n_c \times |C|}$ that map corresponding one-hot representations into embedded spaces of lower dimensions. With the introduced matrices, the user vector in the embedded space is calculated as $\mathbf{u} = \mathbf{E}_U \mathbf{u}_O$. Similarly, the item and the context are represented as embedded vectors $\mathbf{p} = \mathbf{E}_P \mathbf{p}_O$ and $\mathbf{c} = \mathbf{E}_C \mathbf{c}_O$. The embedding matrices are randomly initialized and jointly learned during the training process.

3.2 RNNs with Modified Gated Recurrent Units

RNNs are a class of deep neural networks specially designed to model sequences of any kind. An RNN operates on a sequence of variable length $\mathbf{x} = (x_1, x_2, \ldots, x_L)$ by updating a hidden state $\mathbf{h}$ and optionally produces output $\mathbf{y}$ over $\mathbf{x}$. Here, we consider the variant of RNNs which outputs the next item at each time step t:

$$\mathbf{y}_t = \sigma_y (\mathbf{W}_y \mathbf{h}_t + \mathbf{b}_y), \tag{1}$$

where $\mathbf{W}_y$ and $\mathbf{b}_y$ by are parameter matrix and vector which are learned, σ_y is the softmax function. The hidden state is updated by a function of the previous hidden state and the current input vector $\mathbf{x}_t$ as follows:

$$\mathbf{h}_t = f(\mathbf{h}_{t-1}, \mathbf{x}_t), \tag{2}$$

where f is a non-linear activation function. Here, as function f we consider Gated Recurrent Units (GRUs) [4], which updates the hidden state as follows:

$$\mathbf{z}_t = \sigma_g (\mathbf{W}_z \mathbf{x}_t + \mathbf{V}_z \mathbf{h}_{t-1} + \mathbf{b}_z) \tag{3}$$

$$\mathbf{r}_t = \sigma_g (\mathbf{W}_r \mathbf{x}_t + \mathbf{V}_r \mathbf{h}_{t-1} + \mathbf{b}_r) \tag{4}$$

$$\mathbf{h}_t = \mathbf{z}_t \circ \mathbf{h}_{t-1} + (1 - \mathbf{z}_t) \circ \sigma_h (\mathbf{W}_h \mathbf{x}_t + \mathbf{V}_h (\mathbf{r}_t \circ \mathbf{h}_{t-1}) + \mathbf{b}_h) \tag{5}$$

where $\circ$ denotes element-wise multiplication, σ_g and σ_h are sigmoid and tangent functions respectively, $\mathbf{W}$, $\mathbf{V}$, and $\mathbf{b}$ are parameter matrices and vectors.

In the pure session-based settings where only item ID is considered the input vector $\mathbf{x}_t$ is the item vector $\mathbf{p}_t$ at step t ($\mathbf{x}_t = \mathbf{p}_t$). With user and context vectors $\mathbf{u}_t$ and $\mathbf{c}_t$ available, we now extend the gated unit to incorporate this information when updating the hidden state. In this work, we consider two ways of integrating user and context into the model.

Simple Integration. The most simple and straightforward way is to concatenate user, item, and context vectors and use the resulting vector as the input to the recurrent unit:

$$\mathbf{x}_t = \begin{bmatrix} \mathbf{u}_t \\ \mathbf{p}_t \\ \mathbf{c}_t \end{bmatrix} \tag{6}$$

Since user and context now become parts of the input, they can influence the computation of the hidden state by regulating the amount of information to be forgotten or updated. Once weight matrices are learned, the importance of each input component on the output becomes fixed and remains the same regardless of changing sessions, which is not desired in many cases and makes this integration method not flexible.

Attentive Integration. Depending on concrete situations, factors influencing the user's decision may vary. For example, movie series are often watched in their order regardless of user taste, or Christmas songs are more frequently listened in the context of Christmas season. Thus, a desired characteristic of the system is the ability to adaptively change the focus between the input components depending on specific situations. To achieve this, we propose a modified version of gated cells, in which an attentive gate is introduced to regulate the contributions of the user, item, and context factors. Specifically, at each time step, the attentive gate computes the following weights:

$$a_i = \sigma_a(\mathbf{v}_i^T \mathbf{d}_i + b_i), \ i = 1, 2, 3 \tag{7}$$

$$\alpha_i = \frac{exp(a_i)}{\sum_i exp(a_i)}, i = 1, 2, 3 \tag{8}$$

where $\mathbf{d}_i(i = 1, 2, 3)$ denote $\mathbf{u}_t, \mathbf{p}_t$, and $\mathbf{c}_t$ respectively, $\alpha_i(i = 1, 2, 3)$ are their importance weights; $\mathbf{v}_i$ and b_i are parameters to be learned; σ_a is the sigmoid function. Here, we measure the importance of each component (user, item, context) as the similarity between its embeddings and a vector $\mathbf{v}_i$. We use the sigmoid as a squashing function and get normalized attention weights through a softmax. Note that the user component is fixed for the whole session so we need to compute a_1 only once. Finally, we form the input vector concatenating user, item, and context vectors multiplied by corresponding weights:

$$\mathbf{x}_t = \begin{bmatrix} \alpha_1 \mathbf{u}_t \\ \alpha_2 \mathbf{p}_t \\ \alpha_3 \mathbf{c}_t \end{bmatrix} \tag{9}$$

In this way, the attentive gate controls the contribution of each input component to the computation of the current state and output. Figure 2 shows graphical depiction of the proposed gated units. The attentive gate not only allows regulating the extents to which each component influences the decision but also provides a nice way to deal with the new user problem. A new user is simply represented by a randomized vector with elements close to zeros and the gate will ignore the user component by giving it small weight, thus focusing on session and context information. It is also easy to add different types of context information by considering them as additional input components and apply the same integration method.

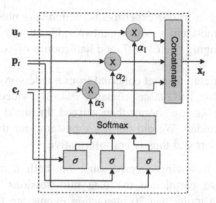

Fig. 2. The attentive gate detached from the complete unit.

3.3 Training

We train our model using standard stochastic gradient descent with Back-Propagation-Through-Time (BPTT) to minimize a chosen loss function. Here we use cross entropy loss between the predicted vector and the one-hot representation of the next clicked item. Note that user embeddings are updated with each click during training process while they are kept fixed throughout the session at prediction time.

To make the input and output of the model similar at the training and prediction times, we prepare training data following the preprocessing procedures described in [17, 18] but with a modification to include user identifiers. Specifically, for a given training session $[e_1, e_2, \ldots, e_T]$ we generate input sequences and corresponding labels $([e_1], e_2), ([e_1, e_2], e_3), \ldots, ([e_1, e_2, \ldots, e_{T-1}], e_T)$ as training samples. We order training sequences by length and user identifiers so that each mini-batch consists of training sequences of the same size but originating from different users. Grouping equal-size sequences together has been widely used in practice as this speeds up training, while distributing users across mini-batches introduces more diversity within a single batch.

4 Experiments

In this section we describe the experiments we performed to evaluate the proposed method and provide an analysis of the results.

4.1 Datasets

We evaluated the proposed method and baselines on the following datasets:

- *Last.fm 1K Users*[1]. This dataset contains music listening history data for 992 users collected by Celma [3] using Last.fm API. The dataset has a total of 19150868 tuples of the form user-timestamp-artist-song. For computational reasons, we used

---

[1] http://www.dtic.upf.edu/~ocelma/MusicRecommendationDataset/.

20% subsample of data in our experiments, which resembles the setting in other works [6]. From a timestamp we extracted two values to use as the context: day-of-week with values ranging from 1 to 7, and half-month-of-year with values ranging from 1 to 24.

– *VOD*. This is a proprietary dataset collected from a video-on-demand service over a period of two months. The dataset contains 7945218 video-viewing events from 749 K users. Events lasting shorter than a fixed threshold (normalized by video length) were not included. We did not use timestamp for this dataset because the time period is too short and thus is not informative.

We performed the following preprocessing on both datasets. First, based on timestamps provided we partitioned the data into sessions with a one-hour idle threshold. Sessions with more than 20 interaction events are further partitioned into session of 20 events with 5-event overlapping. Then, we removed sessions with only one event since those are too short and not informative, and removed users having less than five sessions to provide enough user-related information across sessions. We also removed items appearing less then 10 times as such cannot be modeled reliably. The statistics of two datasets are summarized in Table 1.

For each user, we used the last session to build the training set and the second last session to build the evaluation set that we used to tune model parameters. The remaining sessions form the training set.

Table 1. Statistics of the datasets

	Last.fm	VOD
Users	767	18,775
Items	107,138	61,335
Sessions	203,869	348,172
Events	4,258,811	4,209,416
Average session length	20,89	12,09
Sessions per user	265,8	18,54

4.2 Experimental Setup

Metrics. We consider a typical setting, in which the system returns top K items as recommendations and used Recall@K and Mean Reciprocal Rank (MRR)@K with $K = 5, 20$ as the evaluation metrics.

Parameter Tuning. We varied the embedding dimensions between 50 and 150 but kept them equal for users and items for simplicity (although they can be different in general). The embedding dimensions of time context were fixed at 5. We evaluated our models with 100 and 200-dimensional GRUs. The models were optimized by using standard SGD, with mini-batch size of 64, for 20 epochs with the training rate of 0.001. We implemented and trained our models in TensorFlow. For training sequences longer than 2, we used item dropout with probability of 25% as described in [17].

Baselines. We compared our method with a general recommendation method (Item-based kNN), a for sequential data TribeFlow), and two RNN session-based methods.

- *Item-based kNN*. The first, and very competitive baseline is Item-based kNN. This is a simple, yet effective method, which is often considered industry standard. Item-based kNN measures the similarity between two items as the number of times they co-occur in sessions divided by a normalization factor.
- *TribeFlow*: This method by [7] is designed to handle sequential data by applying a semi-Markov random walk model on short fragments generated from sequences of user activities. TribeFlow was chosen as a baseline based on its reported good performance compared with other sequential recommendation methods.
- *Ses RNN*. We refer to the model proposed in [10] as Ses RNN. This model uses RNNs with standard GRUs to model clicks; no user information is used. We used the implementation provided by the method's authors. We kept all default parameter values with the only exception of GRU size set to 100 for a fair comparison with other RNN models.
- *Improved Ses RNN*. This is an improved version of Ses RNN by [17], in which a number of modifications including data augmentation and adaptation for shifts over time have been applied to improve the prediction accuracy. We used the provided implementation with all parameters set to the default values.
- *User-Ses GRU*: This is our GRU model with simple integration of user and item vectors.
- *User-Ses AGRU*: This is our GRU model that uses attentive gate to combine user, item, and context vectors.

4.3   Results

In the first experiment we compared the performance of our models with embedding and GRU size set to 100, which are default GRU size for other RNN-based baselines. The Recall@K and MRR@K values for each model on Last.fm and VOD datasets are summarized in Tables 2 and 3, respectively. Surprisingly, TribeFlow consistently performed the worst over different evaluation metrics and datasets, but was still able to correctly place nearly 27% of the next items on the top 20 for both datasets. Simple Item-based kNN achieved substantially higher Recall@K and MRR@K values than TribeFlow, but performed worse than most RNN based methods. The superiority of RNN methods over TribeFlow and Item-based KNN may come from the use of session information and modeling power of the RNN models. Our simple GRU model (User-Ses GRU) achieved better results than Ses RNN but much worse results than Improved Ses RNN, suggesting that simple integration of user factor may not give the best results. The results show that User-Ses AGRU consistently achieved strong performance, outperforming other methods on all metrics over both datasets. The only exception is MRR@20 on VOD, where User-Ses AGRU achieved a slightly lower value than Improved RNN.

Two important parameters that influence the performance of RNN based models are the embedding and GRU sizes. We varied embedding size from 50 to 150 (embeddings of higher dimensions were not experimented due to high memory requirements for

Table 2. Recall@K and MRR@K values of the baselines and our models on Last.fm dataset

	Recall@5	Recall@20	MRR@5	MRR@20
Item-based kNN	0.165	0.317	0.104	0.122
TribeFlow	0.067	0.269	0.023	0.1
Ses RNN	0.266	0.295	0.238	0.241
Improved Ses RNN	0.287	0.339	0.241	0.246
User-Ses GRU	0.271	0.303	0.232	0.238
User-Ses AGRU	**0.325**	**0.384**	**0.286**	**0.292**

Table 3. Recall@K and MRR@K values of the baselines and our models on VOD dataset

	Recall@5	Recall@20	MRR@5	MRR@20
Item-based kNN	0.277	0.497	0.184	0.209
TribeFlow	0.057	0.268	0.019	0.099
Ses RNN	0.25	0.512	0.168	0.197
Improved Ses RNN	0.339	0.555	0.209	**0.244**
User-Ses GRU	0.298	0.508	0.180	0.204
User-Ses AGRU	**0.359**	**0.585**	**0.221**	0.242

Table 4. Performance of User-Ses AGRU for different GRU and embedding sizes

GRU size	Embedding size	Last.fm		VOD	
		Recall@20	MRR@20	Recall@20	MRR@20
100	50	0.371	0.276	0.564	0.232
100	100	0.384	0.292	0.585	0.242
100	150	0.391	0.301	0.589	0.248
200	50	0.374	0.289	0.545	0.227
200	100	0.388	0.294	0.565	0.242
200	150	0.393	0.306	0.570	0.243

Last.fm dataset), in combination with GRU size of 100 and 200. Results are summarized in Table 4. As can be seen, increasing GRU size from 100 to 200 does not always yield better performance: GRU models with 200 dimensions just slightly improved Recall on Last.fm while achieved lower Recall on VOD. This result is consistent with those reported in [17]. On the other side, higher embedding sizes resulted in better results. This effect is more significant when moving from 50 to 100. Further increase to 200 resulted in small improvements, suggesting that embedding size of 100 is a good tradeoff between accuracy and computational efficiency.

5   Conclusion

We have presented a method for incorporating user and context factor into session-based recommendation. By extending the GRU architecture with an attentive gate that operated on the embeddings of users and items, the proposed method allows adaptively combining inputs of different types by shifting focus between user, item, and context factors. The model is easy extended to incorporate new type of context conditions. Experimental results show that our method achieved better performance in terms of Recall and MRR metrics on two real-world datasets.

Acknowledgement. We would like to thank FPT for financial support, which made this work possible.

References

1. Bach, N.X., Hai, N.D., Phuong, T.M.: Personalized recommendation of stories for commenting in forum-based social media. Inf. Sci. **352–353**, 48–60 (2016)
2. Bobadilla, J., Ortega, F., Hernando, A., Gutiérrez, A.: Recommender systems survey. Knowl.-Based Syst. **46**, 109–132 (2013)
3. Celma, O.: Music Recommendation and Discovery: The Long Tail, Long Fail, and Long Play in the Digital Music Space. Springer, Heidelberg (2010). https://doi.org/10.1007/978-3-642-13287-2
4. Cho, K., et al.: Learning phrase representations using RNN encoder–decoder for statistical machine translation. In: Proceedings of the 2014 Conference on Empirical Methods in Natural Language Processing (EMNLP), pp. 1724–1734 (2014)
5. Deshpande, M., Karypis, G.: Item-based top-n recommendation algorithms. ACM Trans. Inf. Syst. (TOIS) **22**(1), 143–177 (2004)
6. Donkers, T., Loepp, B., Ziegler, J.: Sequential user-based recurrent neural network recommendations. In: Proceedings of the Eleventh ACM Conference on Recommender Systems (RecSys), pp. 152–160 (2017)
7. Figueiredo, F., Ribeiro, B., Almeida, J.M., Faloutsos, C.: TribeFlow: mining & predicting user trajectories. In: Proceedings of the 25th International Conference on World Wide Web (WWW), pp. 695–706 (2016)
8. Goldberg, D., Nichols, D., Oki, B.M., Terry, D.: Using collaborative filtering to weave an information tapestry. Commun. ACM **35**(12), 61–70 (1992)
9. Hidasi, B., et al.: Parallel recurrent neural network architectures for feature-rich session-based recommendations. In: Proceedings of the 10th ACM Conference on Recommender Systems (RecSys), pp. 241–248 (2016)
10. Hidasi, B., Karatzoglou, A., Baltrunas, L., Tikk, D.: Session-based recommendations with recurrent neural networks. In: Proceedings of the International Conference on Learning Representations (ICLR) (2016)
11. Koren, Y., Bell, R., Volinsky, C.: Matrix factorization techniques for recommender systems. Computer **42**(8), 30–37 (2009)
12. Li, J., Ren, P., Chen, Z., Ren, Z., Lian, T., Ma, J.: Neural attentive session-based recommendation. In: Proceedings of the 2017 ACM Conference on Information and Knowledge Management (CIKM), pp. 1419–1428 (2017)
13. Phuong, N.D., Thang, L.Q., Phuong, T.M.: A graph-based method for combining collaborative and content-based filtering. In: Ho, T.-B., Zhou, Z.-H. (eds.) PRICAI 2008. LNCS (LNAI), vol. 5351, pp. 859–869. Springer, Heidelberg (2008). https://doi.org/10.1007/978-3-540-89197-0_80
14. Quadrana, M., Karatzoglou, A., Hidasi, B., Cremonesi, P.: Personalizing session-based recommendations with hierarchical recurrent neural networks. In: Proceedings of the Eleventh ACM Conference on Recommender Systems (RecSys), pp. 130–137 (2017)

15. Rendle, S., Freudenthaler, C., Schmidt-Thieme, L.: Factorizing personalized markov chains for next-basket recommendation. In: Proceedings of the 19th International Conference on World Wide Web (WWW), pp. 811–820 (2010)
16. Sarwar, B., Karypis, G., Konstan, J., Riedl, J.: Item-based collaborative filtering recommendation algorithms. In: Proceedings of the 10th International Conference on World Wide Web (WWW), pp. 285–295 (2001)
17. Tan, Y.K., Xu, X., Liu, Y.: Improved recurrent neural networks for session-based recommendations. In: Proceedings of the 1st Workshop on Deep Learning for Recommender Systems, pp. 17–22 (2016)
18. Tuan, T.X., Phuong, T.M.: 3D convolutional networks for session-based recommendation with content features. In: Proceedings of the Eleventh ACM Conference on Recommender Systems (RecSys), pp. 138–146 (2017)
19. Vasile, F., Smirnova, E., Conneau, A.: Meta-Prod2Vec: product embeddings using side-information for recommendation. In: Proceedings of the 10th ACM Conference on Recommender Systems (RecSys), pp. 225–232 (2016)

EMD-Based Recurrent Neural Network with Adaptive Regrouping for Port Cargo Throughput Prediction

Yan Li[1], Ryan Wen Liu[1,2(✉)], Quandang Ma[1(✉)], and Jingxian Liu[1]

[1] Hubei Key Laboratory of Inland Shipping Technology, School of Navigation, Wuhan University of Technology, Wuhan, China
{wenliu,qdma}@whut.edu.cn
[2] Hubei Key Laboratory of Transportation Internet of Things, School of Computer Science and Technology, Wuhan University of Technology, Wuhan, China

Abstract. Accurate prediction of port cargo throughput (PCT) plays an important role in economic investment, transportation planning, port planning and design, etc. PCT time series have the properties of non-linearity and complexity. To guarantee high-quality prediction performance, we propose to first adopt the empirical mode decomposition (EMD) to decompose the original PCT time series into high and low frequency components. It is more difficult to predict some components due to their properties of weak mathematical regularity. To take advantage of the selfsimilarities within components, each component will be divided into several small parts which are adaptive regrouped (ARG) via the standardized euclidean distance (SED)-based similarity measure. The regrouped parts are then selected to form the training dataset for long short-term memory (LSTM) to enhance the prediction accuracy of each component. The final prediction result can be obtained by integrating the predicted components. Our proposed three-step prediction framework (called EMD-ARG-LSTM) benefits from the property decomposition and adaptive similarity regrouping. Experimental results have illustrated the superior performance of the proposed method in terms of both prediction accuracy and robustness.

Keywords: Port cargo throughput · Prediction
Long short-term memory · Empirical mode decomposition
Similarity regrouping

1 Introduction

Ports can be considered to be important nodes in the water transportation network. It is well known that port cargo throughput (PCT) has the basic function of economic investment, port planning and transportation planning, etc. [1]. Therefore, accurate prediction of PCT time series play an important role in port development and economic investment.

© Springer Nature Switzerland AG 2018
L. Cheng et al. (Eds.): ICONIP 2018, LNCS 11301, pp. 499–510, 2018.
https://doi.org/10.1007/978-3-030-04167-0_45

In the literature, numerous methods have been applied to predict PCT. Prediction methods usually can be summarized into two categories, which consist of qualitative and quantitative prediction methods [2]. Qualitative prediction methods rely on personal knowledge and experience to predict the future development. Due to the complex and stochastic properties of PCT time series, qualitative prediction methods can not generate high-accuracy prediction. Conventional quantitative prediction methods can be roughly summarized into three categories, i.e., numerical simulation-based methods [3], parametric methods [4] and nonparametric methods [5]. The simulation-based prediction methods easily suffers from randomness in practical applications. Thus, it is difficult to generate satisfactory prediction results by using numerical simulation.

The parametric methods, such as linear regression, exponential smoothing, grey theory-based models (GM) and autoregressive integrated moving average (ARIMA), etc., have been developed for data prediction. Linear regression [6] and exponential smoothing [7] are often applied to predict simple linear data. GM [8] has also been widely adopted to predict time series. However, these mentioned methods are only available for short-term prediction. It is difficult to generate high-accuracy PCT prediction in practical applications. ARIMA [9] can be considered as one of the most popular prediction methods. This method is simple in calculation and has some advantages over other models in the case of short-term prediction. However, due to the complex, unstable and nonlinear properties, ARIMA model could not capture the regular pattern behind PCT time series. The corresponding prediction accuracy will be easily limited. Therefore, parametric prediction methods can not generate high-accuracy prediction for complex, unstable and nonlinear time series.

To further enhance prediction performance, much attention has been paid to the nonparametric methods. Neural network (NN) model is one typical nonparametric prediction method, which could be considered as one of the most famous and successful versions in the field of time series prediction. This model has strong self-learning ability and the ability to seek optimal solution at high speed [10]. There is a great potentiality to accurately predict complex PCT time series by using advanced NN models. Back propagation neural network (BPNN) [11], which has been widely used, is an early method for time series prediction. With the rapid development of big traffic data, traditional NN have been replaced by deep learning (DL) [12] to improve prediction performance. Recent studies have shown that DL methods are able to generate higher prediction accuracy than traditional NN. It is well known that, recurrent neural network (RNN) [13] has a strong self-learning ability for complex and nonlinear time series. The RNN-based prediction framework is essentially a scalable and effective computation method. However, RNN easily suffers from the common problem of gradient vanish from the theoretical point of view. In contrast, long short-term memory (LSTM) [14] has the ability to mitigate this problem to guarantee robust prediction. Therefore, we propose to select LSTM as the basic research framework in our work.

It is generally thought that original PCT time series are commonly composed of several high and low frequency components. Thus, it is difficult to generate high-accuracy prediction results by directly using LSTM. Fortunately, the empirical mode decomposition (EMD) [15] could decompose the complex and nonlinear time series into high and low frequency components. In the literature, Liu *et al.* [16] used EMD with BPNN to predict water temperature. Yu *et al.* [17] proposed to combine EMD with NN to predict crude oil price. Wang *et al.* [18] tended to develop EMD-based ENN to predict wind speed, etc. Although the EMD-based NN methods have been widely studied, to the best of our knowledge, no research has been conducted on PCT time series prediction. The original PCT time series could be decomposed into the sum of several components. It is more difficult to predict some components due to their properties of weak mathematical regularity. To take advantage of the selfsimilarities within components, each component will be divided into several small segments which are adaptively regrouped via the standardized euclidean distance (SED)-based similarity measure [19]. The regrouped parts are then selected to form the training dataset for LSTM to enhance the prediction accuracy of each component. The final prediction result can be obtained by integrating these predicted components. Numerous experiments have shown that our proposed three-step prediction framework could generate more accurate and stable prediction results compared with traditional methods.

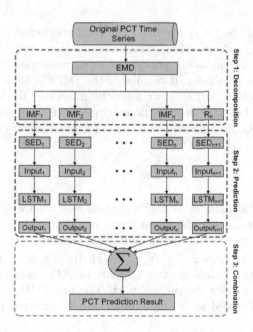

Fig. 1. Flowchart of the proposed three-step prediction framework EMD-ARG-LSTM.

2   EMD-Based LSTM with Adaptive Regrouping

This section will detailedly illustrate our proposed three-step prediction framework, shown in Fig. 1. In the first step, EMD will be introduced to decompose the original PCT time series into high and low frequency components. In the second step, each component is divided into several continuous small (overlapping) segments. Several small segments with high SED-based similarities are adaptive regrouped to form proper training dataset for LSTM to improve prediction accuracy. Each component prediction is combined to form the final prediction results in the final step.

2.1   Empirical Mode Decomposition (EMD)

EMD has been widely applied to decompose the nonlinear and non-stationary signals into a series of different scaled data sequences. Each sequence is designated as an Intrinsic Mode Function (IMF), which must meet the following two conditions [15,20]

1. The number of zero-crossing points and extrema points (local maxima and local minima) in the entire signal should be equal or differ by at most one.
2. At any point in the signal, the mean values of lower and upper envelopes which are represented by the local minima and local maxima should be equal to zero.

Essentially, EMD highly depends on the local characteristics of original signal, such as local minima, local maxima and zero-crossings. The decomposition of original signal was commonly performed using an iterative procedure called sifting algorithm. In particular, it calculated the IMFs at each scale from fine to coarse. For more details on EMD, we refer the interested reader to see [21] and references therein. Let $x(t)$ denote the original PCT time series (i.e., original signal), which can be decomposed as a sum of finite number of IMFs, i.e.,

$$x(t) = \sum_{i=1}^{N} c_i(t) + r_N(t), \tag{1}$$

where N is the total number of IMFs, $c_i(t)$ is the ith IMF, and $r_N(t)$ is the residual signal representing the mean trend of original data $x(t)$. In practice, $r_N(t)$ can be also considered as an IMF and denoted as $c_{N+1}(t)$. Therefore, Eq. (1) can be rewritten as $x(t) = \sum_{i=1}^{N+1} c_i(t)$. In this decomposition, the first IMF is related to the fastest fluctuating part of $x(t)$; meanwhile the last IMF corresponds to the slowest fluctuating part. The first IMF and other IMFs can be respectively considered as the high and low frequency components.

2.2   Standardized Euclidean Distance (SED)

SED is an effective and popular technique for measuring distance (or similarity) between two sequences of time series, which is an improved version of euclidean distance [19]. This method is proposed based on the standard deviation of observed samples. The differences between the values of observed samples have also been considered to enhance the computational robustness. In addition, SED is able to suppress the disadvantage (e.g., the influence of the dimension of observation index) existed in traditional euclidean distance.

Let $X = \{x_1, x_2, x_3, \cdots, x_n\}$ and $Y = \{y_1, y_2, y_3, \cdots, y_n\}$ respectively denote two time sequences. As a consequence, the corresponding computing formula of SED can be recursively calculated by

$$SED(X,Y) = \sqrt{\sum_{i=1}^{n}(\frac{X_i - Y_i}{S_i})^2}, \qquad (2)$$

where S_i is the standard deviation of ith component. Each component is divided into several small segments. The small segments with high similarity are merged into the same cluster. The clustered segments of time series are exploited to form the training dataset for LSTM to enhance the prediction results.

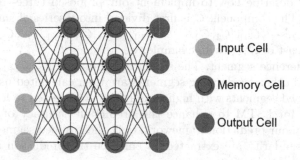

Fig. 2. The architectural diagrams of LSTM prediction method adopted in this paper.

2.3 Long Short-Term Memory (LSTM)

LSTM is the main framework of DL, which is one of the variations of RNN to mitigate the gradient vanish problem of RNN [14]. The structural framework of LSTM used in this work is visually shown in Fig. 2.

Let $v = \{v_1, v_2, v_3, \cdots, v_{n-1}, v_n\}$ with $v_t \in R^d$ $(1 \le t \le n)$ denote the PCT time series. At each position t, there is a series of vectors, which include one input cell ic_t, forget gate fg_t, output cell oc_t and memory cell mc_t. To generate the l-dimensional hidden vector h_t, we obtain the following mathematical formulas

$$ic_t = \sigma(W^{ic}v_t + U^{ic}h_{t-1} + b^{ic}), \qquad (3)$$

$$fg_t = \sigma(W^{fg}v_t + U^{fg}h_{t-1} + b^{fg}), \qquad (4)$$

$$oc_t = \sigma(W^{oc}v_t + U^{oc}h_{t-1} + b^{oc}), \qquad (5)$$

$$mc_t = fg_t \odot mc_{t-1} + ic_t \odot tanh\sigma(W^{mc}v_t + U^{mc}h_{t-1} + b^{mc}), \qquad (6)$$

$$h_t = oc_t \odot tanh(mc_t),\tag{7}$$

where σ represents the sigmoid function, $\odot$ represents elementwise multiplication of two vectors, and W^*, U^* and b^* are the weight vector or matrix to be learned (i.e., $W^* \in R^{l \times d}$, $U^* \in R^{l \times l}$, $b^* \in R^l$).

2.4   Three-Step Prediction Framework

Our proposed EMD-ARG-LSTM framework for PCT time series prediction is illustrated in Fig. 1. The detailed calculation steps are described as follows

Step 1. EMD first decomposes the original PCT time series into high and low frequency components.

Step 2. It is more difficult to predict some components due to their properties of weak mathematical regularity. To take advantage of the selfsimilarities within components, each component will be divided into several small segments which are adaptively regrouped via standardized euclidean distance (SED)-based similarity measure. Taking a decomposed component C as an example, we will describe how to implement our proposed three-step prediction framework. The component C is first divided into a series of small segments, i.e., $\{[C_1, C_2, \cdots, C_n], [C_2, C_3, \cdots, C_{n+1}], \cdots, [C_i, C_{i+1}, \cdots, C_{n+i-1}]\}$. Taking the forecast of C_{n+i} as an example, we consider $[C_i, C_{i+1}, \cdots, C_{n+i-1}]$ as an input reference segment. The similarities between $[C_i, C_{i+1}, \cdots, C_{n+i-1}]$ segment and other overlapping segments are then calculated using the SED. The regrouped segments with high similarities are selected to form the training dataset for LSTM to enhance the prediction accuracy of each component. Within our prediction framework, only the grouped segments with high degrees of similarity are extracted to enhance the prediction accuracy and robustness.

Step 3. Finally, the prediction result can be obtained by integrating the predicted components.

Through the proposed three-step prediction framework EMD-ARG-LSTM, we can generate more accurate and stable PCT prediction performance.

3   Experimental Results and Discussion

To evaluate the prediction accuracy and robustness, our proposed three-step prediction framework will be compared with several state-of-the-art prediction methods on realistic dataset from 18 different ports in China.

3.1   Experiment Settings

In this work, we select the realistic PCT time series from 18 major ports in China from 1996 to 2016. The growing trends of PCT time series are visually illustrated in Fig. 3. It shows that different ports have different properties of growing trends

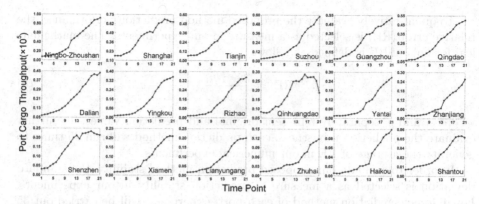

Fig. 3. The growing trends of PCT time series at 18 major ports in China from 1996 to 2016 (i.e., *Time Point from* 1 *to* 21 *in the horizontal axis*). It can be found that different ports have different properties of growing trends.

due to several influence factors, such as geographic location, type of goods, port scale, service level, and management level, etc. This means that the proposed prediction method should have the universality and robustness properties to enhance prediction performance. Our proposed three-step prediction framework EMD-ARG-LSTM will be compared with LSTM and EMD-LSTM.

LSTM framework belongs to DL. Traditional NN can be considered as the foundation of DL. Thus, we trend to select the typical NN method BPNN as competing method in our numerical experiments. Furthermore, BPNN will be also incorporated into our proposed three-step prediction framework (i.e., EMD-ARG-BPNN) to evaluate the prediction performance. In addition, the prediction framework will be compared to traditional BPNN and EMD-BPNN.

- **NN:** Widely-used NN, such as BPNN and LSTM, are introduced to predict PCT time series. Both EMD-NN and EMD-AGR-NN frameworks are developed based on BPNN and LSTM to future enhance prediction performance.
- **EMD-NN:** EMD-NN is two-step prediction framework that combines EMD with NN. Original PCT time series could be first decomposed into several high and low frequency components by using the EMD. The NN methods are then exploited to directly predict each component. The final prediction results are obtained by combining the each predicted component. EMD-NN prediction framework has the capacity of guaranteeing more robust performance compared with single NN.
- **EMD-ARG-NN:** This three-step prediction framework can be considered as an extension of EMD-NN prediction framework. Under this framework, each component will be divided into several small parts which are adaptive regrouped via SED-based similarity measure. The regrouped parts are selected to form the training dataset for NN to enhance the accuracy of each component prediction. Each predicted component is also combined to generate the final prediction results.

To quantitatively evaluate the proposed method prediction performance, the relative error (RE) is selected as a measure in our experiments. The mathematical formula of RE is defined as follows

$$RE_i = \frac{|y_i - \widehat{y_i}|}{y_i}, \tag{8}$$

where y_i and $\widehat{y_i}$ separately represent original and prediction values. To further evaluate the prediction robustness, each prediction method will run 35 times to generate the average of RE in our numerical experiments.

In order to evaluate our proposed prediction method stability, the standard deviation is selected as a measure of prediction stability in our experiments. For different prediction method of each port, experiment will be carried out 35 times. Then, prediction experiment result of standard deviation will be carried out. The corresponding computing formula of standard deviation is given by

$$std = \sqrt{\frac{1}{T} \sum_{t=1}^{T} (X_t - \bar{X})^2}, \tag{9}$$

where $\bar{X}$ represents the mean value of prediction results, X_t denotes one specific prediction value, and T is the total number of prediction results for one method for one port.

3.2   Quantitative Performance Evaluation

To implement the prediction experiments, the different PCT time series extracted from 1996 to 2015 are used to train NN, EMD-NN and our proposed method. The prediction accuracy will be evaluated by data from each port in 2016. We empirically decompose PCT time series into 4 components since numerous experiments have demonstrated that this selection is able to generate satisfactory performance. We take Xiamen Port as an example, shown in Fig. 4.

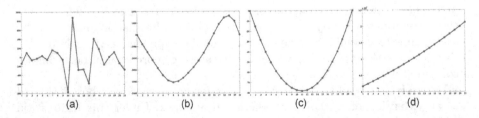

Fig. 4. The EMD-based decomposition of PCT time series for an example of Xiamen Port. From left to right: (a) IMF1, (b) IMF2, (c) IMF3 and (d) IMF4. In particular, from IMF1 to IMF4, respectively, are from high-frequency components to low-frequency components.

Table 1. PCT prediction results [mean $(\times 10^4) \pm$ std $(\times 10^3)$] for 18 major ports in China in 2016. In the table, the adopted prediction methods, from left to right, are NN (i.e., BPNN, LSTM), EMD-NN (i.e., EMD-BPNN, EMD-LSTM), and EMD-ARG-NN (i.e., EMD-ARG-BPNN, EMD-ARG-LSTM), respectively. The right side denotes the value of actual data $(\times 10^4)$.

Methods	NN		EMD-NN		EMD-ARG-NN		Actual data
	BPNN	LSTM	BPNN	LSTM	BPNN	LSTM	
Ningbo-Zhoushan	9.11 ± 5.58	9.35 ± 4.16	9.36 ± 3.83	9.33 ± 3.11	9.37 ± 2.47	9.32 ± 1.78	9.22
Shanghai	6.66 ± 3.27	6.52 ± 2.95	6.53 ± 2.99	6.57 ± 2.24	6.47 ± 2.77	6.47 ± 1.89	6.45
Tianjin	5.66 ± 2.65	5.60 ± 2.86	5.63 ± 2.39	5.63 ± 2.05	5.61 ± 2.36	5.56 ± 2.40	5.51
Suzhou	5.48 ± 5.46	5.93 ± 4.95	5.89 ± 3.67	5.86 ± 3.19	5.76 ± 2.83	5.84 ± 2.79	5.79
Guangzhou	4.92 ± 2.39	5.32 ± 1.71	5.09 ± 2.13	5.35 ± 1.15	5.12 ± 1.79	5.29 ± 0.95	5.22
Qingdao	4.93 ± 2.53	5.11 ± 2.46	5.11 ± 2.87	5.12 ± 2.84	4.93 ± 2.22	5.07 ± 2.24	5.00
Dalian	4.21 ± 3.21	4.51 ± 2.87	4.46 ± 2.20	4.42 ± 2.40	4.45 ± 2.38	4.41 ± 2.03	4.37
Yingkou	3.39 ± 1.89	3.64 ± 1.78	3.57 ± 2.15	3.56 ± 2.04	3.51 ± 1.87	3.59 ± 2.07	3.52
Rizhao	3.39 ± 3.54	3.59 ± 3.95	3.42 ± 2.74	3.57 ± 2.87	3.45 ± 2.36	3.56 ± 2.48	3.50
Qinhuangdao	2.68 ± 1.72	2.66 ± 1.21	2.07 ± 2.15	2.05 ± 1.01	2.03 ± 1.38	2.01 ± 1.21	1.87
Yantai	2.56 ± 1.37	2.71 ± 0.94	2.60 ± 1.35	2.68 ± 1.10	2.60 ± 1.09	2.72 ± 1.07	2.65
Zhanjiang	2.30 ± 2.87	2.78 ± 3.09	2.46 ± 1.39	2.63 ± 1.54	2.52 ± 1.41	2.60 ± 0.72	2.56
Shenzhen	2.21 ± 1.15	2.25 ± 1.01	2.08 ± 1.31	2.18 ± 1.55	2.19 ± 0.91	2.18 ± 0.80	2.14
Xiamen	2.15 ± 1.01	2.16 ± 0.95	2.13 ± 0.63	2.13 ± 0.56	2.13 ± 0.57	2.12 ± 0.31	2.09
Lianyungang	1.96 ± 1.17	2.04 ± 0.98	2.05 ± 1.46	2.05 ± 1.19	2.04 ± 0.81	2.03 ± 0.66	2.01
Zhuhai	1.13 ± 1.33	1.21 ± 1.50	1.22 ± 1.18	1.20 ± 1.39	1.21 ± 0.65	1.20 ± 0.58	1.18
Haikou	0.93 ± 0.54	1.04 ± 0.48	0.95 ± 0.31	1.03 ± 0.38	0.95 ± 0.68	1.02 ± 0.11	1.00
Shantou	0.53 ± 0.31	0.52 ± 0.34	0.50 ± 0.27	0.52 ± 0.27	0.50 ± 0.19	0.51 ± 0.16	0.50

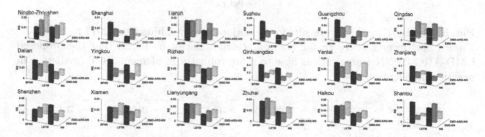

Fig. 5. Visual illustration of RE for different prediction frameworks at 18 major ports in China. Results demonstrate that our proposed three-step prediction framework yields the lowest RE under almost all conditions.

The quantitative results from our proposed prediction framework EMD-ARG-NN and other competing techniques (i.e., BPNN, LSTM and EMD-BPNN, EMD-LSTM) are detailedly depicted in Fig. 5. The mean RE results illustrate that three-step prediction method generates the most accurate prediction under most conditions. EMD-NN methods have constrained the further improvement in prediction performance. Single NN methods generate the lowest prediction accuracy under consideration in most of the cases due to the PCT time series have the predominantly complex, nonlinear and nonstationary properties.

The superior performance of our framework is further confirmed by the statistical results for prediction robustness shown in Table 1. It illustrates the standard deviation of different prediction values adopted to evaluate the prediction robustness. It can be found that our prediction framework generates the most robust prediction. This good performance mainly benefits from the EMD-based property decomposition and SED-based similarity adaptive regrouping.

3.3   Prediction Results and Discussion

In this subsection, our proposed three-step prediction framework was compared with the NN and EMD-NN methods on PCT prediction from 2017 to 2026. The

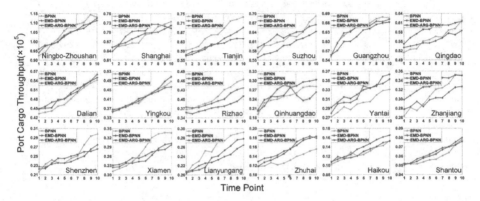

Fig. 6. The PCT prediction results for 18 major ports in China from 2017 to 2026 (i.e., *Time Point from 1 to 10 in the horizontal axis*). It can be found that the proposed EMD-ARG-BPNN framework is able to more robustly implement PCT prediction in practical applications.

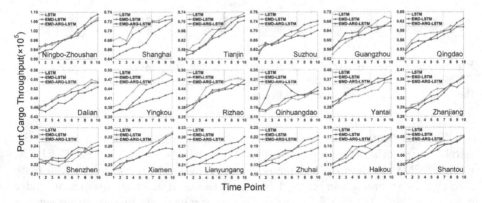

Fig. 7. The PCT prediction results for 18 major ports in China from 2017 to 2026 (i.e., *Time Point from 1 to 10 in the horizontal axis*). It can be found that the proposed EMD-ARG-LSTM framework is able to more robustly implement PCT prediction in practical applications.

prediction results are displayed in Figs. 6 and 7. It can be easily found that our framework has the capacity of robustly implementing the prediction of PCT time series, and our framework could be applied to both traditional NN and DL methods. Figure 3 shows that the trends of PCT in most ports from 1996 to 2016. From the theoretical point of view, the predicted results from 2017 to 2026 should also grow with a similar increase. Figures 6 and 7 show that our proposed framework generates the predicted PCT time series which are more similar to the theoretical results. However, NN and EMD-NN prediction methods have low-quality prediction performance.

4   Conclusions

The reliable PCT time series prediction is of fundamental importance in economic investment, transportation planning, port planning and design, etc. In this paper, we proposed a three-step calculation framework, called EMD-ARG-LSTM, to guarantee the accuracy and robustness of PCT prediction for 18 main ports in China. In particular, the original PCT time series were first decomposed into high and low frequency components using the EMD. Then, each component was divided into several continuous small (overlapping) segments. Several segments with high SED-based similarities are adaptive regrouped to form proper training dataset for LSTM to improve prediction accuracy. The final prediction results were consequently obtained by combining the predicted component. The experimental results for 18 different ports have demonstrated the superior performance of our proposed three-step prediction framework in terms of both prediction accuracy and robustness. According to our experiment, our proposed three-step prediction framework can be also applied to traditional NN methods, such as BPNN. That is to say, EMD-ARG-BPNN prediction framework also has certain superiority in prediction accuracy and robustness compared with conventional BPNN and EMD-BPNN.

Acknowledgment. This work was supported by National Natural Science Foundation of China (Nos.: 51609195 and 51479156), Fund of Hubei Key Laboratory of Transportation Internet of Things (No.: WHUTIOT-2017B003), and Independent Innovation Research Funding for Undergraduates (No.: 2018-HY-A1-01).

References

1. Zhang, C., Huang, L., Zhao, Z.: Research on combination forecast of port cargo throughput based on time series and causality analysis. J. Ind. Eng. Manag. **6**(1), 124–134 (2013)
2. Fjodorova, N., et al.: Quantitative and qualitative models for carcinogenicity prediction for non-congeneric chemicals using CP ANN method for regulatory uses. Mol. Divers. **14**(3), 581–594 (2010)
3. Hindmarsh, M., Huber, S.J., Rummukainen, K., Weir, D.J.: Numerical simulations of acoustically generated gravitational waves at a first order phase transition. Phys. Rev. D **92**(12), 24–30 (2015)

4. Dixon, W.E., Walker, I.D., Dawson, D.M., Hartranft, J.P.: Fault detection for robot manipulators with parametric uncertainty: a prediction-error-based approach. IEEE Trans. Robotic. Autom. **16**(6), 689–699 (2000)
5. Khosravi, A., Nahavandi, S.: Combined nonparametric prediction intervals for wind power generation. IEEE Trans. Sustain. Energy **4**(4), 849–856 (2013)
6. Preacher, K.J., Curran, P.J., Bauer, D.J.: Computational tools for probing interactions in multiple linear regression, multilevel modeling, and latent curve analysis. J. Educ. Behav. Stat. **31**(4), 437–448 (2006)
7. Hyndman, R.J., Koehler, A.B., Snyder, R.D., Grose, S.: A state space framework for automatic forecasting using exponential smoothing methods. Int. J. Forecast. **18**(3), 439–454 (2002)
8. Wang, C.N., Phan, V.T.: An improvement the accuracy of grey forecasting model for cargo throughput in international commercial ports of Kaohsiung. Int. J. Bus. Econ. Res. **3**(1), 1–5 (2014)
9. Liu, R.W., Chen, J., Liu, Z., Li, Y., Liu, Y., Liu, J.: Vessel traffic flow separation-prediction using low-rank and sparse decomposition. In: IEEE ITSC, pp. 1–6 (2017)
10. Odom, M.D., Sharda, R.: Stock market prediction system with modular neural networks. In: IEEE IJCNN, pp. 1–6 (1990)
11. Ping, F.F., Fang, X.F.: Multivariant forecasting mode of Guangdong province port throughput with genetic algorithms and back propagation neural network. Procedia Soc. Behav. **96**, 1165–1174 (2013)
12. Lv, Y., Duan, Y., Kang, W., Li, Z., Wang, F.Y.: Traffic flow prediction with big data: a deep learning approach. IEEE Trans. Intell. Transp. Syst. **16**(2), 865–873 (2015)
13. Connor, J.T., Martin, R.D., Atlas, L.E.: Recurrent neural networks and robust time series prediction. IEEE Trans. Nerual Netw. **5**(2), 240–254 (1994)
14. Ma, X., Tao, Z., Wang, Y., Yu, H., Wang, Y.: Long short-term memory neural network for traffic speed prediction using remote microwave sensor data. Transport. Res. C-Emer. **54**, 187–197 (2015)
15. Naik, J., Satapathy, P., Dash, P.K.: Short-term wind speed and wind power prediction using hybrid empirical mode decomposition and kernel ridge regression. Appl. Soft Comput. **70**, 1167–1188 (2018)
16. Liu, S., Xu, L., Li, D.: Multi-scale prediction of water temperature using empirical mode decomposition with back-propagation neural networks. Comput. Electr. Eng. **49**, 1–8 (2016)
17. Yu, L., Wang, S., Lai, K.K.: Forecasting crude oil price with an EMD-based neural network ensemble learning paradigm. Energ. Econ. **30**(5), 2623–2635 (2008)
18. Wang, J., Zhang, W., Li, Y., Wang, J., Dang, Z.: Forecasting wind speed using empirical mode decomposition and Elman neural network. Appl. Soft Comput. **23**, 452–459 (2014)
19. Bianconi, F., Fernández, A.: Evaluation of the effects of Gabor filter parameters on texture classification. Pattern Recogn. **40**(12), 3325–3335 (2007)
20. Chen, P.Y., Lai, Y.C., Zheng, J.Y.: Hardware design and implementation for empirical mode decomposition. IEEE Trans. Ind. Electron. **63**(6), 3686–3694 (2016)
21. Flandrin, P., Rilling, G., Goncalves, P.: Empirical mode decomposition as a filter bank. IEEE Sig. Process. Lett. **11**(2), 112–114 (2004)

Enhancing the Recurrent Neural Networks with Positional Gates for Sentence Representation

Yang Song, Wenxin Hu$^{(\boxtimes)}$, Qin Chen, Qinmin Hu, and Liang He

School of Computer Science and Software Engineering,
East China Normal University, Shanghai 200241, China
{ysong,qchen}@ica.stc.sh.cn, wxhu@cc.ecnu.edu.cn,
{qmhu,lhe}@cs.ecnu.edu.cn

Abstract. The recurrent neural networks (RNN) with attention mechanism have shown good performance for answer selection in recent years. Most previous attention mechanisms focus on generating the attentive weights after obtaining all the hidden states, while the contextual information from the other sentence is not well studied during the internal hidden state generation. In this paper, we propose a position gated RNN (PG-RNN) model, which merges the positional contextual information of the question words for the inner hidden state generation. Specifically, we first design a positional interaction monitor to detect and measure the positional influence of question word within answer sentence. Then we present a positional gating mechanism and embed it into RNN to automatically absorb the positional contextual information for the hidden state update. Experiments on two benchmark datasets, namely TREC-QA and WikiQA, show the great advantages of our proposed model. In particular, we achieve the new state-of-the-art performance on TREC-QA and WikiQA.

Keywords: Position · Gate · Attention · Recurrent neural network

1 Introduction

Recurrent neural networks (RNNs) have been widely used for natural language processing due to its good performance [1,2], especially in question selection task [3–5]. Given a question and a list of candidate answers, the answer selection task is to rank the candidate answers according to their similarities with the question. In the RNN based answer selection models, hidden layers firstly learn a hidden state for each word in question or answer. Then, the question or answer is represented by aggregating all the hidden states. Finally, the best relevant answer is selected from candidate answers according to the question-answer similarities.

How to generate the hidden states for sentence representation is one of the major challenges in RNNs. Recently, attention mechanism has shown its great

© Springer Nature Switzerland AG 2018
L. Cheng et al. (Eds.): ICONIP 2018, LNCS 11301, pp. 511–521, 2018.
https://doi.org/10.1007/978-3-030-04167-0_46

advantages in answer selection with an attentive sentence representation [3, 6, 7]. To be specific, an attentive weight is firstly learned for each word in the sentence, and then the whole sentence is represented through the sum of the attentive weighted hidden states in RNN.

Most previous attention mechanisms concentrate on learning the attentive weights for all the generated hidden states [3, 6, 8]. However, how to incorporate the contextual information from another sentence for hidden state generation is not well studied. Wang et al. [4] highlighted that the inner activation units in RNN controls the information flow over a sentence, and they proposed an IARNN-GATE model to incorporate the question representation in the active gates which influence the hidden state generation for the answer. However, it adopted information of all question words, which would bring noise and lead to a low performance if the question words were not relevant to the answer current hidden state. Tracing back to the traditional information retrieval (IR) models [9, 10], positional context has been widely adopted to indicate the relevance between the query word (i.e. question word) and the candidate document (i.e. answer). It assumed that an occurrence of a query word within the document had an impact towards its neighboring texts, and the accumulation of influence measured the relevance between the query word and the given document. In addition, work in [11] has proved the effectiveness of position features in neural networks for answer selection. Whereas, it is still unknown how to integrate the positional context into active gate of RNN to influence the hidden state generation for the answer.

In this paper, we proposed a **Position Gated RNN (PG-RNN)** model, where the positional context information of the question word is incorporated into the active gate as an attention. To be specific, we first proposed a positional interaction monitor to detect and measure the relevance of question word and answer word at each position within the answer sentence. Then, a positional gating mechanism is presented and embedded into our model, in which the answer representation is generated in hidden state with a positional attention. We conduct experiments on two widely used benchmark collections, namely TREC-QA and WikiQA. The experimental results demonstrate that our proposed positional gating mechanism can significantly outperform those classical attention mechanisms that do not involve positional gate. Moreover, the performance of the proposed PG-RNN model achieves new state-of-the-art performance in answer selection task.

The main contributions of this paper are summarized as follows: (1) to the best of our knowledge, it is the first attempt to optimize the active gate with the words' positional information as an attention; (2) a positional gating mechanism is presented and nicely embedded into RNN, which can reduce the noise for generating a hidden state by detecting and measuring relevance of question word and answer word; (3) we conduct elaborate analyses of the experimental results on two answer selection datasets, which provides a better understanding of the effectiveness of our model.

2   Related Work

Recently, the deep neural networks have been widely used in answer selection task due to their good performance [5,12], especially the recurrent neural networks (RNN) because of their capacity in modeling the sentence with variable length. [13] and [14] applied the long short-term memory (LSTM) [15] based RNN model to obtain the semantic relevance between question-answer pairs for the community based question selection.

To capture the salient information for better sentence representations, the attention mechanism was introduced into the neural networks [4,6]. [16] proposed an attentive interactive neural network, which focused on the interactions between text segments for answer selection. In addition, the interactions in sentence level or word level are incorporated for the attentive weight generation within the RNN framework. In [3], the attentive weights for an answer sentence relied on the interactions with the question sentence. In [6], the word-by-word interactions were utilized for the attentive sentence representations. Whereas, these attention based RNN models mainly focus on the attentive weight generation for all the learned hidden states, while the inner interactions between to sentences are not well studied during the hidden state generation. [4] proposed an IARNN-GATE model, where the question information was added to the active gates in RNN to influence the hidden state generation for answers. However, all the information of the question sentence is utilized in their model, which brings some noise for modeling the hidden states that are not relevant to the question.

To reduce the noise introduced into the matching model, we trace back to the traditional proximity-based information retrieval (IR) models. In [9,10], positional context was adopted to indicate the relevance between the query word and the candidate document. In particular, with the assumption that an occurrence of a question word within the document had an impact towards its neighboring texts, the positional contextual information was also taken into consideration in those models. In addition, work in [11] had proved the effectiveness of position features on neural networks for answer selection. Whereas, it is still unknown how to integrate the positional context into active gate of RNN to influence the hidden state generation. In this paper, we embed the positional information of question word into RNN active gate as an attention, which induces little attentive noise and enhances the performance of the RNN model in answer selection task.

3   Position Gated Recurrent Neural Network

3.1   The Framework of PG-RNN

Figure 1 shows the proposed framework for sentence modeling, including the position gated RNN for answer sentence modeling (Fig. 1(b)) and the classical RNN model for question sentence modeling (Fig. 1(a)). To be specific, we adopt the gated recurrent unit (GRU) as the basic cell for RNN since it has shown great advantages in many neural language processing tasks and has less parameters compared with long short-term memory (LSTM) [17]. The detailed formulation of GRU is presented as follows:

$$\mathbf{z}_t = \sigma(\mathbf{W}_{xz}\mathbf{x}_t + \mathbf{W}_{hz}\mathbf{h}_{t-1}),$$
$$\mathbf{r}_t = \sigma(\mathbf{W}_{xr}\mathbf{x}_t + \mathbf{W}_{hr}\mathbf{h}_{t-1}),$$
$$\tilde{\mathbf{h}}_t = tanh(\mathbf{W}_{xh}\mathbf{x}_t + \mathbf{W}_{hh}(\mathbf{r}_t \odot \mathbf{h}_{t-1})),$$
$$\mathbf{h}_t = (1 - \mathbf{z}_t) \odot \mathbf{h}_{t-1} + \mathbf{z}_t \odot \tilde{\mathbf{h}}_t,$$

(1)

where $\mathbf{z}_t$ and $\mathbf{r}_t$ stand for update gate and reset gate respectively, $\mathbf{W}_{xz}$, $\mathbf{W}_{hz}$, $\mathbf{W}_{xr}$, $\mathbf{W}_{hr}$, $\mathbf{W}_{xh}$ and $\mathbf{W}_{hh}$ are weight matrices, h_{t-1} and h_t are hidden states, and $\odot$ implies the element-wise multiplication.

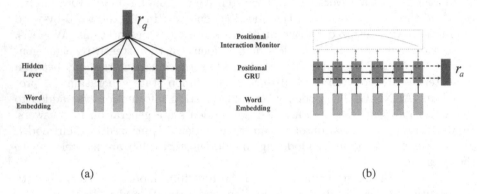

(a)   (b)

Fig. 1. (a) The classical attention based RNN for question representation; (b) the position gated RNN (PG-RNN) for answer representation.

In our framework, question is represented by the classical RNN model introduced in (Fig. 1(a)), and is denoted as r_q. Regarding to the answer, we first design a positional interaction monitor to detect and measure the influence of question word over each hidden state, in which we assume that if a question word has occurrence in the answer sentence, then it has impact on its neighboring text and the impact decreases with the increase of the distance away from the position where the question word has occurrence. After that, positional interaction monitors and related hidden states are assembled in the GRU block as a positional gate to generate answer representation, which is denoted as r_a. In particular, the generation of hidden state h_t not only relies on the directly connected hidden state h_{t-1}, but also absorbs contextual information delivered by positional interaction monitors. Finally, the Manhattan distance with L1 norm (Eq. 2) is adopted to calculate the similarity between question and answer, which is proved to be better for similarity modeling compared with other alternatives such as cosine in [18].

$$sim(r_q, r_a) = exp(-||r_q - r_a||_1)$$

(2)

3.2   Positional Interaction Monitor

Based on the proximity-based information retrieval theory [9,10], we assume that the question word has an impact on its neighboring terms if it has occurrence within the answer sentence. We do believe this impact should act on each hidden state in the networks. The positional interaction monitor is designed to detect and measure this impact. Obviously, the influence can be detected by identifying whether the question word has occurrence within the answer or not. As for the influence measurement, we propose an influence propagation strategy as follows.

Influence Propagation Strategy. According to the traditional proximity-based IR models [9,10], the positional influence tends to decrease with the increase of the distance to the place where the question word has occurrence. In this paper, we use the Gaussian density function $Kernel(u)$ to estimate the word's influence propagation, which is indicated to be effective in position modeling [9,10]:

$$Kernel(u) = exp(\frac{-u^2}{2\delta^2}), \tag{3}$$

where u is the distance between the current word and given question word, δ restricts the influence propagation scope of a question word. Therefore, the value of $Kernel(u)$ implies the impact strength regarding to the positional distance u.

Suppose $\mathbf{q} = \{q_1, q_2, \ldots, q_m\}$ stands for question, $\mathbf{a} = \{x_1, x_2, \ldots, x_n\}$ stands for answer, and m, n indicate the length of question and answer respectively. For $x_i \in \mathbf{a}$, $i = 1, 2, \ldots, n$, the value of its **positional interaction monitor** over the j-*th* hidden state h_j, denoted as $p_j^{x_i}$, is defined by:

$$p_j^{x_i} = \begin{cases} Kernel(|i - j|), & if x_i \in \mathbf{q}, \\ 0, & if x_i \notin \mathbf{q}. \end{cases} \tag{4}$$

3.3   Positional Gating Mechanism

Inspired by the previous work of IARNN-GATE [4], in which the question representation is incorporated into the active gates for hidden state generation, instead of embedding attention information to the hidden vectors, we incorporate positional context attention to the gate inner activation (i.e. $\mathbf{z}_t$ and $\mathbf{r}_t$ in Eq. 1). Since these inner activation units decide the which kind of information and how much information is required to removed or reserved for hidden state resetting and updating, we embed positional information to these active gates as the attention to enrich the hidden representations as follows:

$$\mathbf{z}_t = \sigma(\mathbf{W}_{xz}\mathbf{x}_t + \mathbf{W}_{hz}\mathbf{h}_{t-1} + \mathbf{W}_{zp} \sum_{i<t, x_i \in \mathbf{a}} p_j^{x_i}\mathbf{h}_i)$$

$$\mathbf{r}_t = \sigma(\mathbf{W}_{xr}\mathbf{x}_t + \mathbf{W}_{hr}\mathbf{h}_{t-1} + \mathbf{W}_{rp} \sum_{i<t, x_i \in \mathbf{a}} p_j^{x_i}\mathbf{h}_i),$$

$$\tilde{\mathbf{h}}_t = tanh(\mathbf{W}_{xh}\mathbf{x}_t + \mathbf{W}_{hh}(\mathbf{r}_t \odot \mathbf{h}_{t-1})),$$

$$\mathbf{h}_t = (1 - \mathbf{z}_t) \odot \mathbf{h}_{t-1} + \mathbf{z}_t \odot \tilde{\mathbf{h}}_t,$$

$$\tag{5}$$

where $\mathbf{W}_{zp}$ and $\mathbf{W}_{rp}$ are weight matrices of positional attention. For a better understanding purpose, we show the positional gating mechanism in Fig. 2. With this positional gate, the GRU units not only record the time step information in a series, but also enhance the inner interaction between question and answer.

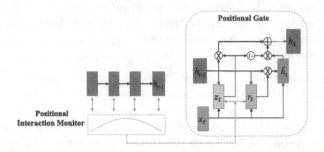

Fig. 2. Positional gating. We show the inner state process of positional gate within the blue box, and highlight the positional context information with the red dotted line. (Color figure online)

With this positional gating mechanism, the answer is represented by the sum of all the hidden vectors:

$$r_a = \sum_{i=1}^{n} h_i \qquad (6)$$

4   Experiments

4.1   Experimental Setup

Datasets and Evaluation Metrics. We perform experiments on two standard answer selection collections, namely TREC-QA and WikiQA. TREC-QA [19] was created based on the data of Text REtrieval Conference QA track from the year of 1998 to 2003. WikiQA [20] is an open domain question-answering dataset in which all answers are collected from the Wikipedia. Both TREC-QA and WikiQA have the train, development and test sets, and each sample is labeled as 1 or 0 to indicate whether the candidate answer is right or wrong for a given question. The statistics of the two data sets are showed in Table 1. In this paper, we adopt the mean average precision (MAP) and the mean reciprocal rank (MRR) to evaluate the effectiveness of the proposed model.

Parameter Settings. We adopt the Bidirectional GRU (BiGRU) [21] model with positional gate for answer representation. The pre-trained 100-dimensional GloVe [22] word vectors[1] are adopted to map the input words to vectors. At the same time, the dimension of the hidden vectors generated by positional gate is set to 50. As for the propagation scope δ (in Eq. 3), we select its value from [5, 55]. Furthermore, the cross-entropy loss function and Adadelta optimizer [23] are adopted model training and parameter updating.

---

[1] http://nlp.stanford.edu/data/glove.6B.zip.

Table 1. Statistics of Datasets. "Question Length" and "Answer Length" denote the average length of questions and answers.

Dataset	# of quetions (train/dev/test)	Question length (train/dev/test)	Answer length (train/dev/test)
TREC-QA	1162/65/68	7.57/8.00/8.63	23.21/24.9/25.61
WikiQA	873/126/243	7.16/7.23/7.26	25.29/24.59/24.59

4.2 Effectiveness of Positional Gating Mechanism

In order to investigate the effectiveness of the proposed positional gate, two basic RNN models which do not involve the positional information are adopted to make comparison in Table 2, including (1) the original GRU model ("G-RNN") with no optimization in the gate unit, (2) a GRU model with an attention gate ("AG-RNN") used in [4]. In particular, we conduct statistical significant test based on the paired t-test at the 0.05 level, where the symbols "*" and "+" indicate the significant improvements over "G-RNN" and "AG-RNN" respectively.

Table 2. Comparison with basic RNN models. "G-RNN" stands for RNN with no optimization in the gate unit, and "AG-RNN" stands for a RNN with an attention gate. "*" and "+" imply significant improvements over "G-RNN" and "AG-RNN" respectively.

Model	TREC-QA		WikiQA	
	MAP	MRR	MAP	MRR
G-RNN	0.6487	0.6991	0.6581	0.6691
AG-RNN	0.7369*	0.8208*	0.7258*	0.7394*
PG-RNN	**0.8121*+**	**0.8813*+**	**0.7397*+**	**0.7485*+**

The experimental results showed that (1) embedding the positional context information of question word into active gates for answer representation can significantly improve the performance of RNN on question selection task; (2) question word' positional information outperforms the general features as an attention in optimizing active gates. "AG-RNN" uses the whole question sentence to implement attention mechanism in active gate. Whereas, it neglects whether the question words are initially relevant to the answer or the surrounding context of the occurred question words, which bring noise into hidden state for sentence representation generation. In contrast, our positional gating mechanism highlighted the inner matching signals between question words and answer words through the positional interaction monitor, which finally enhances the semantic representation of sentences.

4.3   Effectiveness of PG-RNN Model

We also make comparisons with the state-of-the-art and up to date models in answer selection task to further evaluate the effectiveness of the proposed PG-RNN model. Tables 3 and 4 show the experimental results on the collections of TREC-QA and WikiQA respectively. Regarding to the TREC-QA dataset, the following five models are selected for comparison, which is considered as the strong baselines in recent studies. (1) A deep neural matching model which integrate the CNN structure on the top of BLSTM [3]. (2) A combination of the stacked BLSTM relevance model and the BM25 overlapping baseline [5]. (3) A BLSTM model with a positional attention on hidden vectors [11]. (4) A convolution neural network which perform ranking with both convolution features and additional statistic features [12]. (5) A word-alignment-based method, which is implement by the leaning-to-ranking model with neural network based similarity and lexical features [24]. As for the dataset WikiQA, besides the models [3] and [11] introduced above, we also adopt another three up to date models as baselines for comparison. (1) A GRU model with an inner attention in the active gates [4]. (2) A bigram CNN model with average pooling [20] (3) A basic CNN model with attention mechanism both on the convolution and pooling layers [25].

Table 3. Comparisons with recent progress on TREC-QA.

System	MAP	MRR
Wang and Nyberg [5]	0.7134	0.7913
Severyn and Moschitti [12]	0.7459	0.8078
Wang and Ittycheriah [24]	0.7460	0.8200
Santos et al. [3]	0.7530	0.8511
Chen et al. [11]	0.7814	0.8513
PG-RNN	**0.8121**	**0.8813**

Table 4. Comparisons with recent progress on WikiQA

System	MAP	MRR
Yang et al. [20]	0.6520	0.6652
Santos et al. [3]	0.6886	0.6957
Yin et al. [25]	0.6921	0.7108
Wang et al. [4]	0.7341	0.7418
Chen et al. [11]	0.7212	0.7312
PG-RNN	**0.7397**	**0.7485**

It can be found that our proposed PG-RNN achieves the new state-of-the-art performance on both TREC-QA and WikiQA in terms of MAP and MRR.

This validates our previous assumption that active gate in RNN should give more attention to question words that have occurrence in answer as well as the context around them. In addition, our approach makes improvements on both TREC-QA and WikiQA in terms of MRR, which indicates the advantages of our proposed model in capturing the top relevant answers with the positional gate.

4.4 Investigation of Propagation Scope δ

In the PG-RNN model, parameter δ (in Eq. 3) restricts the influence propagation scope of a given word. Given a constant distance u, the value of the positional interaction monitor grows with the increase of δ. We plot the influence of δ on the evaluation metrics MAP and MRR in Fig. 3, where the value of δ is selected in $[5, 15, 25, 35, 45, 55]$. The figures shows that MAP and MRR have the same evolution trend. To be specific, they decrease before $\delta = 15$, and then increase, finally tend to be stable when $\delta > 45$. Therefore, it is recommended to assign δ with the value in the interval $[25, 45]$ to achieve a reliable performance in our experiments.

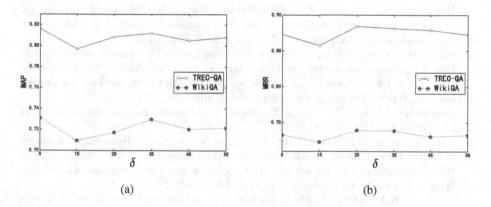

(a)   (b)

Fig. 3. Impact of the propagation scope δ

5   Conclusion and Future Work

In this paper, we propose a novel position gated RNN (PG-RNN) model for question selection task. To be specific, a positional gating mechanism is presented and well embedded into our model, which can automatically absorb the positional contextual information of the question words for better hidden state generation. In particular, a positional interaction monitor is designed to detect and measure the positional contextual information. The experimental results on two benchmark datasets, namely TREC-QA and WikiQA, show the great advantages of our proposed PG-RNN model. It is notable that, we achieve the new state-of-the-art performance on both of the above answer selection datasets. In the future, we will investigate the effect of our model on more tasks, such as paraphrase identification and textual entailment.

Acknowledgements. We thank all viewers who provided the thoughtful and constructive comments on this paper. The second author is the corresponding author. This research is funded by the National Natural Science Foundation of China (No. 61572193). The computation is performed in the Supercomputer Center of East China Normal University.

References

1. Tang, D., Qin, B., Liu, T.: Document modeling with gated recurrent neural network for sentiment classification. In: Proceedings of the 2015 Conference on Empirical Methods in Natural Language Processing, pp. 1422–1432 (2015)
2. Sutskever, I., Vinyals, O., Le, Q.V.: Sequence to sequence learning with neural networks. In: Advances in Neural Information Processing Systems, pp. 3104–3112 (2014)
3. Tan, M., dos Santos, C., Xiang, B., Zhou, B.: LSTM-based deep learning models for non-factoid answer selection. arXiv preprint arXiv:1511.04108 (2015)
4. Wang, B., Liu, K., Zhao, J.: Inner attention based recurrent neural networks for answer selection. In: Proceedings of the 54th Annual Meeting of the Association for Computational Linguistics, vol. 1, pp. 1288–1297 (2016)
5. Wang, D., Nyberg, E.: A long short-term memory model for answer sentence selection in question answering. In: Proceedings of the 53rd Annual Meeting of the Association for Computational Linguistics and the 7th International Joint Conference on Natural Language Processing, vol. 2, pp. 707–712 (2015)
6. dos Santos, C., Tan, M., Xiang, B., Zhou, B.: Attentive pooling networks. arXiv preprint arXiv:1602.03609 (2016)
7. Yang, Z., Yang, D., Dyer, C., He, X., Smola, A., Hovy, E.: Hierarchical attention networks for document classification. In: Proceedings of the 2016 Conference of the North American Chapter of the Association for Computational Linguistics: Human Language Technologies, pp. 1480–1489 (2016)
8. Hermann, K.M., et al.: Teaching machines to read and comprehend. In: Advances in Neural Information Processing Systems, pp. 1693–1701 (2015)
9. Lv, Y., Zhai, C.X.: Positional language models for information retrieval. In: Proceedings of the 32nd International ACM SIGIR Conference on Research and Development in Information Retrieval, pp. 299–306 (2009)
10. Zhao, J., Huang, J.X., He, B.: CRTER: using cross terms to enhance probabilistic information retrieval. In: Proceedings of the 34th International ACM SIGIR Conference on Research and Development in Information Retrieval, pp. 155–164 (2011)
11. Chen, Q., Hu, Q., Huang, J.X., He, L., An, W.: Enhancing recurrent neural networks with positional attention for question answering. In: Proceedings of the 40th International ACM SIGIR Conference on Research and Development in Information Retrieval, pp. 993–996. ACM (2017)
12. Severyn, A., Moschitti, A.: Learning to rank short text pairs with convolutional deep neural networks. In: Proceedings of the 38th International ACM SIGIR Conference on Research and Development in Information Retrieval, pp. 373–382 (2015)

13. Zhao, Z., Lu, H., Zheng, V.W., Cai, D., He, X., Zhuang, Y.: Community-based question answering via asymmetric multi-faceted ranking network learning. In: AAAI, pp. 3532–3539 (2017)
14. Fang, H., Wu, F., Zhao, Z., Duan, X., Zhuang, Y., Ester, M.: Community-based question answering via heterogeneous social network learning. In: Thirtieth AAAI Conference on Artificial Intelligence (2016)
15. Hochreiter, S., Schmidhuber, J.: Long short-term memory. Neural Comput. **9**(8), 1735–1780 (1997)
16. Zhang, X., Li, S., Sha, L., Wang, H.: Attentive interactive neural networks for answer selection in community question answering. In: AAAI, pp. 3525–3531 (2017)
17. Jozefowicz, R., Zaremba, W., Sutskever, I.: An empirical exploration of recurrent network architectures. In: International Conference on Machine Learning, pp. 2342–2350 (2015)
18. Mueller, J., Thyagarajan, A.: Siamese recurrent architectures for learning sentence similarity. In: AAAI, pp. 2786–2792 (2016)
19. Wang, M., Smith, N.A., Mitamura, T.: What is the jeopardy model? A quasi-synchronous grammar for QA. In: Proceedings of the 2007 Joint Conference on Empirical Methods in Natural Language Processing and Computational Natural Language Learning (EMNLP-CoNLL), pp. 22–32 (2007)
20. Yang, Y., Yih, W., Meek, C.: Wikiqa: a challenge dataset for open-domain question answering. In: Proceedings of the 2015 Conference on Empirical Methods in Natural Language Processing, pp. 2013–2018 (2015)
21. Schuster, M., Paliwal, K.K.: Bidirectional recurrent neural networks. IEEE Trans. Sig. Process. **45**(11), 2673–2681 (1997)
22. Pennington, J., Socher, R., Manning, C.D.: Glove: global vectors for word representation. In: Proceedings of the 2014 Conference on Empirical Methods in Natural Language Processing (EMNLP), pp. 1532–1543 (2014)
23. Zeiler, M.D.: ADADELTA: an adaptive learning rate method. arXiv preprint arXiv:1212.5701 (2012)
24. Wang, Z., Ittycheriah, A.: FAQ-based question answering via word alignment. arXiv preprint arXiv:1507.02628 (2015)
25. Yin, W., Schütze, H., Xiang, B., Zhou, B.: ABCNN: attention-based convolutional neural network for modeling sentence pairs. arXiv preprint arXiv:1512.05193 (2015)

Spiking Neural Networks

A Visual Recognition Model Based on Hierarchical Feature Extraction and Multi-layer SNN

Xiaoliang Xu, Wensi Lu, Qiming Fang(✉), and Yixing Xia

School of Computer, Hangzhou Dianzi University, Hangzhou, China
{xxl,fangqiming,yixingx}@hdu.edu.cn, 1018473669@qq.com

Abstract. In this paper, a visual pattern recognition model is proposed, which effectively combines hierarchical feature extraction model and coding method on multi-layer SNN. This paper takes HMAX model as feature extraction model and adopts independent component analysis (ICA) to improve it, so that the model can satisfy the sparsity of information extraction and the output result is more suitable for SNN processing. Multi-layer SNN is used as classifier and the firing of spikes is not limited in the learning process. We use valid phase coding to connect these two parts. Through the experiments on the MNIST and Caltech101 datasets, it can be found that the model has good classification performance.

Keywords: Visual pattern recognition · Multi-layer SNN
HMAX · Phase coding

1 Introduction

In recent years, the bio-inspired object recognition frameworks have become an increasingly active field of research [1]. HMAX model [2] as a biologically based feedforward hierarchical architecture for object recognition, has many extended models [3]. Original HMAX model applies template matching and MAX pooling methods to achieve selectivity and invariance. In order to learn higher features, some researchers have made several adjustments for the model [4,5]. There are two main extensions: one is to reduce calculation cost, the other is to change architecture of the HMAX model. Some researches use sparse-based technique to reduce calculation cost. Mutch et al. [6] adopt sparsity constrain in a similar model like HMAX model. Hu et al. [7] combine independent component analysis and the standard sparse coding with HMAX model. For changing the architecture of the HMAX model, one method is adjusting the layer number of the model. Hu et al. introduced a high-layer architecture including six layers. In [8], using only the earlier stages (one S layer and one C layer) of HMAX model, yet achieves good or better performance. Another method is carrying out the feedback operation into the HMAX model [9] to replace the feedforward architecture.

© Springer Nature Switzerland AG 2018
L. Cheng et al. (Eds.): ICONIP 2018, LNCS 11301, pp. 525–534, 2018.
https://doi.org/10.1007/978-3-030-04167-0_47

When combining the HMAX model with the multiclass linear SVM for good classification performance [10], some researchers have focused on more biologically realistic classifier. Spiking neural networks (SNNs) [11] have the same capability of processing spikes as biological neural system, especially adopting temporal encoding mechanism [12]. Various learning algorithms for SNNs have been proposed, mainly can be divided into two types: single-layer learning algorithms and multi-layer learning algorithms. For single-layer algorithms, the research can be classified into three types: based on convolution of spike train, gradient descent and synaptic plasticity mechanism, respectively. Typical algorithms of these three types are PSD [13], SpikeProp [14] and STDP [15]. Compared with single-layer learning algorithms, there are few researches on multi-layer learning algorithms. Effective multi-layer learning algorithms include the SpikeProp [14], Multi-ReSuMe [16] and the recurrent network learning rules [17]. The main reason for the lack of the typical multi-layer learning algorithm is that the neuronal spike-timing in the SNN is discontinuous, which leads to the difficulty of information feedback.

A lot of work has been done on HMAX model and SNNs in their respective fields. However, there is litter work to combine HMAX model with biological classifiers. In the HMAX+SVM model, the classifier is a mathematical method and does not have biological interpretability. Inspired by [5, 7, 18], in this paper, we propose a hierarchical computation architecture for visual pattern recognition which integrates HMAX and SNN into an integrated model that is biologically interpretable. In this model, we use multi-layer SNN to learn, FastICA to improve HMAX to better integrate with SNN, and Adam [22] to optimize the learning rate.

In the next section we will introduce the methods we use, including hierarchical feature extraction, phase encoding, and the learning algorithm used for multi-layer SNNs. Section 3 mainly discusses the experimental simulation results. The conclusion is given in Sect. 4.

2   Methods

2.1   Hierarchical Model for Feature Extraction

We use a four-layer model for feature extraction similar to HMAX, and make improvements in S2, C2 layers. The overall structure is shown in Fig. 1 and the specific method is described as follows.

(a) *S1 layer*: The results of the S1 layer are obtained by convolving the input image with Gabor filter on four directions $(0°, 45°, 90°, 135°)$.

(b) *C1 layer*: We use max pooling over local neighborhood which increases the tolerance to 2D transformations from layer S1 to C1. The specific operation is to use the $n \times n$ sliding window to select the strongest response value as the feature map of C1 in the window, and the window slides with 1/2 window overlaps.

(c) *S2 layer*: In order to extract more information and not limited by handcraft features [7], we adopt FastICA to learn filters, which is different from the operation of the S1 layer. Computational neuroscience indicates that filters used by the S1 layer can be considered as the result of sparse coding, while ICA is also closely related to sparse coding [20]. The process description is shown below:

$$X = AS \tag{1}$$

$$minimize \ \|X - AS\|_F^2 + \lambda \sum_{i=1}^{k} \|s_i\|_1 \tag{2}$$

$$subject \ to \ \|a_i\|^2 \leqslant 1, \forall i = 1, \ldots, m$$

Where each column of X is a patch x_i, each column of A is a basis a_i and each column of S is a vector $s_i \in \Re^m$ consisting of coefficients of the m bases for reconstructing x_i. $\|\cdot\|_F$ is the Frobenius norm and λ is a positive constant. For the sake of simplicity, we set A as an invertible matrix, then Eq. 2 can be solved by ICA algorithm.

$$maximize \sum_{i=1}^{k} \sum_{j=1}^{m} \log \ f_j \left(w_j^T x_i \right) + k \log |det \ W| \tag{3}$$

where $W = A^{-1}$, w_j^T is the $j - th$ row of W, x_i is the $i - th$ column of X, and $f_j(\cdot)$ denotes a sparse probability distribution function.

(d) *C2 layer*: ICA is a linear model that can extract linear statistical features from input. But in fact, many images contain nonlinear statistical information. Therefore, we apply max pooling on the result of the S2 layer so that nonlinear statistical information can be extracted.

2.2   Phase Encoding

When using SNN for image recognition, coding becomes an important bridge to establish the relationship between images and SNN. We make some minor changes to phase encoding algorithm [19] to make it more suitable for our simulation experiments. The precise time of each pixel in the receptive field (RF) depends not only on the intensity value but also on the position, which can be calculated as

$$
\begin{aligned}
step1: \quad & t_i = (i - 1) * t\_step + \frac{j-1}{n} t_{max}, \quad x_i \in j^{th} \ encoding \ neuron \\
step2: \quad & if \quad t_i > t_{max} \\
& t_i = t_i - t_{max}
\end{aligned} \tag{4}
$$

where x_i is the intensity value of pixel i, t_step is calculation interval, n is number of encoding neuron, and t_{max} is the maximum value of the time window. From

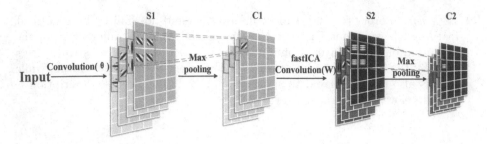

Fig. 1. Hierarchical model for feature extraction: The S layers show the results of the input image under different filtering effects. The S1 represents the result of Gabor filtering with different orientation $(0°, 45°, 90°, 135°)$, and S2 shows the result of processing with FastICA. The C layers represent the results of the S layers after max pooling. The difference is that the sliding windows of the C1 layer overlap, while the C2 layer do not.

Eq. 4 we can find that each encoding neuron approximates the periodic oscillation of a cosine function.

The encoding process is shown in Fig. 2. For the sake of simplicity, we just consider excitatory and inhibitory encoding neurons in the RF. And in the alignment step, time is adjusted to its nearest time in the time list of corresponding neurons. Finally, the firing time collected by the sampling window is mapped to the input neurons.

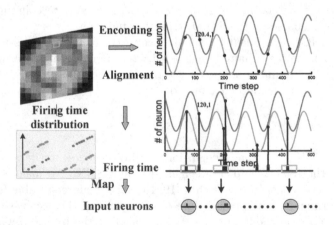

Fig. 2. Phase encoding: (upper left corner) input image is the result of C2 layer; (right) encoding process of the phase encoding algorithm, different color curves represent different encoding neurons; (lower left corner) the distribution of encoded spike trains onto the 2D axis. (Color figure online)

2.3   Learning Algorithm

In order to explore the learning performance of SNNs, we choose the multi-layer SNN. The learning algorithm [18] uses the maximum probability of generating the target spike trains as an objective function, and makes it suitable for multi-layer SNN by combining error back propagation algorithm and STDP. The object function is

$$P\left(z^{ref}|y\right) = exp\left(\int_0^T log\left(\rho_0\left(t|y,z_o\right)\right)z_o^{ref}\left(t\right) - \rho\left(t|y,z_o\right)dt\right) \quad (5)$$

where y is the spike train generated by the hidden layer, z_o is the actual output spike train and z_o^{ref} represents the target spike train. T is the time window and $\rho\left(t\right)$ denotes the stochastic intensity of generating spikes at t. $P\left(z|y\right)$ represents the probability of producing z under y conditions.

3   Simulation Results

In the real world, object recognition is a complex issue because external stimuli are diverse. Here we mainly experiment with the MNIST and Caltech101 datasets, and all images are processed into grayscale images.

3.1   Experimental Setup of Image Coding

Each image has an appropriate conversion process before it enters the spiking neural network. We collectively refer to feature extraction and phase encoding as the encoding process. In S1 layer we choose Gabor filters in four different orientations $(0^o, 45^o, 90^o, 135^o)$. And the other parameters are set to $\gamma = 0.3, \sigma = 2.8, \lambda = 3.5$ and the filter size is 7×7 pixels. After getting the result of S1 layer, we perform max pooling operation on it. The sliding window size is 6×6 pixels and the overlap window is set to be 3 pixels in one axis (x or y). In S2 layer, fastICA is used to autonomously learn 4 bases from the C1 results, and the max pooling used in C2 is to make the linear result become nonlinear. The window size is 4×4 pixels without overlap. After the hierarchical feature extraction operation, we use phase encoding which plays an important role in our model. In the process of phase encoding, the parameters are set to: the type of encoding neurons $encoding\_neuron = 2$, time window $t_{max} = 500$ ms, time step $t\_step = 1$ ms.

3.2   Performance of Multi-layer Learning Rule

In order to improve network performance, we use the multi-layer SNN learning algorithm mentioned in [18]. In this learning algorithm, "escape noise" [11] is used to solve the problem of the discontinuous nature of neuronal spike-timing. We specify some parameters: the threshold $V_{thr} = 15$ mV and the reset kernel

$V_{rest} = -15\,\mathrm{mV}$. We set escape rates at different layers to $\Delta u_h = 0.5\,\mathrm{mV}$, $\Delta u_o = 5\,\mathrm{mV}$. For the sake of simplicity, we use Poisson spike trains with $firing\_rate = 0.006$ to verify the learning ability of this algorithm. From Fig. 3 we can find that the spike trains generated by the hidden layer around the target output spike trains help to increase the possibility of the output layer accurately outputting the target spike trains.

Since using a stochastic neuron model, there is a gap between the actual output spike trains and the target output spike trains when using the van Rossum Distance [21] method to measure their distance. In order to narrow down the gap, we use the Adam [22] method to adjust the learning rate during the learning process and we can see from Fig. 3(d) that the convergence rate has increased.

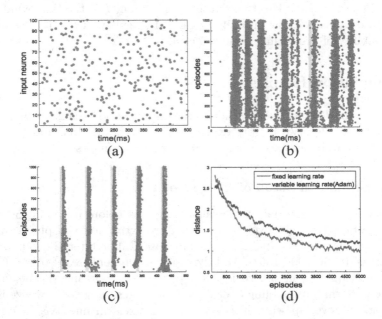

Fig. 3. Learning algorithm learns a specified target output spike train. The network contains 100 input neurons, 10 hidden neurons and one output neuron. The target output spike train is [85, 170, 255, 340, 425] ms. (a) shows the distribution of the generated Poisson spiking time among the time window. (b) denotes the spiking time distribution of the hidden neuron. (c) represents the spiking time distribution of the output neuron. (d) shows the distance between the actual output spike train and the target output spike train in different strategies' learning rate.

3.3   Recognition Performance

In this section, the multi-layer spiking neural network combined with the hierarchical feature extraction is used to deal with the actual classification problems. We choose two datasets (MNIST and Caltech101) for verification.

(1) The effectiveness of coding strategies: Phase encoding plays an important role in our approach. We compare three coding methods, one is the linear mapping that converts pixels directly into corresponding spikes, the second is to convert the pixels into Gaussian distributed spikes and the third is phase encoding used in this paper. We select three categories $(0, 1, 2)$ of images to verify the performance of each encoding method. Each method is run 10 times during the training, and each run contains 100 episodes. During the testing process, each test set consists of 300 images randomly selected from three categories $(0, 1, 2)$ of MNIST. As can be seen from Fig. 4, the training/testing accuracy of phase encoding is higher than the other two methods in the same situation. The main reason for this phenomenon is that the spikes obtained by image2spike and GRF methods are very narrowly distributed in the time window, making less time information available for the learning process. The phase encoding makes spikes more widely distributed in the time window so that the time information of each pixel can be fully utilized.

(2) The effect of different output forms of the feature extraction layer: In different HMAX methods, there are two different output forms, one is an image and the other is a vector, as shown in Fig. 5. The hierarchical feature extraction method used in this paper and the HMAX method in [5] correspond to these two different output forms. Ten types $(0–9)$ of handwritten digits are selected, the training set size is 1000 images/type and the test set is 100 images/type. These two forms are used as input to the multi-layer SNN to obtain two sets of test results. The test accuracy obtained by our method is 89.6%, and the accuracy of using vector as input is 10.2%. The main reason for this gap in accuracy is that when the vector is used as an input to the SNN, the effective information it provides is limited and cannot be fully used by the SNN.

(3) The classification accuracy of different methods: Table 1 shows the classification accuracy of different methods on the MNIST dataset. Method I and method II represent different HMAX enhancement methods and have the same form of output as our method. Method I mainly considers the architecture and uses only the S1 and C1 layers of the hierarchical feature extraction [8]. Method II considers from internal operation which uses sparse HMAX [7] to extract features. From Table 1, it can be seen that although our method and method II both use the four-layer feature extraction model, the accuracy of our method is higher, and the accuracy of method II isn't higher than only SNN. According to the experimental results, it can be concluded that a naive combination of SNN and feature extraction methods may not necessarily improve the classification performance. And the key to make the model more effective is that feature extraction results should highlight the unique features of each category.

To verify our method's ability to handle different datasets, we did experiments on Caltech101. The four methods chosen are the combined model of HMAX+PSD proposed by Xu et al. [23] which uses single-layer SNN, and the three methods that are gradually extended from the structure mentioned in this paper. From Fig. 6, it can be seen that our method has advantages over the

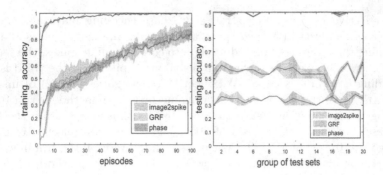

Fig. 4. The training/testing accuracy obtained by different coding schemes, GRF represents gaussian receptive field.

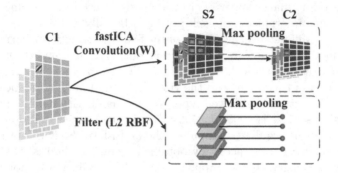

Fig. 5. Feature extraction model with different output forms.

Table 1. The classification accuracy of different methods.

Accuracy (%)	MNIST	
	Training accuracy (%)	Testing accuracy (%)
SNN only	84.45 ± 2.1	78.5 ± 3.1
Method I + SNN	91.8 ± 0.9	84.1 ± 1.2
Method II + SNN	90.05 ± 0.75	72.5 ± 4.5
Our method	93.25 ± 0.85	89.7 ± 1.1

other three methods in dealing with complex datasets, that accuracy obtained on all four datasets is higher, especially Caltech101-8. Caltech101-8 indicates that eight categories were selected from the Caltech101 dataset. Comparing Xu's method with the method using multi-layer SNN only, it can be clearly seen that in complex datasets, the processing capacity of multi-layer SNN is better than single-layer SNN.

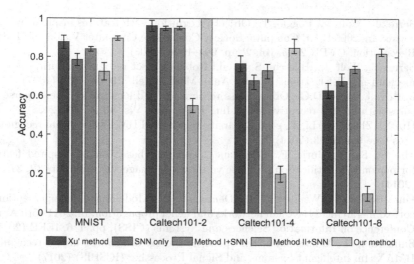

Fig. 6. Classification accuracy of different methods on different datasets.

4 Conclusion

This paper presents a multi-layer SNN combined with hierarchical feature extraction model. The model mimics the processing mechanism of biological vision, which combines simple features into complex features layer by layer, and uses spiking neural network as a classifier, so that the overall structure of the model can approximate biological visual information processing. And the model optimizes the feature extraction layer and extends the neural network learning layer, which not only satisfies different feature processing requirements, but also handles complex visual classification problems with multi-layered spiking neural networks. The model provides an optimizable idea that effectively combines multiple biologically responsive processing methods to reduce the gap between machine vision and biological vision.

Acknowledgments. This work was supported by the National Natural Science Foundation of China under Grant No. 61603119 and Zhejiang Provincial Natural Science Foundation of China under Grant No. LY17F020028.

References

1. Mély, D.A., Serre, T.: Towards a theory of computation in the visual cortex. In: Zhao, Q. (ed.) Computational and Cognitive Neuroscience of Vision. CST, pp. 59–84. Springer, Singapore (2017). https://doi.org/10.1007/978-981-10-0213-7_4
2. Riesenhuber, M., Poggio, T.: Hierarchical models of object recognition in cortex. Nat. Neurosci. **2**, 1019 (1999)
3. Liu, C., Sun, F.: HMAX model: a survey. In: Neural Networks (IJCNN), pp. 1–7. IEEE (2015)

 4. Serre, T., Wolf, L., Poggio, T.: Object recognition with features inspired by visual cortex. In: 2005 IEEE Computer Society Conference Computer Vision and Pattern Recognition. CVPR 2005, vol. 2, pp. 994–1000 (2005)
 5. Serre, T., Wolf, L., Bileschi, S., et al.: Robust object recognition with cortex-like mechanisms. IEEE Trans. Pattern Anal. Mach. Intell. **29**, 411–426 (2007)
 6. Mutch, J., Lowe, D.G.: Object class recognition and localization using sparse features with limited receptive fields. Int. J. Comput. Vis. **80**, 45–57 (2008)
 7. Hu, X., Zhang, J., Li, J., et al.: Sparsity-regularized HMAX for visual recognition. Plos One **9**, e81813 (2014)
 8. Ma, B., Su, Y., Jurie, F.: Covariance descriptor based on bio-inspired features for person re-identification and face verification. Image Vis. Comput. **32**, 379–390 (2014)
 9. Dura-Bernal, S., Wennekers, T., Denham, S.L.: Modelling object perception in cortex: hierarchical Bayesian networks and belief propagation. In: 2011 45th Annual Conference on Information Sciences and Systems (CISS), pp. 1–6. IEEE (2011)
10. Sufikarimi, H., Mohammadi, K.: Speed up biological inspired object recognition, HMAX. In: Intelligent Systems and Signal Processing (ICSPIS) (2017)
11. Gerstner, W., Kistler, W.M.: Spiking Neuron Models: Single Neurons, Populations, Plasticity. Cambridge University Press, Cambridge (2002)
12. Zheng, Y., et al.: Sparse temporal encoding of visual features for robust object recognition by spiking neurons. IEEE Trans. Neural Netw. Learn. Syst. 1–11 (2018). https://doi.org/10.1109/TNNLS.2018.2812811
13. Yu, Q., Tang, H., Tan, K.C., et al.: Precise-spike-driven synaptic plasticity: learning hetero-association of spatiotemporal spike patterns. Plos One **8**, e78318 (2013)
14. Bohte, S.M., Kok, J.N., La Poutre, H.: Error-backpropagation in temporally encoded networks of spiking neurons. Neurocomputing **48**, 17–37 (2002)
15. Caporale, N., Dan, Y.: Spike timing-dependent plasticity: a Hebbian learning rule. Annu. Rev. Neurosci. **31**, 25–46 (2008)
16. Sporea, I., Grüning, A.: Supervised learning in multilayer spiking neural networks. Neural Comput. **25**, 473–509 (2013)
17. Pyle, R., Rosenbaum, R.: Spatiotemporal dynamics and reliable computations in recurrent spiking neural networks. Phys. Rev. Lett. **118**(1), 018103 (2017)
18. Gardner, B., Sporea, I., Grüning, A.: Learning spatiotemporally encoded pattern transformations in structured spiking neural networks. Neural Comput. **27**, 2548–2586 (2015)
19. Nadasdy, Z.: Information encoding and reconstruction from the phase of action potentials. Front. Syst. Neurosci. **3**, 6 (2009)
20. Olshausen, B.A., Field, D.J.: Sparse coding with an overcomplete basis set: a strategy employed by V1? Vis. Res. **37**, 3311–3325 (1997)
21. Rossum, M.V.: A novel spike distance. Neural Comput. **13**, 751–763 (2001)
22. Kingma, D.P., Ba, J.: Adam: a method for stochastic optimization. Computer Science (2014)
23. Xu, X., Jin, X., Yan, R., et al.: Visual pattern recognition using enhanced visual features and PSD-based learning rule. IEEE Trans. Cogn. Dev. Syst. **10**, 205–212 (2017)

Skewed and Long-Tailed Distributions of Spiking Activity in Coupled Network Modules with Log-Normal Synaptic Weight Distribution

Sou Nobukawa[1](✉), Haruhiko Nishimura[2], and Teruya Yamanishi[3]

[1] Department of Computer Science, Chiba Institute of Technology,
2-17-1 Tsudanuma, Narashino, Chiba 275-0016, Japan
nobukawa@cs.it-chiba.ac.jp
[2] Graduate School of Applied Informatics, University of Hyogo,
7-1-28 Chuo-ku, Kobe, Hyogo 650-8588, Japan
[3] Department of Management Information Science, Fukui University of Technology,
3-6-1 Gakuen, Fukui, Fukui 910-8505, Japan

Abstract. Recent studies with neuroimaging modalities have been elucidating a structure of a whole network of the brain and its functional activity. The characteristics of various functional neural activities and network structures exhibit skewed and long-tailed distributions. However, it remains unclear how heavy-tailed structural distribution affects functional distribution. In this study, we constructed spiking neural networks composed of two modules with excitatory post-synaptic potential (EPSP) following log-normal distribution. Through the evaluation of multi-scale entropy analysis and its surrogate data analysis, we reveal that the long-tailed synaptic weight distribution enhances the complexity of spiking activity at large temporal scales and emerges non-linear dynamics. Furthermore, we compared distribution of residence time in each spiking pattern between cases with/without large EPSPs. The results show that strong synapses are crucial in the heavy-tailed distribution of residence time.

Keywords: Spiking neural network · Log-normal distribution
Pattern alternation

1 Introduction

Recent studies with neuroimaging modalities have been elucidating a structure of a whole network of the brain, which is composed of individual regions and connections, called connectome [10,11]. The brain network possesses feedback loops at multiple hierarchical [9] and complex network characteristics, such as a high degree, high clustering, short path length, and high centrality [8,12–15,29]. Recent findings from network studies of the human brain indicate that brain

© Springer Nature Switzerland AG 2018
L. Cheng et al. (Eds.): ICONIP 2018, LNCS 11301, pp. 535–544, 2018.
https://doi.org/10.1007/978-3-030-04167-0_48

function emerges from its topological features in a whole brain network, not merely divided individual regions [1, 15, 31].

In neural systems, characteristics of various kinds of network structure and neural activity exhibit skewed and long-tailed distributions [5, 6]. In the structural connectivity, synaptic connections of the cerebral cortex, the spikes of presynaptic neurons increase the membrane potential of post-synaptic neurons; this is known as excitatory post-synaptic potential (EPSP). Song *et al.* and Lefort *et al.* showed that EPSPs in most synapses exhibit sub-mV potential, while a small number of synapses exhibit large EPSPs ($\gtrsim$1.0 [mV]) [19, 30] (review in [6]). They reported that this distribution of EPSP can be fitted by a log-normal distribution. In the neural activity, the occurrence probabilities of spiking rate [2, 16, 17, 22, 25, 26, 28] and size of synchronous spiking population [22, 23] exhibit log-normal distribution (reviewed in [6]). In visual perception, at the level of the cognitive processes, when an ambiguous figure is presented that has two different interpretations or two different stimuli, the period of dominant perception, which involves unilateral interpretation of stimuli, follows a heavy-tailed, unimodal, maximum peak distribution, such as a gamma or log-normal distribution [4, 20, 21, 33] (reviewed in [3]). However, it remains unclear how heavy-tailed structural distribution affects functional distribution.

To answer this question, in this study, we constructed spiking neural networks composed of two modules with EPSPs following log-normal distribution. We compared the spiking activity and spiking pattern alternation between cases with/without large EPSPs.

2 Materials and Methods

2.1 Spiking Neural Network

As shown in Fig. 1, we use two coupled spiking neural modules composed of excitatory and inhibitory neurons. The excitatory-to-excitatory connections in the internal modules have the synaptic weights following log-normal synaptic weight distribution. The excitatory-to-excitatory and excitatory-to-inhibitory connections join two modules. We assume that the spiking activity of each module suppresses the other's spiking activity under the condition for adequately large strength of excitatory-to-inhibitory connections.

Terame *et al.* proposed the spiking neural network exhibiting spontaneous activity by the log-normal distribution of EPSP [32]. In this study, we used this spiking neural network. In each neuron of this network, the membrane potential $v(t)$ is described by the conductance, based on a leaky-integrate-and-fire neuron model:

$$\frac{dv}{dt} = -\frac{1}{\tau_m}(v - V_L) - g_E(v - V_E) - g_I(v - V_I) + I_{\text{ex}}, \tag{1}$$

$$\text{if } v \geq V_{\text{thr}} \text{ [mV], then } v(t) \rightarrow V_r, \tag{2}$$

where τ_m, and V_E, V_I, and V_L are the membrane decay constant, and the reversal potentials of the α-amino-3-hydroxy-5-methyl-4-isoxazolepropionic acid

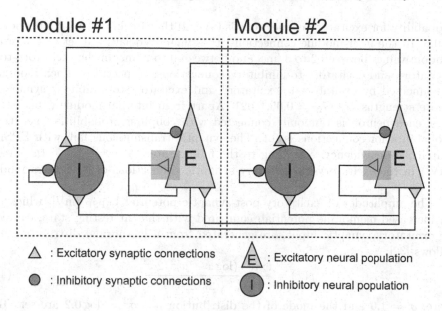

Fig. 1. Topology of spiking neural network. In the red shaded synaptic connections, excitatory post-synaptic potential follows log-normal distribution. (Color figure online)

(AMPA)-receptor-mediated excitatory synaptic current, inhibitory synaptic current, and leak current, respectively. I_{ex} is an external input given by $21 \cdot \delta(t - t_{ex})$ [mV], where input time t_{ex} is drawn from a Poisson process with the input rate Λ_j [Hz] (suffix indicates the module number $j = 1, 2$), for the trigger to produce the spiking activity. The excitatory/inhibitory synaptic conductances $g_E(t)/g_I(t)$ [ms^{-1}] are given by

$$\frac{dg_X}{dt} = -\frac{g_X}{\tau_s} + \sum_j G_{X,j} \sum_{s_j} \delta(t - s_j - d_j), \quad X = E, I, \qquad (3)$$

where, τ_s is the decay constant of the excitatory and inhibitory synaptic conductances. s_j, d_j, $G_{E,j}$, and $G_{I,j}$ are spike times of synaptic input from j-th neuron, synaptic delays, and weights of excitatory and inhibitory synapses, respectively. As the parameter sets, we used $V_I = -80$ [mV], $V_L = -70$ [mV], $V_r = -60$ [mV], $V_{thr} = -50$ [mV], $V_E = 0$ [mV], $\tau_m = 20$ [ms] (excitatory neuron), $\tau_m = 10$ [ms] (inhibitory neuron), and $\tau_s = 2$ [ms]. In this study, Eq. (3) is solved by the Euler method, with the size of time step $\Delta t = 0.1$ [ms]. The refractory period is set to 1 [ms]. In excitatory-to-excitatory connections and other internal module connections, the synaptic delays are set to uniform random values between 1 to 3 [ms] and between 0 to 2 [ms], respectively. The synaptic weights for excitatory-to-inhibitory, inhibitory-to-excitatory, and inhibitory-to-inhibitory connections are set to constant values of $0.018, 0.0019$, and 0.0025, respectively. The size of the module is $N_E = 5000$ for excitatory neurons and $N_I = 1000$ for inhibitory neurons. Each neuron is randomly connected with coupling probabilities; briefly, the probability for excitatory connections is 0.1 and that for inhibitory connections is 0.5. In the inter-module connections, the synaptic delays are set to uniform random values between 2 to 3 [ms] and between 0 to 2 [ms] in the excitatory-to-excitatory and excitatory-to-inhibitory connections, respectively. Each module is connected by excitatory-to-excitatory and excitatory-to-inhibitory synapses whose strengths are $G_E = 0.05, 0.021$. Apart from internal module connectivities, each neuron is randomly connected with coupling probabilities, i.e., the probability for connections is 0.1. The synaptic transmissions fail, with EPSP amplitude dependency, according to the failing rate: $P_E = \frac{a}{a+V_{EPSP}}$ ($a = 0.1$ [mV]) in the excitatory-to-excitatory synaptic connections of inter/intra modules.

The amplitudes of excitatory post-synaptic potential V_{EPSP} [mV], which is an increased membrane potential, compared with that at resting state, caused by excitatory synaptic input, are generated from a log-normal distribution as follows [32]:

$$p(x) = \frac{\exp[-(\log x - \mu)^2 / 2\sigma^2]}{\sqrt{2\pi}\sigma x}. \tag{4}$$

Here, $\sigma = 1.0$ and the mode of the distribution $\mu - \sigma^2 = \log 0.2$ are set. To remove unrealistic values of V_{EPSP} that exceed 14 [mV], a new value is drawn from the distribution. We use this EPSP distribution in the internal module excitatory-to-excitatory connections. We compare the spiking activity between the case with strong synaptic weights ($V_{EPSP} > 2$ [mV]) and the case eliminating them, to confirm the effect of long-tail distribution.

To translate V_{EPSP} as an observable value into synaptic weight G_E, we derive the relationship between V_{EPSP} and G_E as follows:

$$G_E = V_{EPSP}/100. \tag{5}$$

2.2 Evaluation Indexes

Spiking Rate. To observe spiking activity, we used spiking rates in the excitatory neural population of #j module r_E^j [Hz] and the inhibitory neural population r_I^j [Hz] as follows:

$$r_X^j(t) = 1000 \frac{S_X^j(t)}{\Delta t \, N_X} \quad X = E, I, \tag{6}$$

where, S_E^j/S_I^j indicate the frequency of spikes in the bin whose width is 0.1 [ms] in excitatory/inhibitory neural populations. Using these spiking rates, we measured the maintenance period of $r_E^1 > r_E^2$ or $r_E^2 > r_E^1$ as residence time.

Multi-scale Entropy. To evaluate the dependency of the complexity in the time series of r_E^j on temporal scale, we used multi-scale entropy (MSE) [7]. A sample entropy (SampEn) for stochastic variable $\{x_1, x_2, \cdots x_N\}$ is defined by

$$h(r, m) = -\log \frac{C_{m+1}(r)}{C_m(r)}, \tag{7}$$

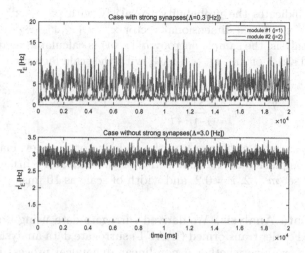

Fig. 2. Time-series of spiking rate r_E^j in the cases with and without strong synapses

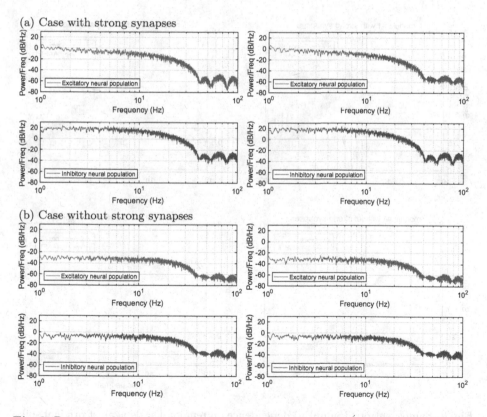

Fig. 3. Power spectrum density for time-series of spiking rate $r_{E,I}^j$ in modules #1 (left part) and #2 (right part)

where $C_m(r)$ indicates the probability to satisfy with $|\mathbf{x}_i^m - \mathbf{x}_j^m| < r$ $(i \neq j,$ $i, j = 1, 2, \cdots)$. $\mathbf{x}_i^m$ is m a dimensional vector $\mathbf{x}_i^m = \{x_i, x_{i+1}, \cdots, x_{i+m-1}\}$.

In MSE analysis, the sample entropy $h^\tau(r, m)$ is calculated against temporal coarse-grained series of $\{x_1, x_2, \cdots, x_N\}$, with the scale factor τ $(\tau = 1, 2, \cdots)$:

$$y_j^{(\tau)} = \frac{1}{\tau} \sum_{i=(j-1)\tau+1}^{j\tau} x_i, \qquad (1 \le j \le N/\tau) \qquad (8)$$

By the dependency of $h^\tau(r, m)$ on the scale factor τ, we evaluated the characteristic of complexity in the time series of the excitatory firing rates r_E^j. In this study, we set $m = 2$, $r = 0.2$, and width of scale as 10 [ms].

Surrogate Data Analysis. We derived surrogate data using iterative amplitude adjusted Fourier transformed (IAAFT) surrogate data analysis for the spiking rate r_E^j to examine whether a non-linear dynamical process is involved in the spiking rates [27]. In this case, iteration number is set to 30.

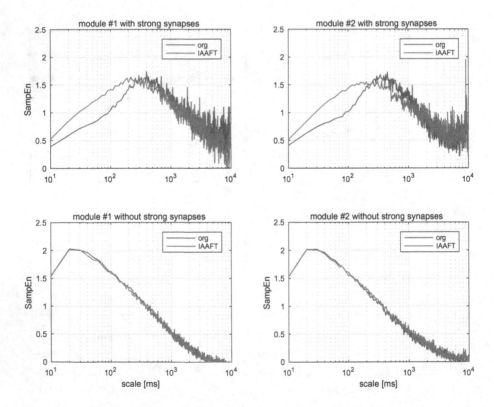

Fig. 4. Sample entropy (SampEn) dependence on temporal scale against original spiking rate r_E^j and its iterative amplitude adjusted Fourier transformed (IAAFT) surrogate data. Upper/lower parts indicate the cases with/without strong synapses.

3   Results

We evaluated the characteristics of spiking activity in two coupled modules with and without strong synapses, respectively. Figures 2 and 3 show the time series of spiking rates of r_E^j and their power spectrum of spiking rates, respectively. In the case with strong synapses, the spiking rates exhibited irregular behavior transiting between $r_E^j \approx 1.0$ and $r_E^j \approx 12.0$ [Hz]. These transitions arise alternately in the modules #1 and #2. Conversely, in the case without strong synapses, the spiking rates exhibit $r_E \approx 3.0$ [Hz], and the explicit alternation was not observed. These power spectra approximately distribute to [1 : 40] [Hz].

Figure 4 shows the dependencies of SampEn on temporal scale. In the case with strong synapses, the profiles of SampEn for both modules exhibit a unimodal maximum peak at approximately 300 [ms]. In comparison with the MSE profile of IAAFT surrogate data, the original SampEn is lower in [10 : 200] [ms]. Conversely, in the case without strong synapses, the peaks are located at shorter periods ($\approx$20 [ms]), and there is no difference between the MSE profiles for the original data and the IAAFT surrogate data. That is, SampEn in the case with strong synapses reflects the inherent non-linear dynamics in the spiking neural networks.

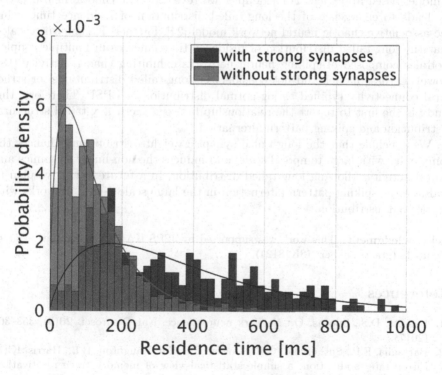

Fig. 5. Residence time distribution in the cases with/without strong synapses

The distribution of residence time in cases with/without strong synapses is represented in Fig. 5. In a case with strong synapses, the peak residence time was located at approximately 200 [ms], and residence time was distributed until 1000 [ms]. Conversely, in a case without strong synapses, the peak residence time was located at approximately 10 [ms], and the distribution of residence time converged to zero at approximately 500 [ms].

4   Discussion and Conclusion

In this study, we constructed two coupled modules of spiking neural networks with excitatory and inhibitory neural populations. Through MSE analysis, we revealed that the long-tailed synaptic weight distribution enhances the complexity of spiking activity at larger temporal scales. Furthermore, the surrogate analysis of spiking activity in cases with/without strong synapses revealed that strong synapses located at the long-tailed parts induce non-linear neural dynamics. Further, through the comparison with residence time between cases with/without strong synapses, we revealed that strong synapses play a crucial role in the emergence of the heavy-tailed distribution of residence time.

So far, the pattern alternation of neural activity has been investigated using a model-based approach. For example, we revealed that chaotic neural activity leads to emergence of the long-tailed distribution of residence time using the associative chaotic neural network model [24]. Further, Kanamaru revealed that the long-tailed distribution of residence time emerges in multiple coupled modules composed of spiking neural networks exhibiting chaotic activity [18]. However, these models did not consider the long-tailed distributions for structural connectivity typified as log-normal distribution of EPSP. Therefore, this study is the first to report the relationship between the heavy-tailed structural distribution and spiking pattern alternation.

We conclude that the long-tailed synaptic weight distribution enhances the complexity with large temporal scale, and induces the non-linear dynamics and neural activity following long-tailed distribution. In a future study, we plan to evaluate the spiking pattern alternation in the large scale neural networks with log-normal distribution.

Acknowledgment. This work was supported by JSPS KAKENHI for Early-Career Scientists (grant number: 18K18124).

References

1. Bassett, D.S., Sporns, O.: Network neuroscience. Nat. Neurosci. **20**(3), 353–364 (2017)
2. Battaglia, F.P., Sutherland, G.R., Cowen, S.L., Mc Naughton, B.L., Harris, K.D.: Firing rate modulation: a simple statistical view of memory trace reactivation. Neural Netw. **18**(9), 1280–1291 (2005)
3. Blake, R., Logothetis, N.K.: Visual competition. Nat. Rev. Neurosci. **3**(1), 13 (2002)

4. Borsellino, A., De Marco, A., Allazetta, A., Rinesi, S., Bartolini, B.: Reversal time distribution in the perception of visual ambiguous stimuli. Kybernetik **10**(3), 139–144 (1972)
5. Bullmore, E., Sporns, O.: Complex brain networks: graph theoretical analysis of structural and functional systems. Nat. Rev. Neurosci. **10**(3), 186–198 (2009)
6. Buzsáki, G., Mizuseki, K.: The log-dynamic brain: how skewed distributions affect network operations. Nat. Rev. Neurosci. **15**(4), 264–278 (2014)
7. Costa, M., Goldberger, A.L., Peng, C.K.: Multiscale entropy analysis of complex physiologic time series. Phys. Rev. Lett. **89**(6), 068102 (2002)
8. Eguiluz, V.M., Chialvo, D.R., Cecchi, G.A., Baliki, M., Apkarian, A.V.: Scale-free brain functional networks. Phys. Rev. Lett. **94**(1), 018102 (2005)
9. Fell, J., Kaplan, A., Darkhovsky, B., Röschke, J.: EEG analysis with nonlinear deterministic and stochastic methods: a combined strategy. Acta Neurobiol. Exp. **60**(1), 87–108 (1999)
10. Glasser, M.F., et al.: A multi-modal parcellation of human cerebral cortex. Nature **536**(7615), 171–178 (2016)
11. Glasser, M.F., et al.: The human connectome project's neuroimaging approach. Nat. Neurosci. **19**(9), 1175–1187 (2016)
12. Hagmann, P., et al.: Mapping the structural core of human cerebral cortex. PLoS Biol. **6**(7), e159 (2008)
13. Hagmann, P., et al.: Mapping human whole-brain structural networks with diffusion MRI. PloS One **2**(7), e597 (2007)
14. van den Heuvel, M., Mandl, R., Luigjes, J., Pol, H.H.: Microstructural organization of the cingulum tract and the level of default mode functional connectivity. J. Neurosci. **28**(43), 10844–10851 (2008)
15. van den Heuvel, M.P., Sporns, O.: Network hubs in the human brain. Trends Cogn. Sci. **17**(12), 683–696 (2013)
16. Hirase, H., Leinekugel, X., Czurkó, A., Csicsvari, J., Buzsáki, G.: Firing rates of hippocampal neurons are preserved during subsequent sleep episodes and modified by novel awake experience. Proc. Natl. Acad. Sci. **98**(16), 9386–9390 (2001)
17. Hromádka, T., DeWeese, M.R., Zador, A.M.: Sparse representation of sounds in the unanesthetized auditory cortex. PLoS Biol. **6**(1), e16 (2008)
18. Kanamaru, T.: Chaotic pattern alternations can reproduce properties of dominance durations in multistable perception. Neural Comput. **29**(6), 1696–1720 (2017)
19. Lefort, S., Tomm, C., Sarria, J.C.F., Petersen, C.C.: The excitatory neuronal network of the C2 barrel column in mouse primary somatosensory cortex. Neuron **61**(2), 301–316 (2009)
20. Lehky, S.R.: Binocular rivalry is not chaotic. Proc. R. Soc. Lond. B **259**(1354), 71–76 (1995)
21. Levelt, W.J.: Note on the distribution of dominance times in binocular rivalry. Br. J. Psychol. **58**(1–2), 143–145 (1967)
22. Mizuseki, K., Buzsáki, G.: Preconfigured, skewed distribution of firing rates in the hippocampus and entorhinal cortex. Cell Rep. **4**(5), 1010–1021 (2013)
23. Mizuseki, K., Buzsaki, G.: Theta oscillations decrease spike synchrony in the hippocampus and entorhinal cortex. Phil. Trans. R. Soc. B **369**(1635), 20120530 (2014)
24. Nagao, N., Nishimura, H., Matsui, N.: A neural chaos model of multistable perception. Neural Process. Lett. **12**(3), 267–276 (2000)
25. O'Connor, D.H., Peron, S.P., Huber, D., Svoboda, K.: Neural activity in barrel cortex underlying vibrissa-based object localization in mice. Neuron **67**(6), 1048–1061 (2010)

26. Peyrache, A., et al.: Spatiotemporal dynamics of neocortical excitation and inhibition during human sleep. Proc. Natl. Acad. Sci. **109**(5), 1731–1736 (2012)
27. Schreiber, T., Schmitz, A.: Improved surrogate data for nonlinearity tests. Phys. Rev. Lett. **77**(4), 635 (1996)
28. Shafi, M., Zhou, Y., Quintana, J., Chow, C., Fuster, J., Bodner, M.: Variability in neuronal activity in primate cortex during working memory tasks. Neuroscience **146**(3), 1082–1108 (2007)
29. She, Q., Chen, G., Chan, R.H.: Evaluating the small-world-ness of a sampled network: functional connectivity of entorhinal-hippocampal circuitry. Sci. Rep. **6**, 21468 (2016)
30. Song, S., Sjöström, P.J., Reigl, M., Nelson, S., Chklovskii, D.B.: Highly nonrandom features of synaptic connectivity in local cortical circuits. PLoS Biol. **3**(3), e68 (2005)
31. Sporns, O.: Contributions and challenges for network models in cognitive neuroscience. Nat. Neurosci. **17**(5), 652–660 (2014)
32. Teramae, J.N., Tsubo, Y., Fukai, T.: Optimal spike-based communication in excitable networks with strong-sparse and weak-dense links. Sci. Rep. **2**, 485 (2012)
33. Walker, P.: Stochastic properties of binocular rivalry alternations. Percept. Psychophys. **18**(6), 467–473 (1975)

Efficient Multi-spike Learning
with Tempotron-Like LTP and PSD-Like
LTD

Qiang Yu(✉), Longbiao Wang, and Jianwu Dang(✉)

Tianjin Key Laboratory of Cognitive Computing and Application,
School of Computer Science and Technology, Tianjin University, Tianjin, China
{yuqiang,longbiao_wang}@tju.edu.cn, jdang@jaist.ac.jp

Abstract. Biological neurons use electrical pulses to transmit and process information in a significantly efficient way. To understand the mysteries of the underlying processing principles of the biological nervous systems, spiking neurons have been proposed to process information in a brain-like way. However, how could neurons learn spikes in an efficient way still remains challenging. In this study, we propose a simple and efficient multi-spike learning rule which could train neurons to associate input spike patterns with different output spike numbers. Our learning algorithm adopts a Tempotron-like LTP and a PSD-like LTD to adapt neuron's efficacies. The results show that the proposed rule is faster than other benchmarks for the given task. A fast running time and simple implementation can largely benefit applied developments in neuromorphic systems. Additionally, we show that neurons with our proposed rule can elicit different output spike numbers in response to input spike patterns. Thus, single neurons are capable of performing the challenging task of multi-category classifications.

Keywords: Spiking neural networks (SNNs) · Multi-spike learning
Neuromorphic computing · Multi-category classification

1 Introduction

Artificial intelligence has been thriving in recent years due to a class of successful techniques called deep learning [1]. The performance of such techniques on certain cognitive tasks such as recognition with respect to accuracy can even outperform human brain. However, deep learning is resource consuming and data hungry as compared to the high efficiency of the brain that can operate with one-shot learning. This gap could be resulted from the way how information is processed and represented in these two systems. In artificial neural networks, information is coded with real values, while spikes are essentially important for information computation and transmission in central nervous systems. In order to reveal the information processing mechanisms of the brain, different spiking models have been introduced to operate on an additional time dimension as

© Springer Nature Switzerland AG 2018
L. Cheng et al. (Eds.): ICONIP 2018, LNCS 11301, pp. 545–554, 2018.
https://doi.org/10.1007/978-3-030-04167-0_49

compared to the traditional neuron models used in deep learning. This makes the spiking neurons more biologically plausible than their counterparts in the traditional neural networks, but more complicated considering the time evolving dynamics. As a result, how to train these neurons to learn spikes still remains challenging.

A train of spikes may convey information according to different schemes such as rate or temporal code [2–6]. In a rate code, spikes are assumed to carray information by their count within a certain time window, while a temporal code postulates that each spike time plays an essential role in information transmission. Different coding schemes have different advantages as well as disadvantages [2,4]. The rate code is more resistant to noise as it utilizes an average over time to code information. However, when the coding efficiency matters, the temporal code is more favorable as it uses fewer number of spikes to transmit information. The sparseness of this temporal code can largely benefit rapid feedforward computation which has been observed in the visual systems [7,8]. Efficiency is one of the major focuses of the neuromorphic computing which could potentially overcome the bottlenecks deep learning currently encounters [9,10]. Therefore, it is of great interest to develop efficient neuromorphic learning algorithms for spiking neurons to process information.

There are various learning rules have been proposed to train spiking neurons to learn precise-timing spike patterns. The tempotron rule, for example, is an efficient rule to train neurons to separate two classes of patterns by firing or not [11]. A single tempotron neuron can only be used for binary classifications unless additional mechanisms are applied in the readout , such as grouping [12,13], to extend its capability to multi-category classifications. Additionally, the binary output of the tempotron neuron constrains its ability to fully utilize the temporal structure of its output. Some other rules have been proposed to train neurons to elicit precise-timing spikes [14–17]. For example, a single neuron with the precise-spike-driven (PSD) learning algorithm can separate multiple categories by assigning different desired times to each category [18]. However, how to design precise timings of the supervisory signal for an optimal performance still remains unclear. This requirement for generating desired teacher spike timings has been discarded in a new family of multi-spike learning rules recently proposed [19–21]. It is shown that neurons with these rules can elicit a desired number of spikes by adapting their critical thresholds [19]. The convergence and robustness of this kind of learning have been proven, and importantly, neurons with this rule can be applied to both rate and temporal spike patterns [20]. Although simplifications have been introduced to improve the performance of this multi-spike learning [20], further efforts are still favorable to make it more efficient such that being more suitable for neuromorphic implementations.

In this study, an improved multi-spike learning algorithm with a high computational efficiency is proposed to associate a spike pattern with a target output spike number rather than their precise timings. Our learning algorithm is developed by combining the advantages of both tempotron [11] and PSD [17] learning rules. We evaluate the learning efficiency of the proposed rule, and show its capability to train single spiking neurons to perform the task of multi-category recognition.

2 Methods

The neuron model used in this paper as well as its dynamics are presented in this section. Then the multi-spike learning algorithm is proposed by combining both the tempotron and PSD rules.

2.1 Neuron Model

The current-based leaky integrate-and-fire spiking neuron model was adopted in our study because of its analytical tractability and simplicity. Neuron continuously integrates input spikes into its membrane potential, and generates output spikes whenever its potential reaches a critical value. To be specific, its membrane potential dynamics, $V(t)$, is given as follows:

$$V(t) = \sum_{i=1}^{N} w_i \sum_{t_i^j < t} K(t - t_i^j) - \vartheta \sum_{t_s^j < t} \exp\left(-\frac{t - t_s^j}{\tau_m}\right). \tag{1}$$

where, t_i^j denotes the j-th spike of the i-th afferent, and t_s^j represents the j-th output spike time of the neuron. ϑ denotes the firing threshold. Every incoming spike will contribute a post-synaptic potential to the neuron. The shape of this contribution is determined by a normalized kernel K while its peak amplitude is controlled by synaptic weight w_i. The chosen K is defined as

$$K(t - t_i^j) = V_0 \left[\exp\left(-\frac{t - t_i^j}{\tau_m}\right) - \exp\left(-\frac{t - t_i^j}{\tau_s}\right) \right], \tag{2}$$

where, τ_s and τ_m are the time constants of the synaptic currents and the membrane potential, respectively. V_0 is a normalization factor that keeps the peak of K to be one. The kernel $K(t - t_i^j)$ is causal and as a result it vanishes for $t < t_i^j$.

Different from the tempotron neuron [11] which will shunt all the afferent input spikes after fire, the neuron model as described by Eq. 1 continuously integrates input spikes through synaptic efficacies. Whenever the neuron's membrane potential reaches the firing threshold, it will elicit an output spike followed by a reset dynamics (see Fig. 1). The output response of the neuron to input spike patterns is thus enriched by the dynamics of continuous integrating-and-firing, which could be significantly important for information transmission, processing and multi-category decision. Neurons with the reset dynamics are free to elicit as many spikes as necessary whenever a firing condition is reached.

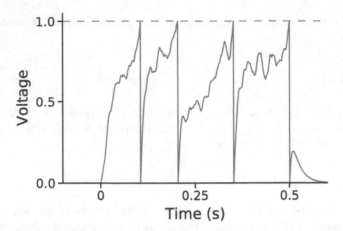

Fig. 1. Demonstration of the membrane potential dynamics of the neuron in response to an input spike pattern. Whenever its membrane potential reaches its firing threshold (dashed horizontal line), the neuron will fire an output spike, followed by a reset dynamics.

2.2   Multi-spike Learning Algorithm

The firing threshold ϑ of a multi-spike neuron can greatly affect its output response to the input spike pattern. A larger ϑ will normally result in a fewer number of output spikes, while a smaller one would make the neuron firing more spikes. The spike-threshold-surface (STS) [19] is introduced to characterize the relation between the output spike number n_{out} and the firing threshold ϑ of the neuron (see Fig. 2). n_{out} normally decreases monotonically with increasing threshold. In another word, a greater threshold would lead to a lower output spike number n_{out}.

STS is characterized by several critical threshold values, ϑ_k^* ($k \in \mathbb{Z}^+$), and n_{out} jumps from $k-1$ to k at every normal critical point (see [20] for more details). If ϑ is greater than the global maximum of membrane potential under the super-high threshold case, i.e. $\vartheta > V_{\mathrm{max}}$, the neuron will never fire an output spike.

The training target for a neuron to elicit a specific number of output spikes can therefore be accomplished by adjusting STS such that a given firing threshold can lead to a desired spike response. The rationale for the training is that long-term potentiation (LTP) and long-term depression (LTD) should operate together to satisfy $\vartheta_{d+1}^* < \vartheta < \vartheta_d^*$ with d representing the desired output spike number.

There could be numerous ways to perform this multi-spike learning task with the above-mentioned rationale. A learning algorithm can be developed by evaluating the gradients of ϑ^* with respect to w [19]. This can be done because each ϑ^* is a function of the weights as reflected by Eq. (1) and thus differentiable with respect to them. In [20], an alternative gradient evaluation was introduced to

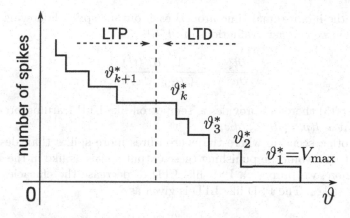

Fig. 2. Demonstration of spike-threshold-surface (STS) which shows the relation between the output spike number n_{out} and the threshold ϑ. This curve is characterized by a number of critical threshold values ϑ_k^*. The location of ϑ with respect to these critical values will determine the neuron's output spike number in response to the input pattern. The LTD process should decrease those critical values that are greater than ϑ, while the LTP increase those that are less than ϑ.

improve the learning efficiency. These approaches with gradient descent could result in a fast adaption of STS, but the gradient evaluation of ϑ^* with respect to $\boldsymbol{w}$ would make the computation complex. Therefore, we seek alternative ways to further simplify the learning to make it more efficient and suitable for neuromorphic computing.

The supervised learning process normally starts whenever there is a difference between the actual and the desired output response. Under the mismatching cases, the actual output spike number o could be either greater or less than the desired one d. If $o < d$, the neuron should increase its efficacies, a process named long-term potentiation (LTP), such that it would fire more spikes. Conversely, a long-term depression (LTD) should occur to decrease the neuron's efficacies when $o > d$.

If the neuron needs to increase o, the most direct way is to enhance its local maximum potential as this point is more closer to firing threshold. We denote this process as Tempotron-like LTP since a similar approach is used in the tempotron rule [11]. In order to derive the learning, we define the maximum of all the peaks of local-continuous sub-threshold voltage as $v_{\text{max}} = V(t_{\text{max}})$ with t_{max} denoting the time of this maximum. The gradients of v_{max} with respect to w_i could thus be evaluated as

$$\frac{dv_{\text{max}}}{dw_i} = \frac{\partial V(t_{\text{max}})}{\partial w_i} + \sum_{j=1}^{m} \frac{\partial V(t_{\text{max}})}{\partial t_{\text{s}}^j} \frac{\partial t_{\text{s}}^j}{\partial w_i}, \tag{3}$$

where, m denotes the total output spike number before t_{max}. Based on the linear threshold crossing assumption [16,22], the internal state of a neuron increases linearly in the infinitesimal time around each output spike. Following the processes in [20], we can get evaluations of $\partial t_s^j/\partial w_i$ as

$$\frac{\partial t_s^j}{\partial w_i} = -\frac{1}{\dot{V}(t_s^j)} \frac{\partial V(t_s^j)}{\partial w_i} . \tag{4}$$

Equation (3) therefore provides a Tempotron-like LTP learning process, and we denote it as $ltp_i = dv_{\max}/dw_i$.

In the other scenario where the neuron fires more spikes than desired, the most direct way would be punishing these output spikes as like in the PSD rule [17]. Therefore, we propose a PSD-like LTD to decrease the efficacies at all of the output spikes. The PSD-like LTD is given as

$$ltd_i = \frac{1}{o} \cdot \sum_{j=1}^{o} \frac{\partial V(t_s^j)}{w_i} . \tag{5}$$

Here, the change amount is normalized by the number of actual output spikes o to balance the contributions of LTP and LTD.

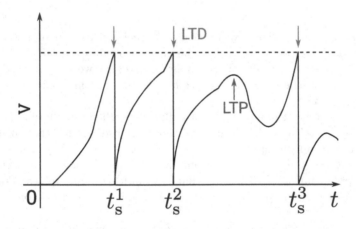

Fig. 3. Demonstration of the synaptic plasticity for multi-spike learning.

Thus, our multi-spike learning rule can be summarized as

$$\Delta w = \begin{cases} -\eta \cdot ltd_i & \text{if } o > d \\ \eta \cdot ltp_i & \text{if } o < d \end{cases} , \tag{6}$$

Here, $\eta > 0$ represents the learning rate that determines the changing size for each synaptic adaptation. Figure 3 illustrates the scheme of our multi-spike learning algorithm. The Tempotron-like LTP is used to increase the output spike number, while the PSD-like LTD decreases it when the neuron spikes more than the target.

3   Experimental Results

In this section, we show our experimental results. First, the learning efficiency of our proposed learning rule is presented, and then we show its capability for training single spiking neurons to perform the task of multi-category recognition.

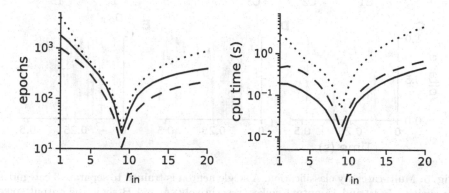

Fig. 4. Convergent speed of different multi-spike learning algorithms versus the afferent firing rate r_{in}. Dashed and dotted lines denote learning algorithms from [20] and [19], respectively. The solid line shows the performance of the proposed learning rule. The learning performance are evaluated with both epochs (Left panel) and cpu time (Right). Simulations were performed on a platform of Intel E5670@2.93 GHz, and data were averaged over 1000 runs.

Following the setups in [20], we set the default experimental parameters as $N = 500$, $\tau_m = 20$ ms, $\tau_s = 5$ ms, $\eta = 10^{-4}$. We initialize the weights with a random Gaussian distribution where both the mean and standard deviation are set as 0.01. Each input spike pattern is generated with every afferent having a Poisson firing rate of r_{in} over a given time window which we define as $T = 500$ ms in this study.

In the first experiment, the neuron is trained to fire a desired spike number of 20 in response to patterns with different afferent firing rate r_{in}. Both the epochs and cpu times were recorded to show the learning speed for convergence. The learning algorithms from both [20] and [19] are also used to perform the same task for a better comparison. As can be seen from Fig. 4, our algorithm needs less epochs than [19], but more epochs than [20], to converge. The algorithm of [20] uses critical threshold values to update efficacies, which can result in a fast change of STS so as the epochs for convergence. Noteworthily, the overall cpu running time of our algorithm is faster than all the other two. This is because that our algorithm takes the efficiency advantage of both the Tempotron and PSD rules, and discards the directive evaluations of critical thresholds with respect to efficacies. A fast running time and simple implementation can largely benefit applied developments in neuromorphic systems.

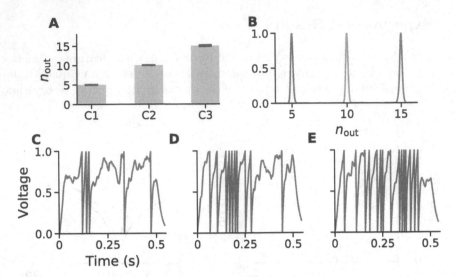

Fig. 5. Multi-category classification. A single neuron is trained to separate 3 categories by eliciting 5, 10 and 15 output spikes, respectively. **A** and **B** show the output spike distribution of the neuron in response to different categories. **C-E** demonstrate membrane potential dynamics of the neuron in response to a sample spike pattern drawn from each category.

In the second experiment, the ability of our proposed learning algorithm to perform a multi-category classification is investigated. In the task, we construct 3 categories by generating a fixed template spike pattern for each. Every new input spike pattern is constructed by adding a 3-ms jitter noise to spikes of its template pattern. Finally, neurons are trained to fire desired output spike number as: 5 (for C1), 10 (C2) and 15 (C3).

Figure 5 shows that our proposed learning algorithm can be successfully applied to train neurons to classify different spike patterns with a different desired output spike number. It has been shown that single spiking neurons with the multi-spike dynamics are capable of performing the multi-category classification task by associating different output spike numbers to each category.

4   Conclusion

In this study, we developed an efficient and simple multi-spike learning algorithm which can be applied to train spiking neurons to fire desired output spike numbers rather than precise spike timings. Our learning algorithm take the efficiency advantage of both the Tempotron and the PSD rule to adapt neuron's efficacies with a Tempotron-like LTP and PSD-like LTD. The results show that our learning rule is faster than its alternatives. Additionally, we show that single neurons with the proposed algorithm are capable of performing the challenging multi-category classification task. In our future works, we will continue to examine the performance of our multi-spike learning algorithm in some practical application tasks.

Acknowledgments. This work was supported by the National Natural Science Foundation of China (No. 61806139, 61771333), and by the Natural Science Foundation of Tianjin (No. 18JCYBJC41700).

References

1. LeCun, Y., Bengio, Y., Hinton, G.: Deep learning. Nature **521**(7553), 436–444 (2015)
2. Gütig, R.: To spike, or when to spike? Curr. Opin. Neurobiol. **25**, 134–139 (2014)
3. Borst, A., Theunissen, F.E.: Information theory and neural coding. Nat. Neurosci. **2**(11), 947–957 (1999)
4. Brette, R.: Philosophy of the spike: rate-based vs. spike-based theories of the brain. Front. Syst. Neurosci. **9**, 151 (2015)
5. Panzeri, S., Brunel, N., Logothetis, N.K., Kayser, C.: Sensory neural codes using multiplexed temporal scales. Trends Neurosci. **33**(3), 111–120 (2010)
6. Yu, Q., Tang, H., Hu, J., Tan, K.C.: Neuromorphic Cognitive Systems: A Learning and Memory Centered Approach, vol. 126, 1st edn. Springer, Cham (2017). https://doi.org/10.1007/978-3-319-55310-8
7. Reinagel, P., Reid, R.C.: Temporal coding of visual information in the thalamus. J. Neurosci. **20**(14), 5392–5400 (2000)
8. Serre, T., Oliva, A., Poggio, T.: A feedforward architecture accounts for rapid categorization. Proc. Natl. Acad. Sci. **104**(15), 6424–6429 (2007)
9. Merolla, P.A., et al.: A million spiking-neuron integrated circuit with a scalable communication network and interface. Science **345**(6197), 668–673 (2014)
10. Yao, P., et al.: Face classification using electronic synapses. Nat. Commun. **8**, 15199 (2017)
11. Gütig, R., Sompolinsky, H.: The tempotron: a neuron that learns spike timing-based decisions. Nat. Neurosci. **9**(3), 420–428 (2006)
12. Yu, Q., Tang, H., Tan, K.C., Li, H.: Rapid feedforward computation by temporal encoding and learning with spiking neurons. IEEE Trans. Neural Netw. Learn. Syst. **24**(10), 1539–1552 (2013)
13. Yu, Q., Tang, H., Tan, K.C., Yu, H.: A brain-inspired spiking neural network model with temporal encoding and learning. Neurocomputing **138**, 3–13 (2014)
14. Ponulak, F., Kasinski, A.J.: Supervised learning in spiking neural networks with ReSuMe: sequence learning, classification, and spike shifting. Neural Comput. **22**(2), 467–510 (2010)
15. Florian, R.V.: The chronotron: a neuron that learns to fire temporally precise spike patterns. PLoS One **7**(8), e40233 (2012)
16. Bohte, S.M., Kok, J.N., La Poutré, J.A.: Error-backpropagation in temporally encoded networks of spiking neurons. Neurocomputing **48**(1–4), 17–37 (2002)
17. Yu, Q., Tang, H., Tan, K.C., Li, H.: Precise-spike-driven synaptic plasticity: learning hetero-association of spatiotemporal spike patterns. PLoS One **8**(11), e78318 (2013)
18. Yu, Q., Yan, R., Tang, H., Tan, K.C., Li, H.: A spiking neural network system for robust sequence recognition. IEEE Trans. Neural Netw. Learn. Syst. **27**(3), 621–635 (2016)

19. Gütig, R.: Spiking neurons can discover predictive features by aggregate-label learning. Science **351**(6277), aab4113 (2016)
20. Yu, Q., Li, H., Tan, K.C.: Spike timing or rate? Neurons learn to make decisions for both through threshold-driven plasticity. IEEE Trans. Cybern. 1–12 (2018). https://doi.org/10.1109/TCYB.2018.2821692
21. Yu, Q., Wang, L., Dang, J.: Neuronal classifier for both rate and timing-based spike patterns. In: Liu, D., Xie, S., Li, Y., Zhao, D., El-Alfy, E.S. (eds.) Neural Information Processing. ICONIP 2017, vol. 10639, pp. 759–766. Springer, Cham (2017). https://doi.org/10.1007/978-3-319-70136-3_80
22. Ghosh-Dastidar, S., Adeli, H.: A new supervised learning algorithm for multiple spiking neural networks with application in epilepsy and seizure detection. Neural Netw. **22**(10), 1419–1431 (2009)

A Ladder-Type Digital Spiking Neural Network

Hiroaki Uchida and Toshimichi Saito[(✉)]

Hosei University, Koganei, Tokyo 184-8584, Japan
tsaito@hosei.ac.jp

Abstract. This paper presents a ladder-type digital spiking neural network and its hardware implementation. Depending on the parameters, the network can exhibit multi-phase synchronization of periodic spike-trains. Applying a time dependent selection switching, the network can output a variety of periodic spike-trains consisting of any combination of desired inter-spike-intervals. The network is a digital dynamical system and is suitable for FPGA based hardware implementation. A test circuit is implemented in a FPGA board by the Verilog and typical phenomena are confirmed experimentally. These results will be developed into several applications including time-series approximation/prediction.

Keywords: Digital spiking neurons · Multi-phase synchronization
FPGA

1 Introduction

Spiking neural networks are analog dynamical systems and can generates various spike-trains. The spike-trains are important to consider information processing function [1–3]. Also, the spike-trains play important roles in engineering applications including image processing, spike-based communication, and central pattern generators [4–7]. Analysis and synthesis of spiking neural networks are important from both fundamental and application viewpoints.

This paper presents a ladder-type digital spiking neural network (LDSN) consisting of digital spiking neurons (DSNs). Repeating integrate-and-fire behavior between periodic base signal and constant threshold, the DSN can output various periodic spike-trains (PSTs) [8–12]. The dynamics of DSN can be visualized by a digital spike map (Dmap) defined on a set of points. Applying spike-based cross-coupling, the body of LDSN is constructed. Depending on parameters, the LDSN can exhibit multi-phase synchronization of various PSTs. Applying a time dependent selection switching, the LDSN can output a variety of PSTs consisting of any combination of desired inter-spike-intervals. Such PSTs are desired in spike-based time series approximation/prediction: the LDSN can be developed into a digital reservoir computing system. Analog reservoir computing systems have remarkable performance in time series approximation/prediction [13,14].

© Springer Nature Switzerland AG 2018
L. Cheng et al. (Eds.): ICONIP 2018, LNCS 11301, pp. 555–562, 2018.
https://doi.org/10.1007/978-3-030-04167-0_50

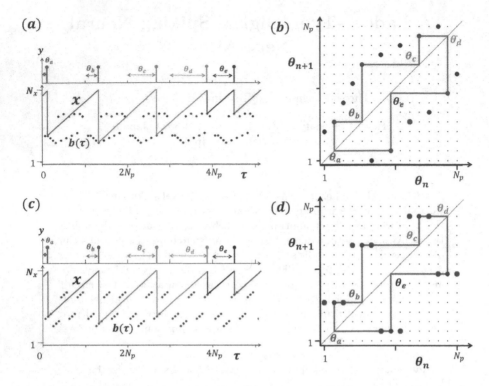

Fig. 1. DSN. (a) PST with period $5N_p$. (b) Dmap and PEO with period 5 for $d = (4, 5, 7, 9, 11, 1, 12, 2, 13, 5, 14, 6, 15, 8, 10)$. (c) Super-stable PST. (d) Dmap and super-stable PEO with period 5 for $d = (5, 5, 5, 11, 11, 11, 2, 2, 2, 14, 14, 14, 8, 8, 8)$.

The LDSN is a digital dynamical system and can be regarded as a digital version of the analog spiking neural networks. The LDSN can be precisely analyzed with computer aid and is suitable for FPGA based hardware implementation. Using shift registers and several kinds of switches, the LDSN can be realized. Using the Verilog, a test circuit of the LDSN is implemented on a FPGA board and typical phenomena are confirmed experimentally.

2   Digital Spiking Neuron and Periodic Spike-Train

Here, we introduce the DSN [10,11], a building block of the LDSN. Let $x(\tau)$ denote a discrete state variable at discrete time τ. Repeating integrate-and-fire behavior between a periodic base signal $b(\tau)$ with period N_p and a constant threshold N_x, the DSN outputs a spike-train $y(\tau)$ as shown in Fig. 1(a):

$$
\begin{aligned}
&\text{Integrating:} \quad x(\tau+1) = x(\tau) + 1, \, y(\tau) = 0 \text{ if } x(\tau) < N_x \\
&\text{Self-firing:} \quad x(\tau+1) = b(\tau), \qquad y(\tau) = 1 \text{ if } x(\tau) = N_x
\end{aligned}
\tag{1}
$$

where $b(\tau)$ is a periodic base signal with period N_p and N_x is a threshold:

$$x(\tau) \in \{0, 1, \cdots, N_x\}, \ N_x \le 2N_p - 1.$$

For simplicity, we assume the following condition for the base signal

$$\tau - 2N_p + 1 \le b(\tau) - N_x \le \tau - N_p \text{ for } \tau \in \{0, \cdots, N_p - 1\}, \ b(\tau + N_p) = b(\tau). \quad (2)$$

In this case, one spike appears per unit interval of N_p points.

$$y(\tau) = \begin{cases} 1 & \text{for } \tau = \tau_n \\ 0 & \text{for } \tau \ne \tau_n \end{cases} \ \tau_n \in [(n-1)N_p + 1, nN_p]. \quad (3)$$

where τ_n denote the n-th spike-position. Let $\theta_n = (\tau_n \bmod N_p)$ denote the n-th spike-phase. A spike-position is given by $\tau_n = \theta_n + N_p(n-1)$ and a spike-train $y(\tau)$ is governed by the digital spike map (Dmap) as shown in Fig. 1(b):

$$\theta_{n+1} = F(\theta_n) \equiv \theta_n - b(\theta_n - 1) + (N_x - N_p), \ \theta_n \in \{1, \cdots, N_p\} \equiv L_{N_p} \quad (4)$$

The Dmap is represented by a characteristic vector of integers:

$$d \equiv (d_1, \cdots, d_{N_p}), \ F(i) = d_i, \ d_i \in \{1, \cdots, N_p\}, \ i \in \{1, \cdots, N_p\} \quad (5)$$

We define periodic orbit and its stability. A point $p \in L_{N_p}$ is said to be a periodic point with period k if $p = F^k(p)$ and $F(p)$ to $F^k(p)$ are all different where F^k is the k-fold composition of F. A sequence of the periodic points $\{F(p), \cdots, F^k(p)\}$ is said to be a periodic orbit (PEO). A PEO with period k is equivalent to a PST with period kN_p. A PEO (and a PST) is said to be super-stable if all the initial points fall directly into the PEO: $p = F(q)$ for either periodic point p of the PEO and all other points q.

Adjusting the base signal under condition (2), the Dmap (DSN) can generate desired PEOs (PSTs) and the PEOs can be super-stable as discussed in [12]. A Dmap with super-stable PEO is shown in Fig. 1(c) and (d).

3 Ladder-Type Digital Spiking Neural Network

Figure 2(a) illustrates the body of the LDSN: M pieces of DSNs are coupled in ladder topology with a common base signal. The dynamics is described by

Integrating: $x_i(\tau + 1) = x_i(\tau) + 1,$ $y_i(\tau) = 0$ if $x_i(\tau) < N_x$
Self-firing: $x_i(\tau + 1) = b(\tau),$ $y_i(\tau) = 1$ if $x_i(\tau) = N_x$
Cross-firing: $x_{j+1}(\tau + 1) = N_x - N_p + 1, \ z_j(\tau) = 1$ if $x_j(\tau) = N_x$ and
$$x_{j+1}(\tau) \le N_x - N_p \quad (6)$$

where $i \in \{1, \cdots, M\}$, $j \in \{1, \cdots, M - 1\}$ and the cross-firing is prior to the integrating. The integrating and self-firing are the same as the single DSN in Eq. (1). The cross-firing connects DSNs in ladder topology: the 1st DSN can apply the cross-firing to the 2nd DSN (see Fig. 3) and the $(j - 1)$-th DSN can apply the cross-firing to the j-th DSN, $j = 2 \sim M$. It can be regarded as a master-slave cascade: the 1st DSN is the master and the j-th DSN is slave of the $(j-1)$-th DSN. The self-firing is characterized by the spike-train $y_i(\tau)$. The cross-firing is characterized by the connection signal:

$$z_i(\tau) = \begin{cases} 1 \text{ if } x_i(\tau) = N_x \text{ and } x_{i+1}(\tau) \leq N_x - N_p \\ 0 \text{ otherwise} \end{cases} \quad (7)$$

As a typical phenomenon of the LDSN, we define the M-phase synchronization (M-SYN) with period MN_p.

$$\begin{aligned}
x_i(\tau) &= x_i(\tau + MN_p), y_i(\tau) = y_i(\tau + MN_p), i \in \{1, \cdots, M\} \\
x_j(\tau) &= x_{j+1}(\tau + N_p), y_j(\tau) = y_{j+1}(\tau + N_p), j \in \{1, \cdots, M-1\} \quad (8) \\
z_i(\tau) &= 1 \text{ for some } \tau \in \{1, \cdots, MN_p\}
\end{aligned}$$

Note that non-zero connection signal $z_i(\tau)$ is required for the M-SYN. If $z_i(\tau) = 0$ for all τ, all the DSNs are isolated without coupling. Figure 2(b) illustrates an M-SYN of PSTs with period MN_p for $M = 5$.

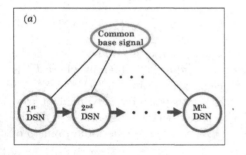
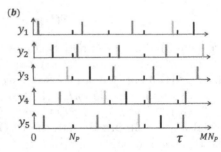

Fig. 2. The body of LDSN. (a) Configuration. (b) M-SYN.

Applying time-dependent selection switches S_i, the output of the LDSN is given as illustrated in Fig. 4. For simplicity, we consider the case where each DSN outputs a PST with period MN_p and the LDSN consists of M pieces of DSNs. We divide the time interval $(0, MN_p]$ into M slots I_1 to I_M:

$$I_1 = [0, N_p), I_2 = [N_p, 2N_p), \cdots, I_M = [(M-1)N_p, MN_p).$$

The i-th time-dependent selection switches S_i, $i = 1 \sim M$, selects either DSN in each time slot and gives the output as the following:

$$\text{Output: } y(\tau) = w_{ij}y_i(\tau), w_{ij} = \begin{cases} 1 \text{ if } y_i \text{ is selected for } \tau \mod MN_p \in I_j \\ 0 \text{ otherwise} \end{cases}$$

where $i = 1 \sim M$. For each time slot I_j, only one DSN can be selected: if $w_{ij} = 1$ then $w_{kj} = 0$ for all $k \neq i$. The output is characterized by selection matrix $W = (w_{ij})$. Figure 4 shows an output for selection matrix

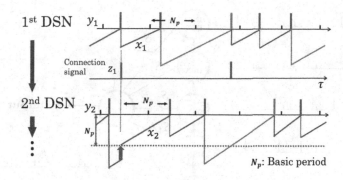

Fig. 3. Cross-firing of the LDSN

$$W = \begin{pmatrix} 1 & 0 & 0 & 0 & 0 \\ 0 & 1 & 0 & 0 & 0 \\ 0 & 0 & 0 & 0 & 1 \\ 0 & 0 & 1 & 1 & 0 \\ 0 & 0 & 0 & 0 & 0 \end{pmatrix} \tag{9}$$

The output PST is characterized by five spike-phases: $(\theta_a, \theta_a, \theta_b, \theta_a, \theta_c)$. Such a PST is impossible in the DSN governed by the Dmap $\theta_{n+1} = F(\theta_n)$: $\theta_a = F(\theta_a)$ and $\theta_b = F(\theta_a)$ contradict to each other. Note that this LDSN can output a variety of PSTs consisting of any combination of five spike-phases $\{\theta_a, \theta_b, \theta_c, \theta_d, \theta_e\}$ which constitute a desired PEO (PST) of a Dmap (DSN). Note that the PST can be stabilized simply by adjusting the base signal [11,12]. In general, if a single DSN outputs a PST with period MN_p consisting of M spike-phases then the LDSN can output a variety of PSTs consisting of any combination of the M spike-phases (or equivalently, M inter-spike intervals).

4   FPGA Based Implementation

The LDSN is a digital dynamical system and is suitable for hardware implementation. The hardware is useful for both observation of nonlinear phenomena and development into engineering applications. As a first step to realize a utility hardware, we consider a simple test circuit for $M = 5$, $N_p = 15$, and $N_x = 29$. Basically, the DSN is realized by two shift registers and a wiring circuit. The basic circuit design can be found in [11]. Applying cross-firing switches and time-dependent selection switches, the LDSN is realized. The LDSN is designed by Verilog and Algorithm 1 shows outline of the Verilog source code. Figure 5 shows a circuit diagram. In the implementation, we have used the following tools:

- Verilog version: Vivado Design Suite 2017.2
- FPGA board: DIGILENT BASYS3, Clock frequency: 100 [MHz].
- Measuring instrument: ANALOG DISCOVERY2, Multi-instrument software: WaveForms 2015

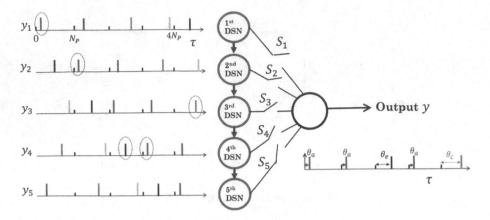

Fig. 4. Output of the LDSN by time-dependent selection switches.

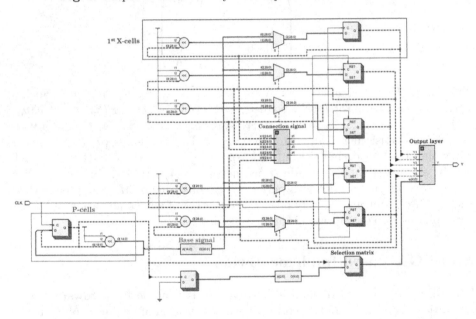

Fig. 5. Circuit diagram of LDSN for FPGA based hardware.

For convenience, we have divided the original CLK frequency 100 [MHz] into 6.25 [MHz]. Using the Verilog, we have implemented the circuit in the FPGA. Figure 6 shows measured PSTs in the FPGA board and simulated PSTs in the Verilog. These laboratory experiments are the first step for implementation of LDSNs with various PSTs and their applications.

Algorithm 1. LDSN

```
module ldsn(outputY, inputCLK)
wire Y1,Y2,Y3,Y4,Y5,Y,z1,z2,z3,z4 //Y1 = X1_{N_x}
reg P,X1,X2,X3,X4,X5,Base,W = initial cindition.
//Instatiate the connection signal part
connection signal(.X1(X1), ..., .X5(X5), .z1(z1), ..., .z4(z4))
//judge each X and output connection signal
//Instatiate the output layer
output layer(.Y1(Y1), ..., .Y5(Y5), .Y(Y), .W(W))
//selection matrix W select either DSN
always @(posedgeCLK)
P <= (P << 1) and if(P_{N_p} = 1), then P_1 <= 1
X1 <= (X1 << 1)
if (X1_{N_x} = 1) then
    X1 is reset to the base signal
end if
if (X2_{N_x} = 1) then
    X2 is reset to the base signal
end if
if (z1 = 1) then
    X2 jumps to N_x - N_p + 1
end if
X3 to X5 are also the same
end always
function Base(P)
correspond p-cells to the base signal
end function
end module
```

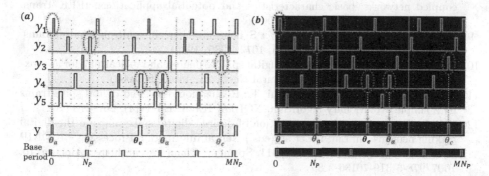

Fig. 6. Multiphase synchronization of PSTs and output of LDSN. (a) Measured waveform in a FPGA board. (b) Verilog simulation.

5   Conclusions

The LDSN is presented and multi-phase synchronization of the PSTs are considered in this paper. Applying the time dependent selection switches, the LDSN can output various PSTs consisting of any combination of desired inter-spike intervals. Using the Verilog, a test circuit was implemented in a FPGA board. Using the test circuit, typical phenomena were confirmed experimentally.

In our future works, we will study detailed stability analysis of multi-phase synchronization phenomena of PSTs and application to spike-based time series approximation/prediction.

References

1. Izhikevich, E.M.: Dynamical Systems in Neuroscience. MIT Press, Cambridge (2006)
2. Izhikevich, E.M.: Simple model of spiking neurons. IEEE Trans. Neural Netw. **14**(6), 1569–1572 (2003)
3. Perez, R., Glass, L.: Bistability, period doubling bifurcations and chaos in a periodically forced oscillator. Phys. Lett. A **90**(9), 441–443 (1982)
4. Campbell, S.R., Wang, D., Jayaprakash, C.: Synchrony and desynchrony in integrate-and-fire oscillators. Neural Comput. **11**, 1595–1619 (1999)
5. Rulkov, N.F., Sushchik, M.M., Tsimring, L.S., Volkovskii, A.R.: Digital communication using chaotic-pulse-position modulation. IEEE Trans. Circuits Syst. **48**(12), 1436–1444 (2001)
6. Iguchi, T., Hirata, A., Torikai, H.: Theoretical and heuristic synthesis of digital spiking neurons for spike-pattern-division multiplexing. IEICE Trans. Fundam. Electron. Commun. Comput. Sci. **E93–A**(8), 1486–1496 (2010)
7. Lozano, A., Rodriguez, M., Roberto Barrio, R.: Control strategies of 3-cell Central Pattern Generator via global stimuli. Sci. Rep. **6**, 23622 (2016)
8. Torikai, H., Hamanaka, H., Saito, T.: Reconfigurable spiking neuron and its pulse-coupled networks: basic characteristics and potential applications. IEEE Trans. Circuits Syst. II **53**(8), 734–738 (2006)
9. Torikai, H., Saito, T., Schwarz, W.: Synchronization via multiplex pulse trains. IEEE Trans. Circuits Syst. I **46**(9), 1072–1085 (1999)
10. Torikai, H., Funew, A., Saito, T.: Digital spiking neuron and its learning for approximation of various spike-trains. Neural Netw. **21**, 140–149 (2008)
11. Saito, T., Yamaoka, K., Hamaguchi, T.: Realization of desired digital spike-trains by a simple evolutionary algorithm. NOLTA, IEICE **E8–N**(4), 267–278 (2017)
12. Uchida, H., Saito, T.: Implementation of desired digital spike maps in the digital spiking neurons. In: Liu, D., et al. (eds.) Neural Information Processing – ICONIP 2017. LNCS, vol. 10639, pp. 804–811. Springer, Cham (2017). https://doi.org/10.1007/978-3-319-70136-3_85
13. Appeltant, L., et al.: Information processing using a single dynamical node as complex system. Nat. Commun. **2**, 468 (2011)
14. Antonik, P., Hermans, M., Haeltermany, M., Massar, S.: Photonic reservoir computer with output feedback for chaotic time series prediction. In: Proceedings of IJCNN, pp. 2407–2413 (2017)

The Effects of Feedback Signals Mediated by NMDA-Type Synapses for Modulating Border-Ownership Selective Neurons in Visual Cortex

Nobuhiko Wagatsuma[1][(✉)] and Hirotoshi Konno[2]

[1] Toho University, Miyama 2-2-1, Funabashi, Chiba 274-8510, Japan
nwagatsuma@is.sci.toho-u.ac.jp
[2] Tokyo Denki University, Ishizaka, Hatoayama-Machi, Hiki-Gun,
Saitama 350-0394, Japan

Abstract. The mean firing rate of a Border-Ownership Selective (BOS) neuron in intermediate-level visual areas represents where a figure direction from its classical receptive field. Recent neurophysiological and computational studies implied that slow modulatory feedback signals mediated by NMDA synaptic currents played an important role for modulating the responses and dynamics of BOS neurons. In order to understand the effects of modulatory feedback signals for modulating the responses of BOS neurons in more detail, we analyzed the simulation data of BOS model neurons through the jitter methods for computing tight synchrony with various ranges of jitter window. In the millisecond range of the jitter window, tight synchrony for the pairs of BOS model neurons was not modulated by selective attention. However, when the jitter window was widened, attention markedly decreased the magnitude of the tight synchrony for BOS model neurons. These behaviors of model neurons were in good agreement with the characteristics of neurophysiological BOS neurons, supporting a critical role of the modulatory feedback signals for modulating the responses of BOS neurons in intermediate-level visual areas. These results provide testable predictions for understanding the neural mechanism of figure-ground segregation and object perception.

Keywords: Border-Ownership · Spike synchrony · Jitter analysis
Modulatory input

1 Introduction

Neural mechanisms for separating a figural object from the background is the most fundamental process to perceive and understand objects and their location [1]. Neurophysiological studies have reported that a majority of neurons in V2 show the selectivity to Border-Ownership: the responses of Border-Ownership Selective (BOS) neurons depended on which side of the contour owned the border [2, 3]. Martin and von der Heydt investigated the characteristics of BOS neuron pairs for responding either to contours of different objects, or contours of the same object [4] (Fig. 1(A)). In

© Springer Nature Switzerland AG 2018
L. Cheng et al. (Eds.): ICONIP 2018, LNCS 11301, pp. 563–570, 2018.
https://doi.org/10.1007/978-3-030-04167-0_51

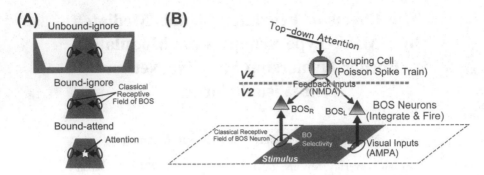

Fig. 1. (A) Examples of conditions for visual stimuli and attention for neurophysiological experiments [4]. Classical receptive fields of BOS neurons are illustrated by black ellipses on the stimulus contours. Arrows mean the figure direction selectivity for BOS neurons. Under the "Unbound" condition, two BOS neurons responded to contours of different white objects. Under the "Bound" condition, the contours of the gray object were given on these BOS receptive fields. On "attend" trials, animals attend the object. In contrast, on "ignore" trials, attention of animals is not directed to the target object. (B) Model architecture proposed by Wagatsuma et al. [5]. Grouping (G) cells are activated for representing a rough sketch of the object shape and attentional on BOS neurons. See Wagatsuma et al. [5] for details of the model.

their experiments, the classical receptive fields of BOS neurons were on the contours of keystone-like stimuli (black ellipses in Fig. 1(A)). Whereas two BOS neurons responded to contours of different objects under the "Unbound" condition, receptive fields of two BOS neurons lie on the contours of the same object under the "Bound" condition. On "attend" trials, attention of animals was directed to the object. Their neurophysiological results implied that presumed modulatory feedback signals due to the grouping structure and attention modulated spike synchrony between BOS neurons [4]. Wagatsuma et al. have developed a computational model to study the neural mechanism for spike synchrony of BOS neurons [5]. In their model, the modulatory common feedback from Grouping cells (G-cells) in V4 through NMDA-type synapses modulated the responses of BOS model neurons (Fig. 1(B)).

The application of jitter method has been utilized for investigating the hypothesis that neurons operate at or below any specific temporal resolution [6, 7]. Martin and von der Heydt [4] computed tight synchrony (correlation on the order of milliseconds) between BOS neurons by using this analysis in which the original spike trains were resampled by distributing the spikes randomly within various jitter window widths (see Methods and Fig. 2). Interestingly, in the small width of the jitter window, the magnitude of tight synchrony under "Bound, attend" condition was similar to that under "Bound, ignore". In contrast, when the jitter window was widened, the tight synchrony was much stronger for the ignored condition than for the attended condition. Wagatsuma et al. also applied the jitter method with jitter window of 20 ms to their simulation data [5]. However, effects of ranges of jitter window on their simulation data are unknown.

To elucidate the neuronal mechanisms of figure-ground segregation, we applied the jitter analysis with various ranges of jitter windows to the simulation data of BOS model neurons [5] for computing tight synchrony and for removing the influence of spikes outside these jitter window widths. Our results for the analyses of simulation data were in good agreement with characteristics of tight synchrony for neurophysiological BOS neurons [4]. These results suggested that the modulation of responses on BOS neurons originates, at least in part, from modulatory feedback signals mediated by NMDA-type synapses.

2 Methods

2.1 Model Architecture

The computational study has reported that the NMDA-synaptic currents played an important role to modulate the responses of BOS neurons [5]. We applied the jitter analysis with various ranges of jitter window [6, 7] to their simulation data [5] for further understanding the mechanism for modulating the responses of BOS neurons. The network architectures proposed by Wagatsuma et al. [5] were illustrated in Fig. 1(B). This model consisted of only the minimum number of neurons and networks for understanding the fundamental mechanism for modulating the responses of BOS neurons. The BOS model neuron whose receptive field was presented by the left (right) circle had right (left) side-of-figure preference (Fig. 1(B)). Therefore, it was called as BOS_R (BOS_L). The BOS model neurons received two types of input: Feedforward inputs represented visual stimuli, whereas feedback inputs from Grouping cells (G-cells) mediated the rough sketch of object shape and attention [8]. These inputs were given by stochastic random processes with Poisson statistics. Feedforward visual inputs were independent process, whereas the feedback signals mediated by G-cells were common to BOS_L and BOS_R neurons. In this model, the mean firing rates of a G-cell ν_G indicated the grouping structures of visual stimuli and attention (Fig. 1(A)). The "Unbound-ignore", "Bound-ignore", and "Bound-attend" conditions were represented by mean rates of a G-cell ν_G of 3, 25, and 45 Hz, respectively. The feedforward inputs for visual stimuli have same statistics in all three conditions, shown by a Poisson spike train with mean rates of 200 Hz.

In the model proposed by Wagatsuma et al. [5], the BOS neurons were modeled by integrate-and-fire neurons. They used $\theta = -50$ mV for the firing threshold, and $V_{reset} = -65$ mV for the reset potential. They selected membrane capacitance $C_m = 0.5$ nF and the membrane time constant $\tau_m = 20$ ms. The dynamics of the subthreshold membrane potential V were computed based on the following equation:

$$\frac{dV(t)}{dt} = -\frac{V(t)}{\tau_m} + \frac{I_{syn}(t)}{C_m}. \tag{1}$$

Here, $I_{syn}(t)$ comprised the synaptic currents for BOS neurons. They were sum of I_{vis} from feedforward visual inputs and I_G from feedback signals from G-cells, as following equation:

$$I_{syn}(t) = I_{vis}(t) + I_G(t). \tag{2}$$

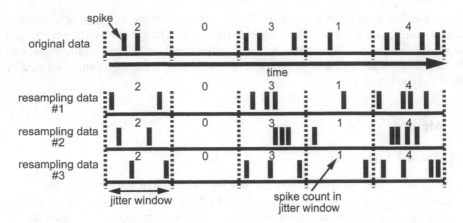

Fig. 2. Diagram of the jitter method [6, 7]. Top row indicates an original spike train consisting of 10 spikes (bar). Jitter windows shown by dashed lines partition time of spike trains. The numbers above the spikes are the spike counts in each jitter window. The original data are resampled. A single resampling spike train (resampling data #1) is generated by uniformly and independently selecting the new spike location that preserves the number of spikes in each jitter window. These processes are repeated for generating additional data (resampling data #2 and #3).

In this model [5], synaptic currents for feedforward visual inputs (I_{vis}) were mediated by the AMPA type receptors. In contrast, G-cells gave common modulatory feedback signals to BOS_L and BOS_R neurons. Modulatory feedback signals (I_G) took the form of currents via NMDA receptors, which played a critical role for modulating the rates of BOS model neurons and spike synchrony between the pairs of these neurons. For a full mathematical description of these synaptic currents, we refer the reader to [5]. The model was simulated by 50 trials of a 200 biological seconds length per stimulus condition.

2.2 Jitter Method for Extracting Tight Synchrony Between BOS Neurons

In jitter analysis, the jitter window separated the data from each neuron. Each spike of each neuron was then independently moved to a new location, selected from the uniform distribution on the jitter window to which it belonged in the original data (Fig. 2). In this way, the number of spikes in each bin for the resampled data was preserved. Smaller jitter windows removed more of the long timescale correlation between neurons (the loose synchrony) while preserving short timescale correlation (the tight synchrony) [7]. We applied the jitter analysis to simulation data of the model proposed by Wagatsuma et al. [5]. In our study, we used jitter windows δ of 20, 40, 60, 80, 100, 150, 200, 250, and 300 ms for computing tight synchrony and for removing the influence of spikes outside these jitter window widths. See also Wagatsuma et al. [5] for more detailed description of the jitter method.

For each spike train of BOS neurons S_{BOS_L} and S_{BOS_R}, spikes were jittered in a uniform distribution. Whereas the original spike trains were binned in 1 ms bins with 1 spike/bin, the jittered spike trains could have as many spikes in a bin as were present in each jitter window width of the original binned spike train. In this method, the firing rate characteristics of each trial were preserved at the resolution of the jitter window width. A sequence of resampling spike trains was produced by repeating this jittering. The cross-correlation between each of the resampling data presented a spike synchrony distribution. I subtracted the mean of this distribution from the mean of synchrony of the original spike trains. The r jittered synchrony were found by taking the trial-wise mean cross-correlation of each jittered resampling spike train $S_{BOS_L}^{i,r*}$ and $S_{BOS_R}^{i,r*}$ in trial i:

$$J^r(\tau) = \frac{1}{t_{end} - t_0} \left\langle \sum_{\mu=t_0-250}^{t_{end}+250} \left(S_{BOS_L}^{i,r*}(\mu+\tau) - f_{BOS_L}^{i,r*}\right) \left(S_{BOS_R}^{i,r*}(\mu) - f_{BOS_R}^{i,r*}\right) \right\rangle_i \quad (3)$$

$$f_j^{i,r*} = \frac{1}{t_{end} - t_0} \sum_{n=t_0}^{t_{end}} S_j^{i,r*}(n) \quad (4)$$

where $< >_i$ denoted the average over trial i. τ signified the time lag between two spike trains $(-250 \leq \tau \leq 250)$. t_0 and t_{end} define the interval of the spike train. $f_j^{i,r*}$ indicated the mean spike count per bin for jittered resampling spike train $S_j^{i,r*}$ between the interval t_0 to t_{end}. In our work, the above step was repeated for 200 times, generating 200 resampling data sets $(r = 200)$ for each jitter window width.

The jittered correlated synchrony (Tight Synchrony, TS) was computed by subtracting the mean jittered synchrony, $<J^r(\tau)>_r$, from the original spike synchrony

$$TS(\tau) = \frac{1}{t_{end} - t_0} \left\langle \sum_{\mu=t_0-250}^{t_{end}+250} \left(S_{BOS_L}^i(\mu+\tau) - f_{BOS_L}^i\right) \left(S_{BOS_R}^i(\mu) - f_{BOS_R}^i\right) \right\rangle_i - \langle J^r(\tau) \rangle_r$$

$$(5)$$

where f_j^i indicated the mean number of spikes per bin for original spike train S_j^i between the interval t_0 to t_{end}.

The integral of the tight synchrony between two BOS model neurons in the range of ± 5 ms was also computed, as shown in following equation:

$$M^* = \sum_{\tau=-5}^{5} TS(\tau) \times binsize \quad (6)$$

M^* shown in Eq. 6 represented the tight synchrony magnitude. In our work, the spike trains have bin size of 1 ms.

3   Results

Recent neurophysiological study has reported that selective attention and grouping structure modulated the spike synchrony between pairs of BOS neurons [4]. Their study also implied that the magnitude of tight synchrony (Eqs. 5 and 6) was markedly dependent on the width of jitter windows. To examine whether the model proposed by Wagatsuma et al. [5] reproduces these characteristics of tight synchrony, we applied the jitter methods [6, 7] with various widths of jitter window (δ = 20, 40, 80, 100, 150, 200, 250 and 300 ms) to their simulation data. In this study, the "Unbound-ignore", "Bound-ignore", and "Bound-attend" conditions were described by mean rates of G-cells ν_G of 3, 25, and 45 Hz, respectively.

Tight synchrony between BOS_L and BOS_R model neurons in three conditions were shown in Fig. 3. Irrespective of mean rates of G-cells ν_G, when large width of jitter window (δ = 300 ms) was applied, the broad peak was evident. In contrast, when the width of jitter window was small (δ = 20 ms), the broad peak was removed and the only short time scale synchrony remained. These results were in good agreement with characteristics of tight synchrony in visual areas [7].

In order to investigate the effects on the width of jitter window for tight synchrony, we computed the magnitude of tight synchrony (Eq. 6) with parametrically varying the width of jitter window δ. Figure 4 summarized the magnitude of tight synchrony as a function of jitter window widths. In all conditions, the magnitude of tight synchrony between BOS_L and BOS_R model neurons monotonically increased with increasing the width of jitter window. Compared with the unbound condition, under bound conditions, simulation data of the model proposed by Wagatsuma et al. [5] exhibited higher magnitude of tight synchrony. When we applied small widths of jitter window ($\delta \leq 80\,\mathrm{ms}$) to simulation data, the magnitudes of tight synchrony for the "Bound, ignore" condition were similar to that for the "Bound, attend". However, under large widths of jitter window ($\delta \geq 100\,\mathrm{ms}$), there was a marked decrease of tight synchrony magnitude from the "Bound, ignore" condition to the "Bound, attend" condition. These results were in good agreement with the neurophysiological findings reported by

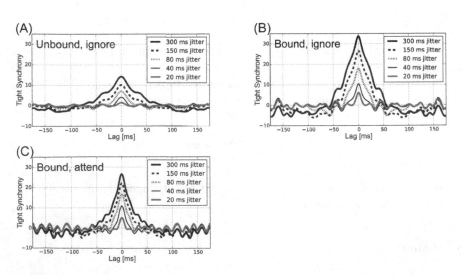

Fig. 3. Tight synchrony between BOS_L and BOS_R model neurons. The modulation patterns of tight synchrony depending on the width of jitter window were similar to neurons in the visual area [7].

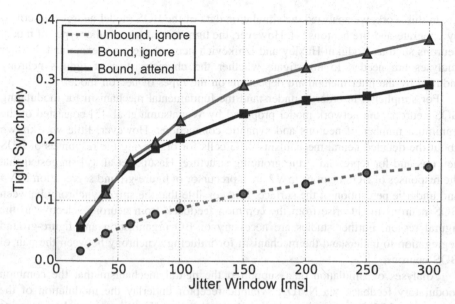

Fig. 4. The magnitude of tight synchrony for simulation data [5] as a function of jitter window widths. When small widths of jitter window ($\delta \leq 80$ ms) were applied to simulation data, the magnitude of tight synchrony for the "Bound, ignore" condition was similar for that for the "Bound, attend". However, under large width of jitter windows ($\delta \geq 100$ ms), there was a marked decrease of tight synchrony from the "Bound, ignore condition" to the "Bound, attend" condition.

Martin and von der Heydt [4] (see also Fig. 5C on [4]), supporting the hypothesis that the modulatory feedback mediated by NMDA synaptic receptors underlay the modulation for the responses of BOS neurons.

4 Discussion and Conclusion

The application of jitter methods is utilized for investigating the hypothesis that neurons operate at or below any specific temporal resolution [6, 7]. In our current study, for computing tight synchrony, we applied various widths of jitter window to the simulation data of the computational model based on the hypothesis that the modulatory feedback signals mediated by NMDA-type synapses modulated the responses of BOS neurons [5]. When small widths of jitter window ($\delta \leq 80$ ms) were applied to the simulation data, the magnitude of tight synchrony for the "Bound, attend" condition was similar to that for the "Bound, ignore". In contrast, under large widths of jitter window ($\delta \geq 100$ ms), the magnitude of tight synchrony markedly decreased from the "Bound, ignore" condition to the "Bound, attend" condition. These results for analyses of simulation data were in good agreement with the neurophysiological findings for the modulation of BOS neurons [4].

In this work, we analyzed the simulation data of the BOS model neurons described by integrate-and-fire neurons [4]. However, the tight synchrony of other types of model neurons such as Hodgkin-Huxley and Izhikevich neurons were not examined. Further analyses are needed to investigate whether the characteristics of tight synchrony induced by the jitter method are dependent on the types of neuron model.

For simplicity, in order to understand the fundamental mechanism for modulating BOS neurons, the network model proposed by Wagatsuma et al. [5] consisted of the minimum number of neurons and synaptic connections. However, little was known about the detailed neural mechanisms of G-cells for organizing the responses of BOS neurons and for representing the grouping structure. Hasuike et al. [9] suggested that the responses of BOS neurons in V2 are a precursor of figure-ground segregation in V4 and underlie perception of the surface. It is possible that the spike synchrony between BOS neurons might arise from the common feedback from neurons selective to the figural region. Further studies are necessary of BO organization and figure-ground segregation to understand the mechanism for inducing synchrony between the pair of BOS neurons.

Analyses of simulation data supported the neural mechanism that the common modulatory feedback via NMDA synaptic receptor underlay the modulation of the responses of BOS neurons in intermediate-level vision [5]. It is expected to examine this prediction from physiological observations.

Acknowledgements. This work was partly supported by KAKENHI (no. 17K12704).

References

1. Lamme, V.A.: The neurophysiology of figure-ground segregation in primary visual cortex. J. Neurosci. **15**, 1605–1615 (1995)
2. Zhou, H., Friedman, H.S., von der Heydt, R.: Coding of border ownership in monkey visual cortex. J. Neurosci. **20**, 6594–6611 (2000)
3. Qiu, F.T., Sugihara, T., von der Heydt, R.: Figure-ground mechanisms provide structure for selective attention. Nat. Neurosci. **10**, 1492–1499 (2007)
4. Martin, A.B., von der Heydt, R.: Spike synchrony reveals emergence of proto-objects in visual cortex. J. Neurosci. **35**, 6860–6870 (2015)
5. Wagatsuma, N., von der Heydt, R., Niebur, E.: Spike synchrony generated by modulatory common input through NMDA-type synapses. J. Neurophysiol. **116**, 1418–1433 (2016)
6. Amarasingham, A., Harrison, M.T., Hastsopoulos, N.G., Geman, S.: Conditional modeling and the jitter method of spike resampling. J. Neurophysiol. **107**, 517–531 (2012)
7. Smith, M.A., Kohn, A.: Spatial and temporal scales of neuronal correlation in primary visual cortex. J. Neurosci. **28**, 12591–12603 (2008)
8. Craft, E., Schutze, H., Niebur, E., von der Heydt, R.: A neural model of figure-ground organization. J. Neurophysiol. **97**, 4310–4326 (2007)
9. Hasuike, M., Yamane, Y., Tamura, H., Sakai, K.: Representation of local figure-ground by a group of V4 cells. In: Hirose, A., Ozawa, S., Doya, K., Ikeda, K., Lee, M., Liu, D. (eds.) ICONIP 2016. LNCS, vol. 9947, pp. 131–137. Springer, Cham (2016). https://doi.org/10.1007/978-3-319-46687-3_14

Modelling and Analysis of Temporal Gene Expression Data Using Spiking Neural Networks

Durgesh Nandini[1]([⊠]), Elisa Capecci[2], Lucien Koefoed[2], Ibai Laña[3], Gautam Kishore Shahi[1], and Nikola Kasabov[2]

[1] Dipartimento di Ingegneria e Scienza dell'Informazione (DISI), University of Trento, via Sommarive 9, 38100 Povo, Trento, TN, Italy
durgeshnandini16@yahoo.in
[2] Knowledge Engineering and Discovery Research Institute (KEDRI), Auckland University of Technology (AUT), AUT Tower, Level 7, cnr Rutland and Wakefield Street, Auckland 1010, New Zealand
[3] OPTIMA Unit. TECNALIA. P. Tecnologico Bizkaia, Ed. 700, 48160 Derio, Spain

Abstract. Analysis of temporal gene expression data poses a significant challenge due to the combination of high dimensionality and low sample size. The purpose of this paper is to present a methodology for classification, modelling, and analysis of short time-series gene expression data using spiking neural networks (SNN) and to uncover temporal expression patterns for knowledge discovery. The classification is based on the NeuCube SNN model. Time-series gene expression data of mouse primary cortical neurons is examined as a case study. The results of the analysis are promising, indicating that SNN methodologies can be effectively used to model and analyse temporal gene expression data with surpassing performance over traditional machine learning algorithms. Additionally, a gene interaction network is constructed from the temporal gene activity modelled using the NeuCube architecture offering a new way of knowledge discovery. Future work will be directed towards using gene interactions networks to help guide pharmacological research for dementia.

Keywords: Spiking neural networks · Gene interaction networks
Gene expression · Microarray · Transcriptome data analysis

1 Introduction

Time-series gene expression data is challenging to analyse and classify as it possesses high dimension and low sample size. Recent advances in next-gen sequencing have lead to an explosion in publicly available gene expression data. However, the methods of processing information from this data, are struggling. The latest literature in the domain of system biology and bioinformatics suggest that

© Springer Nature Switzerland AG 2018
L. Cheng et al. (Eds.): ICONIP 2018, LNCS 11301, pp. 571–581, 2018.
https://doi.org/10.1007/978-3-030-04167-0_52

learning and extracting meaningful information from time-series data is better addressed by artificial neural network techniques, especially Spiking Neural Networks (SNN) [1].

This work makes use of the potentials of the NeuCube [2] SNN architecture. The NeuCube is a development environment and a computational framework for the creation of brain-like artificial intelligent systems based on deep-learning in an evolving SNN machine [2–6].

The NeuCube has demonstrated superior classification performance in several domains [7], such as a better analysis of spatial and temporal activities modelled in the network [8], and an improved understanding of the molecular processes measured through the data [9, 10].

Our work accounts for the problems of high dimensional space and low sample size by taking advantage of the NeuCube SNN model to introduce a new methodology for classification of short time-series data. Additionally, we extract temporal gene expression information through a Gene Interaction Network (GIN).

As a case study, our methodology is applied on gene expression data, measured while studying the molecular mechanisms involved in primary cortical neurons death mediated by hydrogen sulphide (H2S) and the role of N-Methyl-D-Aspartate Acid receptor (NMDAR) antagonists. Cellular signalling can be used to plan pharmacological treatment of NMDAR antagonists towards endogenous H2S expression, which is the cause of neural death in people affected by Alzheimer's Disease (AD).

By combining the proposed SNN methodology with GIN analysis, we aim to study the molecular interaction amongst genes related with cellular apoptosis mediated by H2S signalling. This could help guide personalised medicine research for monitoring pharmaceutical treatments of neurological disorders.

The paper is organised as follows: Sect. 2 describes in details the materials and methods used to carry out the study; Sect. 3 presents and discusses experimental results; Sect. 4 shows the comparison between the performance of SNN with other machine learning algorithms; and Sect. 5 closes with conclusions and future works.

2   Materials and Methods

2.1   Data Description

Data has been retrieved from the Gene Expression Omnibus (GEO) repository of functional genomics data of the National Center for Biotechnology Information (NCBI) (NCBI GEO [11,12] accession GSE16035; [13]). In the study, gene expression data was collected over time to study the effects of NMDAR antagonists to attenuate H2S apoptosis in rodents primary cortical neurons. In total, 24 samples were analysed under two different conditions: administration of H2S or NMDAR treatment. Three replicas were collected for each time-point (5 h, 15 h and 24 h). Gene expression levels were analysed using microarray technology that measured transcript levels in biopsies.

2.2 Gene Expression Data Augmentation and Temporal Feature Selection

To develop a methodology for GIN analysis in a SNN classifier, we need to tackle two initial problems: augmenting the number of time-points to encode sufficient spike activity to activate the network, and identifying the genes, which expression results to be codependent, when describing the problem.

To handle the first issue, synthetic time-points are generated using linear interpolation techniques. To handle the second problem, we need to establish which features are of high significance with respect to the molecular process of study.

First, signature genes were selected, as derived from the original study [13]. Then, a number of feature selection methods were applied to these signature genes. Even though there are several feature selection algorithms, only a few of them ensure to select features for time-series data with reliability and redundancy. Here, we selected the latest proposed methods for temporal feature selection: the (TMRMR-C and TMRMR-M) [14], the Maximum Relevance Minimum Redundancy (mRMR) [15], F-statistic (ANOVA), and Relief-F [16].

2.3 SNN Model

NeuCube represents a set of tools for spatio-temporal data modelling and understanding. The SNN architecture can be divided into four main modules:

- An input module for gene expression data modelling and encoding into spike trains that represent a temporal change in gene expression;
- A SNN module for unsupervised learning of the data in a 3D cube of mapped neurons;
- An output module for supervised learning and classification of the data;
- A GIN module for data analysis and knowledge discovery.

Input Module and Gene Expression Data Encoding. The real-valued gene expression data is first encoded into train of spikes using a pre-selected encoding algorithm, such as the Bens Spiker Algorithm (BSA) [17,18].

Spike encoding is an essential step of the process, as every gene expression time-series is transformed so that only relevant changes of gene expression over time are encoded as binary events. The algorithm used for encoding needs to be optimised, taking into account the case study problem and the model performance in terms of classification accuracy (Fig. 1).

Unsupervised Learning in the Cube. Using a graph matching algorithm [19], the input genes are mapped into a network of $10 \times 10 \times 10$ leaky integrate-and-fire (LIF) neurons, where the encoded information is entered. A Small-World (SW) connectivity rule is used to initialise the cube and randomly generated weights are assigned to the connected neurons. Unsupervised learning is performed throughout the network to modify these weights using Spike-Timing-Dependent Plasticity (STDP) principles [20,21].

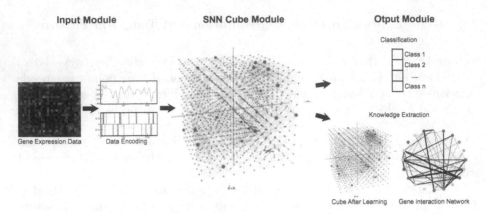

Fig. 1. The proposed SNN methodology for time-series gene expression data modelling with its four main modules: input data encoding module; a 3D SNN cube module; an output module for classification; and knowledge extraction from the data by means of GIN analysis.

Output Module for Supervised Learning and Data Classification. A one-pass supervised learning algorithm, the Dynamic Evolving Spiking Neural Network (deSNN) classifier [22], is used to classify the data into the respective classes.

GIN Module for Knowledge Discovery. The spiking activity and connectivity generated during learning can be analysed and used for knowledge discovery from the data. This is a unique feature of the Cube that is not implemented in traditional machine learning techniques. In a GIN, the input genes represent the nodes of a weighted connectivity graph that describe the temporal information exchanges during learning. The spiking activity and connectivity is clustered after training by assigning each neuron of the network to the corresponding gene with which it exchanges more information. The information represented in a GIN can be used to uncover the complex temporal regulation of gene expression data.

3   Results and Discussion

For our case study, the BSA algorithm was used to encode the data into spike trains. The patterns of temporal activity generated were then classified and analysed to show the effectiveness of the SNN methodology proposed.

3.1   Gene Expression Profiling Data Modelling

The initial dataset was divided into two classes: H2S, and NMDAR. Originally, there were nine samples for each class with three time-points each (5 h, 15 h and 24 h). Data was augmented (as mentioned in Sect. 2.2) to increase the temporal component - up to six time-points - and the sample number - up to a total of 18 samples for each class. This was done via linear interpolation technique assuming a linear combination between the past expression values of a selected set of genes. As reported in the literature, experimental results demonstrated that during short intervals single pre- and post- synaptic spikes pair in hippocampal neurons [20, 23]. Additionally, interpolation methods have been used extensively to estimate unobserved gene expression measurements, as there are considerable literature that deals with this problem through statistical techniques [24–26].

A subset of signature genes was selected, as suggested by the original study [13], to reduce the high number of features. Then, the TMRMR-C, TMRMR-M, mRMR, F-statistic, and Relief-F features selection methods were applied to the selected signature genes to determine the number of genes that influenced the molecular process over time. As a result, we reduced the number of features/genes up to 21. Information about these genes is summarised in Table 1.

Table 1. The table reports the corresponding names of the selected gene features that are used to construct the GIN. Information about the genes has been retrieved from GeneCards (http://www.genecards.org).

Genes	Description
mt3	Metallothionein 3
gstm1	Glutathione Synthetase
apoe	Apolipoprotein E
ube2n	Ubiquitin Conjugating Enzyme E2 N
Gsta1	Glutathione S-Transferase Alpha 1
usp10	Ubiquitin Specific Peptidase 10
ube1x	Ubiquitin Like Modifier Activating Enzyme 1
casp6	Caspase 6
psmc5	Proteasome 26S Subunit, ATPase 5
notch1	Notch 1
prdx4	Peroxiredoxin 4
gsta4	Glutathione S-Transferase Alpha 4
gpx1	Glutathione Peroxidase 1
hmox1	Heme Oxygenase 1
ddit3	DNA Damage Inducible Transcript 3
usp20	Ubiquitin Specific Peptidase 20
bag3	BCL2 Associated Athanogene 3
psmc2	Proteasome 26S Subunit, ATPase 2
usp29	Ubiquitin Specific Peptidase 29
psmb5	Proteasome Subunit Beta 5
psmb10	Proteasome Subunit Beta 10

3.2 Classification

The dataset was classified into two classes: H2S and the NMDAR. A grid search method was used to find the combination of 24 parameters that resulted in the highest classification accuracy. The Monte-Carlo method was used to statistically evaluate the results with leave-one-out cross-validation (LOOCV) method. LOOCV was chosen as it possesses the ability to discriminate variables importance over a random set of entities, when only a small dataset is available. The optimised parameter were:

- The BSA algorithm (with a spike threshold of 0.1);
- The Small World connectivity (with a radius of 2.5);
- The LIF model parameters (with a threshold of firing of 0.5; a refractory time of 6; and a potential leak rate of 0.002);
- The STDP parameter (with a learning rate of 0.1);
- The deSNN parameters (with a mod and drift of 0.8 and 0.005 respectively).

Table 2 reports the accuracy of classification obtained from the combination of parameters given above.

Table 2. Classification accuracy obtained after training the SNN system with the gene expression data available.

Measure	Overall accuracy %
Mean	87
Standard deviation	3

4 Comparative Analysis with Traditional Machine Learning Algorithms

To analyse the efficacy of the SNN methodology proposed, we compared our results with other machine learning algorithms, such as Support Vector Machines (SVM) and Deep Convolutional Neural Networks (CNN).

SVM is a popular method in bioinformatics data analysis, due to its effectiveness in handling high dimensional data with a low sample number [27]. Missing values were handled using linear interpolation techniques.

CNN can select their own features in the course of classification, and that represents the primary advantage of using a deep neural network, as no preprocessing of features is needed. Here, CNN was used with TensorFlow implementation.

The classification accuracy obtained with each of these two algorithms is reported in the Table 3.

Table 3. The classification accuracy of each of the algorithm used.

Algorithm	Overall accuracy %
Support vector machines	73.21
Deep convolutional neural networks	79.64

4.1 GIN Analysis

After classification, the initial mapping of the input features and the optimised parameter values are retained. Then, each class data is used separately to train the Cube in order to carry out a comparative analysis of the network. This process was repeated for 50 times to allow the system to learn accurately from the small dataset. The neural activity generated per class is clustered according to the number of gene features. Each neuron of the network is assigned to its corresponding cluster by evaluating the activity exchanged with this feature during learning. The proportion of activity generated in the Cube for each class along with the combination of both classes is shown in the pie charts Fig. 2. This can be used as an extension to the feature ranking method applied assuming that the more active genes are more important for the data of the corresponding class. The GIN obtained for each class and from the combination of both the classes is shown in Fig. 3.

The total interaction between clusters of neuron for each class along with the combination of both classes is shown in Fig. 4. Each cluster corresponds to a gene highlighted by a different colour.

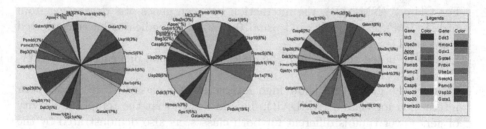

Fig. 2. The proportion of activity generated in the entire 3D cube is illustrated in the pie chart. (left) Class1 (H2S); (centre) Class 2 (NMDAR); (right) combination of both classes.

The GINs formed for individual class show significant differences. Some information derived from the GINs are:

- For Class 1 (H2S), the strongest neuron interaction in percentage was held by gsta4 (17%), psmb (10%), usp29 (8%), gstm1 (8%). A strong interaction between genes implies a concerted regulation in gene expression that changes over time.

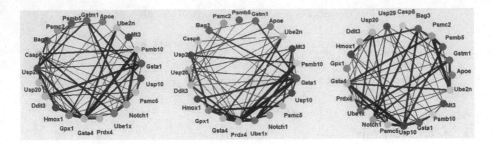

Fig. 3. Gene Interaction Network of the input gene features. (left) Class1 (H2S); (centre) Class 2 (NMDAR); (right) combination of both classes.

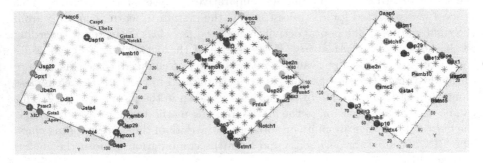

Fig. 4. Interaction of Neuron Clusters. The figure shows the x-y plane of the 3D Cube only. (left) Class 1 (H2S); (centre) Class 2 (NMDAR); (right) combination of both classes. (Color figure online)

- For Class 2 (NMDAR), the interaction between genes was decreasing from prdx4 (19%), gsta1 (9%), and psmb10 (8%). This indicates lower concerted expression levels.
- It is to be noted that some of the genes showed equal or nearly equal interaction in both classes.
- Most of the genes that show high interaction for class 1 (H2S) show poor interaction for class 2 (NMDAR). This indicates that the temporal activity of genes over time is different for both classes, as also demonstrated by the classification accuracy results.

5   Conclusion and Future Work

The analysis of patterns generated by time-series gene expression data constitutes a major goal for the area of bioinformatics and system biology. In this study, we have modelled and classified time-series gene expression data and revealed new realisations from the genes expressed over time by employing a GIN analysis. From our case study, we conclude that the best method to evaluate short time-series gene expression data is through the use of SNN, as its performance surpasses other machine learning algorithms. Additionally, with the use of the NeuCube architecture, we can study the interaction of genes via GIN analysis.

Future work includes:

- Studying the possible implication of directionality between temporal interaction of genes in the network;
- Applying the proposed model with the neuroreceptor dependent plasticity learning rule proposed in [28];
- Integrating gene expression temporal data with brain data (EEG, fMRI) in humans for the understanding and the prediction of brain diseases, such as dementia and AD [29].

Acknowledgment. The SRIF 2017–2018 INTERACT project of the Auckland University of Technology supports the presented study. Several people have contributed to the research that resulted in this paper, especially: Dr Y.Chen, Dr J.Hu, L.Zhou, Dr E. Tu and Maryam Gholami-Doborjeh. A free for research and teaching version of the NeuCube SNN system can be found from the KEDRI web site: https://kedri.aut.ac.nz/R-and-D-Systems/neucube.

References

1. Kasabov, N.K.: Neucube: a spiking neural network architecture for mapping, learning and understanding of spatio-temporal brain data. Neural Netw. **52**, 62–76 (2014)
2. Kasabov, N., et al.: Evolving spatio-temporal data machines based on the neucube neuromorphic framework: design methodology and selected applications. Neural Netw. **78**, 1–14 (2016)
3. Kasabov, N.: Neucube evospike architecture for spatio-temporal modelling and pattern recognition of brain signals. In: Mana, N., Schwenker, F., Trentin, E. (eds.) ANNPR 2012. LNCS, vol. 7477, pp. 225–243. Springer, Berlin Heidelberg (2012). https://doi.org/10.1007/978-3-642-33212-8_21
4. Chen, Y., Hu, J., Kasabov, N., Hou, Z., Cheng, L.: NeuCubeRehab: a pilot study for eeg classification in rehabilitation practice based on spiking neural networks. In: Lee, M., Hirose, A., Hou, Z.-G., Kil, R.M. (eds.) ICONIP 2013. LNCS, vol. 8228, pp. 70–77. Springer, Heidelberg (2013). https://doi.org/10.1007/978-3-642-42051-1_10
5. Kasabov, N.: Neucube: a spiking neural network architecture for mapping, learning and understanding of spatio-temporal brain data. Neural Netw. **52**, 62–76 (2014)
6. Tu, E., et al.: NeuCube(ST) for spatio-temporal data predictive modelling with a case study on ecological data. In: 2014 International Joint Conference on Neural Networks (IJCNN), pp. 638–645, July 2014
7. Kasabov, N., Capecci, E.: Spiking neural network methodology for modelling, recognition and understanding of eeg spatio-temporal data measuring cognitive processes during mental tasks. Inf. Sci. **294**, 565–575 (2015)
8. Marks, S.: Immersive visualisation of 3-dimensional spiking neural networks. Evol. Syst. **8**(3), 193–201 (2017). https://doi.org/10.1007/s12530-016-9170-8
9. Capecci, E., Kasabov, N., Wang, G.Y.: Analysis of connectivity in neucube spiking neural network models trained on eeg data for the understanding of functional changes in the brain: A case study on opiate dependence treatment. Neural Netw. **68**, 62–77 (2015)

10. Espinosa-Ramos, J.I., Capecci, E., Kasabov, N.: A computational model of neuroreceptor dependent plasticity (NRDP) based on spiking neural networks. IEEE Trans. Cogn. Dev. Syst. (2017). https://doi.org/10.1109/TCDS.2017.2776863. ISSN: 2379-8920

11. Edgar, R., Domrachev, M., Lash, A.E.: Gene expression omnibus: NCBI gene expression and hybridization array data repository. Nucl. Acids Res. **30**(1), 207–210 (2002)

12. Barrett, T., et al.: NCBI GEO: archive for functional genomics data sets - update. Nucl. Acids Res. **41**(D1), D991–D995 (2012)

13. Chen, M.J., et al.: Gene profiling reveals hydrogen sulphide recruits death signaling via the N-methyl-D-aspartate receptor identifying commonalities with excitotoxicity. J. Cell. Physiol. **226**(5), 1308–1322 (2011)

14. Radovic, M., Ghalwash, M., Filipovic, N., Obradovic, Z.: Minimum redundancy maximum relevance feature selection approach for temporal gene expression data. BMC Bioinform. **18**(1), 9 (2017)

15. Ding, C., Peng, H.: Minimum redundancy feature selection from microarray gene expression data. J. Bioinform. Comput. Biol. **3**(02), 185–205 (2005)

16. Kononenko, I., Šimec, E., Robnik-Šikonja, M.: Overcoming the myopia of inductive learning algorithms with relieff. Appl. Intell. **7**(1), 39–55 (1997)

17. Schrauwen, B., Van Campenhout, J.: BSA, a fast and accurate spike train encoding scheme. In: Proceedings of the International Joint Conference on Neural Networks, Piscataway, NJ, vol. 4, pp. 2825–2830. IEEE (2003)

18. Nuntalid, N., Dhoble, K., Kasabov, N.: EEG classification with BSA spike encoding algorithm and evolving probabilistic spiking neural network. In: Lu, B.-L., Zhang, L., Kwok, J. (eds.) ICONIP 2011. LNCS, vol. 7062, pp. 451–460. Springer, Heidelberg (2011). https://doi.org/10.1007/978-3-642-24955-6_54

19. Tu, E., Kasabov, N., Yang, J.: Mapping temporal variables into the neucube for improved pattern recognition, predictive modeling, and understanding of stream data. IEEE Trans. Neural Netw. Learn. Syst. **28**(6), 1305–1317 (2017)

20. Song, S., Miller, K.D., Abbott, L.F.: Competitive Hebbian learning through spike-timing-dependent synaptic plasticity. Nat. Neurosci. **3**(9), 919–926 (2000)

21. Hebb, D.O.: The Organization of Behavior: A Neuropsychological Approach. Wiley, Hoboken (1949)

22. Kasabov, N., Dhoble, K., Nuntalid, N., Indiveri, G.: Dynamic evolving spiking neural networks for on-line spatio- and spectro-temporal pattern recognition. Neural Netw. **41**, 188–201 (2013)

23. Caporale, N., Dan, Y.: Spike timing-dependent plasticity: a Hebbian learning rule. Annu. Rev. Neurosci. **31**(1), 25–46 (2008). https://doi.org/10.1146/annurev.neuro.31.060407.125639. pMID: 18275283

24. D'haeseleer, P., Wen, X., Fuhrman, S., Somogyi, R.: Linear modeling of mRNA expression levels during CNS development and injury. In: Biocomputing 1999, pp. 41–52. World Scientific (1999)

25. Aach, J., Church, G.M.: Aligning gene expression time series with time warping algorithms. Bioinformatics **17**(6), 495–508 (2001)

26. Troyanskaya, O., et al.: Missing value estimation methods for dna microarrays. Bioinformatics **17**(6), 520–525 (2001)

27. Huang, S., Cai, N., Pacheco, P.P., Narandes, S., Wang, Y., Xu, W.: Applications of support vector machine (SVM) learning in cancer genomics. Cancer Genomics-Proteomics **15**(1), 41–51 (2018)

28. Espinosa-Ramos, J.I., Capecci, E., Kasabov, N.: A computational model of neuroreceptor dependent plasticity (NRDP) based on spiking neural networks (2017, accepted)
29. Kasabov, N.: Time-Space, Spiking Neural Networks and Brain-Inspired Artificial Intelligence, vol. 7. Springer, Heidelberg (2019). https://doi.org/10.1007/978-3-662-57715-8. https://www.springer.com/gp/book/9783662577134

A Gesture Recognition Method Based on Spiking Neural Networks for Cognition Development

Dong Niu, Dengju Li, Rui Yan, and Huajin Tang(✉)

Neuromorphic Computing Research Center, College of Computer Science,
Sichuan University, Chengdu, China
{2017223045166,2017223045147}@stu.scu.edu.cn, {ryan,htang}@scu.edu.cn

Abstract. This paper proposes a gesture recognition method based on spiking neural network (SNN). The method can be used to develop the cognition behavior by associating the recognition results with semantic information from the observed target. Firstly, a single shot multi-box detector (SSD) is used to recognize the target object and locate it. Then two SNNs based on Izhikevich model are used to record trajectories of plane motion and depth motion. After projecting and translating the data extracted from the SNN, self-organizing mapping (SOM) and support vector machine (SVM) are applied to realize the gesture recognition. Finally, the associative memory model is used to associate gestures with semantics to achieve cognition. The experiment results show that SNN can well memorize the spatial-temporal information of various gestures. Furthermore, based on the spiking trains from the Izhikevich model, we can realize good results from the clustering and classification.

Keywords: Cognitive development · Gesture recognition
Spiking neural network

1 Introduction

As robots begin to blend into people's lives, it becomes more and more important of how ordinary people can interact with robots in a more friendly way and even cooperate. This interaction is widely discussed in the fields of sociology, developmental psychology, relevance theory, and cognitive science [1]. The relevance theory proposed a development direction for robots interaction improvement [2]. It shows that people are not simply exchanging information in the process of interaction, but they are sharing their cognitive environment. The cognitive environment may be related to, for example, who the interaction partner is, what he is talking about, and his body postures or hand gestures. By this way, we can consider developing a robot's own cognitive environment to enhance the interaction experience with people. In recent years, there have also been many studies that enhance interactions through the development of robotic cognition.

© Springer Nature Switzerland AG 2018
L. Cheng et al. (Eds.): ICONIP 2018, LNCS 11301, pp. 582–593, 2018.
https://doi.org/10.1007/978-3-030-04167-0_53

For example, [1] extracts object properties and associate it with object motion gesture to create a cognitive environment for the robot to enhance interaction. [3] proposed a method to recognize human body movements and combine it with contextual knowledge of human-robot collaboration scenarios. The task of robot navigation in the maze environment was studied by using some neurobiologically-inspired cognitive system in [4, 5]. There are also many application scenarios for this interaction research, such as communications robots are used to offer support to the elderly [6] and a conversational robot is used in nursing homes [7].

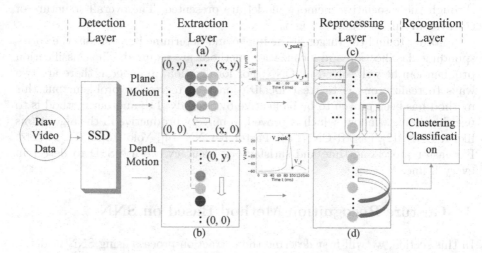

Fig. 1. System structure. The system structure includes: Detection layer; Extraction layer (Extract trajectories of plane motion and depth motion); Reprocessing layer (Projecting and translating data); Recognition layer (Recognize data by clustering or classifying). Introduction of modules: (a) PlaneNet module (b) DepthNet module (c) Projection module (d) Translation module.

This paper focuses on improving the robot's interactive experience by using gestures combined with semantics to support interaction. Many methods can be used for gesture recognition, such as [8] used deep learning to recognize hand gesture. A method based on localist attractor networks (LAN) for online gesture recognition was proposed in [9]. [10] introduced a method for the efficient detection and classification of hand gestures in the complex background. The SNN model is more biologically plausible and suitable for managing spatial-temporal data. In this paper, similar to the method in [1], we choose SNN to record gesture data. Different from some traditional methods which recognize the gestures by extracting the motion of hand parts, the proposed method record the motion trajectory of the moving object in the hand to achieve gesture recognition. Finally, the cognition behavior is realized by associating recognition results with the semantics of the target object.

2   Overview of System

In this section, we will illustrate an overview of the gesture recognition system. We use SSD to detect the object. Izhikevich neurons are organized for recording trajectories of the plane motion and depth motion of the moving object. To memorize gesture, spike timing dependent plasticity (STDP) is used to train SNN. For the data generated by SNN, the projection and translation processing are performed and then we conducted clustering and classification experiments for the processed data. Finally, the results of combining gestures with semantics through the associative memory model are presented. The overall structure of the system can be found in Fig. 1.

In order to find the target in the image and determine its position, the corresponding classification and localization methods are required. The classification problem can be well solved by CNN. For localization, in general, there are two ways to realize: one is to treat localization as a regression problem, but this method has been proven to be ineffective by Szegedy [11]; another method is to use the sliding window, which is proved to be very productive in the algorithms like R-CNN [12], Fast R-CNN [13], Faster R-CNN [14], Yolo [15], SSD [16], etc. For faster processing time and satisfactory accuracy, we use SSD to detect in every frame.

3   Gesture Recognition Method Based on SNN

In this section, we will first describe the extraction process using SNN in detail and then reveal the way of how to reprocess data extracted from SNN.

3.1   SNN for Gesture Extraction

Various types of artificial neural networks (ANNs) have emerged in recent years, such as CNN for processing pictures and RNN for processing time series data. Basically, ANNs are classified into pulse-coded neural networks and rate-coded neural networks, from the viewpoint of their level of abstraction [17].

The pulse coded networks are closer to the neuro-dynamic mechanism of biological neurons than the rate-coded networks. Pulse-coded neural networks have many neuron models, such as one of the most classic model Hodgkin-Huxley (HH) [18]. Unfortunately, because of its intrinsic complexity, HH model is usually difficult to analyze and is computationally expensive in numerical implementations. Therefore some simplified models are proposed such as integrate-and-fire (IF) [19]. But a remaining shortcoming of IF model is that it only implements time-independent memory. The better leaky integrate-and-fire (LIF) [20] model solves the memory problem by introducing "leaky". However, IF and LIF are not so good because they remove the simulation of ion channels in HH. The Izhikevich neuron model is both simplified relative to HH and retains its simulation of biological ion channels.

Compared to [1], we use the Izhikevich model to organize the spike neural network for gesture extraction. Izhikevich model can be represented by the following equations:

$$\begin{cases} dv/dt = 0.04v^2 + 5v + 140 - u + I \\ du/dt = a(bv - u) \\ if\ v \geq 30mV, \\ then\ v \leftarrow c, u \leftarrow u + d \end{cases} \tag{1}$$

where v represents the membrane potential. u is the membrane recovery variable. The I is the external input of the neuron. The parameter $a = 0.02$, $b = 0.2$, $c = -65$ and $d = 8$ are chosen for neuron dynamics control. When membrane potential surpass the threshold potential, a spike will be released according to the following:

$$\begin{cases} if\ v \geq 30mV, \\ then\ p(t) = 0,\ p(t-1) = 1, \\ else\ p(t) = 0 \end{cases} \tag{2}$$

where $p(t)$ means whether the neuron fires a spike at time t. The input of the neuron consists of two parts, one part is the external current input and another part comes from other neurons outputs. The calculative equation is as shown in the following:

$$I_i(t) = I_i^{ext}(t) + I_i^{syn}(t) \tag{3}$$

where $I_i(t)$ represents the input of the ith neuron at time t. $I_i^{ext}(t)$ indicates the external current input. $I_i^{syn}(t)$ stands for the other neurons outputs which can be calculated by the following equation:

$$I_i^{syn}(t) = \sum_{j=1, j \neq i}^{N} w_{j,i} O_j^{EPSP}(t) \tag{4}$$

where $w_{j,i}$ represents the connection weight from jth neuron to ith neuron. N is the number of neurons. The presynaptic spike output is transmitted to the connected neuron according to the excitatory postsynaptic potential (EPSP) in which value can be calculated by the following:

$$O_j^{EPSP}(t) = \sum_{n=0}^{T} \alpha^n q p_j(t - n) \tag{5}$$

where $O_j^{EPSP}(t)$ is the EPSP output of jth neuron at t time. The parameter α is a discount rate ($0 < \alpha < 1$) we use 0.6. T is the current time and q is the total charge injected into the synapse which we take 50 pA. Two SNNs called PlaneNet and DepthNet will be introduced to record the motion series data extracted by using SSD. Firstly, for plane motion, PlaneNet is established as a matrix plane to map the location of the target object in the image (see Fig. 1(a)). The coordinates of the moving object in the image at time t are represented by $C(t) = (c_x(t), c_y(t))$. The ith neuron's coordinates in PlaneNet are $S_i = (s_{x,i}, s_{y,i})$. The ranges of $c_x(t)$ and $s_{x,i}$ are $[0, m-1]$. The ranges of $c_y(t)$ and $s_{y,i}$ are $[0, n-1]$. Where m and n represent the number of neurons in the x-axis and y-axis. Secondly, for depth motion, DepthNet is created to map the size of the target object (see Fig. 1(b)). The object's depth coordinates are represented by $C(t) = (0, max(r_w, r_h)N_d)$. Which r_w and r_h are the ratios of the length and width of the object to the length and width of the image. N_d indicates the number of neurons above the axis. The coordinates of the ith neuron in DepthNet are $S_i = (0, s_i)$. Then the external current input of ith neuron in PlaneNet and DepthNet are calculated by the following:

$$I_i^{ext}(t) = exp(-||a(S_i - C(t))||_2)\gamma^{(c-1)}b \tag{6}$$

In order to avoid the whole SNN being activated, we use a which is the parameter that controls the size of the difference value generated at different positions of the same gesture mapping. γ is the degenerative parameter. c is the number of consecutive activation of the same neuron. b controls the range of external output values. We take a bigger b to allow neurons to be activated in a short time. For the parameters mentioned before, we take $a = 4$, $\gamma = 0.5$, $b = 400pA$. To make the same gesture can be more easily activated, we use STDP to train the SNN. The weight update method can be seen as following:

$$\Delta w_{ji} = \sum_{f=1}^{M} \sum_{n=1}^{N} W(t_i^n - t_j^f) \tag{7}$$

$$W(x) = \begin{cases} A \cdot exp(-x/t_+) & for \ x > 0 \\ -A \cdot exp(x/t_-) & for \ x < 0 \end{cases} \tag{8}$$

where t_i^n and t_j^f represent the time of the nth spike of the ith neuron and the time of the fth spike of the jth neuron. A is the connection weight from neuron j to neuron i. Where t_+ and t_- are time constants we all take 10 ms.

3.2   Feature Reprocess

Based on all previous calculations, we obtain two vectors $V_p = (v_{p1}, v_{p2}, \ldots, v_{pN_p})$ and $V_d = (v_{d1}, v_{d2}, \ldots, v_{dN_d})$, where v_{pi} represents EPSP output of the ith neuron in PlaneNet, v_{dj} is EPSP output of the jth neuron in DepthNet. By 2D projection and translation, the data V_p can be reprocessed (see Fig. 1(c) and (d)). Firstly, we resize V_p to V_{pm}. The shape of V_{pm} is (n, m). A row of V_{pm} is represented by $V_{pm}[i][:]$ and a column of V_{pm} is represented by $V_{pm}[:][j]$, where the range of i is $[0, n-1]$ and the range of j is $[0, m-1]$. The projection result V_{mapped} is calculated as follows:

$$V_{mapped} = \sum_{i=0}^{n-1} V_{pm}[i][:] + (\sum_{j=0}^{m-1} V_{pm}[:][j])^T \tag{9}$$

The V_{mapped} has the shape $(1, m + n)$. T stands for transpose. Secondly, our goal is to identify the gesture itself. In order to avoid the mapping difference of the same gesture made in different spatial locations, we perform a left translation on the V_{mapped} data. For V_d, We only use the left translation to reduce the mapping difference of the same gesture.

4   Experimental Results

In this section, we will first show part of the gesture extraction results by using SNN. Then we further demonstrate clustering and classification experiments. Finally, the results of associating gestures with semantics are presented.

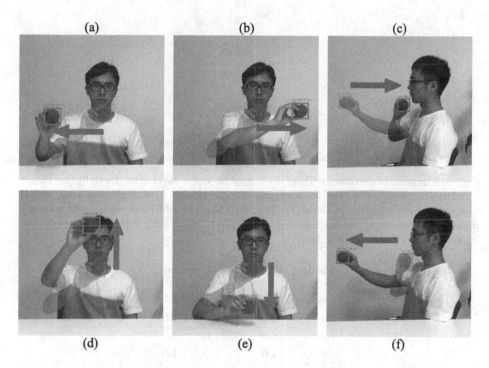

Fig. 2. Basic gestures. (a) Moving horizontally to the right (l2r gesture) (b) Moving horizontally to the left (r2l gesture) (c) Moving horizontally to the near (f2n gesture) (d) Moving vertically to the up (b2t gesture) (e) Moving vertically to the down (t2b gesture) (f) Moving horizontally to the far (n2f gesture).

4.1   Gesture Extraction

This section shows part of the process of extracting hand gesture by using SNN. The size of PlaneNet and DepthNet are 15×15 and 1×20. The basic gestures

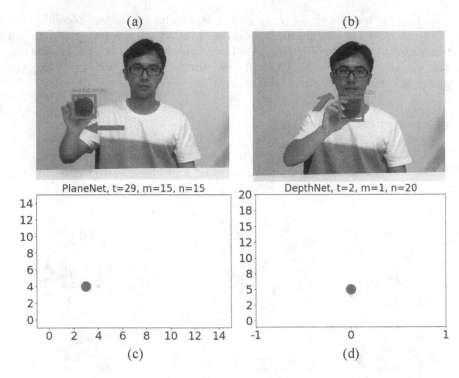

Fig. 3. Mapping of gestures in SNN. (a) Target object is moving to the right (b) Target object is moving to the near (c) Mapping of (a)'s motion in PlaneNet at time 29 (d) Mapping of (b)'s motion in DepthNet at time 2. In (c) and (d), m and n represent the number of neurons in the horizontal axis and the vertical axis. Where t represents the current time. (Color figure online)

Table 1. Gestures classification

	Gesture	Precision	Recall	F1-Score	Support
Plane motion	l2r	1.0	1.0	1.0	17
	r2l	1.0	1.0	1.0	6
	t2b	1.0	1.0	1.0	8
	b2t	1.0	1.0	1.0	13
	avg/total	1.0	1.0	1.0	44
Depth motion	f2n	1.0	1.0	1.0	13
	n2f	1.0	1.0	1.0	10
	avg/total	1.0	1.0	1.0	23
Mixed motion	lr	1.0	1.0	1.0	23
	tb	1.0	1.0	1.0	17
	nf	1.0	1.0	1.0	26
	avg/total	1.0	1.0	1.0	66

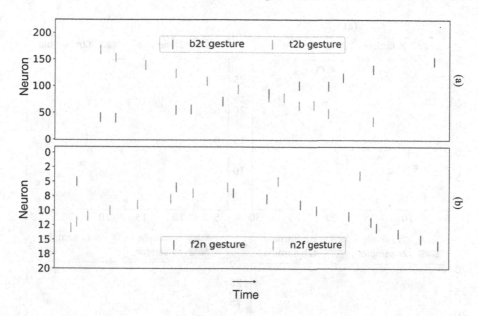

Fig. 4. Pulse distributions. (a) Pulse distributions of the b2t gesture and t2b gesture (see Fig. 2(d) and (e)) generated from PlaneNet (b) Pulse distributions of the f2n gesture and n2f gesture (see Fig. 2(c) and (f)) generated from DepthNet.

Table 2. Associate gestures with semantics

	l2r	r2l	t2b	b2t	n2f	f2n	Behavior
Cup	0	0	0	1	0	1	Drinking
Brush	1	1	0	0	0	0	Brushing

we selected for experiments are shown in Fig. 2. A total of 66 training videos are recorded using 6 basic gestures mixed in different spatial positions. Figure 3 shows the mapping of two gestures in SNN, for the dots in the Fig. 3(c) and (d), the bluer and larger the point is, the closer the neuron in this position is going to fire. The red dot in the figure indicates that the position neuron is fired. Figure 4 shows pulse distributions of four gestures. This implies SNN can well preserve the temporal and spatial features of gesture motion.

4.2   Gesture Clustering

This section demonstrates the experiments of clustering. SOM is chosen for clustering because it is more suitable for incremental learning. The size of SOM network is 30×30. The ratio of training data and testing data is 7 to 3. Firstly, we use the training data to train the SOM, then employ these data to identify multiple clustering centers, and finally, the classification is performed by comparing the distance (e.g. the Euclidean distance) between the test data and the

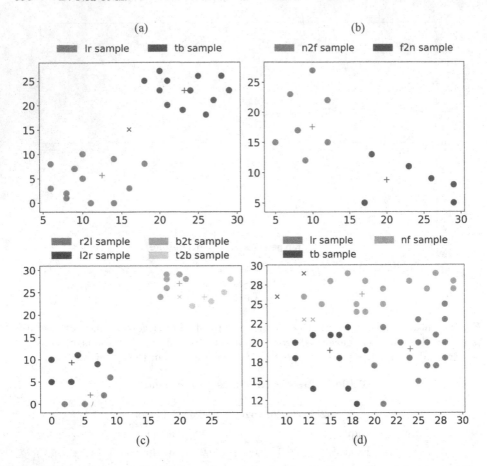

Fig. 5. Clustering results. The sample represented by circle means sample is correctly classified; The sample represented by X mark means sample is incorrectly classified; The sample represented by cross means center of a cluster. The label name of the gesture sample is similar to Fig. 2, and the label of the sample like ab means a a2b gesture sample or a b2a gesture sample.

clustering centers. In the beginning, we only use the data comes from PlaneNet for two clustering and four clustering (see Fig. 5(a) and (c)). Then we use data extracted from DepthNet for two clustering (see Fig. 5(b)). Mixed three clustering is performed using data generated by directly splicing plane motion data and depth motion data (see Fig. 5(d)). Because of the simple direct splicing, from experiments, we can see that it will get worse clustering results when mixing plane motion data and depth motion data together rather than separating them.

4.3   Gesture Classification

This section shows classification experiments. A linear SVM is used for the classifier. The ratio of training data and testing data is 7 to 3. The classification results are shown in Table 1. The range of parameters (Precision, Recall, F1-score, defined in [21]) in the table is [0, 1]. The metric from 0 to 1 represents from poor to excellent performance. Parameter support means the number of testing gestures. From experiments, we find that the classification works well on extracted data.

4.4   Semantic Association

By using clustering or classifying, the trajectory of a particular gesture can be recognized. And when we associate the semantics of the object with its motion information, it should be more straightforward to understand the target behavior. For example, we can extract trajectory of a moving cup or toothbrush, then associate gestures with semantics of the cup or toothbrush. When a toothbrush is moving horizontally, we can identify that it may be in the brushing state. While when the cup is moving from the far-bottom to the near-up, it may be in the state of drinking. An associative memory model is used to establish the connection between gestures and semantics, where Table 2 shows the performance of association.

5   Conclusion

This paper discusses a method for developing cognition behavior. Firstly, we get sequence motion data of the target object by using SSD, then two SNNs based on the Izhikevich model are organized to record two-dimensional plane motion and one-dimensional depth motion. After projection and translation, the reprocessed data is clustered and classified by SOM and SVM. Finally, the recognized gesture is combined with the semantics of the object by using the associative memory model to achieve cognitive ability. From experiments, we find that the data extracted from SNN can well preserve the temporal and spatial features of its motion and further achieve good recognition results after clustering and classification and also it can help to develop cognition to some extent by associating gestures with semantics. This method can be extended to study the relationship between the continuous motion of a single object or the motion of multiple objects.

Acknowledgments. This work was supported by the National Natural Science Foundation of China under grant number 61773271.

References

1. Yorita, A., Kubota, N.: Cognitive development in partner robots for information support to elderly people. IEEE Trans. Auton. Ment. Dev. **3**(1), 64–73 (2011)
2. Sperber, D., Wilson, D.: Relevance: Communication and Cognition, 2nd edn. Blackwell Press, Oxford (1995)
3. Saponaro, G., Salvi, G., Bernardino, A.: Robot anticipation of human intentions through continuous gesture recognition. In: 2013 International Conference on Collaboration Technologies and Systems (CTS), pp. 218–225. IEEE Press, New York (2013)
4. Tang, H., Tian, B., Shim, V.A., Tan, K.C.: A neuro-cognitive system and its application in robotics. In: 10th IEEE International Conference on Control and Automation, pp. 406–411. IEEE Press, New York (2013)
5. Tang, H., Huang, W., Narayanamoorthy, A., Yan, R.: Cognitive memory and mapping in a brain-like system for robotic navigation. Neural Netw. **87**, 27–37 (2017)
6. Roy, N., et al.: Towards personal service robots for the elderly. Carnegie Mellon University (2000)
7. Kanoh, M., Kato, S., Itoh, H.: Facial expressions using emotional space in sensitivity communication robot "ifbot". In: International Conference on Intelligent Robots and Systems, pp. 1586–1591. IEEE Press, New York (2004)
8. Hussain, S., Saxena, R., Han, X., Khan, J.A., Shin, H.: Hand gesture recognition using deep learning. In: International SoC Design Conference (ISOCC), pp. 48–49. IEEE Press, New York (2017)
9. Yan, R., Tee, K.P., Chua, Y., Li, H., Tang, H.: Gesture recognition based on localist attractor networks with application to robot control. IEEE Comput. Intell. Mag. **7**(1), 64–74 (2012)
10. Vishwakarma, D.K.: Hand gesture recognition using shape and texture evidences in complex background. In: International Conference on Inventive Computing and Informatics (ICICI), pp. 278–283. IEEE Press, New York (2017)
11. Szegedy, C., Toshev, A., Erhan, D.: Deep neural networks for object detection. In: Advances in Neural Information Processing Systems, vol. 26, pp. 2553–2561 (2013)
12. Girshick, R., Donahue, J., Darrell, T., Malik, J.: Rich feature hierarchies for accurate object detection and semantic segmentation. In: IEEE Conference on Computer Vision and Pattern Recognition, pp. 580–587. IEEE Press, New York (2014)
13. Girshick, R.: Fast R-CNN. In: IEEE International Conference on Computer Vision, pp. 1440–1448. IEEE Press, New York (2015)
14. Ren, S., He, K., Girshick, R., Sun, J.: Faster R-CNN: towards real-time object detection with region proposal networks. IEEE Trans. Pattern Anal. Mach. Intell. **39**(6), 1137–1149 (2017)
15. Redmon, J., Farhadi, A.: Yolov3: an incremental improvement. arXiv preprint arXiv:1804.02767 (2018)
16. Liu, W., et al.: SSD: single shot multibox detector. In: Leibe, B., Matas, J., Sebe, N., Welling, M. (eds.) ECCV 2016. LNCS, vol. 9905, pp. 21–37. Springer, Cham (2016). https://doi.org/10.1007/978-3-319-46448-0_2
17. Maass, W., Bishop, C.M.: Pulsed Neural Networks, 1st edn. MIT Press, Cambridge (2001)
18. Hodgkin, A.L., Huxley, A.F.: A quantitative description of membrane current and its application to conduction and excitation in nerve. J. Physiol. **117**(4), 500–544 (1952)

19. Burkitt, A.N.: A review of the integrate-and-fire neuron model: I. Homogeneous synaptic input. Biol. Cybern. **95**(1), 1–19 (2006)
20. Koch, C., Segev, I.: Methods in Neuronal Modeling: From Ions to Networks. MIT Press, Cambridge (1998)
21. Goutte, C., Gaussier, E.: A probabilistic interpretation of precision, recall and F-score, with implication for evaluation. In: Losada, D.E., Fernández-Luna, J.M. (eds.) ECIR 2005. LNCS, vol. 3408, pp. 345–359. Springer, Heidelberg (2005). https://doi.org/10.1007/978-3-540-31865-1_25

Delayed Feedback Reservoir Computing with VCSEL

Jean Benoit Héroux[✉], Naoki Kanazawa, and Daiju Nakano

IBM Research - Tokyo, Kawasaki, Kanagawa 212-0032, Japan
{heroux,knzwnao,dnakano}@jp.ibm.com

Abstract. A reservoir computing device built with directly modulated VCSEL chips and multi-mode fiber couplers is described, experimentally realized and tested for a distorted signal recovery task. Numerical and experimental results show little disparity, with a small error count in both cases. The successful realization of this low power system with an all-optical time-delay feedback and electro-optical gain located inside the physical node is promising for the realization of a high speed, board-integrated reservoir or reservoir cluster architecture.

Keywords: Vertical cavity surface emitting laser
Opto-electronic device · Neural network hardware
Reservoir computing

1 Introduction

Impressive progress has been made on the application of deep neural networks to the realization of highly complex computational tasks such as speech or image recognition. However, decreasing the amount of time, energy and computational resources required to train a neural network would be highly desirable to gain more flexibility in machine learning and for environments in which power limitation is a predominant constraint, such as wireless IOT devices. The term Reservoir Computing refers to a kind of recurrent network that may be appropriate to address this pain point, as it is easily trainable: only the weights between the neurons and the external layer are adjusted for a specific goal, while the connections between the input layer and the reservoir nodes, as well as inside the reservoir, are random and fixed [1].

While the implementation of this paradigm as an algorithm in large, high performance computers has led to state-of-the-art results for time series-based tasks such as speech word classification or non-linear pattern prediction, there is also increasing interest in the exploration of new physical media to realize compact, ultrafast and low power reservoir computer devices to fully exploit the advantage of random and fixed internal network connectivity. As the need for local, deterministic weight tuning in the material is avoided, the architecture design is greatly simplified. Reservoir computing has been investigated for

© Springer Nature Switzerland AG 2018
L. Cheng et al. (Eds.): ICONIP 2018, LNCS 11301, pp. 594–602, 2018.
https://doi.org/10.1007/978-3-030-04167-0_54

example in dedicated integrated circuits, memristive and spintronics materials, and biological environments. Photonic implementations have received particular attention due to inherently fast, low power operation and the commercial availability of a wide range of powerful tools and components primarily developed for the telecommunication market that can be adapted to this neuromorphic application.

An attractive scheme was proposed a few years ago in which the input weights are applied to a signal by a time multiplexing process to realize a delayed feedback reservoir [2]. A single node with a feedback loop is then sufficient to perform a task, further simplifying the physical realization. The output weights and final task result are obtained by a de-multiplexing calculation. In the first optical demonstrations [3,4], light from a continously emitting laser was externally modulated, sent into a fiber loop and converted to an electrical signal by a photodiode. This feedback was electronically combined with the weighted input signal for the modulation. In another variant, the laser was directly modulated and an electronic multiple delay line was included to enhance connectivity [5]. All-optical loops with a semiconductor optical amplifier [6] or fiber amplifier [7] were also reported. The potential for high speed processing was recently demonstrated in the realization of a setup with an opto-electronic feedback capable of performing a million words per second classification task [8].

Recently, we introduced an alternative implementation of time-delay photonic reservoir computer adapted from multi-mode optical interconnect technology developed for high performance computers [9,10]. Whereas in previous designs, discrete components were assembled to form a tabletop setup, in the scheme that we propose the optical signal comes from directly modulated Vertical Cavity Surface Emitting Lasers (VCSEL) that are low power, available in multi-channel arrays and could be flip-chip mounted on a board with polymer waveguide layers to be fully integrated on a substrate. The multi-mode waveguide technology that we intend to use is of a scale suitable for a light path length of tens of centimeters, allowing us to potentially obtain several virtual nodes with a delayed feedback at a 25 GS/s sampling rate. In contrast, with a reservoir that is fully integrated on a chip with single mode waveguides [11], realizing a time-delay feedback would be extremely challenging due to the millimeter-scale size and higher propagation loss.

There is currently a research trend toward the exploration of new architectures in which several reservoir computers are linked into clusters or other configurations to solve new problems more efficiently [12,13]. The technology that we describe would be suitable as a platform for the integration of multiple reservoirs. A physical neuron is formed of a photodiode connected back-to-back to the VCSEL with an electrical gain. The opto-electronic conversion is performed inside the neuron as opposed to the feedback loop, so that a node array could eventually be realized with several loops to mix physical and virtual nodes to form a complex system of reservoir clusters. A basic example of such building blocks with two reservoirs made of a neuron and loop, which has not been simulated yet and could be further scaled up, is shown in Fig. 1 to illustrate the possibilities.

The goal of this contribution is to present the first experimental results on a VCSEL-based reservoir with an optoelectronic node having integrated gain, and an all-optical time-delay feedback that has the potential advantage of scalability as described above. For this initial demonstration, the waveguides are graded-index optical fibers of a meter scale and the signal rate is 1 GS/s. At a faster input signal, the waveguide loop could be scaled down and integrated on a board. Apart from the advantages mentioned above, an integrated time delay architecture is of interest in light of recent work where back-propagation was physically implemented in a tabletop experiment to train not only the output but also the input signal weights. The system then is not categorized as a reservoir, but excellent performance for speech recognition has been reported [14]. The importance of the results presented here comes from the rich variety of potential architectures that could eventually be integrated with the technology, including hardware-based back-propagation training on a substrate.

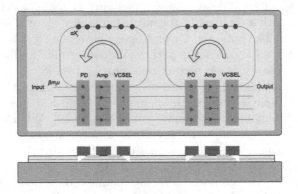

Fig. 1. Example concept of a possible inter-connected multi-reservoir architecture built with optical multi-chip module technology including polymer waveguides. The top and bottom parts show a top and cross section view, respectively. The chip arrays are drawn with four channels for illustrative purposes. The gold lines on the side view are electrical pads. The red circles illustrate virtual nodes. PD, Amp and VCSEL refer to photodiode, electrical amplification and VCSEL chip arrays flip-chip mounted on the substrate respectively. Light is coupled to and from the waveguide cores with micro-mirrors shown on the side view. The figure is not drawn to scale. The total size is in the centimeter range. This optical interconnect technology is described more in details in ref. [15] (Color figure online)

2   Experimental Setup

The time-delay experimental setup that we built is described in Fig. 2. The waveform is input using a computer program with a 2^{16} sample memory into a Field Programmable Gate Array (FPGA), whose output is amplified to obtain a peak-to-peak amplitude around 1 V suitable to modulate the first laser VCSEL1. A bias tee and high speed finger probe (not shown) provide a DC current bias around 7 mA added to the modulated signal. The VCSEL chip array is mounted on a metal holder acting as a heat sink. The output light from the VCSEL is coupled via a lens fiber attached to a probe arm with a loss around 2 dB. This optical signal is sent into one end of the first coupling fiber (brand AFOP)for optical combination with the feedback. Coupling to the active region of the photodiode PD1 is done with another lens fiber with negligible loss. The VCSEL and photodiode chip arrays operate at a 850 nm wavelength, are commercially available products (Phillips) and are designed to operate at 25 and 14 Gb/s respectively.

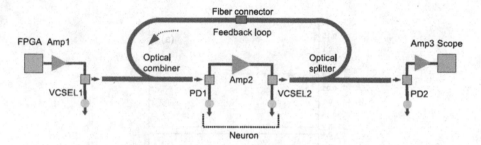

Fig. 2. Schematic description of the experimental setup used to demonstrate a VCSEL-based reservoir with all-optical feedback. Electrical lines are drawn in grey, and optical fibers in blue. The two coupling fibers are linked together by an FPC connector as shown in the loop. Circles below the VCSEL and photodiodes indicate a DC bias added to the modulated signals with bias tees. A 2 V reverse bias is applied to the photodiode PD1. (Color figure online)

The neuron is composed of the photodiode PD1, the amplifier Amp2 and the second laser VCSEL2. The amplifier is commercially available (SHF 450B), provides a gain of approximately 20 dB and has a limited output voltage amplitude, i.e., a built-in saturation which acts as a non-linear response inside the neuron. The amplified response feeds the source VCSEL2 via a high speed probe and a bias tee. The neuron output is sent into a second, identical coupling fiber via another lensed fiber with a 2 dB loss. The signal is split into two parts, 60% of the optical power going to the feedback loop and 40% to the second photodiode PD2 for readout with a time-sampling oscilloscope after photocurrent amplification with a third amplifier AMP3. For the combination PD2 and AMP3 we use a New Focus high speed photo-receiver. The total length of the loop made of the two fiber couplers connected back-to-back is 2 m approximately.

To verify the feedback delay and strength, a pulse waveform was programmed into the FPGA and input into the fibers as shown in Fig. 3. The delay between the electrical and the first output optical pulse is due to the propagation time through the first amplifier and VCSEL, optical and node delay in the main fiber arm, as well as the second photodiode and third amplifier. The amplitude of the lower portion of the optical signal is slightly lower than the upper portion, as the driving current reaches the VCSEL thresholds. This provides further non-linearity to the link.

The secondary optical pulses in Fig. 3(b) are due to the return signal of the system re-entering the node. As the setup has a low optical loss and gain in the node, care must be taken to avoid a feedback that is too strong, for which case the system becomes unstable and self-oscillates at low speed due to noise. The strength of the feedback is simply adjusted by deliberately increasing the optical loss from VCSEL2 to the lens fiber by misaligning the fiber tip position. In the figure, it is adjusted just to the edge of stability, and up to three echo pulses are observed.

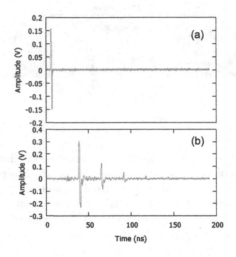

Fig. 3. (a) Experimental electrical waveform from the FPGA directly read from the oscilloscope without optical conversion. (b) Experimental optical waveform detected by the second photodiode as observed with the oscilloscope. As the oscilloscope reading is the greatest source of noise in the setup, the averaging function was used (16 times) for clarity. The negative values indicate a current below the 7 mA VCSEL DC bias, i.e., a modulated signal that decreases the optical power. The vertical scale is the voltage reading after the amplified signal from AMP3. (Color figure online)

3   Task Description and Results

For this demonstration we select a task that is a widely used benchmark in the reservoir community, namely the recovery of a distorted signal that has a practical application in the detection of a wireless signal the undergoes multiple reflections and reaches the detecting element several times with various delays and amplitudes [16]. A series of random integers d in the range $\{-3,-1,1,3\}$ is generated for a number of steps n_{max}. The distorted signal $u(n)$ having non-linear and inter-symbol interference components is computed according to the relation

$$
\begin{aligned}
q(n) = & \ 0.08d(n+2) - 0.12d(n+1) + d(n) + 0.18d(n-1) - \\
& \ 0.1d(n-2) + 0.091d(n-3) - 0.05d(n-4) + 0.04d(n-5) + \\
& \ 0.03d(n-6) + 0.01d(n-7) \\
u(n) = & \ q(n) + 0.036q^2(n) - 0.011q^3(n) + \nu(n)
\end{aligned}
\tag{1}
$$

where $\nu(n)$ is a random noise with a Gaussian distribution that we adjust to obtain a relatively low signal-to-noise ratio of 20 dB. This signal is then normalized to obtain a suitable voltage input into the VCSEL.

The reservoir response can be described in matrix form as

$$
x_i(n+1) = \begin{cases} F_{NL}[\alpha x_{i-1}(n) + \beta m_i u(n+1)] & 2 \le i \le i_{max} \\ F_{NL}[\alpha x_{N+i-1}(n-1) + \beta m_i u(n+1)] & i = 1 \end{cases}
\tag{2}
$$

where $u(n)$ is the temporal step signal input, m_i is the weight mask randomly chosen in the interval [-1:1] and $x_i(n)$ is the optical signal reaching the second photodiode PD2 corresponding to the non-linear node response F_{NL}. The input and output relative signal strengths, determined by the optical losses and gain in the system, are related by the factors β and α. The amount of data steps that can be processed at once is limited by the FPGA program and oscilloscope memory. For this demonstration we set the number of symbols to $n_{max} = 800$, and the number of mask steps (virtual nodes) to $i_{max} = 25$ for a total of 20000 input steps per measurement batch. Three data points of the FPGA are assigned to each step for a total of 60000 data points, and the step rate is 1 GHz. The readout $x_i(n)$ is obtained by numerical averaging of the oscilloscope data with 20 data points for each signal step for conversion into a matrix format.

Figure 4 shows part of a typical signal data stream as input and processed by the system. On top is a section of a calculated, distorted symbol series $u(n)$ for a given interval. In Fig. 4(b), two symbol steps are represented on which the weight mask is applied with a normalization procedure. Figure 4(c) is the experimental electronic reservoir input as measured by the oscilloscope. Finally, Fig. 4(d) shows the experimental output with optical conversion. As the delayed feedback and non-linearity are included, this signal, corresponding to Eq. (2), is different from the input and contains rich information that is essential for reservoir functionality.

To evaluate the reservoir performance, six batches of 800 input symbols were used, one for training and five for testing. Figure 5 shows the number of errors detected for each test batch with the experimental data along with a numerical result with the same input symbols for a similar system with VCSEL laser equations [10]. The important point to emphasize in this figure is that a finite number of errors is found even for the calculated result due mostly to the relatively low input signal to noise ratio of 20 dB as well as the small sample size used for

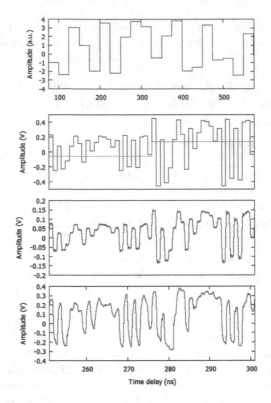

Fig. 4. (a) Section of the ideal signal input $u(n)$. The time origin corresponds to the beginning of the first step of the data stream. This particular waveform is from the second test trial of the task accomplishment. (b) Calculated curve showing two symbol steps with the weight mask applied. The red line indicates the two unmodulated symbols $u(n)$, on a different y axis scale. (c) Experimental FPGA output (reservoir input) for the same waveform. (d) Experimental reservoir output after optical processing. (Color figure online)

training. While the number of errors is slightly larger for the experimental data, it is still in the low single digits and the reservoir is functioning as intended.

Experimental results are promising considering that there is much room for more adjustment and improvement of the system, and the reasons for the disparity between the calculated and experimental performance, while not straightforward to quantify exactly, are understood. Currently, part of the experimental node non-linearity may come from the modulated input current reaching the threshold of the VCSELs which is similar to a cutoff and not an optimal non-linear response. With further tuning of the bias and amplitude, this situation could be avoided and the non-linearity could be entirely provided by the second amplifier AMP2 with a shape close to an ideal hyperbolic tangent, as in the calculations. This adjustment will be required for faster processing at 25 GS/s. Moreover, the experimental input data stream is not ideal due the use of an

FPGA as analog waveform generator. Migrating to better input and readout equipment, as well as finer tuning of the VCSEL modulation and bias, will allow us to reach similar performance for the calculated and measured waveforms. The signal feedback strength could also be increased to obtain an even larger memory effect with more optimization of the system stability and amplifier gain adjustment.

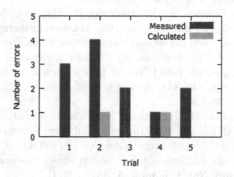

Fig. 5. Number of errors for each random test batch as found experimentally (blue) and numerically (red). (Color figure online)

4 Conclusion

In summary, we present the first experimental demonstration of a time-delay reservoir system driven by directly modulated, low power VCSEL chips having an all-optical feedback and electro-optical gain conversion within the node. With this architecture, reservoir clusters in which virtual and physical neurons could coexist could be realized and integrated on a substrate with polymer waveguides. Much simulation work will be required to identify the best configuration for a given task. Moreover, the speed limitation currently comes from the analog waveform generator equipment, and the reservoir itself would be capable of operating at a much higher data rate. An optical spike processing system could also be considered, as electronic integrating function could be integrated in the node.

References

1. Jaeger, H.: The 'echo state' approach to analyzing and training recurrent neural networks - with an erratum note, Technical report, GMD Report Number 148, Fraunhofer Institute for Autonomous Intelligent Systems (2011)
2. Appeltant, L., et al.: Information processing using a single dynamical node as complex system. Nat. Commun. **2**, 468 (2011)
3. Larger, L., et al.: Photonic information processing beyond turing: an optoelectronic implementation of reservoir computing. Opt. Expr. **20**, 3241–3249 (2012)

4. Paquot, Y., et al.: Optoelectronic reservoir computing. Sci. Rep. **2**, 287 (2012)
5. Martinenghi, R., Rybalko, S., Jacquot, M., Chembo, Y.K., Larger, L.: Photonic nonlinear transient computing with multiple-delay wavelength dynamics. Phys. Rev. Lett. **108**, 244101 (2012)
6. Duport, F., Schneider, B., Smerieri, A., Haelterman, M., Massar, S.: All-optical reservoir computing. Opt. Expr. **20**, 22783–22795 (2012)
7. Dejonckheere, A., et al.: All-optical reservoir computer based on saturation of absorption. Opt. Expr. **22**, 10868–10881 (2014)
8. Larger, L., Baylon-Fuentes, A., Martinenghi, R., Udaltsov, V.S., Chembo, Y.K., Jacquot, M.: High-Speed photonic reservoir computing using a time-delay-based architecture: million words per second classification. Phys. Rev. X **7**, 011015 (2017)
9. Héroux, J.B., Numata, H., Nakano, D.: Polymer waveguide-based reservoir computing. In: Liu, D., et al. (eds.) Neural Information Processing. ICONIP 2017. LNCS, vol. 10639, pp. 840–848. Springer, Heidelberg (2017). https://doi.org/10.1007/978-3-319-70136-3_89
10. Héroux, J.B., Numata, H., Kanazawa, N., Nakano, D.: Optoelectronic reservoir computing with VCSEL. In: 2018 International Joint Conference on Neural Networks (IJCNN). IEEE (2018). https://doi.org/10.1109/IJCNN.2018.8489757
11. Vandoorne, K., et al.: Experimental demonstration of reservoir computing on a silicon photonics chip. Nat. Commun. **5**, 3541 (2014)
12. Keuninckx, L., Danckaert, J., Van der Sande, G.: Real-time audio processing with a cascade of discrete-time delay line-based reservoir computers. Cogn. Comput. **9**, 315–326 (2017)
13. Ortın, S., Pesquera, L.: Reservoir computing with an ensemble of time-delay reservoirs. Cogn. Comput. **9**, 327–336 (2017)
14. Hermans, M., Antonik, P., Haelterman, M., Massar, S.: Embodiment of learning in electro-optical signal processors. Phys. Rev. Lett. **117**, 128301 (2016)
15. Tokunari, M., Hsu, H.H., Toriyama, K., Noma, H., Nakagawa, S.: High-bandwidth density and low-power optical MCM using waveguide-integrated organic substrate. J. Light. Technol. **32**, 1207–1212 (2014)
16. Jaeger, H., Haas, H.: Harnessing nonlinearity: predicting chaotic systems and saving energy in wireless communication. Science **304**, 78–80 (2004)

Modeling the Respiratory Central Pattern Generator with Resonate-and-Fire Izhikevich-Neurons

Pavel Tolmachev[1]([✉]) [iD], Rishi R. Dhingra[2] [iD], Michael Pauley[1] [iD],
Mathias Dutschmann[2] [iD], and Jonathan H. Manton[1] [iD]

[1] Department of Electrical and Electronic Engineering, University of Melbourne,
Parkville, VIC 3010, Australia
ptolmachev@student.unimelb.edu.au
[2] Florey Institute of Neuroscience and Mental Health, 30 Royal Parade,
Parkville, VIC 3052, Australia

Abstract. Computational models of the respiratory central pattern generator (rCPG) are usually based on biologically-plausible Hodgkin Huxley neuron models. Such models require numerous parameters and thus are prone to overfitting. The HH approach is motivated by the assumption that the biophysical properties of neurons determine the network dynamics. Here, we implement the rCPG using simpler Izhikevich resonate-and-fire neurons. Our rCPG model generates a 3-phase respiratory motor pattern based on established connectivities and can reproduce previous experimental and theoretical observations. Further, we demonstrate the flexibility of the model by testing whether intrinsic bursting properties are necessary for rhythmogenesis. Our simulations demonstrate that replacing predicted mandatory bursting properties of pre-inspiratory neurons with spike adapting properties yields a model that generates comparable respiratory activity patterns. The latter supports our view that the importance of the exact modeling parameters of specific respiratory neurons is overestimated.

Keywords: Respiratory central pattern generator · Rhythm generation
Resonate-and-fire neurons · Brainstem

1 Introduction

Respiration is one of the vital processes of life. While in simple single cell organisms respiration is driven by passive diffusion, complex organisms have developed complex breathing organs for the uptake of atmospheric oxygen and excretion of CO_2 (e.g. gills and lungs). In mammals, airflow in and out of the lungs is generated by a variety of respiratory thoracic and abdominal muscles [1], while the strength and duration of pulmonary airflow is regulated by valving muscles in the upper airways [2, 3]. The former include the diaphragm, the primary inspiratory muscle, expiratory and intercostal and finally expiratory abdominal muscles. The latter include laryngeal adductor and abductor muscles, as well as the tongue and various muscles of the soft palate and pharynx. Besides the bronchomotor muscles, all respiratory muscles are skeletal and

L. Cheng et al. (Eds.): ICONIP 2018, LNCS 11301, pp. 603–615, 2018.
https://doi.org/10.1007/978-3-030-04167-0_55

are therefore controlled by the brain. The brain breathing centers of mammals are organized in neuronal columns that span the medulla oblongata and the pons, which form the anatomical substrate for the central respiratory pattern generator (rCPG). Specific compartments of the rCPG are seen to serve a specific function in respiratory rhythm generation and formation of a three-phase sequential motor pattern compromising inspiration, postinspiration (stage I expiration) and expiration (stage II expiration) [4–11].

Over the last century, experimental data accumulated that identified the basic behavior of respiratory neurons that are distributed within specific compartments of rCPG. The class of neurons traditionally encodes the phase of neuronal activity in relation to the inspiratory activity of the diaphragm or the phrenic nerve. In addition, augmenting and decrementing discharge frequencies of these neurons are considered for classification. There is a general consensus that 5 classic respiratory neuron types form the core of the neural circuit that generates the respiratory rhythm and motor pattern: (1) rhythmogenic pre-Inspiratory (pre-I), (2) early-Inspiratory (early-I) with a decrementing discharge pattern (thus also called I-Dec), (3) Inspiratory neurons with augmenting discharge pattern (I-Aug, or ramp-I), post-Inspiratory neurons (post-I) with decrementing discharge pattern, which are active during the first part of expiration (thus also called E-Dec) and finally expiratory neurons that show augmenting discharge (E-Aug) pattern during the second phase of the expiratory interval. These neuron types form the basis for a substantial number of computational models that describe the putative function of the r-CPG. Mathematical models have focused on several dynamical mechanisms including the biophysical bursting properties of rhythmogenic pre-I that are seen to initiate the respiratory cycle [12–14], network oscillation based on reciprocal synaptic inhibition [15] and hybrid models based on excitatory rhythmogenic cell properties and inhibitory synaptic inhibition [7, 16]. The latter sometimes even implement sensory feedback loops [17, 18]. The contemporary hybrid models are complex Hodgkin-Huxley-based models. The main driver for Hodgkin-Huxley-based modeling has largely arisen from the finding in the early nineties that a specific subset of neurons located in the pre-Bötzinger complex (pre-BötC) remain rhythmogenic when isolated from the larger network [19]. Electrophysiological studies of pre-BötC neurons revealed the biophysical basis of pacemaker pre-I neurons and the excitatory synaptic coupling underlying group pacemaker mechanisms [20]. However, the biophysical properties of neurons outside the pre-BötC remain largely unexplored. Thus, the biophysical properties of non-pre-BötC neurons in modeling approaches are often based on speculation. Even more compelling is that fact that biophysical properties (ion channel composition) of bursting neurons in the somatogastric ganglion, an invertebrate model system of a CPG network, are extremely diverse in functionally homogenous neurons and therefore do not define the function of neurons in a rhythmogenic circuit [21] that shares significant similarity with the rCPG.

To simplify the complex and difficult to assess Hodgkin-Huxley-based models of the respiratory CPG, we implement the rCPG using Izhikevich resonate-and-fire neurons. These neurons are simple enough to be computationally efficient and tractable for bifurcation analysis and at the same time can account for various neural activity patterns observed within the respiratory circuit *in vivo*, including intrinsic bursting and spike adaptation [22].

The remainder of the manuscript is organized as follows. In Sect. 2, we present a description of the model. In Sect. 3, we present our results regarding how the model can reproduce previous experimental and theoretical observations, and how the model can be used to test the hypothesis that intrinsically bursting neurons are necessary for the expression of the three-phase respiratory rhythm. Finally, in Sect. 4, we present our overall conclusions and discuss the utility of the model for future research on respiratory rhythm and pattern generation.

2   Modeling Description

2.1   Izhikevich Neurons

To construct a model of the rCPG, which is flexible and more transparent than models utilizing Hodgkin-Huxley neurons, we employed Izhikevich neurons. The model of an Izhikevich neuron consists of two differential equations, resulting in a 2-dimensional system. The phase plane trajectory of these neurons is thereby easily visualized, making it tractable for bifurcation analysis. This simplifies the task of finding parameters to achieve a desired neuronal behavior.

The general form of the equations is presented below:

$$\frac{dv}{dt} = \alpha(v - v_0)^2 + V_b - xu \tag{1}$$

$$\frac{du}{dt} = a(bv - u) \tag{2}$$

with two additional resetting conditions as the spike occurs:

$$if \ V = V_{threshold} : V \rightarrow V_{reset}; \ u \rightarrow u + d \tag{3}$$

Here v is the membrane potential in mV, and u is an adaptation variable. The greater the parameter u, the slower the voltage rises after resetting. $\alpha, v_0, V_b, a, b, x$ are the parameters of the system whose geometric meaning can be seen in Fig. 1. The sensitivity of the membrane potential to adaptation is governed by parameter x. Parameter α is proportional to the width of the parabolic-shaped v-nullcline. v_0 and V_b define the coordinates of the peak of the v-nullcline. The slope of u-nullcline is set by a parameter b. The parameter a is responsible for the rate of relaxation of the adaptation variable.

Our model of the rCPG includes only two sets of parameters for a single neuron to reproduce the two different neural behaviors critical to the established connectivity of the rCPG: spike adaptation—a gradual decline of the frequency given constant tonic input—and intrinsic bursting—the alternation of spiking and quiescent periods due to an oscillation in the slow-subsystem. By visualizing the phase plane of Izhikevich neurons, we were able to find suitable sets of parameters to account for these properties (see Appendix). We also ensured that these parameter sets were associated with biologically plausible firing frequencies.

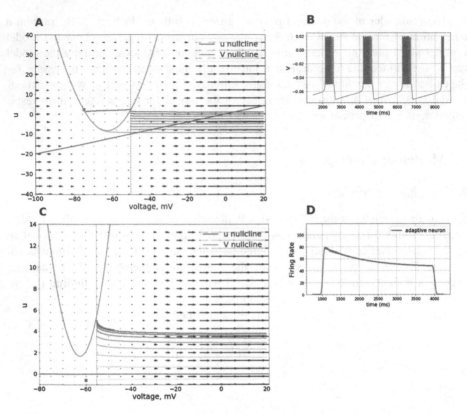

Fig. 1. Examples of a phase plane diagram and the corresponding behaviour of the Izhikevich model for a bursting (top: A, with v plotted against the time in B) and adaptively spiking neuron (bottom: C, with firing rate plotted against time in D). V-nullcline from the expression (1) has a parabolic shape. The u-nullcline is a straight line. The arrows depict the vector field of $\left(\frac{dv}{dt}, \frac{du}{dt}\right)$, where derivatives are the right-hand side of the expressions (1) and (2). The green dashed line indicates the resetting voltage in Eq. (3). The red line indicates the phase trajectory of the neuron: the brighter the line the more recently the system was in that point of the phase-plane. The cross marks indicate the initial conditions that the system evolved from. The adaptation of firing rate (C) occurs because initially a neuron's state is in the region where the time derivative of voltage is larger. The decline of the firing rate continues until the neuron reaches a dynamic equilibrium: after each spike the adaptation parameter is incremented by value d, but during the spike initiation it decays back by the same amount. The derivative of voltage along the resulting stationary trajectory is less than in the regions with smaller u, resulting in the decreased firing rate. In case of bursting (A), the dynamical equilibrium is not reached: eventually, after some spike, the system's state goes into a region inside the parabola. That moment in time marks the end of bursting. The state of the neuron evolves in the phase-space according to a vector field, and the neuron's voltage drops to the minimum when the state reaches the left branch of the parabola. The neuron initiates a new bursting after the adaptation parameter relaxes below the parabola's peak. Our rCPG model is composed strictly of neurons with one of these behaviors. (Color figure online)

To account for the synaptic interactions between the neurons, one has to introduce an extra term into Eq. (1):

$$I_{syn} = g(v - E_{syn}) \tag{4}$$

where g is the conductance of the synapses of a particular neuron and is comprised of two parts:

$$g = g_{net} + g_{tonic} \tag{5}$$

The parameter g_{net} is the network contribution into the overall conductance of the synapses, g_{tonic} is the baseline conductance due to activity of the tonic drive populations. E_{syn} is the synaptic reversal potential: for excitatory synapses, E_{syn} is around -10 mV, for inhibitory synapses, E_{syn} is approximately -70 mV. The synaptic conductance g_{net}, in turn, is subject to the following differential equation with an extra condition on the arrival of the spike from the presynaptic neuron:

$$\frac{dg_{net}}{dt} = \frac{-g_{net}}{\tau} \tag{6}$$

$$if\ spike \in presynaptic\ neuron : g_{net} \to g_{net} + \Delta \tag{7}$$

Thus, the network contribution to the synaptic conductance exponentially decays back to 0, unless the spike arrives from an afferent neuron. The arrival of the spike causes the conductance to swiftly rise by a level defined by parameter Δ (which represents the synaptic efficacy).

All simulations were implemented using the *Brian2* simulation package.

2.2 Model Description

The baseline synaptic connectivity in our model implemented the conceptual model described in [1] (Fig. 2).

The effects of synaptic plasticity are considered to be negligible in CPGs, and in our models, the synaptic strengths for all synapses were set equal. The probabilities of connection between the neuronal populations are listed in Table 1 (see Appendix). The tonic drives for the various groups of neurons are shown in Table 2. In our simulations, each neural group represents a population of 100 neurons with identical parameters, which are described in Table 3 (see Appendix). In the simulations, we have set random initial conditions for the neurons in the model. To account for the heterogeneity, we have also introduced some variability of the synapses' responses to spikes. The parameters d, an increment of the parameter u after the arrival of a presynaptic spike, and Δ, or the synaptic efficacy, are normally distributed random variables with the variance equal 10% of the mean.

In the model, we set the activity of HN identical to the activity of pre-I neurons. The PN activity is set to equal to the activity of ramp-I neurons. The activity of VN was

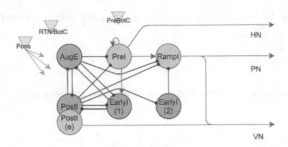

Fig. 2. Network connectivity between six populations of neurons of the rCPG. Blue circles denote inhibitory populations, whereas red circles represent excitatory populations. Post-I neurons have both inhibitory and excitatory subpopulations, which receive almost identical inputs from the rest of the network. Inhibitory connections are depicted by blue lines with a filled circle. The red arrows signify excitation between neural groups. All of the groups receive tonic input from three major sources: PreBötC, RTN/BötC and the pontine compartments. These tonic drives modify the conductance of the neurons by setting a baseline conductance g_{tonic}. The parameters for the tonic drives for each neuronal group are described in the Table 2 (see Appendix). The activities of respiratory neurons are projected to three nerve fibres: Hypoglossal nerve (HN), Phrenic nerve (PN) and Vagal nerve (VN). (Color figure online)

described as the combination of activities of post-I(e) neurons in conjunction with ramp-I neurons with the weighted coefficients 0.75 and 0.25, respectively.

3   Results

3.1   rCPG Replicates the Properties of Hodgkin-Huxley-Based Models

Under normal metabolic conditions, the circuit operates in the three-phase mode described in [16]. Our model demonstrates results consistent with both experimental observations and the model based on the Hodgkin-Huxley neurons [16]. The results of our model simulations are presented in Fig. 3 and are summarized in the next paragraph.

Post-I neurons escape from inhibition by early-I (1) neurons and start to fire. Their firing rate declines through the E_1 phase and reaches the point when the post-I activity is no longer sufficient to fully inhibit aug-E neurons. Aug-E neurons exhibit an incrementing discharge pattern during the E_2 stage, while the activity of the post-I neurons continues to decline. Eventually, inhibition of pre-I neurons by aug-E and post-I neural groups can no longer be maintained, and pre-I neurons start to fire. Since pre-I neurons excite the early-I neurons, the latter also escape from inhibition arising from aug-E and post-I neurons. Early-I neurons then inhibit aug-E and post-I groups and fire throughout the inspiratory phase. An intrinsic adaptation of early-I neurons results in decrementing firing rate until early-I activity is no longer sufficient to inhibit the post-I population. Then, the cycle repeats.

Our model of the rCPG, although lacking its biophysical properties, robustly reproduces the modeling results of similar Hodgkin-Huxley-based models. The discharge patterns of the neurons are also consistent with experimental data, except for the shape of activity of the pre-I neurons, for which the experimental data suggests that the firing rate of pre-I neurons is incrementing. A possible solution to reconcile this discrepancy is also purely mechanistic and not dependent on a neuron's biophysics: one may introduce into the model a new group of decrementing inhibitory neurons, which fire in the inspiratory phase and inhibit the pre-I neurons (data not shown).

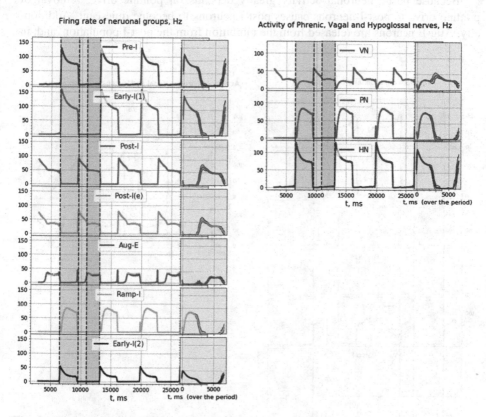

Fig. 3. The simulated activities of the respiratory neurons (left) and nerves which drive the muscles (right: vagal nerve – VN, phrenic nerve – PN and hypoglossal nerve – HN) during normal metabolic conditions. The horizontal axis depicts simulated time in milliseconds. The model robustly reproduces the basic three-phase rhythm: on this and the following figures the example of the inspiratory phase is highlighted by red. E1 and E2 phases of expiration are denoted by the green and the blue shadings respectively. The gray shaded insets on the right represent the firing rate averaged consecutively (1) over the period and (2) over 50 different trials. The variability of a signal within one standard deviation is depicted on each inset by a shading between two black lines with mean laying in-between. The variability of the signal comes from (1) randomness in initial conditions and (2) stochastic character of the connectivity. The average period T = (6372 ± 675) ms, CV = 0.106. The firing patterns of the neurons are consistent with those described in [16]. (Color figure online)

3.2 Simulations of Pontine Transection Are Consistent with Experimental Data and Previous Models

To further test the validity of our model, we have simulated pontine transection, which experimentally causes apneusis—a breathing pattern characterized by prolonged inspiration and an absence of the post-inspiratory phase activity. A representative example of this simulation is presented in Fig. 4.

Because post-I neuronal activity greatly depends on pontine drive, removal of pontine drive to post-I neurons causes post-I neurons to become quiescent. Additionally, Aug-E neurons are released from the inhibition from the post-I population, and, in

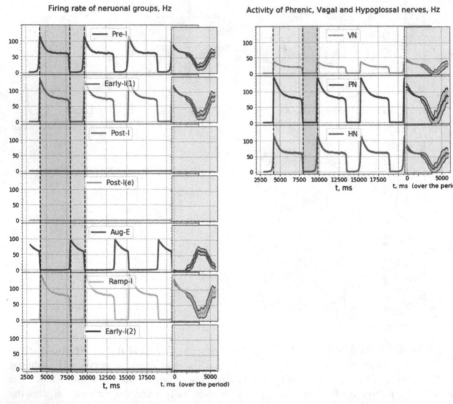

Fig. 4. Removal of pontine drive causes apneusis. The average period of breathing activity is (5620 ± 819) ms, CV = 0.146. The increased variability of the period is consistent with the existing experimental data [26]. The removal of the pontine drive causes the network to produce oscillations with increased inspiratory period and decreased expiratory period in comparison to the inspiratory and expiratory periods of the network operating in intact conditions (see Fig. 3). The apneustic breathing cycle is comprised only of two phases: inspiration (depicted by red shading) and expiration (denoted by green shading). The insets with grey shading depict one period of the activity and show mean and the standard deviation of simulations averaged over 50 trials. This apneustic breathing pattern is consistent with the experimental data. (Color figure online)

conjunction with early-I neurons, these two groups form a half-center oscillator. The oscillator produces a two-phase apneustic respiratory rhythm. Our simplified model produces neuronal and motor outputs which have the same patterns as earlier models and is also consistent with the data obtained from experiments.

3.3 Intrinsically Bursting Properties Are Not Essential to Generate the Three-Phase Respiratory Rhythm

Using the simplified model, we were able to test whether intrinsic bursting properties were necessary for the generation of the three-phase rhythm by introducing only minor changes to the model. In this section, we discuss the effect of switching pre-I neurons' dynamics from a pacemaker to an adaptively spiking parameter set (Fig. 5). To adjust

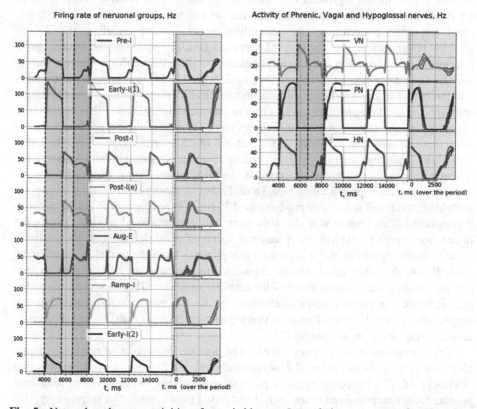

Fig. 5. Network and motor activities after switching pre-I population parameters from bursting to adaptively spiking neurons. To adjust the period of inspiration we have also decreased tonic drive to pre-I neurons while making them adaptively spiking. While there was a minor change in the respiratory period, the discharge patterns of rCPG populations and motor nerves were qualitatively similar to the original pacemaker-dependent model (compare with Fig. 3). The three phases - Inspiration, postinspiration (E1) and expiration (E2) are denoted by red, green and blue shadings respectively. The average period of a respiratory cycle T = (4380 ± 227) ms, CV = 0.052. (Color figure online)

the period of inspiration we have also decreased tonic drive to pre-I neurons. Our simulations show that the network may produce the three-phased rhythm even with non-bursting pre-I neurons. According to our model of the respiratory CPG, switching the dynamics of pre-I neurons did not result in qualitative changes. Even with adaptively spiking pre-I neurons, the network still produced a the three-phase rhythm similar to that produced by the unperturbed model and described in the Sect. 3.1. Our results therefore suggest that the role of the pacemaker neurons in the network under normal metabolic conditions may be overestimated.

4   Conclusion

We conclude that (1) the simplified model reproduces the three-phase respiratory pattern, (2) the model reproduces the effect of the removal of pontine components of the rCPG, and (3) intrinsic bursting properties are not necessary to generate the three-phase respiratory pattern given the currently accepted connectivity of the rCPG.

The respiratory cycle consists of three sequential 'phases': inspiration, post-inspiration and expiration. Neuronal populations within the rCPG are active during specific respiratory phases. Like previous models [7, 16, 17], our simplified resonate-and-fire model largely reproduces this distinct respiratory motor pattern. Employing the resonate-and-fire neurons allowed us to focus on connectivity, rather than biophysics.

However, the current model also shows some instability during the phase transitions that are reflected by short bursts that are quickly terminated when the phase irreversibly changes (see Fig. 3). This is likely due to a competition between the three reciprocally-connected inhibitory populations that arises from the variability of synaptic connection probabilities (see Table 1). For a short period of time during the switching, combined activity of pre-I and post-I groups is not enough to inhibit the aug-E population. This feature was also seen in similar computational models [25]. In the future, we expect to explore the parameter space of possible connectivities using a genetic search algorithm and will remove this property from the model to simplify the analysis. On the other hand, similar transition instabilities can be observed experimentally when pontine components of the network are absent. Previous work suggested that such aberrant phrenic nerve discharges may be due to the widespread excitatory projections of medullary pre-I neurons that trigger synchronous bursting in functionally different respiratory motor output [23].

The current model reproduces the effect of pontine transection or pontine synaptic blockade as it has been observed experimentally [8, 16, 24] and in the corresponding models [7, 16, 17]. In accordance with previous models, this was achieved by removing pontine tonic synaptic input to the early-I and post-I populations. Not surprisingly, this eliminates the synaptic instability during phase transitions. In the future, we hope to use this model to explicitly consider alternative connectivities between medullary and pontine respiratory neurons, which are currently oversimplified.

Interestingly, we observed that replacing intrinsically bursting pre-I neurons with adaptively spiking pre-I neurons did not qualitatively change the three-phase rhythm. This supports our view that biophysical properties of a specific subset of respiratory neurons might be overstated. Thus, we aim to utilize the current model for further investigation of the dependence of respiratory rhythm on respiratory circuit connectivity. While part of our motivation for the current model was to reduce the parameter space of the model, because our model includes many neurons, it ultimately is still prone to overfitting. Others have accomplished this task more effectively [25] by developing a squashing function to reduce a population of identical Hodgkin-Huxley neurons with slow synaptic transmission into a single Hodgkin-Huxley neuron's firing rate. This allowed for bifurcation analysis at a network level to understand the dynamic regulation of phase transition, the hallmark mechanism for respiratory pattern formation that defines breathing.

Appendix

See Tables 1, 2 and 3.

Table 1. Synaptic connectivities.

Synapse	Probability of connection
Pre-I → Pre-I	0.125
Pre-I → Early-I(1)	0.8
Aug-E → Pre-I	0.06
Aug-E → Early-I(1)	0.5
Early-I(1) → Aug-E	0.5
Aug-E → Early-I(1)	0.5
Post-I → Early-I(1)	0.5
Post-I → Aug-E	0.7
Early-I(1) → Post-I	0.5
Aug-E → Post-I	0.1
Post-I → Pre-I	0.15
Aug-E → Post-I(e)	0.13
Early-I(1) → Post-I(e)	0.5
Pre-I → Ramp-I	0.625
Early-I(1) → Ramp-I	0.625
Aug-E → Ramp-I	0.5
Post-I → Ramp-I	0.2
Early-I(2) → Ramp-I	0.8
Aug-E → Early-I(2)	0.2
Post-I → Early-I(2)	0.2

Table 2. Tonic drives to neuronal groups.

Drive → Population ↓	From PreBötC	From RTN/BötC	From Pons
Pre-I	0.1	0.2	0.3
Early-I(1)	0	0.6	0.5
Aug-E	0	1	0.8
Post-I	0	0	0.9
Post-I(e)	0	0	0.6
Ramp-I	0	0	0
Early-I(2)	0	0	0.2

Table 3. Parameters for a single neuron.

Population	Bursting	Adaptation
α	0.004	0.004
v_0	−62.5	−62.5
V_b	−1.6	0.0
a	0.001	0.0005
b	0.2	0.0
E_{synE}	−10	−10
τ_E	10	10
g_{netE}	0.1	0.33
g_{tonicE}	0.1	0.1
E_{synI}	−75	−75
τ_I	15	15
g_{netI}	0.1	1.0
Δ	0.08	0.08
V_{reset}	−50	−55
$V_{threshold}$	20	20
d	0.3	0.5
x	0.06	0.06

References

1. Feldman, J.L.: Neurophysiology of breathing in mammals. Handb. Physiol. Nerv. Syst. Am. Physiol. Soc. Sect. **1**, 463–524 (1986)
2. Dutschmann, M., Paton, J.F.: Inhibitory synaptic mechanisms regulating upper airway patency. Respir. Physiol. Neurobiol. **131**(1-2), 57–63 (2002)
3. Dutschmann, M., Jones, S.E., Subramanian, H.H., Stanic, D., Bautista, T.G.: The physiological significance of postinspiration in respiratory control. In: Progress in brain research, vol. 212, pp. 113–130. Elsevier (2014)
4. Richter, D.W.: Generation and maintenance of the respiratory rhythm. J. Exp. Biol. **100**(1), 93–107 (1982)
5. Richter, D.W., Spyer, K.M.: Studying rhythmogenesis of breathing: comparison of in vivo and in vitro models. Trends Neurosci. **24**(8), 464–472 (2001)
6. Feldman, J.L., Del Negro, C.A.: Looking for inspiration: new perspectives on respiratory rhythm. Nature Rev. Neurosci. **7**(3), 232 (2006)
7. Rybak, I.A., Abdala, A.P., Markin, S.N., Paton, J.F., Smith, J.C.: Spatial organization and state-dependent mechanisms for respiratory rhythm and pattern generation. Prog. Brain Res. **165**, 201–220 (2007)
8. Dutschmann, M., Dick, T.E.: Pontine mechanisms of respiratory control. Compr. Physiol. **2**(4), 2443 (2012)
9. Smith, J.C., Abdala, A.P., Borgmann, A., Rybak, I.A., Paton, J.F.: Brainstem respiratory networks: building blocks and microcircuits. Trends Neurosci. **36**(3), 152–162 (2013)
10. Anderson, T.M., Ramirez, J.M.: Respiratory rhythm generation: triple oscillator hypothesis. F1000Research **6**, 139 (2017)
11. Del Negro, C.A., Funk, G.D., Feldman, J.L.: Breathing matters. Nat. Rev. Neurosci. **19**, 351–367 (2018)
12. Butera Jr., R.J., Rinzel, J., Smith, J.C.: Models of respiratory rhythm generation in the pre-Botzinger complex. I. Bursting pacemaker neurons. J. Neurophysiol. **82**(1), 382–397 (1999)
13. Butera Jr., R.J., Rinzel, J., Smith, J.C.: Models of respiratory rhythm generation in the pre-Botzinger complex. II. Populations of coupled pacemaker neurons. J. Neurophysiol. **82**(1), 398–415 (1999)
14. Del Negro, C.A., Johnson, S.M., Butera, R.J., Smith, J.C.: Models of respiratory rhythm generation in the pre-Botzinger complex. III. Experimental tests of model predictions. J. Neurophysiol. **86**(1), 59–74 (2001)
15. Ogilvie, M.D., Gottschalk, A., Anders, K., Richter, D.W., Pack, A.I.: A network model of respiratory rhythmogenesis. Am. J. Physiol. Regul. Integr. Comp. Physiol. **263**(4), R962–R975 (1992)
16. Smith, J.C., Abdala, A.P.L., Koizumi, H., Rybak, I.A., Paton, J.F.: Spatial and functional architecture of the mammalian brain stem respiratory network: a hierarchy of three oscillatory mechanisms. J. Neurophysiol. **98**(6), 3370–3387 (2007)
17. Rybak, I.A., et al.: Modeling the ponto-medullary respiratory network. Respir. Physiol. Neurobiol. **143**(2–3), 307–319 (2004)
18. Molkov, Y.I., Bacak, B.J., Dick, T.E., Rybak, I.A.: Control of breathing by interacting pontine and pulmonary feedback loops. Front. Neural Circ. **7**, 16 (2013)
19. Smith, J.C., Ellenberger, H.H., Ballanyi, K., Richter, D.W., Feldman, J.L.: Pre-Botzinger complex: a brainstem region that may generate respiratory rhythm in mammals. Science **254**(5032), 726–729 (1991)
20. Del Negro, C.A., Morgado-Valle, C., Feldman, J.L.: Respiratory rhythm: an emergent network property? Neuron **34**(5), 821–830 (2002)

21. Schulz, D.J., Goaillard, J.M., Marder, E.: Variable channel expression in identified single and electrically coupled neurons in different animals. Nat. Neurosci. **9**(3), 356 (2006)
22. Izhikevich, E.M.: Simple model of spiking neurons. IEEE Trans. Neural Netw. **14**(6), 1569–1572 (2003)
23. Jones, S.E., Dutschmann, M.: Testing the hypothesis of neurodegeneracy in respiratory network function with a priori transected arterially perfused brain stem preparation of rat. J. Neurophysiol. **115**(5), 2593–2607 (2016)
24. Dhingra, R.R., Jacono, F.J., Fishman, M., Loparo, K.A., Rybak, I.A., Dick, T.E.: Vagal-dependent nonlinear variability in the respiratory pattern of anesthetized, spontaneously breathing rats. J. Appl. Physiol. **111**(1), 272–284 (2011)
25. Rubin, J.E., Shevtsova, N.A., Ermentrout, G.B., Smith, J.C., Rybak, I.A.: Multiple rhythmic states in a model of the respiratory central pattern generator. J. Neurophysiol. **101**(4), 2146–2165 (2009)
26. Dhingra, R.R., Dutschmann, M., Galán, R.F., Dick, T.E.: Kölliker-Fuse nuclei regulate respiratory rhythm variability via a gain-control mechanism. Am. J. Physiol. Regul. Integr. Comp. Physiol. **312**(2), R172–R188 (2016)

Proposal of Carrier-Wave Reservoir Computing

Akira Hirose[1(✉)], Gouhei Tanaka[1], Seiji Takeda[2], Toshiyuki Yamane[2],
Hidetoshi Numata[2], Naoki Kanazawa[2], Jean Benoit Heroux[2], Daiju Nakano[2],
and Ryosho Nakane[1]

[1] The University of Tokyo, Tokyo, Japan
ahirose@ee.t.u-tokyo.ac.jp
[2] IBM Research - Tokyo, Kawasaki, Japan
http://eeip.t.u-tokyo.ac.jp/

Abstract. Reservoir computing is highly compatible with physical waves such as optical wave and spin wave. In such wave-realized reservoir computing, signals are not limited to the raw signals, or baseband signals, but also expressed as frequency-shifted signals having a carrier. This paper proposes such a reservoir computing architecture, namely, carrier-wave reservoir computing. We present its construction and suitable learning dynamics with which we deal with the phase information explicitly. The merits of the proposed carrier-wave reservoir computing are its frequency-dependent processing functions, frequency-domain multiplexing ability and explicit phase-information utilization. It is also useful for material evaluation from the viewpoint of computational ability in the reservoir computing.

Keywords: Reservoir computing · Complex-valued neural networks
Phase-sensitive learning

1 Introduction

Reservoir computing attracts many researchers as a promising neural network architecture to deal with time-sequential data in the near future mainly because of its low energy consumption and high speed learning [1,10,18,28,30]. It includes echo state neural networks and other networks based on similar ideas [19]. Theoretical analysis and simulations have revealed its various natures such as the powerfulness in the operation at the so-called edge of stability [34] and the general relationship between the embedding dimensions and operation errors [33]. Its attractive characteristics in hardware architecture have also been reported, for example, on how to construct a reservoir computer by using memristors [26], and the fact that variability works favorable for efficient learning [25]. Demonstrations on practical tasks are also reported recently, namely, the distinction of sinusoidal time-sequential signals from triangular waves [27] and simple speech recognition [29].

© Springer Nature Switzerland AG 2018
L. Cheng et al. (Eds.): ICONIP 2018, LNCS 11301, pp. 616–624, 2018.
https://doi.org/10.1007/978-3-030-04167-0_56

The use of waves is also emphasized in these years. For example, synchronization of connected oscillations works as a reservoir in total [32]. Lightwave reservoirs are analyzed on its online training methods [2] and on the optimization of external feedback parameters [24].

Another wave reservoir computing employs spin wave [22,23]. The paper reported a successful estimation of square-wave duration when a baseband rectangular wave is fed to the reservoir. This spin-wave reservoir computing hardware also utilizes wave phenomenon, in particular its hysteresis, nonlinearity and asymmetry. Here, hysteresis means that input-signal effects depend on the front-end states of the reservoir. Nonlinearity includes the nonlinear interference generating various harmonics. Asymmetry is the anisotropy observed in spin wave propagation in two- or three-dimension.

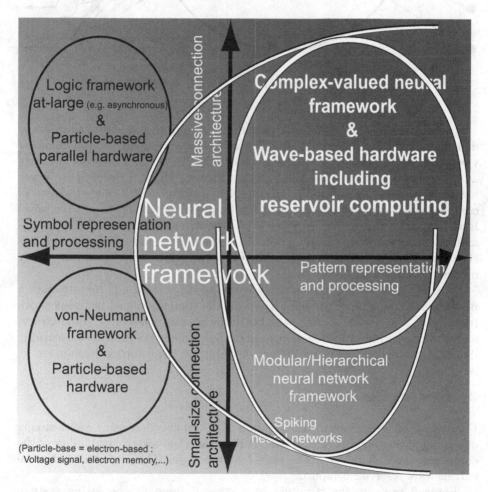

Fig. 1. Positions and relationship of various computations in the coordinate of connection scale and information representation [8].

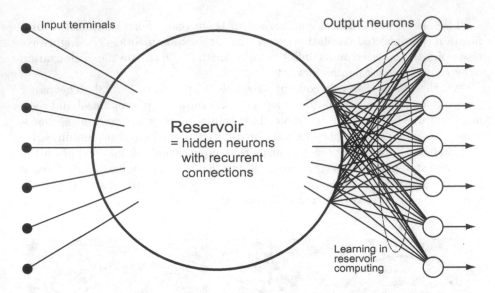

Fig. 2. Basic construction of a reservoir neural network.

In parallel, dispersion is also a potentially useful physics. Dispersion is the dependence of wave speed on its frequency. It is directly related to the following viewpoint for physical/material reservoirs [3,20,21]. We can analyze, evaluate and discuss the nature of the material that forms a physical reservoir by focusing on the functions realized in the physical reservoir. The characteristics of computation in a material depends on the frequency of the signal, i.e., the wave. If we can assign the carrier frequency explicitly, we can elucidate the computing-material characteristics as a function of the frequency, and can even modulate the characteristics to extend the freedom of computation and to realize frequency multiplexing [11–15,17].

In this paper, we propose carrier-wave reservoir computing. We introduce a phase sensitive learning dynamics based on the complex-valued neural-network framework [8,9]. We focus on the weights in the neural output layer, that extracts and synthesizes desirable time-sequential output signals. We present also learning dynamics to realize a good generalization ability in the complex domain.

2   System Construction

Figure 1 is a diagram showing the position of neural networks using waves among other computing architectures widely. It illustrates the wide extendability of the complex-valued neural networks placed at top right. This region includes the carrier-wave reservoir computing.

Figure 2 presents the structure of a general reservoir computing system. Time-sequential signals are fed to the reservoir module to be transformed into high-dimensional time-sequential data. The data is selected and/or weighted and

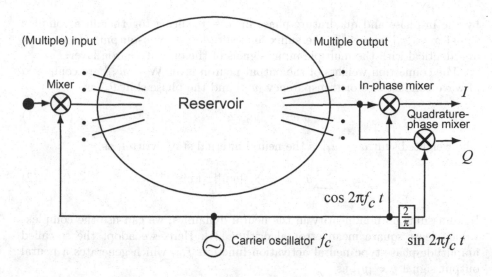

Fig. 3. System construction of the proposed carrier-wave reservoir computing.

extracted by the neural output layer to generate desired time-sequential signals. This paper gives a supervised learning dynamics to be used in this layer later. Though, in some cases, the activation function may be linear in this output layer, we assume a nonlinear activation function for a wider availability in this paper.

Figure 3 shows the construction of the proposed carrier-wave reservoir computing system by focusing on the specific input feeding part and the output readout part including the relationship with the carrier wave. We prepare a carrier oscillator, such as an electrical oscillator or a diode laser, oscillating at frequency f_c. A signal having frequency f_{sig}, which is usually distributed, is mixed by a mixer, or an optical modulator such as acoust-optical modulator (AOM), to become a carrier signal of frequency $f_{sig} + f_c$. Unlike ordinary baseband signals, the carrier signal can realize a carrier-frequency dependent reservoir function. This fact results also in the possibility of frequency-domain multiplexing [13–15]. We can also utilize the phase information in information representation and processing.

The carrier signal is converted down to a baseband signal before it is fed to an analog-to-digital converter (ADC) and to the output neuron layer. We use mixers again, or the ADC directly for this down-conversion. The next section derives phase-sensitive learning dynamics.

3   Learning Dynamics in Complex Domian

In this section, we consider the complex-valued steepest descent dynamics in a single-layered output neural network in a carrier-wave neural network shown in Fig. 3. We employ an approach similar to coherent lightwave computing system [4–7, 15, 16]. The carrier-wave signal generated in the reservoir module is detected by the in-phase and quadrature-phase mixers, resulting together in a complex signal $\boldsymbol{x} = [x_i]$ where i is the suffix for multiple reservoir-output signals, that are identical with the multiple input signals of the output neuron layer.

The connection weights of the output neuron layer $\mathbf{W} = [w_{ji}]$ are composed of wave conductance or transparency $|w_{ji}|$ and the phase shift θ_{ji} as

$$w_{ji} = |w_{ji}| \ e^{i\theta_{ji}} \tag{1}$$

The weighted sum $\boldsymbol{u} = [u_j]$ is the neural internal state written as

$$u_j = \sum_i w_{ji} x_i = \sum_i |w_{ji}||x_i| \ e^{i(\theta_{ji} + \theta_i)} \tag{2}$$

If we assume no nonlinearity in the neural dynamics, we can use the complex-valued least square mean error algorithm [31]. Here, we adopt the so-called amplitude-phase-type neural activation function f_{ap}, which generates a neural output signal $\boldsymbol{y} = [y_j]$ as

$$y_j = f_{\mathrm{ap}}(u_j) = \tanh(|u_j|) \ \exp(i \arg(u_j)) \tag{3}$$

That is, the function makes the amplitude saturate, but no change in the phase. We define an error function E as the sum of the squared difference between the teacher signals $\hat{\boldsymbol{y}} = [\hat{y}_j]$ and the output signals $\boldsymbol{y}$ under learning as

$$
\begin{aligned}
E_l &\equiv \frac{1}{2}|\boldsymbol{y} - \hat{\boldsymbol{y}}|^2 = \frac{1}{2}\sum_j |y_j - \hat{y}_j|^2 \\
&= \frac{1}{2}\sum_j \Big(\tanh^2(|u_j|) + \tanh^2(|\hat{u}_j|) \\
&\qquad\qquad - 2\tanh(|u_j|)\tanh(|\hat{u}_j|)\cos(\theta_j - \hat{\theta}_j) \Big)
\end{aligned} \tag{4}
$$

Here, $\hat{\boldsymbol{u}} = [|\hat{u}_j|e^{i\hat{\theta}_j}]$ is a variable that is defined as the equivalent internal state to generate the output teacher signal $[\hat{y}_j]$.

Then, we derive the amplitude-phase-type steepest descent change in terms of w_{ji} as follows. Note in general that the changes in amplitude $|u_j|$ and phase θ_j of the internal state are orthogonal to each other. We sum separately-obtained partial derivatives in the amplitude and phase directions. Since the derivative of the nonlinear function $\tanh(|u|)$ gives $(1 - \tanh^2(|u|))$, we obtain

$$
\begin{aligned}
\frac{\partial E}{\partial(|u_j|)} &= \frac{1}{2}\Big(2\tanh(|u_j|)(1 - \tanh^2(|u_j|)) \\
&\qquad - 2(1 - \tanh^2(|u_j|))\tanh(|\hat{u}_j|)\cos(\theta_j - \hat{\theta}_j) \Big) \\
&= (1 - |y_j|^2)(|y_j| - |\hat{y}_j|\cos(\theta_j - \hat{\theta}_j)) \tag{5}
\end{aligned}
$$

$$\frac{\partial E}{\partial \theta_j} = |y_j||\hat{y}_j|\sin(\theta_j - \hat{\theta}_j) \tag{6}$$

This is the same case as the coherent lightwave neural networks [14]. Then, we obtain the learning rule in the two directions in the complex domain for the amplitude $|w_{ji}|$ and the phase θ_{ji} of the complex-valued weight as

$$\tau\frac{d(|w_{ji}|)}{dt} = -\frac{\partial E}{\partial(|w_{ji}|)} = -\Big((1 - |y_j|^2)(|y_j| - |\hat{y}_j|\cos(\theta_j - \hat{\theta}_j))\, |x_i|\, \cos\theta_{ji}^{\mathrm{rot}}$$

$$-|y_j||\hat{y}_j|\sin(\theta_j - \hat{\theta}_j)\frac{|x_i|}{|u_j|}\sin\theta_{ji}^{\mathrm{rot}}\Big) \tag{7}$$

$$\tau\frac{d\theta_{ji}}{dt} = -\frac{1}{|w_{ji}|}\frac{\partial E}{\partial\theta_{ji}} = -\Big((1 - |y_j|^2)(|y_j| - |\hat{y}_j|\cos(\theta_j - \hat{\theta}_j))\, |x_i|\, \sin\theta_{ji}^{\mathrm{rot}}$$

$$+|y_j||\hat{y}_j|\sin(\theta_j - \hat{\theta}_j)\frac{|x_i|}{|u_j|}\cos\theta_{ji}^{\mathrm{rot}}\Big) \tag{8}$$

where

$$\theta_{ji}^{\mathrm{rot}} \equiv \theta_j - \theta_i - \theta_{ji} \tag{9}$$

We can find that this learning process works in a distributed manner since the learning rule uses only the variables obtainable locally. The rotation angle $\theta_{ji}^{\mathrm{rot}}$ defined in (9) is the difference between the temporary output phase θ_j and the phase of the input we focus on $\theta_i + \theta_{ji}$, which is weighted but not summed yet. But this phase value is also locally available.

We also obtain the complex-valued gradient descent dynamics in the discrete time. We modify (7) and (8) as

$$|w_{ji}|(t + 1) = |w_{ji}|(t) - K\Big((1 - |y_j|^2)(|y_j| - |\hat{y}_j|\cos(\theta_j - \hat{\theta}_j))\, |x_i|\, \cos\theta_{ji}^{\mathrm{rot}}$$

$$-|y_j||\hat{y}_j|\sin(\theta_j - \hat{\theta}_j)\frac{|x_i|}{|u_j|}\sin\theta_{ji}^{\mathrm{rot}}\Big) \tag{10}$$

$$\theta_{ji}(t + 1) = \theta_{ji}(t) - K\Big((1 - |y_j|^2)(|y_j| - |\hat{y}_j|\cos(\theta_j - \hat{\theta}_j))\, |x_i|\, \sin\theta_{ji}^{\mathrm{rot}}$$

$$+|y_j||\hat{y}_j|\sin(\theta_j - \hat{\theta}_j)\frac{|x_i|}{|u_j|}\cos\theta_{ji}^{\mathrm{rot}}\Big) \tag{11}$$

where K determines the learning time constant.

Accordingly, we have derived the learning dynamics in the output layer by referring to the complex-valued neural-network framework. Then we can utilize not only the amplitude but also the phase information explicitly. This leads to the reservoir computing based on the signal representation utilizing the carrier wave, that is, the carrier-wave reservoir computing.

4   Conclusion

We proposed the carrier-wave reservoir computing. We presented its construction, in particular the input/output signal conversions. Then we derived the phase-sensitive learning dynamics suitable for this reservoir signals. The carrier-wave reservoir computing realizes the utilization of the phase information as well as the amplitude, reservoir functions dependent on the carrier-wave frequency, and frequency multiplexing. It is useful for evaluation of materials from the viewpoint of computational ability.

References

1. Special issue: Reservoir computing. J. IEICE (to appear)
2. Antonik, P., Duport, F., Smerieri, A., Hermans, M., Haelterman, M., Massar, S.: Online training of an opto-electronic reservoir computer. In: Arik, S., Huang, T., Lai, W.K., Liu, Q. (eds.) ICONIP 2015. LNCS, vol. 9490, pp. 233–240. Springer, Cham (2015). https://doi.org/10.1007/978-3-319-26535-3_27
3. Fernando, C., Sojakka, S.: Pattern recognition in a bucket. In: Banzhaf, W., Ziegler, J., Christaller, T., Dittrich, P., Kim, J.T. (eds.) ECAL 2003. LNCS (LNAI), vol. 2801, pp. 588–597. Springer, Heidelberg (2003). https://doi.org/10.1007/978-3-540-39432-7_63
4. Georgiou, G.M., Koutsougeras, C.: Complex domain backpropagation. IEEE Trans. Circ. Syst. II **39**(5), 330–334 (1992)
5. Hirose, A.: Continuous complex-valued back-propagation learning. Electron. Lett. **28**(20), 1854–1855 (1992)
6. Hirose, A., Eckmiller, R.: Behavior control of coherent-type neural networks by carrier-frequency modulation. IEEE Trans. Neural Netw. **7**(4), 1032–1034 (1996)
7. Hirose, A., Eckmiller, R.: Coherent optical neural networks that have optical-frequency-controlled behavior and generalization ability in the frequency domain. Appl. Optics. **35**(5), 836–843 (1996)
8. Hirose, A., et al.: Complex-valued neural networks for wave-based realization of reservoir computing. In: Liu, D., Xie, S., Li, Y., Zhao, D., El-Alfy, E.S. (eds.) ICONIP 2017. LNCS, vol. 10637, pp. 449–456. Springer, Cham (2017). https://doi.org/10.1007/978-3-319-70093-9_47
9. Hirose, A., et al.: Complex-valued neural networks to realize energy-efficient neural networks including reservoir computing. In: International Symposium on Nonlinear Theory and Its Applications (NOLTA), Cancun. A2L-E-3 (2017)
10. Jaeger, H., Haars, H.: Harnessing nonlinearity: predicting chaotic systems and saving energy in wireless communication. Science **304**, 78–80 (2004)
11. Kawata, S., Hirose, A.: Coherent lightwave neural network systems: use of frequency domain. World Scientific Publishing Co., Singapore (2003)
12. Kawata, S., Hirose, A.: Coherent optical adaptive filter that has an arbitrary generalization characteristic in frequency-domain by using plural path difference. In: International Conference on Artificial Neural Networks (ICANN) 2003 / International Conference on Neural Information Processing (ICONIP), pp. 426–429, June 2003
13. Kawata, S., Hirose, A.: Coherent optical neural network that learns desirable phase values in frequency domain by using multiple optical-path differences. Opt. Lett. **28**(24), 2524–2526 (2003)
14. Kawata, S., Hirose, A.: Frequency-multiplexed logic circuit based on a coherent optical neural network. Appl. Opt. **44**(19), 4053–4059 (2005)
15. Kawata, S., Hirose, A.: Frequency-multiplexing ability of complex-valued Hebbian learning in logic gates. Int. J. Neural Syst. **12**(1), 43–51 (2008)

16. Leung, H., Haykin, S.: The complex backpropagation algorithm. IEEE Trans. Signal Process. **39**, 2101–2104 (1991)
17. Limmanee, A., Kawata, S., Hirose, A.: Phase signal embedment in densely frequency-multiplexed coherent neural networks. In: OSA Topical Meeting on Information Photonics (OSA-IP), Charlotte. No. ITuA2, June 2005
18. Lukosevicius, M., Jaeger, H.: Reservoir computing approaches to recurrent neural network training. Comput. Sci. Rev. **3**(3), 127–149 (2009)
19. Mori, R., Tanaka, G., Nakane, R., Hirose, A., Aihara, K.: Computational performance of echo state networks with dynamic synapses. In: Hirose, A., Ozawa, S., Doya, K., Ikeda, K., Lee, M., Liu, D. (eds.) ICONIP 2016. LNCS, vol. 9947, pp. 264–271. Springer, Cham (2016). https://doi.org/10.1007/978-3-319-46687-3_29
20. Nakajima, K., Hauser, H., Li, T., Pfeifer, R.: Information processing via physical soft body. Sci. Rep. **5**, 10487 (2015)
21. Nakajima, K., Li, T., Hauser, H., Pfeifer, R.: Exploiting short-term memory in soft body dynamics as a computational resource. J. Roy. Soc. Interface **11**, 20140437 (2018)
22. Nakane, R., Tanaka, G., Hirose, A.: Demonstration of spin-wave-based reservoir computing for next-generation machine-learning devices. In: International Conference on Magnetism (ICM), San Francisco, pp. 26–27, July 2018
23. Nakane, R., Tanaka, G., Hirose, A.: Reservoir computing with spin waves excited in a garnet film. IEEE Access **6**, 4462–4469 (2018)
24. Takeda, S., et al.: Photonic reservoir computing based on laser dynamics with external feedback. In: Hirose, A., Ozawa, S., Doya, K., Ikeda, K., Lee, M., Liu, D. (eds.) ICONIP 2016. LNCS, vol. 9947, pp. 222–230. Springer, Cham (2016). https://doi.org/10.1007/978-3-319-46687-3_24
25. Tanaka, G., et al.: Exploiting heterogeneous units for reservoir computing with simple architecture. In: Hirose, A., Ozawa, S., Doya, K., Ikeda, K., Lee, M., Liu, D. (eds.) ICONIP 2016. LNCS, vol. 9947, pp. 187–194. Springer, Cham (2016). https://doi.org/10.1007/978-3-319-46687-3_20
26. Tanaka, G., et al.: Nonlinear dynamiccomdynamic of memristive networks and its application to reservoir computing. In: International Symposium on Nonlinear Theory and Its Applications (NOLTA), Cancun. A2L-E-2 (2017)
27. Tanaka, G., et al.: Waveform classification by memristive reservoir computing. In: Liu, D., Xie, S., Li, Y., Zhao, D., El-Alfy, E.S. (eds.) ICONIP 2017. LNCS, vol. 10637, pp. 457–465. Springer, Cham (2017). https://doi.org/10.1007/978-3-319-70093-9_48
28. Tanaka, G., et al.: Recent advances in physical reservoir computing: a review. arXiv preprint. arXiv:1808.04962 (2018)
29. Torrejon, J., et al.: Neuromorphic computing with nanoscale spintronic oscillators. Nature **547**, 428–431 (2017)
30. Verstraeten, D., Schrauwen, B., D'Haene, M., Stroobandt, D.: An experimental unification of reservoir computing methods. Neural Netw. **20**(3), 391–403 (2007)
31. Widrow, B., McCool, J., Ball, M.: The complex LMS algorithm. Proc. IEEE **63**, 719–720 (1975)
32. Yamane, T., Katayama, Y., Nakane, R., Tanaka, G., Nakano, D.: Wave-based reservoir computing by synchronization of coupled oscillators. In: Arik, S., Huang, T., Lai, W.K., Liu, Q. (eds.) ICONIP 2015. LNCS, vol. 9491, pp. 198–205. Springer, Cham (2015). https://doi.org/10.1007/978-3-319-26555-1_23

33. Yamane, T., et al.: Simulation study of physical reservoir computing by nonlinear deterministic time series analysis. In: Liu, D., Xie, S., Li, Y., Zhao, D., El-Alfy, E.S. (eds.) ICONIP 2017. LNCS, vol. 10634, pp. 639–647. Springer, Cham (2017). https://doi.org/10.1007/978-3-319-70087-8_66

34. Yamane, T., et al.: Dynamics of reservoir computing at the edge of stability. In: Hirose, A., Ozawa, S., Doya, K., Ikeda, K., Lee, M., Liu, D. (eds.) ICONIP 2016. LNCS, vol. 9947, pp. 205–212. Springer, Cham (2016). https://doi.org/10.1007/978-3-319-46687-3_22

Spiking Neural Networks for Cancer Gene Expression Time Series Modelling and Analysis

Jack Dray[✉], Elisa Capecci, and Nikola Kasabov

Knowledge Engineering and Discovery Research Institute (KEDRI), Auckland University of Technology (AUT), AUT Tower, Level 7, Cnr Rutland and Wakefield Street, Auckland 1010, New Zealand
xrj3564@autuni.ac.nz, elisa.capecci@gmail.com, nkasabov@aut.ac.nz

Abstract. Gene expression can be used for profiling of cancer cell state and classification of disease. Some cancer variants have been attributed to one or few significant gene expression features. This paper investigates the combination of novel features selection methods - Minimum-Redundancy, Maximum-Relevance - and artificial neural networks - the spiking neural network NeuCube architecture - for genomic data classification and analysis. A NeuCube model performs not only a better classification than other machine learning methods, but most importantly contributes to the feature extraction and marker discovery along with providing gene interaction network analysis for selected genes. Results demonstrated that the methodology proposed could contribute to bioinformatics data analysis for the treatment of disease by discovery of new biomarkers from gene expression data.

Keywords: Spiking Neural Networks · Gene expression Bioinformatics

1 Introduction

Due to recent advances in technology there has been a large influx of biological data, such as gene expression data. Gene expression data is the measurement of the activity of up to several thousands of genes at one time, giving a large overview of cellular processes. These genetic profiles can provide information enabling the distinction between disease and healthy state cells, as well as how these cells react in altered environments. With this and newer techniques such as full genome sequencing, there has been a rising challenge on how to process and analyse this data in a meaningful way [1]. With such a large output of genomic data, current analytical software and methods are playing catch up. Genetic data presents a unique problem of high feature counts with a low sample number and traditional methods are unsatisfactory in solving these biological problems. An exploration of different techniques is necessary to broaden the applications of bioinformatics in the treatment of disease.

© Springer Nature Switzerland AG 2018
L. Cheng et al. (Eds.): ICONIP 2018, LNCS 11301, pp. 625–634, 2018.
https://doi.org/10.1007/978-3-030-04167-0_57

This paper aims to investigate machine learning techniques, specifically, Spiking Neural Networks (SNN), to interpret and understand the aforementioned data. In doing so, we aim at identifying potential new biomarkers, which remained undiscovered by a previous study using more conventional methodology, and to determine a Gene Interaction Network (GIN) between these markers. A classifier to distinguish between tissue experiencing differing dosage levels will also be derived. This will act as a proof of concept to alternative computational methods for genomic data analysis in the area of genetic cancer research.

The gene expression data for this research was acquired from a study published in 2009 [2]. Cancer patients were given the cancer therapy drug R547 at different concentrations and blood samples taken at timed intervals post-drug. The gene expression levels were analysed with a DNA microarray, which is capable of recording genome-wide gene expression [3]. The study then undertook several genomics-based approaches in testing the antiproliferative effect of the drug R547 to identify relevant pharmacodynamic biomarkers. Significant genes were discovered using linear modelling and were then ranked using the bioinformatics tools Ingenuity Pathway Analysis provided by QIAGEN. The study found 26 potential biomarkers and then compared results with two other similar studies to propose eight genes as suitable dose-responsive pharmacodynamic biomarkers for further study [2].

Minimum-Redundancy, Maximum-Relevance (mRMR) uses a heuristic approach of measuring maximal statistical dependency based on mutual information as well determining redundancy within the features. The mRMR approach has shown to be better than other feature selection methods for support vector machine analysis and NCI cancer cell lines [4]. Proficient feature selection is an integral step in minimizing classification error and we have examined the same data using an mRMR algorithm.

Following feature selection, a machine learning approach was applied with the use of SNN, which incorporates the neuronal and synaptic states of artificial neural network (ANN). This allows for learning important patterns of changes in temporal data [5–7]. For this purpose we used the SNN architecture and software NeuCube [7–9], which provided learning, classification and the analysis of a GIN.

2   Materials and Methods

2.1   Data Description

The dataset was obtained from the Gene Expression Omnibus (GEO), which is part of the National Centre for Biotechnology Information (NCBI). GEO query data set number GSE15395 [2]. *Homo sapiens* biopsies were used for the analysis of HCT116 cancer cells that were treated with the CDK inhibitor R547 at three doses. DNA microarray data (Affymetrix Human Genome U133 Plus 2.0 Array) was measured over time up to 24 h from the drug intake (1, 2, 4, 6, 24 h). The drug causes an anti-proliferative effect on various cancer lines and the results were compared to DU145 cells (GDS5267) to identify pharmacodynamic biomarkers for the R547 drug [2].

The data contains 74 samples, replicates A, B, C, D, E, F, a feature count of 54,675 and four sample groups - the control and three different concentrations of R547, namely ID50 (LOW), IC90 (MED) and 3xIC90 (HIGH) that the cells were treated with.

2.2   Feature Selection

The high-dimensionality of the data was handled by an mRMR feature selection method [10]. This feature selection allowed for any desired amount of ranked features to be evaluated.

Features chosen were then compared to the original study to look for concordance of results as well as a search of relevant literature of identified genes for potential roles in cancer genetics.

2.3   SNN Model: The NeuCube Architecture

The ANN method used is the NeuCube architecture as depicted in Fig. 1 [7, 8, 11]. NeuCube takes inspiration from the brain as a spatio-temporal information processor and can be used to analyse different types of data, especially spatio-temporal data. The NeuCube is based on a 3D evolving SNN that approximates a map of structuring and functions features of significance. NeuCube learns from the time-series data to modify the initial set of connection weights assigned. This is done through the several functional modules within the NeuCube. The input data encoding module encode continuous value input information into trains of spikes. Encoding can be done by several algorithms such as the Threshold-Based (TB) algorithm [11] used in this study. Then, a 3D SNN reservoir module is used to construct and train an unsupervised SNN with an algorithm such as Spike-Timing Dependant Plasticity (STDP) [12], used in this study, to learn spike sequences that represent individual input patterns. A classification module constructs and trains in a supervised mode an evolving classifier to differentiate SNN reservoir activities that represent different input patterns of classes. The algorithm used for classification is Dynamic Evolving SNN (deSNN) [13], which combines rank-order and temporal learning rules. Finally, the NeuCube is optimized by changing the parameters of these modules during several iterations until maximum classification accuracy is achieved. This model is then able to be recalled with new data. Once learning is done, the NeuCube can be used to classify and predict functional pathways of temporal activity and for new knowledge discovery from the data. This allows for visualisation of the temporal activity generation after learning, which can be used for network structure analysis. This is used extensively in this paper, as it can be useful for clinical predictions of cells states [6]. The NeuCube was specifically chosen for its capability of processing temporal data which is vital for gene analysis expression over time and building a GIN.

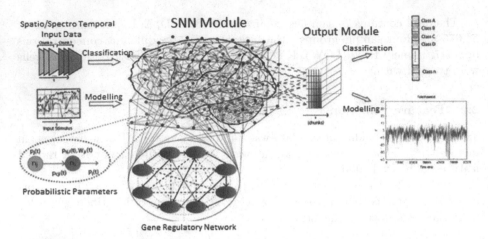

Fig. 1. A diagram of the NeuCube architecture illustrating the modules: input data encoding module; 3D SNN reservoir module; classification module and an optional GRN module.

3   Results and Discussion

3.1   Data Pre-processing and Augmentation

The data went through a series of pre-processing before analysis. Not all of the data from the original paper was used for analysis. The control data only had one reading per sample and was not formatted temporally so was excluded. Any other samples that only had one-time point reading were also removed. The data was then arranged into each class of low (class one), med (class two) and high dosage (class three) according to the time of collection.

Before analysis in the NeuCube the data was also augmented via linear interpolation to increase the time points from 5 to 17. This was done as encoding spikes for SNN is more successful with higher temporal counts and due to the nature of human sample collection a low temporal count is common.

3.2   Feature Selection

The top 10 significant genes resulting from the mRMR selection method were chosen for further analysis, as reported in Table 1.

A cut-off of 10 was chosen, as 10 novel significant candidates for biomarkers in cancer cell response to R547 were detected.

Out of the 10 features, seven have known functions or known roles in pathways according to Gene Ontology Consortium. Four have known functions in protein binding (*HYLS1, COL4A, EFNA4, UNC13D*), one plays a role in negative regulation in cell growth cycle (*CTDSPL*), two have a role in GTPase activity (*ARHGEF33, UNC13D*), one is an integral part of membrane function (*MIR6748*) and finally there are three with no known function (*BE672168, CCDC160, BC038291*).

Table 1. Features identified by mRMR and their main function. Features are ranked in order of significance.

Gene ID	Function
HYLS1	Protein binding
BE672168	Unknown
CTDSPL	Negative regulation of G1
COL4A1	Protein binding
EFNA4	Protein binding
MIR6748	Integral component of membrane
ARHGEF33	Positive regulation of GTPase
UNC13D	GTPase binding/protein binding
CCDC160	Unknown
BC038291	Unknown

3.3   Data Classification with the NeuCube

The numerous parameters of the three main algorithms of the NeuCube need to be optimised to find the best combination that leads to high classification accuracy. To achieve that, we ran a greedy-search that evaluated the model that resulted in the highest classification accuracy using Leave-One-Out Cross Validation (LOOCV) method. The optimised parameters and their corresponding values were: the threshold of the TB encoding algorithm (set at 0.01); the parameters of the leaking-integrate and fire neurons, such as the threshold of firing (set at 0.5), the refractory time (set at 6) and the potential leak rate (set at 0.002); the STDP rate parameter of the unsupervised learning algorithm (set at 0.006); and the variables mod and drift of the deSNN classifier (set at 0.5 and 0.008 respectively).

As reported in Table 2, our classifier obtained an overall accuracy of 76.9%. This is a reasonable score and it indicates that using different dosage of the same drug results in clearly different gene expression temporal profiles. The NeuCube classifier is scalable and accuracy can be increased with the introduction of more data and improved optimization of its parameters.

Being able to classify the state of a cancer cell under different treatment and conditions can have beneficial implications for the design of cancer treatment. A high accuracy classifier can be used as a diagnostic and prognostic tool to identify the response to treatment under different dosage of a drug.

Class three reported the lowest accuracy of classification. This is most likely due to the small sample size. Having more samples for class three would improve this accuracy result and in turn improve the overall accuracy.

Table 2. NeuCube classification accuracy results.

Class	Accuracy
Overall	76.90%
Class 1 (LOW)	80%
Class 2 (MED)	80%
Class 3 (HIGH)	66.60%

3.4   Comparison with Support Vector Machine Classifier

Support Vector Machine (SVM) is another popular method for classification of gene expression data that has been widely proposed in the literature for the analysis of cancer genomic data [14]. They are based on finding a hyperplane that best divides dataset into two classes. It is accurate when classifying small, clean datasets. It is not the best suited for larger and noisier datasets which biological data can be. The multiclass variant was used to compare classification accuracy with the SNN method (Table 3). Here, we have utilised the error-correcting output codes (ECOC) model [15] with one-versus-one coding design for a multiclass SVM classifiers, as this model is able to improve classification accuracy, when compared to other multiclass models [16]. Parameters of this model have been optimised by using automatic hyperparameter optimization, which minimise five-fold cross-validation loss. As reported in Table 3, the proposed method showed superior result, when compared with traditional statistical and machine learning methods, especially in the following aspects:

1. Better data analysis and classification accuracy of temporal data;
2. Better understanding of the data and the processes that are measured through network analysis, which is not possible with other methods, such as SVM;
3. Enabling new information and knowledge discovery through meaningful interpretation of the models and the processes that regulate the data.

SNN having a better performance than SVM may be attributed to several factors. SNNs have a much higher flexibility for types of data sets they can support. Due to the size and noisiness of gene expression data the SNN model is better able to accommodate for this. Also, as the data used was temporal, SNN is uniquely designed to incorporate time which would improve its classification learning over SVM. These attributes of the SNN make it more accurate for biological processes as seen in our results.

3.5   Connectivity Analysis of the NeuCube SNN Model

As shown in Fig. 2, the NeuCube learns from gene expression temporal data to create connections between neurons, each connection representing temporal association between the activity of the connected neurons. NeuCube uses spike-time depended plasticity learning rule (STDP). It can be seen that the input

Table 3. Comparison between the classification accuracy obtained with SNNs and SVM method.

Class	SNN	SVM
Overall	76.90%	44%
Class 1 (LOW)	80%	45%
Class 2 (MED)	80%	43%
Class 3 (HIGH)	66.60%	50%

features (the selected 10 genes) are connected between each other that makes a pattern of their interaction in time.

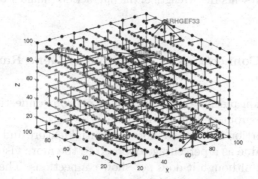

Fig. 2. A trained NeuCube showing neural connections between input features (i.e. the selected 10 genes). Genes are represented as input nodes of the network and are highlighted in different colours. Positive connections between neurons are shown in blue, while negative connections are shown in red. (Color figure online)

3.6 GIN Extraction from a Trained NeuCube Model

A GIN of the 10 input gene variables was extracted from a trained NeuCube model on the 3 classes of data, as shown in Fig. 3. The lines between nodes indicate an interaction between genes. The thickness of the lines shows how much neural information has been exchanged between the clusters of neurons centred around the input variables (the genes). The genes with the strongest interactions between them are the most affected during the administration of the different dosages of the drug.

As seen in Fig. 3, genes *BE672168* and *CCDC160* seem to be the most affected. Both of these genes are of unknown function yet seem to be playing a major role in the genetic expression state of the cancer cells in response to treatment. Future research could be undertaken to define their function. By visualizing the interaction between the genes it can be seen that gene *CCDC160* has been affected significantly, while on the ranking in Table 1 it appears as No.9.

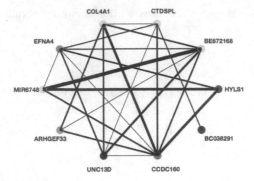

Fig. 3. GIN extracted from a trained NeuCube model on the three classes of gene expression time series. Black lines indicate interaction between features and the thickness of the lines represents the strength of the interaction (number of exchanged spikes during training).

3.7   Neuronal Connections as Indicators of Gene Ranking by Importance

A neuron proportion pie shows the amount of neuron connections attributed to each feature as an input neuron to the model.

A higher proportion indicates a more influential feature and vice versa. It also gives a concrete value as a percentage opposed to the more visual representation of the GIN, Fig. 4, although it does not show connections. The feature with the highest proportion value was *CCDC160* (24%), having the largest effect in the regulation of the other genes. This was previously seen in the GIN. This gene should be investigated further. The second highest is *BE672168* (20%), another gene with unknown function also observed in the GIN. The lowest proportion is

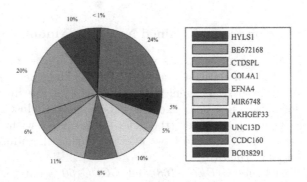

Fig. 4. Proportion of connectivity of input variables (genes) that represent their activity during the training procedure. Each gene is highlighted in a different colour and is used to construct the activity with other genes. The *CCDC160* gene is ranked on the top, followed by *BE672168*. (Color figure online)

BC038291 with 1%. Increasing the number of features would have located genes with greater similarity and therefore have had a redundant effect on classification. A point of interest is that a total of 44% of the neuronal connections can be attributed to only two features. Even in the top 10 significant features there is still two having a disproportionate influence over the rest.

4   Conclusion and Future Direction

We have proposed a method for modelling gene expression time series data and have demonstrated the method on benchmark gene expression data of genes measured over time as a result of the cancer cells treatment with a drug. We have used the mRMR feature selection method and the NeuCube SNN system. The results differed from those of the original study whilst using the same data set. None of the 10 significant genes selected in this study matched those previously identified. Three of the most significant features were found to have no known function according to Gene Ontology Consortium.

Post-feature selection analysis with the NeuCube provided new information on which features interact with each other and to what degree. Two of the genes with unknown functions, *BE672168* and *CCDC160*, showed the most gene interaction and had the highest proportions of neuronal connectivity in the model. A route for further study could be to use these two as candidates for gene function and expression studies to further understand their role of gene regulation in cancer cells.

This new approach to gene expression time series modelling and knowledge discovery can be further studied for cancer temporal data analysis.

More information about the NeuCube modelling system can be obtained from https://kedri.aut.ac.nz/R-and-D-Systems/neucube. More information about SNN methods, systems and applications can be found in [9].

Acknowledgments. EC has been funded by the Auckland University of Technology (AUT) SRIF INTERACT project 2017-18 and by the Knowledge Engineering and Discovery Research Institute (KEDRI, www.kedri.aut.ac.nz). Many thanks to Amanda Dixon-McIver of IGENZ Ltd, Auckland, New Zealand for contributing to the paper. Several people have contributed to the research that resulted in this paper, especially: Dr Y. Chen, Dr J. Hu, L. Zhou and Dr E. Tu. A free for research and teaching version of the NeuCube SNN system can be found from the KEDRI web site: https://kedri. aut.ac.nz/R-and-D-Systems/neucube.

References

1. Alberts, B., Johnson, A., Lewis, J., Walter, P., Raff, M., Roberts, K.: Molecular Biology of the Cell, 4th edn. Garland Science, New York (2002). International Student Edition
2. Berkofsky-Fessler, W., et al.: Preclinical biomarkers for a cyclin-dependent kinase inhibitor translate to candidate pharmacodynamic biomarkers in phase I patients. Mol. Cancer Ther. **8**(9), 2517–2525 (2009)

3. Tarca, A.L., Romero, R., Draghici, S.: Analysis of microarray experiments of gene expression profiling. Am. J. Obstet. Gynecol. **195**(2), 373–388 (2006)
4. Peng, H., Long, F., Ding, C.: Feature selection based on mutual information criteria of max-dependency, max-relevance, and min-redundancy. IEEE Trans. Pattern Anal. Mach. Intell. **27**(8), 1226–1238 (2005)
5. Medler, D.A.: A brief history of connectionism. Neural Comput. Surv. **1**, 18–72 (1998)
6. Maass, W.: Networks of spiking neurons: the third generation of neural network models. Neural Netw. **10**(9), 1659–1671 (1997)
7. Kasabov, N.: NeuCube: a spiking neural network architecture for mapping, learning and understanding of spatio-temporal brain data. Neural Netw. **52**, 62–76 (2014)
8. Kasabov, N., et al.: Evolving spatio-temporal data machines based on the neucube neuromorphic framework: design methodology and selected applications. Neural Netw. **78**, 1–14 (2016)
9. Kasabov, N.: Time-Space, Spiking Neural Networks and Brain-Inspired Artificial Intelligence. Springer, Heidelberg (2018). https://doi.org/10.1007/978-3-662-57715-8. https://www.springer.com/gp/book/9783662577134
10. Ding, C., Peng, H.: Minimum redundancy feature selection from microarray gene expression data. J. Bioinform. Comput. Biol. **3**(2), 185–205 (2005)
11. Tu, E., Kasabov, N., Yang, J.: Mapping temporal variables into the neucube for improved pattern recognition, predictive modeling, and understanding of stream data. IEEE Trans. Neural Netw. Learn. Syst. **28**(6), 1305–1317 (2017)
12. Song, S., Miller, K.D., Abbott, L.F.: Competitive Hebbian learning through spike-timing-dependent synaptic plasticity. Nat. Neurosci. **3**(9), 919–926 (2000)
13. Kasabov, N., Dhoble, K., Nuntalid, N., Indiveri, G.: Dynamic evolving spiking neural networks for on-line spatio- and spectro-temporal pattern recognition. Neural Netw. **41**, 188–201 (2013)
14. Huang, S., Cai, N., Pacheco, P.P., Narandes, S., Wang, Y., Xu, W.: Applications of support vector machine (SVM) learning in cancer genomics. Cancer Genomics Proteomics **15**(1), 41–51 (2018)
15. The MathWorks Inc.: Statistics and machine learning toolbox: User's guide (r2012b) 2012–2018. https://au.mathworks.com/help/stats/fitcecoc.html# References
16. Fürnkranz, J.: Round robin classification. J. Mach. Learn. Res. **2**, 721–747 (2002)

Dimensionality Reduction by Reservoir Computing and Its Application to IoT Edge Computing

Toshiyuki Yamane[1(✉)], Hidetoshi Numata[1], Jean Benoit Héroux[1],
Naoki Kanazawa[1], Seiji Takeda[1], Gouhei Tanaka[2], Ryosho Nakane[2],
Akira Hirose[2], and Daiju Nakano[1]

[1] IBM Research - Tokyo, Kawasaki, Kanagawa 212-0032, Japan
{tyamane,hnumata,heroux,knzwnao,seijitkd,dnakano}@jp.ibm.com
[2] Department of Electrical Engineering and Information Systems,
The University of Tokyo, Tokyo 113-8656, Japan
gouhei@sat.t.u-tokyo.ac.jp, nakane@cryst.t.u-tokyo.ac.jp,
ahirose@ee.t.u-tokyo.ac.jp

Abstract. We propose a method of dimension reduction of high dimensional time series data by reservoir computing. The proposed method is a generalization of random projection techniques to time series, which uses a reservoir smaller than input time series. We demonstrate the method by echo state networks for artificially generated time series data. We also discuss an implementation as physical reservoirs and its application of the proposed method to IoT edge computing, which is the first proposal for industry application of physical reservoir computing beyond standard benchmark tasks.

Keywords: Dimensionality reduction · Random projection
Reservoir computing · Echo state network · Internet of Things
Edge computing

1 Introduction

In recent years, the importance of cognitive computing for business is rapidly increasing which handles unstructured big data related to human cognition such as voices, images, videos and texts. For example, a number of deep learning neural networks have been proposed recently based large and deeply layered neural networks and have been successfully applied to handling such big cognitive data. The recent remarkable trend of such cognitive big data is that huge amount of them are generated by IoT (Internet of Things). Therefore, it is becoming unfeasible to apply cloud computing approach which transfers all data to cloud datacenters and gives feedback to IoT environments based on the analytics since heavy network traffic, high power consumption and response latency are major bottleneck of overall system.

© Springer Nature Switzerland AG 2018
L. Cheng et al. (Eds.): ICONIP 2018, LNCS 11301, pp. 635–643, 2018.
https://doi.org/10.1007/978-3-030-04167-0_58

Edge computing is proposed as a promising way of overcoming these issues. So far, the term edge generally refers to the end points of the internet, and with the advent of the era of IoT, the edge means all the sensors, equipment, mobile devices, cars, robots and everything mutually interconnected and deployed all around the world. The basic principle of edge computing is to move computations close to data sources, instead of moving data to computation at remote datacenters, and thus achieve more efficiency of IoT systems. However, edge of the IoT systems have very different performance requirements from cloud datacenters such as strict power constraints, high speed and real time operations. Considering these requirements, implementation of deep learning algorithms by traditional multi-purpose CPU and software approach have proved to be so power-hungry and computer-intensive that they are not necessary suitable for operation near the edge devices. In fact, the famous cat recognition by Google implemented a 9-layered locally connected sparse auto-encoders on distributed 1,000 machines (equivalent to 16,000 cores) and achieved hierarchical abstraction of features in input images after three days of training [5].

Reservoir computing (RC) can be an attractive solution for these issues of edge computing. RC is a new architecture of recurrent neural network [7], which is composed of an input layer, a reservoir and a readout (output) layer. From the viewpoint of machine learning, the function of reservoirs is kernel functions which is nonlinear mapping of input signal to high dimensional feature space. The remarkable feature is that the internal weight of the reservoir and interconnection weights between input and reservoir are initialized randomly and fixed throughout their operation and only interconnection weights in the readout layer are updated by an linear adaptive filter. Therefore, RC takes much less learning cost than current large scale deep neural networks which need hard optimization of huge number of parameters. In addition, the reservoir layers can be implemented by nonlinear physical dynamics since the reservoir layers are random network of nonlinear activation function. The physical implementation of reservoirs can achieve significant performance gain such as real time, higher speed and lower power operation compared to software implementation, which makes RC suitable for IoT edge computing near sensor devices.

In this work, we apply RC to dimension reduction of sensor data collected from huge number of spatially deployed edge sensors. As the number of edge sensors increases exponentially, it is becoming infeasible to send all of sensor data to cloud datacenters. However, the real dimension of the sensor data can be relatively low since the sensor data are often spatially redundant if they are located close to each other. Therefore, making use of this latent low dimensionality, we can compress the sensor data before sending them to cloud datacenters. Though the size of reservoirs are usually larger than the dimension input signal for classification tasks, we show that small size reservoirs can be used for dimension reduction. We also show that physical reservoir computing can be implemented effectively and consistently with the architecture of IoT edge systems.

2   Relation to Prior Work on Dimensionality Reduction Techniques

Dimensionality reduction which finds lower-dimensional structure in higher dimensional data is of fundamental value for a number of analytics such as feature selection for classification and visualization, and saving of resources for communication and storage. In this section, we shall briefly review prior work on such dimensionality reduction techniques and discuss the relation to our reservoir computing based approach. The readers are also referred to the thorough survey by [6] in general and the references therein.

Principal Component Analysis (PCA) would be the most well-known technique for statistical dimension reduction. PCA tries to find the linear projection subspace that maximize spread of observed data and minimize reconstruction error. This problem can be transformed into singular value decomposition of covariance matrix. However, the computational cost of singular value decomposition becomes prohibitive for the high-dimensional data. *Random projection* uses randomly selected linear projection subspace instead of calculating the optimal projection subspace. In random projection, given D dimensional data $\mathbf{u}$ is projected onto N dimensional subspace such that $N \ll D$ by calculation of matrix multiplication $\mathbf{x} = W\mathbf{u}$, where W is a random $N \times D$ matrix which is constructed by

$$W_{i,j} = \begin{cases} +\sqrt{3} & \text{with probability } \frac{1}{6} \\ 0 & \text{with probability } \frac{2}{3} \\ -\sqrt{3} & \text{with probability } \frac{1}{6} \end{cases} \tag{1}$$

In spite of the random construction of the projection matrix, random projection guarantees the preservation of positional relation of two points in high dimensional space after projection to lower dimensional space.

PCA and random projection is a linear method for dimensionality reduction and they often fail to catch nonlinear structure of data. *Kernel PCA* uses predesigned nonlinear transformations of data (called kernel function) and apply PCA for transformed data. Generally, kernel PCA avoids costly explicit computation of inner product in high dimensional space by kernel trick. *Autoencoder* is a three-layer (input-hidden-output) feed-forward neural network whose hidden layer is intentionally designed with smaller number of neurons than input dimension. The size of output layer is the same as input layers and the output layer is so trained that the output reproduce the input. Therefore, the hidden layer learns compressed representation of the input feature. *Manifold Learning* (ML) also deals with mapping of high dimensional data to lower spaces, but assumes that the original data lie on a low dimensional surface (more generally, manifold). ML tries map the original data on the surface onto lower dimensional space while keeping the spatial relations of the proximity data as precisely as possible.

As for the extension to dynamic time series data, *Karhunen-Loéve expansion* (KL expansion) is known as a time series version of PCA. KL expansion is based on the singular value decomposition of autocorrelation function $R(s,t) =$

$\text{Cov}[X(s), X(t)]$ of stochastic process $\{X(t); 0 \leq t < \infty\}$. Since $R(s, t)$ defines Hermite positive definite kernel, one can apply eigen function decomposition of $R(s, t)$ and obtain eigen functions $\{\psi_i(t)\}$. One can achieve dimension reduction of original time series data by taking a few number of eigen functions from $\{\psi_i(t)\}$.

These dimension reduction techniques can be classified by two viewpoints, linearity (linear or nonlinear) and dynamics (static or dynamic) as the technology mapping shown in Fig. 1. Following the classification in Fig. 1, our approach can be characterized as

- dynamic: the input data is time series,
- nonlinear: the model is based on nonlinear recurrent neural network,
- random: the model exploits random structure for projection of input data in input and reservoir layer.

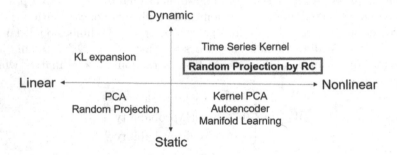

Fig. 1. Technology mapping of dimensionality reduction

3   Random Projection by Reservoir Computing

Consider we have a time series data $\{\mathbf{u}(t) \in \mathbb{R}^D\}$, which is very high dimensional but its real dimension is relatively low due to its spatial redundancy. If we apply random projection straightforward to $\mathbf{u}(t)$ by random matrix $W_{\text{in}} \in R^{N \times D}, N \ll D$ defined by (1), we have lower dimensional time series $\{\mathbf{x}(t) \in \mathbb{R}^N\}$ as

$$\mathbf{x}(t) = W_{\text{in}}\mathbf{u}(t). \tag{2}$$

However, the random projection at each time step cannot take the internal nonlinear dynamics $\{\mathbf{u}(t)\}$ into account. We model the internal dynamics by echo state network (ESN), which is a variant of reservoir computing, defined by the following equations

$$\mathbf{x}(t + 1) = \tanh(W_{\text{res}}\mathbf{x}(t) + W_{\text{in}}\mathbf{u}(t + 1)), \tag{3}$$

where $\mathbf{x}(t)$ is the N-dimensional reservoir state, W_{res} is the $(N \times N)$ reservoir internal weight matrix, W_{in} is the $N \times D$ input weight matrix to the reservoir and random projection matrix to lower dimensional linear space. Both the matrix W_{res} and W_{in} are initialized randomly and fixed throughout its operation.

The readout layer is typically defined as linear filter as follows:

$$\mathbf{y}(t) = W_{\text{out}}\mathbf{x}(t). \tag{4}$$

At training phase, the readout layer is trained by least mean square error principle, that is, it tries to minimize the mean square error of reconstruction of the input $\{\mathbf{u}(t), t = 1, \ldots, T\}$ from the reservoir state $\{\mathbf{x}(t), t = 1, \ldots, T\}$. The optimization is formulated as minimization problem as

$$W_{\text{out}} = \arg\min_{W} \left\{ \frac{1}{T} \sum_{t=1}^{T} \|\mathbf{u}(t) - W\mathbf{x}(t)\|^2 + \lambda\|W\| \right\}, \tag{5}$$

where the parameter $\lambda > 0$ is a regularization constant. The optimum readout weight W_{out} is given by

$$W_{\text{out}} = UX^+, \tag{6}$$

where $X = (\mathbf{x}(1), \ldots, \mathbf{x}(T)) \in \mathbb{R}^{N \times T}$ is the reservoir state collection matrix, $U = (\mathbf{u}(1), \ldots, \mathbf{u}(T)) \in \mathbb{R}^{D \times T}$ is the teacher signal collection matrix, and $X^+ = X^T(XX^T)^{-1}$ is the Moore-Penrose pseudo inverse of X.

In order for reservoirs to work for dimension reduction, the dynamics of reservoirs should have separation property which requires that reservoirs are dynamic so that significant difference of the input must remain distinguishable over time after being mapped to reservoir states. In addition, the reservoirs should have fading memory property which means that reservoir should be stable enough to forget the effect of past input eventually and be driven only by recent input. Without fading memory, reservoirs might have highly sensitive dynamics like chaotic motion and enhance even minor difference of input. The separation property and fading memory property are conflicting requirements to reservoirs and there is trade-off between them. It is often claimed that tuning the dynamics toward edge of chaos or edge of stability provides good compromise of two properties. As an empirical criteria to meet this condition, the spectral radius of W_{res}, denoted by $\rho(W_{\text{res}})$, is often set to be close to one but slightly less than one.

The separation property was investigated experimentally in [3] using separation ratio metric and was evaluated for different echo state networks with a wide variety of parameters. On the other hand, fading memory property is not evident from spectral radius condition $\rho(W_{\text{res}}) \lesssim 1$ since $\rho(W_{\text{res}}) = 1$ means that the dynamics of ESN is at a critical bifurcation point [10]. Instead, we see how reservoir states can be bounded from above in the sense of ℓ^2 norm under the condition $\rho(W_{\text{res}}) \leq 1$. Hereafter, we use the operator norm $\|A\|$ for matrix A defined by $\|A\| \equiv \sup_{\|x\| \neq 0} \frac{\|Ax\|_2}{\|x\|_2} = \sigma_{\max}(A)$, where $\sigma_{\max}(A)$ means maximum of singular values of A. We assume that W_{res} is symmetric and therefore $\|W_{\text{res}}\| = \rho(W_{\text{res}})$. The differences of reservoir states $\mathbf{x}(t)$ and $\mathbf{x}'(t)$ for two different input $\mathbf{u}(t)$ and $\mathbf{u}'(t)$ can be bounded from above as follows Let $f(x) = \tanh(x), \delta(t) = \|\mathbf{u}(t) - \mathbf{u}'(t)\|, \varepsilon(t) = \|\mathbf{x}(t) - \mathbf{x}'(t)\|$ and $1 \leq t \leq T$, then we have

$$
\begin{aligned}
\varepsilon(t) &= \|f(W_{\mathrm{res}}\mathbf{x}(t-1) + W_{\mathrm{in}}\mathbf{u}(t+1)) - f(W_{\mathrm{res}}\mathbf{x}'(t-1) + W_{\mathrm{in}}\mathbf{u}'(t))\| \\
&\leq \|W_{\mathrm{res}}(\mathbf{x}(t-1) - \mathbf{x}'(t-1))\| + \|W_{\mathrm{in}}(\mathbf{u}(t) - \mathbf{u}'(t))\| \\
&\quad \text{(due to the Lipschitz continuity of } f, \|f(x) - f(y)\| \leq \|x - y\|) \\
&\leq \rho(W_{\mathrm{res}})\varepsilon(t-1) + \|W_{\mathrm{in}}\|\delta(t) \leq \cdots \\
&\leq \|W_{\mathrm{in}}\| \sum_{r=0}^{t} \rho(W_{\mathrm{res}})^r \delta(t-r) \leq \|W_{\mathrm{in}}\| \sum_{r=0}^{t} \delta(r)
\end{aligned}
\tag{7}
$$

$$
\frac{1}{T}\sum_{t=0}^{T} \varepsilon(t) \leq \frac{\|W_{\mathrm{in}}\|}{T} \sum_{t=0}^{T}\sum_{r=0}^{t} \delta(r)
\tag{8}
$$

If $\mathbf{u}(t)$ is finite energy signal in the sense that $\mathbf{u}(t) \in \ell^2$, then $s(t) = \sum_{r=0}^{t}\delta(r)$ converges as $t \to \infty$ and $\frac{1}{T}\sum_{t=0}^{T} s(t)$ also converges as $T \to \infty$. This means that the echo state networks with spectral radius smaller than one have stability in the sense that they do not enhance the difference of the input signal and preserve the original structure, which is essential for dimensionality reduction.

We apply the random projection by ESN to artificial nonlinear time series model NARMA (nonlinear auto-regression moving average) given by

$$
\begin{aligned}
w_1(t) &= w_2(t-5)v_1(t-10) + v_2(t-2)^2, \\
w_2(t) &= w_2(t-1)v_2(t-3) + v_2(t-2)w_1(t-2),
\end{aligned}
$$

where $v_1(t)$ and $v_2(t)$ are random number sequences over $[0, 1]$. The two dimensional NARMA time series $\{\mathbf{w}(t) = (w_1(t), w_2(t))\}$ are transformed to 10 dimensional time series data $\{\mathbf{u}(t)\}$ by $\mathbf{u}(t) = A\mathbf{w}(t)$, where A is a random $(2, 10)$ matrix. We set the size of the reservoir to be $N = 2$ and use an internal connection matrix W_{res} with the spectral radius $\rho(W_{\mathrm{res}}) = 0.99$ and regularization constant $\lambda = 0.001$. We generated 2,000 length NARMA(10) time series data, 1,000 of which were used for training and the rest 1,000 were used for testing. We evaluated the reconstruction performance in terms of normalized root mean squared error (NRMSE) defined by $NRMSE = \sqrt{\frac{\langle\|\mathbf{y}(t)-\mathbf{u}(t)\|^2\rangle}{\langle\|\mathbf{u}(t)-\langle\mathbf{u}(t)\rangle\|^2\rangle}}$ and We obtained NRMSE $= 0.2736$ for training and NRMSE $= 0.2968$ for testing. Figure 2 shows the reconstruction results for the first 5 dimensions out of 10 dimensional NARMA(10) time series.

4   Application to IoT Edge Computing

The remarkable recent trend of RC is that some interesting physical implementations of have been reported so far, for example, photonic systems [9], electronic

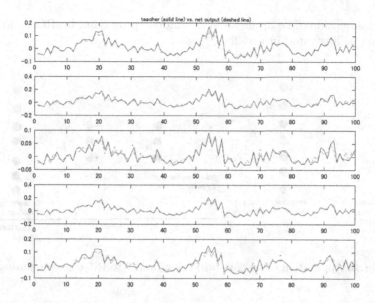

Fig. 2. Reconstruction results for NARMA(10) time series. The first 5 dimensions out of 10 dimensions of time series data are plotted.

circuits [1], nonlinear feedback laser systems [8], polymer wave guide reservoir [4]. These devices perform information processing such as kernel methods and construction of basis functions by directly making use of nonlinearity and high dimensionality of physical dynamics and can achieve higher speed and more power efficiency than software implementation of RC. However, in spite of these advantages, the physical RC is applied only to standard benchmark tasks such as spoken digit recognition, but not applied to any industrial scale tasks so far. Considering the advantages of physical RC, we propose operation of RC with edge sensors as powerful accelerators for edge computing in IoT systems under severe performance constraints.

Specifically, we apply dimensionality reduction of sensor data generated by huge number of spatially deployed edge sensors. The compression of sensor data is not an entirely new problem, for example, Brillinger [2] dealt with transmission of multivariate time series data through limited number of communication lines, where encoding and decoding was performed by linear FIR filters. Now it is worth rethinking this problem in view of RC and IoT edge computing since more and more connected devices result in communication cost in recent IoT systems. More specifically, the reservoir layer in RC and the readout layer are separated completely. Therefore, we can compute dimension reduction near the sensor devices at edge by power efficient physical reservoir layers and send the compressed data through the internet. Then, we can reconstruct the original data by applying readout adaptive filter at cloud data centers. The readout layers are usually implemented as inexpensive adaptive filters which do not cause serious power consumption. The application of random projection by RC to IoT edge computing is illustrated in Fig. 3.

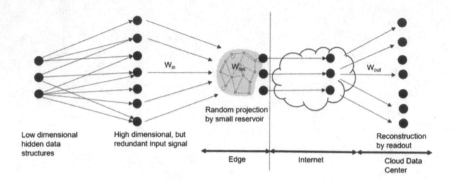

Fig. 3. Application of random projection by RC to IoT edge computing

5   Conclusion

We have discussed the challenges and benefits of IoT edge computing. We have shown that physical reservoir computing can be very attractive accelerator for edge computing since it can achieve power efficiency and high speed operation and enables near-sensor computing. In particular, we have proposed the dimension reduction by physical reservoir computing and how it can fit in the context of IoT edge computing. The implementation of the proposed architecture on a real IoT platform is left to future work.

References

1. Appeltant, L., et al.: Information processing using a single dynamical node as complex system. Nat. Commun. **2**, 468 (2011)
2. Brillinger, D.R.: Time Series: Data Analysis and Theory (Expanded Edition). Holden-Day, San Francisco (1981)
3. Gibbons, T.: Unifying quality metrics for reservoir networks. In: International Joint Conference on Neural Networks (IJCNN) (2010)
4. Héroux, J.B., Numata, H., Nakano, D.: Polymer waveguide-based reservoir computing. In: Liu, D., Xie, S., Li, Y., Zhao, D., El-Alfy, E.S. (eds.) Neural Information Processing, vol. 10639, pp. 840–848. Springer, Cham (2017). https://doi.org/10.1007/978-3-319-70136-3_89
5. Le, Q.V., et al.: Building high-level features using large scale unsupervised learning. In: Proceedings of the 29th International Conference on Machine Learning (2012)
6. Lee, J.A., Verleysen, M.: Nonlinear Dimensionality Reduction. Springer, New York (2007). https://doi.org/10.1007/978-0-387-39351-3
7. Lukoševičius, M., Jaeger, H.: Reservoir computing approaches to recurrent neural network training. Comput. Sci. Rev. **3**(3), 127–149 (2009)

8. Takeda, S., et al.: Photonic reservoir computing based on laser dynamics with external feedback. In: Hirose, A., Ozawa, S., Doya, K., Ikeda, K., Lee, M., Liu, D. (eds.) ICONIP 2016, Part I. LNCS, vol. 9947, pp. 222–230. Springer, Cham (2016). https://doi.org/10.1007/978-3-319-46687-3_24
9. Vandoorne, K., et al.: Toward optical signal processing using photonics reservoir computing. Opt. Express **1615**, 11182–11192 (2008)
10. Yamane, T., et al.: Dynamics of reservoir computing at the edge of stability. In: Hirose, A., Ozawa, S., Doya, K., Ikeda, K., Lee, M., Liu, D. (eds.) ICONIP 2016, Part I. LNCS, vol. 9947, pp. 205–212. Springer, Cham (2016). https://doi.org/10.1007/978-3-319-46687-3_22

Author Index

Ahmed, Saad Bin 307
Alizadehsani, Roohallah 172

Bach, Ngo Xuan 487
Bhattacharyya, Pushpak 214
Bian, Lina 416
Bresson, Xavier 362

Cai, Congbo 162
Cao, Longbing 12
Capecci, Elisa 571, 625
Chang, Liang 275
Chen, Chengbo 297
Chen, Cong 319
Chen, Qin 511
Chen, Wuya 297
Chen, Xingyu 12
Chen, Xu 140
Chen, Yefei 226
Cheng, Jun 33
Cheng, Li 439
Cheng, Zhengxin 275

Dang, Jianwu 545
Defferrard, Michaël 362
Dhingra, Rishi R. 603
Ding, Xinghao 162
Dray, Jack 625
Duan, Fuqing 275
Dutschmann, Mathias 603

Fang, Qiming 525
Fang, Ting 319
Fu, Xuzhou 406
Fuh, Chiou-Shann 263

Gao, Yang 350
Gu, Chaochen 21
Gu, Shenshen 238
Guan, Xinping 21
Guo, Jinrong 107

Hagiwara, Katsuyuki 59
Han, Jiaming 319
Han, Jizhong 107

Hao, Pengyi 33
He, Liang 511
He, Yuqing 385
Héroux, Jean Benoit 594, 616, 635
Hirose, Akira 340, 616, 635
Hong, Xiaowei 152
Hossain, Ibrahim 172
Hu, Bo 284
Hu, Guoxiong 319
Hu, Qinmin 511
Hu, Wenxin 511
Huang, Kaizhu 182
Huang, Yue 162
Hussain, Amir 182

Ide, Hidenori 332
Inuzuka, Nobuhiro 204

Jia, Xiaodong 439
Jiang, Han 406
Jiao, Wencong 394
Jing, Xiao-Yuan 439
John Lewis, D. 3

Kabir, H. M. Dipu 172
Kanazawa, Naoki 594, 616, 635
Kasabov, Nikola 571, 625
Kebria, Parham M. 172
Khosravi, Abbas 172
Koefoed, Lucien 571
Konno, Hirotoshi 563
Koohestani, Afsaneh 172
Kumar, Roneel 476
Kurita, Takio 332

Laña, Ibai 571
Lee, You-Hsien 263
Li, Dengju 582
Li, Ruixuan 425
Li, Ting 284
Li, Xiang 120
Li, Xu 70, 140
Li, Xuewei 406
Li, Yan 499

Li, Yantao   83
Li, Yuhua   425
Lin, Bing-Jhang   263
Liu, Fanghui   12
Liu, Jingxian   499
Liu, Jinhai   152
Liu, Qiuyang   130
Liu, Ryan Wen   499
Liu, Wantao   107
Liu, Wei   46
Liu, Xiao   251
Lu, Changsheng   21
Lu, Gang   33
Lu, Qu   107
Lu, Wenhuan   385
Lu, Wensi   525
Luo, Sihui   96

Ma, Chao   120
Ma, Fei   439
Ma, Quandang   499
Ma, Xiaohu   416
Majumdar, Angshul   3
Manton, Jonathan H.   603
Matsui, Tohgoroh   204
Moriyama, Koichi   204
Murakami, Kazuhiro   204
Mutoh, Atsuko   204

Nahavandi, Saeid   172
Nakane, Ryosho   616, 635
Nakano, Daiju   594, 616, 635
Nandini, Durgesh   571
Natsuaki, Ryo   340
Naz, Saeeda   307
Ngo, Son   452
Nguyen, Anthony   452
Ni, Zhiwen   416
Nishimura, Haruhiko   535
Niu, Dong   582
Nobukawa, Sou   535
Numata, Hidetoshi   616, 635

Pang, Yongheng   284
Pauley, Michael   603
Phuong, Tu Minh   487
Prasad, Vishal   476

Qi, Fumin   439
Qian, Yang   192
Qiao, Hong   192

Qiao, Xuejun   275
Qiao, Yu   120
Qin, Zheng   70, 140
Qu, Fuming   152
Quan, Xiaojun   297

Razzak, Muhammad Imran   307

Saha, Sriparna   214
Saha, Tulika   214
Saito, Toshimichi   555
Salaken, Syed Moshfeq   172
Seo, Youngjoo   362
Sha, Chunlin   463
Shahi, Gautam Kishore   571
Sharma, Adarsh Karan   476
Sharma, Anuraganand   476
Shen, Chengchao   96
Shen, Hongqian   406
Singhal, Vanika   3
Song, Jiarong   319
Song, Jinhua   350
Song, Mingli   96
Song, Yang   511
Su, Jianbo   226
Sun, Liyan   162
Sun, Shiliang   130
Sun, Xiao   416
Sunaga, Yuki   340

Takeda, Seiji   616, 635
Tanaka, Gouhei   616, 635
Tang, Huajin   582
Thanh, Tran Cong   487
Tian, Yanling   33
Tolmachev, Pavel   603
Tran, Son N.   452
Tsan, Ting-Chen   263
Tu, Enmei   12
Tung, Tzu-Chia   263

Uchida, Hiroaki   555

Vandergheynst, Pierre   362
Verma, Sunny   46
Vu, Xuan-Son   452

Wagatsuma, Nobuhiko   563
Wang, Chen   46
Wang, Chenhan   406

Wang, Chunhe 439
Wang, Haizhen 394
Wang, Hao 284, 350
Wang, Haotian 21
Wang, Longbiao 545
Wang, Shuiquan 275
Wang, Zhe 385
Wei, Jianguo 385
Wei, Xuhong 226
Wu, Huafeng 162
Wu, Kaijie 21
Wu, Yawen 162

Xia, Yixing 525
Xiao, Yi 374
Xie, Tianying 83
Xiong, Mengting 425
Xiong, Naixue 385
Xu, Jing 192
Xu, Xiaoliang 525

Yamane, Toshiyuki 616, 635
Yamanishi, Teruya 535
Yan, Rui 582
Yang, Dongsheng 284
Yang, Jie 12, 120
Yang, Liang 439

Yang, Qi 425
Yang, Xudong 251
Yang, Yezhou 96
Yang, Yue 238
Yang, Zhong 319
Yao, Kai 182
Ye, Wenwen 70
Yin, Yanling 96
Ying, Xiang 406
Yu, Qiang 545
Yu, Ruiguo 406
Yusof, Rubiyah 307

Zhang, Qieshi 33
Zhang, Qing 452
Zhang, Qiuyan 319
Zhang, Rui 182
Zhang, Weitong 33
Zhang, Yu 152
Zhao, Hongyong 463
Zhao, Mankun 406
Zhao, Ya 96
Zheng, Yan 374
Zhong, Guoqiang 394
Zhu, Liming 46
Zhu, Xianyi 374
Zhu, Xiaoke 439